Springer Series in Optical Sciences Volume 58
Edited by Arthur L. Schawlow

Springer Series in Optical Sciences

42 **Principles of Phase Conjugation**
By B. Ya. Zel'dovich, N.F. Pilipetsky, and V.V. Shkunov

43 **X-Ray Microscopy**
Editors: G. Schmahl and D. Rudolph

44 **Introduction to Laser Physics**
By K. Shimoda 2nd Edition

45 **Scanning Electron Microscopy**
Physics of Image Formation and Microanalysis
By L. Reimer

46 **Holography and Deformation Analysis**
By W. Schumann, J.-P. Zürcher, and D. Cuche

47 **Tunable Solid State Lasers**
Editors: P. Hammerling, A.B. Budgor, and A. Pinto

48 **Integrated Optics**
Editors: H.P. Nolting and R. Ulrich

49 **Laser Spectroscopy VII**
Editors: T.W. Hänsch and Y.R. Shen

50 **Laser-Induced Dynamic Gratings**
By H.J. Eichler, P. Günter, and D.W. Pohl

51 **Tunable Solid State Lasers for Remote Sensing**
Editors: R.L. Byer, E.K. Gustafson, and R. Trebino

52 **Tunable Solid-State Lasers II**
Editors: A.B. Budgor, L. Esterowitz, and L.G. DeShazer

53 **The CO_2 Laser** By W.J. Witteman

54 **Lasers, Spectroscopy and New Ideas**
A Tribute to Arthur L. Schawlow
Editors: W.M. Yen and M.D. Levenson

55 **Laser Spectroscopy VIII**
Editors: W. Persson and S. Svanberg

56 **Atomic and Molecular Spectroscopy**
By S. Svanberg

57 **Single-Mode Fibers I – Fundamentals**
By E.-G. Neumann

Volumes 1–41 are listed on the back inside cover

P. Hess J. Pelzl (Eds.)

Photoacoustic and Photothermal Phenomena

Proceedings of the 5th International
Topical Meeting,
Heidelberg, Fed. Rep. of Germany,
July 27–30, 1987

With 419 Figures

Springer-Verlag Berlin Heidelberg New York
London Paris Tokyo

Professor Dr. Peter Hess

Institute of Physical Chemistry, University of Heidelberg,
Im Neuenheimer Feld 253, D-6900 Heidelberg, Fed. Rep. of Germany

Professor Dr. Josef Pelzl

Ruhr University, Institute of Experimental Physics,
Universitätsstr. 150, D-4630 Bochum, Fed. Rep. of Germany

ISBN 3-540-18782-0 Springer-Verlag Berlin Heidelberg New York
ISBN 0-387-18782-0 Springer-Verlag New York Berlin Heidelberg

Printing: Druckhaus Beltz, 6944 Hemsbach/Bergstr.
Binding: J. Schäffer GmbH & Co. KG, 6718 Grünstadt.
2153/3150-543210

Preface

This volume comprises the proceedings of the Fifth International Topical Meeting on Photoacoustic and Photothermal Phenomena, which was held at the University of Heidelberg (FRG), July 27–30, 1987. The proceedings contain twelve invited talks, six progress reports and a large number of selected contributed papers providing latest results on the fundamental investigation of photothermal and photoacoustic processes and on their applications in physics, chemistry, biology, medicine and materials science.

The conference was truly international with 230 scientists representing the following countries: Australia, Belgium, Brazil, Bulgaria, Canada, People's Republic of China, Czechoslovakia, Denmark, England, Finland, France, Federal Republic of Germany, Hungary, India, Ireland, Israel, Italy, Japan, Yugoslavia, Republic of Korea, The Netherlands, Poland, Portugal, Romania, Scotland, Spain, Switzerland, Tunisia, the United States, and the USSR.

Spectroscopy traditionally plays an important role in this area, and therefore the first part of the proceedings is devoted to this subject. This is followed by papers on dynamical processes, which are collected in the second part. High sensitivity makes photoacoustic methods suitable for trace analysis; measurements on gases and on condensed matter are reported in Part III. A considerable number of contributions are concerned with surfaces or thin films and ablation, adsorption and desorption processes (Part IV). Accounts of investigations of optical and electronic properties in semiconductors can be found in Part V. Ultrasonic detection of the thermal response offers good time and spatial resolution and is increasingly applied in microscopy (Part VI). Various procedures have been used to investigate heat and mass diffusion in different materials (Part VII). One of the most important areas of application of thermal waves is the nondestructive evaluation of subsurface thermal, optical and magnetic properties. A large variety of applications are reported in Part VIII. Contributions to Part IX deal with new trends in experimental techniques. The last part of this volume is devoted to applications of the photothermal and photoacoustic techniques in medicine and biology (Part X).

These proceedings make reports of the current activities in this field available to a wider audience. The review articles may be very useful for newcomers, and the contributed papers provide a timely survey on the large variety of activities in the field and the numerous applications of photothermal and photoacoustic techniques.

The conference could not have been held without the dedicated help of many people. The gratitude of the conference committee of the Fifth International Topical Meeting on Photoacoustic and Photothermal Phenomena (S.E. Braslavsky, G. Busse, P. Hess, P. Korpiun, J. Pelzl, M.W. Sigrist, R. Tilgner, and F. Träger), therefore, goes to many individuals and organizations. We gratefully acknowledge financial support from the Deutsche Forschungsgemeinschaft (DFG), the Ministry of Science and Arts of the state Baden-Württemberg, and the University of Heidelberg. Without their help the conference and the proceedings would not have been possible. We are particularly grateful to Bärbel Hess for her tireless efforts during all phases of the conference organization.

The editors of this volume are indebted to all members of the conference committee, the international advisory board, the invited speakers and session chairmen, who diligently reviewed papers, thereby contributing to the high scientific level of these proceedings. We thank Springer-Verlag for rapid publication of this volume. Especially, we would like to express our sincere appreciation to Deborah Hollis for her assistance in the preparation of these proceedings.

November 1987 *Peter Hess*
Heidelberg, Bochum *Josef Pelzl*

Contents

Part I Spectroscopy

Part II **Kinetics and Relaxation**

Part III **Trace Analysis**

Part IV **Surfaces and Thin Films**

Part V **Applications to Semiconductors**

Part VI Ultrasonic Detection and Characterization

Part VII **Mass and Heat Transfer**

Part I

Spectroscopy

Thermal Wave Spectroscopies – Where Do They Win?

R.M. Miller

Unilever Research Port Sunlight Laboratory, Quarry Road East, Bebington, Wirral, L633JW, United Kingdom

The questions to be addressed in this paper are: where do thermal wave spectroscopies offer real advantages over alternative methods, and how can they escape into general application from being the province of a limited number of specialised practitioners?

Photoacoustic and photothermal science are clearly thriving at the present time. This Conference is the 5th in the series, now spanning 10 years of research. There are over 170 contributed papers in this meeting, of which over 50 are on spectroscopic applications of thermal waves. There are over 220 delegates at the meeting. All this indicates a very healthy level of interest in this field of research. Publications in the literature at large show a continuing high level of interest in photoacoustic and photothermal spectroscopy. Figure 1 shows the total number of publications between 1971 and the middle of 1987. These totals were obtained by a computer search of Chemical Abstracts, and may therefore have missed a number of publications, but they clearly show the trend. After climbing steadily through the 1970's, the total number of publications peaked in 1983 and has since dropped back slightly to a more steady level. The total for 1987 is of course as yet incomplete.

This levelling off of the rate of publication may be seen as evidence of a maturing science. However, the level of publication suggests that despite its apparent advantages, photoacoustic and photothermal spectroscopy is still a specialised minority interest. A comparison with the number of publications concerning another mature, generic spectroscopic tool such as fluorescence will make this clear. Despite this, the publication rate is still some 200-250 papers per annum,

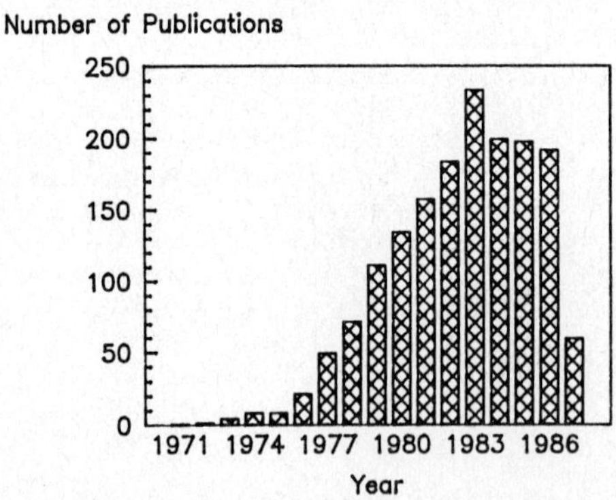

Fig. 1 Number of publications in photoacoustic or photothermal spectroscopy cited in Chemical Abstracts between 1971 and 1987

equivalent to one paper per working day. Further, these publications are not in obscure journals, but find their way into the mainstream of chemistry and physics periodicals. The evidence of the usefulness of the techniques is there, but is it being taken up by a broad range of users, and is it still dominated by developments in the methodology rather than applications?

Hard evidence is difficult to come by, but there are indications that photoacoustic and photothermal spectroscopy have still not made the transition from a field of experimental study to a routinely available tool. There is still only one manufacturer of dedicated UV/vis and NIR spectrometers[1], and the total size of the market at present is small considering the relatively low cost of the instrumentation. The bulk of photoacoustic cells are probably sold as FTIR accessories, and it is not clear that they are actually being used. There is anecdotal evidence that many purchasers of FTIR spectrometers, particularly in the industrial sector, purchase a full spread of detection and sampling systems at the same time as the spectrometer, but then only develop an expertise in a few methods. Photoacoustic detectors for FTIR are unforgiving of poor experimental technique, and require considerable care to obtain the best results from them. It is probable that many users of FTIR systems have conducted a few preliminary experiments with their new photoacoustic cell, and promptly concluded that the signal to noise ratio was unacceptable for their needs. This feeling that the signal to noise ratio of photoacoustic detectors is a problem can be found in the literature, as can be seen in the following quotation from a recent text book on Fourier transform infra-red spectrometry by GRIFFITHS and DE HASETH[2].

> "As a general rule, it can be stated that if the spectrum of a given sample can be measured using either photothermal spectrometry or an alternative "conventional" sampling technique, it is very likely that the spectra measured using the alternative technique will have a higher SNR."

Despite the above statement, these authors clearly have an ambivalent feeling towards photoacoustic and photothermal methods for FTIR detection. A sense that in specific applications, with careful preparation, the methods do offer distinct advantages to the user. After an analysis of the difficulties associated with these detection schemes, the authors conclude:

> "Nevertheless, we believe that no analytical laboratory in which a wide range of samples is to be characterised using a FT-IR spectrometer can afford to be without a photoacoustic accessory".

Apart from the limited number of systems sold as integrated photoacoustic spectrometers, and the photoacoustic accessories sold for FTIR spectrometers, all the other workers are constructing their own systems. This sort of approach is typically found in the very early stages of a technology, but after 15 years of development, it might be anticipated that more integrated systems would now be available, particularly considering the stress which has been laid by the practitioners on the suitability of these techniques for routine and continuous monitoring.

Clearly there are some difficulties in converting the enthusiasm and work which have been done on photoacoustic and photothermal spectroscopies into routine tools, but what are they? One difficulty is that the apparent advantages of thermal wave spectroscopies are not sufficient to justify the necessary investment. For example, in an industrial context it is necessary to justify the investment in both capital and human resource required to achieve the capability of making new kinds of measurements by the value of the information so obtained. In the case of thermal wave spectroscopies the investment is relatively high; not so much in the capital costs, but in the manpower costs of putting together a system to meet a specific need, and in developing the necessary expertise. This high level of investment is acceptable when the information obtained is of sufficient importance, but is thermal wave spectroscopy capable of giving unique information?

Thermal wave spectroscopy is usually represented as having particular advantages because of its capability of dealing with:

1. Samples of very low optical density.

2. Samples of very high optical density.

3. Light scattering samples.

4. Specularly reflecting samples.

These advantages may be considered as sampling advantages. The ability to work with materials in a wide variety of physical forms, and therefore to reduce sample handling and preparation. This is an undoubted gain, but suffers from the problem that it suggests that the same information can be obtained using existing instrumentation, with techniques the users are already familiar with, by appropriate sample preparation methods. There will therefore always be a strong pressure, both from managers and industrial spectroscopists, to continue with the existing methods. The advantages are real, and there are circumstances where they will be able to justify the necessary investment; for example the work of McCLELLAND in the infra-red spectroscopy of micro-samples[3], and of KIM and others in the photoacoustic spectroscopy of trace actinide solutions[4,5]. However, it seems unlikely that thermal wave spectroscopies will ever find routine applications if they are used to generate spectroscopic information which is essentially available from other sources.

An alternative approach is to look for applications of thermal wave spectroscopy where it yields unique information, unobtainable by other techniques, and where the value of the information justifies the expenditure. It is undoubtedly a unique capability of thermal wave spectroscopy to track the fate in both space and time of energy introduced into the sample by absorption. There are two active areas of research at the moment which exploit this capability. Both are spectroscopic, and provide information of a different kind to that obtained with other types of spectroscopy; these are studies of energy transport and transfer, and of chromophore distribution.

Within the area of energy transport and transfer, examples of significant applications of thermal wave spectroscopy can be found in the photo-synthesis work of LEBLANC *et al*[6,7], the photo-chemistry and photo-biology of BRASLAVSKY *et al*[8,9], the gas phase work of HESS, TAM and others[10-11], and the work on semi-conductors and photo-electrochemistry by MANDELIS and others[12-14]. There are major sessions at this conference, reported in this volume, on kinetics and relaxation, biological and medical applications, and semi-conductors, which all contain examples of the use of this unique property of thermal wave spectroscopies. There is a significant increase in activity in these areas over the previous meeting at Montreal[15], and there is clearly a growing interest in its applications.

The other area which appears to be successfully exploiting the unique characteristics of thermal wave spectroscopies is the study of chromophore distributions in condensed phases. The depth profiling capabilities of photoacoustic spectroscopy through the control of thermal diffusion length has been implicit from the earliest days of its renaissance, but has been relatively little exploited. More recently there has been considerable interest in the use of pulse and multi-frequency methods to obtain the thermal wave impulse response of the sample, from which various properties can be estimated. Examples of this are found in the work of TAM[16] and COUFAL[17,18] using pulsed lasers and piezo-electric or pyro-electric detectors. Spectroscopic applications are found in the work of IMHOF on opto-thermal transient emission radiometry (OTTER)[19,20], the studies by SUGITANI on correlation photoacoustic spectroscopy (CPAS)[21,22] using continuously modulated light sources, the extension of the CPAS technique by MANDELIS into the frequency modulated method of time delay spectrometry (TDS) [23-25], and the work of our own group on impulse response photoacoustic spectroscopy (IMPAS) and the study of systems where the distribution of the chromophore varies with time[26-30].

As an example of this type of work, there follows a brief description of the current state of our studies in time varying chromophore distributions.

This work has been based upon the proposition that since the photoacoustic impulse response of a sample gives information about the chromophore distribution, and the impulse response of a sample can be measured with good signal to noise ratio in 1-10 seconds, then it is feasible to study systems where the chromophore distribution is slowly varying with time. As an example, we have studied the diffusion of chromophore through a polymer film. The rear surface of a polymer film 80 microns thick is placed in contact with a reservoir containing a solution of a dye in an organic solvent. The dye gradually diffuses from the rear surface of the film to the front surface, and the photoacoustic impulse response is measured at regular intervals using the cross-correlation method with psuedo-random binary sequence modulation and gas microphone detection. Typical results of this experiment are shown in Fig. 2[29]. A set of 10 impulse responses measured at 3 minute intervals are overlayed, and it is clear that the strong signal from the chromophore gradually moves to shorter time delays, increases in intensity, and becomes narrower. These changes are all consistent with studies of static systems where the depth below the surface of a narrow band of chromophore is varied[26]. Relative changes in peak height and peak delay during the experiment are plotted in Fig. 3. The apparent straight line relationship in the variation of peak height with time appears to be preserved over a range of diffusion rates and time scales.

The information obtained from such experiments is useful, but the lack of a suitable theoretical model limits it to essentially comparative studies. An analytical solution for the relevant equations is difficult to obtain because of the continuously changing chromophore distributions, and the variable boundary conditions which must be imposed. Accordingly, we have used a finite difference simulation of both the chromophore

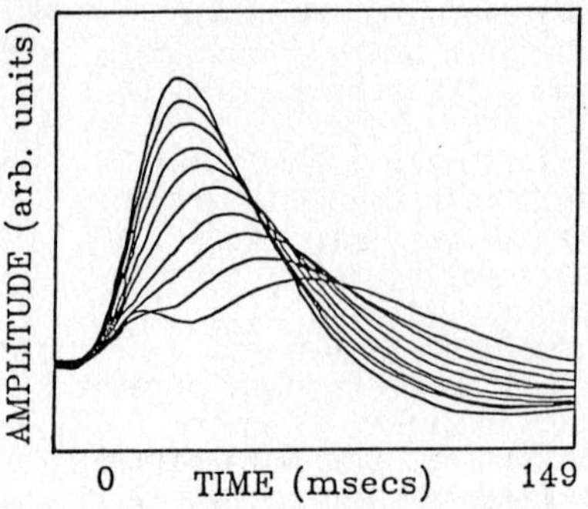

Fig. 2 Impulse response recorded at 3 minute intervals during diffusion of a dye through a polymer film

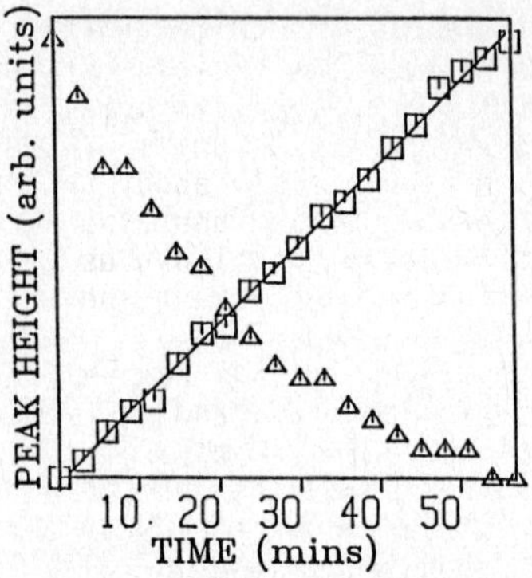

Fig. 3 Plot of peak height (□) and peak delay (Δ) against time for the data in Fig. 2

diffusion and the heat diffusion equations[30]. Similar formalisations have found considerable success in interpreting the data from other thermal wave experiments[31]. To test the suitability of the model we evaluated the predicted responses for two different hypotheses about a diffusing chromophore. In one case we assumed that the chromophore distribution was a step function boundary moving at a constant velocity through the film, and in the other we assumed that Fick's first law applied. These two hypotheses are physically completely different, and it should be possible to discriminate between them using the experimental data. Figures 4 and 5 show the calculated variations in peak height and peak amplitude with time for these two different hypotheses. It is immediately evident by comparison with Fig. 3 that the shape of the curves obtained assuming a moving boundary are completely different; whereas the curves obtained assuming Fickian diffusion have a very similar form to the experimental curves.

This demonstrates that the results obtained from the simulation and from the experiment are physically reasonable, but there are still limitations.

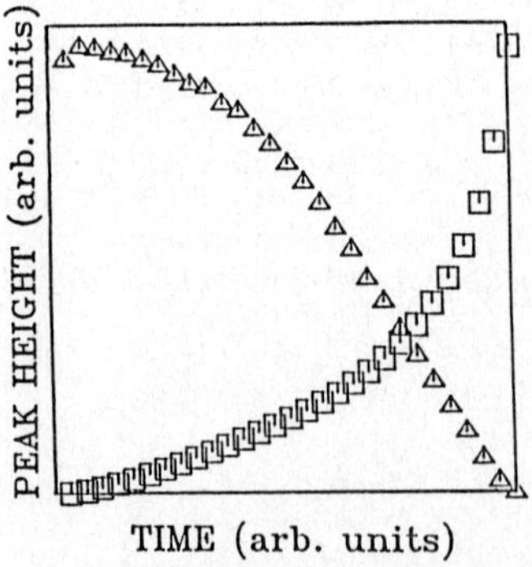

Fig. 4 Simulated peak height (□) and peak delay (Δ) for dye diffusion assuming a moving boundary

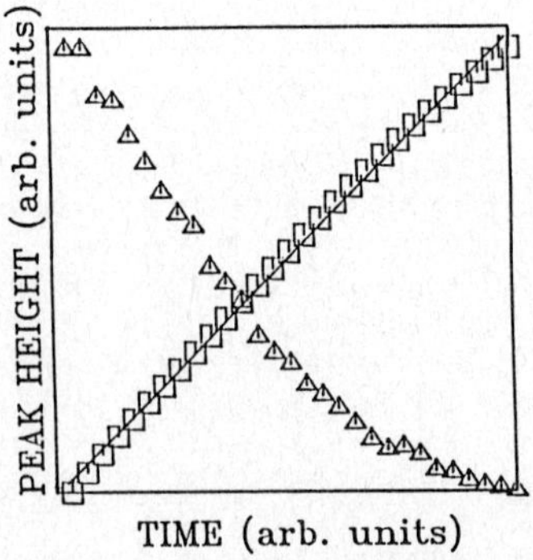

Fig. 5 Simulated peak height (□) and peak delay (Δ) for dye diffusion assuming Fick's Law

It is not possible to tell from the experimental data, or from the simulations, what the optical absorption coefficient of the chromophore is, nor to provide exact information on the spatial distribution of the chromophore in absolute units, unless a considerable amount of information is available about both chromophore and diffusing medium. However, the simulation does show good agreement with the observed experimental data, and allows us to place certain upper and lower limits on the diffusion coefficient and the optical extinction coefficient.

Work is continuing on the analysis and physical interpretation of data from diffusing systems, and on extending the method to the study of more complex biological systems[32]. The information obtained using the thermal wave impulse response method is unique in that it can be gained non-destructively from intact systems. Interpretation of the data is difficult, but in combination with information from other techniques can provide a valuable insight into dynamic systems. As more groups become involved in this type of study, we can expect to see the technique applied to an increasing range of problems, and a growing understanding of the relationship between the subtle changes in the thermal wave response of the sample, and the underlying dynamics of the system.

Thermal wave spectroscopy is unlikely to enjoy much success if it presents itself as an alternative in competition with conventional spectroscopies. However, as a unique source of information which can be used in combination with other experiments to characterise complex physico-chemical systems, it will flourish.

References

1. Model OAS 400, EDT Research Ltd, London, UK

2. P.R. Griffiths and J.A. de Haseth, Fourier Transform Infra-Red Spectrometry. (Wiley, New York, 1986), Chap. 9.

3. J.F. McClelland, L.M. Seaverson and S.M. Kauslink, Paper WA13 This Symposium, also Submitted for Publication.

4. R. Stumpe, J.I. Kim, W. Schrepp and H. Walther, Appl. Phys, B34, 203 (1984).

5. G. Buckau, R. Stumpe and J.I. Kim, J. Less Common Met., 122, 565 (1986).

6. R. Carpentier, H.C.P. Matthijs, R.M. Leblanc and G. Hind, Can. J, Phys., 64, 1136 (1986).

7. R. Popovic, M. Beauregard and R.M. Leblanc, Biochem. Biophys. Res. Commun, 144, 198 (1987).

8. K. Heihoff and S.E. Braslavsky, Chem. Phys. Lett., 131, 183 (1986).

9. S.E. Braslavsky, Photochem. Photobiol., 43, 667 (1986).

10. H. Sontag, A.C. Tam and P. Hess, J. Chem. Phys., 86, 3950 (1987).

11. D. Kumer, L. Klasinc, P.C. Clancy, R.V. Nauman and S.P. McGlynn, Can. J. Phys, 64, 1107 (1986).

12. R.E. Wagner, V.K.T. Wong, A. Mandelis, Analyst (London), 111, 299 (1986).

13. A. Mandelis, E.K.M. Siu, Phys. Rev. B : Condens. Matter, 34, 7209 (1986).

14. A.F.S. Penna, J. Shah, A.E. Di Giovanni, A.Y. Cho and A.C. Gossard, Appl. Phys. Lett, 47, 591 (1985).

15. Can. J. Phys, 64 (9), 1021-1344 (1986).

16. A.C. Tam and H. Sontag, Appl. Phys. Lett, 49 1761 (1986).

17. H. Coufal, IEEE Trans on UFFC, 33, 507 (1986).

18. H. Coufal and P Hefferle, Appl. Phy. A, 38, 213 (1985).

19. R.E. Imhof, D.J.S. Birch, F.R. Thornley, J.R. Gilchrist and T.A. Strivens, J. Phys D : Appl. Phys, 18, L103, (1985).

20. R.E. Imhof, D.J.S. Birch, F.R. Thornley and J.R. Gilchrist, Anal. Proc, 24, 17 (1987).

21. A. Uejimi, Y. Sugitani and N. Kozo, Anal. Sci, 1, 5 (1985).

22. M. Habiro and Y Sugitani, Anal. Sci, 3 3 (1987).

23. A. Mandelis, Rev. Sci. Instrumen, 57, 617 (1986).

24. A. Mandelis, L.M.L. Borm and J. Tiessinga, Rev. Sci. Instrumen, 57, 622 (1986).

25. A. Mandelis, L.M.L. Borm and J. Tiessinga, Rev.Sci. Instrumen, 57, 630 (1986).

26. G.F. Kirkbright and R.M. Miller, Anal. Chem, 55, 502 (1983).

27. G.F. Kirkbright, R.M. Miller, D.E.M. Spillane and Y. Sugitani, Anal. Chem, 56, 2043 (1984).

28. G.F. Kirkbright, R.M. Miller, D.E.M. Spillane and I.P. Vickery, Analyst (London), 109, 1443 (1984).

29. R.M. Miller, G.R. Surtees and C.T. Tye, Analyst (London), In Press.

30. R.M. Miller, Spectrochmica Acta B, In Press.

31. D. Levesque, G. Rousset and L. Bertrand, Can. J. Phys, 64, 1030 (1986).

32. D.A. Adesida, R.M. Miller and M.L. Waller, Submitted for publication.

Fourier Transform Photoacoustic Spectroscopy of Solids

B.S.H. Royce and J. Alexander

Applied Physics and Materials Laboratory, Princeton University, Princeton, NJ 08544, USA

1. Introduction

Fourier transform photoacoustic spectroscopy is now a well established technique for measuring the optical properties of materials in the visible and infra-red spectral regions [1,2]. Both commercial constant scan-rate spectrometers and custom step-and-integrate spectrometers have been used to provide the modulated radiation required for the photoacoustic method. The measurements have employed gas-microphone, photothermal deflection, and piezoelectric detection of the PAS signal, and have been made in both the absorption and emission modes. This paper discusses the generation of the complex photoacoustic signal using a constant scan-rate spectrometer, and its relationship to the optical properties of the sample, the transfer function of the spectrometer, and the characteristics of the photoacoustic cell.

2. Signal Generation

The central component of an FTIR-PAS spectrometer is a Michelson interferometer. The optical configuration of an ideal interferometer is shown in Fig.1, and indicates that the interferometer is symmetric with respect to the input and output ports. Radiation of wavenumber, w, entering the interferometer is sent along two different optical paths and then recombined at the beam splitter to generate an exit beam. This beam has an intensity that depends upon the spectral content of the source and the optical path difference between the two arms of the interferometer. At the mirror position corresponding to zero path difference (ZPD), the transmitted intensity will be a maximum. Displacement of one of the

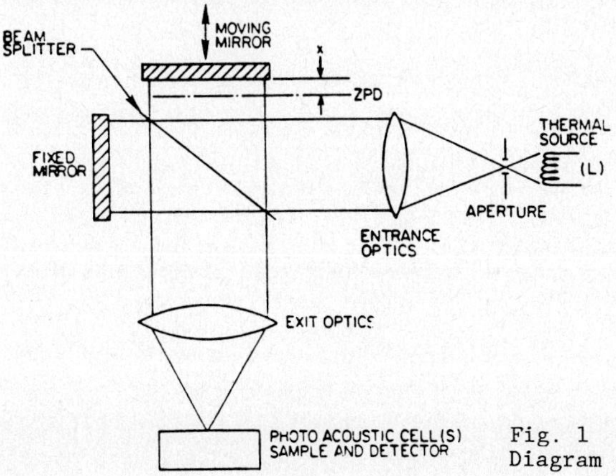

Fig. 1
Diagram of the FTIR-PAS configuration

mirrors by an amount, x, with respect to this ZPD location will introduce a retardation, $2\pi wx$, between the two beams, and the transmitted intensity from a source with a spectral power, $B(w)dw$, in the wavenumber interval between w and $(w + dw)$ has the form (1)

$$I(x) = \int_{-\infty}^{\infty} B(w)\, dw + \int_{-\infty}^{\infty} B(w) \exp(2\pi iwx)\, dw, \quad (1)$$

where $I(x)$ contains a contribution from each spectral element of the source, and it has been assumed that $B(w)$ is a symmetric function (although negative values of w have no physical meaning). The first term in (1) is independent of the mirror location, and has a value equal to half the transmitted intensity, $I(0)$, at the ZPD. The second term is the interferogram that is used in conjunction with the photoacoustic detection method to obtain information about the sample. For this ideal case, the interferogram is perfectly symmetric about $x = 0$.

In a real instrument using photoacoustic detection, wavenumber dependent phase delays will occur, and an asymmetric interferogram will be obtained. The phase delays arise from such sources as systematic sampling errors in digitizing the interferogram, dispersion in the optical components of the interferometer, unbalanced electronic delays in the sampling electronics, and signal delays from the photoacoustic process. In general, all of these terms may be written in the form (2)

$$P(w) = A(w) \exp(i\theta w), \quad (2)$$

and the interferogram can then be expressed in the complex form (3):

$$I(x) = \int_{-\infty}^{\infty} A(w)\, B(w) \exp(i\theta w) \exp(2\pi iwx) dw. \quad (3)$$

The real interferogram is thus the inverse Fourier transform of a complex "spectral function," and the Fourier transform of the interferogram will contain lumped amplitude and phase information from each of the signal handling components.

The Bomem DA3 FTIR spectrometer used in our laboratory employs the Forman method for evaluating the phase contribution to the overall signal. A double-sided interferogram is collected at circa 8 cm^{-1} spectral resolution, and the phase spectrum, $\theta(w) = \arctan\{Im(w)/Re(w)\}$, is computed from the complex Fourier transform of this apodized, double-sided interferogram. This information is used to compute the impulse response function, $f(x)$, of a filter function, $F(w)$, via an inverse Fourier transform of the filter (4):

$$f(x) = \int_{-\infty}^{\infty} \exp(-i\theta(w)) \exp(2\pi iwx)\, dw \quad . \quad (4)$$

Apodization and spectral range information may also be included in this impulse response function. By convolving the asymmetric interferogram, $I(x)$, with the impulse function, $f(x)$, in the interferogram domain, a symmetric interferogram is obtained. The real (cosine) Fourier transformation of this apodized, phase-corrected interferogram then yields the product, $A(w)B(w)$, from which the amplitude response of the photoacoustic sample may be determined, if the optical throughput of the spectrometer and non-sample related phase contributions are measured independently.

In order to obtain accurate PA spectral data, it is necessary to correct the sample and reference PAS signals with phase information collected under conditions identical to those employed for collecting the spectral data. The volume of a typical gas-microphone PA cell is small, and the signal is sensitive to changes in sample volume. The stored phase data used to compute the impulse function for the sample should, therefore, be collected on the sample itself using the same mirror speed and cell gas as for the data acquisition. Similarly, the reference sample, which is frequently a black absorber such as graphite powder, should be spectrally corrected employing an impulse function measured with this reference sample prior to the collection of the reference spectrum under the same conditions of mirror velocity and cell gas as those used for the sample phase and amplitude measurements. These precautions are particularly important for the case of powder samples, where the porosity of the powder strongly influences contributions to the complex PA signal arising from the thermal expansion of the inter-particle gas, and may influence the acoustic resonance conditions of the PA cell.

The response of the black graphite powder is a reasonable approximation to that of a saturated PA sample over the 4000 to 400 cm^{-1} spectral range, making it a repeatable reference sample that does not contain residual hydrocarbon absorption peaks. Taking the ratio of the sample amplitude spectrum to the amplitude spectrum for this reference sample will yield PAS data corrected for the spectral characteristics of the spectrometer system. Since the various phase contributions to $\theta(w)$ are additive, subtraction of the reference sample phase from that of the sample under study will provide an indication of the PA phase contributions of the sample corrected for spectrometer effects.

2.1 Wavenumber Dependence of Photoacoustic Amplitude and Phase

Because the PA signal is dependent upon both the optical absorption, $\beta(w)$, and the thermal diffusion length of the sample and the cell gas, it is necessary to take the wavenumber dependence of these quantities into account when interpreting the PA spectral data. With a constant scan-rate spectrometer, the modulation frequency, f, at any wavenumber is mirror velocity, V, dependent such that: $f = 2Vw$. At a particular mirror velocity the modulation frequency varies linearly across the spectral region, and consequently, the thermal diffusion length (which depends upon the inverse square root of the modulation frequency) will also vary across the spectral range. The effect of this on the PA amplitude and phase is most easily seen for a thermally thick, homogeneous sample with a negligible thermal expansion. For such a sample the PA amplitude is given by (5)

$$q(w) = (pI(w)\mu_s\mu_g/4TL)\{\beta\mu_s/[(\beta\mu_s)^2 + 2(\beta\mu_s) + 2]^{1/2}\}, \tag{5}$$

where p, and T are the gas temperature and pressure, respectively, L is the length of the gas above the sample, I(w) is the wavenumber dependent light energy incident on the sample, and μ_i is the thermal diffusion length in the solid, i=s, or cell gas, i=g. I(w) depends on the wavenumber of the radiation through the photon energy, which is linear in w. Substituting for the wavenumber dependent modulation frequency in terms of the mirror velocity, and for I(w), (5) becomes

$$q(w)/n(w) = (D/V)\{1 + (2/\beta)[2\pi\rho cVw/K]^{1/2} + (2/\beta^2)[2\pi\rho cVw/K]\}^{-1/2}, \tag{6}$$

where D is a constant, and n(w) is the number of photons exiting the interferometer with a wavenumber, w, ρ is the sample density, K the sample thermal conductivity, and c its heat capacity. The first term in (6) is independent of wavenumber, but inversely proportional to the interferometer

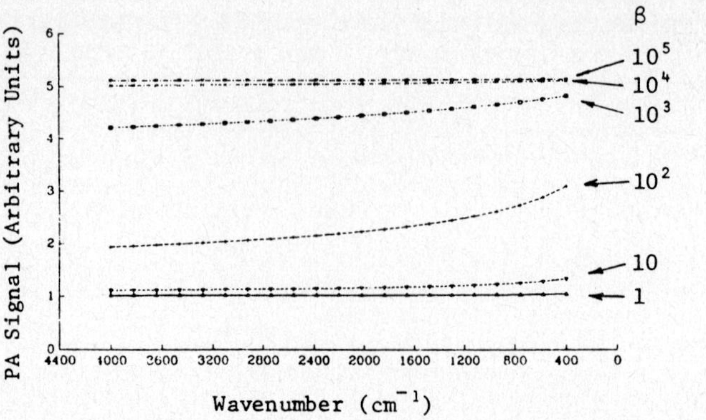

Fig. 2 Wavenumber dependence of the PA signal per photon for a grey body

mirror velocity. The wavenumber dependence of the amplitude response at constant mirror velocity is contained in the dimensionless second term, directly through $\beta(w)$ and indirectly through the modulation frequency. As $\beta(w)$ becomes large and the sample response goes to PA saturation, the second term tends to unity, and there is no wavenumber dependence of the PA amplitude except that due to the number of incident photons. The response of a grey body, for which β is independent of w, is shown in Fig. 2 for several values of β. In the range $1 < \beta < 1000$ cm^{-1}, the wavenumber dependence of the PA amplitude per photon causes the signal at constant (β, V) to vary by less than a factor of two across the 4000 to 400 cm^{-1} spectral region.

The PA phase for this sample type also has a simple form that is wavenumber dependent as shown in (7).

$$\theta(w) = -\pi/4 - \arctan(1 + 2/\beta\mu_s) = \arctan(\beta G/(Vw)^{1/2} + 1), \quad (7)$$

where G is a constant. For a sample in PA saturation, the PA phase is constant, whereas for a grey body, the phase will vary with wavenumber at a constant mirror velocity and with mirror velocity at a constant wavenumber.

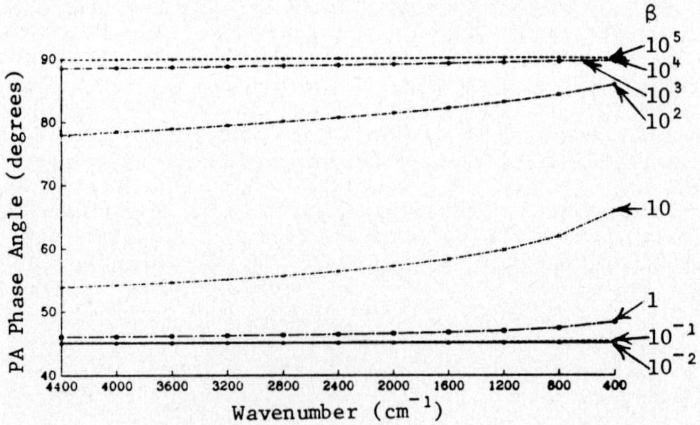

Fig. 3 Wavenumber dependence of the PA phase for grey bodies with different values of β

This again points to the importance of making phase collection and spectral data collection measurements under identical photoacoustic conditions. Fig. 3 shows the wavenumber dependence of the PA phase for a series of grey bodies with absorption coefficients between 10^{-2} and 10^{+5} cm^{-1}. As expected, the phase angle changes between approximately 45 and 90 degrees for this range of absorption coefficients. A wavenumber dependence of the phase is also seen for intermediate values of the absorption coefficient, and for β=1, this contributes approximately a three-degree phase shift across the spectral range. For very low or high values of the absorption coefficient, the phase is "saturated," and the grey body response is wavenumber independent.

TENG and ROYCE [3] have suggested a method for correcting the PA spectral response of thermally thick, homogeneous samples for the wavenumber dependence of the signal amplitude. This technique requires the collection of sample and reference data at two different spectrometer mirror velocities. The method can be used with several types of commercial FTIR spectrometer equipped with scan mirror velocity control.

2.2 Signal Generation with Powder Samples

When powder samples are being studied, the assumption that the thermal expansion can be neglected may no longer be valid. McGOVERN et al. [4] and MONCHALIN et al. [5] have extended the composite piston model of McDONALD and WETSELL [6] to include the expansion of the inter-particle gas in the powder sample. McGOVERN et al. point out that the inter-particle gas has a much larger thermal expansion coefficient, and a lower thermal conductivity, than the solid component of the composite system. This will influence the effective thermal diffusion length in the composite sample, altering the thermal diffusion contribution to the PA signal. In addition, heat transferred to this gas within an optical absorption length of the illuminated powder surface will cause a pressure change that will be transmitted to the gas above the composite sample at the velocity of sound. For a high porosity powder, this pressure wave suffers little attenuation and will add to the pressure change produced by the periodic heat transfer from a thermal diffusion depth below the powder surface.

The pressure and thermal terms of this composite material model are characterized by different length scales. The thermal term is associated with the thermal diffusion length in the composite sample, whereas the pressure term is associated with the optical absorption length of the sample. For a low optical absorption coefficient, the deposition length is large, and the pressure term may dominate the total signal. At high optical absorption coefficients, the main contribution to the PAS signal comes from the thermal term. The phase contributions made by these two modes are also different, with the pressure term exhibiting an essentially zero phase delay over the complete range of optical absorption coefficients. Figure 4 shows the computed amplitude and phase responses as a function of absorption coefficient for a powder-gas sample with the thermal properties of silica and nitrogen. It can be seen that the pressure term increases the PA sensitivity at low absorption coefficients by about an order of magnitude and, because of its essentially instantaneous response time, significantly modifies the PA phase. The phase is no longer a single valued function of the absorption coefficient. For the example shown, it is seen that the phase of the PA signal has a minimum value at an absorption coefficient of about 1000 cm^{-1}. This will influence the phase structure associated with strong absorption peaks in the IR, and the phase reversal may give rise to a local phase minimum at an

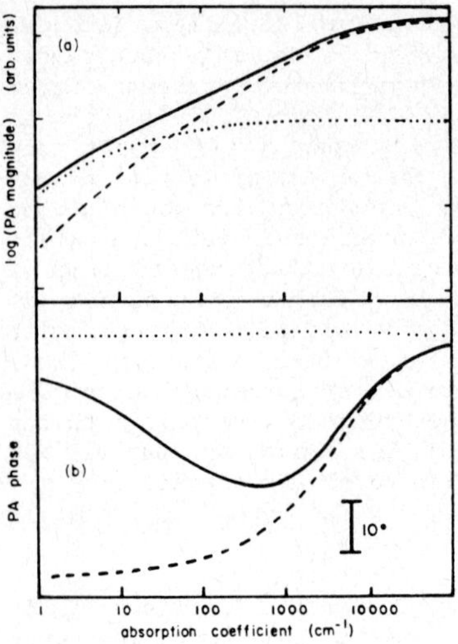

Fig. 4 PA amplitude (a) and phase (b) for a powder sample as a function of $\beta(w)$. Total signal (solid line), pressure contribution (dotted line), thermal contribution (dashed line). From [4]

absorption peak rather than a phase maximum. This behavior again points to the necessity of collecting the FTIR-PA phase on the sample, or on the reference material used to determine the spectrometer throughput.

The presence of this additional contribution to the PA signal complicates the quantitative evaluation of peak heights across the spectral range. Further difficulties will be encountered if photo- or thermal-desorption processes increase the number of gas phase molecules in the cell and add further terms to both the amplitude and phase of the total signal, as has been discussed by KORPIUN et al. [7]. Under these conditions, PAS will give the correct wavenumber location of spectral features, but care must be taken in interpreting peak heights in terms of the number of chromophores present in the sample.

2.3 Emission Photoacoustic Spectroscopy

Figure 1 indicates that an FTIR spectrometer has optical symmetry with respect to the light source and the sample. As was pointed out by TANNER and McCALL [8], for a given location of the moving mirror, radiation emitted by the sample at a given wavenumber experiences the same retardation by the interferometer as radiation emitted by the light source. Energy conservation considerations show that, for an ideal interferometer, radiation from the source or sample that does not leave the interferometer by the opposite arm is returned to its point of origin. The interferometer is, therefore, a device that switches energy between its entrance and exit ports. It either permits energy at a given wavenumber to pass through and exit at the opposite arm or returns some percentage of it to the sample and to the source for the same mirror position.

In PA measurements, the sample is also the radiation detector, and the temperature change that gives rise to the PA signal depends upon the net radiation gain or loss by the sample. The total signal is, therefore, a measure of net effect of the temperature increase that results from the absorption of radiation from the light source, and the temperature decrease that results from the emission of radiation by the sample. These two terms in the overall signal have phases that differ by 180 degrees. With a sample at room temperature, emission effects are normally unimportant over the spectral range between 4000 and 400 cm^{-1}. Raising the sample temperature above 100°C progressively increases the emission contribution to the PA signal, modifying the apparent absorption. This occurs first at the low wavenumber end of the spectrum for a low temperature sample, and then at higher wavenumbers as the sample temperature is increased. McGOVERN et al. [9] have shown that PA emission spectra may be collected from a sample in a gas microphone cell by turning off the light source and causing the radiation from the sample to exit the interferometer into a water-cooled absorbing enclosure. These effects are illustrated in Fig. 5. LOW [10] has also reported the observation of PA emission effects, with these measurements being made using a photothermal beam deflection (PDS) technique in conjunction with a liquid nitrogen temperature receiver at the opposite end of the interferometer. The PDS method is not sensitive to the thermal expansion of the inter-particle gas in a powder sample, and therefore only measures the thermal contribution to the PA signal.

3. Experimental

Measurements have been made using a Bomem DA3 FTIR spectrometer in conjunction with a custom photoacoustic cell having its first resonance frequency significantly above the modulation frequency range generated in the FTIR. For illustrative purposes, the reference sample was a graphite powder and the sample a silicalite powder. Nitrogen gas was employed to purge the cell and act as the transducer fluid. The cell employed a B&K electret microphone with a sensitivity of 50mV/Pa that was directly coupled to a preamplifier with a gain of circa 20. The output signal from the cell was further amplified by a PAR 113 low noise-variable gain amplifier before being returned to the spectrometer for signal processing and Fourier transformation. The amplifier gains were kept constant over the complete retardation range employed in the measurements. The DA3 permits spectral data to be acquired at a resolution of about 8cm^{-1}, either using a double-sided interferogram (the PHASE mode) or a phase corrected single-sided interferogram (the SCAN mode). Measurements were made on both the sample and the reference materials under identical conditions of mirror velocity, resolution, and electronic gain, employing both of these techniques.

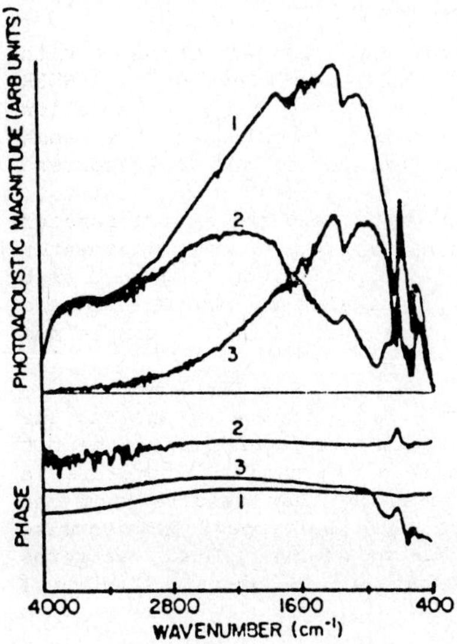

Fig. 5 Amplitude and phase effects due to the emission of radiation by a hot PA sample in a gas-microphone cell. From [9]. 1. Room Temperature Absorption Spectrum, 2. Absorption/Emission Spectrum at 350 C, 3. Emission Spectrum at 350°C

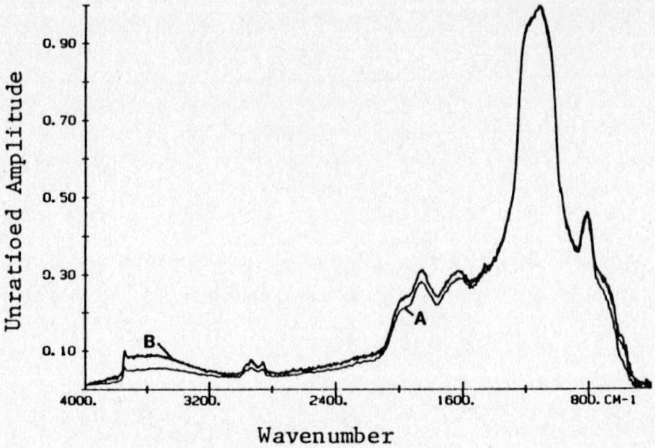

Fig. 6 Unnormalized spectra for a silicon dioxide powder sample: A. Double-sided interferogram; B. Phase corrected single-sided interferogram

Figure 6 shows the unratioed spectra of the silicon dioxide sample obtained using these two methods. It is seen that if the two spectra are normalized to agree at the peak absorption due to the Si-O lattice vibrations, significant peak height differences exist in the overtone and surface hydroxyl regions of the spectrum. Presumably, these differences arise from the algorithms used in the two different Fourier transformations, a complex Fourier transformation for the double-sided interferogram, and a real Fourier transformation for the single-sided interferogram that was phase corrected using a phase file taken with the sample itself.

Measurements made on the graphite reference sample under essentially identical conditions indicate a similar discrepancy between the unnormalized spectral response curves obtained by these two methods. In order to correct the PA-amplitude signal for the spectral throughput of the spectrometer, it is necessary to divide the sample spectrum by the reference spectrum. When this is done it is seen, Fig. 7, that the two

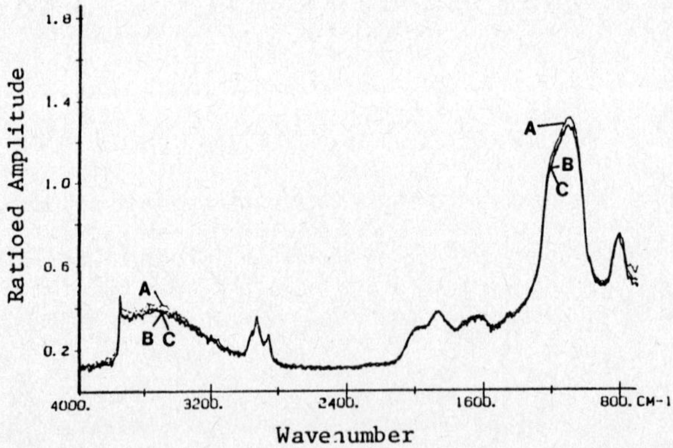

Fig. 7 (Silica Amplitude)/(Reference Amplitude) for: A. Double-sided interferograms; B. Phase corrected single-sided interferograms; C. Blackbody phase corrected single-sided interferograms.

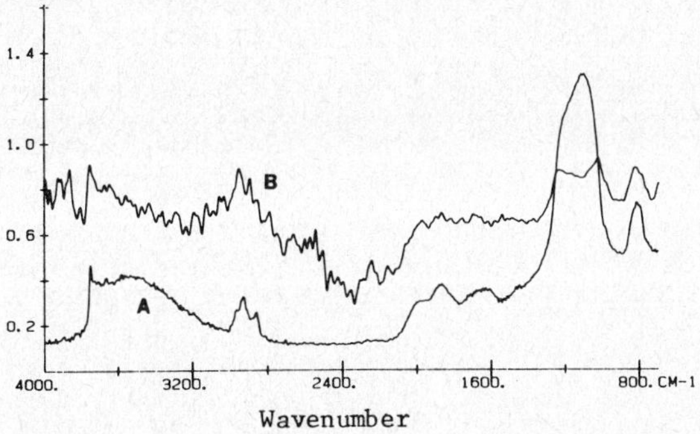

Fig. 8 Amplitude and phase spectra for a silicon dioxide powder sample. A. Ratioed amplitude spectrum; B. PA-Phase spectrum

methods, applied self-consistently, yield spectra that are essentially the same within experimental accuracy. Spectrum C in this figure shows the effect of using a black body phase to correct both the sample and the reference spectra obtained using single-sided interferograms. This also agrees with the other spectra within experimental error.

Subtraction of the phase files for the sample and the reference material should yield the photoacoustic phase response of the sample. This data is shown in Fig. 8 superimposed on the ratioed amplitude spectrum. Both the phase and the amplitude spectrum must be interpreted using the information shown in Fig. 4. In the 4000 to 2400 cm^{-1} range, the pressure contribution to the amplitude signal increases the magnitude of the peaks above the value expected from the thermal term alone. In the region of the lattice modes, the thermal contribution to the signal tends to dominate the spectral amplitudes. The phase signal shows a peak structure that follows that of the amplitude spectrum and, as shown, the phase peaks have the same sign as the amplitude peaks. Phase reversal due to the minimum in the phase angle dependence on the absorption coefficient occurs in the region of the peak of the Si-O vibrational absorption. The diagram indicates the complexity of the phase response of powder systems and the significant modification of peak amplitudes that occurs with these materials as the balance between the pressure and thermal contributions to the total signal changes with wavenumber.

4. Conclusions

This paper has discussed some of the parameters that are important in interpreting FTIR-Photoacoustic data. The amplitude and phase of the signal from a thermally-thick homogeneous sample have been shown to be wavenumber dependent, and for a grey body this has a simple algebraic form. Additional phase and amplitude contributions arise from the thermal expansion of the inter-particle gas if a powder sample is employed, and these make a quantitative evaluation of the PA-spectrum difficult. Measurements using both double-sided and phase corrected single-sided interferograms have shown that these two techniques are equivalent, provided data is consistently acquired in one mode or the other. These experiments also indicated that correction of the single-sided interferogram by a black body phase file, rather than a phase file collected on the sample, produced spectral differences that were close to the uncertainty in the data. Sample emission effects further complicate both the amplitude and phase response of samples at high temperature. However, PA-emission measurements are of utility in providing spectral information about hot samples in the low wavenumber region.

5. Acknowledgements

The authors thank Professor H. Lam for useful discussions. Grants from the State of New Jersey and RCA are gratefully acknowledged.

References

1. B.S.H. Royce and J.B. Benziger: IEEE Transactions UFFC-33, 561 (1986)
2. B.S.H. Royce: In Photoacoustic and Thermal Wave Phenomena in Semiconductors, ed. by A. Mandelis, (Elsevier, New York, 1987, in press)
3. Y.C. Teng and B.S.H. Royce: Applied Optics 21, 77 (1982)
4. S.J McGovern, B.S.H. Royce and J.B. Benziger: J. Appl. Phys. 57, 1710 (1985)
5. J-P Monchalin, L. Bertrand, G. Rousset and F. Lepoutre: J. Appl. Phys. 56, 190 (1984)
6. A. McDonald and G. Wetsell: J. Appl. Phys. 47, 64 (1978)
7. P. Korpiun, W. Herrmann, A. Kindermann, M. Rothmeyer and B. Buchner: Canadian J. Phys. 64, 1042 (1986)
8. D.B. Tanner and R.P. McCall: Applied Optics 23, 2363 (1984)
9. S.J. McGovern, B.S.H. Royce and J.B. Benziger: Applied Optics 24, 1512 (1985)
10. M.J.D. Low: Spectroscopy Letters 17, 279 (1984)

Photoacoustic X-Ray Absorption Spectroscopy

T. Masujima

Institute of Pharmaceutical Sciences, Hiroshima University, School of Medicine, Kasumi 1-2-3, Hiroshima 734, Japan

X-Ray Photoacoustic Effect

When materials are irradiated by X-rays, several effects occur, e.g. absorption, scattering, fluorescence, and electron release, as seen in Fig.1. Heat generation should also be observed on absorption of electromagnetic waves even for the X-ray absorption process. However, it has not been reported that transient heat generation by X-ray absorption has been detected, and almost no attention has been paid in a scientific sense to the heat produced by X-ray irradiation.

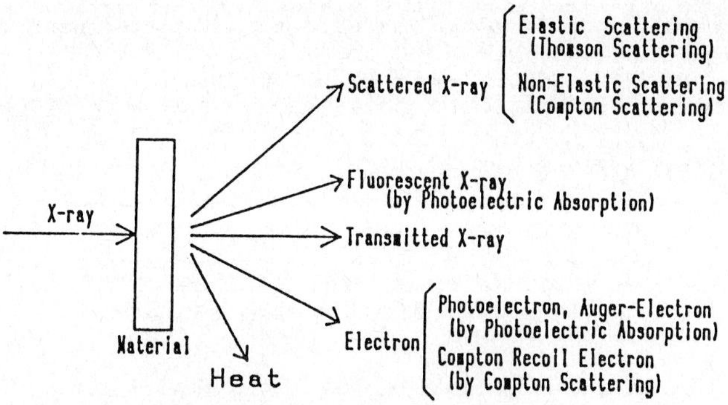

Fig. 1 X-ray absorption and scattering

Since the photoacoustic method, especially microphonic detection, is one of the most sensitive methods of detecting the heat generation from the solid surface, a study of the photoacoustic effect of solid materials on X-ray irradiation was started by this method, using a synchrotron orbital X-ray source. The photoacoustic effect on X-ray irradiation was very hard to detect initially, however, after several improvements of the cell[1,2], it was finally found that transient heat is generated in the materials by X-ray absorption. Figure 2 shows a cross-sectional view of the cell for solid materials[1]. The components and design were more or less the same as our conventional cell for UV and IR light, except the windows are of beryllium. However, electrical noise should be carefully avoided in order to get a high S/N ratio. The microphone was an electret type that is commercially available. The electronics and the general concept of this type of cell are detailed elsewhere [3]. The studies were performed using the synchrotron storage ring at the Photon Factory (PF) of the National Laboratory for High Energy Physics (KEK), Japan.

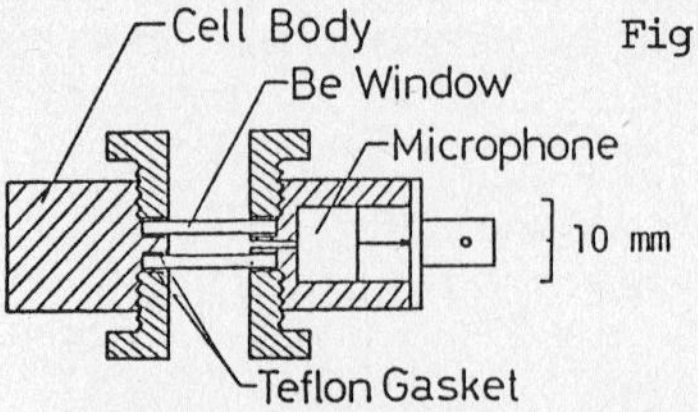

Fig. 2 Cross-sectional view of PAXAS cell

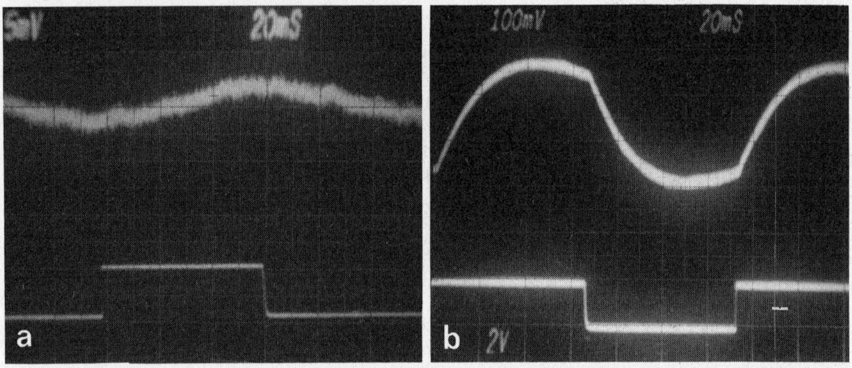

Fig. 3 X-ray photoacoustic signals of a) glass plate (8.1 Hz, ring current (R.C.) 91.1 mA) and b) Ar gas (7.5 Hz, R.C. 197.0 mA at cell length 10 cm).

Figure 3a shows the signal for solid materials (glass plate 100 µm thick) detected with 1.56 Å monochromatic X-rays. The triangular shape of the waveform shows an integrated temperature rise of the solid material. The signal intensity for the lead sample (50 µm) was proportional to the storage ring current for white X-rays as seen in Fig. 4, and also to the ion chamber current for monochromatic light, both of which are the measure of the photon flux. This shows that this method can be applied to dosimetry of X-rays, and further studies are in progress[4]. The signal amplitude decreased when the metal sample thickness was large. This result seems to be due to the higher heat capacity of the thicker samples. Metal samples have a high heat diffusivity, and thus the signal intensity did not always depend just on the linear absorption coefficient but also on the heat capacity and heat diffusivity of the materials. Figure 3b shows the signal for argon gas[5]. It differs from that of the solids, the shorter rise

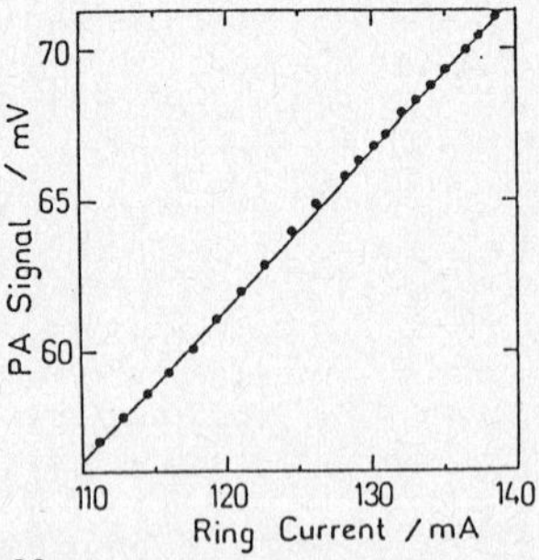

Fig. 4 Dependence of photoacoustic signal on ring current.

time and the following saturation in the waveform reflect a large heat expansion coefficient and low heat capacity. Precise computer simulation is in progress.

Photoacoustic X-Ray Absorption Spectroscopy(PAXAS)

Initially, the heat production by X-ray absorption had been thought to be a secondary or multi-process result and so not to contain any spectroscopic information.

Figure 5 shows an experimental setup for spectroscopic study. Since the intensity of X-rays from a synchrotron radiation source decreases with time, the ion chamber (5 cm) was set upstream in the X-ray line in order to monitor the X-ray dose and to correct the photoacoustic signal amplitude. As seen in Fig. 6, the sudden increase in the photoacoustic signal at 1.38 Å corresponds to the K-edge absorption of copper[6]. This shows that PAXAS spectra primarily show the variation of the X-ray absorption coefficient and reflect the heat generation on the release of a K-shell electron and its following process. The increase in the high energy region seems to show that the total heat energy also depends on the absolute value of the X-ray energy.

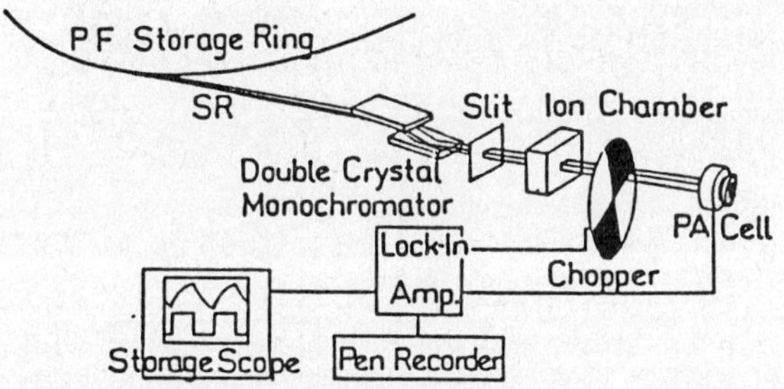

Fig. 5 Schematic diagram of the experimental set-up for PAXAS.

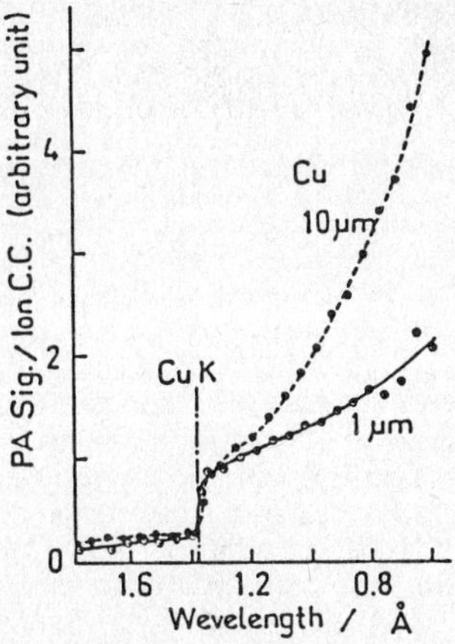

Fig. 6 PAXAS ratio spectra of copper (1 and 10 μm). ▶

Photoacoustic EXAFS

The fine structure near the edge absorption region is known as EXAFS (Extended X-ray Absorption Fine Structure) which gives us information on the number of surrounding atoms and the distance between the central and surrounding atoms.

Figure 7 shows the spectrum of copper foil (5 μm) in comparison with the absorption spectrum detected by ion chambers which were set at both sides of the photoacoustic cell. The corresponding feature, which was detected simultaneously, confirms that the EXAFS effect is also included in the PAXAS spectra[7]. The difference betwen these two spectra was the monotonically increasing trend of the PAXAS spectrum at high photon energy, which is due to the energy dependence of heat generation as noted previously. We can conclude now that the X-ray photoacoustic effect reflects the absorption coefficient of the materials and the absorbed X-ray energy; the intensity of the signal also depends sensitively on the thermal property

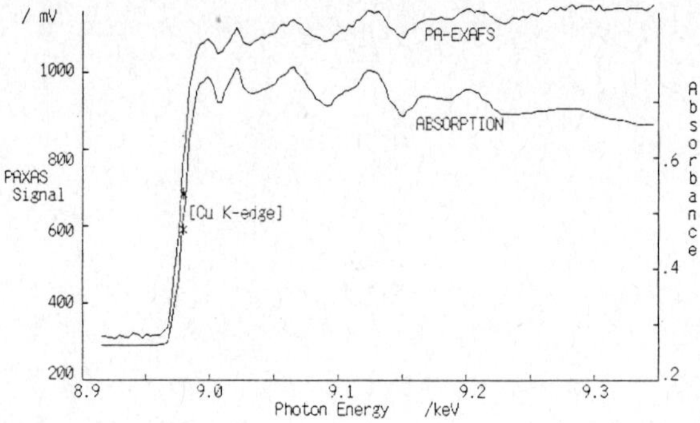

Fig. 7 Photoacoustic EXAFS spectrum for copper foil (5 μm) in comparison with its absorption spectrum.

of the sample and is the result of heat propagation in the materials. This conclusion is the same in principle as for the photoacoustic effect at UV, visible and IR wavelengths, while the mechanism of heat production is different.

The heat generation process seems to include many phenomena in the case of X-ray absorption. The final process should be lattice vibration; however, the primary process should be an energy-dissipating interaction between X-rays and electrons or nuclei. Non-elastic scattering process, ionization of atoms, and the photoelectron process should be studied precisely from this viewpoint.

Applications

Several applications were studied. Using a focused beam or using a pencil beam coupled with computer tomography(CT), imaging of an artificial sample was performed. Figure 8a shows the PAXAS image of a model sample which was made with several metal foils attached to the paper as seen in Fig. 8b[8]. Focused monochromatic X-rays (1.48 Å at PF beam line

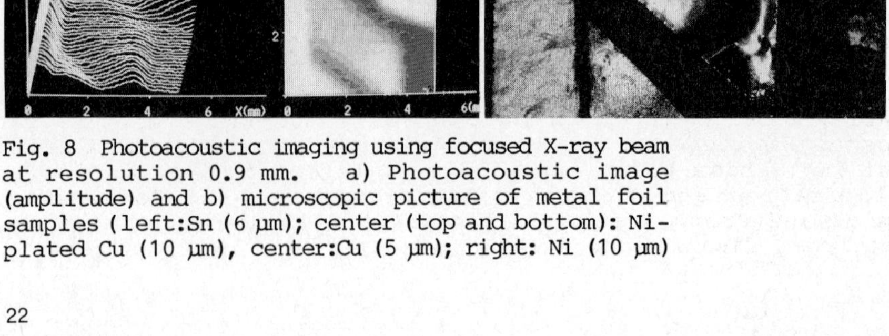

Fig. 8 Photoacoustic imaging using focused X-ray beam at resolution 0.9 mm. a) Photoacoustic image (amplitude) and b) microscopic picture of metal foil samples (left:Sn (6 μm); center (top and bottom): Ni-plated Cu (10 μm), center:Cu (5 μm); right: Ni (10 μm)

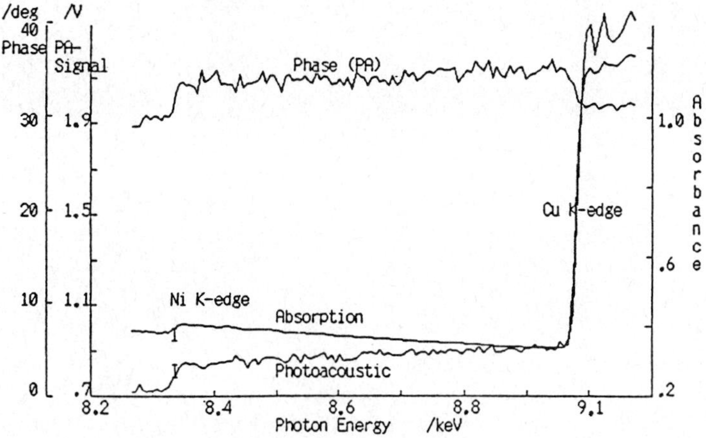

Fig. 9 Spectra of PAXAS signal amplitude and phase for Ni-plated Cu (10 µm) foil in comparison with absorption.

15A) were used with an aperture of 0.9 mmφ which was set in front of the PA cell. A strong PA signal was observed in the region where the tin (6 µm) (left rectangle) and nickel (10 µm) (right triangle) foils were located. Even in the same material region, the signal shows differences due to the thermal condition, for example, the difference in the contact with backing paper. In the spectroscopic study, the nickel-plated copper foil showed a sudden change not only in the amplitude (K-edges of Ni and Cu) but also in the phase of the signal depending on the photon energy, which reflected the edge absorption of the sub-surface material, as seen in Fig. 9 [7]. This edge absorption effect shows that the PAXAS method is quite useful for depth profiling of solid materials. Transparency to X-rays is another advantage for depth profiling. Studies with other detection methods, and other applications, are in progress.

Conclusion

An X-ray photoacoustic effect has recently been found, as reported in this paper, and its study is still at a primitive stage; however, this finding has expanded photoacoustic spectroscopy to the X-ray region and, hopefully, also opened a spectroscopic method which has the potential to give us new information about materials.

Acknowledgment

After the discovery of this effect, this study has proceeded as a cooperative research project (KEK proposal No.86-020 (Chairman: T.M.)) (now 28 persons) as reflected in the authors of our papers. The author would like to express his sincere gratitude to Dr. Hiroshi Kawata(KEK), Dr. Masaharu Hoshi (Hiroshima Univ.) and other project members for their collaboration and for discussion, and to Dr. Yoshiyuki Amemiya and Prof. Masami Ando (KEK) for their support and directional discussion.

The author is also grateful for financial support in the form of a Grant-in-Aid for Scientific Research from the Ministry of Education, Science and Culture, from the Research Foundation for Pharmaceutical Sciences, and from the Yamada Scientific Research Foundation, Japan.

References

1. T.Masujima, H.Kawata, Y.Amemiya, N.Kamiya, T.Katsura, T.Iwamoto, H.Yoshida, H.Imai, M.Ando, Photon Factory Activity Report, 4,314(1986).
2. T.Masujima, H.Kawata, Y.Amemiya, N.Kamiya, T.Katsura, T.Iwamoto, H.Yoshida, H.Imai, M.Ando, Chemistry Letters, 1987,973.
3. E.M.Eyring, S.J.Komorowski, T.Masujima, "Analytical Instrumentation Handbook" Ed. by G.Ewing, Marcel Dekker Inc., New York, in press.
4. M.Hoshi, T.Masujima, C.Nagoshi, Y.Sugitani, T.Sano, S.Sawada, H.Kawata, Y.Amemiya, M.Ando, Photon Factory Activity Report, 4,313(1986).
5. T.Masujima, Y.Amemiya, H.Kawata, S.Uehara, S.Sakura, H.Imai, M.Hoshi, T.Ikeda, M.Ando, Photon Factory Activity Report, 5,(1987) in press.
6. T.Masujima, M.Hoshi, Y.Sugitani, T.Sano, C.Nagoshi, H.Yoshida, H.Imai, H.Kawata, M.Ando, Photon Factory Activity Report, 4,315(1986).
7. T.Masujima, H.Kawata, M.Kataoka, M.Nomura, S.Sakura, H.Imai, M.Hoshi, C.Nagoshi, S.Uehara, H.Yoshida, M.Ando, Photon Factory Activity Report, 5, (1987) in press.
8. T.Masujima, H.Imai, H.Shiwaku, H.Kawata, Y.Amemiya, A.Iida, M.Hoshi, C.Nagoshi, S.Uehara, M.Kataoka, M.Danno, T.Ikeda, H.Makihara, H.Yoshida, M.Ando, Photon Factory Activity Report, 5,(1987) in press.

X-Ray Absorption Spectroscopy of Thin Films with a Pyroelectric Thermal Wave Detector

*H. Coufal, J. Stöhr, and K. Baberschke**

IBM Almaden Research Center, 650 Harry Road,
San Jose, CA 95120, USA

Due to the availability of a tunable source of X-rays in the form of synchrotron radiation, X-ray spectroscopy has been thriving in the last decade. A number of detection techniques have been developed to monitor the absorption of X-rays by thin films and surface layers [1]. All these techniques measure a signal (e.g., electrons, ions or fluorescence X-rays) which is linked to the excitation probability of a core electron. Since the absorption of X-rays in a sample results in the generation of heat, photoacoustic or photothermal spectroscopy should provide another method of studying X-ray absorption. The characteristics of X-ray induced ultrasonic waves have been studied, and their use for nondestructive ultrasonic imaging has been explored [2,3]. Recently, microphones have been utilized to monitor X-ray induced photoacoustic effect [4,5] to use it for dosimetry [6] and to obtain the first X-ray absorption spectra with acoustic detection [7]. The present paper presents first results of a feasibility study exploring the potential of pyroelectric detection of X-ray induced thermal waves.

Very few photothermal techniques are compatible with the vacuum requirements of most Synchrotron radiation experiment. Among these, a recently developed pyroelectric calorimeter has been shown to combine ultrahigh vacuum compatibility with utmost sensitivity [8]. It is particularly suitable for thin film samples. Details of the calorimeter and the signal generation and detection process are described in full detail elsewhere [9,10].

An amorphous carbon film, thickness 500 nm, was sputtered on top of the 120 nm thick silver electrode of the β-poly(vinylidene fluoride) (β-PVDF) calorimeter (20 μm). The thermal diffusivity of the carbon film was determined with conventional photothermal techniques [11] to be of the order of 8.0×10^{-4} cm^2/s. The sample was then cleaned by ion bombardment and analyzed with Auger spectroscopy to ensure a well-defined sample. Subsequently, the sample was mounted in a vacuum system on beam line I-1 at the Stanford Synchrotron Radiation Laboratory (SSRL). Using the grasshopper monochromator, the sample was irradiated with soft X-rays in the range 275-300 eV.

In a first series of experiments, an attempt was made to use the pulsed structure of Synchrotron radiation to excite a corresponding thermal wave with a frequency of approximately 4 MHz. Due to the tremendously large coherent radio frequency background at the storage ring and the associated shielding and grounding problems, no thermal wave signal was observed.

In a second series of experiments, the beam was, therefore, modulated at a frequency of approximately 3 Hz with a mechanical chopper inside the vacuum system. The resulting photothermal signal from the pyroelectric calorimeter was amplified with a high input

*Permanent address: Institut für Atom- und Festkörperphysik, Freie Universität Berlin, D-1000 Berlin, Germany

impedance, low noise preamplifier (Ithaco 1201) and detected with a two-phase lock-in-amplifier (PAR 5208). The observed signal was then normalized by the incident X-ray intensity, which is determined by measuring the total electron yield signal from a metal grid (80% transmission) reference monitor. With this arrangement, a considerable improvement of signal/noise ratio was achieved. Coherent noise due to modulation of the ion beam from ion pumps used in the system and modulation of radio frequencies still limited the sensitivity. Figure 1 demonstrates signal and noise levels by illuminating the sample with 275 eV photons and then interrupting the beam. The noise sources will be eliminated in an upcoming experiment. In the present experiment, the signal/noise ratio was, however, sufficient to demonstrate the utility of this detection scheme. A scan across the spectral range of interest from 275 to 300 eV photon energy was feasible in 5 minutes with a signal/noise ratio of 10 using a 10 sec time constant of the lock-in amplifier.

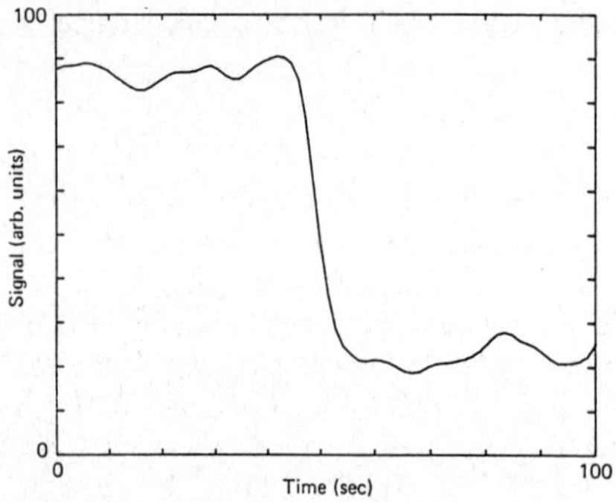

Fig. 1. Change in X-ray induced thermal signal when blocking the beam

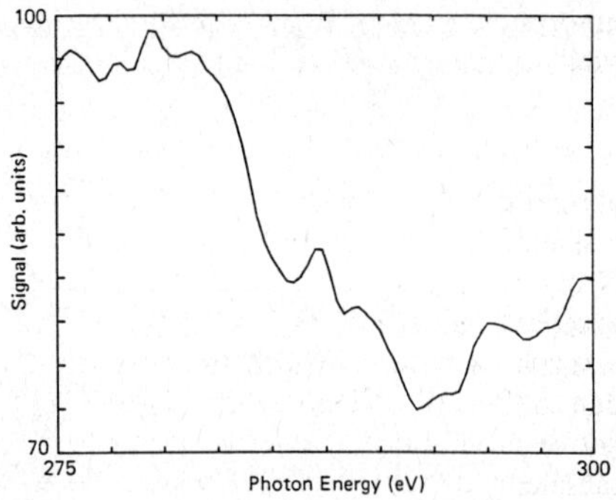

Fig. 2. X-ray absorption spectrum of an amorphous carbon film recorded with a pyroelectric calorimeter

The sample and the detector both contain carbon. They are separated by the silver electrode. Below the carbon K-edge (285 eV), the sample is partially transparent, above that edge, opaque. Characteristic thermal and X-ray data are summarized in Table 1.

The sample-detector assembly can hence be looked at as two samples separated by a thermal wave phase shifter, and the techniques developed for such samples can be applied [12]. Figure 2 shows the normalized photothermally detected X-ray transmission spectrum of the amorphous carbon film. With in-phase detection only, energy generated by X-rays that are

Table 1. Geometric and thermal characteristics (thickness ℓ, thermal diffusion length μ and phase lag Δ) of the sample detector arrangement at 3 Hz and X-ray absorption length β at 275 eV and 300 eV.

Layer μm	ℓ μm	μ μm	Δ degree	β_{275} μm	β_{300}
C	0.5	9.0	0.1	1.2	0.06
Ag	0.012	4.1×10^3	0	0.1	0.1
Al	0.0015	2.9×10^3	0	0.1	0.1
PVDF	20	70		0.6	0.06

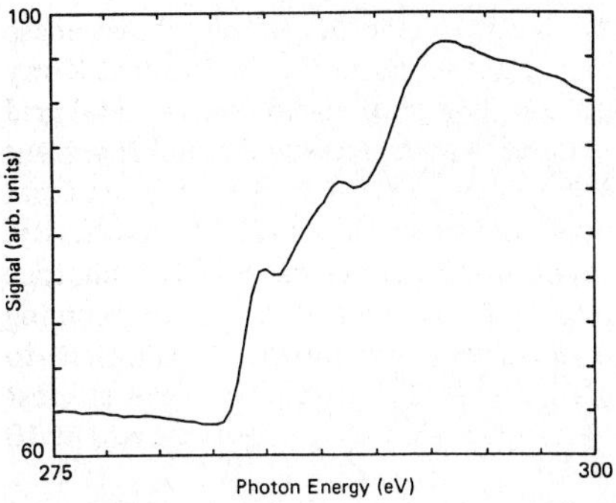

Fig. 3. Total electron yield spectrum of the amorphous carbon film. The total electron yield is proportional to X-ray absorption

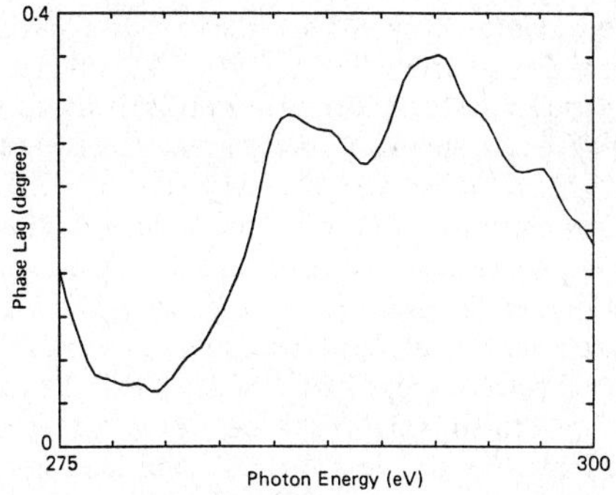

Fig. 4. Phase angle of the X-ray induced thermal signal as a function of photon energy

absorbed in the pyroelectric β-PVDF, i.e., are transmitted through the sample, gives rise to a signal. The spectrum hence resembles a transmission spectrum. A comparison with the X-ray absorption spectrum recorded with total electron yield detection (Fig. 3) shows excellent agreement between both techniques. Heat generated by X-rays absorbed in the amorphous carbon film will diffuse across the sample into the calorimeter. This time delay due to thermal diffusion will be reflected by a phase lag. This phase lag is evident in Fig. 4 between the signal above the absorption edge, when the majority of X-rays is absorbed in the sample, and below the edge, where the X-rays are absorbed in the calorimeter. This feature emphasizes the depth-profiling capability of thermal detection schemes. Experiments with other materials, i.e., samples that do not contain carbon on β-PVDF calorimeters and carbon samples on other pyroelectric detector materials are in progress.

We would like to thank E. Matthias for stimulating discussions and for drawing our attention at this interesting application of photothermal detection. Support by R. Grygier, C. Heino, C. Troxel and the staff at SSRL is gratefully acknowledged. The experiments were done at the SSRL which is supported by the Office of Basic Energy Sciences of DOE and the Division of Materials Research of NSF.

References

1. J. Stöhr: In X-Ray Absorption: Principles, Applications, Techniques of EXAFS, SEXAFS and XANES, ed. by D. C Koningsberger and R. Prins (Wiley and Sons, New York, 1987)
2. K. Y. Kim and W. Sachse: Appl. Phys. Lett. 43, 1099 (1983)
3. W. Sachse, K. Y. Kim and W. F. Pierce, IEEE Trans. Ultrason., Ferroelectrics, and Freq. Contr. UFFC-33, 546 (1986)
4. T. Masujima, H. Kawata, Y. Amemiya, N. Kamiya, T. Katsura, T. Iwamoto, H. Yoshida, H. Imai and M. Ando, Chem. Lett. (5), 973 (1987)
5. T. Masujima, H. Kawata, Y. Amemiya, N. Kamiya, K. Katsura, T. Iwamoto, H. Yoshida, H. Imai and M. Ando, Photon Factory Activity Report 1985/86 (1987)

6. M. Hoshi, T. Masujima, C. Nagoshi, Y. Sugitani, T. Sano, S. Sawada, H. Kawata, Y. Amemiya, and M. Ando, Photon Factory Activity Report 1985/86 (1987)
7. T. Masujima, M. Hoshi, Y. Sugitani, T. Sano, C. Nagoshi, H. Yoshida, H. Imai, H. Kawata, Y. Amemiya, and M. Ando, Photon Factory Activity Report 1985/86 (1987)
8. H. Coufal, F. Träger, T. Chuang and A. Tam, Surf. Sci. 145, L504-508 (1984).
9. H. Coufal, IEEE Trans. Ultrason., Ferroelectrics, and Freq. Contr., UFFC-33, 507-512 (1986).
10. H. Coufal, R. Grygier, D. Horne and J. Fromm, J. Vac. Sci. Technol., in print.
11. H. Coufal and P. Hefferle, Appl. Phys. A38, 213 (1985)
12. H. Coufal, Appl. Phys. Lett. 45, 516-518 (1984)

Vacuum Ultraviolet Photoacoustic Spectroscopy of Organic Solids Using Synchrotron Radiation as a Light Source

T. Inagaki[1], *Y. Furusawa*[2], *K. Hieda*[3], *H. Maezawa*[4], *and T. Ito*[5]

[1]Physics Department, Osaka Kyoiku University, Tennoji, Osaka 543, Japan
[2]Department of Molecular Biology, School of Medicine, Tokai University, Isehara, Kanagawa 259-11, Japan
[3]Biophysics Laboratory, Rikkyo (St. Paul's) University, Toshima, Tokyo 171, Japan
[4]Radiology Department, School of Medicine, Tokai University, Isehara, Kanagawa 259-11, Japan
[5]Institute of Physics, College of Arts and Sciences, University of Tokyo, Meguro, Tokyo 153, Japan

The vacuum ultraviolet (UV) region below 190nm has been one of the most difficult portions of the photon spectrum on which to perform spectroscopic studies. It is well known, however, that the major part of the oscillator strength of organic substances due to the valence electrons is localized in this spectral region, where photon energies are high enough to produce a variety of higher excited states. It is an intriguing subject to study relaxations of these higher excited states by photoacoustic (PA) method.

There are two major experimental difficulties in performing spectroscopic PA measurements of solids in the vacuum UV. For PA measurements, it is necessary to use a light source with a high flux density. Conventional light sources such as a cold-cathode discharge and a condensed spark source were found insufficient in their photon flux to generate a good PA signal from solids. In the present study, this was overcome by exploiting synchrotron radiation as a light source. The second difficulty arises from strong absorption exhibited by most substances in the vacuum UV. For organic solids, the penetration depths of photons in this region are of the order of 10^{-4}-10^{-6}cm, which are much shorter than the thermal diffusion length usually attainable in PA experiments. In order to obtain spectral information for such substances from PA measurements, it is necessary to use samples in the form of thin films which are uniform in thickness and optically semitransparent. In the present study, we have used self-supporting films and films supported by a thin film substrate of collodion, Formvar and polyethylene.

We have employed [1,2] the gas-microphone technique to explore primarily the region above 105nm, below which no transparent window material is available. A nonresonant type of PA cell with a magnesium fluoride window was set on an adaptor flange of an ultrahigh vacuum chamber installed on the beamline No. 5 of the electron storage ring of the Synchrotron Radiation Laboratory of the University of Tokyo. The total volume of the cell cavity was variable and normally set at $0.17cm^3$, and the sensitivity of the microphone was 0.5mV/μbar. Air inside the PA cell was purged by dry helium gas after mounting a sample in the PA cell. The monochromator used was of a 2m modified Wadsworth type and used without entrance and exit slits. A beam of monochromatic synchrotron radiation was chopped mechanically by a specially designed light chopper hung from the top flange of the vacuum chamber. The intensity spectrum of the incident light at the sample position was measured photoacoustically using a silver foil as a "black" absorber and used to normalize the observed PA signals. At a ring current of 150mA, the typical operating condition of the electron storage ring, the incident flux was estimated to be roughly 10μW/cm^2 with a spectral resolution (the full spectral width at half maximum) 6.2nm. In the region below 150nm, the flux density decreased gradually due to decreasing throughput of the monochromator-mirror system, which consisted of two aluminum mirrors and an aluminum diffraction grating.

An example of the PA amplitude spectra obtained for a series of organic solids is presented in Fig. 1, where the PA spectrum for an evaporated film of thymine is compared with the 1-T spectrum of the same material measured by the conventional transmission measurement. Spectral features appearing in these two absorption spectra correlate well with one another. Discrepancies found in the relative heights of the absorption peaks are due to reflection loss at the sample surfaces which are not included in the 1-T spectrum. A rapid decline of the PA absorption curve below 140nm may be indicative of involvement of nonthermal relaxations.

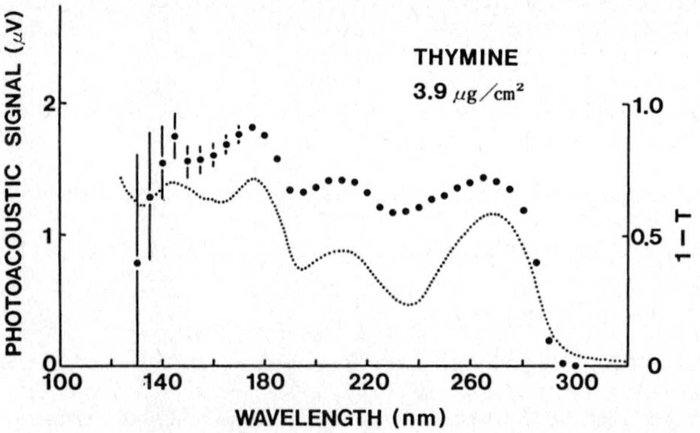

Fig. 1. Photoacoustic amplitude spectrum of an evaporated film of thymine on a polyethylene film. The mass density of the evaporated film was 3.9μg/cm^2. Dotted curve (small dots) is the 1-T spectrum derived from transmission data, where T is the transmittance.

After considerable efforts to improve the signal to noise ratio of the PA signal, it was found that the observed PA spectra were appreciably affected by desorption of water molecules adsorbed on the sample surfaces, unless the samples were dried thoroughly before mounting in the PA cell. In contrast to the case of the well-known effect of thermal desorption of adsorbates on the PA signal [3], this effect was found significant only in the region below 185nm, where water molecules absorb photons. This indicates that the desorption is not due to thermal agitation but is caused by electronic excitations. A detailed experimental study on this photodesorption of water molecules by the vacuum UV photons is now under way.

This research was supported by a Grant-in-Aid for Scientific Research from the Ministry of Education, Science and Culture. We gratefully acknowledge the cooperation of the SOR-RING staff members of the Synchrotron Radiation Laboratory, the University of Tokyo.

1. T. Inagaki, A. Ito, M. Motosuga, K. Hieda, K. Kobayashi, H. Maezawa, and T. Ito: Photochem. Photobiol. 41, 527 (1985).
2. T. Inagaki, A. Ito, K. Hieda, and T. Ito: Photochem. Photobiol. 44, 303 (1986).
3. P. Ganguly and T. Somasundaram: Appl. Phys. Lett. 43, 160 (1983).

Simultaneous Detection of Photoacoustic and Fluorescence Signals by X-Ray Excitation

Y. Sugitani[1], *S. Aoki*[1], *K. Kato*[2], *and Y. Kobayashi*[2]

[1]Institute of Chemistry, Institute of Applied Physics, University of Tsukuba, Sakura-mura, Ibaraki 305, Japan

[2]National Chemical Laboratory for Industry, Yatabe-machi, Ibaraki 305, Japan

The X-ray fluorescence method is advantageous in terms of non-destructive elemental analysis as well as convenience of operation, while the results of analysis are strongly dependent on the specimen. Photoacoustic measurements provide information about the thermal properties of bulk and layered materials. So far, photoacoustic measurements have been limited to UV through IR spectral regions. Recently, photoacoustic measurements in the X-ray region were successfully conducted using conventional X-ray tubes (1,2) and synchrotron radiation (3,4) as exciting sources. The former experiment using an X-ray tube can be easily combined with X-ray fluorescence measurements, and will provide a new analytical tool for materials science. On the basis of these findings, the simultaneous measurement of photoacoustic and fluorescence signals has been attempted on an X-ray fluorescence spectrometer equipped with a photoacoustic cell designed for this purpose.

Experimental

A built-in type X-ray tube with a molybdenum target was used as an exciting source. The X-ray beam was narrowed and collimated by the molybdenum collimator with a diameter of 1mm. A molybdenum spinning rod with a through hole (Fig. 1) was used to chop the exciting beam. Figure 1 shows the photoacoustic cell used for the experiment. The thicknesses of the beryllium windows were 0.5mm (upper) and 0.25mm (lower). The sample room was designed to be close to the detector to avoid fluorescence from the wall material (brass and aluminum) of the cell. Detection of X-ray

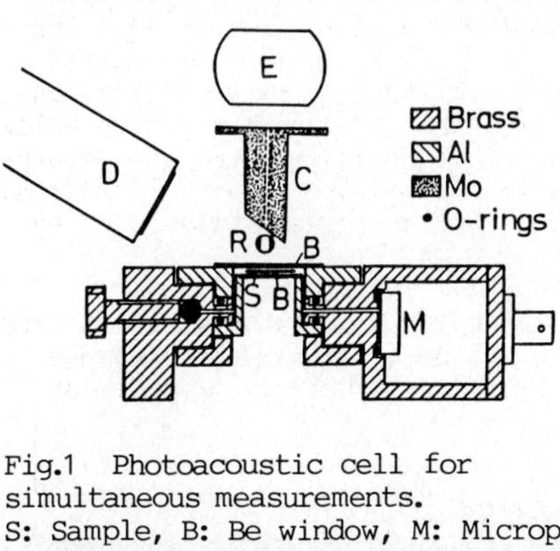

Fig.1 Photoacoustic cell for simultaneous measurements.
S: Sample, B: Be window, M: Microphone, D: Si(Li) detector, E: Exciting source, C: 1mm ϕ collimator, R: Rod chopper.

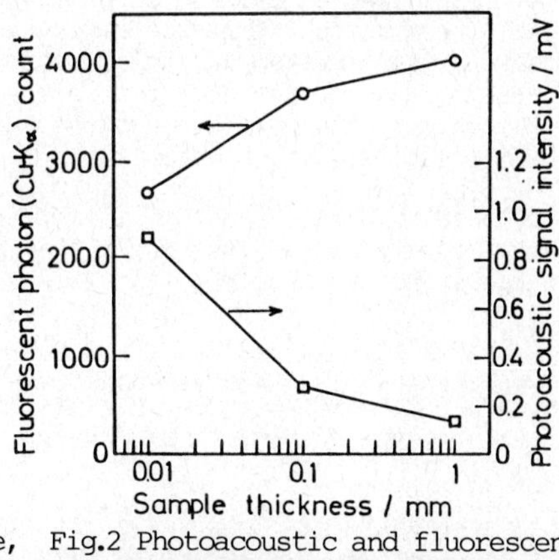

Fig.2 Photoacoustic and fluorescence signal intensities for different thicknesses of copper foil.

fluorescence was by a solid state detector Si(Li). The photoacoustic signal was detected by a condensor microphone and was amplified by a lock-in amplifier to be read on a digital voltmeter.

Results and Discussion

Photoacoustic and fluorescence signals were simultaneously detected for the conditions of 40mA and 40kV in the X-ray tube. The fluorescence signal, however, was very strong in these conditions, therefore the X-ray tube current was reduced to 2mA when only the fluorescence signal was measured. The chopping frequency was 16Hz. Figure 2 shows the change of photoacoustic signal intensity plotted as a function of the thickness of copper foil. The reason for the decrease of photoacoustic intensity with the increase of sample thickness is considered to be due to the heat dissipation in the foil material. Under these conditions the thermal penetration lengths for the samples used were longer than the thickness of the sample, so that the heat generated by X-ray absorption was considered to be uniformly spread through the sample. Results of the simultaneous measurements for various samples are presented in Table 1.

Table 1. Results of simultaneous measurements of the photoacoustic signal and fluorescence spectra.

Sample	Z	Thickness [μm]	PA intensity [a.u.]	Fluorescence peaks and intensities [cps]			
Al	19	200	173	-		-	
Fe	26	200	620	Fe-Kα	78.6	Fe-Kβ	13.2
Ni	28	200	474	Ni-Kα	76.3	Ni-Kβ	11.6
Cu	29	100	554	Cu-Kα	52.0	Cu-Kβ	10.1
Mo	42	100	363	Mo-Kα	over	Mo-Kβ	73.4
Sn	50	600	460	Sn-Kα	13.9	Sn-Kβ	3.1
Pb	82	100	1392	Pb-Lα	20.6	Pb-Lβ	18.8

References

(1) H. Vargas and C. L. Cesar, Med. Phys. 11,73(1984).
(2) K. Kato, Y. Kobayashi, S. Aoki and Y.Sugitani, Anal. Sci.3,275(1987).
(3) T. Masujima, H. Kawata, Y. Amemiya, N. Kamiya, T. Katsura, T. Iwamoto, H. Yoshida, H. Imai and M. Ando, Chem. Lett. 973(1987).
(4) Photon Factory Report 4,313;4,314;4,315(1986).

Study of the Photoacoustic Signal of Transparent Double-Layer Polymer Films Using FTIR

J. Philippaerts, E. Vanderheyden, and E.F. Vansant

Department of Chemistry, University of Antwerp (UIA), Universiteitsplein 1, B-2610 Wilrijk, Belgium

Abstract : FTIR-PAS measurements of a packaging material containing aluminum foil, polyamide and polyethylene (P.E.) have been carried out. An attempt has been made to obtain information about the thickness of the layers and about the surface characteristics. The influence of the reflection of transmitted IR light by the aluminum foil is investigated using a total absorber (carbon foil) as backing material.

Introduction : For the study of polymers, the PA detection technique in FTIR has a lot of advantages over conventional detection methods since optically thick samples can be examined without sample preparation, thus without changing the sample characteristics. A lot of work has already been done in this field : studies of curing processes [1], polymer coatings on fibers [2], layered polymers [3], etc. In this work the PA signal of a polymer-coated aluminum foil is examined.

Experimental : The polymer films were delivered by Exxon Chemicals (Belgium). The spectra were recorded on a Nicolet 5DXB equipped with a MTEC 100 photoacoustic detector at 4 cm^{-1} resolution. The different mirror velocities (10, 20, 30 and 40) are respectively 0.08, 0.16, 0.32 and 0.65 cm/sec. All spectra were ratioed against carbon black spectra, taken at the same scan speed.

Results and discussion : The packaging foil consists of a thin aluminum foil with a 14 µm thick layer of polyamide and a 16 µm thick layer of P.E. on top of it. Figure 1a shows the PAS spectrum at scan speed 10 of the material without the P.E. coating, giving the normal vibration spectrum of polyamide. Figures 1b, c and d are the spectra of the foil with polyamide and P.E., respectively, taken at velocities 10, 20 and 30. In the high frequency region, the NH polyamide vibrations gradually disappear and they even show inverted vibration bands at the highest scan speeds. This is caused by a decrease of the thermal diffusion length (μ). At 3400 cm^{-1} μ is calculated as 8.7, 6.2 and 4.4 µm respectively for velocities 10, 20 and 30 with the thermal diffusivity of P.E. (α_{PE}) = 1.3×10^{-3} cm^2/sec according to TOULOUKIAN et al. [4]. Since, at low scan speeds, polyamide vibrations are still visible (which means that μ should be at least 16 µm), a more exact value of α_{PE} is necessary for the determination of the thickness of the layer. The inverted peaks are the result of a time delay between the detec-

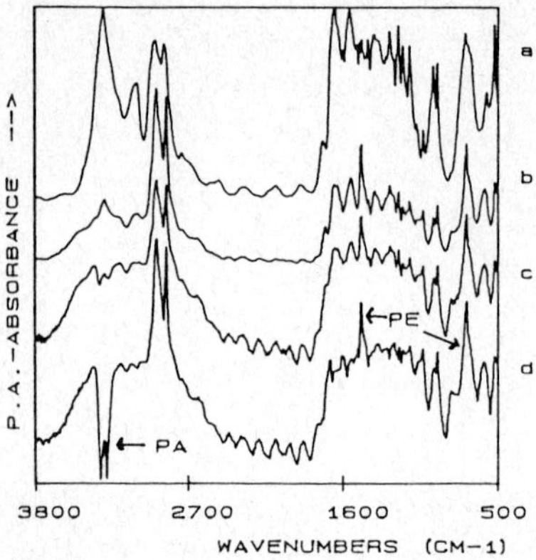

Figure 1 : a) Al-foil with polyamide at speed 10, Al-foil with polyamide and P.E. at speed b) 10, c) 20 and d) 30

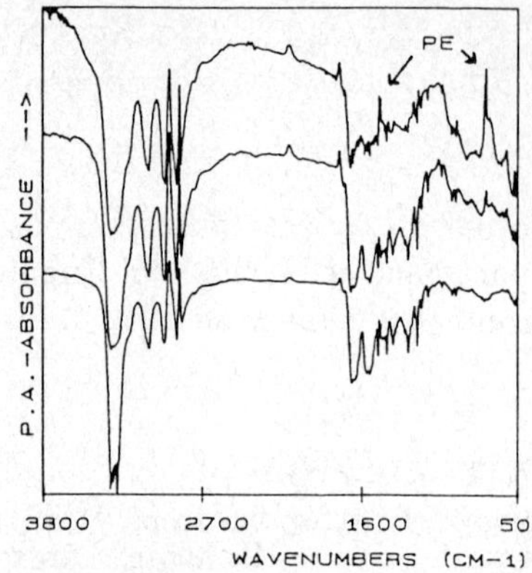

Figure 2 : Carbon foil with polyamide and P.E. at speed a) 10, b) 20 and c) 30

tion of the heat generated at the surface (P.E.) and the heat generated in the bulk of the material. The reflection of the transmitted light or the generated heat by the aluminum might also have a contribution to these phase shifts [5]. When the aluminum is replaced by a total absorber, carbon foil, the spectra are the result of the photoacoustic response of the polymer films and the response of the signal of the carbon foil. The latter is, in principle, a transmittance spectrum of the double layer polymer film. For scan speed 10 (Fig. 2a) the PAS-signal of polyethylene is still visible in the low frequency region. With scan speed 30 the transmittance signal compensates for the PA signal except in the region of CH stretching vibrations. Since the carbon black signal compensates for the bulk PA signal, an accentuation of surface compound vibrations is observed like the C=O stretching vibrations of corona-treated polyamide and polyethylene films.

References :

1) J.A. Graham, W.M. Grimm III and W.G. Fately in "FTIR-spectroscopy Vol.4 Appl. to Chemical Systems", Ed. J.R. Ferraro and L.J. Basile, 370 (Academic Press 1985)
2) C.Q. Yang, T.J. Ellis, R.R. Bresee and W.G. Fately, Proc. ACS division of Polymer. Material Science and Eng. 53, 169
3) M.W. Urban and J.L. Koenig, Appl. Spectr. 40, 994 (1986)
4) Y.S. Touloukian, R.W. Powell, C.Y. Ho and M.C. Nicolasu, Thermal Diffusivity, IFI/Plenum, New York (1973)
5) L. Bertrand, Personal Communications (1987)

Photopyroelectric Spectroscopy (P^2ES) of Electronic Defect Centers in Crystalline n-CdS

A. Mandelis, W. Lo, and R.E. Wagner

Photoacoustic and Photothermal Sciences Laboratory,
Department of Mechanical Engineering, University of Toronto,
Toronto, Ontario M5S1A4, Canada

I. INTRODUCTION

Photopyroelectric spectroscopy (P^2ES) has recently emerged as a simple, powerful photothermal technique for the spectroscopic study of condensed phase materials [1-6]. Pyroelectric thin-film sensors, such as polyvinylidene difluoride (PVDF) [7], have proven to be capable of providing high SNR, high quality spectra of thin-film materials [1,4,8], as well as spectra of condensed phase samples in the thickness range above 200 μm [3].

In this work we exploited the non-acoustic nature of the detection mechanism of thin-film P^2ES to obtain high quality, high SNR spectra of n-CdS single crystals pyroelectrically for the first time. The back-detection character of P^2ES proved to yield extreme sensitivity to subbandgap defect centers, superior to microphonic PAS. The overall simplicity and open-cell geometry of the detection method further point to P^2ES as an optimum photothermal spectroscopy for non-radiative defect studies in semiconductors.

II. MATERIALS AND EXPERIMENTAL

CdS samples used in this work were high purity n-type single crystals from Eagle-Pitcher, Inc., Miami, Oklahoma with nominal resistivity $\rho = 1.3 - 1.9 \times 10^5$ ohm-cm. The nominal mobilities were 190-215 $cm^2 V^{-1} s^{-1}$. The crystal growth details were published elsewhere [9]. The CdS samples were electrically isolated from the PVDF Ni-Al metallization by a thin Teflon membrane stretched over the pyroelectric film which was housed in an Inficon quartz microbalance sputtering sensor body assembly (Model 007-048) with a crystal holder (Model 007-049) and a ceramic retainer (Model 007-023). This assembly, conventionally used as a quartz crystal housing, was readily adapted to mount PVDF thin films, offering mechanical support, RF-shielding and electrical contacts. The Teflon membrane, besides electrically decoupling the samples from the detector, served as a barrier to direct PVDF film irradiation at subbandgap wavelengths where CdS is optically transparent.

Electrical connections were made using an In-Ga mixture at opposite crystal edges with surface normals perpendicular to the direction of incident radiation. This metal-semiconductor interface was shown to exhibit approximately ohmic characteristics in the voltage range between -10V and +20V, with a slight change in the slope of the I-V curve upon reversal of the sign of the field [9].

III. RESULTS AND DISCUSSION

Three kinds of experiments were performed at room temperature: (a) open-circuit P^2ES; (b) P^2ES and PCS with an applied transverse DC bias; and (c) P^2ES and PCS with an applied transverse AC bias.

A. Open-circuit P^2E Spectra

Previous photopyroelectric spectroscopic measurements [2,4] indicate, and the developed theory [5] predicts, that there occurs an inversion in the spectral character of P^2E spectra with increasing modulation frequency above a transition frequency f_o such that

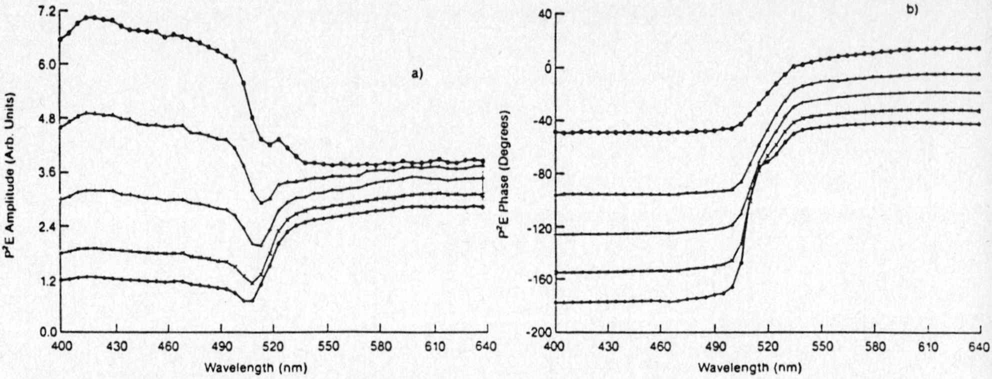

Fig. 1: Modulation-frequency dependence of photopyroelectric spectra of CdS crystal (1 mm thickness) -•-•-: 3 Hz; –Δ–Δ–: 6 Hz; +-+-: 9 Hz; x-x-: 12 Hz; ♦-♦-: 15 Hz (a) Amplitude; (b) Phase (Resolution: 2 nm)

$$\mu_s(f_o) \approx L \quad , \tag{1}$$

where L is the sample thickness. As the material becomes thermally thick, the absorption spectrum gradually becomes a transmission spectrum. For our 1 mm thick CdS single crystal, $f_o \approx 5$ Hz. Figure 1a shows the onset of spectral inversion at ca. f = 6Hz (amplitude spectrum). The P^2E phase, however, does *not* exhibit inversion characteristics, Fig. 1b, in agreement with the photopyroelectric theory [5].

A thin n-CdS crystal exhibiting the low frequency spectrum of Fig. 1 was further etched to thickness ca. 0.4 mm, so that the condition

$$\mu_s(2\,\text{Hz}) = 1.55\ \text{mm} \gg L \tag{2}$$

was valid and the sample was definitely thermally thin. When deliberate mechanical damage was introduced to the crystal in the form of a narrowline fracture along the c-axis, which did not entirely sever the crystal but introduced a damage network below the surface and outward from the fracture line, the 2 Hz P^2E spectrum of the crystal exhibited strong inversion of features with respect to the 3 Hz curve of Fig. 1a, with the sub-bandgap signal increasing approximately tenfold over its value before the fracture, Fig. 2a, while the superbandgap signal remained essentially unchanged. The P^2E phase of the intact crystal at 3 Hz (Fig. 1b) shows a gradual increase (indicated as a more positive value in this and later figures, and henceforth designated as a "phase lead"; the opposite effect will be designated as a "phase lag") over ca. 100 nm span before saturating in the region above λ_g (= hc/E_g), as expected from the shift of the heat centroid from the remote front crystal surface (opaque superbandgap region; $\lambda < 490$ nm) into the bulk and closer to the back PVDF detector with decreasing optical absorption coefficient. This trend has been interpreted theoretically previously [5]. The P^2E phase of the fractured crystal, however, (Fig. 2b) is characterized by a much more abrupt increase above λ_g of ca. 10 nm span before saturation. Below λ_g, the phase exhibits a minimum at ca. 513 nm (which corresponds to a maximum in phase lag) in lieu of the monotonic behavior shown in Fig. 1b. This phase minimum becomes more pronounced with increasing modulation frequency as was verified with experiments performed at 6, 10 and 75 Hz. The P^2E phase at 75 Hz showed that the spectral position of the maximum lag remained unchanged with frequency at ca. 513 nm. A spectral behavior similar to that of Fig. 1b of the phase of an intact CdS single crystal in a microphone gas-coupled PA cell with back surface detection has been observed by TAKAUE et al. [10] at frequencies ranging from 8 Hz to 80 Hz, with a flat superbandgap profile and a broad subbandgap increase with onset around 520 nm and a span of ca. 100-120 nm before saturation. The frequency-

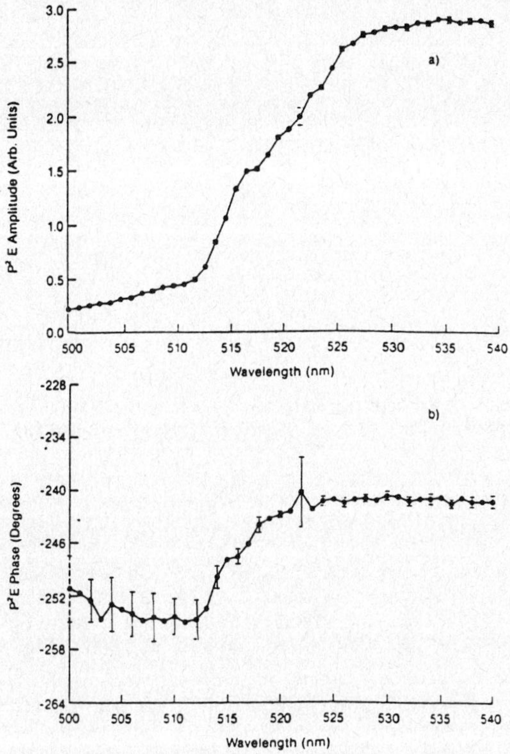

Fig. 2: Photopyroelectric spectrum of n-CdS single crystal with narrowline fracture. Thickness ca. 0.4 mm; modulation frequency: 2 Hz; a) Amplitude, b) Phase. (Resolution: 2nm).

dependent P^2E phase increase below 513 nm in Fig. 2b is most likely related to the perturbed heat centroid location after the introduction of damage. In the presence of large densities of non-radiative defects the heat centroid may no longer be dominated entirely by the optical absorption length in the material, but rather by bulk defect centers via their contributions to the wavelength dependent non-radiative quantum efficiency, $\eta_{NR}(\lambda)$.

B. P^2E and PC Spectra: Modulated Optical Excitation with DC Electric Field

To test further some of these ideas, several experiments were performed with a DC voltage applied transversely across the CdS crystal. The magnitude of the applied bias was varied between 5 and 30 V. Both photopyroelectric and photocurrent spectra from the fractured crystal were obtained simultaneously and are shown in Fig. 3. In Fig. 3a the spectrum corresponding to the V_{DC} = 0V case, when expanded, is described by that of Fig. 2a for the fractured crystal. The P^2E and PC amplitudes, Fig. 3a and b, respectively, show large increases with increased bias and a maximum at ca. 529 nm, in agreement with previous photoacoustic data [9,11]. No local maximum at shorter wavelengths can be identified in the P^2E spectrum, unlike the PA spectra of crystals from the same batch reported in Ref. [11]. We attribute the absence of the high energy peak to complete spectral domination by the large increase of the subbandgap P^2E signal due to the enhanced Joule effect [9,11,12] in the presence of non-radiative scattering centers, perhaps in conjunction with the suppression of interband transition features due to the back surface detection, as observed previously by HATA et al. [13] with a piezoelectric transducer.

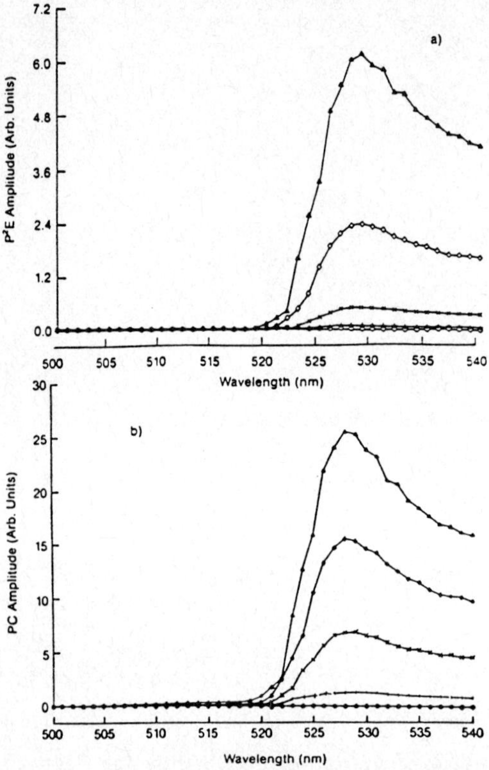

Fig. 3: a) P^2E amplitude spectra of n-CdS single crystal (narrowline fracture) with an applied transverse DC bias of 0 V (-o-o-); 5 V (+-+-); 10 V (x-x-); 20 V (♦-♦-); 30 V (Δ–Δ–); b) PC spectra corresponding to (a). (Resolution: 2nm).

The P^2E and PC amplitude decay to very small values below λ_g in Figs. 3a and b is consistent with surface excitation and the absence of bulk carrier contributions to either signal, in agreement with earlier observations [9,11,12]. The strong peaks in the P^2E spectra of Fig. 3a and PC spectra of Fig. 3b were identified photoacoustically [14] to be consistent with the spectral variation of the non-radiative quantum efficiency $\eta_{NR}(\lambda)$ in the presence of large defect densities. The PC phases for all applied biases essentially coincided below λ_g and exhibited a slight downward trend (lag) at long wavelengths similar to that reported in Ref. [14].

C. P^2E and PC Spectra: Unmodulated Optical Excitation with AC Electric Field

The main difference between this mode of experimentation and the preceding one is that this mode is *only* sensitive to thermal energy released due to the Joule effect in the crystal (P^2E spectra), while PC spectra are dominated by electronic transport of carriers which survive Joule effect scattering and decay. This is so, because the photogenerated carrier densities are DC (unmodulated), and therefore thermal energy from nonradiative carrier recombination is also not temporally modulated and thus will not be lock-in detected by the P^2E and PC probes. The only AC dependence in the sample is that due to the alternating electric field E(t) contributing a Joule heat density proportional to $E^2(t)$

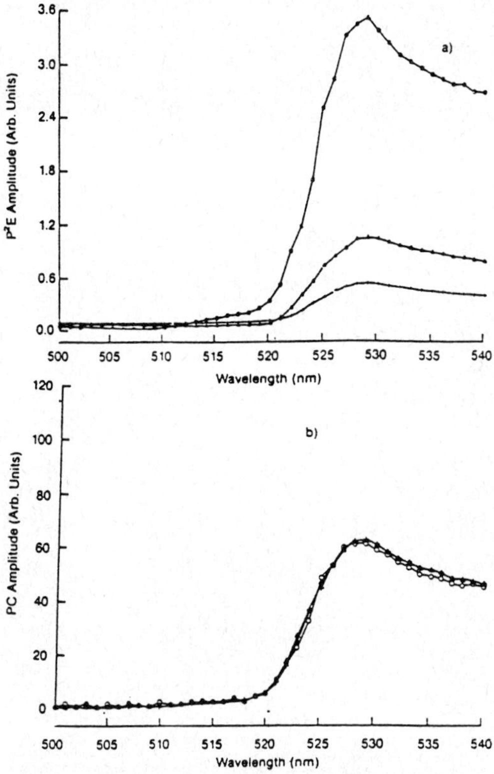

Fig. 4: a) Magnitude spectra of P^2E responses from defect-rich CdS crystal with an AC transverse electric field of 20 V peak-to-peak. (-o-o-): 20 Hz; (Δ–Δ–): 40 Hz; (-+-+-): 55 Hz; b) PC spectra corresponding to (a). (Resolution: 2 nm).

and to the DC photogenerated carrier density [11,12,14]. The magnitudes of the P^2E and PC spectra with a 20-V peak-to-peak AC electric field applied transversely across the defect-rich CdS crystal are shown in Figs. 4a and b. The P^2E maximum, in Fig. 4a occurs at 529 nm for all modulation frequencies, i.e. in the same spectral location as the maxima in Fig. 3a; so do the PC maxima in Fig. 4b, similar to those in Fig. 3b. The constancy of the spectral position of the P^2E 528-529 nm maximum with mode of experimentation (DC or AC electric field applied) and under different modulation frequencies was previously observed photoacoustically [9] and interpreted as related to the wavelength-dependent non-radiative quantum efficiency. The independence of the PC magnitude spectrum, Fig. 4b, from modulation frequency is also expected, as the free carrier density within the crystal (the quantity directly related to the PC magnitude) is not affected by the low modulation frequencies employed here, compared to the carrier response rate. The P^2E magnitude response is, of course, thermal diffusion-limited and exhibits large attenuation with increasing frequency, Fig. 4a.

Figures 4 a, b indicate that the band-to-band recombination effects are essentially entirely dominated by the increase in the non-radiative quantum efficiency which saturates at ca. 530 nm [14] in agreement with the externally introduced defect network presence.

IV. CONCLUSIONS

The present work has established the use of P^2ES as a spectroscopic technique sensitive to the non-radiative energetics of electronic defect centers in the subbandgap region of defect-rich semiconductors such as n-CdS single crystals. The spectral domination of the subbandgap region by defect-related heat release was shown to be complete, leading to spectral feature inversions between the intact and the defective CdS crystals. This behavior shows that P^2E spectra are much more sensitive to defect centers than similar PA spectra [11], where both interband transition and defect-related spectral peaks of similar magnitudes were found to co-exist. The high non-radiative defect center sensitivity of P^2ES, coupled with the extreme simplicity of the experimental system may possibly establish this technique as the leading candidate for spectroscopic studies of defect physics in crystalline and non-crystalline semiconductors.

REFERENCES

1. H. Coufal: Appl. Phys. Lett. **44,** 59 (1984)
2. A. Mandelis: Chem. Phys. Lett. **108,** 388 (1984)
3. D. Dadarlat, M. Chirtoc, R.M. Candea and I. Bratu: Infrared Phys., **24,** 469 (1984)
4. H. Coufal: Appl. Phys. Lett., **45,** 516 (1984)
5. A. Mandelis and M.M. Zver: J. Appl. Phys., **57,** 4421 (1985)
6. H. Coufal and A. Mandelis: In *Photoacoustic and Thermal Wave Phenomena in Semiconductors,* ed. by A. Mandelis, (North-Holland, New York 1987), Chapter 7.
7. Data concerning PVDF KYNAR pyroelectric film can be found in "KYNAR^{TM} Piezo Film Technical Manual", Pennwalt Corp. (1983); Information on PVDF can be found in: Ferroelectrics, **32** (1981)
8. K. Tanaka, K. Sindoh and A. Odajima: Rept. Progr. Polym. Phys. Jpn., **29,** 367 (1986).
9. A. Mandelis and E.K.M. Siu: Phys. Rev., **B34,** 7209 (1986)
10. R. Takaue, M. Matsunaga and K. Hosokawa: J. Appl. Phys., **56,** 1543 (1984)
11. T. Dioszeghy and A. Mandelis: J. Phys. Chem. Solids, **47,** 1115 (1986)
12. I.N. Bandeira, H. Closs and C.C. Ghizoni: J. Photoacoust., **1,** 275 (1982)
13. T. Hata, Y. Sato and M. Kurebayashi: Jpn. J. Appl. Phys. **22,** Suppl. **22-3,** 205 (1983)
14. E.K.M. Siu and A. Mandelis: Phys. Rev., **B34,** 7222 (1986)

Characterization of n-CdS/Polysulfide Photoelectrochemical Cells via Photothermal Deflection and Photoaction Spectroscopies

R.E. Wagner and A. Mandelis

Photoacoustic and Photothermal Sciences Laboratory,
Department of Mechanical Engineering, University of Toronto,
Toronto, Ontario M5S 1A4, Canada

I. INTRODUCTION

Photothermal deflection spectroscopy (PDS), like the other thermal-wave methods, can be used to obtain the absorption spectra of materials. Several groups have obtained PDS spectra of semiconductor electrodes in photoelectrochemical (PEC) cells [1-3]. Subsequent to our original work [1], we have carried out an extensive study involving the measurement of both PDS and photoaction spectra for several different n-CdS electrodes in polysulfide electrolytes. The information obtained by the two combined spectroscopic techniques can be qualitatively compared taking into account both the magnitude and phase variations of the spectra. The phase information is especially useful in helping one to decide whether the weak subbandgap signals are due to bulk absorptions, surface absorptions, or a combination of both.

II. EXPERIMENTAL

Figs. 1 to 3 show data for a system consisting of a high resistivity n-CdS single crystal (polished and etched in HCl) immersed in a (1,0.01,1) NaOH, S, Na_2S electrolyte. Figure 1 depicts the PD and photovoltage (PV) spectra at 25 Hz. The PD spectrum was normalized by a blackbody PD spectrum, while the PV spectrum was normalized by the square root of the blackbody PD spectrum, as it was found that the PV signal was roughly dependent upon the square root of the excitation intensity for superbandgap wavelengths. Fig. 2(3) presents the PD (PV) phase spectra at 25 and 79 Hz.

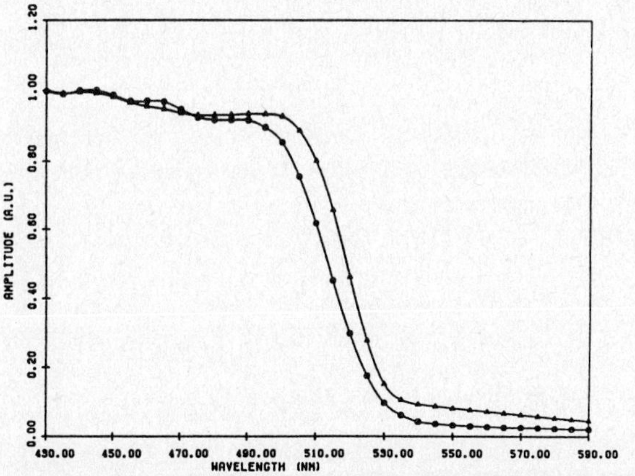

Fig. 1: High resistivity CdS spectra at 25 Hz. –●–●–: PDS; –Δ–Δ–: PVS

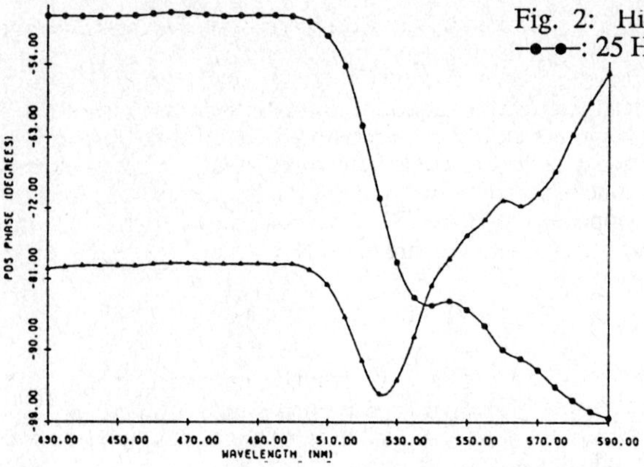

Fig. 2: High resistivity CdS; PD spectra. –●–●–: 25 Hz; –Δ–Δ–: 79 Hz

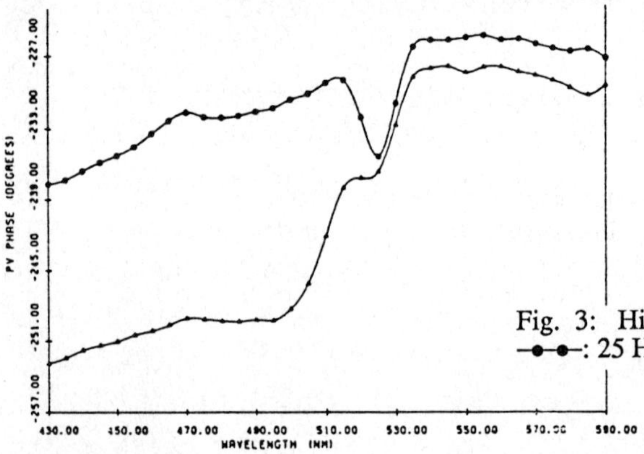

Fig. 3: High resistivity CdS; PV spectra. –●–●–: 25 Hz; –Δ–Δ–: 79 Hz

III. DISCUSSION

The PDS magnitude and phase spectra may be explained as follows: The magnitude spectrum (Fig. 1) consists of three regions: one, a high energy plateau which drops gently from 430 to 490 nm, and corresponds to the fundamental interband absorption across the CdS bandgap. This absorption is strong, the absorption length of about 10^{-5} cm [4] being considerably shorter than the thermal diffusion length at 25 Hz ($\approx 4.4 \times 10^{-2}$ cm). The second region is the "band edge" ($\approx$ 490-530 nm) where the absorption coefficient begins to decrease towards longer wavelengths, but electronic transitions are still mainly interband. The third region ($\lambda \geq 530$ nm) corresponds to a weak signal with heating due to weak bulk absorptions between a variety of localized and extended electronic states (both in the gap and the valence/conduction bands), in addition to surface state transitions. Figure 2, the PD phase spectra at 25 and 79 Hz, provides important information regarding the origin of the PDS signal: At 25 Hz, the phase lag is constant for strong fundamental absorption and then the lag increases for $\lambda \geq 500$ nm. Thus, the 25 Hz subbandgap signal appears to be derived mainly from weak bulk absorption. On the other hand, at 79 Hz the phase behavior mimics that at 25 Hz for $\lambda \leq 525$ nm, but then the lag decreases considerably for longer wavelengths as the PD phase becomes dominated by surface absorption. At 79 Hz, the PD signal is expected to be more sensitive to surface absorption, when compared to the 25 Hz signal. This is so, due to the large contribution of a surface layer to the optical absorption profile in the composite solid (surface layer + bulk) at subbandgap wavelengths, where surface layer absorption can dominate the PD phase by strongly affecting the heat centroid position at high f (79 Hz) and less so at longer thermal diffusion lengths (25 Hz) [5].

The photovoltage spectrum (Fig. 1) is similar to the PDS spectrum The photovoltage phase spectra (Fig. 3) can be explained as follows: both spectra have two distinct regions, one from 430 to 525 nm, the second from 525 to 590 nm. The high energy region corresponds to a fundamental bandgap excitation, and is quite sensitive to chopping frequency. The low energy region probably corresponds to surface state absorption, since weak bulk absorptions are not expected to contribute significantly to the photovoltage. The phase of this region is much less sensitive to chopping frequency than the fundamental absorption region, the opposite from the PD phase spectra of Fig. 2. On the other hand, PD and PV magnitude spectral features are relatively insensitive to the value of chopping frequency used.

IV. CONCLUSION

We can conclude that both the amplitude and phase information contained in the PDS and photoaction signals are useful in interpreting the phenomena occurring at the PEC cell working electrode, with the phase containing easier to interpret spatial information concerning the origin of the observed signals.

REFERENCES

1. R.E. Wagner, V.K.T. Wong, A. Mandelis: Analyst **111,** 299 (1985).
2. J.P. Roger, D. Fournier, A.C. Boccara, R. Noufi, D. Cahen: Thin Solid Films **121,** 11 (1985).
3. M.A. Tamor, R.E. Hetrick: Appl. Phys. Lett. **46,** 460 (1985).
4. D. Dutton: Phys. Rev. **112,** 785 (1985).
5. A. Mandelis, Y.C. Teng and B.S.H. Royce: J. Appl. Phys. **50**, 7138 (1979).

Photoacoustic Detection of Light-Shifts in Molecules

A. Di Lieto[1,2], *P. Minguzzi*[1], *S. Profeti*[1], *and M. Tonelli*[1]

[1]Dipartimento di Fisica, Piazza Torricelli 2, I-56100 Pisa, Italy
[2]Scuola Normale Superiore, Piazza dei Cavalieri 7, I-56100 Pisa, Italy

It is well known that an intense laser radiation detuned off the resonant frequency of an atomic or molecular transition causes a change in the energy of the ground and excited states of the system. A laser tuned below resonance shifts the ground-state energy to lower values and raises the energy of the excited state of the same amount. The existence of such an effect is a key element in the achievement of optical trapping of atoms (as recently demonstrated by CHU et al. /1/) and therefore a renewed interest in it has arisen.

The modification of energy levels is usually detected by observing how the frequency of a transition coupled to either level is affected by the non-resonant light. These kinds of experiments in three-level systems were performed even before the advent of the laser (see e.g. BARRAT and COHEN TANNOUDJI /2/) and are well documented in different regions of the spectrum.

Experiments on two-level systems (i.e. where the perturbing laser has a frequency close to the monitored transition and only two energy states effectively interact with radiation) are more difficult, and they have been seldom attempted. In these cases, for strongly driven systems, they may involve multi-photon effects and complicated lineshapes /3/. Thus, a direct observation is more difficult, but photoacoustic detection is of great help in solving the problems of nonlinear spectroscopy in small samples /4/. In fact, when using a photoacoustic apparatus, the power on the detector is negligible outside the resonance. This is a remarkable advantage in all cases of nonlinear effects where a strong intensity of the laser source is required. Optoacoustic detection has an additional unique advantage: its sensitivity is larger when the sample volume is small. This allows using small cells and obtaining high power density with laser sources of moderate intensity.

In this work we describe a two-level experiment performed in the infrared region on a rotovibrational transition of the v_2 band of ammonia. By means of Stark effect we select two non-degenerate sublevels and we measure the transition frequency by using a tunable CO_2 laser (probe) and a nonlinear Doppler-free technique /5/. A second CO_2 laser (pump) is used as a fixed-frequency perturbing source. Different conditions of pump detuning are easily achieved by adjusting the static electric field. The details of the experimental apparatus are shown in Fig. 1, which is largely

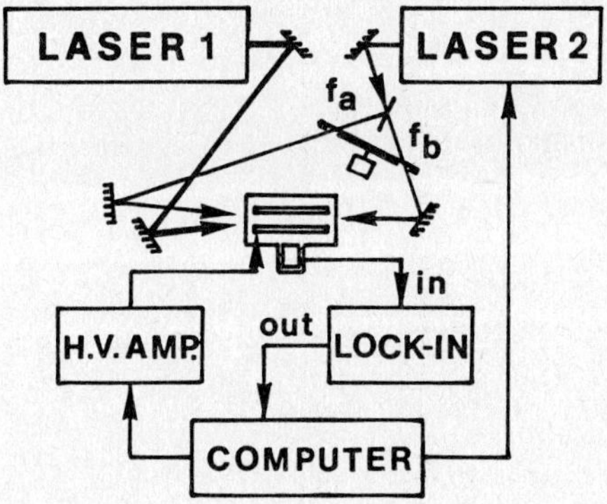

Fig. 1. Schematic diagram of the experimental apparatus

self-explanatory. The pump laser is frequency-stabilized by means of the saturated-fluorescence technique /6/ and it is used also as the reference for the frequency-offset locking of the tunable probe laser.

In order to obtain Doppler-free lineshapes we adopt an intermodulated-detection scheme /7/ : The probe beam is split into two counterrunning parts that intersect in the centre of the optoacoustic cell. These two beams are modulated at two different frequencies by a single mechanical chopper, and a lock-in amplifier, tuned to the sum frequency, processes the microphone signal. The frequency of the waveguide laser is digitally scanned under control of a microcomputer, which also collects data from the lock-in amplifier (signal) and from a digital voltmeter (electric field). The perturbing beam is focused to the centre of the cell at the region sampled by the probe laser; its direction is almost collinear with one of the probe beams.

A typical experimental run consists of two steps. First the absorption (acoustic) lineshape is sampled at several discrete frequencies of the probe laser, in the absence of the perturbing radiation. Then the pump laser is switched on and the scan is repeated. Each data set is recorded at a fixed electric field corresponding to a given detuning of the pump frequency. The example shown in Fig. 2 was obtained for the asQ(5,3) M=5

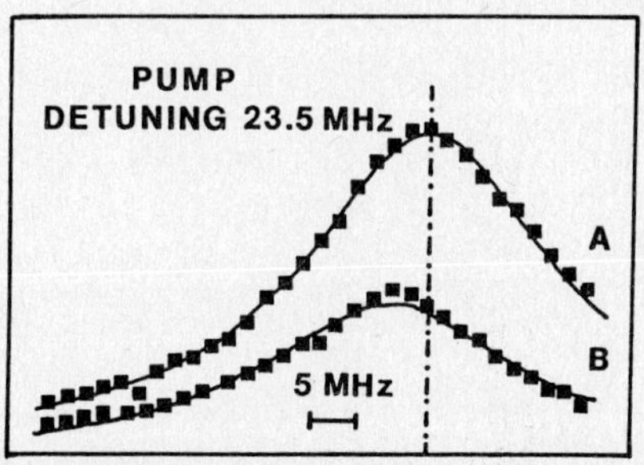

Fig. 2. Intermodulated lineshapes recorded at a pressure of 62 mTorr. (A): the perturbing laser is absent; (B): the power of the perturbing laser is 5.8 Watt

$\Delta M=0$ line of $^{14}NH_3$;the emission line of the CO_2 lasers was the P(32) of the 10 μm band and the Stark field was about 12.3 kV/cm. We note that the signal-to-noise ratio is not very good because the powerful nearly resonant laser markedly depletes the population difference between the molecular levels.

When the detuning is sufficiently large (several MHz) the lineshape may be naively described as a Lorentzian profile. On the contrary, if the detuning is small, the lineshape is more complicated because multi-photon effects, similar to those described in Ref. /3/, are no longer negligible. In particular, at the saturation levels achieved in our experiment, twin-peak curves are expected and observed. These complicated cases will not be described here and we report only measurements recorded at large detunings. Effective Lorentzian lineshapes are fitted to the data (the solid lines in Fig. 2) and the shift in the centre is measured at different Stark fields, i.e. at different pump detunings Δ. Within the validity of this approximate description, the centre shift s is given by

$$s = - \frac{\Omega_R^2}{2\Delta} , \qquad (1)$$

where Ω_R is the Rabi frequency for the pump laser, $\Delta = \omega - \omega_0$ the detuning between the frequency of the pump laser and the absorption line.

The results for the shift measurements are plotted in Fig. 3 as a function of pump detuning; the power of the perturbing laser is about 5.8 Watt and the gas pressure is 62 mTorr. The solid line is the best fit of (1) to the data and the estimated Ω_R is about 15 MHz, in reasonable agreement with the expected value.

As a conclusion we may state that we have obtained clear evidence of a light-shift effect in the infrared region for a two-level molecular system.

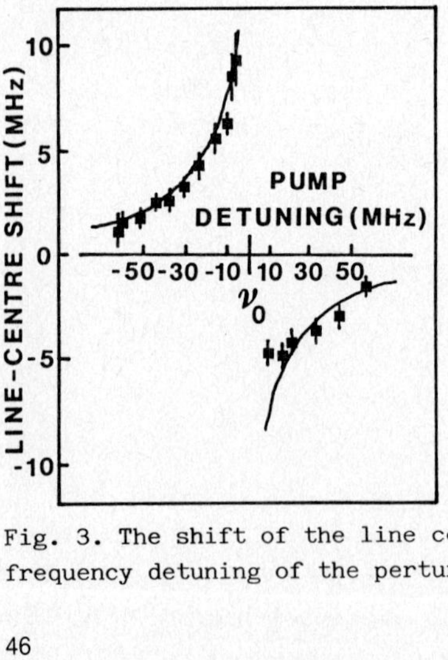

Fig. 3. The shift of the line centre is plotted as a function of the frequency detuning of the perturbing laser

The case of large detuning is explained in a very simple intuitive approach, and it is well in agreement with expectations. The case of small detuning requires a careful analysis and work is in progress to compare experimental results with theory.

References

1. S. Chu, J.E. Bjorkholm, A. Ashkin, and A. Cable: Phys. Rev. Lett. 57, 314 (1986)
2. J.P. Barrat and C. Cohen Tannoudji: J. Phys. Rad. 22, 329 (1961)
3. F.Y. Wu, S. Ezekiel, M. Ducloy, and B.R. Mollow: Phys. Rev. Lett. 38, 1077 (1977)
4. P. Minguzzi, S. Profeti, M. Tonelli, and A. Di Lieto: J. Opt. Soc. Am. B 3, 1075 (1986)
5. M. Tonelli, P. Minguzzi, and A. Di Lieto: J. Phys. (Paris) 44, C6-553 (1983)
6. C. Freed and A. Javan: Appl. Phys. Lett. 17, 53 (1970)
7. P. Minguzzi, M. Tonelli, A. Carrozzi, and A. Di Lieto: J. Mol. Spectrosc. 96, 294 (1982)

Spectroscopic Parameters of the DODCI Photoisomer from Laser-Induced Optoacoustic Spectroscopy

G.M. Bilmes[1], *J.O. Tocho*[1], *and S.E. Braslavsky*[2]

[1]Centro de Investigaciones Opticas, C.C. 124, 1900 La Plata, Argentina
[2]Max-Planck-Institut für Strahlenchemie, Stiftstraße 34–36, D-4330 Mülheim a.d. Ruhr, Fed. Rep. of Germany

1. Introduction

The photoisomerization of DODCI starting from the first excited singlet state of the stable form (N) as well as the reverse, thermal isomerization of the ground state of the less stable photoisomer (P) are well studied processes. In addition, at wavelengths where P absorbs, a photoisomerization of P to N has been suggested. Using laser-induced optoacoustic spectroscopy (LIOAS) we have recently determined that the rate constant for that process is less important than the internal conversion taking excited P back to its ground state [1]. However, several spectroscopic parameters of P had to be assumed since they were not well determined, including the values for the fluorescence quantum yield, $\Phi_f(P)$, and for the absorption cross sections of P, $\sigma_P(\lambda)$, which we have now measured by LIOAS under conditions of steady-state population of both species.

2. Experimental

Optoacoustic signals were obtained using a N_2-laser pumped dye laser tuned between 570 and 655 nm in a set-up with 1 μs time resolution [2]. The photoequilibrium between N and P was achieved by illuminating the 1 cm sample cell (10^{-6} M DODCI in ethanol) with a CW Ar-ion laser at 514.5 nm. The ratio of the amplitudes of the first acoustic signal with the CW laser on and off was studied as a function of the CW photon fluence. In this manner the signal from N was used as reference. No alteration of the steady-state populations was introduced by the low-energy laser pulse ($E_0 < 1$ μJ).

3. Results and Discussion

A strong dependence of the signals was observed with λ of the pulsed laser and with the CW laser fluence. An isooptoacoustic point was found at ca. 595 nm. At this wavelength the signal did not depend on the CW-laser power. The results are interpreted in terms of a model with no absorption by P at the CW laser λ and negligible excited states populations. The triplet state population is also negligible, due to the small intersystem crossing yield ($\Phi_{isc} < 0.005$)[3], and efficient triplet quenching by O_2 in the air saturated solutions. Under these conditions the steady-state populations of both N and P ground states, N_0 and N_2, respectively, are given by:

$N_0 = N/(1 + a\ \sigma_N\ \Phi_{NP}/k\ )$, $N_2 = N - N_0$, where N is the total number of molecules, a the CW photon fluence, σ_N the absorption cross section of N at 514.5 nm, Φ_{NP} the quantum yield of the photoisomerization and k the rate constant for the process P⟶N on the ground states potential energy surface. Since the pulsed dye laser is absorbed by N and P, the signal arises from the radiationless processes undergone by both species. The amplitude of the first deflection H can be written as :
$H = K(\alpha_N\ E_N + \alpha_P\ E_P)$, where K includes instrumental factors and thermoelastic properties of the sample, E_N and E_P are the energies absorbed by N and P, and α_N and α_P are the fractions of heat dissipated promptly by N and P, respectively. Eq. (1) results for H:

$$H = K\ E_0\ \left[\frac{\alpha_N\ +\ B(\lambda)\ \Omega\ \alpha_P}{1 + B(\lambda)\ \Omega}\right]\left[\ 1\ -\ 10^{-A\left(\frac{1\ +\ B(\lambda)\ \Omega}{1\ +\ \Omega}\right)}\right], \qquad (1)$$

where $B(\lambda) = \sigma_P(\lambda)/\sigma_N(\lambda)$ is the ratio of the absorption cross sections of P and N at the pulsed laser wavelength, A is the absorbance of the sample at low fluence, and $\Omega = a\ \sigma_N\ \Phi_{NP}/k$. Under the same conditions the signal H_0 was measured without the CW laser background. In this case $H_0 = k\ E_0\ \alpha_N(1 - 10^{-A})$ and the ratio H/H_0 is given by eq. (2) used to adjust the results:

$$\frac{H}{H_0} = \frac{(\ \alpha_N\ +\ B(\lambda)\ \Omega\ \alpha_P\ )}{(1\ +\ B(\lambda)\ \Omega)\ \alpha_N\ \ (1\ -\ 10^{-A})}\left[\ 1\ -\ 10^{-A\left(\frac{1\ +\ B(\lambda)\ \Omega}{1\ +\ \Omega}\right)}\right]. \qquad (2)$$

At the isosbestic point ($B(\lambda) = 1$) eq. (2) is reduced to $H/H_0 = (1 + \Omega\ \alpha_N/\alpha_N)/(1 + \Omega)$, which is independent of the absorbance. This expression was used to fit the results at 600 nm resulting in $\alpha_P = 0.96 \pm 0.02$. $\Phi_f(P) = 0.05$ and the quantum yield of reverse photoisomerization, $\Phi_{PN} = 0.02$, were calculated combining $\alpha_P = (1 - \Phi_f(P) + \Delta E\ \Phi_{PN}\ \lambda_1/hc)$, with the molar energy content of the ground state of P, $\Delta E = 1.3$ eV [2], and data derived from fluorescence measurements [4]. hc/λ_1 is the energy of the laser pulse.

H/H_0 was measured between 579 and 630 nm. From the fitting with eq. (2) the $B(\lambda)$ values and hence the $\sigma_P(\lambda)$ values resulted. Over 630 nm, the small absorption of N rendered the use of the reference (negligible H_0) difficult. LIOAS were performed at $\lambda > 630$ nm and σ_P values calculated for 630-655 nm, relative to $\sigma_P(630)$. This yields a precise spectrum of P. The parameters determined in this work support our previous conclusion [1,4] that the main radiationless process of excited P is internal conversion without passing through the twisted state.

4. References

[1] G.M. Bilmes, J.O. Tocho and S.E. Braslavsky: Chem. Phys. Lett. 134, 355 (1987)

[2] G.M. Bilmes: Ph.D. Thesis, University of La Plata, Argentina, 1987

[3] D.N. Dempster, T. Morrow, R. Rankin and G.F. Thompson: J. Chem. Soc. Faraday Trans. II 68, 1479 (1972)

[4] L. Scaffardi, G.M. Bilmes, D. Schinca and J.O. Tocho: Chem. Phys. Lett. in press (1987)

Step-and-Integrate Interferometry in the Mid-Infrared with Photothermal Beam Deflection and Sample-Gas-Microphone Detection

R.A. Palmer[1], *M.J. Smith*[1], *C.J. Manning*[1], *J.L. Chao*[2], *A.C. Boccara*[3], *and D. Fournier*[3]

[1]Department of Chemistry, Duke University, Durham, NC 27706, USA
[2]IBM, Research Triangle Park, NC 27713, USA
[3]Laboratoire d'Optique Physique, ESPCI, 10, rue Vauquelin, F-75231 Paris, Cedex 05, France

The utility of photothermal (PT) detection in Fourier transform infrared spectrometry (FTIR-PTS) has been amply illustrated during the past several years. However, the rapid scanning mode, which is characteristic of all current commercial FTIR instruments, has several inherent disadvantages for PT detection, which have prevented the full realization of the potential of the method. The source of the problem is the temporal modulation of the spectrometer beam intensity, which also naturally creates the PT modulation in rapid scanning FTIR-PTS. The fact that the Fourier frequencies are different for each wavelength and that, for even the lowest practical mirror velocities, these frequencies are relatively high for mid-infrared wavelengths, has three negative consequences for PT detection: (1) lock-in amplification is not possible, (2) the thermal diffusion depth is different for all wavelengths and (3) the modulation frequencies are generally higher than desirable for optimum PT signal strength. These problems are shared by both common types of photothermal detection, sample-gas-microphone and PT beam deflection (the mirage effect). However, the lack of lock-in amplification is particularly serious for mirage detection because of its generally greater sensitivity to external vibration.[1]

The solution to this dilemma is to replace the rapid scanning mode with step-and-integrate scanning. This, in fact, is the obvious choice of scanning mode for all Fourier transform interferometry with PT detection since it provides for low-, single-frequency modulation and thus enhanced S/N with lock-in detection, and a thermal diffusion depth constant with respect to wavelength.[2] Although these advantages have been used with great effect in the NIR and VIS,[3] they have not heretofore been realized in conjunction with modern commercial FTIR instrumentation. In this paper we report progress in the development of such an instrument and illustrate its use with both sample-gas-microphone and mirage effect detection.

An IBM IR/44 optical bench and interferometer (with airbearing supported mirror) has been used for the instrument. The step-and-integrate drive control electronics are modeled after the circuits described by Debarre, Boccara and Fournier,[4] with necessary modifications required by the different mode of moving mirror support. Computer control and data manipulation programs include the applicable parts of the IR/44 software package running on an IBM PC/AT, supplemented by code specific to the step-and-integrate drive and data collection. Sample-gas-microphone data were collected using a PAR FTIR-PAS cell, and mirage data were measured using a compact beam deflection device similar to that described by Charbonnier and Fournier.[5]

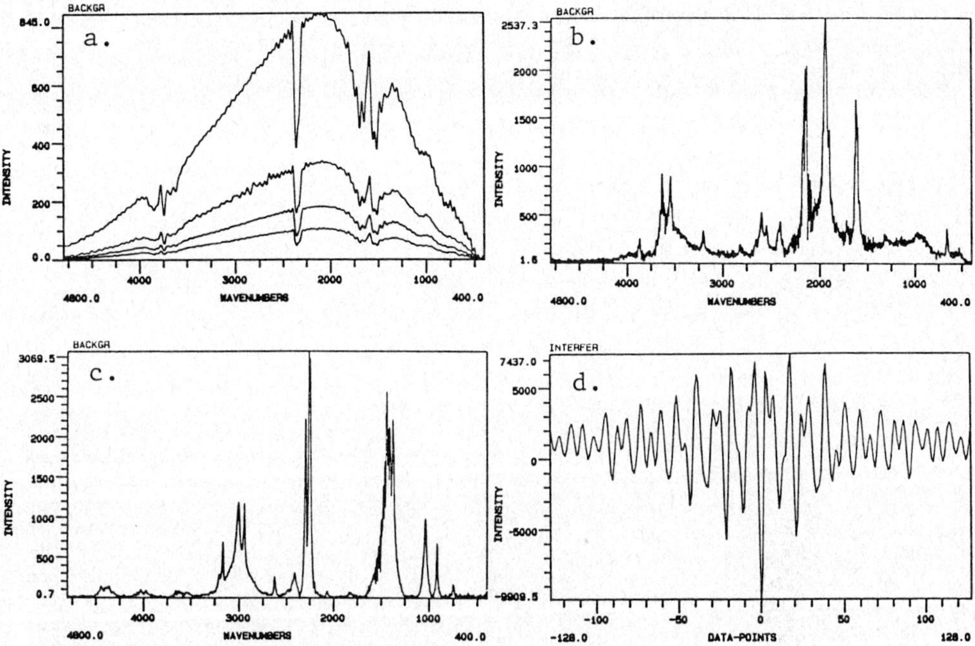

Fig. 1a. Carbon Black PAS at 50, 75, 100, 125 Hz (from top) modulation (35 cm^{-1}). b. PAS spectrum of sodium nitroprusside (8 cm^{-1}, 100 Hz). c. Mirage spectrum at Si-CH_3CN interface (8 cm^{-1}, 200 Hz). d. Interferogram of c.

Similar to the instrument described previously,[4] the interferometer moving mirror is controlled by an electronic circuit which monitors a HeNe laser interference pattern. The circuit consists of a signal generator, two lock-in amplifiers and a signal processing board, configured as a feedback loop. The loop includes the laser interference detector, alternately one or the other of the amplifiers, the signal processing board and the moving mirror via its linear motor drive. The mirror position is sensed by sinusoidally modulating it, and detecting the resulting modulation of the laser interference signal. Any error is converted to a current which corrects the mirror's position. Pulses from the computer cause the mirror to step by switching the signal path and phase in the signal processing board. The feedback then drives the mirror to the next position. The step is one-eighth of the HeNe wavelength (79 nm), corresponding to $(1/4)\lambda$ retardation. The stepping frequency is adjusted so as to allow the desired data collection time at each position. The $(1/4)\lambda$ step creates a free spectral range of 31,600 cm^{-1}, although the optics of the instrument as presently configured limit operation to the 400-4800 cm^{-1} range.

Although the mirror jitter can be used to generate the PT modulation [4], the amplitude of mirror jitter appropriate for stepping control is too small for optimum phase modulation of IR wavelengths. Modifications are in progress to implement internal (phase) modulation with PT detection. However, in the initial experiments reported here, external chopper (amplitude) modulation has been used.

The preliminary data shown below indicate the potential of step-and-integrate FTIR coupled with PT detection. The 8 cm^{-1} resolution data were acquired in 1.25 hr. Rapid scan data collected in the same time typically show significantly better S/N. However, modifications to the instrument, including implementation of phase modulations[6], will improve this comparison significantly.

1. See for example: R.A. Palmer and M.J. Smith, Can. J. Phys., 64, 1086. (1986) and M.J.D. Low and M. Lacroix, Infrared Phys., 22, 139 (1982).
2. L.B. Lloyd, S.M. Riseman, R.K. Burnham, E.M. Eyring and M.M. Farrow, Rev. Sci. Inst., 51, 1488 (1980).
3. W.B. Jackson, N. Amer, A.C. Boccara and D. Fournier, Appl. Opt., 20, 1333 (1980).
4. D. Dèbarre, A.C. Boccara and D. Fournier, Appl. Opt., 20, 4281 (1981).
5. F. Charbonnier and D. Fournier, Rev. Sci. Inst., 57, 1129 (1986).
6. J. Chamberlain, Infrared Phys., 11, 25 (1971).

Photoacoustic UV Spectra of Some Environmental Chemicals with Low Volatility Adsorbed on SiO_2 and Al_2O_3 Powders

H. Scharping and C. Zetzsch

Fraunhofer-Institut für Toxikologie und Aerosolforschung,
Nikolai-Fuchs-Str. 1, D-3000 Hannover 61, Fed. Rep. of Germany

1 Introduction

Adsorption of molecules on the surface of solid substances often causes perturbations of their electronic spectra. These perturbations may result in shifts of the absorption spectra, and in some cases large red shifts have been observed /1/. Under these circumstances, chemicals normally transparent to sunlight in the gas phase and in solution may become susceptible to the absorption of tropospheric sunlight at $\lambda > 300$ nm, when adsorbed on materials like sand or even silica /2, 3/, leading to heterogeneous photodecomposition in the presence of sunlight at ground level. This might be a dominant sink for these chemicals in the environment, especially for substances with low volatility, occurring predominantly adsorbed on atmospheric aerosols.

2 Experimental Details

Spectra of the substances under investigation were taken at 3 nm resolution in the wavelength region from 200 to 380 nm using a single beam apparatus consisting of a 450 W high pressure Xe arc lamp, a 25 cm Czerny-Turner-monochromator, a mechanical chopper operating at 65.5 Hz, and a photoacoustic cell (MTEC model 100, with suprasil window). The signal from the microphone was detected with a lock-in amplifier (Ithaco).

The substances used as model aerosols are three different kinds of SiO_2, namely Aerosil 200 (Degussa), silica gel 60 (15 to 40 µm diameter), and silica gel 60 reinst (63 to 200 µm) (both Merck), and aluminium oxide C (Degussa). The compounds investigated are di(2-ethylhexyl-)phthalate (Fluka), pentachlorobenzene (Janssen), and hexachlorobenzene and pentachlorophenol (Riedel de Haen).

The substances are adsorbed on the model aerosols by mixing the compound and the aerosol in a solution of frigen 113 and evaporating the solvent. The degree of surface covering, estimated from the BET surface and the organic content of the solutions, varied from less than 0.01 to 1 monolayer.

3 Results and Discussion

Spectra of all compounds adsorbed on the different aerosols are taken with different concentrations. In table 1 the absorption onsets of the lowest electronic transition (indicating the 0-0-band if allowed) are summarized for all compounds together with the data in solution.

In the adsorbed state, pentachlorobenzene shows only slight shifts which do not differ much for the different aerosols compared to the data in solu-

Table 1 Onsets of UV absorption of the aromatics in solution and adsorbed on the various powders

powder / compound	aerosil	silica gel 60 r	silica gel 60	aluminium oxide C	solution
di(2-ethylhexyl-)-phthalate	285	284	283	282	283[a]
hexachlorobenzene	305	303	308	305	302[a]
pentachlorobenzene	299	296	296	300	298[a]
pentachlorophenol	305	316	317	322	320[b]
					303[c]

[a] n-hexane solution [b] alkaline aqueous solution [c] acidic aqueous solution

tion. For hexachlorobenzene the shifts to longer wavelengths are more pronounced, probably because the high symmetry of this molecule is disturbed in the adsorbed state. The spectra of di(2-ethylhexyl-)phthalate are shifted slightly to the blue as expected for a $\pi^* \leftarrow n$-transition in contrast to $\pi^* \leftarrow \pi$-transitions like in chlorinated aromatic compounds.

Pentachlorophenol is the compound that is disturbed in the strongest way. This may be caused by the acidic phenol group interacting with surface OH-groups or adsorbed water. This is indicated as well by the different absorption onsets in acidic and alkaline solution, see table 1. Aerosil and aluminium oxide C have acidic surfaces, nevertheless aluminium oxide appears to adsorb pentachlorophenol as phenolate.

When comparing the absorption spectra with the spectrum of the sun at ground level, which has its cut-off near 300 nm, direct photolysis can be neglected for the phthalate and for pentachlorobenzene, adsorbed on inert aerosols, while for hexachlorobenzene the potential is not very large. For pentachlorophenol on the other hand the possibility of direct photolysis cannot be excluded. The present study was restricted to model aerosols that are transparent in the wavelength region under investigation. Other, natural aerosols might influence the spectra in a stronger way than the above-mentioned aerosols do.

4 References

1 P.A. Leermakers, H.T. Thomas, L.D. Weis, and F.C. James: J. Am. Chem. Soc. 88, 5075 - 5083 (1966)

2 P. Ausloos, R.E. Rebbert, and L. Glasgow: J. Res. Natl. Bur. Standards 82, 1 - 8 (1977)

3 S. Gäb, S. Nitz, H. Parlar, and F. Korte: Chemosphere 4, 251 - 256 (1975)

Photopyroelectric Spectroscopy of Water in the Near Infrared

M. Chirtoc[1], *D. Dadârlat*[1], *I. Chirtoc*[1], *and D.D. Bicanic*[2]

[1]Institute of Isotopic and Molecular Technology, R-3400 Cluj-Napoca 5, P.O. Box 700, Romania
[2]Department of Physics and Meteorology, Wageningen Agricultural University, Duivendaal 2, NL-6701 AP Wageningen, The Netherlands

1 Introduction

The advantages of frequency domain photopyroelectric spectroscopy (PPES) of the condensed phase over conventional photoacoustic spectroscopy (PAS) has been discussed by Mandelis [1]. Measurements of the absorption coefficient β for strongly absorbing liquids ($\sim 400\ cm^{-1}$) by means of CO_2 laser transverse photothermal deflection method have been reported recently [2]. This paper is concerned with a new photopyroelectric method based on the reflection mode of operation. Essentially, the technique is based on detecting the heat changes in the medium due to the periodic absorption using a pyroelectric detector. Clearly, the detector should only sense the heat developed in the medium and not the exciting radiation itself [3].

2 Experimental

The experimental assembly used in this study consists of the chopped (12 Hz) quartz tungsten source (2410 K) the radiation of which is focussed (NaCl lens) into a Carl-Zeiss SPM-2 prism monochromator and the measuring PPE cell-detector assembly. This latter, shown in Fig.1 is a lead zirconate titanate PZT ceramic disc (10 mm diameter and 1 mm thick) provided with metalized Al electrodes. Highly reflecting Al foil F, 10 µm thick, is glued to the detector by means of the silicone adhesive. Two Mylar spacers (strips) S having thickness of 60 µm, separate the reflecting surface from the 150 µm thick glass window W (transmittance T is 0.91 for the wavelength range from 0.8 to 2.7 µm) and define at the same time the volume that accommodates the injected liquid sample. Due to the presence of Al-electrodes the radiation leaving the monochromator traverses the liquid twice, generating the periodic heat and the proportional PPE signal when the sample is thermally thin [4]. Additional heat is developed upon the reflection at the Al-electrodes and in the window while the higher order reflections can be neglected [3].

3 Results

Figure 2 displays the signals $S(\lambda)$ and $S_R(\lambda)$ obtained at a given wavelength λ with the PPE cell filled with distilled water and with the empty cell respectively. These signals were normalized versus another pyroelectric detector with flat spectral response. The absorption peaks of water are observed at 1.43 and 1.92 µm. The absorption peak centered around 3 µm in Fig.2 appears saturated due to a very large β value. The absorption coefficient β of the sample can be found from a total heat balance for the cell and Beer's law as

$$\beta = (2\ d)^{-1} \ln \left[\frac{T^2(\lambda)\ R(\lambda)}{1 - S(\lambda)}\right], \qquad (1)$$

where $R(\lambda)$ represents the reflectance of the electrode and is proportional to $S_R(\lambda)$ [3]. The calculated values for the absorption coefficient at the two

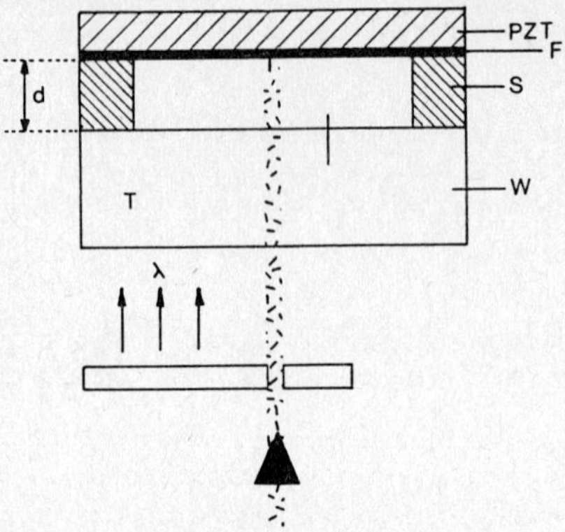

Figure 1: The photopyroelectric cell used in this study.

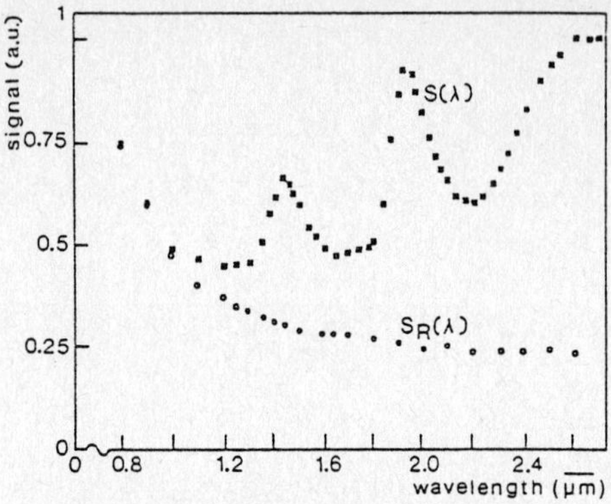

Figure 2. The photopyroelectric signal of distilled water S(λ) and the residual cell absorption $S_R(\lambda)$ in the near infrared region.

wavelengths mentioned above are 48.6 cm^{-1} and 175 cm^{-1} respectively, which is in good agreement with the expected literature values [5].

For a given value of spacer thickness d this technique provides a simple means to evaluate the absorption coefficient β ranging anywhere from the optically thin region ($2\,d\,\beta < 1$) to the optically thick region ($2\,d\,\beta > 1$).

According to the above relationship ($\beta \sim \frac{1}{2}d$) the use of a thinner spacer allows the studies of material samples with larger β values. In the absence of photothermal saturation, the practical limit of β measurability is actually determined by the experimentalist's capability to produce thin spacers of satisfactory mechanical strength.

4 References

1. A.Mandelis: Chem. Phys. Letts. 108, 388 (1984).
2. D.D.Bicanic et al.: submitted for publication to Appl. Phys. A.
3. M.Chirtoc, I.Chirtoc, D.Dadarlat, D.Bicanic: will be submitted to Appl. Spectr.
4. A.Mandelis, M.Zver: Jour. Appl. Phys. 57, 4421 (1984).
5. W.Pepperhoff: Wissensch. Forschungsberichte Naturwiss. Reihe 65 (1956).

Chalcopyrite Semiconductors – A PAS Study

S. Venkataraman and A.K. Bhatnagar

School of Physics, University of Hyderabad,
Central University P.O. Hyderabad-500 134, India

The four chalcopyrite semiconductors $CuInSe_2$, $CuGaSe_2$ and $CuGaS_2$ belong to the I-III-VI_2 group and are of technological importance as terrestrial photovoltaic and non-linear optical materials. Study of optical properties of polycrystalline bulk materials by conventional spectroscopic methods is rather difficult and cumbersome, especially for highly absorbing powder samples. Photoacoustic spectroscopy (PAS) can be conveniently used to study optical and thermal properties of such compounds.

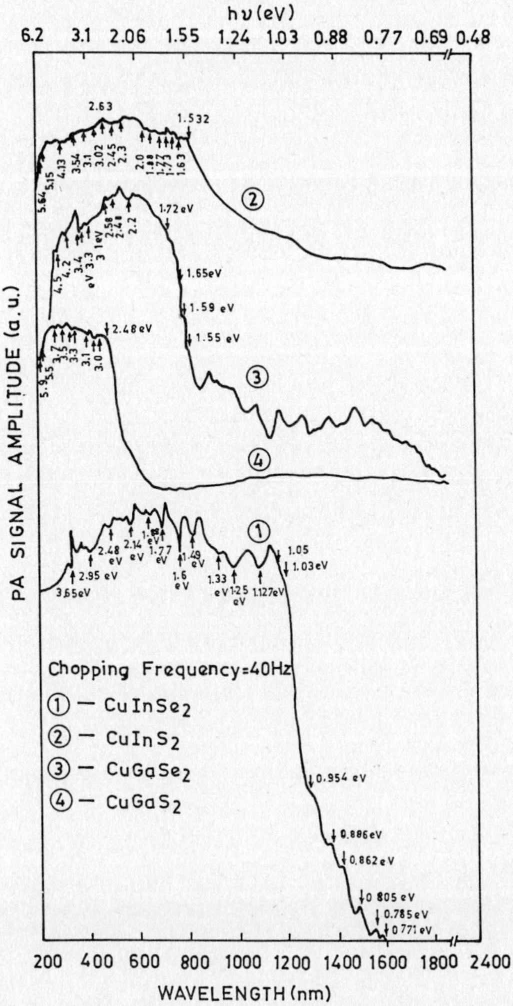

Fig.1: Typical PAS amplitude spectra of $CuInSe_2$, $CuInS_2$, $CuGaSe_2$, $CuGaS_2$ polycrystalline semiconductors. All spectra are not of the same magification

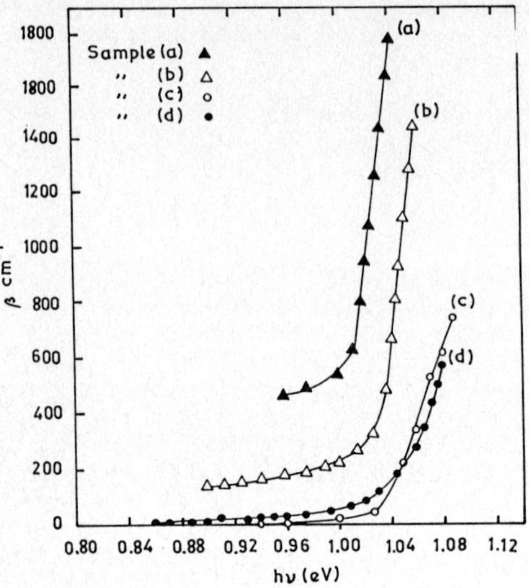

Fig.2: $\beta(cm^{-1})$ vs. hν of $CuInSe_2$ bulk samples

Fig.3: $\beta(cm^{-1})$ vs. hν of $CuInS_2$ bulk ▶

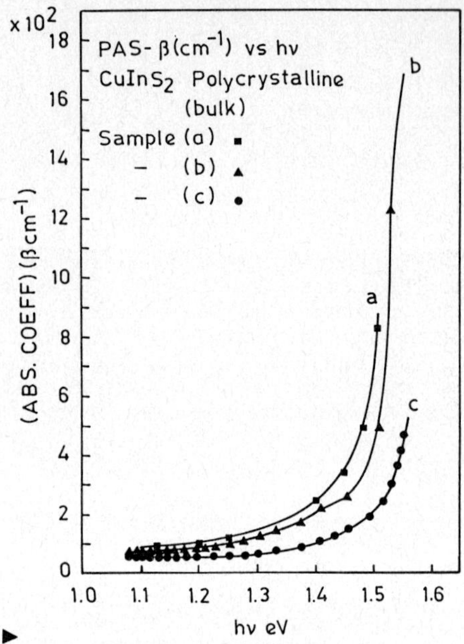

$CuMX_2$ (M=In/Ga, X=S/Se) compounds are prepared by direct fusion methods starting with 5N pure elements and are characterised by XRD and AES techniques for identification of chalcopyrite phase and composition. All PASs (amplitude) are recorded with a μ-p controlled PAR 6001 photoacoustic spectrometer from 200 to 2400 nm using carbon black spectra as the reference. The compounds are optically opaque and thermally thick and hence there is no dependence of the spectra on the sample cup design [1]. Figure 1 shows the amplitude spectra of the four chalcopyrites as a function of incident energy (hν). The absorption coefficient (β) vs. hν and $(\beta)^{\frac{1}{2}}$ vs. hν calculated using the relations of Fesquet et al. [2], and the thermal properties of $CuInSe_2$, $CuInS_2$ and $CuGaS_2$ samples are shown in Figs. 2-7. The observation of large band tails below the absorption edge as shown in Figs. 5-7 are due to Dow-Redfield effects [3] caused by the electric fields in the grain boundaries of polycrystalline materials. The refractive index n at E_g of the chalcopyrites is computed using the Moss relation [4]. Table I summarises the results of E_g, nature of band transition, refractive index n at E_g and the intrinsic defect level transitions of the four chalcopyrites. The values 1.03, 0.94, 0.88 eV for $CuInSe_2$ are identified as the free-exciton transition, and donor-acceptor transitions between Se vacancy donors and Cu or In antisite acceptors, respectively, in accordance with Lange et al.[5]. In $CuGaSe_2$ the peak at 1.65 eV is a transition from the free electron to the bound hole (Cu or Se vacancies). The peak at 1.59 eV can be related to the donor-acceptor transitions in $CuGaSe_2$. No defect transitions have been observed in $CuInS_2$ and $CuGaS_2$ from the room

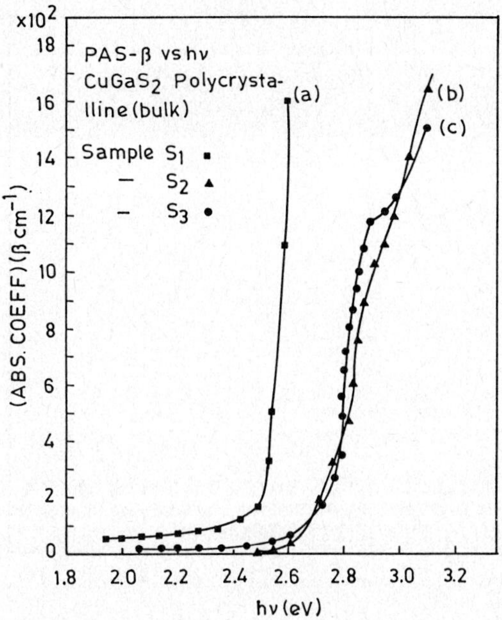

Fig.4: $\beta(cm^{-1})$ vs. hν of $CuGaS_2$ bulk sample

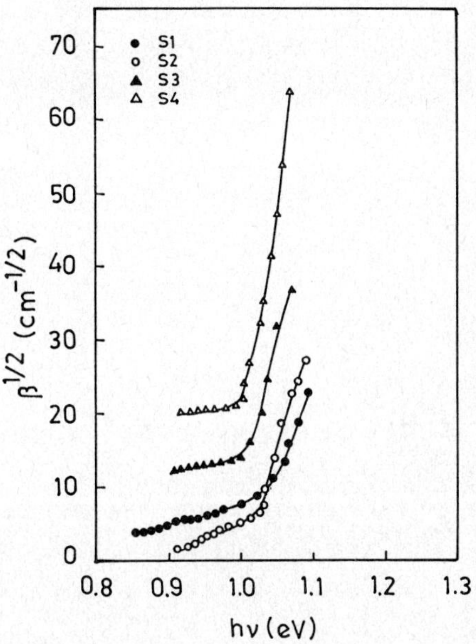

Fig.5: $\beta^{1/2}$ vs. hν of four $CuInSe_2$ samples

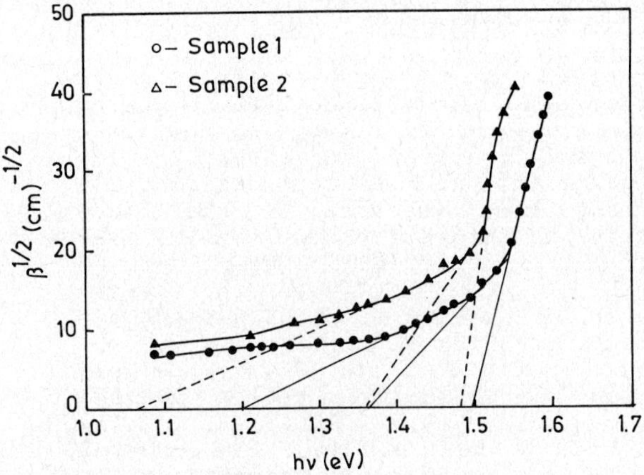

Fig.6: $\beta^{1/2}$ vs h of $CuInS_2$

TABLE I

Material	E_g at 300 K ± 0.02 eV [eV]	Band transition	n at E_g	Defect level transition ± 0.02 eV [eV]
$CuInSe_2$	1.05	Direct	3.1	1.03,0.94,0.89
$CuInS_2$	1.53	Direct	2.8	--
$CuGaSe_2$	1.72	Direct	2.73	1.65,1.59,1.55
$CuGaS_2$	2.48	Direct	2.48	--

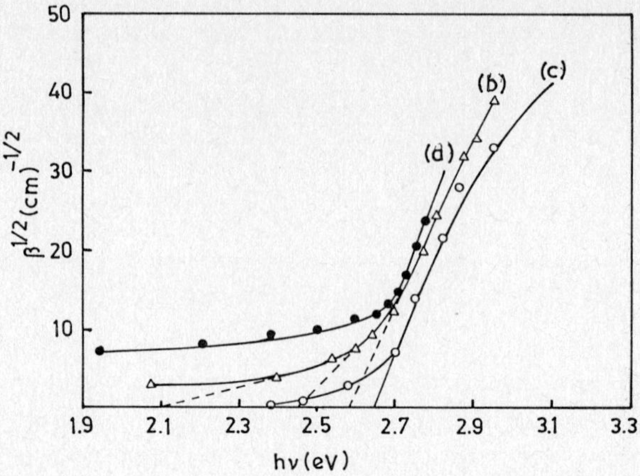

Fig.7: $\beta^{1/2}$ vs. hν of $CuGaS_2$

temperature spectra. In Fig. 1 the high energy transitions above the E_g transition correspond to the reflectivity spectrum of the chalcopyrites. The values between 2.5 to 4.0 eV can be assigned to transitions from Cu d-levels to the bottom of the conduction band [6] and below that to the polycrystallinity of the films.

REFERENCES

1. L.C. Aamodt, J.C.Murphy and J.C.Parker, J.Appl. Phys. Vol:48, No:3, 1977, pp: 927-933.

2. J.Fesquet, B.Girault and M.D.M.Razafindrandriatsimaniry, Applied Optics, Vol:23, No:16, 1984, 2784-2787.

3. L.L.Kazmerski, M. Hallerdt and P.J.Ireland, J.Vac. Sci. Technol. A1(2) 1984, 395-398.

4. T.S.Moss, Phys. Stat. Sol.(b). 131, 415 (1983).

5. P. Lange, H. Neft, M. Fearheiley and K.J. Bachmann, Phys. Rev. B, Vol:31, No:6, 1985, 4974-4976.

6. W. Horig and H.Neumann and I.Godmanis, Solid State Communication, Vol:36, pp. 181-183, 1980.

Optical Properties of Cholesteric Liquid Crystals by Photoacoustics

A. Hadj-Sahraoui[1,2], *G. Louis*[1], *P. Peretti*[1,2], *and J. Billard*[3]

[1]D.R.P. Université Pierre et Marie Curie, CNRS UA. 71, Place Jussieu, Tour 22, F-75252 Paris Cedex 05, France
[2]LPPC Université, F-75252 Paris Cedex 05, France
[3]LPMC, Collège de France, CNRS UA. 542, Place Marcellin Berthelot, F-75231 Paris Cedex 05, France

In a recent paper /1/, we have used the photoacoustic technique to study the second order smectic A-nematic phase transition in cyanobiphenyl liquid crystals. The aim of the present work is to detect selective reflection of light with a cholesteric mesogen.

The cholesteric liquid crystals structure is formed by the helical arrangement of thermotropic mesogen molecules. Several optical properties, such as very strong apparent rotatory power, arise from this spatially periodic structure and depend upon the cholesteric pitch.

When a linearly polarized light beam is sent parallel to the helical axis, polarization features of reflected and transmitted waves are spectacular. If we analyze the incident light into two components of opposite circular polarizations, only the component with the instantaneous spatial field pattern matching the spiraling cholesteric direction, is strongly reflected. The second component is transmitted without significant reflection loss. For a given value of the pitch, a right handed cholesteric mesogen reflects right circularly polarized light /2/ and transmits left circularly polarized light. The cholesteric pitch is a function of the temperature. Any variation of the sample temperature shows different colours in reflection. This property is used for medical and industrial temperature control.

Our experiments were performed on a 100 µm thick chiral optically transparent sample. To minimize the Lehmann effect /3/, we have chosen a mesogen having a cholesteric mesophase with small temperature dependence of the pitch : the 2-methyl-butyl-4-benzal-amino-4'- cinnamate /4/, /5/. This sample exhibits a cholestericNd-liquid L phase transition at 108.5°C. The selective reflection wavelength variation as a function of temperature is weak and anomalous. During the cooling from the liquid phase, this wavelength varies from red to yellow.

The experimental set-up is a standard photoacoustic configuration /1/. The light source is an Argon ion laser operating at 488 nm and 514.5 nm. The laser of 130 mW is intensity-modulated at 160 Hz by an acousto-optic modulator. The light beam is defocused to avoid thermal gradients into the sample which is thermostated.

Both amplitude and phase of photoacoustic signal have been plotted as a function of temperature. Figure 1 shows the results obtained starting from the crystalline phase. The left or right circular polarization of the incident light is obtained by insertion of a quarter wave plate. We observe in a small temperature range, a peak corresponding to the selective reflection of the light. The peak position increases 10°C when the wavelength

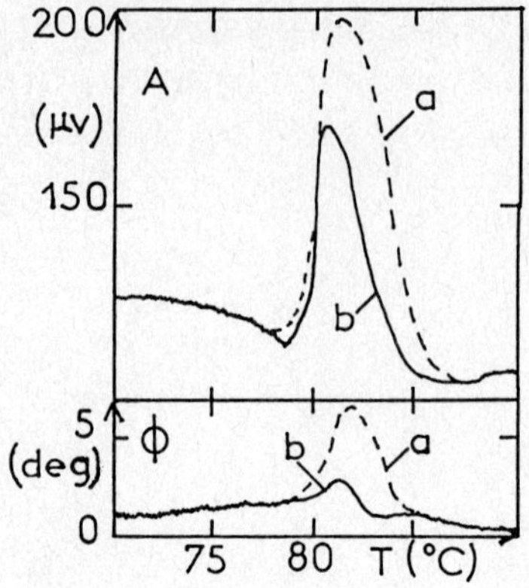

Fig. 1. Temperature dependence of amplitude A and phase angle Ø as a function of temperature. The temperature change rate is 1.5°C min^{-1}. (a) and (b) : right and left circularly polarized light (488 nm) respectively.

excitation increases from 488 nm to 514.5 nm. If the chiral phase is reached through cooling, the peak deviates - 6°C from the precedent value. This hysteresis may be caused by long time scale molecular rearrangements. The left to right ratio has a finite value different from zero indicating that our sample-droplet does not have a uniform orientation.

The signal enhancement observed at the selective reflection temperature may be explained as follows : the Bragg reflection increases the optical path in a zone adjacent to the free surface. The depth of this zone is equal to the thermal diffusion length (about 22 μm).

These preliminary results show that the photoacoustic method is a relevant technique to study optical properties of twisted anisotropic media.

References

1. G. Louis, P. Peretti, J. Billard and B. Mangeot: Mol. Cryst. Liq. Cryst. 122, 261 (1985)
2. F. Giesel: Phys. Z. 11, 192-3 (1910)
3. P.G. De Gennes: The Physics of Liquid Crystals, Clarendon Press, Oxford (1974) p. 258
4. H.Stolzenberg: Dissert. Halle (1911)
5. M. Leclercq, J. Billard and J. Jacques: Mol. Cryst. Liq. Cryst. 8, 367 (1969)

High Accuracy Measurements of Reflectivity of Laser Mirrors by a Novel Photoacoustic Technique

H.G. Walther, E. Welsch, and P. Sladky

Friedrich-Schiller-Universität Jena, Sektion Physik, Max-Wien-Platz 1, DDR-6900 Jena, GDR

The measurement of mirror reflectivities R of about 1 is of current interest in modern optics. Conventional optical methods such as R/R_o - ratio measurements fail in the determination of $R \gtrsim 0.999$. The determination of R by measuring the finesse of a passive resonator is possible but complicated due to the expensive experimental set-up /1/.

The promising PA-technique suggested by us /2/ enables a separate and absolute estimation of the optical losses transmittance T, absorptance A and light scattering in the backward and forward directions S_B and S_F. R has to be calculated by

$$R = 1-(T + A + S_B + S_F). \qquad (1)$$

The experimental procedure is carried out by means of a specially designed gas-cell-microphone set-up as shown in Fig. 1.

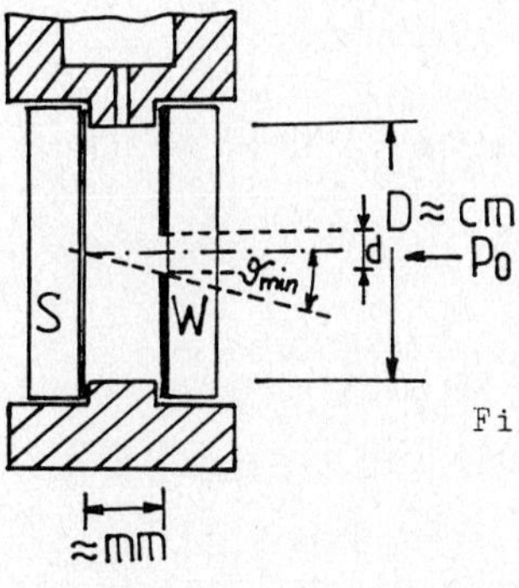

Fig. 1: Cross section of the PA-cell

Here the entrance window W is made of transparent material with one of its two surfaces almost fully overcoated by a black absorber film with the exception of only a small aperture of diameter d. By simply turning W inside the cell we can perform two states of operation characterized by the different complex-valued sensitivities ζ_1 (transparent surface inside) and ζ_2 (blackened surface inside).

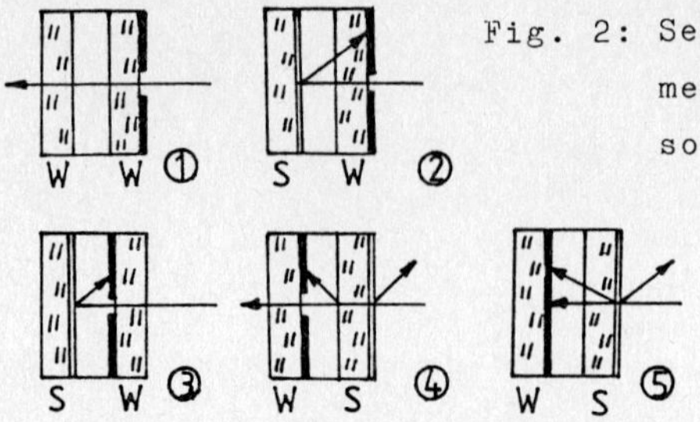

Fig. 2: Sequence of PA-measurements for solving (2)-(6)

Corresponding to the various positions of sample (S) and window as shown in Fig. 2 we find the complex-valued PA-signals

$$\mathfrak{M}_1 = 2\, P_1^W A_W \zeta_1 \tag{2}$$
$$\mathfrak{M}_2 = (P_2^W A_W + P_2^S A)\, \zeta_1 \tag{3}$$
$$\mathfrak{M}_3 = \mathfrak{M}_2 + P_3^S S_B \zeta_2 \tag{4}$$
$$\mathfrak{M}_4 = T\, \mathfrak{M}_1 + P_4^S S_F \zeta_2 \tag{5}$$
$$\mathfrak{M}_5 = \mathfrak{M}_4 + T\, P_o \zeta_2 \tag{6}$$

where P^x is the actual light power density at cell locus x.

Rearranging (2)-(6) a separate determination of the optical mirror losses becomes possible.

In order to test the method described above, we investigated dielectric argon-ion laser mirrors illuminated within the PA-cell by an 1-W laser beam at 515 nm. The experimental results yielded are summarized in Table 1.

Table 1: Measured optical losses and reflectance of a dielectric Ar-ion laser mirror

A (%)	S_B (%)	S_F (%)	T (%)	R (%)
0.016	0.0179	0.004	0.02	99.79 ± 0.04

The error limit in R- measurement was estimated to be within the range $\delta R \approx 2 \times 10^{-4}$ for R = 0.999 and $\delta R \approx 8 \times 10^{-4}$ for R = 0.996.

/1/ J. M. Herbelin, J. M. McKay; Appl. Opt. 20, 3341 (1981)
/2/ H.G. Walther, E. Welsch; Patentschrift DD 240 955 A1

Part II

Kinetics and Relaxation

Optoacoustic Measurement of Isotope Effects in the H_2–Cl_2 Reaction

S.G. Wyers and G.J. Diebold

Department of Chemistry, Brown University, Providence, RI02912, USA

ABSTRACT

The progress of photochemical chain reactions can be monitored with the chemically amplified optoacoustic effect. Measurement of the acoustic amplitude as a function of time in $H_2 - Cl_2$ and $D_2 - Cl_2$ mixtures irradiated with the $488nm$ output of an Ar ion laser gives a profile of reaction rate versus time. Measurement of the relative rates of reaction for H_2 and D_2 with Cl_2 obtained from the optoacoustic data along with thermochemical data permits relative rate parameters for these two reactions to be determined.

I. INTRODUCTION

When exothermic chain reactions are initiated by absorption of radiation, the release of energy through the consumption of reactants causes a heating of gas which gives rise to the optoacoustic effect. An example of this is the two-center chain reaction of H_2 with Cl_2 through the sequence

$$Cl_2 + h\nu \rightarrow 2Cl \tag{1}$$

$$Cl + H_2 \underset{k_{-2}}{\overset{k_2}{\rightleftarrows}} HCl + H \tag{2}$$

$$H + Cl_2 \overset{k_3}{\rightarrow} HCl + Cl \tag{3}$$

$$2Cl + M \underset{k_{-4}}{\overset{k_4}{\rightleftarrows}} Cl_2 + M\,, \tag{4}$$

where $h\nu$ represents the incident photon energy. When H_2 is not present and only reactions 1 and 4 describe the chemical processes in the gas then the optoacoustic effect comes as a result of the deposition of the bond energy of the Cl_2. The presence of H_2 changes the mechanism of sound generation altogether. The evolution of heat through numerous cycles of reaction 2 followed by reaction 3 dominates the liberation of energy in the system so that the optoacoustic signal[2] becomes (in the steady state) proportional to $k_2[H_2][Cl]$. The exothermicity of the $H_2 - Cl_2$ reaction combined with the long chain length of the reaction (defined as the number of HCl molecules produced per Cl radical introduced) gives rise to a pronounced chemical amplification[3,4] of the optoacoustic signal. Note however that the kinetics are still determined by the three-body recombination reaction 4. That is, when $d[H]/dt$ is zero, reactions 2 and 3 taken together leave the Cl concentration unchanged so that the overall Cl concentration is determined by reactions 1 and 4.

As a consequence it is possible to compare the magnitude of the optoacoustic signals in $H_2 - Cl_2$ and $D_2 - Cl_2$ mixtures under identical excitation conditions and to determine from the experimental data relative kinetic parameters for the two reactions. The theory for such a quantitative comparison is given in Section II; experiments are reported in Section III. It is shown that, along with thermodynamic data, reliable values of the rate constant for reaction 2 can be obtained.

II. THEORY

The time rate of change of the chlorine concentration has contributions from all of reactions $1-4$ and contains 6 terms. The corresponding equation for $d[H]/dt$ has three terms, which, given the steady state assumption for $[H]$ permits reduction of the Cl rate equation to
2

$$\frac{dx}{dt} + a\,x^2 = b + \delta \sin \omega t\,, \tag{5}$$

where

$$x = [Cl]/[Cl_2]\,, \qquad a = 2k_4[M][Cl_2]\,,$$

$$b = 2(\rho B + k_{-4}[M]),\ \delta = 2\rho B d\ ;$$

ρ is the radiation density, B is the Einstein coefficient for absorption and photodissociation, d is the modulation depth of the radiation, ω is the modulation frequency, and t is the time. This equation is a nonlinear, first order differential equation of the form of a generalized Riccati equation, whose solution is not apparent. Provided that the modulation depth of the radiation ($d \cong \delta/b$) is small compared with unity, then a perturbation solution to Eq 5 can be obtained[2,5] for x. The rate of energy release into translational heating of the gas is given by

$$\frac{d\epsilon}{dt} = \frac{k_2[H_2][Cl]}{1 + k_{-2}[H\ Cl]/k_3[Cl_2]}\Delta U\quad , \tag{6}$$

where ϵ is the energy per unit volume of the gas, and ΔU is the reaction exothermicity for reactions 2 and 3. Substitution of the perturbation solution to Eq 5 into Eq 6 along with the ideal gas law ($dp/dt = (\frac{2}{5})\frac{d\epsilon}{dt}$) gives the fundamental frequency component of the pressure as

$$p(t) = \frac{-(\frac{1}{5})k_2[H_2][Cl_2]\Delta U}{1 + k_{-2}[HCl]/k_3[Cl_2]}\left[\frac{(\delta/b)(b/a)^{\frac{1}{2}}}{\omega(1+\lambda^2)^{\frac{1}{2}}}\right] cos(\omega t - \phi) \quad , \qquad (7)$$

where $\lambda = \omega/2(ab)^{\frac{1}{2}}$ and $\phi = tan^{-1}\lambda$. This solution is valid when the vibrational relaxation rate of HCl is much faster than the modulation frequency, and when the sound wavelength is much larger than the dimensions of the cell. In principle, the magnitude of the acoustic signal could be determined as the reaction proceeds given values for the time dependent H_2, Cl_2 and HCl concentrations and the various rate constants for reactions 1– 4.

This solution shows the usual ω^{-1} dependence on frequency when $\lambda >> 1$. The signal falls off more rapidly when $\lambda >> 1$ (at high frequency or very low light intensity). A linear dependence on modulation depth δ/b is also evident. In contrast with the optoacoustic effect in the infrared (arising from vibrational relaxation), Eq 7 predicts a square root dependence of the signal on radiation intensity for $\lambda << 1$. On the other hand, for $\lambda >> 1$ (low light intensity or high modulation frequency) the signal is linear in intensity for a fixed modulation depth. It is noteworthy that the acoustic signal can be written in terms of the photochemical chain length, which is defined[6,7] as

$$\Phi_{ph} = \frac{2k_2[H_2]}{1 + k_{-2}[HCl]/k_3[Cl_2]}(ab_{ph})^{-\frac{1}{2}} \quad , \qquad (8)$$

where $b_{ph} = 2\rho B$. (This is valid provided $\rho B/k_{-4}[M] >> 1$.) Substitution of Eq 8 into Eq 7 gives

$$p(t) = \frac{-[Cl_2]\rho B\Phi_{ph}\Delta U}{5\omega(1+\lambda^2)^{\frac{1}{2}}}\left(\frac{\delta}{b_{ph}}\right)\cos(\omega t - \phi) \quad , \qquad (9)$$

which shows the acoustic signal to be directly proportional to the photochemical chain length. The chain length is, of course, dependent on the light intensity.

Consider now comparison of the optoacoustic signals in stoichiometric mixtures of $H_2 - Cl_2$ and $D_2 - Cl_2$ irradiated at the same pressure with the same laser power. Provided that the three-body recombination rates for Cl in the presence of H_2 and D_2 are the same[8], and that the data are taken when $[HCl] = [DCl] \cong 0$, then according to Eq 7, the pressure ratio for the deuterium reaction (denoted D) relative to the hydrogen reaction (denoted H) is given by

$$\frac{p_H}{p_D} = \frac{k_2^H}{k_2^D} \frac{\Delta U_H}{\Delta U_D} . \tag{10}$$

Thus, from a knowledge of thermodynamic parameters and optoacoustic signal amplitudes it is possible, in principle, to determine relative rate constants.

III. EXPERIMENTS

Optoacoustic signal amplitude measurements were taken in a cylindrical Teflon cell (1.8 cm diameter by 3.8 cm long) fitted with Pyrex windows (see Fig 1). The acoustic signal was detected through a thin Teflon membrane by an electret microphone whose output was fed to a lock-in amplifier. The membrane protected the microphone from the corrosive gases in the cell and yet resulted in only a small loss in signal amplitude. The 488nm output from an Ar ion laser beam was modulated by an acousto-optic light modulator at 20Hz and expanded to irradiate the cell volume uniformly. Radiation at this wavelength is known to produce two ground state Cl atoms through predissociation of the B state of Cl_2. The zeroth order beam was taken from the modulator so that the beam had a modulation depth of 0.26. Great care was taken in maintaining gas purity. Molecular oxygen is known to act as a catalyst for the removal of Cl atoms thus altering the kinetic mechanism given above[9].

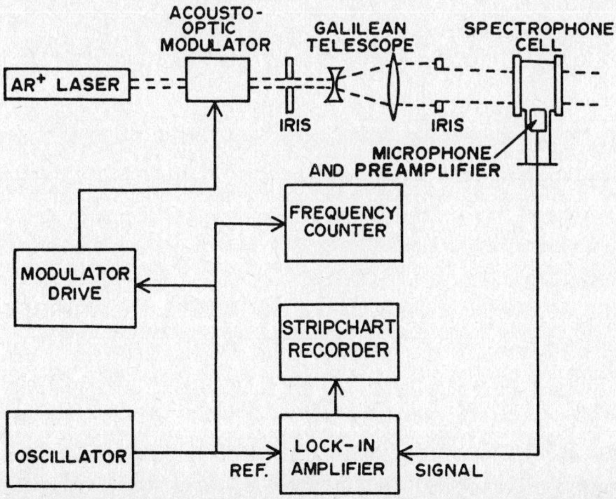

Figure 1: Diagram of the experimental apparatus

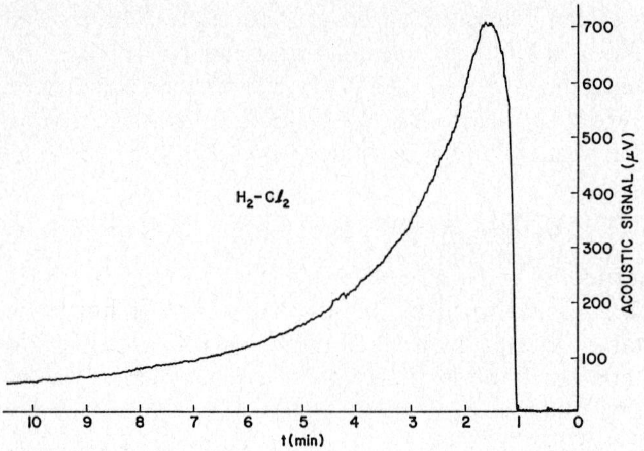

Figure 2: Lock-in amplifier signal versus time in a stoichiometric $H_2 - Cl_2$ mixture at one atm and 300 K. The lock-in time constant is $1s$.

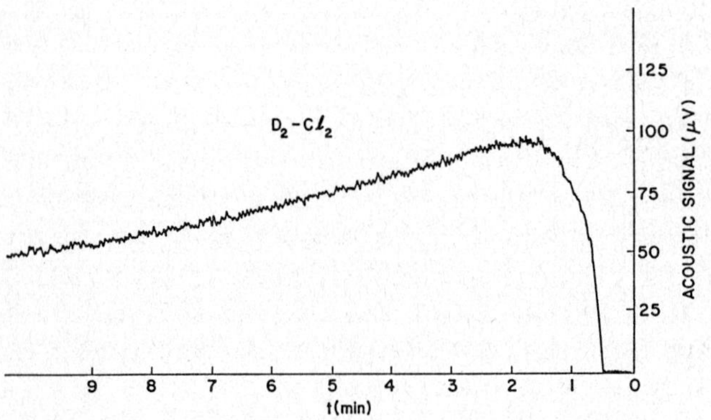

Figure 3: Lock-in amplifier signal versus time as in Fig. 2 except with $D_2 - Cl_2$.

The traces in Figs. 2 and 3 show typical optoacoustic signals as a function of time for H_2 and D_2 respectively. For the same incident radiation intensity ($7mW$) the $H_2 - Cl_2$ reaction can be seen to give a much larger signal amplitude and a faster decay than the $D_2 - Cl_2$ reaction. Although the absorption coefficient is only $4x10^{-3}\,cm^{-1}$ at $488nm$, a high signal-to-noise ratio in either trace, characteristic of the chemically amplified optoacoustic effect, can be seen. Of further note is the slow rise to the peak signal amplitude in both traces. Some loss of response is caused by the $1s$ time constant on the lock-in amplifier, however, the slow rise of the signal is clear evidence of an induction period for the reaction. Data were taken immediately after the end of the induction period. Using previous values for k_2^H/k_2^D of 9.55 (Ref. 10) or 8.98 (Ref. 11) along with[12] a value of $\Delta U_H/\Delta U_D = 1.03$, the value of p_H/p_D should be 9.8 or 9.2 for these two references, respectively. The optoacoustic experiments give a value of 7.2 ± 0.3 in fair agreement with these figures. There is, of course, some question as to when during the course of the reaction the amplitude should be recorded.

IV. DISCUSSION

An important feature of the $H_2 - Cl_2$ reaction is that it leaves the overall mole number unchanged in the spectrophone. It has been shown previously[1] that a change in mole number results in an additional mechanism for acoustic wave generation that considerably complicates the analysis.

A serious problem in studying chain reactions arises when certain impurities are present in the reaction vessel. It is known, for instance, that O_2 acts in competition with reaction 4 for removal of Cl radicals. The rate is sufficiently fast compared with the slow termolecular recombination reaction 4 that a complete change in mechanism results when small amounts of O_2 are added. It has been reported[12] that the addition of only 1% O_2 to a $H_2 - Cl_2$ mixture can change the rate of reaction by a factor of 10^3. Such effects are not surprising given the remarkably large chain length of this reaction. Under the conditions of the above experiments, Φ_{ph} for the $H_2 - Cl_2$ reaction at a laser power of $7mW$ is calculated [13] from Eq. 8 to be 10^4. This figure increases as the laser power is reduced and ultimately reaches[2] the thermal chain length of 10^7. It is thus clear that extreme care must be exercised in eliminating impurities from the reaction vessel.

The optoacoustic method of determining rate constants described here is shown to give rate constants that are in agreement with more direct measurements. Rather than providing a means for accurate measurement of rate parameters, it would appear that the salient feature of optoacoustic measurements in chemically reacting systems is that the method permits *in situ* studies of the influence of slight changes in parameters such as light intensity, chemical composition, or impurities on the overall reaction rate on a time scale that is fast compared with the duration of the reaction, and at the same time with high sensitivity.

ACKNOWLEDGMENT

The authors are grateful for support of this work by the U. S. Department of Energy.

REFERENCES

1. M. T. O'Connor and G. J. Diebold, J. Chem. Phys. **81**, 812 (1984).
2. G. J. Diebold and J. S. Hayden, Chem. Phys. **49**, 429 (1980).
3. M. T. O'Connor and G. J. Diebold, Nature (London) **301**, 32 (1983).
4. J. G. Choi and G. J. Diebold, Anal. Chem. **57**, 2989 (1985).
5. L. A. Pipes *Operational Methods in Nonlinear Mechanics* (Dover, New York, 1965).
6. G. J. Diebold, J. Phys. Chem. **84**, 2213 (1980).
7. S. Benson, *The Foundations of Chemical Kinetics* (McGraw-Hill, New York, 1960).
8. In the absence of measured values of k_4, it can be assumed that this constant changes only insignificantly from H_2 to D_2. See Ref 7.
9. See M. T. O'Connor, R. B. Stewart and G. J. Diebold, J. Phys. Chem. **90**, 711 (1986) and references therein.
10. A. Persky and F. S. Klein, J. Chem. Phys. **44**, 3617 (1966).
11. J. C. Miller and R. J. Gordon, J. Chem. Phys. **79**, 1252 (1983); See also J. Weston, J. Phys. Chem. **83**, 61 (1979).
12. See Ref. 7 p 341. This value is computed from spectroscopic data for the zero point energies of the various species and a value of ΔU_H from Ref. 7.
13. The value of $2(ab)^{\frac{1}{2}}$ from Ref 1 is scaled to $7mW$.

Measurement of the Reaction Rate of BrNO + Br → Br_2 + NO by the Thermal Lens Method

K. Koseki, M. Koshi, and H. Matsui

Department of Reaction Chemistry, The University of Tokyo, Hongo, Bunkyo-ku, Tokyo 113, Japan

1. Introduction

The photochemistry of nitrosyl halides has been intensively investigated by many workers; especial interest has been concentrated on the mechanisms of vibrationally excited NO production. Direct formation of vibrationally excited NO is now confirmed by the UV photolysis studies of ClNO by many investigators, including Basco and Norrish [1] and Moser et al. [2]. However, the photolysis of BrNO at visible wavelengths has not been completely established. Grimley and Houston [3] reported that vibrationally excited NO (v=2-3) was produced in the photolysis of BrNO for wavelengths greater than 480 nm. They assigned the mechanism of the formation of NO(v) as

$$Br + BrNO \rightarrow Br_2 + NO \quad (\Delta H = -17.4 kcal/mol) \qquad (1)$$

by monitoring IR emissions in the 5.2 μm region through a cold gas filter of NO. The rate constant of this process evaluated by them was $k_1 = 5.16 \times 10^{-12}$ [$cm^3 molecule^{-1} s^{-1}$], which was consistent with that measured by flash photolysis/mass spectrometry by Price and Ratajczak [4], but about two orders of magnitude smaller than that reported by Hippler et al. [5] who monitored Br_2 absorption.

Recently, Cedlacek and Wight [6] observed IR fluorescence in the 5.3 μm region in the 500 nm laser flash photolysis of excess Br_2 + NO systems and suggested the important contribution of Br*. They concluded that the observed IR emission was due to the vibrationally excited BrNO which was produced via the E-V energy transfer from Br* to BrNO. They also concluded that neither the E-V energy transfer in Br*-NO collisions nor the production of highly excited BrNO($v \geq 2$) had important roles in the IR fluorescence.

The present study was carried out in order to reinvestigate the rate constant k_1 and make an examination of the unclarified problems regarding the formation of NO(v), as discussed above, in the visible photolysis of BrNO.

2. Experimental System

The thermal lens technique was employed in this study because of its simplicity and high sensitivity to the chemical energy release to the translational degree of freedom. Also, the fast response of the photomultiplier used as a detector of the TLE signals was advantageous compared with IR detectors which were used in the previous works for monitoring very fast reactions such as (1). From TLE signals, on the other hand, one can only get information associated with the time dependence of the total heat release, so it is generally hard to examine the detailed mechanisms of the reaction processes. Nevertheless, the information directly related to the energy transfer rate to the translational degree of freedom is often very useful because it is utilized as a different source for discussing the reaction

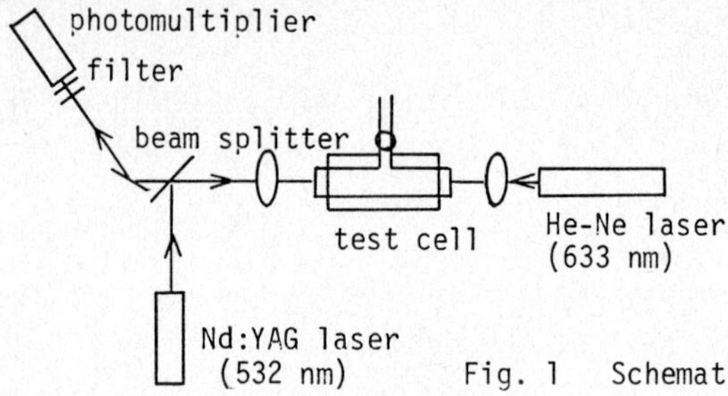

Fig. 1 Schematic of the experimental apparatus

mechanisms. A schematic of the experimental system used in this study is shown in Fig. 1. A Nd:YAG laser (SHG:532 nm) was used to initiate the photodecomposition of BrNO in a Pyrex glass cell of 0.8 cm i.d. and 26 cm length. The consecutive reactions were studied by monitoring the TLE signals of a probe He-Ne laser (633 nm). Several cutoff filters were placed in front of a photomultiplier to eliminate the stray light of the pumping laser. The optical alignment was carefully adjusted so that the output signal of the photomultiplier was maximized. The TLE signals were recorded in a high speed digitizer and averaged over 256 to 1024 times. As the sensitivity of the detection system was found to be sufficiently high, the present experiment was carried out at a low level of the photolyzed Br atom concentrations, i.e., the absorption intensities for various BrNO-Ar mixtures for the 532 nm laser beam were measured by using 87 cm long Pyrex glass cell, and the fraction of photolyzed BrNO was estimated to be about 0.1 % or less. Test gases used in this study were Ar (Nihon Sanso:99.999%), NO (Matheson:CP grade), and Br_2 (Wako:special grade). They were purified by a trap-to-trap procedure, then mixed and stored in 10 liter glass bulbs for more than 24 hours before use. The initial composition of the test gases was evaluated by using the reported equilibrium constants [7],[8].

3. Experimental Results

The experiment was carried out for the test gases with excess NO in order to depress the effect of the photolysis of Br_2 existing in the samples. Except in the initial submicrosecond range, the TLE signals showed a single exponential time dependence with a time constant of less than a few microseconds, followed by slower decay, which was attributed to thermal diffusion. We tentatively assumed that this initial rise of the TLE signals was due to the heat released by the process (1); thus an estimation of the first order rate constant k_1 was tried. The Stern-Volmer plot of this rate constant was very well expressed by a straight line, as is shown in Fig. 2, against the partial pressure of BrNO. From the slope of this straight line, the rate constant for the process (1) was estimated as $k_1 = (5.3 \pm 0.7) \times 10^{-12}$ [$cm^3 molecule^{-1} s^{-1}$] at 298 K. It was confirmed that the observed rate constant was not affected by the mole fraction of excess NO. As is summarized in Table 1, the present result for k_1 is consistent with the previous ones reported by Grimley and Houston [3] and Price and Ratajczak [4].

The temperature dependence of the rate constant was also measured by using a heat/ cold bath in the temperature range 273-310 K. The upper and lower limits of this temperature range arose from the low fraction of BrNO equilibrated in $BrNO \rightleftarrows 1/2Br_2 + NO$, and that of Br_2 (gas) $\rightleftarrows$ (liquid bromine), respectively. The Arrhenius plot of the rate constant k_1 is shown in Fig. 3 for the two systems of the premixed samples. Both results were consistent

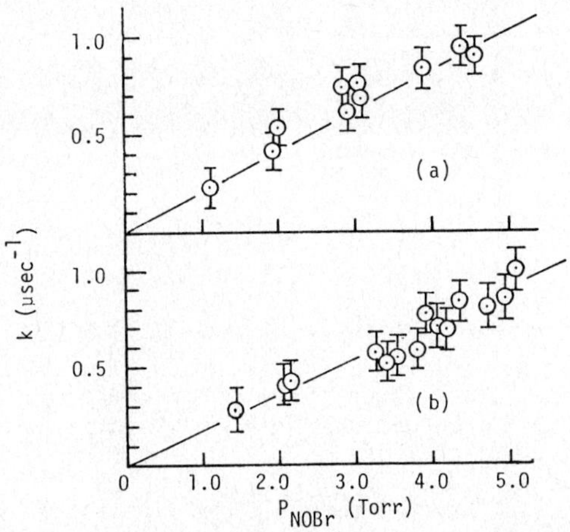

Fig. 2 A Stern-Volmer plot of the rate constant of the process (1) (a) $[(NO)_0/(Br_2)_0 = 4/1$, T = 298.0 K], (b) $[(NO)_0/(Br_2)_0 = 11.6/1$, T = 298.2 K]

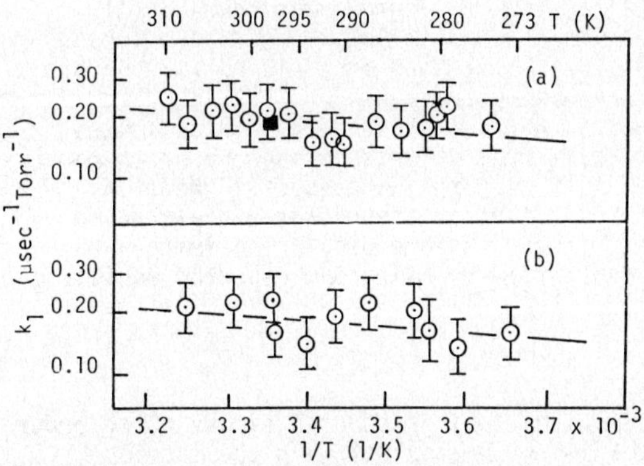

Fig. 3 The Arrhenius plot of the rate constant of the process (1). Sample gas: (a): 2.10 torr BrNO + 7.50 torr NO + 0.54 torr Br_2 + 39.9 torr Ar at 298 K, (b): 1.40 torr BrNO + 5.39 torr NO + 0.34 torr Br_2 + 20.7 torr Ar at 297 K.

with each other and showed a very small temperature dependence. The least squares fit of these data yields the activation energy, E_a = 1.3 ± 0.4 kcal/mol.

4. Discussion

Examination of the chemical and energy transfer processes competing with (1) is evidently very important.

If vibrationally excited NO is produced, as was mentioned by Grimley and Houston [3], the vibrational relaxation of NO has to be examined in the analyses of this study. Based on the relaxation rate constants that have been published for NO-Br [9] and NO-NO [10], the relaxation time of NO for the present experimental conditions was evaluated to be more than 40 μs, so this process can be excluded in the analyses of TLE signals.

The other reactions such as

$$Br_2 + 2\ NO \rightarrow 2\ BrNO \tag{2}$$

or

$$Br + NO + M \rightarrow BrNO + M \tag{3}$$

can also be neglected on the time scale of this study [11]. Small fractions of Br_2 were contained in the samples of this study. The partial pressure of it was dependent on the total pressure of the test gas and was in the range 0.8-1.5 torr for the mixture of $(NO)_0/(Br_2)_0=4/1$. As 532 nm radiation can produce $Br^*(^2P_{1/2})$ [12], the contributions of the following processes should be examined:

$$Br^* + BrNO \rightarrow Br + BrNO(v=1) \quad \text{[E-V transfer]} \tag{4}$$

$$Br^* + BrNO \rightarrow Br + BrNO \quad \text{[collisional quenching]} \tag{5}$$

$$BrNO(v=1) + M \rightarrow BrNO(v=0) + M \quad \text{[T-V transfer]} . \tag{6}$$

As reported by Sedlacek and Wight, these energy transfer processes have very high efficiencies, so they could affect the observed TLE signals to some extent. No experimental data for the vibrational relaxation of BrNO (6) is available, however, the decay rate of the IR emission of BrNO observed by Sedlacek and Wight in a test gas of 205 mtorr Br_2 + 40 mtorr BrNO + 8 mtorr NO supplies information on this, i.e., $k_6 \sim 10^{-12}$ [$cm^3 molecule^{-1} s^{-1}$]. If the collisional efficiency of Br_2 for the T-V relaxation of BrNO is not much different from that of Ar, the addition of 10 $\sim$ 100 torr Ar to the samples of this study would make the relaxation time of BrNO comparable to that of the observed TLE signals. Figure 4 shows the experimental result of the effect of the addition of Ar on the observed rate constant. Evidently, excess Ar showed no significant contributions.

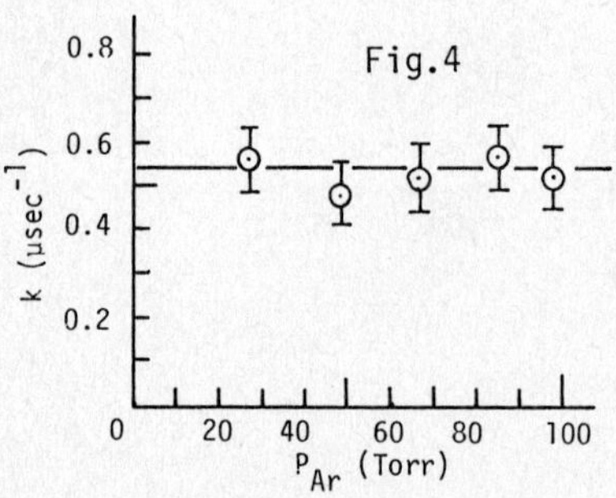

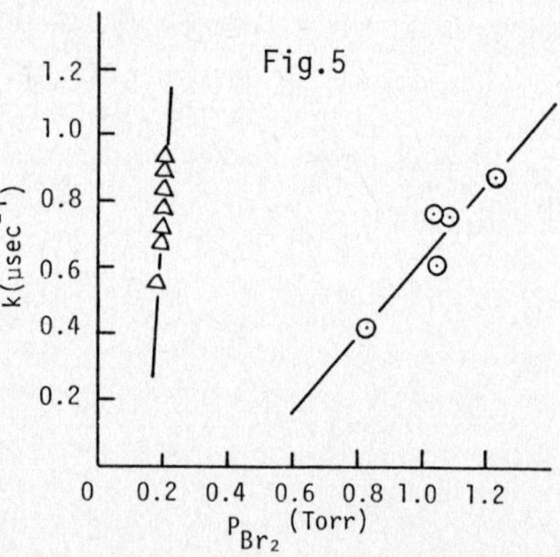

Fig. 4 The dependence of the first order rate constant of the TLE signals on the addition of Ar [sample gas at 298.2 K: 3.83 torr BrNO + 22.8 torr NO + 0.19 torr Br_2 + Ar]

Fig. 5 The dependence of the rate constant of the TLE signals on the partial pressure of Br_2.
[O: $(NO)_0/(Br_2)_0$ = 4/1, T = 298 K, Δ: $(NO)_0/(Br_2)_0$ = 11.6/1, T = 298.2 K]

Table 1 Summary of the rate constant of (1)

	$k_1[cm^3molecule^{-1}s^{-1}]$	Ref.
Hippler, Luu, Teitelbaum and Troe	3.6×10^{-10}	5
Grimley and Houston	5.16×10^{-12}	3
Price and Ratajczak	2.2×10^{-12}	4
This work	5.3×10^{-12} (298 K)	

The rate constant shown in Fig. 2 is plotted vs. the partial pressure of Br_2 in Fig. 5. If the formation of Br* and the consecutive energy transfer processes (4)-(6) are dominant in the TLE signals, such a plot should be a single straight line. The Stern-Volmer plot vs. the partial pressure of Br_2 shows evidence that the rate constant measured in this study was independent of the concentration of Br*; thus it may be concluded that the observed rate constant summarized in Table 1 corresponds to the process (1).

Burns and Dainton [13] studied the photochemical processes in phosgene and evaluated a rate constant of the process

$$Cl + ClNO \rightarrow Cl_2 + NO. \quad (7)$$

According to their results, the process (7) is about three times faster than (1) but the two processes have almost the same activation energy. They also discussed the mechanism of (7) based on the potential energy of Cl_2NO estimated by the London equation, where NO was treated as an atom. The calculation suggested that Cl_2NO was very unstable with very low entrance and exit barriers. In spite of such a simple treatment of the potential energy, the rate constant evaluated using the transition state theory was in good agreement with the experimental one.

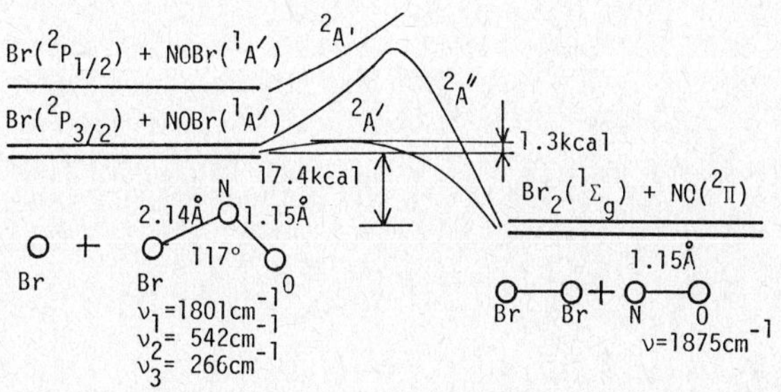

Fig. 6 Schematic of the potential energy of Br + BrNO system

A schematic energy diagram for the process (1) is shown in Fig. 6. The difference between the fission energy of N-Br bond in BrNO and that of the product Br_2 molecule just corresponds to the exothermicity of the process (1). Because of the very low activation energy of this process and the small change of the N-O bond length from BrNO to NO, it is suggested that the chemical energy is released as that of recoil motion between Br_2 and NO without causing a large distortion of the NO bond; thus the probability of direct formation of vibrationally excited NO in (1) may not be too large.

A more quantitative examination of the mechanism of NO(v) formation has to be carried out. According to the method proposed by Sorbie and Murrell [14], the potential energy of Br_2NO is now being calculated in order to allow a more detailed discussion.

References

1. N. Basco and R.G.W. Norrish: Proc. R. Soc. A 268, 291 (1962)
2. M.D. Moser, E. Weitz, and G.C. Schatz: J. Chem. Phys. 78, 757 (1983)
3. A.G. Grimley and P.L. Houston: J. Chem. Phys. 72, 1471 (1980)
4. D. Price and E. Ratajczak: Dyn. Mass Spectro. 5, 176 (1978)
5. H. Hippler, S.H. Luu, H. Teitelbaum, and J. Troe: Int. J. Chem. Kinet. 10, 155 (1978)
6. A.J. Cedlacek and C.A. Wight: J. Chem. Phys. 86, 2787 (1987)
7. C.M. Blair, Jr., P.D. Brass, and D.M. Yost: J. Am. Chem. Soc. 56, 1916 (1934)
8. N. Houel and H. van den Bergh: Int. J. Chem. Kinet. 9, 867 (1977)
9. R.P. Fernando and I.W.M. Smith: J. Chem. Soc., Faraday Trans. 2, 77, 459 (1981)
10. J. Kosanetzky, U. List, W. Urban, and H. Vormann: Chem. Phys. 50, 361 (1980):; J.C. Stephenson: J. Chem. Phys. 60, 4289 (1974)
11. I.C. Hisatsune and L. Zafonte: J. Phys. Chem. 73, 2980 (1969)
12. H.K. Hougen, E. Weitz, and S.R. Leone: J. Chem. Phys. 83, 3402 (1985)
13. W.G. Burns and F.S. Dainton: Trans. Faraday Soc. 48, 52 (1952)
14. K.S. Sorbie and J.N. Murrell: Molec. Phys. 29, 1387 (1975)

Determination of Dynamic Properties of the System $N_2O_4 = 2NO_2$ by Photoacoustic Resonance Spectroscopy

M. Fiedler and P. Hess

Institute of Physical Chemistry, University of Heidelberg, Im Neuenheimer Feld 253, D-6900 Heidelberg 1, Fed. Rep. of Germany

Chemical relaxation in the system $N_2O_4 = 2NO_2$ has been included in the theory of acoustical resonances in a cylindrical cavity. The acoustic resonance profiles of the first radial mode were measured employing a computer controlled photoacoustic resonance spectrometer. For optical excitation of the acoustic modes we used a waveguide CO_2 laser and SF_6 as sensitizer with SF_6 concentrations of 50 to 250 Pa and total pressures in the region of 100 - 10000 Pa. The theory has been fitted to the data and values for the rate constant of dissociation k_2, the dissociation energy D and the mean relaxation time for vibrational and rotational relaxation ($p\tau$) were obtained.

The fit yields $k_2 = (1.47 \pm 0.07) \cdot 10^5\ s^{-1}$ at 1 bar , $D = (53.0 \pm 1.0)$ kJ/mol and $p\tau = (1.71 \pm 0.07) \cdot 10^{-8}$ bar·sec at 300 K. No evidence for vibrational relaxation was found in the f/p region of chemical relaxation .

1. INTRODUCTION

The effect of chemical relaxation, i.e. the influence of the chemical reaction on the sound velocity in a reacting gas mixture, has been known for a long time [1] and a theoretical model has been developed [2][3][4] for a simple type of reaction, namely the dissociation of a molecule into two identical parts $A_2 = 2A$. Later, the theory has been extended to a larger number of different types of reactions [5].

Early attempts to measure the predicted effect in the $N_2O_4 = 2NO_2$ system failed [1][6]. Ultrasonic methods using electrostatic transducers could detect the effect of chemical relaxation and determine the rate constant [7]. Recently the $N_2O_4 = 2NO_2$ system was investigated using the temperature jump relaxation method employing a pulsed laser [8].

In this work photoacoustic resonance spectroscopy was used as a powerful tool to determine precisely the sound velocity and sound absorption via resonance frequency and halfwidth of the resonance profile in the $N_2O_4 = 2NO_2$ mixture in the pressure range of 100 - 10000 Pa at frequencies of about 3 kHz. So far acoustical resonators have been applied to such different problems as precision measurement of sound velocity [9], measurement of vibrational relaxation times [10] and detection of gaseous air pollution in small concentrations [11]. A quantitative theory of acoustical resonators has been developed that describes the photoacoustic signal [12], the dispersion of resonance frequencies and the broadening of the resonance profile [10]. The photoacoustic resonance spectroscopy method has now been extended to the investigation of binary reacting gaseous mixtures giving information about the kinetics on time scales of about 10^4 gas kinetic collisions.

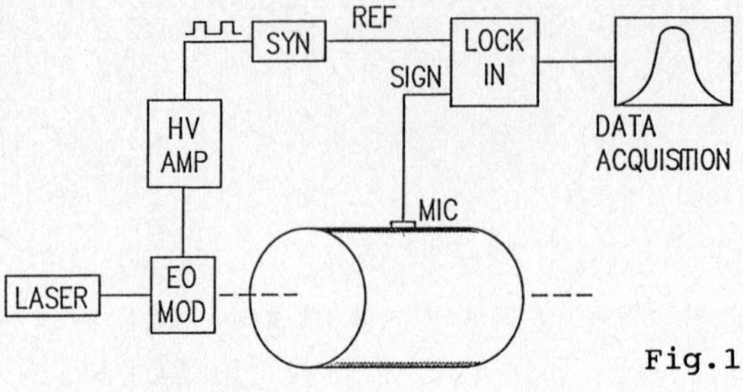

Fig.1 Experimental setup

2. EXPERIMENTAL

Figure 1 shows the setup of a photoacoustic resonance spectrometer. The cylindrical resonator cavity in which standing pressure waves are excited is machined from stainless steel with an inside radius of 0.05 m and an inside length of 0.1 m. Into the front and back plate of the resonator NaCl windows are glued to allow the laser beam to traverse the cylinder along the axis. An Edinburgh Inst. waveguide CO_2 laser with an output power of about 3 W tuned to the 10 P20 transition served as the light source. The $1/e^2$ beam diameter was 1.3 mm and the divergence was 10 mrad. The beam was modulated by a Pockels cell supplied by II-VI Inc., Pennsylvania, USA, with a modulation depth of 80 % driven by a computer controlled Philips frequency synthesizer model PM 5190 via a linear high voltage amplifier.

A standard electret microphone mounted flat with the interior resonator wall at half the length of the cell transformed the pressure waves into an electronic signal recorded by a computer controlled Ithaco lock-in amplifier model 3961 with a sampling time of 1 sec. A Hewlett Packard Vectra personal computer was used for data processing. More details on the photoacoustic resonance technique can be found elsewhere [13].

The temperature was measured by the piezo quartz oscillation method described in [14]. Pressure measurements were performed by MKS baratron capacitance manometers.

The N_2O_4 gas was produced by mixing oxygen and nitrogen oxide both supplied by Messer Griesheim with stated purities of 99.995 % (O_2) and 99.8 % (NO). The contamination of the gases were stated to be < 5 vpm H_2O in O_2 and < 200 vpm N_2O in NO. The N_2O_4 was frozen at 77 K with a surplus of oxygen present. Allowing slow melting of the crystals the middle fraction was kept and refrozen. This procedure was repeated until the crystals showed a white colour. The whole vacuum system was evacuated to about 1 Pa and became inert against the aggressive gases after filling with N_2O_4 for some time.

Since neither N_2O_4 nor NO_2 possess any absorption in the 10 μm region SF_6 was added in concentrations of 50 to 250 Pa as a sensitizer. The 10 P20 transition of the CO_2 laser was used for excitation.

3. RESULTS

3.1 Experimental data

The resonance curves for the first radial mode were recorded point by point for various total pressures and nine different SF_6

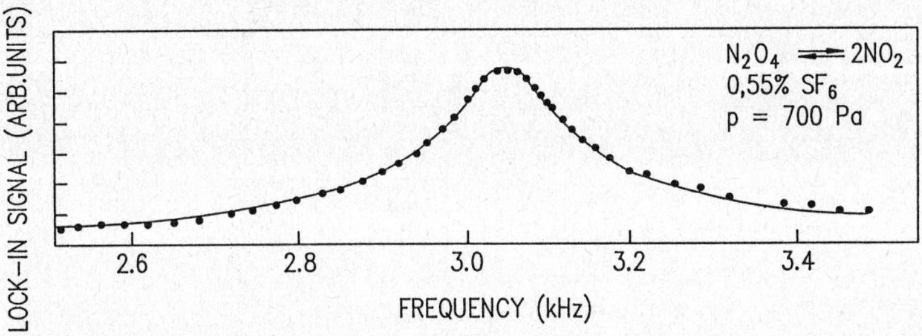

Fig.2 Resonance curve of the first radial mode at 700 Pa total pressure and 0.55% SF_6. The line is the fitted Lorentzian profile

concentrations. An example of such an acoustic resonance is given in Fig. 2, where the full circles represent the data, the solid line shows the fit of a Lorentzian profile to the data from which center frequency, halfwidth and amplitude of the resonance profile are extracted.

The dispersion of the resonance frequencies and the broadening of the resonance profiles as a function of total pressure are shown in Fig. 3 and Fig. 4, respectively.

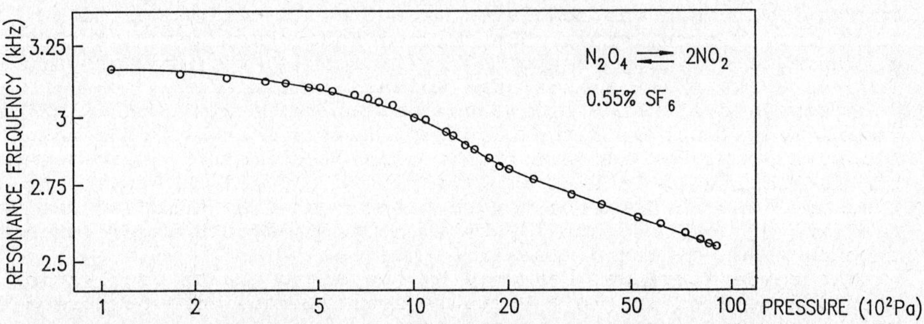

Fig.3 Pressure dependence of the resonance frequency of the first radial mode (0.55% SF_6). Circles: experimental data, line: theoretical fit

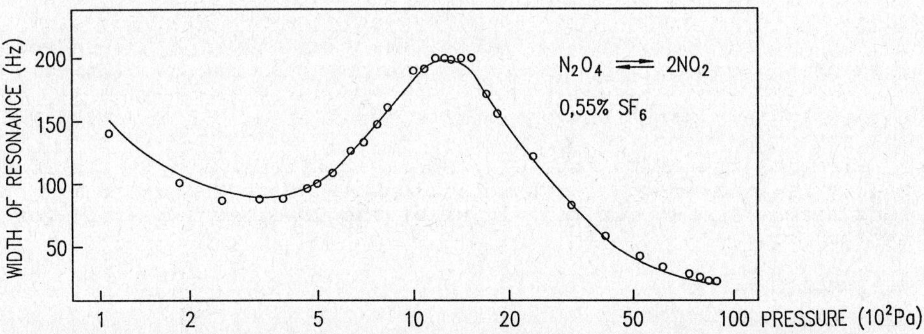

Fig.4 Pressure dependence of the halfwidth of the resonance profile of the first radial mode (0.55% SF_6). Circles: experimental data, line: theoretical fit

3.2 Theoretical Description

The pressure dependence of the measured quantities were compared with a theoretical model including a mean relaxation time for vibrational and rotational energy transfer, losses at the resonator walls and chemical relaxation. The theory of the first two effects is described in detail in [10]. Chemical relaxation of a binary mixture has been included in the model on the basis of a formula derived by Einstein [2] and Luck [3] assuming ideal gas behavior. The formula gives the complex sound velocity as a function of the rate constant and dissociation energy for a given thermodynamical state. The real part of this complex number equals the physical sound velocity, the imaginary part represents the absorption of sound. The kinetical model neglects preceding vibrational excitation of the N_2O_4 molecule before dissociation. Real gas effects are not considered here. The specific heat was calculated from spectroscopic data [15], the transport coefficients were taken from [16][17].

The complete model was fitted simultaneously to the resonance frequency and to the halfwidth data, see solid line in Fig. 3 and Fig. 4 . The rate of dissociation k_2, the dissociation energy D, the mean relaxation time for rotational and vibrational relaxation $(p\tau)$ and the radius of the resonator r were taken as free parameters. Values were determined and extrapolated to zero sensitizer concentration. The fit yields for a temperature of 300 K and a pressure of 1 bar $k_2 = (1.47 \pm 0.07)\cdot 10^5\ s^{-1}$, $D = (53.0 \pm 1.0)$ kJ/mol and $(p\tau) = (1.71 \pm 0.07)\cdot 10^{-8}$ bar·sec. The value obtained for the radius of the resonator that includes corrections for geometrical and systematical imperfections in radial direction was $r = (5.11 \pm 0.01)$ cm.

4. DISCUSSION

In order to compare the results of the different experiments carried out at different temperatures the rate constant has to be recalculated. Assuming the activation enery of the dissociation to equal $E_A = (58.1 \pm 3.7)$ kJ / mol [18] we find for a temperature of 293 K $k_2 = (0.85 \pm 0.07)\cdot 10^5\ s^{-1}$.

Recently temperature jump experiments [8] were performed in argon at 60 mbar and 780 mbar with SiF_4 as sensitizer. Using the formula for the temperature dependence [8] of the rate constant of dissociation for 780 mbar results in $k_2 = 0.55\cdot 10^5\ s^{-1}$ at 293 K and 1 bar. This value agrees with our result within the errors due to uncertainties concerning the maximal rate constant and the activation energy. The value presented for the mean vibrational and rotational relaxation time is in agreement with estimates based on the Lambert-Salter plot [16] assuming fast V-V energy transfer and V-T transfer from the lowest N_2O_4 level at 70 cm^{-1}, which belongs to the torsion of the molecule around the N-N bonding. Uncertainties in our measurements result from the temperature variations during one series of measurements. The temperature enters the exponential expression and cannot easily be corrected for. Systematic errors occur if the halfwidth of resonance profiles are larger than about 100 Hz due to interference with resonances of neighbouring acoustic modes.

The advantages of photoacoustic resonance spectroscopy for studying the kinetics in the $N_2O_4 = NO_2$ system are the large amount of information obtained, the simple principle of the method, the high sensitivity and the possibility of quantitative description of the measurements. The large amount of experimental information obtained by photoacoustic resonance spectroscopy allows a sensitive test of the internal consistency of spectroscopic, thermodynamic, transport and kinetic properties of the system used to model the data.

5. ACKNOWLEDGEMENTS

We wish to thank the Deutsche Forschungsgemeinschaft and the Fonds der Chemischen Industrie for support of this work.

6. LITERATURE

[1] F. Keutel: Inaugural Dissertation,Berlin 1910
[2] A. Einstein: Sitz.-Ber. Berlin. Akad. Wiss. (1920),380
[3] D. Luck: Phys. Rev. (II) 40, 440 (1932)
[4] H. Kneser,O. Gauler: Phys. Z. 37, 677 (1936)
[5] M. Eigen, L. de Mayer: In Techniques of Organic Chemistry, Vol. VIII, part II, 895, ed by A. Weissberg A. Weissberg (Interscience N.Y. 1963)
[6] W. Richards,J. Reid: J. Chem. Phys. 1 ,114 and 737 (1933)
[7] G. Sessler: Acoustica 10, 44 (1960)
[8] P. Gozel,B.Calpini,H.v.d.Bergh: Israel J. Chem. 24,210 (1984)
[9] J. Mehl,M. Moldover: J. Chem. Phys. 74, 4062 (1981)
[10] A. Karbach,P. Hess: J. Chem. Phys. 83, 1075 (1985)
[11] M. Sigrist: J. Appl. Phys. 60, 83 (1986)
[12] A. Karbach,P. Hess: J. Chem. Phys. 84, 2945 (1986)
[13] A. Karbach,J. Roper,P. Hess: Rev. Sci. Inst. 55, 892 (1984)
[14] W. Perkins,M. Wilson: Phys. Rev. 85, 755 (1952)
[15] R. Brokaw,R. Svehla: J. Chem. Phys. 44, 4643 (1966)
[16] H. Sontag,A. Tam,P. Hess: J. Chem. Phys. 86, 3950 (1987)
[17] J. Lambert: In Vibrational and Rotational Relaxation in Gases p 66 (Clarendon, Oxford,1977)
[18] W. Richards,J. Reid: J. Chem. Phys. 1, 144 (1933)

Study of Intracavity Absorption and Laser Behaviour by Photoacoustic Resonance Spectroscopy

A. Neubrand, J. Röper, and P. Hess

Institute of Physical Chemistry, University of Heidelberg, Im Neuenheimer Feld 253, D-6900 Heidelberg 1, Fed. Rep. of Germany

Operation of a resonant photoacoustic cell inside the cavity of a HeNe laser is described and the pressure dependence of the photoacoustic signal is investigated. Three basic types of behaviour are observed for weak, medium and strong absorbers. A quantitative description of the signal is possible up to medium losses. Laser behaviour near threshold is investigated for the first time using photoacoustic resonance spectroscopy.

1. INTRODUCTION

In photoacoustic resonance spectroscopy modulated electromagnetic radiation is absorbed by a gas contained in symmetric resonators such as cylinders or spheres. If the absorbed light is converted into translational energy and this energy is released with the resonance frequency of an acoustical mode of the resonator, this mode is excited and the resonance frequency, halfwidth and mode amplitude of the resonance profile can be recorded.

Photoacoustic resonance spectroscopy in the gas phase allows the determination of molecular constants such as relaxation times [1] and rate constants [2]. An important quantity is the so-called photoacoustic signal, which is defined as the resonance amplitude of an acoustic mode divided by the Q-factor. The photoacoustic signal of the first radial mode of a cylinder can be described quantitatively [3]. It is a function of the laser power and the optical absorption coefficient and, therefore, provides a way of determining the laser power absorbed in the gas. Hence ,a photoacoustic cell operated inside the cavity of a laser yields important information on the properties of intracavity radiation.

2. EXPERIMENTAL

For our experiments the apparatus shown in Fig.1 was employed. The light source was a HeNe laser set up for operation at 3.39 μm with a typical output of 10 mW for this laser line. A mechanical chopper and a cylindrical photoacoustic cell of 10 cm length and diameter were placed inside the laser resonator. The signal was detected by an electret microphone and amplified by a lock-in amplifier. The resonance curves of the first radial mode were recorded employing a microcomputer with interface which served to control the chopping frequency via a frequency synthesizer with subsequent frequency to voltage converter. The computer also accomplished the integration of the signal from the lock-in amplifier. The resonance profiles recorded point by point by measuring at discrete modulation frequencies were fitted to a Lorentzian profile by a least squares procedure yielding values of

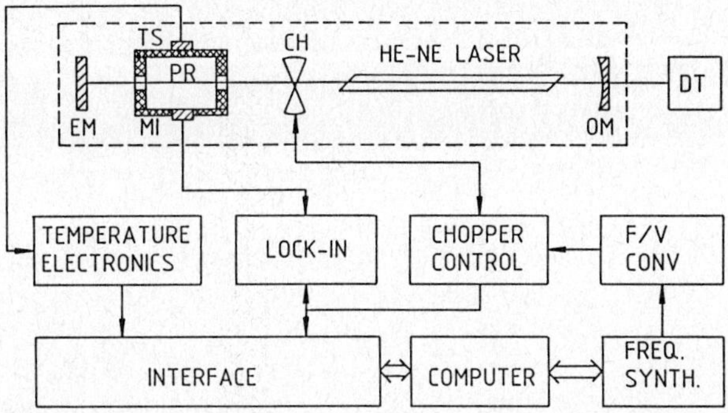

Fig.1. Schematic diagram of the experimental setup

the photoacoustic signal, halfwidth and resonance frequency of the resonance curve.

The output of the HeNe laser at 3.39 μm was monitored extracavity using a power meter. The signal of the first radial mode of the cylindrical resonator was recorded as a function of concentration for the gases C_2H_4, CH_3Cl and CH_4-N_2. The gases were supplied by Messer Griesheim with a stated purity of 99.9995% for CH_4, 99.996% for N_2, 99.95% for C_2H_4, 99.8% for CH_3Cl and were used without further purification.

3. RESULTS

There are at least 3 basic types of laser behaviour which can be observed for different types of absorbing gases. The first type are weakly absorbing gases such as C_2H_4, where threshold was not reached below 760 torr [4]. The second type are medium absorbers, which allow the threshold for laser oscillation to be reached, and the third case are strong absorbers with lines coincident with the laser transition. Representatives of the second and third types of behaviour are CH_3Cl and CH_4, respectively, which we have chosen for the present investigation.

3.1. Results for Chloromethane

Fig.2 shows the photoacoustic signal measured for the first radial mode of the cylinder filled with CH_3Cl as a function of pressure. CH_3Cl is only a moderate absorber because there is no coincidence of a rovibrational transition of this molecule with the HeNe laser line at 2947.90 cm^{-1} (Fig.3). In a cell of 10 cm length more than 250 torr are necessary to absorb 99% of the laser radiation. For the pressures investigated, collisional broadening is the dominant broadening mechanism and thus the absorption coefficient increases strongly between 10 and 100 torr. This is reflected in the photoacoustic signal which shows a drastic increase, but then levels off at about 80 torr. Around this pressure the increase in absorption is compensated by the decrease in intracavity power available as the laser is quenched. Above 200 torr the threshold for laser operation is approached and the photoacoustic signal suddenly drops. However, even at the highest pressures investigated, a signal is observed which does not disappear when the highly reflecting mirror of the optical cavity is blocked.

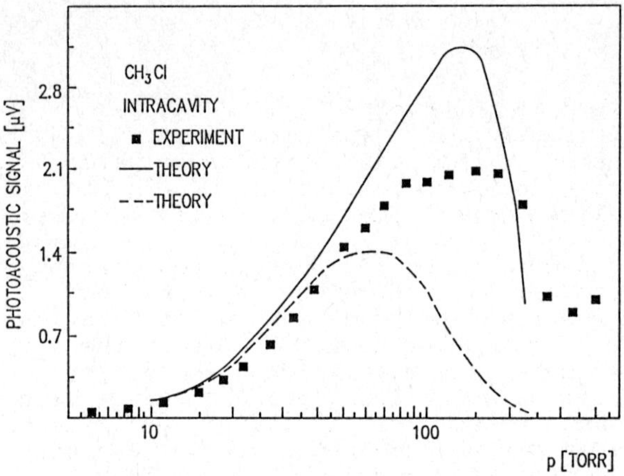

Fig.2. Photoacoustic signal versus pressure for CH_3Cl. Squares: experimental data; solid line: theory for inhomogeneous intracavity power and no gain saturation; the dashed line is for strong gain saturation.

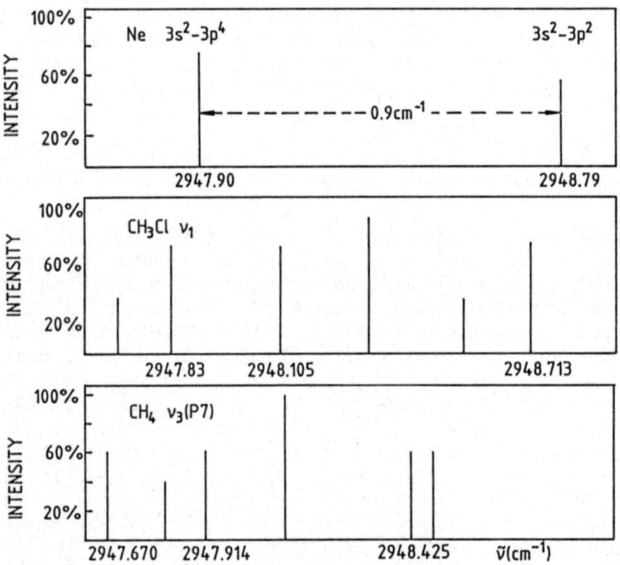

Fig.3. Simplified spectra in the 3.39μm region; upper scheme: transitions of Ne; middle scheme: CH_3Cl spectrum; lower scheme: CH_4 spectrum

This signal may be attributed to amplified spontaneous emission, a phenomenon commonly encountered in high gain lasers.

The theoretical values of the photoacoustic signal were calculated as follows : If the microphone is mounted midway between the cell ends and the laser beam is coaxial with the cylinder axis, the photoacoustic signal PAS of the first radial mode of the cylinder is given by [3]

$$\mathrm{PAS} = \frac{(c_p/c_v)-1}{2\pi f_R\, V\, I_o(x_{01})}\, F_T(f_R)\, P_d, \qquad (1)$$

where c_p/c_v is the specific heat ratio, f_R is the resonance frequency, $F_T(f_R)$ is the fraction of translational energy which contributes to the formation of the standing pressure wave, V is the resonator volume, I_o is the Bessel function, x_{01} is the first zero of the derivative of the Bessel function, P_d is the power deposited in the cell.

The power deposited depends on the intracavity power and the optical absorption coefficient α. In order to estimate the circulating intracavity power we used the measured extracavity output multiplied by 1/T, T being the transmittance of the output mirror. This procedure is generally used to describe the intracavity enhancement inside a low loss optical cavity. With absorption coefficients determined from extracavity experiments, this simple model describes the photoacoustic signal very well in the case of low optical losses [4]. For the higher losses introduced in the optical resonator by CH_3Cl the signal calculated this way is too small. Due to the fact that at higher losses a light wave experiences substantial gain and loss on travelling through the optical resonator, the power of this light wave strongly depends on its position in the optical cavity. To account for this effect, the variation of the light wave travelling from the output mirror to the acoustical cell was calculated on the basis that the losses per round trip must be equal to the gain. The power deposited in the cell is given by

$$P_d = \sum_m \Big[[1-\exp(-\alpha(m)d)][(1/T)-1]\; P_t(m) \times [(1-L_b)(R_2/R_1)^{1/2} + (R_1R_2(1-L_b)^2\exp(-2\alpha(m)d))^{-1/2}]\Big], \qquad (2)$$

where R_1, R_2 are the reflectivities of the output mirror and the highly reflecting mirror, respectively, T is the transmittance of the output mirror, $\alpha(m)$ is the optical absorption coefficient of the laser mode m, $P_t(m)$ is the laser output of this mode, d is the length of the acoustical resonator and L_b is the fractional loss of the back window of the acoustical cell. If R_1, R_2, T, L_b and α are identical for all laser modes m, the deposited power and, as a consequence, the photoacoustic signal can be calculated from the total output of the laser $\Sigma P_t(m)$. The result is shown as solid line in Fig.2. Good agreement between theory and experiment is obtained up to medium losses, but, at high losses the calculated signal is too large. This leads to the conclusion that one of the assumptions made to derive (2) must be incorrect. It was assumed that the amplification of two light waves travelling through the laser amplifier in opposite directions is the same. In fact, this does not have to be the case in an inhomogeneously broadened laser amplifier with a laser mode spacing larger than the homogeneous line width. If, at any point in the laser resonator, all laser modes have intensities much larger than the saturation intensity, the growth of a light wave passing through the laser amplifier is linear [5]. The photoacoustic signal calculated for this condition is represented by the dashed curve in Fig.2. Indeed, up to medium losses, this model predicts the experimental values of the photoacoustic signal. At about 60 torr the theoretical values suddenly become smaller than the experimental ones. This probably indicates that the losses introduced by the absorbing gas at higher pressures reduce the intensity of most laser modes to values below the saturation intensity.

3.2. Results for Methane

Methane is a strong absorber for the 3.39μm radiation of the HeNe laser, because it possesses a rovibrational transition almost coincident with the laser line (Fig.3). The dependence of the photoacoustic signal on the partial pressure of methane in a mixture with N_2 of 720 torr total pressure is completely different compared to CH_3Cl. Two distinct maxima are observed in this case (Fig.4). The decrease of the photoacoustic signal above 10 torr methane partial pressure in Fig.4 corresponds to an almost constant laser output (see Fig.5). The explanation for this

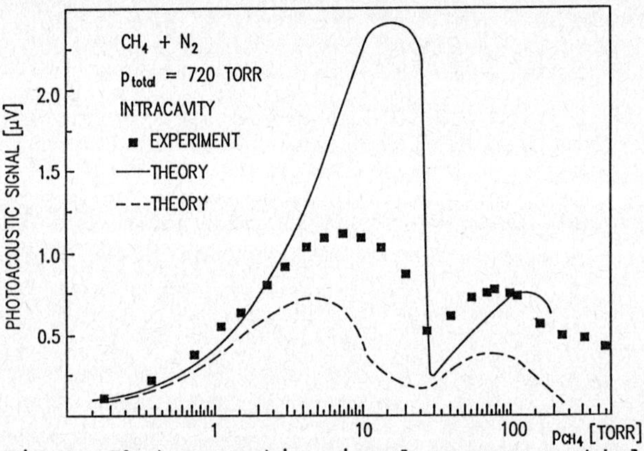

Fig.4. Photoacoustic signal versus partial pressure of CH_4 at fixed total pressure of N_2; squares: experimental data; solid line: theory for inhomogeneous intracavity power and no gain saturation; the dashed line is for strong gain saturation.

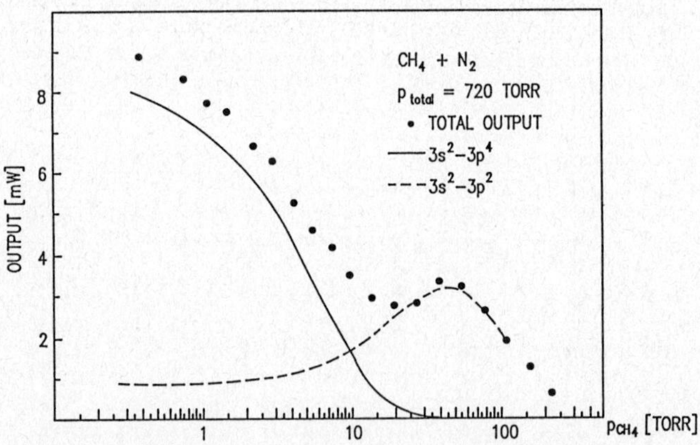

Fig.5. Output of HeNe laser near 3.39μm as a function of CH_4 partial pressure in an intracavity cell. The solid and the dashed lines are laser outputs of the two Ne transitions calculated from the respective absorption coefficients determined by extracavity experiments.

behaviour is laser line hopping. As the partial pressure of methane is raised, the losses for the $3s^2$-$3p^4$ transition at 2947.90 cm^{-1} grow drastically, whereas the losses for the $3s^2$-$3p^2$ transition at 2948.79 cm^{-1} are scarcely increased, because there is no methane line in this region (Fig.3). As a consequence, the population of the $3s^2$ Ne level increases and more and more modes of the $3s^2$-$3p^2$ transition reach threshold. As the absorption coefficients for both laser lines can be determined, it is possible to calculate the portion of $3s^2$-$3p^4$ and $3s^2$-$3p^2$ radiation emitted by the laser. The result is shown by the solid and dashed lines in Fig.5. At 10 torr the output of the two transitions is equal. With the knowledge of the output power from the two laser transitions and the appropriate absorption coefficients we were able to calculate the photoacoustic signal for the two lines separately. The change of the laser oscillation from the 2947.90 cm^{-1} line to the 2948.79 cm^{-1} line also explains the minimum observed for the photoacoustic signal around 30 torr (see Fig.4).The two models discussed in detail before are also applied to the results obtained for the CH_4-N_2 system (Fig.4). At low losses a reasonable description of the experimental data is found as before. At higher losses the deviations are qualitatively similar to those observed in the case of CH_3Cl. Again, at the highest concentrations threshold for stimulated emission is reached and amplified spontaneous emission is observed.

4. DISCUSSION

Intracavity photoacoustic resonance spectroscopy is able to reveal important features of laser behaviour such as mode and line hopping, behaviour at threshold and amplified spontaneous emission. A quantitative evaluation of the photoacoustic signal is possible for low cavity losses, but becomes difficult as threshold is approached. Near threshold, the correlation between absorption coefficient and intracavity power is very strong. This, and the occurrence of amplified spontaneous emission, which is not included in our model, because the gain per cavity round trip is not equal to the losses in this case, complicates the situation.

An important question is under which conditions it is possible to split the measured deposited power into the two quantities of interest, namely, intracavity power and absorption coefficient. For weak broadband absorbers the relationship between intracavity and extracavity laser power is simply 1/T and therefore α can easily be deduced from P_d. Thus, very small optical absorption coefficients as encountered in trace gas analysis or overtone spectroscopy can be measured [6],[7]. If narrow band absorbers are introduced into the cavity of homogeneously broadened lasers, mode hopping is likely to occur [8]. This effect complicates the evaluation of absorption coefficients.

We conclude that laser properties can be investigated by introducing strong or narrowband absorbers into the laser cavity by resonant photoacoustic spectroscopy.

5. ACKNOWLEDGEMENT

We wish to thank the Deutsche Forschungsgemeinschaft and the Fonds der Chemischen Industrie for support of this work.

6. REFERENCES

1. A. Karbach, P. Hess: J. Chem. Phys. 83, 1075 (1985)
2. M. Fiedler, P. Hess: to be published
3. A. Karbach, P. Hess: J. Chem. Phys. 84, 2945 (1986)
4. J. Röper, G. Chen, P. Hess: Appl. Phys. B43, 57 (1987)
5. L. W. Casperson, A. Yariv: IEEE J. Quant. Electron. QE-8, 80 (1972)
6. S. Shtrikman, M. Slatkine: Appl. Phys. Lett. 31, 830 (1977)
7. K. V. Reddy, D. F. Heller, M. J. Berry: J. Chem. Phys. 76, 2814 (1982)
8. K. R. German: Proceedings of the International Conference on Lasers 1980, 128

Studies of Vibrational Relaxation of Silane and Its Fluorine Derivatives by a Time Resolved Photoacoustic Technique

M. Koshi, M. Yoshimura, K. Koseki, and H. Matsui

Department of Reaction Chemistry, The University of Tokyo, Hongo, Bunkyo-ku, Tokyo 113, Japan

1. Introduction

Chemical kinetics and energy transfer processes in silane and its fluorine derivatives are interesting because of their importance in practical applications such as CVD processes. However, little is known thus far about the elementary chemical/energy transfer processes of these molecules, in contrast to the case of methane and its fluorine derivatives.

Owing to the vast experimental and theoretical efforts in the past, the vibrational relaxation rates of many polyatomic molecules have been accumulated and a universal rule predicting their relaxation rates has been searched for. An empirical, but simple correlation rule for the energy transfer rate against the vibrational energy of the lowest fundamental mode seems to be successful for many polyatomic molecules[1]. However, this empirical correlation holds only for x=4-2 for CH_xF_{4-x}(x=0-4) molecules. It may be important to compare the vibrational relaxation times of SiH_xF_{4-x} (x=0-4) with those of CH_xF_{4-x} or the empirical predictions. Such information is available only for SiH_4 at present (to the best of our knowledge), thus the measurements of V-T rate constants of the species in this group are necessary for understanding the fundamentals of molecular interactions.

Time resolved photoacoustic techniques [2-6] have been widely used in the studies of molecular energy transfer, because of their simplicity and high sensitivity to the energy transferred to translational degrees of freedom. Using this technique, the present study was carried out to find the relaxation rates of SiH_4, SiH_2F_2 and SiF_4. The observed rate constants were compared with the theory proposed by MOORE [7] to test the validity of the proposed simple V-R mechanism.

2. Experimental

In our experiments sample gases were excited in a cylindrical stainless-steel cell (10 mm diameter and 10 cm long) with a small condenser microphone. The ends of the cell were terminated by KCl windows at the Brewster angle. The cell was evacuated to less than $2x10^{-5}$ Torr before charging it with a sample gas. The rise time of the microphone (NICHIKON ESM-10) was measured to be less than 75 μs. A grating tunable, mechanically Q-switched CO_2 laser was used for the vibrational excitation. To excite $SiH_4(\nu_4)$, $SiH_2F_2(\nu_2)$ and $SiF_4(\nu_3)$, the CO_2 laser was operated on the 10.6μm P(20), 10.6μm R(16) and 9.6μm P(30) lines, respectively. Microphone signals were amplified before transmission to a transient digitizer (AUTNICS S121) and a signal averager (AUTNICS F601).

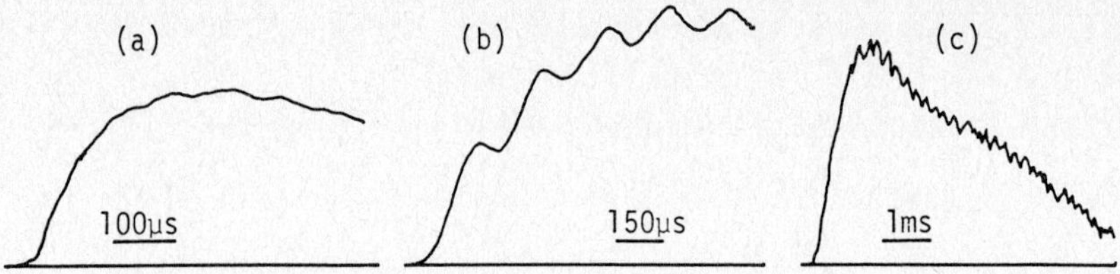

Fig.1 Photoacoustic signals observed from gas mixtures of SiH_4 and Ar. (a) 0.14 Torr SiH_4 in 83 Torr Ar,(b) 0.62 Torr SiH_4 in 21 Torr Ar, (c) 0.12 Torr SiH_4 in 12 Torr Ar

3. Results and discussion

The signal detected by the microphone in these experiments is the superposition of many acoustic waves resonating within the cell [8]. The rise rate of the envelope of this interference pattern reflects the V-T relaxation process. Typical examples of the observed signals are presented in Fig.1. When the concentration of absorber gas was high, the standing wave caused by the longitudinal acoustic resonance of the cell was observed (Fig.1b). This behaviour is well explained by the theory proposed by WROBEL and VALA [8]. Although this resonance prevented the precise determination of the rise rate, the effect of this acoustic wave could be eliminated by considering that inflection points of the wave represented net values of the pressure change [2]. It can also be seen from Fig.1c that at a later time the signal drops back towards the base line. This could be attributed to the effects of thermal conduction to the cell walls which should occur with the time constant $R^2/(4\kappa \ln 2)$ [9], where R is the radius of the cell and κ is the thermal conductivity. This time constant is estimated to be 370 μs for Ar at 1 Torr. Because of this effect the envelope of the acoustic signal,I(t), is given by the approximate expression [10]

$$I(t) = I_0 \exp(-t/\tau_d)[1-\exp(-t/\tau_v)], \quad (1)$$

where τ_d is the time constant for the thermal conduction and τ_v is the vibrational relaxation time. The observed time profiles of the pressure waves were fitted to this functional form to extract the vibrational relaxation times.

The Stern-Volmer plots of the relaxation rates,$k_v=\tau_v^{-1}$, for pure SiF_4 and mixtures of SiF_4 and Ar are shown in Fig.2. Similar plots were also obtained for SiH_2F_2. In the case of self-relaxation of SiH_4, it was found that its relaxation rates were faster than the response time of the microphone at pressures above 4 Torr. At lower pressure, rapid decay of the signals caused by heat conduction prevented precise determination of the relaxation time for pure SiH_4. Therefore self-relaxation rates of SiH_4 were obtained from the extrapolation of the Stern-Volmer plots for various mixtures of SiH_4 (0.1-0.7 Torr) and Ar (10-150 Torr). The resulting V-T relaxation times and the collision numbers for the vibrational energy transfer,Z_{10}, are summarized in Table 1. Collision number Z_{10} was calculated from the expression

$$Z_{10} = Z\tau_v(C_\ell/C_s)[1-\exp(-h\nu_\ell/kT)], \quad (2)$$

where Z is the hard-sphere collision number, C_ℓ is the vibrational heat capacity of the lowest mode whose vibrational frequency is ν_ℓ and C_s is the total vibrational heat capacity.

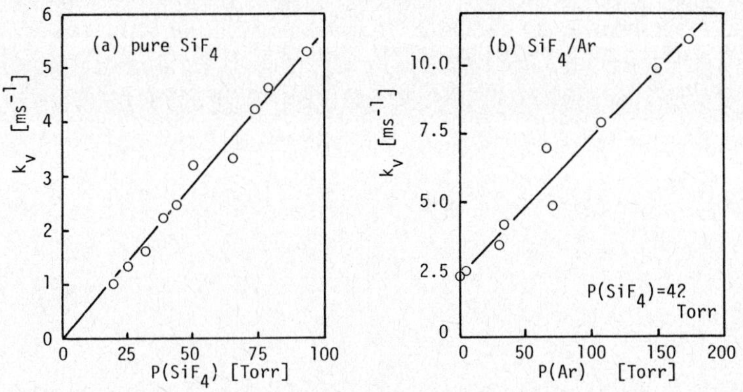

Fig.2 Pressure dependence of relaxation rates for (a) pure SiF_4 and (b) mixtures of SiF_4 and Ar

Table 1. Vibrational relaxation times for fluorinated silane and methane

		ν_ℓ[cm^{-1}]	I [amu.A^2]	τ_V[ms.Torr]	Z_{10}	Reference
SiH_4	$-SiH_4$	910	5.88	0.22±0.1	2184	This Work
				0.087	570	(11)
				0.162	1608	(12)
	-Ar			19±3	355500	This Work
				0.49	5160	(12)
SiH_2F_2	$-SiH_2F_2$	332	20.1	6.6±1.5	17100	This Work
	-Ar			25±3	68500	This Work
SiF_4	$-SiF_4$	264	122.0	17±2	35870	This Work
	-Ar			20±5	63140	This Work
CH_4	$-CH_4$	1306	3.19	1.43	20735	(13)
	-Ar			15.4	193841	(13)
CH_2F_2	$-CH_2F_2$	529	10.33	0.024	105	(14)
	-Ar			0.36	2196	(14)
CF_4	$-CF_4$	437	88.55	1.1	2194	(5)
	-Ar			2.4	6250	(5)

The vibrational relaxation rates of SiH_4 have been reported by two groups. The present data for self-relaxation agrees with that obtained by CHESNOKOV and PANFILOV [12], however, our relaxation rate of SiH_4-Ar is much slower compared with their value. The reason for this discrepancy is not clear. The vibrational relaxation times for CH_4, CH_2F_2 and CF_4 are also included in Table 1 for comparison. Irrespective of its lower frequency of the lowest vibrational mode, the relaxation times for SiH_2F_2 and SiF_4 obtained in this work are longer than those of CH_2F_2 and CF_4. On the other hand, the self-relaxation of SiH_4 is faster than that of CH_4 and the relaxation time for SiH_4-Ar collisions is the same order of magnitude as that for CH_4-Ar. It was suggested that V-R energy transfer was more important than V-T transfer for the molecules containing light atoms. According to a simple theory of V-R transfer [7], Z_{10} is given by the following equations:

$$1/Z_{10} = (Y/Z_0)\exp(-1.78\, X\, \alpha^{-2/3}), \qquad (3)$$

where $Y = [17.1\, I^{13/6} \nu_\ell^{4/3} / d^{13/3}\, T^{1/6}\, M\, \alpha^{7/3}] \exp(0.719 \nu_\ell / T)$,

and $X^3 = [I\, \nu_\ell^2 / d^2\, T]$.

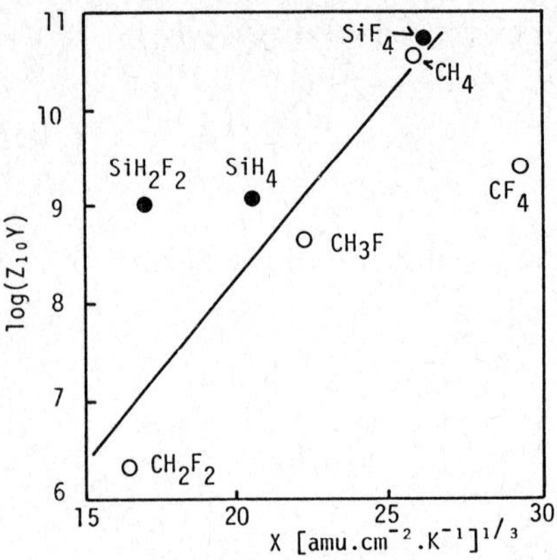

Fig.3 Plots of $\log(Z_{10}Y)$ versus X. Definitions of Y and X are given in equation (3). Line is obtained with $\alpha = 3\ A^{-1}$ and $Z_0 = 6.75$.

Here I is the moment of inertia, ν_ℓ is the vibrational frequency of the lowest mode, d is the distance from the axis of rotation to the rotating atom. M is the reduced mass of the vibrator, Z_0 is a steric factor and α is the range parameter for the intermolecular potential. As can be seen in Fig.3, where values of $\log(Z_{10}Y)$ are plotted against values of X, equation (3) cannot be applicable to the energy transfer of Si-containing species listed in Table 1. Further work is needed to understand the very slow relaxation of these species.

References

1. J.T.Lambert,R.Slater: Proc.Roy.Soc. A253, 277 (1959)
2. T.Aoki,M.Katayama: Japan J.Appl.Phys. 10, 1303 (1971)
3. J.Schurman,G.W.Wegdem: Chem.Phys.Lett. 73, 429 (1980)
4. E.A.Rohlfing,J.Gelfand,R.B.Miles,H.Rabitz: J.Chem.Phys. 75, 4893 (1981)
5. N.J.G.Smith,C.C.Davis,I.W.M.Smith: J.Chem.Phys. 80, 1922 (1985)
6. K.M.Beck,A.Ringwelski,R.J.Gordon: Chem.Phys.Lett. 121, 259 (1985)
7. C.B.Moore: J.Chem.Phys. 43, 2979 (1965)
8. J.Wrobel,M.Vala: Chem.Phys. 33, 93 (1978)
9. C.C.Davis,S.J.Petuchowski: Appl.Opt. 20, 2539 (1981)
10. L.G.Rosengren: Appl.Opt. 14, 1960 (1975)
11. T.L.Cottrel,A.J.Matheson: Trans.Faraday Soc. 58, 2336 (1962)
12. E.N.Chesnokov,V.N.Panfilov: Teor.Eksp.Khimiya 17, 699 (1981)
13. J.T.Yardley,C.B.Moore: J.Chem.Phys. 45, 1066 (1964)
14. L.A.Gamss,A.M.Ronn: Chem.Phys. 9, 319 (1975)

A Time-Resolved Thermal-Lensing Investigation of Photosensitization of Singlet Oxygen

R.W. Redmond and S.E. Braslavsky

Max-Planck-Institut für Strahlenchemie,
Stiftstraße 34–36, D-4330 Mülheim a.d. Ruhr, Fed. Rep. of Germany

1. Introduction

Dual beam time-resolved thermal-lensing (TRTL) is a sensitive calorimetric tool capable of resolving heat evolution due to radiationless deactivations as either fast or slow with respect to the acoustic transit time τ_a in the system (in our case using focused beams, $\tau_a \approx 100$ ns). This enables us to obtain information regarding quantum yields and lifetimes of energy storing species. We have previously outlined the characteristics and capabilities of our system with a number of examples [1] and more recently shown its application in evaluating the photosensitizing properties of hematoporphyrin [2], a molecule attracting interest as a sensitizer in photodynamic therapy of cancers. In the present study we have investigated the production of singlet molecular oxygen, $O_2(^1\Delta_g)$, formed by energy transfer from a variety of triplet sensitizers in oxygen saturated solutions, in an attempt to gain an insight into the properties determining the efficiency of photosensitization.

$$^3Sens^* + O_2(^3\Sigma_g^-) \text{ ----------> } {}^1Sens + O_2(^1\Delta_g) \qquad (1)$$

Radiationless deactivation occurs on time-scales shorter than τ_a and thus the heat released in the course of the these processes is responsible for the 'fast' amplitude (U_1) of the TRTL signal. $O_2(^1\Delta_g)$ in our system acts as the only energy storing species, and its radiationless decay back to the ground state is solely responsible for the slow heat component of the TRTL signal such that its quantum yield of formation (Φ_Δ) may be calculated, using (2).

$$\Delta U/U_{tot} = \Phi_\Delta \cdot E_\Delta / N_A \cdot h \cdot (\nu_1 - \Phi_f \cdot \nu_f) \qquad (2)$$

where N_A is the Avogadro number; h is the Planck constant; ν_1 is the frequency of the excitation laser; Φ_f and ν_f are the quantum yield and integrated average frequency of sensitizer fluorescence in our system; E_Δ is 22.4 kcal/mol, the energy content of $O_2(\Delta_g)$; ΔU and U_{tot} are amplitudes of signal due to slow heat dissipation and total signal amplitude, respectively. A typical example of a TRTL signal obtained from our experiments is shown in fig. 1.

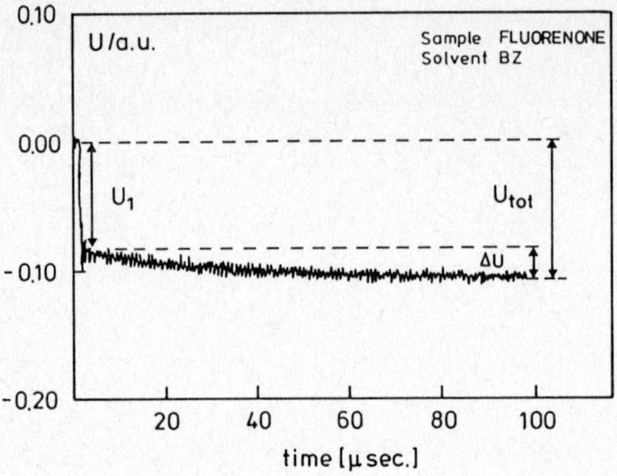

Figure 1 Typical TRTL signal showing resolution of fast and slow heat components

2. Experimental

The basis of our dual beam TRTL apparatus has been described previously [1,2]. In this work, 354 nm emission from a Nd/YAG laser (JK Lasers System 2000) and 820 nm emission from a laser diode [Telefunken Electronic TXCK 1203 Laserpen with collimating optics, supplied by a stabilized laser diode power supply (Canadian Instrumentation and Research Ltd.) and cooled by a Peltier device] were used as excitation and probe beams, respectively. The molecules studied are representatives of various classes of sensitizers. These were the aromatic hydrocarbons anthracene, phenanthrene, pyrene and chrysene; the ketones acetophenone, benzophenone, fluorenone and 2-acetonaphthone; the heterocyclic anthracene anologues acridine and phenazine, and the quinone anthraquinone.

3. Results and Discussion

TRTL signals were treated by exponential analysis of the slow decaying component, giving the lifetime of the energy storing species, and the amplitudes ΔU and U_{tot} evaluated. For each sample the energy of the irradiation pulse was varied such that from the resulting signals a plot of ΔU vs. U_{tot} could be obtained. These plots were subjected to linear regressional analysis to obtain the value of the slope, as depicted in fig. 2. Values were inserted into the TRTL equation, along with the measured fluorescence characteristics, to yield Φ_Δ for each sensitizer.

In oxygen-saturated benzene solutions all sensitizer triplet molecules are quenched by oxygen. Thus, the efficiency of $O_2(^1\Delta_g)$ formation [S_Δ, the fraction of sensitizer triplet quenching interac-

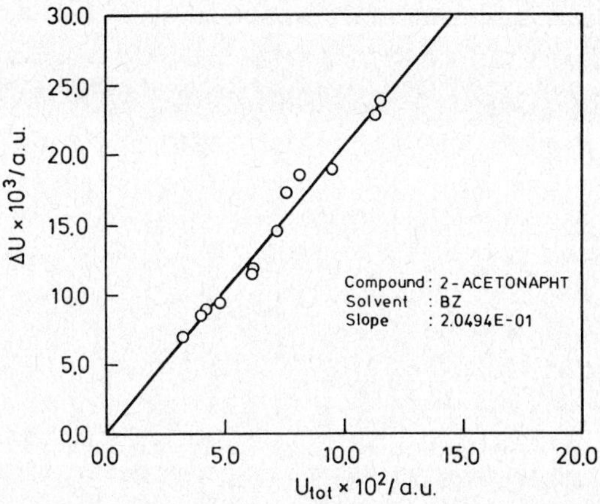

Figure 2 Regressional analysis of ΔU vs. U_{tot} plot.

tions resulting in the formation of $O_2(^1\Delta_g)$] may be evaluated as shown in (3).

$$S_\Delta = \Phi_\Delta / \Phi_{isc} \tag{3}$$

Table 1 shows Φ_Δ values obtained by TRTL, known intersystem crossing yields, Φ_{isc}, and some other properties of the respective sensitizers. The literature values of Φ_{isc} are for the most part obtained under conditions where oxygen is excluded from the system to avoid quenching. Oxygen being paramagnetic enhances the intersystem crossing process and it may well be that some of the Φ_{isc} values are lower under our conditions of oxygen saturation. This may be the case for acridine, for example, for which S_Δ apparently exceeds unity. Support for this argument is found in the case of the

Table 1 Measured Φ_Δ values and calculated sensitizing efficiencies

Sensitizer	E_T kcal/mol	Triplet Character	Φ_Δ	Φ_{isc}	S_Δ
Acetophenone	73.7	nπ*	0.31	1.00	0.31
Benzophenone	69.1	nπ*	0.35	1.00	0.35
Anthraquinone	62.4	nπ*	0.17	0.90	0.19
Phenanthrene	61.9	ππ*	0.57	0.70	0.81
2-Acetonaphthone	59.3	ππ*	0.73	0.84	0.87
Chrysene	56.6	ππ*	0.68	0.81	0.84
Fluorenone	53.3	ππ*	0.82	0.93	0.88
Pyrene	48.7	ππ*	0.74	0.96	0.77
Acridine	45.2	ππ*	0.84	0.76	1.00
Phenazine	44.4	ππ*	0.84	0.85	1.00
Anthracene	42.0	ππ*	0.68	0.78	0.87

related phenazine, where S_Δ is unity when calculated using Φ_{isc} measured in the presence of oxygen. The difference in energy between triplet sensitizer and $O_2(^1\Delta_g)$ (energy content = 22.4 kcal/mol) has no influence on the sensitizing efficiency. The aromatic hydrocarbons all have relatively high S_Δ values as have the heterocyclics acridine and phenazine, whereas the ketones benzophenone and acetophenone along with anthraquinone have much lower efficiencies. Within the ketone family there is a distinction between the ππ* and nπ* triplets as 2-acetonaphthone and fluorenone have efficiencies corresponding to that of the other ππ* aromatics. It thus appears that the ππ* or nπ* nature of the sensitizer may determine the efficiency of singlet oxygen sensitization.

4. Acknowledgement - We are grateful for the support of Professor Kurt Schaffner.

5. Literature

1. G. R. Rossbroich, N. A. Garcia and S. E. Braslavsky: J. Photochem. 31, 37 (1985).
2. R. W. Redmond, K. Heihoff, S. E. Braslavsky and T. G. Truscott: Photochem. Photobiol. 45, 209 (1987)

The Thermal Lens Applied to Photoacoustics

R.T. Bailey, F.R. Cruickshank, and D. Pugh

Department of Pure and Applied Chemistry, University of Strathclyde, 295 Cathedral Street, Glasgow G1 1XL, United Kingdom

1. Introduction

The pressure wave constituting the ultrasonic pulse, which is the photacoustic signal, is created, in the gas phase, by transfer of energy from the energy manifold responsible for the absorption of the laser photon to translational energy. Such energy transfer can occur at a wide variety of rates and, as has been demonstrated [1] previously, in both directions. Equations already developed [2] for quantifying the thermal lens have been extended to quantify the pressure pulse under these conditions. The consequent pressure relationships can be converted to describe the optical signals observed superimposed on the thermal conductivity-controlled decay of the thermal lens signal.

2. Theory

The pressure disturbance is governed by the equation

$$\partial^2\delta p/\partial t^2 - c^2\nabla^2\delta p =$$

$$\omega_s Q_o \exp l[-(R/R_g)^2 - \omega_s t].[-\omega_s H(t) + \delta(t)] \qquad (1)$$

(equation 8 [2]), where R_g is the 1/e radius of the Gaussian, TEM_∞ laser beam, $\omega_s = 1/\tau_{V-T}$, H(t) is the Heaviside unit function and Q_0 is the strength of the laser pulse.

To modify the previous treatment [2] for a finite cell radius, only the time - frequency Fourier transform is taken and the spatial variation is represented by a Fourier - Bessel expansion in which the boundary condition,

$$(\partial\delta p/\partial R)_{R=Ro} = 0 \qquad (2)$$

is satisfied. This ensures that the radial flow velocity is zero at the boundary, where R_0 represents the cell radius.

The formula for δp is,

$$\delta p\ (R,t) = 2\pi\,\omega_s\, Q_o\, (R_g/R_o)^2 \sum_n 1/(\omega^2 n + \omega^2)[-\omega_s e^{-\omega_s t} + \omega_n \sin\omega_n t + \omega_s \cos\omega_n t] \times [\exp - (\gamma_{1n} R_g/2R_o)^2]\, J_o\,(\gamma_{1n}\, R/R_o) / [J_o\,(\gamma_{1n})]^2, \qquad (3)$$

where $\omega_n = \gamma_{1n} c/R_0$ and γ_{1n} is the nth root of $J_1(X)$. In deriving the formula, the fact that in all our experiments $R_g \ll R_0$ has been used to replace the upper limit in the integrals,

$$\int_0^{R_o} (R/R_o)\ [\exp - (R/R_g)^2] J_o(\gamma_{1n}\ R/R_o) dR\ , \qquad (4)$$

by infinity, means they can be evaluated analytically. In this respect our treatment is simpler than that of previous groups [3]. Other terms (thermal diffusivity and diffusion, for example) can be included in the same way, but merely lead to more complex time dependent factors. Results for outward bound waves are not affected by the above departure from the infinite cell model, it being required for treatment of the observed reflected waves only. The density perturbation can be obtained from $p\delta(R,t)$ as described in [2] and an analysis of reflected thermal lens signals can then be made. Future analysis will include the important effect of laser beams which are not truly axial in the cell.

3. Equipment

The thermal lens system used was similar to that detailed previously [4]. However, the cell was 40mm diameter, with 10mm thick stainless steel walls. The microphone (Syrinx, Edinburgh) was a wideband (> 10 MHz) polyvinylidene fluoride peizoelectric film applied to the bottom of a 13mm diameter stainless steel rod 60mm long. This microphone was positioned along a radius with the electret 10mm from the cell axis and 50mm (midway) from the end windows of the cell.

4. Results

The above derivation of acoustic profiles assumes energy release in a single exponential decay. Further sums of exponentials can be incorporated in the future, however, in order to examine the simplest profile, SO_2 was selected, being capable of V-T relaxation in one acoustic lifetime at 600 Torr. This system was also used to examine the effect of positioning the laser beams off-axis. The effect of this latter misalignment is to increase the amplitude of the acoustic pulse returning from the cell wall after the first pass and similar signals in the second pass. In a well-centred configuration the first pulse on the microphone and its reflection from the opposite wall of the cell are of far larger amplitude than all other signals, which are reduced to the noise level. Such a centred system was used for all further work. With 603.4 Torr SO_2 there was a 3μs delay before the thermal lens signal rose, but thereafter the lifetime of the relaxation was equal to $\tau_{acoustic}$ (2.2μs). The observed acoustic pulse was qualitatively correct, but rather narrower than predicted (on a normalised R/R_g timescale). More significantly, the rarefaction was much more pronounced than predicted, even for a relaxation ten times faster than observed. The wake was of the correct duration, although distorted by the reflected pulse from the end of the steel rod supporting the microphone.

One of the less obvious energy transfer kinetic schemes, which acts as a photoacoustic source function, is that giving rise to an endothermic lens signal. As demonstrated earlier [4] such schemes do not require any endothermic transition between levels in the manifold, but merely require the appropriate balance of rate constants to favour, initially at least, abstraction of energy from the translational bath. A few systems have already been quantified [4], but the CO_2 + He system has recently been analysed to yield the unusual values, with the notation of [4], $k_{x3} + k_{x4} = 45$ s^{-1} $Torr^{-1}$ and $k_{010} = 1500$ s^{-1} $Torr^{-1}$. These are based on the generally accepted 387 and 255 s^{-1} $Torr^{-1}$ respectively for CO_2. A typical signal, together with the fitted theoretical curve, is shown in Figure 1. The fit is equally good over the pressure range of 80 to 500 Torr He. The extremely high V-T rate constant, k_{010} compared to the global V-V rate constant k_{x3} + k_{x4}, is most unusual, but is essential to account for the, initially very fast, endotherm. Helium is not alone in exhibiting such high V-T rate

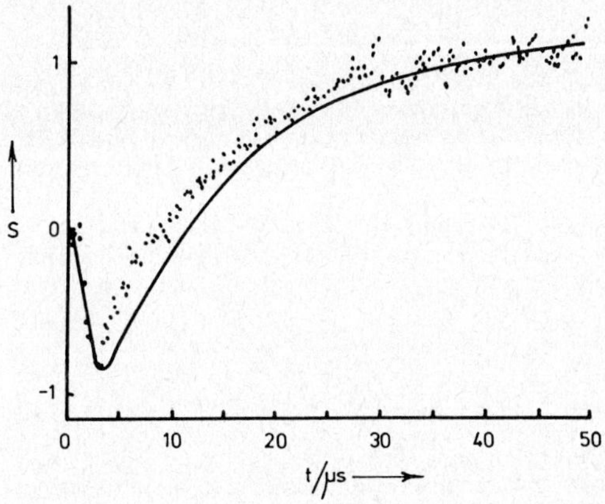

Fig. 1
Thermal lens signal for 141 Torr CO_2 + 431 Torr He.

constants. Hydrogen + CO_2 has also been examined and the thermal lens signal risetime is found to be acoustically limited, i.e. the energy relaxations are fast compared with the time required for the pressure pulse to traverse the pulsed laser R_g distance.

This requires that: $k_{x3} + k_{x4} > 4100\ s^{-1}\ Torr^{-1}$ and $k_{010} > 98000\ s^{-1}\ Torr^{-1}$. These values compare with e.g. $1.6 \times 10^5\ s^{-1}\ Torr^{-1}$ of previous work [5] where it was stated that the 001 level relaxation was measured, but, in fact, this level must relax more slowly than the 010 level to produce the sharp minimum (endotherm) observed.

The effect of such endotherms on the photoacoustic signal has been examined for energy relaxation with a lifetime 0.04 times the acoustic time (59.5 Torr CO_2). The outward bound wave at the mid point of a radius (total length 40 x R_g i.e. the dimensions of our cell) is shown in Figure 2.

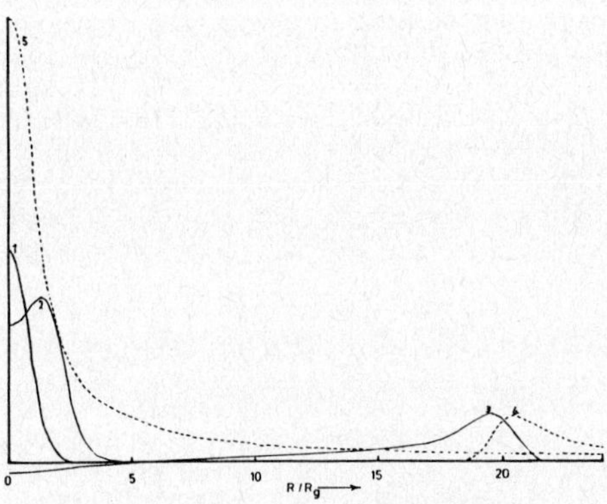

Fig. 2
Calculation of acoustic wave profiles.

1, t = 0.5 x $\tau_{acoustic}$
2, t = 2 x $\tau_{acoustic}$
3, t = 20 x $\tau_{acoustic}$
4, t = 60 x $\tau_{acoustic}$
5, t = 80 x $\tau_{acoustic}$
——— = outward
------ = inward

Microphone signals from 59.5 Torr CO_2 have been recorded digitally and do not have the shape predicted by the simple theory above. The observed pulse is far broader (again on a normalised R/R_g timescale) than predicted and the rarefaction very pronounced, whereas none was predicted. Clearly, the correct source function will have to be included in future calculations, as expected. Larger rarefactions than expected from this initial treatment have also been reported on work with SF_6 when relaxation times were short [6], and we have also seen discrepancies in this direction when modelling thermal lens signals under such conditions [2]. As stated previously, density oscillations off axis may be responsible for these. These effects may also cause the deviations from simple acoustic pulse profiles in the SO_2 case.

Acoustic signals superimposed on thermal lens signals of SF_6 systems have also been digitised and may be interpreted to yield V-T rate constants in approximate agreement with our earlier data [7].

Simultaneous records of the microphone signal and thermal lens signal show that the lens signal is typically 1000 times larger than the acoustic signal for the 59.5 Torr CO_2 relaxing with a global lifetime of .04x $\tau_{acoustic}$. This augurs well for the thermal lens as a sensitive spectrometric detector. Relaxation kinetics have a dramatic effect on these amplitudes.

5. Conclusions

Although the microphone bandwidth is quite sufficient to record faithfully the signals presented to it, these are not as predicted even for the simplest cases. The rarefaction seems to be regularly larger than predicted, and this probably stems from insufficiently detailed modelling of the source function. Examination of the CO_2 signals shows that a source function of known complexity and endothermicity has a dramatic effect on the rarefaction wave in particular.

6. Literature

1. R.T. Bailey, F.R. Cruickshank, K. M. Middleton, D. Pugh : J. Radioanalytical and Nuclear Chem. 101, 383 (1986).
2. R.T. Bailey, F.R. Cruickshank, D. Pugh, R. Guthrie, I.J.M. Weir : Mol. Phys. 43, 81 (1983).
3. J.R. Barker, T. Rothem : Chem. Phys. 68, 331 (1982).
4. R.T. Bailey, F.R. Cruickshank, D. Pugh, K. Middleton : J. Chem. Soc. Trans. Faraday II, 81, 255 (1985).
5. R.R. Jacobs, K.J. Pettipiece, S.J. Thomas : Appl. Phys. Lett. 24, 375 (1974).
6. K.M. Beck, A. Ringwelski, R.J. Gordon : Chem. Phys. Lett. 121, 529 (1985).
7. R.T. Bailey, F.R. Cruickshank, D. Pugh, R. Guthrie, I.J.M. Weir : Chem. Phys., in press.

Optical Detection of the Pulsed Photoacoustic Signal Combined with Thermal Lens Measurement: A Powerful New Tool for Radiationless Relaxation Studies

S.J. Komorowski[1], *S.J. Isak*[2], *and E.M. Eyring*[2]

[1]Polish Academy of Sciences, Institute of Physical Chemistry, PL-Warsaw, Poland

[2]Department of Chemistry, University of Utah, Salt Lake City, UT 84112, USA

A powerful combination of the fast optical detection of photoacoustic (OD PA) pulses with a conventional thermal lens (TL) measurement is depicted in Fig. 1. Each method covers its own characteristic time range. The TL method measures heat released by radiationless relaxation processes with time constants between ∼ 1 μs and ∼ 200 ms. Faster heat release processes are inaccessible to TL measurements. In contrast, OD PA spectroscopy with a fast photodiode can measure radiationless relaxation processes with time constants ranging from a few nanoseconds to a few hundreds of nanoseconds [1]. Unfortunately, the required geometry for the excitation and probe laser beams is quite different in these two detection schemes. In TL experiments the probe beam is located inside the sample region illuminated by the excitation beam. In OD PAS the probe beam must lie outside this illuminated region. We have resolved these conflicting requirements by placing both the excitation and probe beams very near to and parallel with the sample cell wall but with the probe beam still inside the excitation beam. In this geometry the deflection of the probe beam caused by a refractive index change in the illuminated region and a very fast deflection caused by the reflection of the acoustic wave at the cell wall are both detectable. To measure the faster signal a small active surface, fast photodiode (FPD) is used with a high frequency AC preamplifier (1 kHz-100 MHz bandwidth.) The very slow heat release in the TL experiment requires a

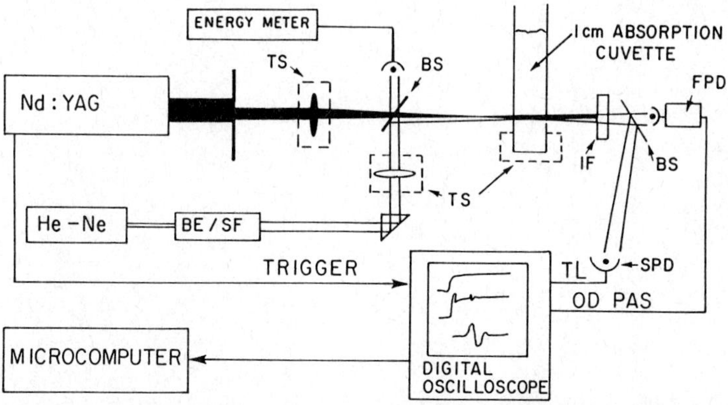

Fig. 1. Schematic of combined optical detection of pulsed photoacoustic signals and thermal lens measurements in a single experimental set-up. BE/SF, beam expander/spatial filter; TS, translation stage; IF, interference filter

DC slow photodiode (SPD) preamplifier set-up. Thus, the probe beam is split in two at the beam splitter (BS), so that the slow TL signal can be observed simultaneously with the combined fast TL and OD PA signals. Thus, the entire nanosecond to hundreds of milliseconds time scale for heat releasing processes is monitored in a single experiment, Fig. 2. A key advantage is the constancy of laser beam geometries in all the measurements.

Usefulness of this new method was verified by measuring the dependence of the TL and OD PA signals on absorption of a $CoCl_2$ in methanol solution and on the focusing of the λ = 532 nm, 200 μJ/pulse excitation beam. For both TL and OD PA signals, the dependence on absorption (from 0.15 to 0.003 cm^{-1}) is linear. The ratio of the amplitude of the TL signal to that of the OD PA signal remains constant as the excitation laser beam is defocused from a diameter of 50 to 100 μm (and both signals diminish.)

In a further test the laser dye 3,3'-diethyl-oxadicarbocyanine iodide (DODCI) was studied in water-methanol solution. A pulsed photoacoustic study of this system limited to time scales longer than a microsecond was reported recently [2]. In the present study signals from DODCI were compared with those from $CoCl_2$. For identical amplitudes of the TL signals from these two solutes there is a significant difference in the amplitude of the fast OD PA signals (Fig. 3). This indicates the existence of a slow (> 70 ns) heat releasing process in DODCI. This conclusion is further supported by the slower increase of the TL signal during the first microsecond after excitation for DODCI compared to $CoCl_2$.

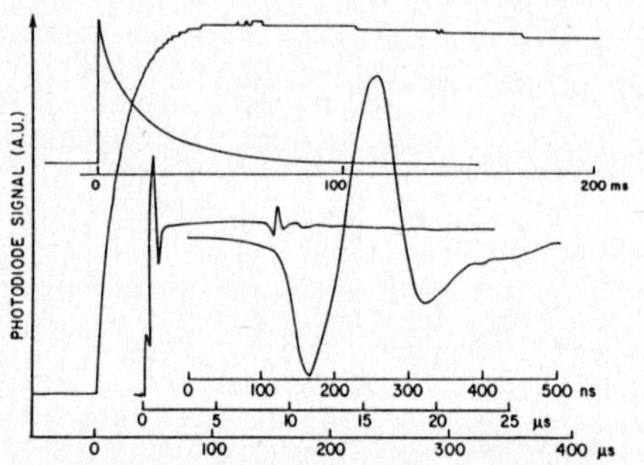

Fig. 2. Example of optical detection of a photoacoustic signal and the thermal lens signal in an experiment with $CoCl_2$ dissolved in methanol

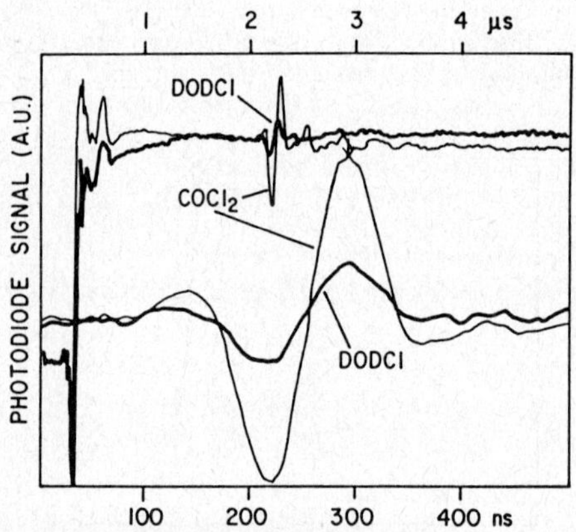

Fig. 3. Comparison of the fast photodiode signals for DODCI and $CoCl_2$ in methanol/30% water (v/v) solution at concentrations giving rise to a 0.01 cm^{-1} absorption in each case

This research was funded by the Department of Energy, Office of Basic Energy Sciences.

1. S. J. Komorowski and E. M. Eyring: the present volume
2. G. M. Bilmes, J. O. Tocho and S. E. Braslavsky; Chem. Phys. Lett. 134, 335 (1987)

Time-Resolved Laser-Induced Optoacoustic Spectroscopy

K. Heihoff and S.E. Braslavsky

Max-Planck-Institut für Strahlenchemie, Stiftstraße 34–36, D-4330 Mülheim a.d. Ruhr, Fed. Rep. of Germany

The use of broad band piezoelectric detectors in optoacoustics permits the determination of yields of production and lifetimes (τ_T) of photoproduced transient states. After the absorption of the 15 ns laser pulse, the fast dissipated heat is related to the amplitude of the first pressure pulse H according to (1):

$$H = K \cdot \alpha \cdot E_0 \cdot (1 - 10^{-A}) \ , \tag{1}$$

where K is an apparative constant, E_0 the pulse energy and A the absorbance of the sample. α is the fraction of absorbed energy dissipated as 'prompt heat', i.e. faster than the instrumental time resolution. This is determined by the transit time τ_a [1], the time that the pressure pulse needs to travel through the illuminated region. For a 'reference sample' ($\tau_T \ll \tau_a$) $\alpha = 1.0$.

Figure 1 shows signals from a reference sample at three different laser beam diameters, resulting in different signal widths.

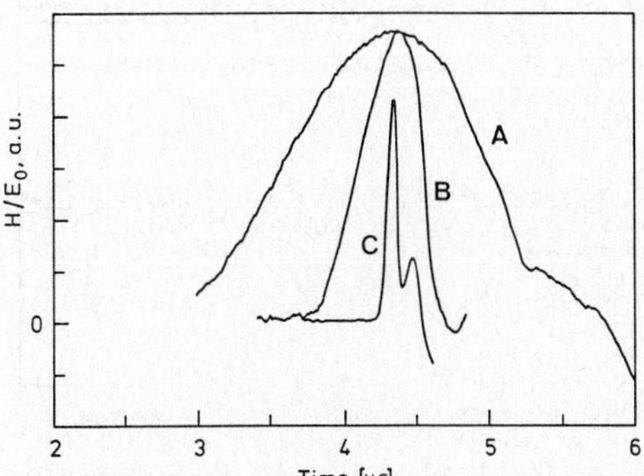

Fig. 1. PVF$_2$-OAS signals with $\tau_a \approx$ 1.7 μs (A), 550 ns (B), 100 ns (C), benzophenone + KI

The radiationless decay of a transient leads to a different amount of 'prompt heat' and thus to different α values, depending on the lifetime τ_T relative to τ_a.

A variation of τ_a allows the determination of τ_T by measuring $\alpha = \alpha(\tau_a)$ [2]. The decay of a transient leads also to a change of the signal form [3]. The measured decay signal (NR) can be described as a convolution of the transient decay function with the instrumental response function (PR), obtained using a reference.

These methods have been tested with benzophenone in acetonitrile. The benzophenone triplet was quenched by KI to lifetimes ranging from 20 ns to 15 μs. Deconvolution gave the best values of lifetimes and permitted the determination of τ_T from 60 ns - 10 μs ($\tau_a \approx$ 450 ns, cf. Fig. 2) [2]. The dynamic range can be estimated from some ns to some μs. Figure 2a shows the OAS signals of benzophenone containing 80 μM KI (NR) and reference sample (PR, [KI] $\geq$ 3 mM, the lifetimes from flash photolysis are indicated). Figure 2b shows one result obtained from deconvolution. We were able to fit the decay by a single exponential function with a shift relative to PR, as expected [3].

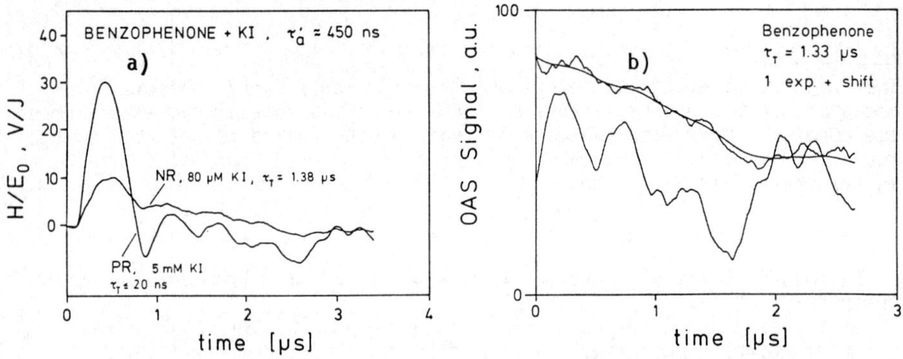

Fig. 2. a) OAS decay signal (NR) and prompt response (PR), b) deconvolution

The variation of τ_a led to lifetimes for the benzophenone triplet from 850 ns to 5 μs, which were in good agreement with those measured by flash photolysis.

This method was used to estimate the quantum yield of the primary photoreaction of the plant pigment phytochrome [4]. With the same method we determined for the first time the lifetime of a photoisomer produced after excitation of the tetrapyrrole phycocyanobilin dimethyl ester (PCBE), a model compound for the phytochrome chromophore. The isomer lifetime is $\tau_T \approx 150 \pm 30$ ns.

This work was supported by a fellowship awarded to K.H. by the Alfried Krupp von Bohlen und Halbach-Stiftung. We thank Prof. K. Schaffner for his support and interest.

1. H.M. Lai, K. Young: *J. Acoust. Soc. Am.* **72** , 2000 (1982).
2. K. Heihoff, S.E. Braslavsky: *Chem. Phys. Lett.* **131** , 183 (1986).
3. C.-Y. Kuo, M.M.F. Vieira, C.K.N. Patel: *J. Appl. Phys.* **55** , 3333 (1984).
4. K. Heihoff, S.E. Braslavsky, K. Schaffner: *Biochem.* **25** , 1422 (1987).

Photoacoustic Investigation of Energy Transfer in Adsorbed Layers

U. Frank and H.D. Breuer

Institut für Physikalische Chemie, Universität des Saarlandes, D-6600 Saarbrücken, Fed. Rep. of Germany

1. Introduction

The transfer of electronic excitation energy from a donor molecule to an acceptor can proceed by various mechanisms. In the trivial radiating process the donor molecule deactivates emitting a photon which is adsorbed by the acceptor. In case of a radiationless process the electronic excitation energy is transferred in a one-step reaction to the acceptor

$$D^* + A \rightarrow D + A^* \quad . \qquad (1)$$

In the dipole-dipole interaction mechanism the rate constant for energy transfer is according to Förster /1/

$$k = (1/\tau) \cdot (R_0/R)^6 ; \qquad (2)$$

τ is the lifetime of the excited state of the donor in absence of an acceptor, R is the distance between the excited donor and acceptor, R_0 is the critical distance, at which dipole-dipole transfer has the same probability as all other relaxation processes and $k = (1/\tau)$.

The intention for our investigations results from the growing interest in understanding the mechanism of energy transfer on surfaces. This mechanism plays an important rule in sensitization and energy conversion processes in biological and technical systems.

2. Experimental

In our experiments we used the Xanthene dyes Acridine Red (ACR), Rhodamine B (RHB) and Lissaminrhodamine B (LIS) as donor molecules. Due to their high fluorescence quantum yield one can expect a poor photoacoustic signal. In all cases the Xanthene derivative Thionine (THI) was the acceptor. Its absorption spectrum shows a good overlap with the donor emission spectra. The samples were prepared by adsorbing the dyes from an aqueous solution onto silica. The total dye concentration of the solution was held constant at $c = 1 \cdot 10^{-4}$ mol/l with a varying donor/acceptor ratio. The surface concentrations were determined by photometric measurements of the dye solutions before and after adsorption. Photoacoustic spectra were recorded with a double beam spectrometer described in /2/. Due to the strong overlap of the donor and the acceptor absorption, the peaks were separated by a computer fit, assuming a Gaussian line shape.

Figure 1 shows the calculated dependence of the PA-signal on the molar fraction of donor in the system ACR/THI. Curve 1 represents the intensity of PA-signal of pure donor dye excited at 530 nm. The concentration of the pure donor corresponds to the donor concentration in the mixture. Curve 2

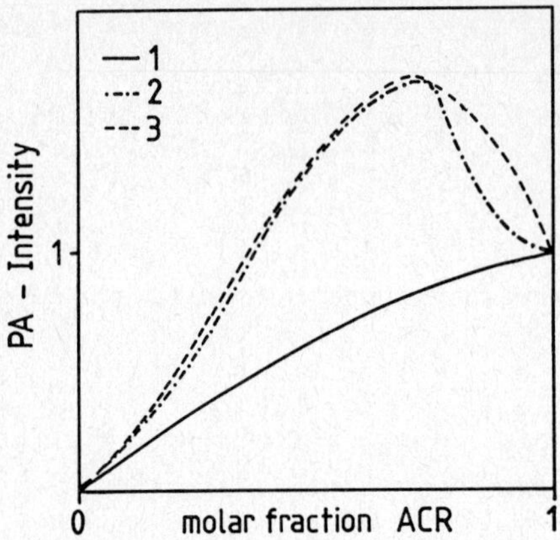

Fig. 1 Photoacoustic signal of Acridine Red 3B

shows the PA-signal of the dye mixture. Adding the acceptor causes a strong rise in donor intensity due to donor deactivation via energy transfer to the acceptor. Assuming that the transfer mechanism is independent of the surface concentration, the absorption intensity can be calculated according to (3):

$$I = q \cdot I_o(c) \cdot X_D \cdot X_A \cdot \qquad (3)$$

The constant q considers the efficiency of transfer and is different for all dye systems. $I_o(c)$ is the PA-intensity for the pure donor at a distinct concentration c. X_D and X_A are the molar fractions of the donor and acceptor dyes. Curve 3 shows the calculated signal. Up to a donor mole fraction of 0.7 there is good agreement between experimental and calculated signal intensities. Since only a fraction of the molecules is excited, the probability of finding an acceptor in a distance less than R_o is reduced at higher donor ratios. Thus at higher donor concentrations, the distance between excited donor molecules and acceptor molecules becomes too large for an effective dipole-dipole interaction and other deactivation paths become important. However, a multistep mechanism including energy migration according to (4) can be excluded since the Förster theory /1/ indicates a critical distance for donor-donor transfer of 18-19 Å in the investigated systems. In all experiments the actual distance between molecules on the surface is about 30 Å.

$$D^* + D \ldots + A \rightarrow D + D^* \ldots + A \rightarrow D + D \ldots + A^* \qquad (4)$$

Critical distances for donor-acceptor transfer on solid samples evaluated from fluorescence lifetime measurements are given in the table. These values are somewhat larger than comparable results which are obtained in solution /4, 5/. Thus, in the adsorbed state other quenching mechanisms seem to be less effective than in solution.

Table 1 Critical distances

Donor	Acceptor	R [Å]
RHB	THI	92
LIS	THI	79
ACR	THI	94

3. Conclusion

Similar as in solution the pure donor dyes deactivate on the surface mainly by fluorescence. It has been shown that, depending on the concentration, radiationless energy transfer is an effective deactivation for donor/acceptor systems in adsorption layers. Furthermore it has been shown that the Förster theory can be applied to describe the observed results.

1. Th. Förster: Fluoreszenz organischer Verbindungen, Verl. Vandenhoeck & Ruprecht, Göttingen (1982)
2. H. Jacob: Examensarbeit, Universität Saarbrücken (1979)
3. C. Bojarski, J. Dudkiewiesz: Z. Naturf., 27a, 1751 (1972)
4. C. Bojarski: Acta Physica Polonia, A63, 251-262 (1983)
5. S. Schneider, H. Coufal: J. Chem. Phys., 76(6), 2919 (1982)

Photothermal Investigation of In Vitro Photochemistry of Psoralens

R. Röben, W. Tuszynski, and E. Strauss

Department of Physics, P.O. Box 2503,
D-2900 Oldenburg, Fed. Rep. of Germany

1) Introduction Psoralens are used as pharmaceutical substances in the UV-A treatment of various skin diseases. The mode of action is a photochemically driven cross-linking of the DNA strands [1,2]. They are also potentially photomutagenic and photocarcinogenic. The lowest triplet state is considered to be mainly involved in the photochemistry, despite the rather low in vitro triplet yields observed [3,4]. These low yields seem to contradict the quite strong photoreactivity in biological material. We have reexamined their in vitro photophysical parameters using time-resolved photothermal spectroscopy [5] and observed a significantly higher triplet yield than reported previously.

2) Materials and Methods The psoralens were used as supplied (Sigma Chemie GmbH). All solutions were prepared from spectrograde solvents, at concentrations of about 10^{-4} mol/l. Samples were bubbled with Ar to avoid quenching of the triplet by oxygen, or with N_2O to scavenge photochemically dissociated electrons in aqueous solutions.

Excitation was with a Q-switched and tripled Nd:YAG laser (355 nm, 15 ns, <0.1 mJ, TEM_{00}-mode) at repetition rates < 20 Hz. The time dependence of the heat release, or rather the subsequent thermal lens, was sensed by collinear photothermal beam deflection. The response time of the detection system was < 150 ns (10 % to 90 %). The deflection signal was recorded with an averaging transient digitizer (8 bit).

With this time resolution, prompt heat released by comparatively "fast" processes originating from the singlet state can be distinguished from delayed heat released by the "slow" triplet relaxation processes. The time dependence of the slow heat gives directly the triplet lifetime (if photochemical intermediates do not interfere). The triplet yield Φ_T can be inferred from the relative signal amplitudes of the slow heat A_S and the total heat A_T. At relatively low pump intensity I and no phosphorescence, Φ_T is given by

$$\Phi_T = \frac{A_S}{A_T} \frac{(\tilde{\nu}_E - \Phi_F \tilde{\nu}_F)}{\tilde{\nu}_T} ,$$

where the subscripts of the frequencies ($\tilde{\nu}$) indicate excitation (E), fluorescence (F) and the triplet state (T). The ratio A_S/A_T is obtained from the slope of the plot of A_S versus A_T at various small pump intensities far from saturation. Note that no deflection amplitude calibration is needed and that only known spectroscopic parameters enter the equation.

3) Results We observed in aqueous solutions of psoralen that, at constant pulse energy, the prompt heat increased at higher repetition rates. This increase can be explained by assuming that long-lived (several 100 ms) photochemical products are produced in the excitation volume which are mainly producing prompt heat upon further excitation. All experiments

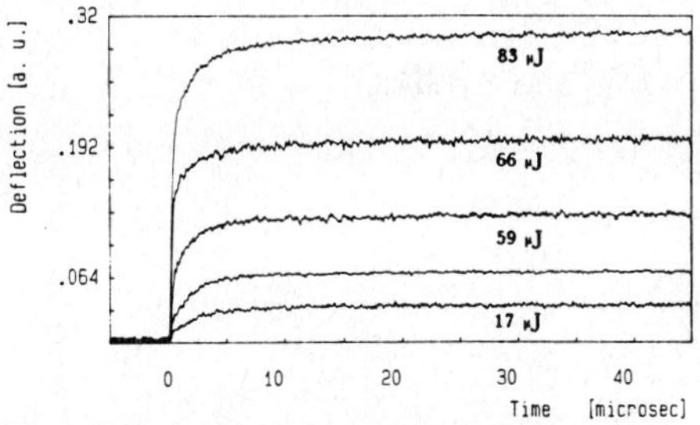

Fig.1 PtS-signal with psoralen/H_2O at various pulse energies.

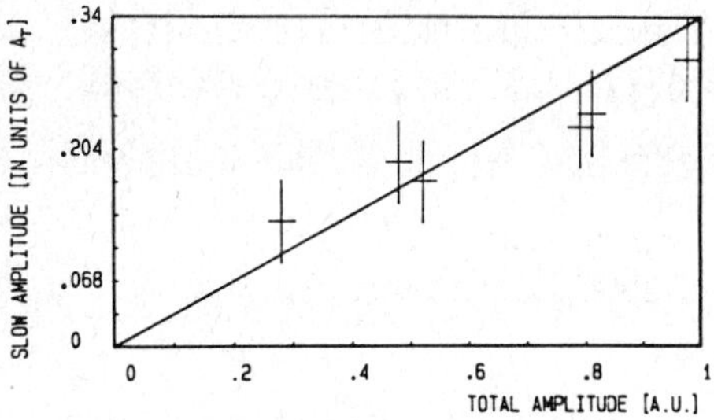

Fig 2. A_S to A_T plot of psoralen in water.

presented were performed at pulse rates < 1 Hz to avoid complications. Figure 1 shows that the photothermal signal depends on excitation intensity as expected from the equations. The rise time of the slow heat, i.e. the triplet lifetime, remains essentially unaffected. The slight decrease in the A_S/A_T ratio with increasing excitation intensity is due to the onset of excited state absorption. At small intensities (pulse energies 5 to 20 µJ) the A_S/A_T ratio is nearly constant, see Fig.2. Thus Φ_T can be determined from the slope of a linear fit. The results obtained for various psoralens are summarized in Table 1. Also referenced are data derived from T-T-absorption measurements.

Table1

Sample	1)Lifetime [microsec]	2)Lifetime [microsec]	3)Triplet Yield	4)Triplet Yield	
$1*10^{-4}$ m Trioxalen/Ethanol + Ar	2.5	2.2 (3)	0.59	0.093	(3)
$7*10^{-5}$ m Psoralen/Water + N_2O	3.8	3.3 (7)	0.60	0.12	(7)
$1*10^{-4}$ m Psoralen/Ethanol + Ar	2.8	2.6 (4)	0.33	0.06	(4)
$1*10^{-4}$ m Bergapten/Benzol + Ar	/	4.0 (7)	<0.05	<0.01	(7)

4) Conclusion The triplet lifetimes are in excellent agreement with the T-T absorption data. However, the triplet yields determined photothermally are systematically 5 to 6 times larger than the yields derived from T-T absorption. This indicates that the relaxation in the psoralens is considerably different from that assumed so far. This result may possibly explain the strong in vivo photobiological activity of the psoralens.

5) References:

(1) J.A.Parrish, T.B. Fitzpatrick, L. Tanenbaum, M.A. Pathak New Engl.J.Med. 291, 1207 (1974)

(2) B.J. Parsons Photochem.Photobiol.32, 813 (1980)

(3) P.C. Beaumont, B.J. Parsons, G.O. Phillips, J.C. Allen Biochim.Biophys. Acta 562, 214 (1979)

(4) M.T. Sa e Melo, D. Averbeck, R.V. Bensasson, E.J. Land, C.Salet Photochem. Photobiol.30, 645 (1979)

(5) R.W. Redmond, K. Heihoff, S.E. Braslavsky, T.G. Truscott Photochem.Photobiol.45, 209 (1987)

(6) W.W. Mantulin, P.S. Song J.Am.Che.Soc. 95, 5122 (1973)

(7) R. Bensasson, T.G. Truscott, E.J. Land, Flashphotolysis and Pulse Radiolysis , Oxford (1983), p.196 ff

Part III

Trace Analysis

Atmospheric Trace Gas Monitoring by Laser Photoacoustic Spectroscopy

M.W. Sigrist

Institute of Quantum Electronics, ETH, CH-8093 Zürich, Switzerland

The main characteristics of laser-photoacoustic (PA) spectroscopy applied to the monitoring of gaseous air pollutants are discussed. A brief review on previous studies is followed by theoretical considerations addressing the problem of analyzing PA spectra of multicomponent mixtures. The discussion on instrumentation emphasizes the design of appropriate PA cells with regard to continuous measurements. Finally, some recent achievements obtained with our stationary CO- and mobile CO_2-laser PA system are presented.

1. Introduction

Today, the pollution of the terrestrial atmosphere is of major concern for various reasons. One distinguishes between primary air pollutants like e.g. sulphur dioxide (SO_2), nitric monoxide (NO), hydrocarbons etc., which are emitted into the troposphere from various sources, and secondary pollutants. The latter include nitric oxides (NO_x), ozone (O_3), aldehydes, peroxiacetylnitrate (PAN) etc., which are produced from the primary products by chemical and photochemical processes occurring in the atmosphere, partly under the influence of the UV sunlight. The concentrations of all these substances range from sub-ppb ($< 10^{-9}$) to ppm (10^{-6}) levels. The specific and sensitive detection of these trace gases is a prerequisite for the understanding of the complex atmospheric chemistry. Since conventional methods like chemiluminescence, flame ionization detection etc. do not meet all the requirements, new methods have to be studied for the continuous and simultaneous recording of numerous pollutants in situ or remote.

In recent years various laser techniques have been applied to air pollution monitoring. These include remote, single-ended schemes like LIDAR [1,2] (in general the DIAL version), long-path absorption measurements either in a cell filled with polluted air [3,4] or as double-ended schemes in the free atmosphere [5,6], as well as the photoacoustic (PA) detection scheme (see e.g. [7-10]). In the lower atmosphere, the (single-ended) LIDAR scheme is in general restricted to emission measurements (i.e. to high concentrations of trace gases), whereas long-path absorption or PA methods can in principle also be applied to immission studies (i.e. to low concentrations). In particular the PA scheme has been demonstrated to permit measurements of minimum absorption coefficients $\alpha_{min} \simeq 10^{-10}$ cm^{-1} for a cell length ℓ of only 10 cm, i.e. a minimum $\alpha\ell$ of 10^{-9}. This value is superior to those which are accessible by the differential absorption technique (minimum $\alpha\ell \simeq 10^{-4}$) or the derivative absorption method (minimum $\alpha\ell \simeq 10^{-8}$).

2. Characteristics of Photoacoustic Schemes

Apart from the high sensitivity, the PA detection offers several advantages:

i) Simple setup, which is important for in situ measurements.

ii) Easy calibration with certified gas mixtures.

iii) Wide dynamic range of over 5 orders of magnitude, i.e. immission and emission measurements can be performed with the same apparatus.

iv) Continuous recording of various trace constituents appears feasible.

In spite of the high sensitivity of the PA method, trace gas concentrations in the ppb range can only be detected if the molecules of interest exhibit reasonable absorption cross sections. Best suited for trace gas detection is therefore the fingerprint region of the spectrum in the IR (≃ 2-15 μm), where most molecules exhibit characteristic absorption lines. At present, continuously tunable intense laser sources throughout this region are not available. A few studies have been performed with a tunable spin-flip Raman (SFR) laser [7,11], a $PbS_{1-x}Se_x$ diode laser [12] and a step-tunable DF laser [13]. However, for most studies, either a step tunable CO_2 laser in the 9-11 μm wavelength range or a CO laser tunable in the 5-6 μm range have been used (see e.g. Refs. in [10]). These lasers offer simple operation and high output powers, yet they are not continuously tunable. Therefore, accidental coincidences between laser transitions and absorption lines of trace gases are mandatory. Fortunately, this does not really reduce the application of these lasers in trace gas detection because numerous gases absorb in their wavelength ranges. For most gases and vapors, detection limits in the ppb concentration range have been achieved. However, most PA studies mentioned above have been performed on certified gas mixtures in the laboratory. The first in-situ PA study was performed by Patel [11] on stratospheric NO with a balloon-borne SFR laser-PA spectrometer. Later, Perlmutter et al. [14] reported on measurements in realistic atmospheres, yet not in situ, with a stationary CO_2 laser PA system. Recently, other research groups including our own have started PA studies on trace gas detection under realistic conditions (see e.g. [15,16] and various reports at this conference).

3. Theoretical considerations

The PA signal generation in a gas by the absorption of modulated laser light has been treated by several authors (see e.g. [8,10,17]). In many cases of PA studies the PA amplitude S is given by the following expression:

$$S \text{ prop. } N\sigma\, I\, \tau A_n\, \Delta E, \qquad (1)$$

where N and σ are the density and absorption cross section of the absorbing molecules, respectively, I is the laser intensity, τ the total lifetime of the excited state including radiative and nonradiative decay, A_n is the nonradiative decay rate due to collisions of the excited molecules, thus τA_n represents the probability for nonradiative relaxation, and ΔE is the heat energy released per deexcitation. Additional factors enter (1) depending on the geometry of the PA cell used. However, it should be noted that the simple relationship of (1) is only valid under the following conditions:

i) The incident laser intensity I has to be small enough not to saturate the absorption of the gas molecules,

ii) $\alpha \ell \ll 1$ where $\alpha = N\sigma$,

iii) $\omega \ll 1/\tau$ where ω is the angular modulation frequency of the laser intensity.

If these conditions are fulfilled, (1) implies that absorption spectroscopy can be performed on weakly absorbing gases as well as on trace gases with large σ yet low concentrations. With respect to trace gas detection, conditions i) and ii) can easily be fulfilled whereas condition iii) has to be considered carefully. For polyatomic molecules at a total pressure of 1 bar and IR excitation to a rotational-vibrational state, the nonradiative V-R,T relaxation usually occurs within less than 10^{-5}-10^{-6} s [17], so that also high modulation frequencies, e.g. for resonant PA cells, can be used. However, in air pollution studies performed with a CO_2 laser PA system attention has to be paid to the absorption by atmospheric CO_2 and also by H_2O vapor. These molecules increase the lifetime of the excited states drastically due to kinetic cooling [17], and cause a phase reversal of the PA signals. Thus, this effect can be taken into account by considering both amplitude and phase of the PA signals [10, 14,19].

In practice, one always deals with mixtures of trace gases and other constituents. We perform detailed analyses of measured PA spectra by using calculated spectra. The latter are deduced on the basis of known PA spectra of single constituents, including phase considerations, according to the following linear system of equations:

$$S_i \cos \phi_i = \sum_j c_j S^0_{ij} \cos \phi^0_{ij} \quad , \tag{2}$$

with $\quad i = 1,\ldots, N$
$\quad\quad j = 1,\ldots, n \qquad (j \leqslant i)$.

S_i and ϕ_i represent the normalized signal amplitude and phase, respectively, at the laser transition i (if a step-tunable laser is used), c_j is the concentration of the air component j giving rise to a normalized and calibrated amplitude S^0_{ij} and phase ϕ^0_{ij} at the laser transition i. A least-squares fit of the experimental data based on (2) yields the individual concentrations of the components taken into account.

Overlapping absorption spectra which result in cross-sensitivities between different gases represent a general problem in spectroscopy. However, for any multicomponent mixture composed of a priori known gases, a cross-sensitivity table can be calculated [20-22]. Thus possible interferences can be taken into account and the appropriate choice of laser transitions for the detection of the individual constituents with maximum sensitivity and minimum interferences can be evaluated.

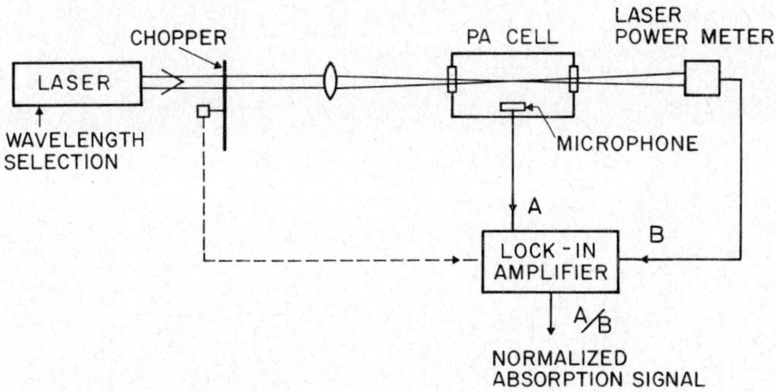

Fig. 1 Experimental setup for photoacoustic spectroscopy on gases

4. Instrumentation

The typical arrangement for PA studies on gases is shown in Fig. 1.

In general, the beam of a tunable cw laser is amplitude-modulated by a mechanical chopper, although other modulation techniques have been introduced as well. The beam is directed through a resonant or non-resonant PA cell which contains the gas sample under study. The laser power is monitored by a power meter which can be placed after the cell in the case of "transparent" gas samples (signal B in Fig. 1). The acoustic signal A is recorded with a gas-phase microphone and processed by a lock-in amplifier. The normalized absorption is obtained as ratio A/B.

In the past, many different versions and modifications of the general scheme presented in Fig. 1 have been introduced (see e.g. [8-10]). A considerable effort has been put into cell design in order to optimize the characteristics of PA detection. With respect to in situ monitoring of air pollution the PA cell has to fulfill the following requirements:

i) It should be applicable also in noisy environments (traffic etc.).

ii) A high sensitivity should be guaranteed also for flowing gas mixtures, i.e. for continuous monitoring.

Both requirements can be met by the use of specially designed resonant PA cells. As an example we developed for our mobile CO_2 laser PA system [10] a resonant cell based on the design by Gerlach and Amer [24], but with additional buffer volumes. It operates on the first radial mode with a Q-factor of ca. 320 at a frequency of ca. 2640 Hz. The microphone is positioned on the axis where the standing acoustic wave exhibits its maximum.

The pressure nodes are located at the entrance and exit windows and at the gas in- and outlets. This design reduces the window heating signal considerably and permits the operation of the cell also with flowing gases at flow rates of ≤ 1.5 ℓ/min without appreciable loss in sensitivity (see also [15]). The high resonance frequency allows measurements in noisy environments, e.g. close to a road with heavy traffic.

A completely different cell geometry was designed for our CO laser PA system [25,26]. Due to the special emission characteristics of the CO laser and the strong interference of water vapor within the CO laser wavelength range, we use a dual-beam arrangement with a sample and reference cell. Detailed calculations resulted in a novel cell which is operated at the second longitudinal mode. Its resonance frequency can be adjusted, which is an important property with regard to the dual-beam setup.

In the following, the performance and some recent results of our CO laser and CO_2 laser PA systems are briefly discussed.

5. CO Laser PA System

Table I summarizes the absorption ranges of some of the gases and vapors of environmental concern which absorb within the emission range of the CO laser in use (5.15 µm - 6.35 µm). The list contains nitric oxides, specific hydrocarbons and also the important smog product PAN. As indicated in the first column the absorption strength decreases from top to bottom from an average value of 75 cm^{-1} atm^{-1} to 1 cm^{-1} atm^{-1}.

The two columns on the right indicate typical sources and immission concentrations for the molecules listed. The strong water vapor absorp-

Table I: List of some gases and vapors absorbing in the CO laser wavelength range

cw sealed off CO laser (Edinburgh Instr Typ PL-3)

5.0 λ [µm] 6.5

Emission range of CO laser

ca. 90 laser transitions

5 mW - 1.5 W cw

Absorption strength [$cm^{-1}atm^{-1}$]	Absorption range of different gases and vapors	Source	Imm. conc. [ppb]
75	Acetaldehyde CH_3CHO	Industry	?
	Nitrogen Dioxide NO_2	Smog	10 - 100
	Peroxiacetylnitrate PAN $CH_3CO_3NO_2$	Smog	< 5
	Acrolein (Propenal) CH_2CHCHO	Industry	?
	Formaldehyde HCHO	Traff/Ind	?
	Hydrogen Nitrate HNO_3	Industry	?
	Nitric Oxide NO	Traffic	10 -100
	Vinyl Chloride C_2H_3Cl	Industry	1 - 100
	Ethylene C_2H_4	Traffic	1 - 100
	Toluene $C_6H_5CH_3$	Traffic	1 - 100
1	Benzene C_6H_6	Traffic	1 - 100
$2*10^{-3}$	Carbon Dioxide CO_2	Nature	350 ppm
10^{-3} - 5	Water Vapor H_2O	Nature	0.1 - 3 %

tion (indicated at the bottom) represents a severe problem. Even with the dual-beam setup, drying or diluting of the air sample is usually necessary in order to detect trace gas concentrations with a resolution of approximately 10 ppb. At present, the performance of the computer-controlled stationary system is tested on exhaust samples from various sources. Examples are discussed in Ref. 21.

6. CO_2 Laser PA System

A list of pollutants that absorb within the wavelength range of the step-tunable CO_2 laser (9.2 - 10.8 µm) is presented in Table II together with the mean absorption strengths, typical sources and concentrations.

Table II: List of some pollutants absorbing in the CO_2 laser wavelength range

cw, sealed off CO_2 laser (Ultra Lasertech)

9µm 10µm 11µm

Emission range of CO_2 laser

ca. 60 laser transitions

2 - 12 W cw

Absorption strength [$cm^{-1}atm^{-1}$]	Absorption range of different gases	Source	Imm. conc. [ppb]
80	Ammonia NH_3	Nature	1 - 50
	Freon 12 Cl_2CF_2	Industry	< 5
	Ethylene C_2H_4	Traffic	1 - 100
	Acrolein CH_2CHCHO	Industry	?
	Ozone O_3	Smog	10 - 200
	Vinyl Chloride C_2H_3Cl	Industry	1 - 100
	Perchloroethylene C_2Cl_4	Industry	1 - 50
	Trichloroethylene C_2HCl_3	Industry	1 - 60
	Freon 13 $CClF_3$	Industry	< 5
	Toluene C_7H_8	Traffic	2 - 200
1	Benzene C_6H_6	Traffic	2 - 100
$3*10^{-3}$	Carbon Dioxide CO_2	Nature	~ 350 ppm
$1*10^{-4}$	Water Vapor H_2O	Nature	0.1 - 3 %

In contrast to the CO laser wavelength range, the absorption by atmospheric water vapor exhibits some weak absorption lines superimposed on a continuum [27], and does not represent an interference problem. The absorption by atmospheric CO_2, however, has to be taken into account by phase considerations as mentioned above.

With respect to the serious forest decline, the measurement of the ambient ethylene (C_2H_4) concentration is of particular interest [28]. We performed C_2H_4 measurements with our mobile CO_2 laser PA system at different locations [22,23]. The minimum detectable C_2H_4 concentration in the real atmosphere is 5 ppb (< 1 ppb with N_2 as buffer gas). In Fig. 2 a comparison between PA and gaschromatography (GC) data of the ambient C_2H_4 concentration vs daytime is presented.

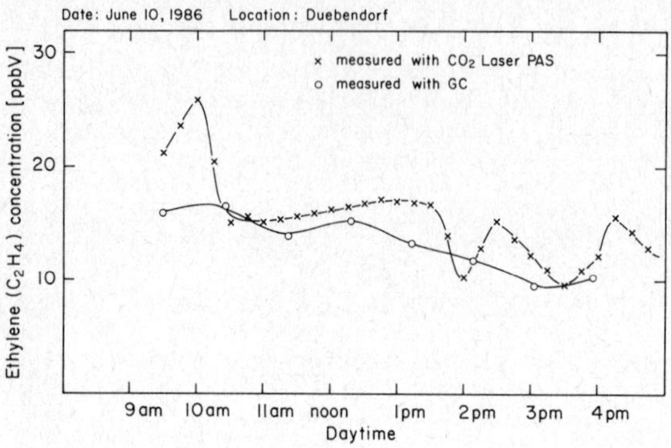

Fig. 2 C_2H_4 concentration profiles measured with PAS and gaschromatography

While GC data could be obtained only every hour, the PA data were taken every 15 minutes. The agreement between the two independent measurements on the same air volumes is remarkable. Figure 2 also demonstrates the advantage of the PA method, which permits almost continuous measurements. Our system, which has been improved and automated in the meantime, can now record concentration variations on a time scale of < 1 min.

At present, multicomponent gas mixtures are investigated with special emphasis put on the problem of cross-sensitivities [22].

Acknowledgement

The author is grateful to S. Bernegger and P. Meyer for their valuable collaboration, and to R. Baumann of the EMPA for the GC data. This work was funded by the National Research Program 14 of the Swiss National Science Foundaton.

References

1. E.D. Hinkley, ed.: Laser Monitoring of the Atmosphere, Topics in Applied Physics, Vol. 14 (Springer, Berlin, Heidelberg, 1976)
2. A.I. Carswell: Can. J. Phys. 61, 378 (1983)
3. A. Mayer, J. Comera, H. Charpentier, C. Jaussaud: Appl. Opt. 17, 391 (1978)
4. M. Khelkhal, F. Herlemont, M. Lyszyk, J. Lemaire: Appl. Phys. B 29, 227 (1982)
5. U. Persson, S. Lundqvist, B. Marthinsson, S.T. Eng: Appl. Opt. 23, 998 (1984)
6. C.R. Webster, R.T. Menzies: Appl. Opt. 23, 1140 (1984)
7. C.K.N. Patel: Science 202, 157 (1978)
8. A.C. Tam: In Ultrasensitive Laser Spectroscopy, ed. by D. Kliger (Academic, New York, 1983), p. 1

9. V.P. Zharov, V.S. Letokhov: Laser Optoacoustic Spectroscopy, Springer Ser. Opt. Sci., Vol. 37 (Springer, Berlin, Heidelberg, 1986).
10. M.W. Sigrist: J. Appl. Phys. 60, R 83 (1986)
11. C.K.N. Patel: Opt. Quant. Electron. 8, 145 (1976)
12. T.H. Vansteenkiste, F.R. Faxvog, D.M. Roessler: Appl. Spectrosc. 35, 194 (1981)
13. T.F. Deaton, D.A. Depatie, T.W. Walker: Appl. Phys. Lett. 26, 300 (1975)
14. P. Perlmutter, S. Shtrikman, M. Slatkine: Appl. Opt. 18, 2267 (1979)
15. G.L. Loper, J.A. Gelbwachs, S.M. Beck: Can. J. Phys. 69, 1124 (1986)
16. M.A. Leugers, G.H. Atkinson: Anal. Chem. 56, 925 (1984)
17. P. Hess: In Topics in Curr. Chem., Vol. 111 (Springer, Berlin, Heidelberg, 1983), p. 1
18. A.D. Wood, M. Camac, E.T. Gerry: Appl. Opt. 10, 1879 (1971)
19. R. Velusamy, M.M. Rao: Appl. Opt. 20, 3828 (1981)
20. P.L. Meyer, St. Bernegger, M.W. Sigrist: Digest 11th Int. Conf. IR and mmWaves, Tirrenia-Pisa (I), Oct 20-24 (1986), p. 507
21. St. Bernegger, P.L. Meyer, M.W. Sigrist: This conference, paper Tu B3
22. P.L. Meyer, St. Bernegger, M.W. Sigrist: This conference, paper Tu B8
23. P.L. Meyer, B. Suter, M.W. Sigrist: Helv. Phys. Acta 59, 1027 (1986)
24. R. Gerlach, N.M. Amer: Appl. Phys. 23, 319 (1980)
25. St. Bernegger, P.L. Meyer, M.W. Sigrist: Digest 11th Int. Conf. IR and mm Waves, Tirrenia-Pisa (I), Oct 20-24 (1986), p. 497
26. St. Bernegger and M.W. Sigrist: Appl. Phys. B 44, 125 (1987)
27. J. Hinderling, M.W. Sigrist, F.K. Kneubühl: Infrared Phys. 27, 63 (1987)
28. H. Mehlhorn, A.R. Wellburn: Nature 327, 417 (1987)

CO-Laser Photoacoustic System for IR Spectroscopy of Gases and Vapors

St. Bernegger, P.L. Meyer, C. Widmer, and M.W. Sigrist

Institute of Quantum Electronics, Physics Department, ETH, CH-8093 Zürich, Switzerland

1. Introduction

Many important gaseous air pollutants, like nitric oxide (NO), nitrogen dioxide (NO_2), ethylene (C_2H_4), propylene (C_3H_6), 1,3-butadiene (C_4H_6), benzene (C_6H_6), toluene ($C_6H_5CH_3$), vinylchloride (C_2H_3Cl), formaldehyde (H_2CO), acetaldehyde (CH_3CHO), acrolein (CH_2CHCO), ammonia (NH_3), etc., show strong and specific absorption bands between 5.0 and 6.5 μm. Therefore, a CO laser, which can be tuned over about 90 lines with output powers between 5 mW and 1 W within the wavelength range of 5.1 and 6.4 μm is of special interest as a source of radiation for photoacoustic spectroscopy (PAS).

The features of the CO laser and the interference of water vapor absorption make it, however, difficult to detect trace gas concentrations in the ppbv range. Therefore, a special experimental setup was introduced [1]. In order to optimize our system, we have developed a new type of spectrophone [1,2]. The cell operates in a longitudinal mode with a resonance frequency of 555 Hz. The quality factor is 43 and the sensitivity 1640 Pa cm/W. For a dried air sample, a minimum detectable absorption coefficient α_{min} of approximately $2 \cdot 10^{-8}$ cm^{-1} is achieved.

2. Experimental

A simplified scheme of the experimental setup for PAS with our CO laser is shown in Fig. 1. Two identical spectrophones, as shown in Fig. 2, are operated in a parallel arrangement. The sample cell contains the gas under study while the reference cell is filled with water vapor diluted in a buffer gas. The cw laser beam is modulated by a mechanical chopper at the resonance frequency of the second longitudinal mode of the sample cell. The resonance frequency of the reference cell can be adjusted by changing its cavity length with the aid of a movable piston, which is mounted at one end of the cell. Changes of the resonance frequency due to changes in the gas temperature are automatically compensated by the computer which controls the experiment.

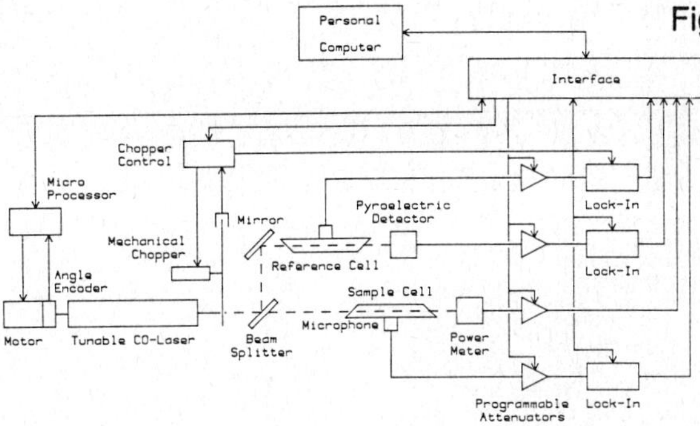

Fig. 1: Experimental setup with reference cell for CO laser PA spectroscopy

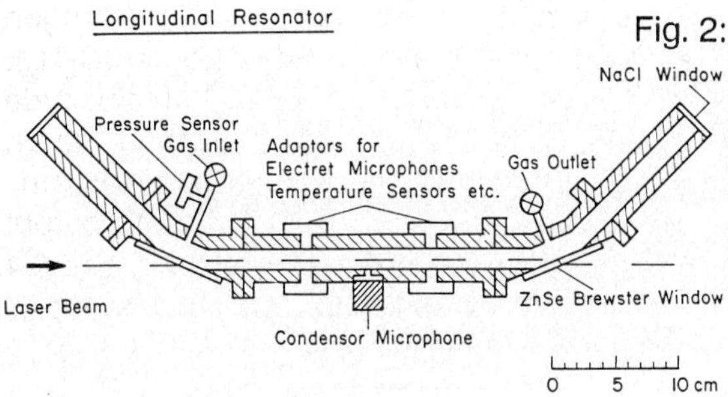

Fig. 2: Longitudinal resonant spectrophone. The Brewster windows are mounted at the pressure nodes of the second longitudinal mode of the cell

The PA signals are detected with condenser microphones. Gain programmable attenuators are used to adapt the signals to the ranges of the dual-phase lock-in amplifiers. The CO laser is tuned with the aid of a grating which is positioned by a microprocessor-controlled motor. A pyroelectric detector is used as a fast feedback to the computer. The P-polarized CO laser beam is divided by a 70% transmission beamsplitter into two paths of identical length, providing the same spectral content of the beams in the two cells. The use of the reference cell enables the determination of the water vapor absorption with an accuracy of 1%.

3. Results

We have used our system for measuring the absorption cross sections of various vapors and gases at the CO laser wavelengths. As an example, the absorption spectrum of 35 ppmv acrolein in 952 mbar N_2 is shown in Fig. 3. The solid lines represent the measured absorption coefficients including the water vapor contribution, while the bars represent the absorption due to acrolein alone. The latter have been calculated by subtracting the appropriate amount of the water vapor spectrum, which has been recorded simultaneously in the reference cell.

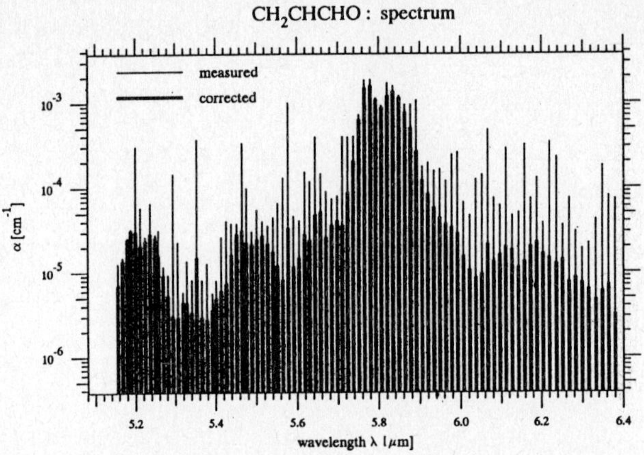

Fig 3. Absorption of 35 ppm acrolein diluted in N_2. The bars represent the spectrum corrected with respect to H_2O absorption

Table I: Minimum detectable concentrations and cross-sensitivities of selected gases and vapors

Substance	H_2O	NO	C_2H_4	C_2H_3Cl	C_3H_6	C_6H_6	C_7H_8	CH_2O	C_2H_4O	C_3H_4O
Water vapor H_2O	40 ppb	-	-	1/100	1/200	1/100	1/50	-	1	1/10
Nitric oxide NO	1/700	15 ppb	1/1000	1/1000	1/300	1/2000	1/400	1/1000	1/30	1/40
Ethylene C_2H_4	1/70	1/8	60 ppb	1/70	1/50	1/125	1/80	1/50	1/5	1/15
Vinylchloride C_2H_3Cl	1/1000	-	-	15 ppb	1/50	1/500	1/10	1/500	1/50	1/50
Propene C_3H_6	1/25	-	-	1/30	60 ppb	1/100	1/30	1/10	1	1/5
Benzene C_6H_6	1/40	1/8	1/100	1/30	1/5	150 ppb	1/10	1/10	2	1/2
Toluene C_7H_8	1/100	-	-	1/40	1/5	1/50	80 ppb	1/20	1	1/2
Formaldehyde CH_2O	1/200	-	-	1/2500	1/5000	1/2000	1/700	2.5 ppb	1/5	1/40
Acetaldehyde C_2H_4O	1/1000	-	-	-	-	-	-	1/50	0.7 ppb	1/100
Acrolein C_3H_4O	1/300	-	-	-	1/2000	-	1/1000	1/100	1/5	1.5 ppb

In multiple-component samples, the minimum detectable concentration c_{min} of a specified substance depends on the contribution to the absorption by all other components. The data for c_{min} and the cross sensitivities for 11 gases and vapors are listed in Table I. The diagonal elements c_{min} correspond to an absorption coefficient α_{min} of $5 \cdot 10^{-8}$ cm^{-1}. The nondiagonal elements represent the smallest ratio of concentration for which the gas in the left column shows for at least one CO laser line the same absorption coefficient as the gas on the top line. This means that, e.g., 1 part of acrolein can still be detected in 100 parts of formaldehyde, while 1 part of the latter can be detected in 40 parts of the former.

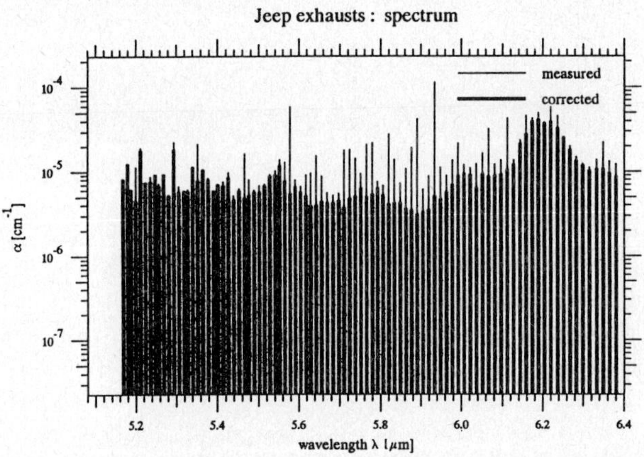

Fig 4. Absorption of 93 mbar Jeep exhausts diluted in N_2

As an application we have used our system for the analysis of car exhausts. As a typical result, Fig. 4 shows the absorption spectrum of 93 mbar Jeep exhausts diluted in N_2 at a total pressure of 957 mbar. Again, the solid lines represent the measured absorption coefficients, while the bars represent the spectrum corrected with respect to the water vapor absorption. The result of the analysis based on the PA calibration spectra of components is shown in Table II.

Table II: Analysis of Jeep exhausts diluted in N_2

Component	Molecule	Concentration
Carbon dioxide	CO_2	19000 ppm
Nitric oxide	NO	2.3 ppm
Ethylene	C_2H_4	24 ppm
Propene	C_3H_6	14 ppm
Benzene	C_6H_6	5.6 ppm
Toluene	C_7H_8	57 ppm

Figure 5 shows a comparison of the measured spectrum with the spectrum calculated on the basis of these concentrations. The bars which are centered around the calculated absorption coefficients represent the uncertainty of the absorption strength, while the solid lines represent the measured values. The agreement is excellent with the exception of the wavelength range around 5.9 μm, where a molecule which has not yet been identified contributes to the spectrum.

An interesting effect is shown in Fig. 6 where the phases of the two photoacoustic signals are plotted. The phase shifts observed in the sample cell for a few lines in the wavelength range between 5.2 and 5.5 μm occur

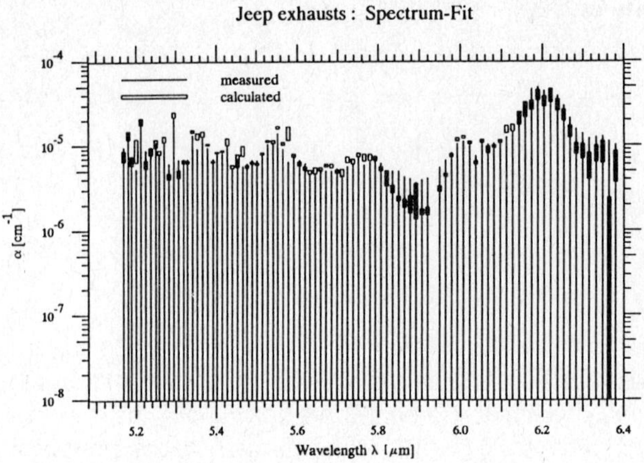

Fig 5. Analysis of the Jeep exhausts. The components listed in Table II yield the calculated spectrum

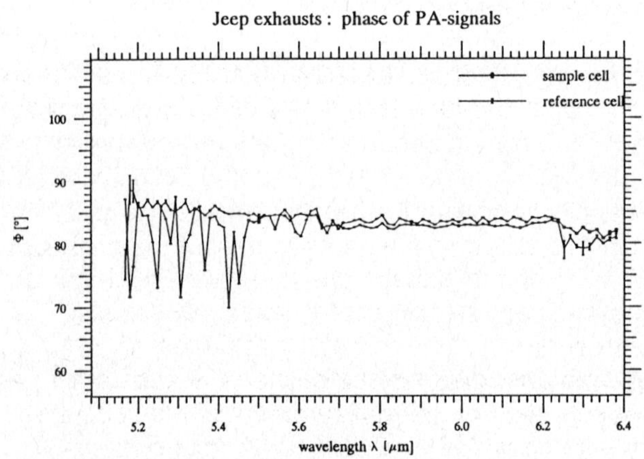

Fig 6. The phases of the PA signals are very stable. The deviations occur at strong NO absorption lines

exactly at the strong absorption lines of nitric oxide. These phase shifts vanish if more water vapor is present in the sample gas. We assume that the mechanism is similar to the kinetic cooling effect observed for carbon dioxide in N_2 at CO_2 laser frequencies [3].

This work was supported by the Swiss Natl. Science Foundation and ETH Zurich.

1. S.Bernegger, M.W.Sigrist: Appl.Phys.B 44,125 (1987)
2. S.Bernegger, P.L.Meyer, M.W.Sigrist: Dig. 11th Int.Conf. on IR and mm Waves, Tirrenia-Pisa (I), Oct 20-24 (1986), p. 497
3. F.G.Gebhardt, D.C.Smith: Appl.Phys.Lett. 20,129 (1972)

On-Line Monitoring of Air Pollutants with a Mobile Computer-Controlled CO_2 Laser Photoacoustic System

P.L. Meyer, St. Bernegger, and M.W. Sigrist

Institute of Quantum Electronics, Physics Department, ETH, CH-8093 Zürich, Switzerland

Hitherto only a few PA studies have considered the cross sensitivities for selected substances with respect to corresponding concentration measurements in outdoor air [1]. Here we present such data for various gases and vapors which have to be taken into account for absorption measurements within the CO_2 laser wavelength range. By considering these cross sensitivities, in situ measurements on multi-component mixtures have been performed. Typical concentration profiles are presented.

Our CO_2 laser PA system which is installed in a trailer has been described previously [2]. The system is now completely computer-controlled in order to improve the time resolution and accuracy of the measurements.

In Table 1 we present a list of important air constituents absorbing in the CO_2 laser wavelength range. The diagonal elements represent the minimum detectable concentrations c_i^{min} of the substances i in pure synthetic air obtained with our PA system at the indicated laser transitions. The nondiagonal elements correspond to the cross sensitivities $Q_{ij} = \sigma_{ij}/\sigma_{jj}$, where σ_{ij} and σ_{jj} represent the absorption cross sections of substance j at laser transitions i and j, respectively. For the measurement of the concentration c_i of substance i one has to consider the influence of an uncertainty Δc_j of the concentration c_j of substance j according to

$$\Delta c_i >= Q_{ij} \cdot \Delta c_j \cdot c_i^{min} / c_j^{min} . \tag{1}$$

If, for example, the vinylchloride (C_2H_3Cl) concentration of an $C_2H_4/C_2H_3Cl/N_2/O_2$-mixture is known with an uncertainty of $\Delta c_j = 20$ ppb then the uncertainty Δc_i of the measured C_2H_4 concentration is raised by

$$\Delta c_i >= 0.15 \cdot 20 \text{ ppb} \cdot 1 \text{ ppb} / 3.2 \text{ ppb} = 0.95 \text{ ppb} . \tag{2}$$

In the 10 μm branch of the CO_2 laser the absorption of ambient air is dominated by H_2O, CO_2 and C_2H_4 because the concentrations of other substances are rather low. Furthermore the influence of other pollutants on

Substance j / Substance i	Water H_2O	Ethylene C_2H_4	Carbon-dioxide CO_2	Ammonia NH_3	Vinyl-chloride C_2H_3Cl	Trichloro-ethylene C_2HCl_3	Perchloro-ethylene C_2Cl_4	Acrolein CH_2CHCHO	Benzene C_6H_6	Toluene C_7H_8
Water 10 R (20)	32 ppm	$4.7*10^{-2}$	1.1	$1.7*10^{-3}$	$1.6*10^{-1}$	$1.6*10^{-2}$	$< 9*10^{-4}$	$7.5*10^{-1}$	$2.8*10^{-3}$	$6.5*10^{-2}$
Ethylene 10 P (14)	$1.4*10^{-1}$	1.0 ppb	$8.8*10^{-1}$	$3.0*10^{-2}$	$1.5*10^{-1}$	$6.1*10^{-1}$	$< 9*10^{-4}$	$6.2*10^{-1}$	$1.2*10^{-3}$	$3.1*10^{-2}$
Carbondioxide 10 P (20)	$1.5*10^{-1}$	$8.3*10^{-2}$	8.7 ppm	$6.6*10^{-3}$	$3.0*10^{-1}$	4.4	$< 9*10^{-4}$	$9.5*10^{-1}$	$1.0*10^{-3}$	$4.3*10^{-2}$
Ammonia 10 R (6)	$1.2*10^{-1}$	$6.9*10^{-2}$	$6.3*10^{-1}$	0.95 ppb		$1.0*10^{-2}$	$< 9*10^{-4}$	$9.7*10^{-1}$	$2.1*10^{-3}$	$2.0*10^{-2}$
Vinylchloride 10 P (22)	$1.2*10^{-1}$	$5.5*10^{-2}$	1.1	$4.2*10^{-3}$	3.2 ppb	1.18	$< 9*10^{-4}$	$6.9*10^{-1}$	$1.2*10^{-3}$	$5.3*10^{-2}$
Trichloroethylene 10 P (28)	$1.1*10^{-1}$	$5.2*10^{-2}$	2.6	$1.3*10^{-2}$	$2.5*10^{-1}$	2.0 ppb	$4.5*10^{-3}$	$4.9*10^{-1}$	$1.3*10^{-3}$	$6.6*10^{-2}$
Perchloroethylene 10 P (42)		$9.1*10^{-2}$	$1.8*10^{-1}$			$1.7*10^{-1}$	1.2 ppb		$1.2*10^{-3}$	$6.9*10^{-2}$
Acrolein 10 R (14)	$1.7*10^{-1}$	$5.6*10^{-2}$	1.8	$3.0*10^{-1}$	$1.7*10^{-1}$	$1.0*10^{-2}$	$< 9*10^{-4}$	10 ppb	$2.0*10^{-3}$	$6.5*10^{-2}$
Benzene 9 P (30)	$8.2*10^{-2}$	$4.0*10^{-2}$	$6.4*10^{-1}$	$2.8*10^{-3}$	$1.5*10^{-1}$	$7.1*10^{-3}$	$< 2.7*10^{-3}$	$1.2*10^{-1}$	13 ppb	$6.9*10^{-1}$
Toluene 9 P (36)	$6.3*10^{-2}$	$2.6*10^{-2}$	$4.1*10^{-1}$	$2.0*10^{-2}$	$1.4*10^{-1}$	$6.5*10^{-3}$	$2.7*10^{-3}$	$1.5*10^{-1}$	$3.7*10^{-1}$	30 ppb

Table 1 : Minimum detctable concentrations c_i and cross sensitivities Q_{ij} of selected gases and vapors the concentration measurement of these three components is very small as Table 1 implies. Therefore accurate concentration profiles for H_2O, CO_2 and C_2H_4 can be obtained by continuous PA measurements at the 10R(20), 10P(20) and 10P(14) laser transitions [3]. In Fig.1 we present typical data measured in Biel (Switzerland) at a location approx. 20 meters distant from a road. The C_2H_4-concentration profile exhibits three peaks of 20 to 35 ppb which coincide with the morning and evening rush hours. The evening peak is shifted to later times due to increasing wind (bottom of Fig.1). The uncertainty is approx. 5 ppb which corresponds to the minimum detectable concentration for C_2H_4 in ambient air.

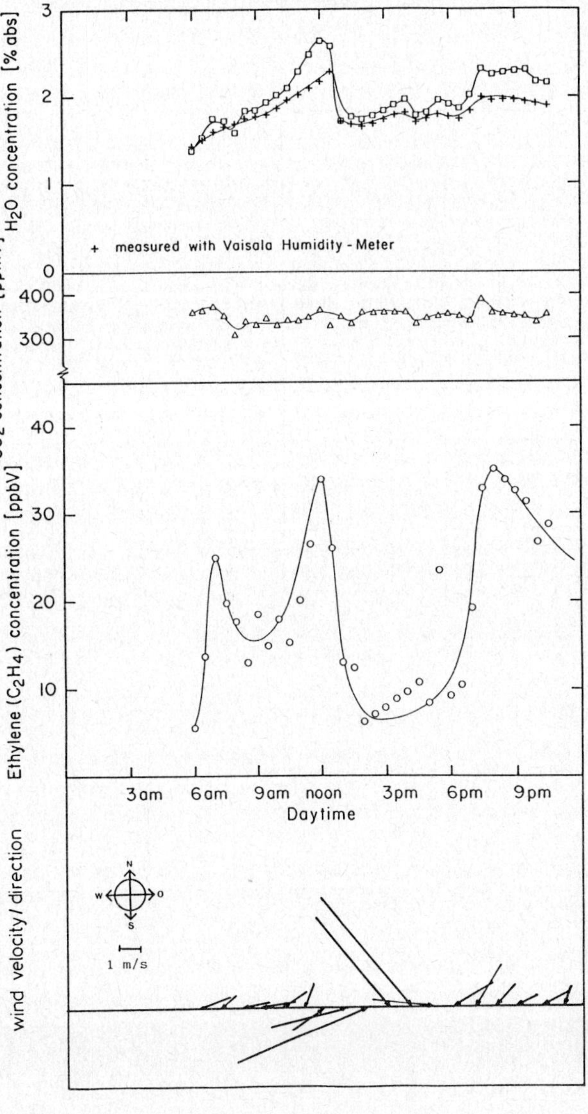

Fig. 1 :

Concentration profiles of H_2O vapor, CO_2 and C_2H_4, and wind data, taken at Biel (Switzerland) on July 3, 1986

Further measurements of additional substances at other locations are in progress and will be discussed later.

This work is supported by the National Research Program NFP 14 of the Swiss National Science Foundation.

References

1. V.P.Zharov, V.S.Lethokov: In Springer Ser. Opt. Sci. Vol. 37 (Springer, Berlin, Heidelberg 1986)
2. P.L.Meyer, St.Bernegger, M.W.Sigrist: Helv.Phys.Acta 58, 833 (1985)
3. P.L.Meyer, St.Bernegger, M.W.Sigrist: Dig. 11th Int. Conf. on IR and mm Waves, Tirrenia-Pisa (I), Oct 20-24 (1986), p.507

Trace Analysis of Freons by Optoacoustic CO_2 Laser Detection

Z. Zelinger[1], *P. Engst*[1], *Z. Papoušková*[2], *and M. Jakoubková*

[1]J. Heyrovský Institute of Physical Chemistry and Electrochemistry, CS-165 02 Prague 6-Suchdol

[2]Institute of Chemical Process Fundamentals, Czechoslovak Academy of Sciences, CS-165 02 Prague 6-Suchdol

Freons are typical pollutants coming mostly from sprays. They also find industrial use as refrigerants. Since they could restrict the Earth's ozone layer, their sensitive and selective detection is important. For this purpose, the laser optoacoustic (OA) method [1,2] may be advantageous, especially when a tunable CO_2 laser of sufficient power is available.

1 Experimental

A classical cylindrical OA cell (i.d. 6 mm) was used for the detection. The cell was equipped with a condenser microphone (sensitivity 100 μV/Pa) as a pressure transducer located in the center of the effective cell length. A focused beam of radiation from a home-made CO_2 laser was modulated by a chopper. The passage of the beam through the cell was defined by internal diaphragms, which act as acoustic suppression elements. The detection scheme was employed over a wide range of modulation frequencies, but mainly at the acoustic resonance of the cell [3] (about 2 kHz). To decrease the background, the cell was completed by buffer spaces. Signals from the microphone and pyrodetector passed through lock-in amplifiers, a comparator and an X-Y recorder.

2 Results and Discussion

To test our apparatus we used permeation tubes with refilling stopcocks filled with Freons 11 or 12 as standards. Tube calibration [4] was carried out gravimetrically (Table I). The permeating substance was diluted by a carrier gas, usually air. The resulting concentration could be determined from the known value of carrier gas flow. Coarse changes of concentration could be obtained by using different permeation tubes made from Teflon or

Table I. Freon 11 losses from permeation tubes and actual concentrations

Tube material	Weight loss [mg/day]	Concentration [ppm] ranges controlled by	
		air flow[a]	temperature[b]
Polyethylene	247.4	11.68 - 93.44	46.72 - 99.92
Teflon	25.0	1.18 - 9.44	4.72 - 10.42
Teflon	5.2	0.25 - 1.96	0.98 - 2.08
Teflon	1.5	0.07 - 0.57	0.28 - 0.68

[a]5 - 40 ml at temperature 298 K, [b]298 - 308 K at air flow 10 ml/s

polyethylene. Using one permeation tube the concentration could be changed by varying the carrier gas flow. The effect of temperature on the permeation rate offers another means of concentration control. Of course, knowledge of the temperature dependence of vapor tension of the filling substance is assumed.

With the method described we have detected Freons (Freon 11: line 9 R(24) and Freon 12: line 10 P(42)) at concentrations down to about 10 ppb. Moreover, concentrations higher than 15 ppm were detected by grating IR spectrometer as well. The results obtained by both OA and IR methods agree well. The OA method has found applications in refrigerator manufacturing plants.

References

1. L.B. Kreuzer: J. Appl. Phys. 42, 2934 (1971)
2. L.-G. Rosengren, E. Max, S.T. Eng: J. Phys. E: Sci. Instrum. 7, 125 (1974)
3. P. Perlmutter, S. Shtrikman, M. Slatkine: Appl. Opt. 18, 2267 (1979)
4. M.L. Stellmack, K.W. Street Jr.: Anal. Lett. 16 (A2), 77 (1983)

Infrared Lasers for Optoacoustic Detection of Vapours

S.A. Johnson[2], *S.A. Bone*[1], *P.B. Davies*[1], *P.G. Cummins*[2], *and E.J. Staples*[2]

[1]Department of Physical Chemistry, University of Cambridge, Lensfield Road, Cambridge, CB21EP, United Kingdom
[2]Unilever Research Port Sunlight Laboratory, Quarry Road East, Bebington, Wirral, Merseyside, L633JW, United Kingdom

1 Introduction

In this study cw Pb-salt diode laser optoacoustic spectroscopy has been applied to the quantification of water vapour in a sample cell. The method could be used for a quick absolute determination of water activity in liquids and solids (by equilibration of sample with a vacuum or gas), for the accurate recording of rovibrational lineshapes, and for possible investigations of the infrared water vapour continuum (caused by dimers or far-wings [1]).

2 Experimental

In general terms the equipment used here has already been described [2]. A diode laser spectroscopy review has recently appeared [3], and its use for water line parameter measurements demonstrated [4 and refs. therein]. A few diode laser optoacoustic experiments have already been reported [2,5-8].

The optoacoustic cell employed here was a non-resonant design of 2 cm^3 volume (including dead space) and consisted of a 4.5 cm long glass tube of diameter 5 mm, closed with LiF windows, and with a Knowles BT-1754 microphone attached. 60 db amplification gave effective oscilloscope display. The cell could be evacuated via a teflon tap. Water vapour and air could also be introduced through this. A glass bulb containing degassed water and surrounded by an ice bath enabled convenient water vapour control in the region of 5-20 torr. Pressures were measured by a combination of an MKS capacitance gauge (0-10.000 torr) attached to the bulb, and a low volume Edwards EPS 10 (0-1000 mbar) transducer directly on the cell.

A Spectra-Physics diode laser was tuned to provide 50 μW operation over an ortho/para doublet at 1942 cm^{-1} belonging to the ν_2 band of water. The radiation was focussed through the cell onto a liquid nitrogen cooled InSb detector for transmission measurements. To achieve high S/N ratios both for transmission and optoacoustic measurements involved the use of a lock-in-amplifier in conjunction with either a light chopper, or laser frequency modulation. Output was fed directly to an X-Y recorder.

3 Results and Discussion

Figure 1 shows the optoacoustic spectrum of the doublet (118 Hz chopper, 10 ms t.c.). In fact, this is the superimposition of two rapidly swept scans, and their coincidence illustrates how high accuracy lineshapes can be recorded.

Under the conditions of Fig.1, three other unabsorbed laser modes of significant power were simultaneously supported, so it proved convenient to calcu-

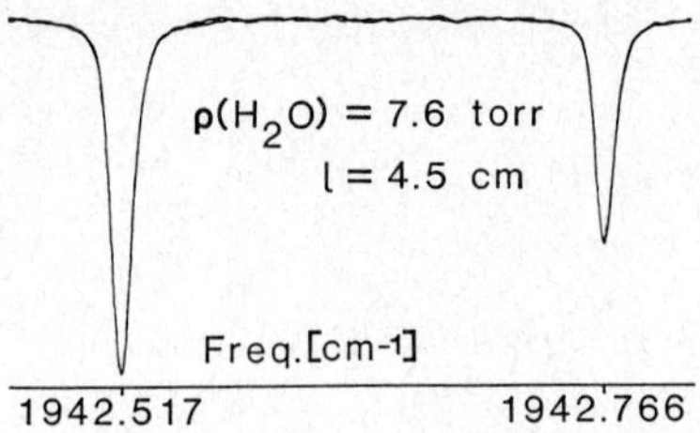

Fig.1. Spectrum of o/p doublet

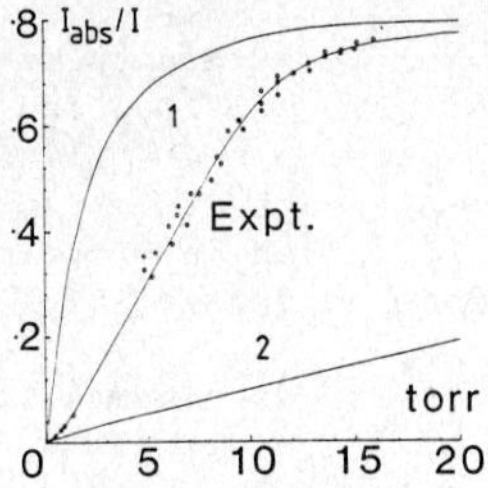

Fig.2. Fractional absorption and opto-acoustic response v $\rho(H_2O)$

late (296 °K) the pressure dependence of line-centre fractional absorption, using existing linestrength and broadening estimates and a Voigt profile, rather than relying on total power transmission measurements.

For the line at 1942.517 cm^{-1} [6,5,2 <- 5,4,1 (J,K_a,K_c)] the linestrength, S, is 1.56×10^{-20} cm^{-1}. $molecules^{-1}$ cm^2 [9]. The Doppler-broadened Gaussian HWHM, b_D, is 0.0028 cm^{-1} (easily calculated [10]). In the absence of high resolution data pertaining to this line, the collision-broadened Lorentz HWHM, b_L, has to be derived as a function of partial pressures (in torr) by theory:- $b_L = 7.63 \times 10^{-5}\ \rho(air) + 4.671 \times 10^{-4}\ \rho(H_2O)\ cm^{-1}$. The air-broadening constant has been calculated [11] by the quantum fourier transform method of DAVIES [12] (0.9 x tabulated H_2O-N_2 value), whilst results of ANDERSON-TSAO-CURNETTE theory [13] used to estimate [14] the self-broadening (with the appropriate temperature factor). Broadening constants from these sources would be expected to be reliable to around ± 15% [11].

Inserting b_D, b_L, S and the pathlength l (= 4.5 cm) in the appropriate equations given in FRIDOVICH [15] allows the line-centre fractional absorption, I_{abs}/I, to be calculated as a function of pressure. This is illustrated in Fig. 2, where the Voigt curves for pure water vapour(1), and for water in 760 torr air(2) are plotted. Also shown is the actual opto-acoustic peak response (several sets of experiments, normalized) for the pure water case, illustrating how the acoustic response to I_{abs} is inefficient at low total pressures.

In Figs. 3 and 4 the experimental FWHM are plotted against $\rho(H_2O)$ and $\rho(air)$. The scatter of points on Fig.3 is around $\pm 1 \times 10^{-3}$ cm^{-1}. This could be improved with more careful pressure measurements. For comparison purposes, calculated $2b_V$ curves are also shown (derived using b_D and b_L [16]-for Fig.4 self-broadening was ignored). In Fig.3 data is presented both for the

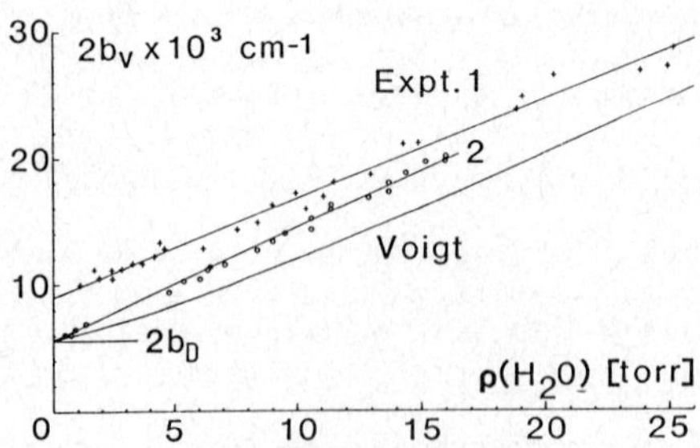

Fig.3. Voigt FWHM v $\rho(H_2O)$

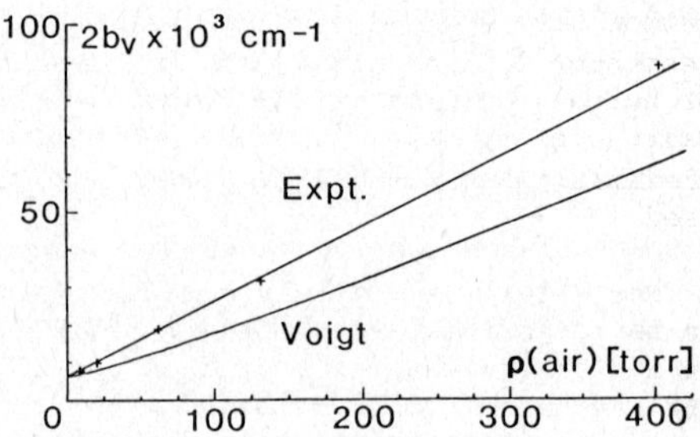

Fig.4. Voigt FWHM v $\rho(air)$

laser mounted in the Laser Analytics two-point supported cooler, as well as in the newer three-point model. The trace for the former(1) shows an instrumental contribution to the linewidth which is absent with that(2) from the latter. Although it is tempting to ascribe this effect to the poorer vibration-isolation in the older device, it could just have arisen from a feedback-induced instability [4].

For Figs.3 and 4 it was assumed that the Voigt width was to be measured at the half height of the optoacoustic signal, M. In fact it should be measured [15] at $[(I^2/I_{abs}^2 - I/I_{abs})^{0.5} + 1 - I/I_{abs}]M$ above the peak response rather than 0.5M. Thus the $2b_v$ values are likely to have been overestimated, which would to some extent account for the differences between experiment and theory. For more accurate results either much shorter path-lengths would be required, such as to validate the 0.5M approximation, or I_{abs}/I would need to be measured. In establishing I_{abs}/I, any curvature in the laser power profile would also usefully be determined.

With the straight transmission method by itself there are problems both with atmospheric absorption, and with determining the 100% transmission line without an iteration step [15]. Here, the 100% line naturally comes out at the zero level of the optoacoustic spectrum, and measurements can be made up to atmospheric pressure if desired.

For measuring vapour in equilibrium with liquid water, the peak amplitude flattens off markedly above 10 torr (Fig.2). Since the system would require rigorous temperature control anyway ($\pm 10^{-1}$°K $\simeq \pm$ 1% change in vapour pressure) for meaningful results it would be most sensible to control this around 10°C to keep the peak response in a relatively fast varying regime. Less attractive would be to make measurements in the presence of buffer gas (decreased line centre response, but could use a high J line exhibiting Dicke collisional narrowing [17]), or by integrating whole lineshapes.

In the monitoring mode, peak amplitude is measured by laser frequency modulation (of depth $\simeq$ linewidth FWHM) and lock-in-detection at 2f. A second lock-in-amplifier is set at 1f to line-lock the laser to the peak maximum. This is a standard diode laser method [18], but is more satisfactory here since the 1f zero crossing is at zero volts. This is rarely true for a transmission measurement where baseline fluctuations can lead to a drifting lock point.

Using this method, the signal to noise on the 2f peak was at the most 500:1 with a 1 second averaging time (100 ms line-lock t.c.) - the noise being microphone noise. In future work a reference water vapour opto-acoustic cell would be necessary, since the laser frequency modulation depth chosen influences the signal amplitude [19]. Although the transmission method offers at present more signal to detector noise (10^6:1 on the 2f response, 1s t.c.), this did not translate into a better measurement because of noise and drift on the laser-induced baseline. Efforts to reduce this using the techniques of REID et al [20] proved only partially successful, presumably because of the complicated spatial mode structure of the laser used.

For detecting water in gas samples the present setup would allow 10 ppm to be determined at an optimum (for air) total pressure of around 20 torr on a stronger water vapour line (e.g. $S = 3 \times 10^{-19}$ cm^{-1}. molecule^{-1} cm^2 at 1684.84 cm^{-1}). This is still quite some way from the ppb detectivity of long-path 2nd harmonic diode laser absorption spectroscopy, or CO_2/CO laser optoacoustic methods.

Conclusions

The merits of various moisture sensors have been recently described [21]. The present laser method offers similar advantages as FTIR with the added potential features of greater specificity, speed and accuracy. Its two main drawbacks are its cost, which would be much reduced using 1.4 or 1.9 μm near-infrared lasers, and the contacting nature of the measurement.

A methodology for water lineshape measurements has been identified that would complement both transmissive diode laser [4] and FTIR [22] efforts in this direction.

1. J. Hinderling, M.W. Sigrist, F.K.Kneubühl: Infrared Phys. 27, 63 (1987)
2. S.A. Johnson: Anal. Proc. 23, 1 (1986)
3. Yu A. Kuritsyn: In Laser Analytical Spectrochemistry, ed. by V.S. Letokhov (Adam Hilger, Bristol 1986)
4. K. B. Thakur, C.P. Rinsland, M.A.H. Smith, D.C. Benner, V. Malathy Devi: J. Mol. Spectry. 120, 239 (1986)
5. E.D. Hinkley: Opto-Electronics 4, 69 (1972)
6. T.H. Vansteenkiste, F.R. Faxfog, D.M. Roessler: Appl. Spectry. 35, 194 (1981)
7. Y. Kawabata, T. Kamikubo, T. Imasaka, N. Ishibashi: Anal. Chem. 55, 1420 (1983)
8. K. Stephan, W. Hurdelbrink: In The Photoacoustic Effect, ed. by E. Luscher, P. Korpiun, H.J. Coufal, R. Tilgner (Vieweg, Braunschweig 1984)
9. J.-M. Flaud, C. Camy-Peyret, R.A. Toth: In Water Vapour Line Parameters from Microwave to Medium Infrared (Pergamon, Oxford 1981)
10. J.A. Mucha: Appl. Spectry. 36, 141 (1982)
11. R.R. Gamache, R.W. Davies: Appl. Optics 22, 4013 (1983)
12. R.W. Davies: Phys. Rev. A12, 927 (1975)
13. C.J. Tsao, B. Curnette: J. Quant. Spectry. Radiat. Transfer 2, 44 (1962)
14. W.S. Benedict, L.D. Kaplan: J. Quant. Spectry. Radiat. Transfer 4, 453 (1964)
15. B. Fridovich: J. Mol. Spectry. 105, 53 (1984)
16. J.J. Olivero, R.L. Longbothum: J. Quant. Spectry. Radiat. Transfer 17, 233 (1977)
17. R.S. Eng, A.R. Calawa, T.C. Harman, P.L. Kelley, A. Javan: Appl. Phys. Lett. 21, 303 (1972)
18. J.A. Mucha: Appl. Spectry. 36, 393 (1982)
19. J.A. Mucha: Appl. Spectry. 38, 68 (1984)
20. J. Reid, R.S. Sinclair, W.B. Grant, R.T. Menzies: Opt. Quant. Electron. 17, 31 (1985)
21. K. Carr-Brion: Moisture Sensors in Process Control (Elsevier, London 1986)
22. J.-Y. Mandin, J.-M. Flaud, C. Camy-Peyret: J. Quant. Spectry. Radiat. Transfer 26, 483 (1981)

Photoacoustic Detection of Formaldehyde as a Minority Component in Gas Mixtures

M. Boutonnat, K.A. Keilbach, D.A. Gilmore, N. Oliphant, P. Vujkovic Cvijin, and G.H. Atkinson

Department of Chemistry and Optical Sciences Center, University of Arizona, Tucson, AZ85721, USA

Abstract

Formaldehyde concentrations as low as 51 ppbv are measured in gaseous samples containing mixtures of H_2CO, NO_2, CH_3CHO, and N_2 using a nanosecond, pulsed laser photoacoustic spectrometer operating in the ultraviolet.

The value of photoacoustic (PA) spectroscopy in the quantitative detection of atmospheric pollutants has been established for a variety of experimental approaches [1], including the recent development of PA detection derived from pulsed lasers and nonlinear optics operating in the ultraviolet [2-4]. Fast (5-7 ns), pulsed laser excitation provides the opportunity to time-resolve PA signals and thereby eliminates the need for reference cells. Quantitative determinations of concentrations in the sub-ppbv range have been demonstrated using this technique [3]. In this paper, an experimental protocol is given for the quantitative detection of H_2CO as minority component in mixtures of H_2CO, NO_2, CH_3CHO, and N_2 at a total pressure of one atmosphere.

A pulsed laser photoacoustic spectrometer consisting of a Nd:YAG pumped dye laser and accompanying nonlinear optics (Quanta Ray, model DCR2A, PDL-1, WEX) is used to initiate PA signals. The second harmonic of the Nd:YAG laser can be used directly (at 532 nm) to generate a PA signal or to pump the dye laser. Nonlinear optics are used with the dye laser to provide tunable ultraviolet radiation in the 300 nm -310 nm range. The cell configuration, the gas sampling system and the procedure for recording PA signals are similar to that used for SO_2 detection [3].

A measurement protocol that allows maximum sensitivity for H_2CO detection while maintaining a high level of discrimination between the three components (H_2CO, NO_2, and CH_3CHO) in a mixture is determined using high resolution (0.5 cm^{-1}) PA spectra in the 302.5 nm - 303.8 nm region (Figure 1). The protocol is: (1) NO_2 concentration is measured using λ_1 = 532 nm (of the three mixture components, only NO_2 absorbs at this wavelength) [5], (2) background due to both NO_2 and CH_3CHO is measured using λ_2 = 302.65 nm, where the PA signal of H_2CO is negligibly small, and (3) λ_3 = 303.59 nm corresponds to the largest PA signal for H_2CO in the 302.5 nm - 303.8 nm region and therefore is chosen to provide the maximum sensitivity for H_2CO detection.

The compounds H_2CO, NO_2, and CH_3CHO are characterized separately in order to determine the calibration curves of each compound at λ_1, λ_2, and λ_3. Plots of PA signal versus the concentration of each compound exhibit a linear dependance. The linear least squares fits to these data have a 99.8% confidence limit over two orders of magnitude of concentration. A detection limit for H_2CO in N_2 at 303.59 nm is found to be 20 ppbv. Excitation in the 302 nm - 304 nm region causes photodissociation in all the three compounds [6] [H_2CO (H + HCO and H_2 + CO), NO_2 (NO + O), and CH_3CHO (CH_3 + HCO and CH_3CO + H)]. The linear dependance of the PA calibration curves for each compound suggests that PA signals are not greatly affected by photochemistry.

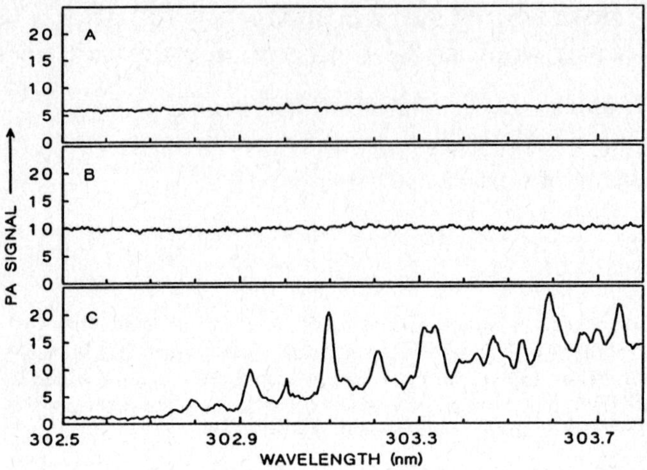

Figure 1. PA spectra of (A) NO_2 (3.86 ppmv), (B) CH_3CHO (4.48 ppmv), and (C) H_2CO (5.70 ppmv) in the 302.50 nm - 303.80 nm region (resolution 0.5 cm^{-1}). Vertical scale is the same in all three spectra [7].

Mixtures containing H_2CO (36 to 887 ppbv), NO_2 (5.45 ppmv), CH_3CHO (2.60 ppmv) and N_2 are studied using the protocol described above. Using the calibration curves, the contribution in the PA signal at 303.59 nm due only to H_2CO is calculated and plotted versus H_2CO concentration . The curve shows a linear dependance with a 99.8% confidence limit over an H_2CO concentration range of 36 - 887 ppbv. A detection limit of 51 ppbv for H_2CO is achieved (instead of 20 ppbv in the case of H_2CO alone with N_2). Futher experiments show that this variation was due to errors inherent in the multiple measurements and calculations necessitated by the protocol rather than photochemical interferences between the mixture components [7].

This example of quantitative PA detection of H_2CO in mixtures demonstrates the value of PA spectroscopy as an experimental method for real-time atmosperic monitoring.

REFERENCES

1. A. C. Tam, Rev. of Modern Phys. 58, 318-431 (1986).
2. M. A. Leugers, G. H. Atkinson, Anal. Chem. 56, 926-929 (1984).
3. P. Vujkovic Cvijin, D. A. Gilmore, M. A. Leugers and G. H. Atkinson, Anal. Chem. 49, 300-304 (1987).
4. P. Vujkovic Cvijin, D. A. Gilmore, and G. H. Atkinson, (submitted for publication).
5. O. Poizat, and G. H. Atkinson, Anal. Chem. 54, 1485-1489 (1982).
6. D. L. Baulch and al, J. Phys. Chem. Ref. Data 11, 327 - 496 (1982).
7. M. Boutonnat, K. A. Keilbach, D. A. Gilmore, and G. H. Atkinson, (submitted for publication).

Laser-Induced Photoacoustic Spectroscopy for the Speciation of Actinides in Submicromol Concentrations

R. Klenze, R. Stumpe, and J.I. Kim

Institut für Radiochemie, TU München,
D-8046 Garching, Fed. Rep. of Germany

Laser-induced Photoacoustic Spectroscopy (LPAS) is developed for the direct speciation of actinides in aquatic solutions. The speciation sensitivity is attained in the submicromol range, e.g. Am^{3+} (ε = 400 L $mol^{-1}cm^{-1}$) down to 10^{-8} mol L^{-1}, or ppb. The method provides spectroscopic information of chemical states of actinides and hence facilitates the study of their hydrolysis-reaction, complexation, redox-reaction and colloid generation. The application of LPAS is demonstrated for the speciation of Am and Pu ions in different solutions.

1. Introduction

Understanding of the chemical behaviour of actinides in the geosphere is an important aspect for the safety analysis of nuclear waste disposal in which actinides comprise the major part of long-lived nuclides. The chemistry of these elements in natural aquatic systems is very complex because of the various coexisting oxidation states, e.g. Pu(III) to Pu(VI), which undergo different extents of chemical reactions like hydrolysis, complexation, redox reaction and colloid generation [1]. Absorption spectroscopy of 5f transitions in the UV-VIS region is a widely used method for the speciation of oxidation and chemical states of actinide ions. Since, in near-neutral solutions, the solubility of actinides is very low due to hydrolysis reactions ($<10^{-6}$ mol L^{-1}), the conventional absorption spectroscopy is not applicable directly. As a very sensitive method, the LPAS has been developed [2,3] to obtain spectroscopic information on actinide ions in the submicromol range directly in natural aquatic systems.

2. Experimental

As a light source, an excimer pumped dye laser (Lamda Physik EMG201E and FL2002) of some mJ pulse energy and about 10 ns pulse length is used. The pulse energy is monitored by a pyroelectric detector (Gentec ED-100) and kept constant over each dye range using a variable attenuator. To overcome problems involved in measuring radioactive or corrosive solutions, the sample, in a 2 cm quartz cuvette, is mounted on an Al-coated quartzplate which is glued to a piezoceramic disc (Vernitron PZT # 4). Such an arrangement makes for easy sample handling, but with some reduction in sensitivity. Amplified piezoelectric signals are integrated by a gated boxcar, normalized to the laser pulse energy and further processed by a personal computer to evaluate averaged final data. The background signal of the solvent is compensated by a differential method, measuring the sample and reference at the same time by two identical detection cells in a tandem alignment. Detailed information on the sample preparation is given elsewhere [4].

3. Results and discussion

The spectroscopic capability of LPAS has been examined for a broad optical range using lanthanide and actinide ions at different concentrations. Typical LPAS spectra of the 1:1 complex of Am(III) with ethylene diamine tetraacetic acid (EDTA) are shown in fig. 1 for the absorption range of the $^7F_0 \rightarrow {}^5L_6'$ transition (λ = 506.5 nm, ε = 580 L $mol^{-1}cm^{-1}$). The EDTA-complex is proved to be well suitable for calibration at such low concentrations because of its high chemical stability against sorption on the vessel surface or formation of colloids. The halfwidth and peakform of the LPAS spectrum are comparable to the absorption spectrum. A detection limit of 7 x 10^{-9} mol L^{-1} for Am-EDTA is obtained, which corresponds to an absorbance of 4 x $10^{-6}cm^{-1}$. This detection limit is determined by the background absorption of water (about $10^{-4}cm^{-1}$ at 505 nm) and by the precision of about 1 % for photoacoustic measurements. In more transparent solutions like CCl_4, which exhibits much more favorable photoacoustic properties, a detection limit is found one order of magnitude lower for the organic complex of Am(III) with bis-2-ethyl-hexyl-pyrophosphoric acid.

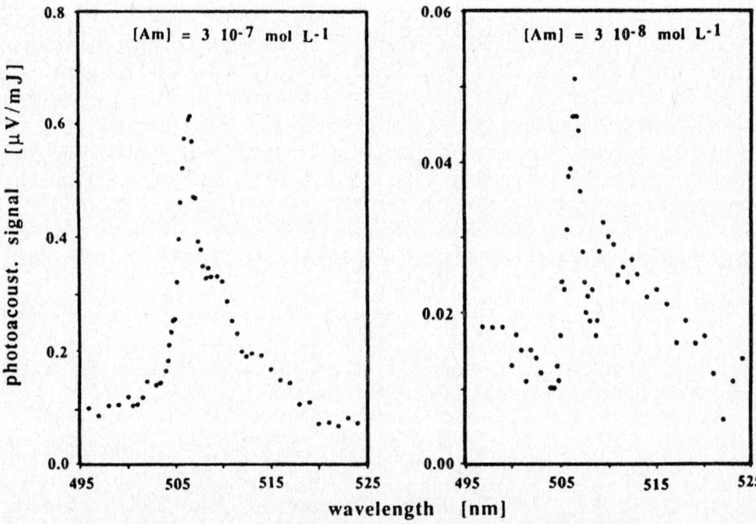

Fig. 1 LPAS spectra of Am(III)-EDTA at different Am concentrations

The photoacoustic signal is found to be proportional to absorbance in the range from 10^{-2} down to 10^{-5} cm^{-1}. Statistical fluctuations, however, are rather large, indicating that the reproducibility of acoustic detection is not the best. This is mainly ascribed to the poor long term stability of the dye laser. The diameter and shape of the laser beam are crucial parameters for short pulse excitation time and it is not easy to keep them constant with time as well as with different dyes. For the absorbance quantification, a calibration standard of known absorbance is necessary. For this purpose, use of a grey standard sealed in a cuvette, which covers the whole spectral range, is found to be convenient. For the present experiment a mixture of different transition metal ions (Cu^{2+}, Cr^{3+}, Co^{2+} and $K_2Cr_2O_7$) in aqueous solution is used throughout. The overlapping of the absorption bands of these ions generates quite a homogenious spectrum for the range of 350 - 900 nm. The standard is chemically and photochemically stable and shows no fluorescence or scattering. The chosen absorbance of about 2 x 10^{-2} cm^{-1}, which can be determined accurately by absorption spectroscopy, yields strong but not saturated LPAS signals.

The actinide chemistry hitherto studied by LPAS involves hydrolysis reactions, complexation with various ligands like carbonate, humic acid, chloride, EDTA etc. [4], redox reactions [5] and colloid generation [6], which are important aspects for understanding the migration behaviour of these elements in the geosphere.

The spectroscopic capability of LPAS is demonstrated by some examples for the speciation of Pu in different oxidation states. Fig. 2 shows the photoacoustic spectrum of Pu in 0.5 M NaCl, which is prepared by dissolution of $^{238}PuO_2$ for 900 days producing a concentration of 2 x 10^{-5} mol L^{-1} (pH = 4.3 and Eh = 532 mV). The main peak at 569 nm with a satellite peak at 555 nm is characteristic for Pu(V) (ε (569 nm) = 20 L $mol^{-1}cm^{-1}$). The intensity of the peak indicates that nearly all of the dissolved Pu is pentavalent state. No significant bands for other oxidation states can be observed. The sharp increase at about 600 nm is due to the H_2O absorption band which is not compensated properly.

At higher salinity (>3 M NaCl), the redox potential is increased by the generation of oxidizing chlorine species induced by α-radiation of ^{238}Pu.

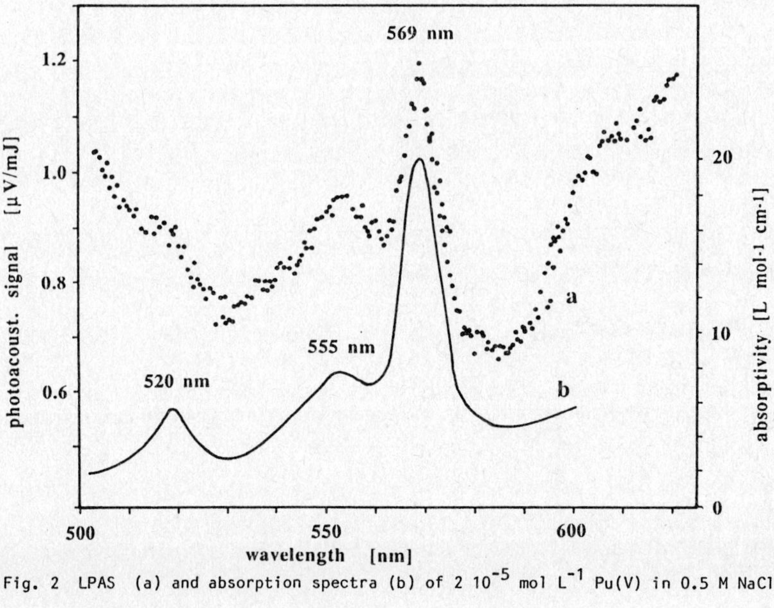

Fig. 2 LPAS (a) and absorption spectra (b) of 2 10^{-5} mol L^{-1} Pu(V) in 0.5 M NaCl

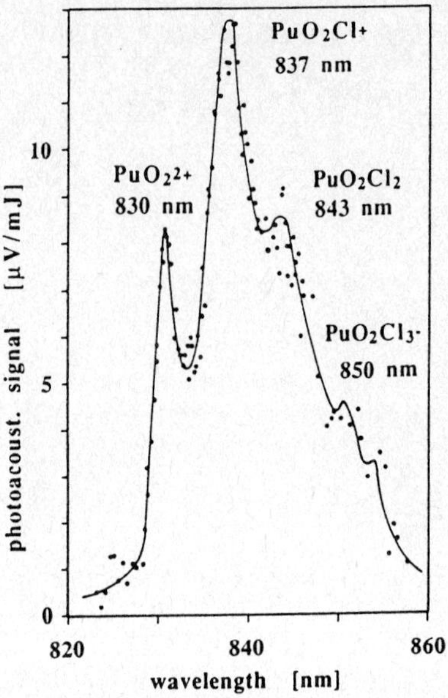

Fig. 3 LPAS spectrum of 1.2 10^{-5} mol L^{-1} Pu(VI) in 3 M NaCl

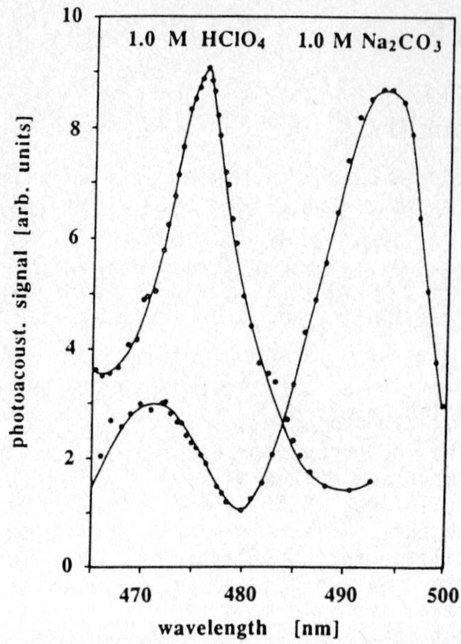

Fig. 4 LPAS spectra of 9 10^{-6} mol L^{-1} Pu(IV) in 1 M $HClO_4$ and 1 M Na_2CO_3 solutions

In such solutions Pu(VI) is well stabilized and hence results in a higher solubility of Pu. In fig. 3 the spectrum of Pu(VI) (1.2 x 10^{-5} mol L^{-1}) in 3 M NaCl is shown. Apart from the absorption of PuO_2^{2+} at 830 nm, (ε = 490 L $mol^{-1}cm^{-1}$) other peaks at 837 nm and 843 nm with a shoulder at 850 nm are observed, which are assigned to be PuO_2Cl^+, PuO_2Cl_2 and $PuO_2Cl_3^-$, respectively.

Contrary to the chloride solution which exhibits oxidizing properties due to radiolysis, in carbonate solution the Pu is stabilized as Pu(IV) carbonate complexes. In fig. 4 the spectrum of 9 x 10^{-6} mol/L Pu(IV) in 1 M Na_2CO_3 is shown in comparison to the ionic peak of Pu^{4+} in 1 M $HClO_4$. A peak shift of about 17 nm enables the differentiation between ionic and carbonate species.

References

1. J.I. Kim: In Handbook on the Physics and Chemistry of the Actinides, Vol. 4, Eds.: A.J. Freeman and C. Keller, Elsevier Science Publishers B.V., 1986, p. 413
2. W. Schrepp, R. Stumpe, J.I. Kim, H. Walther: Appl. Phys. B32, 207 (1983)
3. R. Stumpe, J.I. Kim, W. Schrepp, H. Walther: Appl. Phys. B34, 203 (1984)
4. R. Stumpe, J.I. Kim: Rep. RCM 02386, 1986 (Institut für Radiochemie, Technische Universität München)
5. K. Büppelmann, S. Magirius, Ch. Lierse, J.I. Kim: J. Less. Comm. Met. 122, 329 (1986)
6. G. Buckau, R. Stumpe, J.I. Kim: J. Less. Comm. Met. 122, 555 (1986)

CO_2 Laser Based N_2O Photoacoustic Monitor

D.D. Bicanic[1], *A. Bizzarri*[1], *B.F.J. Zuidberg*[1], *G.P.A. Bot*[1], *T. de Jong*[1], *and H.C.P. Wegh*[2]

[1]Department of Physics and Meteorology, Wageningen Agricultural University, Duivendaal 2, NL-6701 AP Wageningen, The Netherlands
[2]Department of Air Pollution, Wageningen Agricultural University, De Dreijen 12, NL-6703 BC Wageningen, The Netherlands

1 Introduction

The indoor/outdoor exchange of air by natural ventilation is one of the most important of the physical processes taking place in greenhouses and livestock buildings, as it directly affects the microclimate within the agricultural objects. This is of vital horticultural and economic importance during crop tillage for optimization of yield, and hence the quantification of the basic greenhouse processes and their incorporation into a dynamic physical model is being studied intensively [1, 2].
Both the decay rate and the constant flow approach require the injection of a tracer gas, followed by the study of concentration evolution in time. The tracer gas, possessing constant natural abundance and similar properties to those of air, should be neither produced nor condensed in the test area. In daily practice, nitrous oxide N_2O is used as a tracer gas to study full-scale greenhouse dynamics. Its concentration at appropriate locations is determined spectroscopically (with a long path absorption cell) using the frequencies of the strong stretching fundamental bands ν_1 and ν_3 centered near 4 and 8 µm respectively.
The CO_2 laser photoacoustic spectroscopy has been shown to have great potential for monitoring the majority of atmospheric (and hence also greenhouse) pollutants and is currently receiving countrywide attention from the agricultural and environmental sector. Despite the fact that values of the absorption cross sections at CO_2 laser wavelengths are not very favourable, the high sensitivity of photoacoustic spectroscopy justified the laboratory study to investigate the prospects of a CO_2 laser based N_2O on-line monitor. As far as is known, no such measurements have been carried out before.

2 Experimental Results

The experimental set-up consisted of a grating tuned CO_2 waveguide laser, the radiation of which is chopped mechanically at 642 Hz corresponding to a first longitudinal resonance of an essentially "organ-pipe" type cell [3]. Four identical miniature ceramic microphones (Knowless 1670/II) mounted flush with the wall at 90° with respect to each other were used. The N_2O samples were prepared from a cylinder of medical grade purity gas by dilution in pure nitrogen. The measurements were carried out at room temperature and atmospheric pressure in the cell.
Figure 1a shows the normalized photoacoustic spectrum of pure N_2O gas for the 10 P band of the CO_2 laser. The appearance of four distinguishing lines P(16), P(7), R(10) and R(12) of the N O (0 0° 1) - (1 0° 0) vibrational combination band is due to a good degree of spectral overlap with the 10P(40), 10P(32), 10P(16) and 10P(14) lines of the CO_2 laser [4]. Frequency detuning varies from 0.01 cm^{-1} at 10P(16) to 0.28 cm^{-1} at 10P(14).

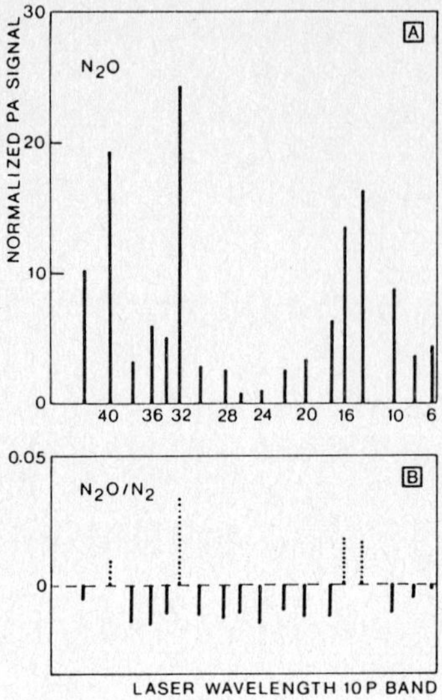

Figure 1: Normalized PA spectrum (in 10P band) of pure N_2O (upper) and a mixture of 400 ppmv in nitrogen (lower) recorded at atmospheric pressure.

Figure 1b displays the spectrum of a mixture containing 400 ppmv of N_2O in nitrogen. Dotted lines represent the N_2O spectrum at four previously mentioned transitions. However, negative signals due to a phase shift caused by the coherent background effects are observed at all other laser lines [5]. The maximum value of the lock-in amplifier signal obtained with 400 ppmv of N_2O in N_2 was 12 μV at the 10P(32) laser line with 0.75 W incident power.
The laboratory results reported here indicate the possibility of using CO_2 laser radiation for photoacoustic monitoring of N_2O concentration. The noise voltage associated with the absorption line and the background was typically of the order of 0.5 μV under the normal operating conditions, from which the detection limit of approximately 20 ppmv N_2O in N_2 can be derived.
This sensitivity is thought to be high enough to allow the practical application in agricultural buildings. Comparative field measurements in realistic greenhouse climatological conditions are scheduled for the future.

3 References

1. T.Takakura, K.Kurata, Y.Honjo: Acta Hortic. 174, 97 (1985).
2. G.P.A.Bot: Ph.D. Thesis, Agricultural University Wageningen (1983).
3. F.Harren, C.Sikkens, D.Bicanic: SPIE Series 701, 525 (1987).
4. R.Kagann: Jour. Molec. Spec. 95, 297 (1985).
5. M.Margottin-Meclou, L.Doyennette, L.Henry: Appl. Opt. 10, 1768 (1971).

Photoacoustic Monitoring of the Time Course of Ammonia Formation During the Spoilage of Inoculated Beef at Room Temperature

D.D. Bicanic[1], *B.F.J. Zuidberg*[1], *R. Friedhoff*[2], *F. Harren*[3], *D.P.v.d. Akker*[1], *and A. Bizzarri*[1]

[1]Department of Physics and Meteorology, Wageningen Agricultural University, Duivendaal 2, NL 6701-AP Wageningen, The Netherlands

[2]Department of Food Technology, Microbiological Division, Wageningen Agricultural University, De Dreijen 12, NL-6703 BC Wageningen, The Netherlands

[3]Department of Laser and Molecular Physics, Catholic University, Toernooiveld, NL-6235 ED Nijmegen, The Netherlands

1 Introduction

The natural aerobic spoilage of meat and that of meat inoculated with pure cultures is a complex process caused by microbial growth associated with the onset of various energy metabolisms. Chemical changes characterizing the end phase of bacterial activity at each of the distinct stages of their growth lead to the production of a well-defined sequence of volatiles [1]. In microbiology, determination of the microbial count number (number of bacteria per unit surface of meat) by elaborate plate counting techniques has become the common way to follow the growth and establish spoilage criteria. In principle predictions of the shelf-life of various meat products can be made if the relationship between the microbial count and the concentration of a specific volatile as the indicator of product deterioration can be found.

2 Experimental

A CO_2 waveguide laser in conjunction with a sensitive PA cell of the "organ-pipe" type [2], shown in Fig.1 was used in this study. There exist a considerable number of spectral coincidences between the CO_2 laser lines and the absorption lines for many of the identified volatiles, and for a number of them the absorption coefficients are also known.
Slices of meat (3 cm x 3 cm x 0.5 cm) were sterilized (10 kGray γ radiation) and then inoculated (initial count 10^4 bacteria/cm^2) with a pure culture of either Pseudomonas putida or Brochotrix thermosphacta strain, bacteria known to dominate the spoilage flora at chill temperatures. Beef, placed in a 100 ml Pyrex flask, was flushed by the environmental air at a constant flow rate of 1.5 cm^3/sec carrying away the volatiles from the head space to the PA

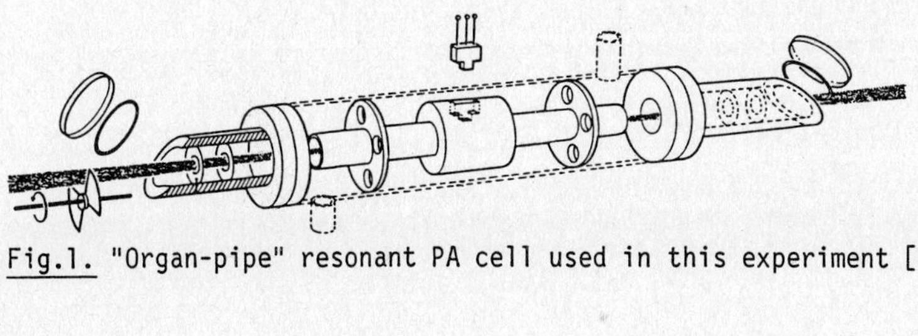

Fig.1. "Organ-pipe" resonant PA cell used in this experiment [2].

cell, thereby passing through the Zeolite molecular sieve and/or KOH traps to reduce water vapour and carbon dioxide content. A three-way valve enabled quick connection to the inoculated control sample whenever required. The measurements were carried out (at 18°C and pressure in the PA cell slightly below the atmospheric value) automatically using a HP 3421 acquisition data unit that samples vector output signal and the input laser power at selected wavelengths.

3 Results

The initial experiment had two goals: i) to find out whether the CO_2 laser can provide specific detection of some species produced in the early stage of spoilage and ii) to study the formation of ammonia known to take place at later time. Acetoin, an aliphatic monoketone is reported to be a compound produced in the earliest phase [1].

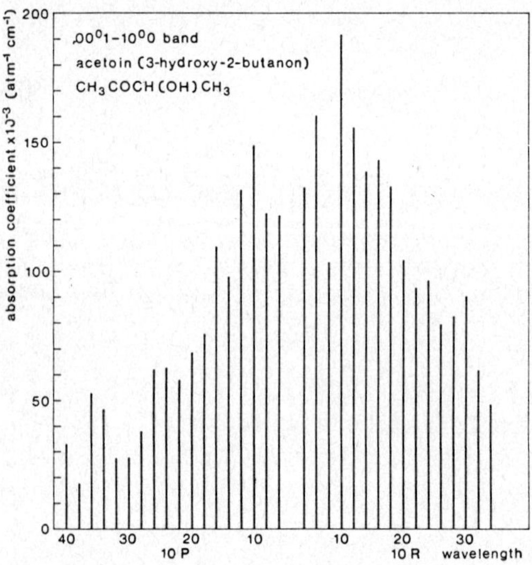

Fig.2. Absorption coefficient of acetoin (in 10 μ band) at 291 K and 1 atmospheric pressure.

The plot of its measured absorption coefficient versus the laser emission in 10 μm region shown in Fig.2 has a "grasslike" appearance. The maximum absorption in the 10R band is about 40 times lower than for example, that of ethyl acetate (another constituent found during the initial phase) in 9P band. The time course of beef spoilage was studied at discrete but regular intervals over a period of several days. The presence of acetoin (at sub-ppmv level) was found in the head space of both samples. A strong ammonia concentration gradient was observed between 30 and 60 hours after the start of the experiment, with the NH_3 concentration reaching a level of several ppmv, which is probably due to glucose depletion and the onset of aminoacid metabolism. Based on data collected in a parallel experiment that related the microbial count to the spoilage duration, bacterial densities of the order of 10^8-10^{10} /cm^2 are expected within the same time interval. Our results suggest that the growth of Pseudomonas exceeded that of Brochothrix under the given experimental conditions. Towards the end of the third day, a decrease of NH_3-concentration was experienced (this finding is confirmed to a satisfactory extent by NEN 6472 (ISO/DIS 7150), a standardized colorimetric test for ammonia) indicating the growth reduction as shown in Fig.3. However, the

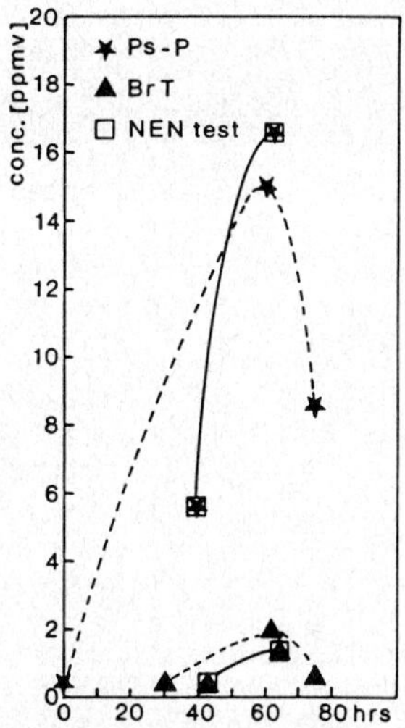

Figure 3: Time course of ammonia concentration for two beef samples inoculated with Ps.P. and Br.T. bacteria.

ultimate sensitivity of CO_2 laser PA method spectroscopy in detecting the ammonia is higher than that of the NEN test.

4 References

1. R.H.Dainty, R.A.Edwards, C.M.Hibbard: Jour. Appl. Bacteriol. 59, 303 (1985).
2. F.Harren, C.Sikkens, D.Bicanic: SPIE Series 701, 525 (1987).

Ethylene Exhalation of a Single Flower Detected by Photoacoustic Methods

F. Harren[1], J. Reuss[1], D.D. Bicanic[2], and E. Woltering[3]

[1]University of Nijmegen, NL-Nijmegen, The Netherlands
[2]Agricultural University of Wageningen, NL-Wageningen, The Netherlands
[3]Sprenger Institute Wageningen, NL-Wageningen, The Netherlands

Introduction

The combination of a powerful CO_2 waveguide laser and a resonant photoacoustic cell provides sufficient sensitivity to investigate the emission of ethylene from a single orchid flower (Cymbidium, Mary Pinchess "Del Rey") in the interval up to 70 hours after emasculation of the flower. Ethylene plays a special role in agriculture. As an air pollutant it inhibits the growth of plants, causes yellowing and abscission of leaves and wilting of flowers /1/. As a gaseous hormone produced by the plant itself, it causes the same effect in unpolluted plants under closed storage.

Orchid flowers show an increase in ethylene production just before or during the wilting process, which can be hastened by removal of the anther cap and pollinia (=emasculation). The flowers were put in a closed volume for accumulation of the ethylene and the concentration was sampled and analysed by a gas chromatograph every few hours /2/. The accumulation of ethylene and CO_2 during the experiment yielded unwanted effects with respect to the ethylene production. Therefore the experiment was repeated under flow conditions with the more sensitive photoacoustic effect. To this end a single orchid flower was placed in a sampling cell containing water.

Experimental set-up

An infrared CO_2 waveguide laser is used as excitation source. In comparison to a conventional CO_2 laser, the waveguide laser has a small plasma bore diameter (ø 3 mm) and is more compact (discharge length 50 cm, FSR 300 MHz). The overall output power amounts to 8 W (70% outcoupling mirror) and line tunability is achieved over more than 80 discrete laser transitions. The chopper consists of a rotating blade. The motor is phase-stabilized to decrease the noise level of the photoacoustic signal.

Cell design and sensitivity

The photoacoustic cell is a single pass resonant type with an inner open resonator /3/ (l=300 mm, d=9 mm) excited in its first longitudinal mode. The resonator is mounted inside a larger volume (L=470 mm, D=60 mm). We have chosen this set-up to minimize the effect of absorption of the laser power in the ZnSe Brewster windows on the microphones, which produces a background signal. The inner wall of the argentan organ tube is polished. In the center part of the tube are four holes (ø 1 mm each) to allow acoustic coupling to the four miniature Knowles microphones (sensitivity 10 mV/Pa). The microphones are mounted in a Teflon ring which is pulled over the tube. The low Q-value for the resonator is mainly caused by surface losses. The experimental and theoretical values are Q=28 and Q=21 respectively /4/.

The noise of the photoacoustic signal at the resonance frequency and at 1 atm pressure is 1 $\mu V/\sqrt{Hz}$. The acoustic background signal with "pure" nitrogen amounts to 20 μV/W at the 10P14 CO_2 laser line. A reduction to 5 μV/W can be achieved by dissociation of the hydrocarbon traces in the nitrogen by means of a catalyst, and by removing the reaction products H_2O and CO_2 with a KOH-scrubber. The overall response of the photoacoustic cell is 156 V/W per unit absorption coefficient ($atm^{-1}cm^{-1}$). With the mentioned noise level, the limiting sensitivity of the photoacoustic cell amounts qo 6.4×10^{-9} cm^{-1} (1 Watt laser power, 1 Hz bandwidth). This gives for ethylene, with its strong absorption coefficient (30.4 $atm^{-1}cm^{-1}$) /5/ on the 10P14 CO_2 laser line, a minimum detection limit of 0.03 parts per billion ($1:10^9$, S/N=1, 7 Watt laser power).

Orchid flower conditions and results

Before entering the sampling cell the air flows through a copper tube (l=6 m) filled with a platinized aluminium oxide catalyst at 350 °C to remove any trace hydrocarbons present. A KOH-scrubber between the sample cell and the resonator eliminates H_2O and CO_2. A flow of 0.9 l/h (time constant 14 minutes) gives no accumulation of gases (C_2H_4, CO_2) in the sampling volume (50 cm^3). However, it gives a measurable ethylene concentration with respect to the production of the flower (<1 nanoliter/gram hour). In Figure 1 the ethylene production is shown as a function of time elapsed after emasculation. The concentration is measured on a strong (10P14) and a weaker (10P12) absorption line of ethylene. A clear peak production of 0.13 nanoliter/gram hour is observed at about 8 hours after emasculation. At 45 hours the actual wilting process starts.

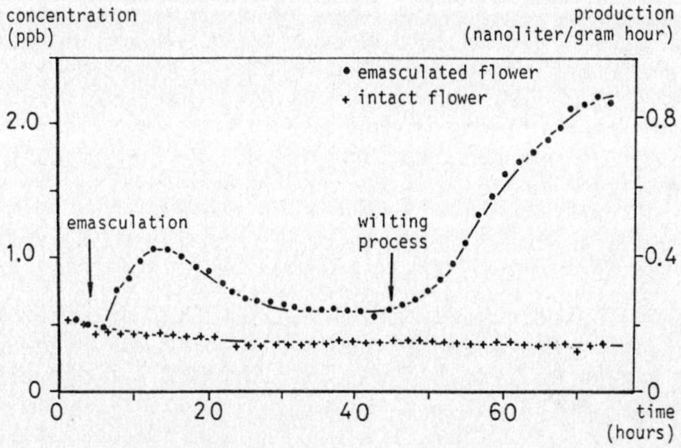

Fig.1. Ethylene production versus time of a single orchid flower. Due to the time scale the measurements were automated

1. E. Woltering and E. Sterling: Acta Horticultura **181** (1986) 483-488
2. C.J. Goh, A.H. Haleny, R. Engel and A.M. Kofranek: Scientia Horticultura **26** (1985) 57-67
3. G.J. Schrerer, K.K. Lehmann and W. Klemperer: J. Chem. Phys. **81** (1984) 5319-5325
4. A. Karbach, J. Röper and P. Hess: Chem. Phys. **82** (1983) 427-434
5. R.J. Brewer, C.W. Bruce and J.L. Mater: Applied Optics **21**, 22 (1982) 4092-4100

Detection of Ultrafine Particles in Liquids Using a Breakdown Acoustic Effect

T. Kitamori[1], *K. Suzuki*[1], *K. Yokose*[1], *T. Sawada*[2], *A. Harada*[2], *and Y. Gohshi*[2]

[1]Energy Research Laboratory, Hitachi, Ltd., 1168 Moriyama, Hitachi, Ibaraki 316, Japan

[2]Department of Industrial Chemistry, Faculty of Engineering, University of Tokyo, 7-3-1 Hongo, Bunkyo, Tokyo 113, Japan

As water and liquid reagents used in semiconductor technology and biochemical engineering are highly purified, analytical and detecting methods for ultratrace and ultrafine particulate impurities are desired. For example, ultrafine particulate matter as small as 10^{-1} μm can break the insulation between wiring on a very large scale integrated circuit (VLSI), so detection, counting or determination methods are required for the water and liquid reagents used in VLSI production. However, the weight concentration of these particulate impurities is estimated to be below 10^{-2} ppt (number density: 10^{1} mL^{-1}), and this is too small for turbidimetric or even photoacoustic determination [1]. On the other hand, the detection limit size of particle counting in liquids using laser scattering is about 0.1 - 0.3 μm, because of the background due to medium Rayleigh scattering. We propose a new method for counting ultrafine particles in liquids using a breakdown acoustic effect.

Figure 1 shows a particle in a focused beam. In the ordinary process of photoacoustic (PA) signal generation, the particle absorbs optical radiation and generates heat. The heat released causes a thermoelastic expansion of the medium, with generation of PA signals. However, when the power density of the optical radiation is too high, and the electric field intensity of the beam waist exceeds the threshold of the particle dielectric breakdown, the particle may become plasma and a strong acoustic emission due to explosive expansion of the plasma is expected. Hence, ultrafine particles can be counted by counting the acoustic pulses induced by this breakdown acoustic effect (BAE).

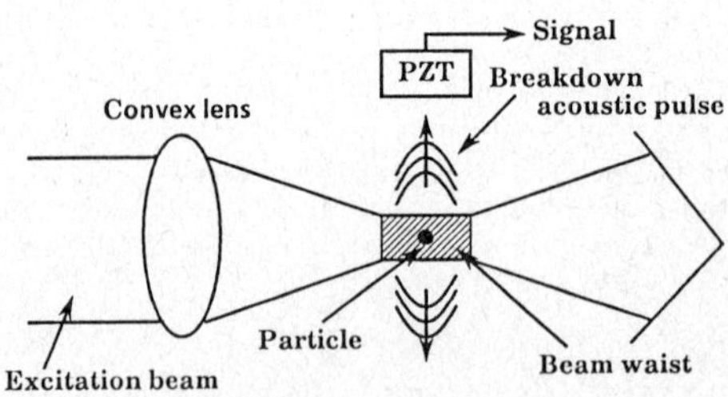

Fig. 1. Optical arrangement

Monodispersed polystyrene latex particles of 85 nm uniform size were used as a sample to verify the principle of this particle counting method. Second harmonic of the pulse YAG laser (532 nm, 0.55 mJ, 6 ns FWHM) was focused with a convex lens (50 mm focal length). As the beam size had a 2.5 mm radius, the spot size of the focused beam waist (Fig. 1) was calculated using waveoptics as 6.4 μm radius, and the length of the approximately cylindrical region of the beam waist was estimated to be 128 μm. Hence, the volume of the beam waist was 1.6 X 10^{-8} mL. The number density of the particles, at which an expectation of the particle number in the beam waist becomes one per beam shot, was obtained as a reciprocal of the volume, and this value was 6.3 X 10^{7} mL^{-1} (22 ppb). Then, concentrations of the polystyrene turbid solutions were prepared at 0.5 - 2.5 ppb. At these concentrations, the excitation beam pulse should hit a particle once per 8.8 - 44 shots, and there would be little possibility of more than two particles being in the beam waist at the same shot.

A cylindrical direct coupling PA cell was used for measurements [2]. The excitation beam was repeated at 10 pps and acoustic pulses were counted for 90 seconds. Hence 900 excitation beam pulses were shot for one sample. First, the concentration dependence of the acoustic pulse counts was measured. The acoustic pulse counts for 0.5 -2.5 ppb samples were 45 - 170, and they were proportional to the concentration. Hence, generation and detection of the acoustic pulse for one particle was confirmed.

A waveform of the acoustic pulse was recorded by a storage oscilloscope and an example is shown in Fig. 2. The first peak P_1 corresponded to the acoustic pulse, and the following peaks P_2 - P_9 were echoes of the acoustic pulse in the cell, because the frequency of the wave was close to the calculated natural frequency of the cell. Hence the peak value of P_1 was the pulse height of the acoustic pulse. The pulse height was 110 mV and it was at least three orders larger than that of the ordinary PA signal. The electric field intensity at the beam waist for the excitation beam of 0.55 mJ and 6 ns pulse width was calculated as 4.1 X 10^{6} V/cm, and this value exceeded the breakdown threshold of polystyrene, 2 - 3 X 10^{6} V/cm. Therefore, the polystyrene particles were considered to be plasma, and strong acoustic emission was induced. As an acoustic pulse of 110 mV could be easily counted with a good signal to noise ratio, application of the present method to counting 85 nm polystyrene ultrafine particles in water was confirmed.

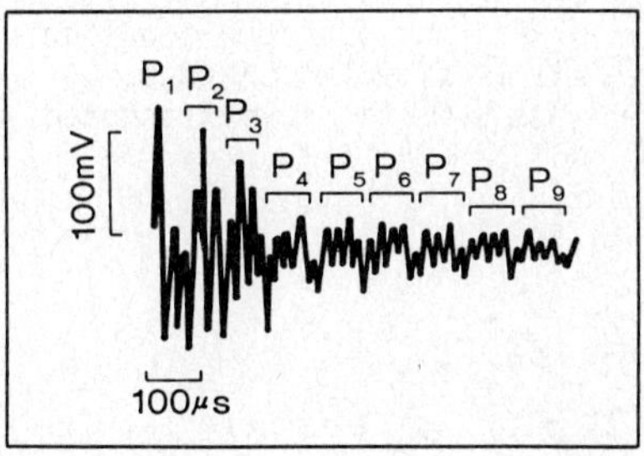

Fig. 2. Waveform of the breakdown acoustic signal from a 85 nm polystyrene particle

References

1 S. Oda, et al., Anal. Chem. 52, 650 (1980)
2 T. Kitamori, et al., J. Spectrosc. Soc. Jpn. 36, 359 (1985)

Optothermal Stabilization of a CO_2 Waveguide Pump Laser

S.T. Kornilov, I.V. Ostreikovski, N.M. Prokopova, and E.D. Protsenko

Moscow Engineering Physics Institute, Kashirskoe sh., 31,
Moscow, 115409, USSR

Gaseous absorption of modulated infrared radiation can increase the temperature of a gas. This is known as the optothermal (OT) effect /1/. A number of works devoted to OT registration of absorption spectra /2,3/ as well as to the study of OT detectors (OTD) basic characteristics and heat transfer mechanisms /4,5/ have been published. These detectors are not large in size, are highly sensitive at low gas pressures (10^{-7} cm^{-1}W Hz$^{-1/2}$ in the present case) and have better vibrational immunity as compared to the optoacoustic ones.

It is very promising to combine an OTD and a waveguide CO_2 laser for FIR laser pumping. In optical pumping of the lines which are considerably removed from the laser transition center the laser stabilization is impeded by either external cell or laser optogalvanic signal /6/. The OT stabilization ensures the laser tuning to the maximum of the power absorbed by the media being pumped. Such a technique was realized with the optoacoustic detector utilization /7/.

The CO_2 waveguide laser (Fig.1) was made up of a 150 l/mm diffraction grating 1, a BeO waveguide 2,and 99.5% dielectric mirror 3 fixed on a PZT. The 5 cm OTD was attached to a waveguide CO_2 laser resonator next to the dielectric mirror. Such a geometry minimized the pumping radiation losses since most of the output power was emitted via the laser diffraction grating. An additional OTD optical alignment while tuning the laser to other transitions became unnecessary. The radiation from the mirror 3 heated gas molecules while passing through the OTD. The temperature variations of the gas were detected by the pyroelement 6. Teflon diaphragms 5 were placed in the OTD to shield the pyroelement from direct illumination by the radiation scattered at the windows 4.

The laser was stabilized by OT tracking of SF_6 and C_2H_4 absorption lines on 10P and 10R branches, and was heterodyned with a stable low pressure CO_2 laser. The modulation of the laser length, i.e. the radiation frequency was obtained by applying dither voltage to the PZT. The signal corresponding to

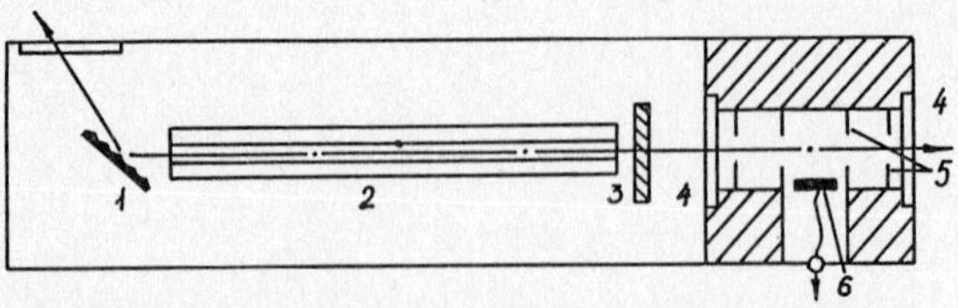

Fig.1 A CO_2 waveguide pump laser with an optothermal detector

the detuning of the radiation frequency from the gas absorption line centre was used to correct the resonator length.

The frequency stability measurements were carried out by observing the beat note on a spectrum analyzer /6/. The laser stabilization at such "strong" lines as SF_6 on 10P(20) laser transition provides the 10^{-8} frequency stability of the beat note position over the time intervals up to 30 min. In this case the 300 Hz voltage applied to PZT revealed 2 MHz excursion of the beat note, the characteristic time of the servo loop being equal to 0.02 s. The laser frequency deviation resulted in less then 0.2% variation of the absorbed power.

The stabilization which employed weak absorption lines with a broad Doppler profile, e.g. for C_2H_4 at 10P(18) laser transition suggested the increase of the operational pressure up to 0.5 Torr in order to get a proper OT signal. The modulation frequency was decreased up to 100 Hz due to an essential dependence of the OTD amplitude-frequency characteristics on pressure /5/. In order to obtain similar frequency stability the peak-to-peak deviation was increased up to 7 MHz.

1. L.-G. Rosengren: Infrared Phys. 13, 173 (1973)
2. C. Hartung, R. Jurgeit, H.-H. Ritze: Appl. Phys. 23, 407 (1980)
3. S. T. Kornilov, I. V. Ostreikovski, E. D. Protsenko: Sov. Quantum Electron. 12, 2379 (1985)
4. C. Hartung, R. Jurgeit: Appl. Phys. B27, 39 (1982)
5. S. T. Kornilov, I. V. Ostreikovski, N. M. Prokopova, E. D. Protsenko: Infrared Phys. in press (1987)
6. M. J. Kavaya, R. T. Mensies, U. P. Oppenheim: IEEE J. Quantum Electron. QE-18, 19 (1982)
7. G. Busse, E. Basel, A. Pfaller: Appl. Phys. 12, 387 (1977)

Part IV

Surfaces and Thin Films

Photoacoustic Studies of Surface Plasmons in Metallic Microstructures

T. Inagaki[1], *J.P. Goudonnet*[2], *and E.T. Arakawa*[3]

[1]Physics Department, Osaka Kyoiku University, Tennoji, Osaka 543, Japan

[2]Laboratoire de Photoelectricité, Université de Dijon, F-21004 Dijon, France

[3]Health and Safety Research Division, Oak Ridge National Laboratory, Oak Ridge, TN 37831, USA

Use of the photothermal method made it possible to explore new aspects of surface plasmon behavior in various types of metallic microstructures.

1. Introduction

Interaction of incident photons with metal surfaces having a topological microstructure, such as a statistical roughness and a periodic corrugation, has been a subject of considerable interest in recent years. Owing to the presence of the microstructure, photons interact with such surfaces more efficiently than with a planar surface. This leads to a significant increase in photoabsorption under a variety of conditions. A class of roughness or corrugation-induced photoabsorption, which has been attracting a great deal of attention, is due to resonant coupling of incident photons with surface plasmons [1]. Surface plasmons are electromagnetic waves propagating along a metal surface with the field strengths decaying away from the surface, and accompanied by a collective oscillation of the conduction electrons in the metal. Since the momenta of surface plasmons are greater than that of photons of the same frequency in the medium with which the metal surface contacts, they are unable to couple with incident photons at a planar surface. However, if the metal surface has a topological microstructure, photons are able to convert into surface plasmons by gaining additional momenta in the direction parallel to the surface provided by the microstructure. Photoabsorptions caused by this extraneous mechanism are of resonance type and their intensities depend strongly on topological details of the surface structures.

The electric field carried by surface plasmons on a metal surface is confined spatially in a narrow region around the surface, producing a strong localization of the electromagnetic energy [1]. This has been found in recent years to play a crucial role in a variety of surface enhancement phenomena in optical interaction at roughened or corrugated metal surfaces [2]. Experimentally, the resonant coupling of incident photon to surface plasmon may be observed most dramatically on metal surfaces having a shallow periodic corrugation. In the past, the conventional spectroscopic techniques used to study this coupling have suffered to a greater or lesser extent [3] because they required measurements of light intensity diffracted into various directions, which varies in a complicated manner as a function of the wavelength and the angle of incidence of photons. The photoacoustic (PA) methods which probe the heat generation occurring inside the sample under study allow us to measure photoabsorption in such a surface in a straightforward manner and, thus, open a new experimental approach to the problem.

Using the gas-microphone method, we have studied [3-9] plasma resonance in silver surfaces having a periodic corrugation under a variety of conditions.

In this paper, we will review the results obtained from these studies together with the results obtained for the microstructures produced by thin organic solid films on a planar silver surface.

2. Experiment

Silver surfaces with a periodic corrugation were produced by the holographic method. A cleaned glass slide was spin-coated with a photoresist film and then exposed to two intersecting beams of 458nm photons from an argon-ion laser. The exposed film was developed and then overcoated with a 120 to 160nm-thick opaque film of silver by vacuum evaporation. Self-supporting silver layers with a periodic corrugation were made by evaporating silver onto a piece of plastic sheet grating which was overcoated with a thin collodion film. The silvered side of the grating was glued onto a tiny aperture of an aluminum frame and then immersed into a pool of amylacetate. The collodion film dissolved away within a few hours, leaving a corrugated silver layer separated from the grating mold. The spatial period d and amplitude h of the corrugation on these surfaces and thin layers were determined [4] by measuring the first-order diffraction angle and the intensity, respectively, of 633nm photons from a He-Ne laser at normal incidence.

The PA cells used were equipped with a glass window and a condenser microphone whose sensitivity was 0.5mV/μbar. The total volume of cell cavity was typically $0.15cm^3$ and filled with air. There are two different methods of observing the resonant coupling of incident photons to surface plasmons. In the first method the resonance absorption is observed by changing the wavelength of photons at a fixed angle of photon incidence (the ω-scan method) and in the second method the resonance absorption is observed by changing the angle of incidence of photon whose wavelength is fixed (the k-scan method). In the present study, the latter was employed with 633nm photons from a He-Ne laser. The photon beam was chopped at 208 or 300Hz by a mechanical light chopper. At these chopping frequencies, the silver films used to form corrugated surfaces and self-supporting layers were thinner than the thermal diffusion length by a factor of 10^3 to 10^4. The PA signals from such thermally thin films may be shown [3] to be proportional to the absorptance, the fraction of the incident photon energy dissipated in the films. Using the response constants of the PA cells, some of the observed PA signal were normalized to the absolute scale of the absorptance for comparison with theoretical calculations.

3. Experimental Results and Discussion

Plasma resonance with a single periodic corrugation. Experimental studies of the plasma resonance in metal surfaces with a single periodic corrugation (periodic in one direction) have been made extensively by using the conventional techniques [1]. Due to the experimental difficulty mentioned above, these studies have been restricted to the case of the planar diffraction geometry, i.e., the grooves perpendicular to the plane of incidence. The present PA method allowed us to study how the incident photons couple with surface plasmons in nonplanar diffraction geometries (see Fig. 1) with the grooves at any orientation θ relative to the plane of photon incidence. In Fig 2, we present the results obtained for a silver surface corrugated with a period d=2186nm and an amplitude h=25nm.

For the planar diffraction geometry θ=90°, the absorption curve of p-polarized photon (with the electric field parallel to the plane of inci-

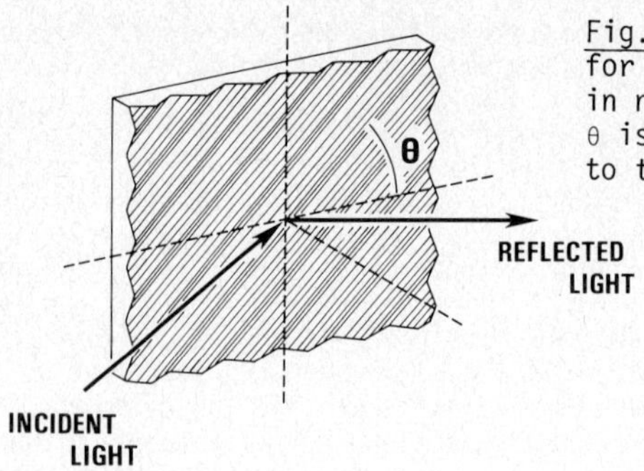

Fig. 1. Experimental configuration for observation of plasma resonance in nonplanar diffraction geometries. θ is the groove orientation relative to the plane of photon incidence

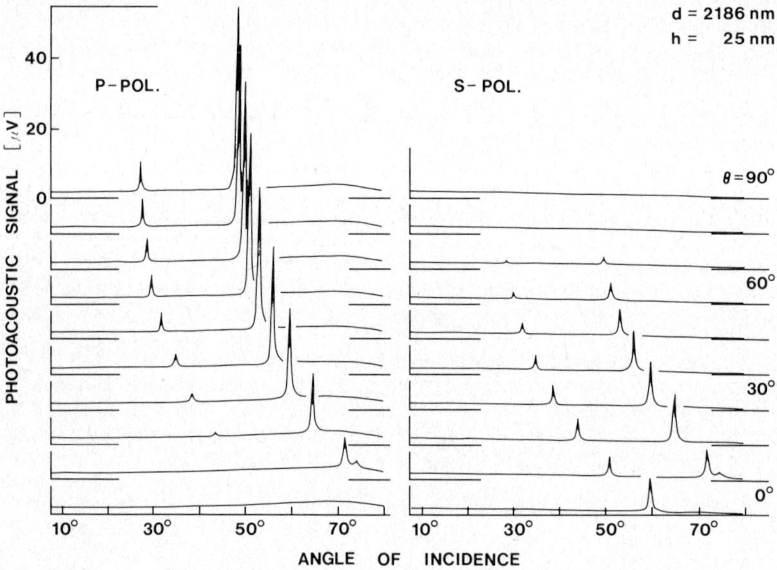

Fig. 2. Plasma resonance absorption in a silver surface having a single periodic corrugation

dence)exhibits two sharp peaks of plasma resonance at 48.4° and 27.5°, which correspond to the n=+1 and the n=+2 resonances, respectively, as given by the resonance condition $\vec{k}_{sp}=\vec{k}'+n\vec{g}$, where $\vec{k}_{sp}$ is the plasmon wavevector, $\vec{k}'$ the incident photon wavevector projected on the surface, $\vec{g}$ the corrugation vector, and n=±1, ±2, ±3, ···. As θ varies from 90° to 0°, the resonance peaks shift to higher angles and their intensities vary systematically with no significant change in the intrinsic absorption in the off-resonance angular regions. The shifts of the resonance angles are due to changes in the direction of $\vec{g}$ relative to $\vec{k}'$. The peak intensities were found to vary, depending primarily on the electric field component of the incident photon perpendicular to the grooves at each resonance. However, their polarization dependence was not explicable by this simple consideration. Experimental results obtained for surfaces having shallower and deeper corrugations indicated [7] that the coupling efficiency of s-polarized photon (with the electric field perpendicular to the plane of incidence) to surface plasmon relative to that of p-polarized photon is strongly dependent on the corrugation amplitude and becomes larger as the corrugation becomes deeper.

Phase dependence of plasma resonance absorption. As is seen in the results in Fig. 2, the resonance peaks are absent from the absorption curves of p-polarized photon for $\theta=0°$ (the conical diffraction) and of s-polarized photon for $\theta=90°$ (the planar diffraction). This is due to lack of an electric field component of the respective photons perpendicular to the grooves at these geometries. Between these two groove orientations, surface plasmons are excited by both p- and s-polarized photons. These plasmons are excited at the same resonance angles, propagate in the same direction and are thus coherent to each other. Phase correlation between them may be observed directly in the resonance absorption intensity of photons having both p- and s-components in a certain phase relation. What one may expect to observe is that the resonance intensity varies depending on the relative phase of the p- and s-components even if their intensities are fixed.

Photoacoustic measurements of the resonance absorption of elliptically polarized incident photons were made at the resonance angle for the n=+1 resonance for $\theta=45°$. The signals were measured as a function of the azimuthal angle of a quarter-wave plate placed between the polarizer and the PA cell which changed the relative phase of p- and s-components of the incident photons. The results presented in Fig. 3 exhibit a striking effect of the phase correlation, demonstrating that the plasmon resonance absorption is phase-dependent. In this experiment, the intensities of p- and s-components were variable. Plotted by a dashed curve in Fig. 3 are the resonance absorptions calculated by assuming no phase correlation. For a planar and isotropic surface, the p- and s-components of incident photons induce the polarization currents inside the surface in the directions perpendicular to each other and, hence, the absorption by no means depends on their relative phase. A quantitative analysis of the results in Fig. 3 showed [4] that the phase lag during the excitation of surface plasmons by the p-component is 11° larger than that by the s-component. This result was found consistent with the prediction from the perturbation theory of corrugation-induced photoabsorption.

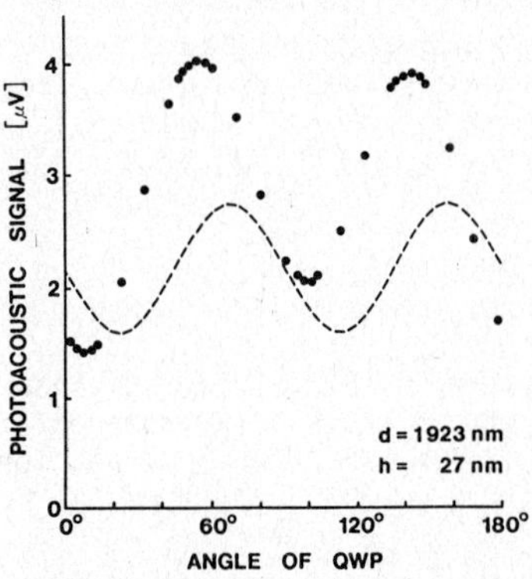

Fig. 3. Plasma resonance absorption of elliptically polarized photons as a funtion of the azimuthal angle of a quarter-wave plate

Plasma resonance with double periodic corrugations. Little experimental study has been made on plasma resonance in metal surfaces with double periodic corrugations (periodic in two directions). The present PA method seems to be almost the only experimental method available for probing the resonance absorption in such surfaces. We have studied silver surfaces having two sets of shallow periodic corrugations crossing perpendicularly. In Fig. 4, we present the results obtained for a surface corrugated with periods $d_1=d_2=2186$nm and amplitudes $h_1=h_2=48$nm.

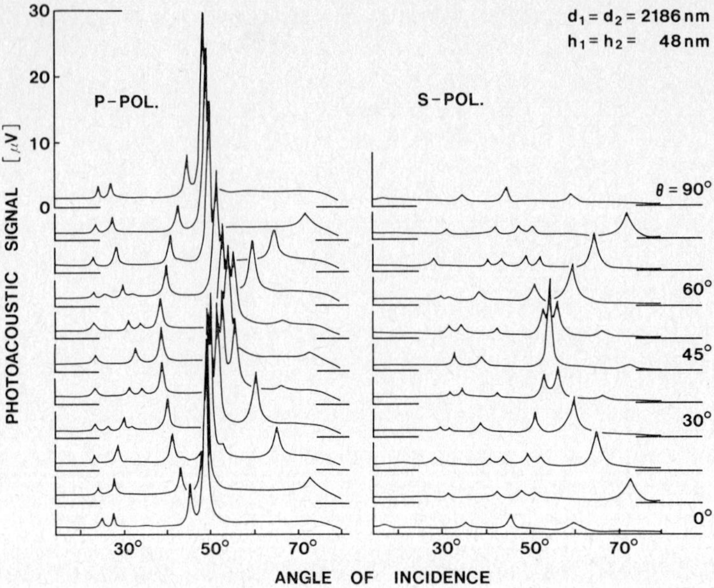

Fig. 4. Plasma resonance absorption in a silver surface having two perpendicularly crossing periodic corrugations

The resonance structures obtained for the various groove orientation θ are symmetric with respect to θ=45°, reflecting the symmetry in the surface topological structure. In contrast to the case of singly corrugated surfaces, the resonance couplings of both p- and s-polarized photons occur for any groove orientations. As θ varies from 90° to 45°, the resonance angles and the peak intensities vary systematically and, at θ=45°, two major peaks at the resonance angles near 55° merge into a single peak. These peaks may be assigned to the (m, n)=(±1, 0) and (m, n)=(0, ±1) resonances using the resonance condition $\vec{k}_{sp}=\vec{k}'+m\vec{g}_1+n\vec{g}_2$. It was found [6] that, upon the mergence into a single peak, the intensities for s-polarized photons increased drastically while those for p-polarized photons decreased in a manner of compensating it. At θ=45°, two plasmons which propagate in two different directions symmetric with respect to the plane of photon incidence were excited simultaneously. These two plasmons may couple to each other through the momentum exchange mediated by the corrugation structure. The redistribution of the resonance intensity between two photon polarizations is a result of this interaction. This has been accounted for by a recent nonperturbative calculation of photoabsorption in doubly corrugated surfaces [10].

Plasma resonance in periodically corrugated thin layers. Experimental results presented above were those obtained for a corrugated single surface of an effectively infinitely thick silver and, hence, the resonant couplings of incident photons observed were to surface plasmons as a single surface mode. It is theoretically well-known [11] that in a thin layer, surface plasmons on the two surfaces interfere with one another and set up a coupled two-surface mode if the layer is thin enough and bounded on both sides by media with similar refractive index. We have studied these coupled modes of surface plasmons in thin silver in the thickness range from 26 to 121nm. The silver layers were self-supporting and corrugated with a period of 1888nm and ampli-

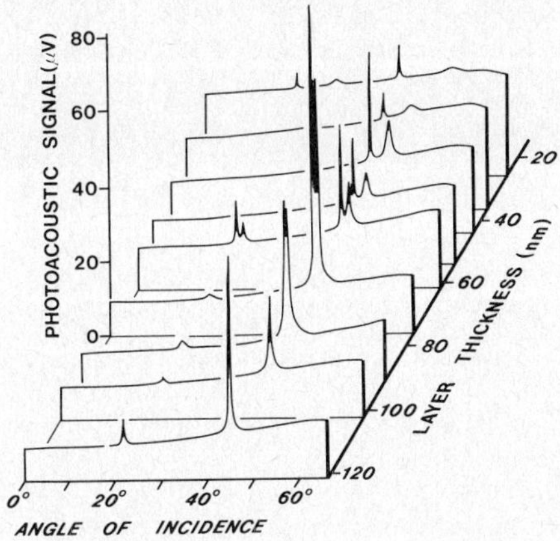

Fig.5. Plasma resonance absorption in periodically corrugated self-supporting thin silver layers

tudes varying from 6 to 12nm, depending on the layer thickness. In Fig. 5 we present the results obtained with p-polarized photons in the planar diffraction geometry.

The absorption curve obtained for the 121nm-thick layer exhibits the resonance peaks at 44.8° and 21.7°, corresponding to the n=+1 and n=+2 resonances, respectively. This feature of the resonance absorption is similar to one due to the single-surface mode observed for a corrugated surface of thick silver shown in Fig. 2. As the layer becomes thinner, it is seen that each peak splits into two peaks, one at a higher angle (the k^+ mode) and the other at a lower angle (the k^- mode), as a result of the interaction of plasmons on the two sides of the layers. The angular split between these modes increases as the layer thickness decreases. Also, the resonance widths of the k^- and k^+ modes become narrower and broader, respectively, with decrease of the layer thickness, exhibiting their long and short-range propagation properties.

These results of the resonance angle and the resonance width observed as a function of layer thickness were compared with theoretical predictions derived from the dispersion relation of the coupled surface plasmon [11] by ignoring the effects of the periodic corrugations on both sides of the layer. It was found [9] that agreements between the experimental and theoretical results were generally satisfactory. The total intensity of the k^+ and k^- resonance was found strongly dependent both on the corrugation amplitude and the layer thickness.

Plasma resonance induced by an overlayer on a planar surface. As was mentioned before, at a planar surface incident photons are unable to couple with surface plasmons. However, if the planar surface is overcoated with a thin layer whose top surface has a statistical roughness or a periodic corrugation, incident photons may couple to surface plasmons [12]. We have studied coupling using a planar silver surface covered with an evaporated layer of organic solids, the results of which are presented in Fig. 6. In this experiment, the resonance absorption was measured as a function of the wavelength of incident photon using a 150W xenon lamp as a light source. The overlayers of acridine orange with three different thicknesses from 5 to 15nm were formed by evaporating it successively onto a planar silver surface.

It is seen that the absorption exhibited by the bare silver is quite weak in the visible region and increases steeply below 350nm. The overlayer of

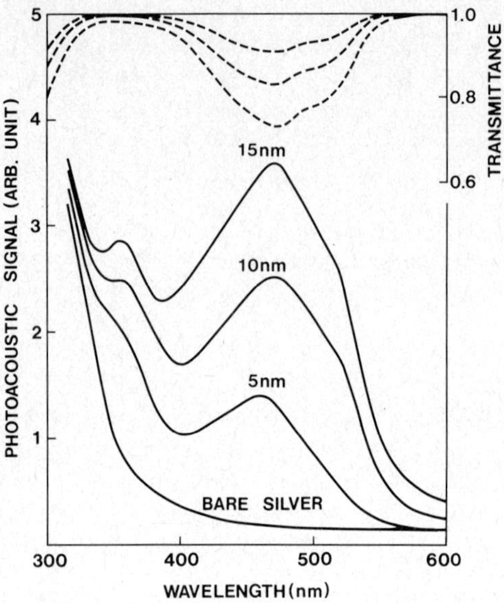

Fig. 6. Absorption spectra of planar silver surfaces covered with acridine orange overlayers of three different thicknesses. Dashed lines are transmission spectra of the acridine orange layers deposited on a quartz plate

acridine orange causes a broad absorption band with a peak at 460nm and a shoulder at 520nm. In addition to this, it is seen that a peak appears at 360nm as the thickness of the overlayer increases. This additional structure at 360nm is absent both from the absorption curves of acridine orange on a quartz plane (dashed curves) and from that of the bare silver but similar to one observed for a statistically roughened bare surface of silver. The overlayers used in the present study were in a form of discontinuous island film and their topological structure formed on the planar silver surface induced this plasma resonance. Results similar to this have been obtained for non-absorbing evaporated layers of stearic acid.

4. Conclusion

Using the gas-microphone method, we have studied plasma resonance in a variety of silver surfaces having topological microstructures. The technique was found to be a quite useful and powerful tool for probing the resonance absorption and made it possible to study the resonance absorption under conditions which have been hitherto difficult to study by the conventional techniques. These studies have revealed new aspects of surface plasmon behavior; plasma resonance of s-polarized photons, phase correlation between incident photons and surface plasmons, simultaneous excitations of two surface plasmons and their interaction. We were also able to study the coupled modes of surface plasmon in self-supporting corrugated thin silver layers. The present experimental technique may be applied to many other studies of surface plasmons and the related phenomena.

Acknowledgment. This research was sponsored in part by the Office of Health and Environmental Research, U.S. Department of Energy, under contract DE-AC05-84OR21400 with Martin Marietta Energy Systems, Inc.

References

1. H. Raether: Excitation of Plasmons and Interband Transition (Springer, Berlin 1980), pp. 116-171
2. R. H. Chang and T. E. Furtak, eds.: Surface Enhanced Raman Scattering (Plenum, New York 1982)
3. T. Inagaki, M. Motosuga, K. Yamamori, and E. T. Arakawa: Phys. Rev. B28, 1740 (1983)
4. T. Inagaki, M. Motosuga, and E. T. Arakawa: Phys. Rev. B28, 4211 (1983)
5. T. Inagaki, M. Motosuga, K. Yamamori, and E. T. Arakawa: Appl. Opt. 22, 3009 (1983)
6. T. Inagaki, J. P. Goudonnet, J. W. Little, and E. T. Arakawa: J. Opt. Soc. Am. B2, 433 (1985)
7. T. Inagaki, J. P. Goudonnet, and E. T. Arakawa: J. Opt. Soc. Am. B3, 992 (1986)
8. T. Inagaki, M. Motosuga, E. T. Arakawa, and J. P. Goudonnet: Phys. Rev. B31, 2548 (1985)
9. T. Inagaki, M. Motosuga, E. T. Arakawa, and J. P. Goudonnet: Phys. Rev. B32, 6238 (1985)
10. N. G. Glass and A. A. Maradudin: Opt. Commun. 56, 339 (1986)
11. E. N. Economou: Phys. Rev. 182, 539 (1968)
12. G. S. Agarwal: Phys. Rev. B31, 3534(1985)

Nanosecond Photoacoustic Studies of UV Laser Ablation of Polymers and Biological Materials

P.E. Dyer

Department of Applied Physics, University of Hull, Hull, HU6 7RX, United Kingdom

1. INTRODUCTION

The ability to precisely ablate organic polymers and biological tissue with high spatial resolution and minimal thermal damage using uv excimer lasers has aroused much interest, both with regard to the basic mechanisms involved and the potential applications [1]. The apparently low thermal damage associated with uv lasers has been attributed to the fact that fast material removal occurs through a direct bond-breaking 'ablative photodecomposition' process [2]. It has, however, alternatively been suggested [3] that removal is due to localized thermal degradation in which the heat affected zone is restricted by the small penetration depth for radiation and short time available for heat flow. In general it is observed that thermal effects become more predominant at longer wavelengths.

This paper describes the application of polyvinylidene fluoride (PVDF) film piezoelectric transducers to study the photoacoustic response of uv laser irradiated polymers and biological materials. The nanosecond resolution furnished by this technique proves useful in both the subablation threshold regime, where information on the thermal relaxation of excited states and optical properties of the material can be deduced from the thermoelastic response, and under ablative conditions where characteristic timescales for the process can be established. Knowledge of the latter can help ascertain the relative contributions of photothermal and photochemical pathways in the ablation.

2. EXPERIMENTAL ARRANGEMENT

2.1 Transducer Design

The photoacoustic transducer used a 9μm thick PVDF film bonded onto a 4mm thick lucite acoustic impedance matching stub, Fig. 1. Thin organic polymer films or biological samples were mounted onto the transducer surface allowing stress waves generated by laser irradiation to be detected. The transducer output was amplified using a 100 MHz bandwidth 1M oscilloscope plug-in giving the time resolved normal force at the transducer, $F(t)$ [4]. For thin samples the stress waves remain planar and the stress is $\sigma(t) = F(t)/A$, where A is the laser irradiated area. In the impedance matched design the overall response time for the transducer-display system was ~ 3.6ns [5].

2.2 Laser Irradiation

Samples were irradiated using XeCl, KrF or ArF laser pulses of 10-20ns (fwhm) duration. The laser output was spatially filtered and imaged onto the samples using a lens to produce a uniform fluence spot [4].

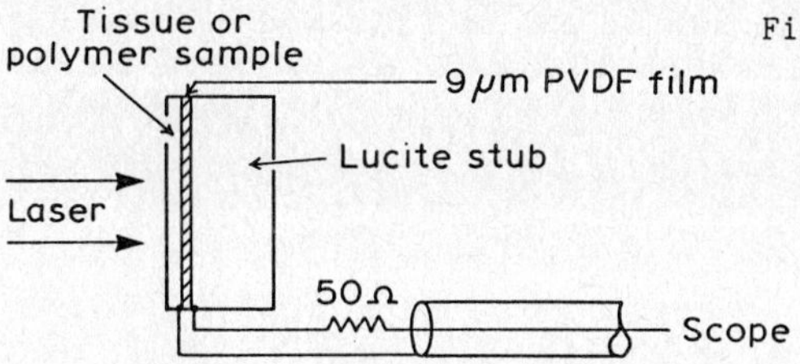

Fig. 1 Photoacoustic transducer

Synchronized recording of the stress wave and laser pulse was achieved using Si photodiodes to trigger the oscilloscope and record the laser pulse. The optical and electrical delay times for each sensor were adjusted to compensate for the $\lesssim$ 4ns acoustic transit-time in the thin ($\leq$ 10μm) organic polymer samples. Synchronization to better than ± 1ns was estimated to be achieved, with acoustic generation referred to the front surface of the film [4]. For biological materials the difficulty of preparing suitable thin films, and the need for optically-thick samples to avoid radiation penetrating to the transducer, prevented such accurate synchronization. However, in most cases the timing of the signals could be judged from the characteristic thermoelastic signal generated in the subthreshold region.

3. STRESS WAVE GENERATION

3.1 Laser Induced Stress Waves

A variety of mechanisms can lead to stress waves in laser irradiated targets [5]. However, in strongly absorbing materials such as polymers and biological tissue in the uv, the dominant sources are the thermoelastic effect [7] and material ablation [8]. Plasma production does not appear to be involved with ablation except at high fluence and can thus be neglected.

3.2 Thermoelastic waves

Stress wave generation by thermal expansion accompanying the rapid laser heating of surfaces has been studied extensively [7,9]. For short pulse lasers it is convenient to model the laser irradiance using the form $P = P_O(1 - \exp - kt) \exp - mt$, where k and m define the rates of rise and fall of the pulse; in this case the stress wave produced for a free surface becomes [10]

$$\sigma = \frac{\alpha \Gamma P_O k e^{at'}}{2(m+a)(k+m+a)} \qquad t' \leq 0$$

$$\sigma = \frac{\alpha\Gamma}{2} P_O \left\{ \frac{2m e^{-mt'}}{(m^2-a^2)} - \frac{2(m+k)e^{-(m+k)t'}}{(m+k)^2 - a^2} - \frac{ke^{-at'}}{(m-a)(k+m-a)} \right\} \qquad t' > 0, \qquad (1)$$

where Γ is the Gruneisen constant, $t' = t - x/v$, where v is the acoustic velocity, and $a = \alpha v$ where α is the optical attenuation coefficient for the laser. Exponential attenuation and instantaneous transformation of absorbed laser energy to heat is assumed in deriving (1).

An important feature of (1) is that in its early stages ($t' \leq 0$) the stress waveform is exponential with a time constant that can be related to the attenuation coefficient in the sample [7]. This proves valuable, particularly for biological materials under conditions where both scattering and absorption can contribute to attenuation. The maximum attenuation coefficient that can be measured is limited by the transducer response time, t_r, which must satisfy $t_r \lesssim (\alpha v)^{-1}$; for biological samples with $v \simeq 1.5 \times 10^5$ cm/s this sets a limit of $\alpha \leq 1800$ cm^{-1} for $t_r \simeq 3.6$ns.

3.3 Ablation Generated Stress Waves

Above a relatively well defined threshold, E_T, which is dependent on the polymer or tissue properties but typically in the range 30-300mJcm^{-2}, rapid material ablation occurs with excimer lasers, producing large amplitude stress waves through the recoil momentum [8]. If ablation is assumed to proceed 'layer-by-layer' [11], and accelerative terms in the flow are neglected, the ablative stress is approximately

$$\sigma_A = v\rho\dot{x},$$

where ρ is the target density, v is the ablation velocity, and $\dot{x}$ the surface recession velocity. This approximation is justifiable if the ablation products are weakly absorbing or transparent or if, in the low density expansion, energy absorbed by the fragments does not couple to the flow because of the low thermal relaxation rate. Then if Beers law, which fits the etch depth - fluence scaling over limited regions [1,3], is taken to apply <u>during</u> the pulse, the stress becomes

$$\sigma_A = 0 \qquad \text{for} \qquad f = \int_0^t P\,dt \leq E_T$$

$$\sigma_A = \frac{v\rho P}{\alpha \int_0^t P\,dt} \qquad f > E_T \;.$$

Here the retention of Beers law implies that the ablation products have the same attenuation as the parent material.

For a square laser pulse of duration 20ns, a fluence $E = 2E_T$, and a typical polymer with $\rho \simeq 1$ grm cm^{-3}, $\alpha \simeq 10^5$ cm^{-1} and $v \simeq 10^5$ cms^{-1} [12], σ_A is calculated to be ~ 10^7 Pa (~ 100 atm). Thus, large amplitude stress waves are anticipated to accompany ablation.

4. ORGANIC POLYMER ABLATION

The ablation of polyethylene terephthalate (PET), polyimide and polymethymethacrylate (PMMA) was investigated using ArF and XeCl lasers. Fig. 2a shows time synchronized laser pulses and ablation impulses for a 2.5µm thick PET film irradiated using the 193nm ArF laser. The positive going impulse corresponds to a compressive wave and signifies that ablation commences within $\lesssim$ 4ns of the start of the laser pulse, the delay appearing to reduce as the fluence increases. The peak stress, shown in Fig. 2b as a function of fluence, approaches 10^7Pa at 100 mJcm^{-2}, in keeping with estimates made in Section 3.3. It is interesting that for PET at 193nm, as with other strongly absorbing polymers (eg. polyimide at 308 and 193nm where $\alpha \simeq 10^5$ cm^{-1}), compressive transients persist at well below the generally accepted threshold for ablation [4].

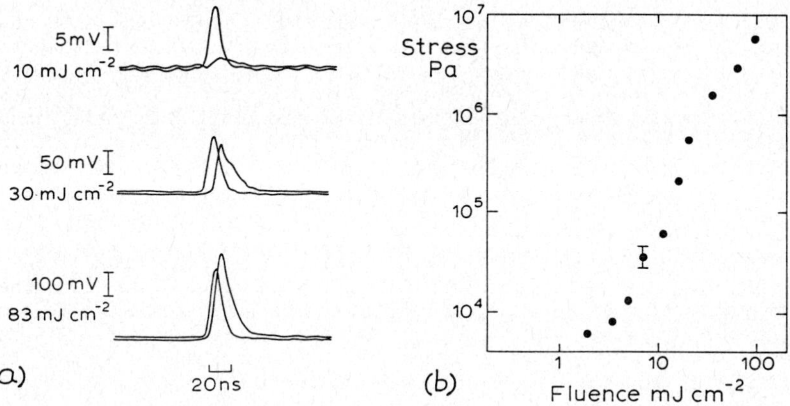

Figure 2 (a) Synchronized laser and ablation impulses for PET at 193nm (the laser is the first pulse). (b) Peak stress versus fluence for PET at 193nm.

In contrast, for less strongly absorbing polymers such as PMMA at 193nm and PET at 308nm, low fluence irradiation produced a bipolar stress wave which, with increasing fluence above threshold, gradually become purely compressive. This is illustrated in Fig. 3a for PMMA irradiated at 193nm. The peak stress increased linearly with fluence up to the threshold at ~ 60 mJcm^{-2} and then more rapidly as ablation commenced (Fig. 3b). This initial linearity is consistent with a thermoelastic generation mechanism in the subthreshold region (equation 1), and the appearance of the thermoelastic signal with the start of the laser pulse indicates that relaxation of photoexcited states to produce heating is rapid. Above threshold the stress appears initially to be due to the thermoelastic component with ablation then commencing within $\lesssim$ 8ns of the start of the laser pulse. From this, the temperature estimated using the fluence delivered just prior to ablation and the effective absorption coefficient for PMMA is $\lesssim$ 800K. At this temperature the thermal degradation rate is

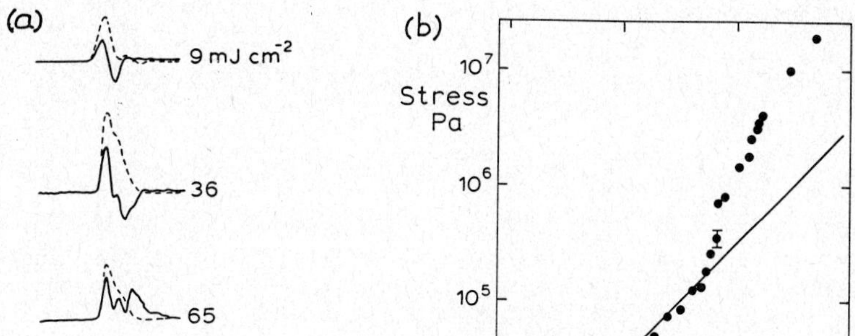

Figure 3 (a) Synchronized laser (dashed) and stress pulses for PMMA at 193nm. (b) Peak stress versus fluence, PMMA at 193nm.

too low to account for the rapidity of ablation and is suggestive that the process is driven by a substantial photochemical component [4]. Similar conclusions can be drawn from the ablation characteristics of PET.

Although the absence of a thermoelastic signal in PET at 193nm and polyimide at 193 and 308nm due to slow thermalization cannot be ruled out, a more likely explanation for this can be given as follows: in strong absorbers stress relaxation by expansion occurs very rapidly because the time, $(\alpha v)^{-1}$, for an acoustic wave to propagate through the absorption depth is much shorter than the laser pulse-width. Consideration of (1) with a >> k, m shows this significantly reduces the peak stress. Under these conditions weak 'subthreshold' ablation probably dominates the thermoelastic signal leading to a purely compressive stress wave [4,12].

5. CORNEAL TISSUE

The very sharply defined incisions that can be produced in corneal tissue by the ArF laser [13] has led to great interest in its potential use in various ophthalmic surgery procedures [14]. To shed light on this interaction, studies of corneal ablation using the acoustic transducers were carried out with ArF, KrF and CO_2 lasers.

For time resolved ablation studies at 193nm, thin corneal sections (~ 30μm) were prepared by dissection and mounted on the 9μm film transducer. Although synchronization to the accuracy achieved with organic polymers was not possible, the results confirmed that ablation commences within the ~ 10ns laser pulse and at a temperature not exceeding ~ 900K [15]. Under these conditions thermal decomposition of the peptide bond, modelled using available data for nylon 66, is too low to explain the ablation rate and suggests direct bond-breaking by the laser plays an important role in fragmentation. Large amplitude stress waves accompany the ablation as can be seen from Fig. 4, where the peak stress as a function of fluence indicates that in excess of 10^7 Pa (~ 100 Atm) is produced above ~ 500 $mJcm^{-2}$. These stress transients may be contributory to cell damage observed in corneal ablation studies [14]. From Fig. 4 the onset of a detectable acoustic transient occurs at ~ 50 $mJcm^{-2}$ which is in keeping with the ablation threshold deduced from material removal experiments [16]. In contrast to findings with the pulsed CO_2 and KrF laser, no thermoelastic signal could be detected below the threshold for the 193nm ArF laser, although calculations based on (1) suggest that the signal should increase at this wavelength because of the relative magnitude of the attenuation coefficient. The slow relaxation of photoexcited states, possibly due to intersystem crossing to triplets, could account for this and would also explain the low degree of thermal damage at 193nm if thermal diffusion was fast in comparison with deactivation. Direct bond-breaking of the organic component of tissue with water being expelled as droplets, as opposed to a high temperature evaporative process which appears to account for the CO_2 and KrF laser ablation, could then be invoked to explain the extremely sharp incisions produced at 193nm.

6. VASCULAR TISSUE

Interest in using excimer lasers to ablate vascular tissue stems from their potential use in laser angioplasty, that is the direct removal of blockages in blood vessels by photovapourisation using an intraluminal fibre [17]. By studying the photoacoustic response, information on the optical properties, ablation thresholds and ablation time-scales for normal and

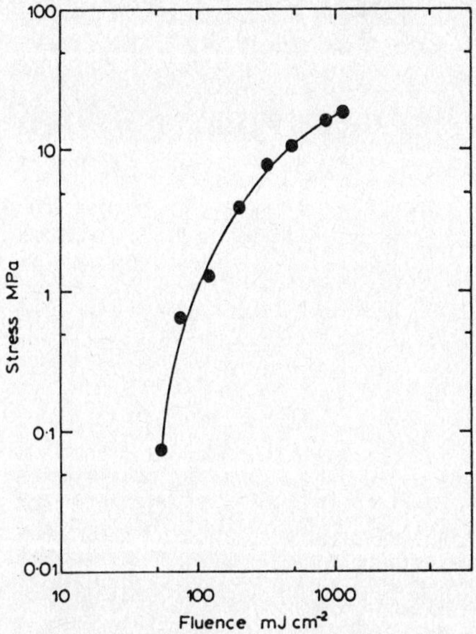

Figure 4 Stress as function of fluence for ArF laser irradiated cornea.

diseased (atheroma) vascular tissue can be obtained at uv and visible wavelengths [18,19].

Human aorta samples $\leq$ 1mm thick were mounted on the transducer in Fig. 1 and irradiated in air or a saline environment using KrF, XeCl or frequency doubled YAG lasers. At low fluence the leading edge of the stress waveform was exponential allowing the attenuation coefficient to be determined (Section 3.2). Figure 5 shows examples of the calculated and experimental response of vascular tissue in air. The attenuation coefficients for normal and atheroma tissue were measured to be 320 cm^{-1} and 310 cm^{-1} at 308nm and 450 cm^{-1} and 420 cm^{-1} at 248nm respectively [18].

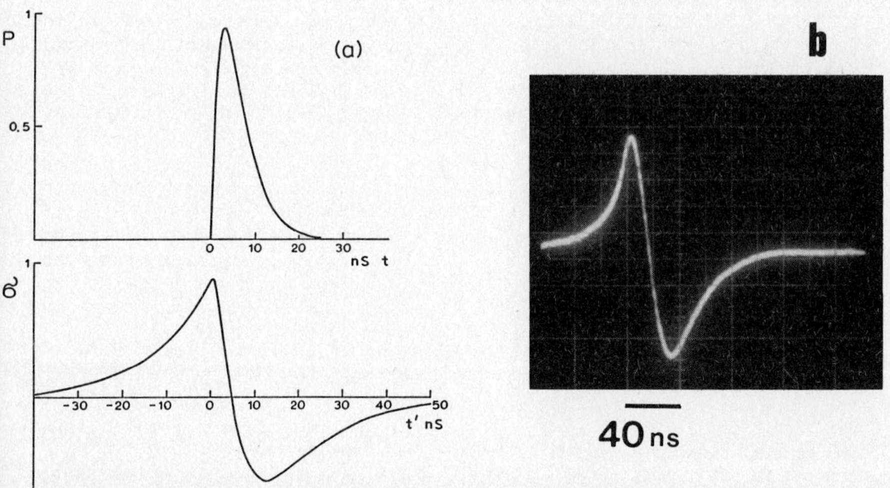

Figure 5 (a) Calculated stress response for vascular tissue. (b) Experimental stress waveform for XeCl laser irradiated normal vascular tissue.

Ablation was observed to start within the timescale of the laser pulse at a threshold of ~ 150-240mJcm^{-2} depending on the uv-wavelength, with little difference between normal and atheroma samples. Lower attenuation coefficients and much higher thresholds were measured at 532nm [18].

It is noteworthy that the effective absorption coefficient deduced from removal rate data in the ablative regime for the uv lasers was typically 20-50 times larger than measured below threshold. This suggests that ablation is intimately coupled with a large change in absorption coefficient, possibly due to the onset of non-linear effects. Similar results were obtained for saline immersed samples, the only major differences being the generation of a purely compressive, rather than bipolar, stress wave below threshold because of the different boundary conditions, and a significant lengthening of the ablative stress at high fluence.

ACKNOWLEDGEMENTS

The important contributions to this work made by Dr. R. Srinivasan of the IBM Research Laboratory (York Town Heights), Mr. R.K. Al-Dhahir of the University of Hull, Dr. F.W. Cross of University College, London, and Dr. A.J. MacRobert of the Royal Institution London are gratefully acknowledged.

REFERENCES

1 R. Srinivasan, Science 234, 559 (1986)
2 R. Srinivasan and W.J. Leigh, J. Am. Chem. Soc., 104, 6784 (1982)
3 J.E. Andrews, P.E. Dyer, D. Forster and P.H. Key, Appl. Phys. Lett. 43, 717 (1983)
4 P.E. Dyer and R. Srinivasan, Appl. Phys. Lett. 48, 445 (1986)
5 P.E. Dyer, B.L. Tait and J. Sidhu, accepted for publication in IEEE Trans. on Ultrasonics, Ferroelectrics and Frequency Control.
6 S.F. Cleary: In Laser Applications in Medicine and Biology, Vol. 3, ed. by M. Wolbarsht (Plenum, New York, 1977)
7 E.F. Carome, N.A. Clark and C.E. Moeller, Appl. Phys. Lett. 4, 95 (1964)
8 G. Gorodetsky, T.G. Kazyaka, R.L. Melcher and R. Srinivasan, Appl. Phys. Lett. 46, 828 (1985)
9 J.C. Bushnell and D.J. McCloskey, J. Appl. Phys. 39, 5541 (1968)
10 P.E. Dyer (unpublished)
11 B.J. Garrison and R. Srinivasan, J. Appl. Phys. 57, 2909 (1985)
12 R. Srinivasan and R.W. Dreyfus; in Laser Spectroscopy VII, ed. by T.W. Hansch (Springer, New York 1985)
13 S.L. Trokel, R. Srinivasan and B. Braren, Am. J. Ophthalmol. 96, 710 (1983)
14 J. Marshall, S. Trokel, S. Rothery and H. Schubert, Ophthalmol. 92, 749 (1985)
15 R. Srinivasan, P.E. Dyer and B. Braren, Laser in Surg. and Med. 6, 514 (1987)
16 R.R. Krueger, S.L. Trokel and H.D. Schubert, Invest. Ophthalmol. and Visual Sci., 26, 1455 (1985)
17 R. Linsker, R. Srinivasan, J.J. Wynne and D.R. Alonso, Lasers in Surg. and Med. 4, 201 (1984)
18 F.W. Cross, R.K. Al Dhahir, P.E. Dyer and A.J. MacRobert, Appl. Phys. Lett. 50, 1019 (1987)
19 D.L. Singleton, G. Paraskevopoulos, G.S. Jolly, R.S. Irwin, D.J. McKenney, W.S. Nip, E.M. Farrell and L.A.J. Higginson, Appl. Phys. Lett. 48, 878 (1986)

Pressure and Velocity Profiles Generated by Laser-Heated Spherical Droplets

G.J. Diebold and X.C. Hu

Department of Chemistry, Brown University, Providence, RI 02912, USA

When a spherical droplet is irradiated by a light beam the heat released by absorption of light causes expansion of the sphere which gives rise to a photoacoustic effect. The acoustic waveforms generated by Gaussian heating functions can be calculated from convolution integrals. The velocity of the sphere can be determined by differentiation of the expressions for the pressure.

I. INTRODUCTION

When radiation is absorbed by a liquid or solid, the resulting deposition of heat causes the medium to expand thereby launching a sound wave. Such generation of acoustic waves by absorption of light is known as the photoacoustic effect. Since this process is contingent on little more than a nonzero absorption and thermal expansion coefficient of the medium, it is not surprising that laser heated particulate matter is also known to generate the photoacoustic effect[1–5]. Recently a detailed calculation of the properties of the acoustic wave generated by uniformly heated, spherical, fluid "particles" has been given[6]. The acoustic wave is found as a solution to a boundary value problem with a monopole source function. The pressure inside the sphere is a sum of driving term and a spherical Bessel function of the form $j_o(kr)$, while the pressure outside is described by an outwardly propagating spherical wave. By equating the pressure and acceleration at the boundary between the sphere and the surrounding fluid a frequency domain expression for the acoustic pressure is found, which, when Fourier transformed can be written as

$$p_f(t) = \frac{\kappa}{4\pi}\int_{-1}^{1} r'\, d\, r' \int_{-\infty}^{\infty} \frac{e^{-iq(\hat{\tau}-r')}}{\left[\left(1-\frac{\rho_s}{\rho_f}\right)\frac{\sin\, q}{q} - \cos\, q + i\,\frac{\rho_s c_s}{\rho_f c_f}\sin\, q\right]}\, dq\ , \tag{1}$$

where κ is defined as

$$\kappa = \frac{\alpha\beta E_0\, c_s^2}{C_p \hat{r}}\ ,$$

and where α, β, c_s, and C_p are the absorption coefficient, sound speed, and heat capacity of the particle, respectively, E_0 is the laser fluence, $\hat{r} = r/a$ where r is the distance from the center of the sphere and a is the radius of the particle, $\hat{\tau} = \frac{c_s}{a}(t - \frac{r-a}{c_f})$ where c_f is the sound speed in the fluid, and ρ_s and ρ_f are the densities of the sphere and fluid, respectively.

From Eq 1 it is possible to derive closed form expressions for the pressure in three limiting cases: when the pulsewidth of the exciting light is much greater than the acoustic transit time across the sphere, when the densities of the spheres and the fluid are the same (but the sound speeds differ), and when the density of the sphere is much smaller than that of the surrounding fluid. It is also possible to obtain an analytic solution for the pressure when the sphere has a higher density than the surrounding medium albeit the mathematics are cumbersome. Waveforms derived from Eq 1 represent the response to a delta function heating of the particle.

In this paper we evaluate several convolution integrals for the pressure giving the acoustic response to light pulses with a Gaussian temporal profile. In addition, expressions are obtained that give the velocity of the sphere at its surface in the limiting cases given above.

II. PRESSURE SIGNAL FROM GAUSSIAN PULSES

When the laser pulsewidth is long compared with the transit time for sound across the sphere, the integrand in Eq 1 can be expanded to lowest order in the frequency parameter $q = \omega a/c_s$, where ω is the modulation frequency giving [6]

$$p_f(t) = \frac{\kappa}{3}\frac{\rho_f}{\rho_s}\frac{d}{d\hat{t}}\left[\frac{1}{2\pi}\int_{-\infty}^{\infty} e^{-iq(\hat{t}-\frac{c_s}{c_f}\hat{r})}\, dq\right] , \tag{2}$$

which can be written in terms of the derivative of a delta function. If the laser intensity is of the form

$$I(t) = \frac{E_0}{\Theta\sqrt{\pi}} e^{-t^2/\Theta^2} , \tag{3}$$

where E_0 is the energy per unit area in the pulse and Θ is the pulse width, then the acoustic signal is given by the convolution integral

$$P_f(t) = \frac{1}{2\pi}\int_{-\infty}^{\infty} p_f(y) I(t-y)dy , \tag{4}$$

which reduces to

$$P_f(t) = \frac{2\kappa\, \rho_f}{3\sqrt{\pi}\, \rho_s \hat{\Theta}^3} \left[(\hat{t} - \frac{c_s}{c_f}\,\hat{r})\, e^{-(\hat{t}-\frac{c_s}{c_f}\hat{r})^2/\hat{\Theta}^2} \right] \tag{5}$$

where $\hat{\Theta} = c_s\Theta/a$. The pressure is thus proportional to the derivative of the heating function which follows directly from the delta function form of Eq 2. The signal is a function of the quantity $(\hat{t} - \frac{c_s}{c_f}\hat{r}) = \frac{c_s}{a}(t - r/c_f)$, which is the retarded time from the center of the sphere, as opposed to the surface of the sphere. This is expected in this long pulse limit since the transit time across the droplet is negligible relative to the length of the exciting pulse.

Now, when the densities of the sphere and the fluid are the same, the delta function pressure response is given by[6]

$$P_f(t) = \frac{\kappa}{(1 + \frac{c_s}{c_f})} \sum_{n=0}^{\infty} (-\xi)^n [\hat{r} - (2n+1)] \tilde{\theta}_{2n,2n+2}(\hat{r}) \tag{6}$$

where $\xi = (1 - c_s/c_f)(1 + c_s/c_f)$ and where the square wave function $\tilde{\theta}$ is defined in terms of Heaviside functions θ as

$$\tilde{\theta}_{t_1,t_2}(\hat{r}) = \theta(\hat{r} - t_1) - \theta(\hat{r} - t_2)\ .$$

Substitution of Eqs 6 and 3 into the convolution integral Eq 4 gives the acoustic pressure as

$$P_f(t) = \frac{-\kappa}{\pi^{\frac{1}{2}}(1 + \frac{c_s}{c_f})\,\hat{\Theta}} \sum_{n=0}^{\infty} (-\xi)^n \int_{2n}^{2n+2} [y - (2n+1)] e^{-(y-\hat{r})^2/\hat{\Theta}^2}\, dy, \tag{7}$$

which can be written in terms of exponential and error functions as

$$P_f(t) = \frac{-\kappa\sqrt{\pi}}{2(1 + \frac{c_s}{c_f})} \sum_{0}^{\infty} (-\xi)^n \Big\{ [\hat{r} - (2n+1)][erf(b) - erf(a)] - \frac{\hat{\Theta}}{2}\left[e^{-b^2} - e^{-a^2}\right] \Big\} \tag{8}$$

where $a = (2n - \hat{r})/\hat{\Theta}$ and $b = (2n + 2 - \hat{r})/\hat{\Theta}$. When the laser pulse width is small compared with the acoustic transit time across the particle ($\hat{\Theta} << a/c_s$) then the pressure signal should resemble the N shaped wave (or series of N shaped waves) described in Ref 6, since the exciting pulse approximates a delta function. At the opposite

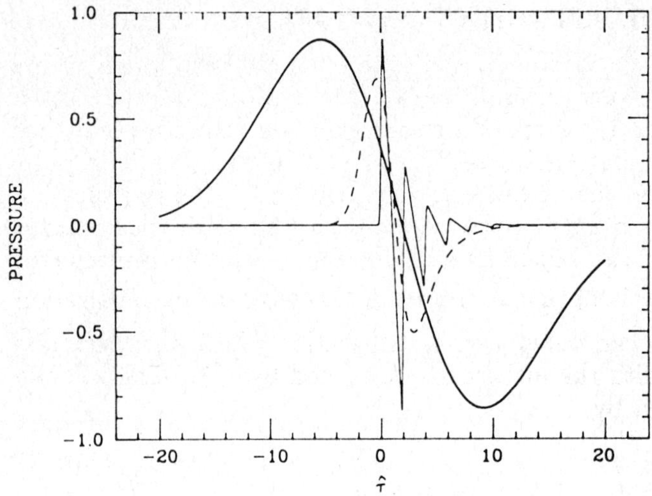

Figure 1: Pressure (dimensionless units) versus $\hat{\tau}$ (in dimensionless units) for $\rho_s/\rho_f = 1$ and $c_s/c_f = 1$ from Eq 8 with $\hat{\Theta} = 0.1$ (—— triangular pulse), $\hat{\Theta} = 2$(− − −) and, $\hat{\Theta} = 10$ (—— broad Gaussian pulse). The parameter κ has been set to unity. For $\hat{\Theta} = 2$ the curve has been multiplied by a factor 2; for $\hat{\Theta} = 10$ the curve has been multiplied by a factor of 10.

extreme, when $\hat{\Theta}$ is much greater than a/c_s, then, according to Eq 5, the pressure should be proportional to the derivative of the heating function. Neither of these limiting responses is immediately evident from Eq 8. Numerical evaluation of this equation gives the curves of pressure versus retarded time shown in Fig 1. The sharp waveform beginning at approximately $\hat{\tau} = 0$ can be seen for $\hat{\Theta} = 0.1$. The other limiting case, where $\hat{\Theta} = 10$ (*i.e.* $\Theta >> a/c_s$), shows the derivative waveform irrespective of the presence of reflections. There is a shift in the zero of the $\hat{\Theta} = 10$ waveform that depends on the ratio c_s/c_f. In fact, the retarded time $\hat{\tau}$ can be written as $\hat{\tau} = \frac{c_s}{a}(t - \frac{r}{c_f}) + \frac{c_s}{c_f}$, while the retarded time in Eq 5 is given as $\frac{c_s}{a}(t - \frac{r}{c_f})$. Thus, there is an artificial constant, c_s/c_f, that is added to the abscissa as a consequence of plotting the pressure as a function of the retarded time from the surface of the sphere instead of the retarded time relative to the origin of the sphere. Of further note is the diminution of the peak amplitude of the long $\hat{\Theta}$ waveform for $c_s/c_f = 2$ relative to its value for $c_s/c_f = 0.5$.

III. VELOCITY WAVEFORMS

The velocity of the surface of the sphere can be calculated from a knowledge of the pressure through use of the relation $\dot{u} = -(1/\rho)\nabla p$. That is, the pressure is differentiated to give the acceleration which is then integrated to determine the velocity. Consider the long pulse approximation, where the pressure is given by Eq 5. The pressure has a dependence on the radius in the factor κ as well as in the retarded time. In the case of the latter, the derivative $\frac{\partial}{\partial r}$ can be replaced by $-\frac{c_s}{c_f a}\frac{\partial}{\partial \hat{t}}$. The velocity is obtained from the acceleration by integration from $-\infty$ to t, which can be written as integration of the dimensionless time from $-\infty$ to $\hat{t}$ with the integrand multiplied by a/c_s. The velocity at $r = a$ is thus given by

$$u(t) = \frac{2\alpha\beta E_0 c_s}{3\sqrt{\pi}\rho_s C_p \hat{\Theta}^3}\int_{-\infty}^{\hat{t}} \left[(t' - \frac{c_s}{c_f a} r) + \frac{\partial}{\partial t'}(t' - \frac{c_s}{c_f})\right] e^{-(t' - \frac{c_s}{c_f})^2/\hat{\Theta}^2} . \tag{9}$$

This can be integrated to give

$$u(t) = \frac{2\kappa_u}{3\sqrt{\pi}\hat{\Theta}}\left[\frac{2c_s}{\hat{\Theta}c_f}(t - \frac{c_s}{c_f}) - 1\right] e^{-(\hat{t} - \frac{c_s}{c_f})^2/\hat{\Theta}^2} , \tag{10}$$

where κ_u is a quantity with dimensions of velocity defined by $\kappa_u = \left(\frac{\alpha\beta E_0}{\rho_s C_p}\right) c_s$.

A plot of this function shows the velocity to peak at the maximum in the heating pulse for large values of $\hat{\Theta}$.

For the case when the densities of the particle and the surrounding fluid are equal, the particle velocity can be found using the same procedure. From the acoustic pressure given by Eq 6, the particle velocity is found to be

$$u(t) = \frac{-\kappa_u}{(1 + \frac{c_s}{c_f})}\sum_{n=0}^{\infty}(-\xi)^n\left\{\frac{\hat{r}^2}{2} + [\frac{c_s}{c_f} - (2n+1)]\,\hat{r} - \frac{c_s}{c_f}(2n+1)\right.$$

$$\left. +2n(n+1)\right\}\tilde{\theta}_{2n,2n+2}(\hat{r}) . \tag{11}$$

A plot of this function is given in Fig 2. The nonlinear dependence of the signal on $\hat{r}^2$ can be dominant, or of no consequence depending on the value of the n and c_s/c_f. Evaluation of this function at $\hat{r} = 0$ gives the velocity as $u(\hat{r} = 0) = (\kappa_u c_s/c_f)/(1 + c_s/c_f)$; at $\hat{r} = 2$ the velocity

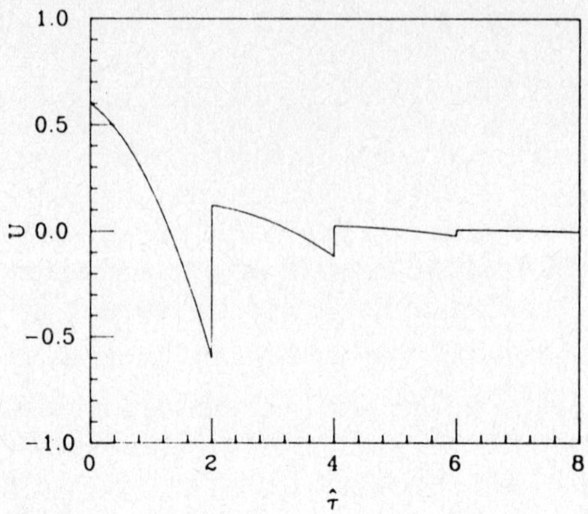

Figure 2: Velocity (dimensionless units) vs $\hat{\tau}$ (dimensionless units) from Eq. 11 with $\kappa_u = 1$ with $c_s/c_f = 1.5$.

is the negative of this. Equal amplitude swings are also found at the beginning and end of each reflection.

When the density of the sphere is higher than that of the surrounding liquid, the initial acoustic wave (*i.e.* for $\hat{\tau}$ between 0 and 2) can be written as[6]

$$p(t) = \frac{\kappa}{(1+\frac{\rho_s c_s}{\rho_f c_f})\lambda}[e^{-\lambda\hat{\tau}}(\lambda+1) - 1]\hat{\theta}_{0,2}(\hat{\tau}) \tag{12}$$

where $\lambda = (\frac{\rho_s}{\rho_f} - 1)/(1+\frac{\rho_s c_s}{\rho_f c_f})$. Using the procedure described above, the velocity of the sphere is found to be

$$u(t) = \frac{\kappa_u}{(1+\frac{\rho_s c_s}{\rho_f c_f})\lambda^2}\left[(\lambda+1)(1-\frac{c_s}{c_f}\lambda)(1-e^{-\lambda\hat{\tau}}) + \frac{c_s}{c_f}\lambda^2 - \lambda\hat{\tau}\right]\tilde{\theta}_{0.2}(\hat{\tau})\ . \tag{13}$$

This function is plotted in Fig 3 for several values of λ. It is possible to find both the pressure and velocity for larger values of $\hat{\tau}$, but the mathematical expressions become complicated.

Evaluation of Eq 13 at $\hat{\tau} = 0$ and $\hat{\tau} = 2$ gives

$$u(\hat{\tau}=0) = \frac{\kappa_u(\rho_s c_s/\rho_f c_f)}{(1+\frac{\rho_s c_s}{\rho_f c_f})} \quad \text{and} \tag{14}$$

$$u(\hat{\tau}=2) = \frac{\kappa_u}{(1+\frac{\rho_s c_s}{\rho_f c_f})\lambda^2}\left[(1-e^{-2\lambda})(\lambda+1) - \lambda(2+\frac{c_s}{c_f}) + \frac{c_s}{c_f}e^{-2\lambda}\lambda(\lambda+1)\right]$$

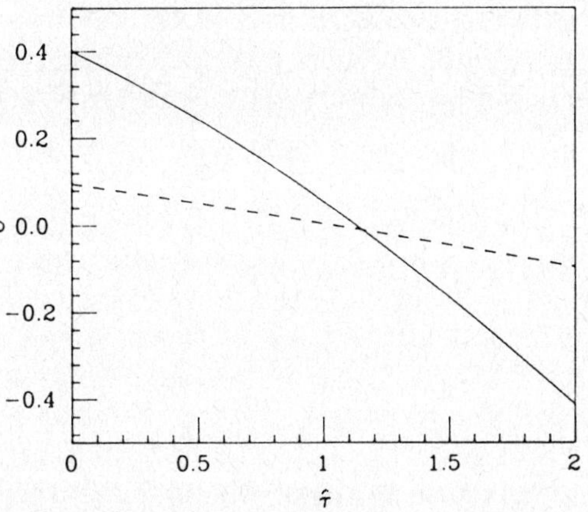

Figure 3: Velocity (dimensionless units) vs $\hat{\tau}$ (dimensionless units) from Eq. 13 with $\lambda = 0.2(- \; -)$ and $\lambda = 0.47(- \;\; - \;\; -)$.

which indicates modification of the velocity over the case for equal densities. For large values of λ (*i.e.* $\rho_s/\rho_f >> 1$), λ equals approximately c_f/c_s; Eqs 14 then reduce to $u(\hat{\tau} = 0) = \kappa_u = -\,u(\hat{\tau} = 2)$ which shows a symmetry in the velocity as in the equal density case. If both ρ_s/ρ_f and c_s/c_f are large compared with one, Eq 13 reduces to

$$u(t) \cong \kappa_u(1 - \hat{\tau})\tilde{\theta}_{0,2}(\hat{\tau}) \,.$$

Of course, when λ is small, Eq 14 reduced to the equal density case described by Eq 11.

DISCUSSION

In a number of places above the factor $\kappa_u = (\alpha\beta E_0/\rho_s C_p)c_s$ appears. It would seem that if the energy fluence is made large enough, then the dimensionless factor in parentheses that multiplies c_s could be made arbitrarily large giving a velocity of the sphere surface that exceeds the sound velocity. However, this dimensionless factor is ordinarily small. In addition the use of the isobaric volume expansion coefficient in Ref 6 for the internal pressure implicitly assumes that the pressure increase within the sphere is small compared with the ambient pressure. Any departure from this assumption requires a completely new derivation of the frequency domain expression for the pressure.

The sharp waveforms with abrupt features predicted here differ qualitatively from the smooth waveform calculated by Hu or Segrist and Kneubuhl[8] for a spherical geometry, or Lai and Young[9] and

Heritier[10] for a heated cylindrical volume. This would appear to be a consequence of the sharply defined droplet-fluid interface where the optical absorption coefficient has a discontinuity in the present calculation as opposed to the gradual dropoff in the absorbed radiation in space assumed by other authors. Of course, the present work assumes a discontinuity in acoustic properties at the droplet-fluid interface whereas the other calculations are valid only for uniform media.

Determination of the velocity of the surface of the sphere could be carried out by scattering light from the sphere and measurement of the Doppler shift with a high resolution interferometer. Alternately, optical heterodyne measurement would permit the determination of the surface velocity as a function of time. The expressions above thus provide a basis for interpretation of data in terms of the physical properties of the sphere.

The support of this research by the U. S. Department of Health and Human Services is gratefully acknowledged.

REFERENCES

1. C. W. Bruce and R. G. Pinnick, Appl. Opt **16**, 1762 (1977).
2. C. W. Bruce, Y. P. Lee, and S. G. Jennings, Appl. Opt. **19**, 1873 (1980).
3. C. W. Bruce and N. M. Richardson, Appl. Opt. **22**, 1051 (1983); **23**, 13 (1984).
4. R. S. Davidson and D. King, Anal. Chem. **56**, 1409 (1984).
5. S. Oda, T. Sarvada, T. Moreguchi, and H. Kamada, Anal. Chem. **54**, 121 (1973).
6. G. J. Diebold and P. Westervelt, "The Photoacoustic Effect Generated by a Spherical Particle", submitted to J. Acoust. Soc. Am.
7. C. L. Hu, J. Acoust. Soc. Am. **46**, 729 (1968).
8. M. Sigrist and F. Kneubuhl, J. Acoust. Soc. Am. **64**,
9. H. M Lai and K. Young, J. Acoust. Soc. Am. **72**, 2000 (1982).
10. J. Heritier, Opt. Commun. **44**, 267 (1983).

Surface Temperature and Thin Film Investigations by Pulsed Laser Photothermal Displacement Spectroscopy

Q. Kong, J. Fischer, and F. Träger

Physikalisches Institut der Universität Heidelberg,
Philosophenweg 12, D-6900 Heidelberg, Fed. Rep. of Germany

1. INTRODUCTION

Recently, photothermal and photoacoustic techniques have been applied in a number of experiments (see e.g. [1-8]) as versatile tools for surface studies. In combination with laser excitation, these methods offer advantages such as high sensitivity, high spectral and high temporal resolution. This makes it possible to investigate adsorbate-substrate as well as adsorbate-adsorbate interactions for coverages as low as a few thousandths of a molecular monolayer [9]. In addition, photothermal laser spectroscopy has been used e.g. to study phase transitions at interfaces [10] or to investigate laser-induced desorption phenomena [11]. Experiments were performed with continuous-wave as well as pulsed lasers and in combination with contact and non-contact surface probing.

The present paper describes new results obtained by nanosecond time-resolved photothermal displacement spectroscopy [3,4]. Experiments were performed with two goals in mind. Firstly, a thermoelastic surface deformation reflects the temperature of the surface and its variation as a function of time. The measurement of surface temperatures is of great importance e.g. for studies of desorption phenomena and laser-induced etching processes, or more generally speaking, for the understanding of surface reaction dynamics. Secondly, a time-resolved photothermal displacement signal reflects directly *both* the optical properties which are obtained from the maximum amplitude of the signal and the thermal properties of the surface that follow from an analysis of the decay of the signal. They are essential for surface characterization and change if adsorbates or thin films are condensed. One would therefore also anticipate that the method can provide an interesting alternative for the existing techniques of thin film analysis.

2. EXPERIMENTAL

In our experiments the beam of an excimer laser pumped dye laser with a pulse duration of 15 ns impinges on the sample surface. Part of the incident photon energy is absorbed and converted into heat by radiationless deexcitation. As a consequence, the temperature rises and the surface expands locally. The resulting thermoelastic surface displacement is probed with a He-Ne laser, whose beam is reflected from the shoulder of the deformation. The change of the reflection angle of the probe beam is measured time-resolved with a position sensitive detector. A second photodiode monitors the time dependence of the intensity of the probe laser beam. This allows us to correct the displacement signal for possible variations of the surface reflectivity induced by the pump pulse.

The experimental arrangement is shown schematically in Fig. 1. It consists of an ultrahigh vacuum system with a base pressure of 10^{-10} mbar. The sample is attached to a manipulator and can be heated and cooled down to liquid nitrogen temperature. A small crucible serves to evaporate metals onto the surface. There is also provision for coverage measurements with a quartz crystal microbalance. In order to facilitate the optical alignment of the laser beams, a precisely machined stainless steel block is installed inside the vacuum chamber. It serves two purposes: Firstly, it carries the lenses for focusing of the laser beams. Secondly, the block defines the position of the surface under investigation. This allows for spatial translations or

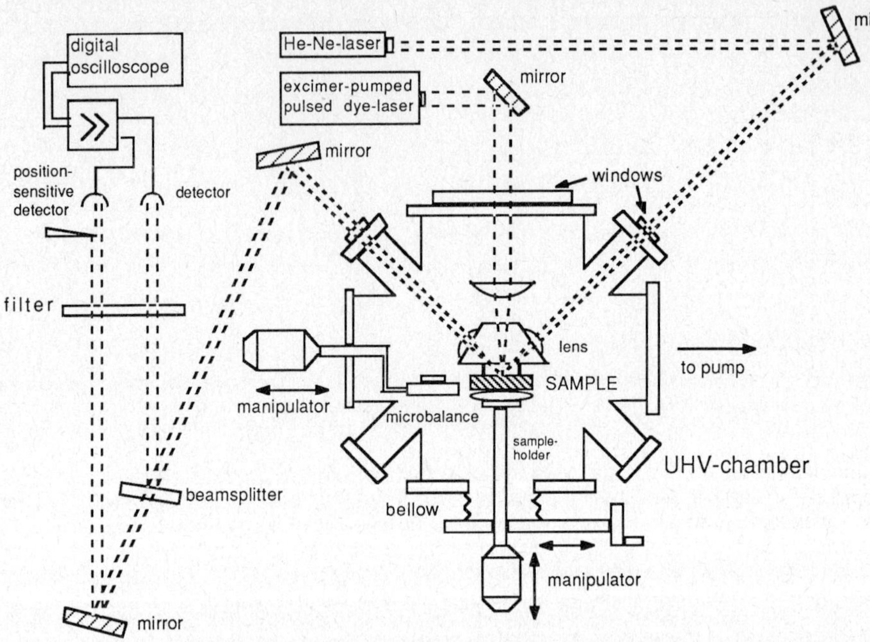

Figure 1: Experimental arrangement for pulsed laser photothermal displacement spectroscopy for surface and thin film studies in ultrahigh vacuum

rotations of the sample with the manipulator in the vacuum system, e.g. for thin film deposition or surface analysis, and simultaneously guarantees a reproducible realignment with respect to the foci of the light beams. The time-resolved displacement signal is either displayed on an oscilloscope or processed by a transient digitizer with a sampling rate of 100 MHz.

3. MEASUREMENTS AND RESULTS

The high sensitivity of the method is demonstrated by the fact that the time-resolved signal can be registered with a *single* laser pulse. The signals (see Fig. 2a) generally exhibit a rapid increase when the laser pulse hits the surface, followed by a slower decrease. The amplitude reaches its maximum value within about 40 ns. The decay typically lasts several microseconds. It depends on the material under investigation and reflects the heat diffusion from the surface layer to the bulk of the material. The time-resolved measurement of the signal makes it possible to observe the change of the laser-induced surface deformation as a function of time. For this purpose data have been taken with different relative positions of the foci of the pump and probe beams. Initially, the shape of the surface displacement is Gaussian, reflecting the Gaussian laser beam profile. As the heat flows to the bulk of the material, however, the shape changes and can only be described by a complicated, non-Gaussian function. For further investigations different samples such as metal surfaces and Si wafers have been studied under well defined conditions in the ultrahigh vacuum environment. Also thin films on different substrates were investigated.

3.1 SURFACE TEMPERATURE INVESTIGATIONS

As already mentioned above, the thermoelastic deformation and its change as a function of time reflect the variation of the surface temperature. The temperature rise can be estimated for the maximum amplitude of the signal, i.e. for the situation where the heat is still confined in that

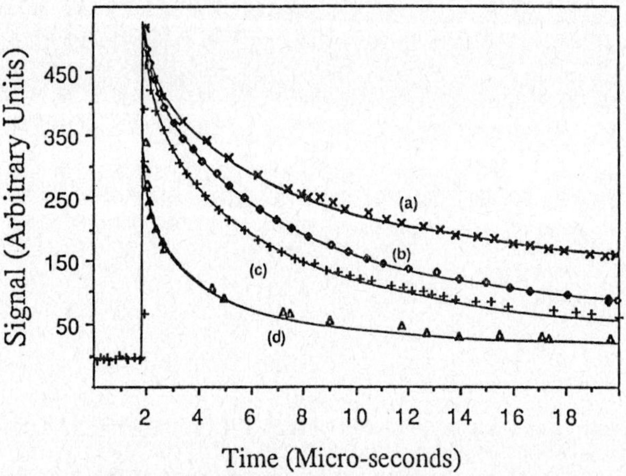

Figure 2: Time dependence of the photothermal displacement signal of a stainless steel sample in ultrahigh vacuum with (a) clean surface, (b) surface covered with a 0.5 μm thick silver film, (c) for a Ag film thickness of 1.0 μm and (d) of 1.5 μm.

layer at the surface in which it is generated with the laser light. One obtains for the height h of the displacement

$$h = \alpha\, L\, \Delta T_{av} \,. \tag{1}$$

α stands for the thermal expansion coefficient, ΔT_{av} for the average temperature rise and L for the thickness of the heated surface layer, which can be approximated by the penetration depth of the light. The temperature change ΔT is proportional to the deposited amount of heat and therefore to the laser pulse energy. Since the amplitude of the measured signal is proportional to the height of the displacement h, a linear relationship between the maximum amplitude of the time-resolved signal and the laser pulse energy is obtained. In our experiments this has been verified for a wide range of temperature changes. For stainless steel, for example, we find that the linear relationship holds from very low laser pulse energies up to the damage threshold of the sample. Preconditions are that the thermal expansion coefficient remains constant in the considered temperature interval and that L is known. One can then correlate the measured signal in mV with the temperature rise in Kelvin averaged over the total thickness of the layer which contributes to the surface deformation. For most surface studies, on the other hand, only the temperature in the first few layers is of relevance. If we again consider only the instant at which the energy is deposited (i.e. maximum amplitude of the signal), the temperature rise in the heated layer decays exponentially as a function of distance from the surface and one can extrapolate the true surface temperature with

$$\Delta T_{surface} = \Delta T_{av} / [1 - 1/e] \,. \tag{2}$$

In certain cases a correction for the different thermal expansion coefficient at the surface as compared to the bulk may also be necessary. It should be emphasized, however, that the temperature determination described here is only valid for the maximum amplitude of the thermoelastic deformation. As soon as the heat starts to diffuse into the bulk, the situation becomes increasingly complicated, particularly because the deformation is no longer Gaussian (see above) and the thickness of the heated layer L is not constant. Therefore, theoretical calculations of the spatial and temporal temperature distribution at the surface are presently carried out for the boundary conditions of our experiment. For this purpose the equations for thermoelastic surface deformation are solved numerically. This should make it possible to calibrate also the temperature variation during the decay of the signal and thus use the technique as an ultrafast thermometer for surface temperatures. A preliminary comparison of the time-dependent signal with calculations of laser induced temperature changes has been made already and shows that the results of Bechtel [12] provide a good description of transient surface temperatures.

An alternative way to calibrate the displacement signal is the observation of a phase transition in the heated surface layer. Melting, for example, occurs at a well known temperature which can be used to find a relation between the signal amplitude and the temperature rise. In our experiments such a calibration could be obtained by measuring the maximum signal amplitude for an aluminum sample as a function of laser pulse energy. In this case, we do not find a linear dependence over the whole temperature range (as for stainless steel), but observe a plateau if the material starts to melt at about 660 ^{0}C. Further details will be given elsewhere [13].

3.2 THIN FILM STUDIES

Thin film studies have been performed e.g. with copper and stainless steel substrates. For this purpose silver was deposited in situ on the surface. The shape of the time-resolved photothermal displacement was monitored as a function of film thickness. Even in the submicrometer range dramatic changes as compared to clean surfaces are observed. For the stainless steel substrate (see Fig. 2), for example, we find that the more the film thickness increases the more rapidly the signal decays. This reflects the higher thermal diffusivity of Ag as compared to the substrate. If the film approaches a thickness in the micrometer range, the thermoelastic signal has converged into that of a bulk silver sample.

4. REFERENCES

1. P.E. Nordal, S.O. Kanstadt, Physica Scripta **20**, 659 (1979)
2. F. Träger, H. Coufal, T.J. Chuang, Phys. Rev. Lett. **49**, 1720 (1982)
3. M.A. Olmstead, N.M. Amer, S. Kohn, D. Fournier, A.C. Boccara, Appl. Phys. **A32**, 141 (1983)
4. C. Karner, A. Mandel, F. Träger, Appl. Phys. **A 38**, 19 (1985)
5. H. Coufal, Appl. Phys. Lett. **44**, 59 (1984)
6. F. Träger, H. Coufal, T.J. Chuang in "Surface Studies with Lasers", Springer Series in Chemical Physics Vol. **33**, p. 110, Springer Verlag, Berlin, Heidelberg, New York, Tokyo, F.R. Aussenegg, A. Leitner, M.E. Lippitsch, eds. (1983)
7. G. Rousset, L. Bertrand, P. Cielo, J. Appl. Phys. **57**, 4396 (1985)
8. A. Rosencwaig, J. Opsal, D.L. Willenborg, Appl. Phys. Lett. **43**, 166 (1983)
9. H. Coufal, F. Träger, T.J. Chuang, A.C. Tam, Surface Science **145**, L504 (1984)
10. H. Coufal, W. Lee in "Laser Processing and Diagnostics", Springer Series in Chemical Physics, Vol. **39**, p. 25, Springer Verlag, Berlin, Heidelberg, New York, Tokyo, D.Bäuerle ed. (1984)
11. I. Hussla, H. Coufal, F. Träger, T.J. Chuang, Ber. Bunsenges. Phys. Chem. **90**, 240 (1986)
12. J. Bechtel, J. Appl. Phys. **46**, 1975 (1975)
13. J. Fischer, Q. Kong, F. Träger, to be published

Laser Energy Deposition at Sapphire Surfaces Studied by Pulsed Photothermal Deformation

R.W. Dreyfus, F.A. McDonald, and R.J. von Gutfeld

IBM T.J. Watson Research Center, P.O. Box 218,
Yorktown Heights, NY 10598, USA

Abstract

A surface deformation technique measures the energy deposited in sapphire surfaces irradiated by focused excimer laser light. The absorbed energy creates $\lesssim$3 electron-hole pairs/oxygen atom, consistent with a photochemical etching mechanism.

Photoacoustic and photothermal techniques have been used for some time to study optical absorption in condensed matter. Recently, the interaction of pulsed and CW laser radiation with solids has been studied through the resulting surface deformation in and near the illuminated region. Detection of the surface displacement has been made by interferometric means; alternatively, the surface slope has been detected through the deflection of a probe beam incident on the sample surface[1].

The ability to measure optical energy absorption makes it possible to resolve questions of energy deposition mechanisms associated with the excimer laser excitation of surfaces just below fluences required for finite etching. The area of interest for the present experiments includes those processes leading to the removal of monolayers with the objective of clarifying processes that occur when microns of material are removed. One of the basic problems in interpreting laser etching has always been knowledge of the surface temperature. For temperatures sufficiently high to generate vapor pressures $\gtrsim$1 bar, the etching can proceed by a purely thermal process. On the other hand, if the temperature during etching barely exceeds room temperature, direct photochemical decomposition is responsible. Ref. 2 describes surface temperatures determined by laser induced fluorescence (LIF), a technique sensitive to diatomic radicals ejected during UV excimer laser etching of the surface. While these results are informative, they do not complete the picture in two areas: collisional effects in the etch plume could only be deduced to be insignificant, and optical energy absorption mechanisms which lead to etching could not be identified from these experiments.

To overcome these limitations, the present pulsed photothermal deformation (PPTD) studies were undertaken, consisting of two parts: first, a rigorous three dimensional, thermoelastic calculation of PPTD is obtained, and second, quantitative experiments on sapphire and other surfaces are described which give values for the energy absorbed in the surface layers. The calculated results were obtained by applying standard thermoelastic theory for isotropic solids to the case of an infinite plate of finite thickness; Laplace and Hankel transform techniques with numerical inversion were used[3]. Detection of photothermal deformation utilized a He-Ne laser beam (0.2 mm dia) focused onto the shoulder of the excimer irradiated area(~0.4mm dia)[4]. Angular deflection of the probe beam was detected by means of a split silicon photodetector. Sensitivity was limited by the high frequency fluctuation in the He-Ne beam intensity and found to be equivalent to ~50 nJ of absorbed energy on the surface.

The theoretical results identify three regimes in the temporal behavior of the near-field surface deformation: 1) local thermal expansion during the pulse, followed by 2) deformation due to propagating surface and bulk elastic pulses, followed or accompanied by 3) "plate motion", that is, bending motion of the sample as a whole (this regime being absent for thick samples) [4]. The importance of these results is the following: The plate motion, 3), gives 10 to 100 times larger signals than the other two modes for thin ($<100\mu$ thick) samples. Furthermore, the plate motion is primarily sensitive to surface energy deposition and hence is the mode of interest; small bulk absorption produces negligible plate motion. Also, quantitative measurement of unknown absorption, for example, on a sapphire surface, can be calibrated in terms of a material with known optical, thermal and mechanical properties, namely a silicon wafer surface.

The primary objective of the present experiments is to obtain a better understanding of the UV energy absorption mechanisms at a sapphire surface just below etching threshold [5]. Our results indicate that 3 x 10^{-2} of the incident energy is absorbed by each surface of the as-received commercial sapphire with 193nm laser pulses. It is also found that the absorption is independent of laser fluence. The absorption drops off to 10^{-2} at 248nm and to 10^{-3} at 351 nm. High temperature annealing followed by hot phosphoric/chromic acid etch reduced the 193nm absorption to approximately 4 x 10^{-3}. The effect of this surface treatment is consistent with a reduction in surface states which give rise to surface absorption and confirms that we are observing a surface rather than bulk absorption. The relatively large absorption at 193 nm presumably accounts for the ease of etching sapphire with excimer pulses at this wavelength. Furthermore, the independence of absorption from fluence indicates that a one photon ($h\nu$) absorption is the controlling step. We find no evidence for a 'true' multiphoton surface effect. Stepwise excitation with two or more photons would agree with the present results as long as one particular step was the controlling absorption. The PPTD results agree well with other observations on UV etching of sapphire. Using a thermal diffusivity of $0.08 cm^2/s$, and hence an effective heated layer of $0.57 \mu m$, we estimate the temperature increase to be ~130K just <u>prior</u> to ablation. This value extrapolates smoothly into the 200 to 300K rises found [2] with LIF measurements just <u>above</u> threshold fluence. These relatively small temperature increases reinforce the concept that short wavelength (193nm) UV etching of sapphire occurs primarily by a photochemical rather than a thermal vaporization process.

Having determined a value for the energy deposited in the surface region permits us to determine an upper limit for the density of electrons excited into the conduction band. The 3% absorption amounts to $18 mJ/cm^2$ at etching threshold. Assuming the aforementioned 2-photon sequential absorption so that ≥ 12.8 eV ($= 2h\nu$) are required to form an electron-hole (or exciton) pair, the above energy density represents $\lesssim 2.2 \times 10^{23}$ e-h pairs/cm^3. The largest unknowns in the calculation are the depth over which the pairs are distributed and the (additional) energy required per pair when their density becomes large. We use a depth which is only an order of magnitude estimate, namely 0.4nm, i.e. about 2 monolayers. This implies (neglecting recombination) $\lesssim$10 electron-hole pairs/ Al_2O_3 molecule or $\lesssim$3 pairs/O atom. Considering that recombination and other losses in the excitation energy undoubtedly occur, a density approaching 3 e-h pairs/anion appears more than sufficient to produce photochemical etching by UV laser light.

In summary, a PPTD technique has allowed us to follow the linear absorption of UV energy on sapphire surfaces. Quantitative comparison using a Si standard allows us to say that the surface temperature rise was $\lesssim$130K, while an implied exciton density is in the range sufficient to produce photochemical etching.

1. S. Ameri, E. A. Ash, V. Neuman and C. R. Petts, Electron Lett. **17**, 337(1981), M. A. Olmstead, A. M. Amer, S. Kohn, D. Fournier and A. C. Boccara, Appl. Phys. **A32**, 141 (1983), and C. Karner, A. Mandel and F. Träger, Appl. Phys. **A38**, 19 (1985).
2. R. W. Dreyfus, Roger Kelly, R. E. Walkup and R. Srinivasan, Proc. Soc. Photo Opt. Instru. Eng. **710**, 46 (1987), and R. W. Dreyfus, R. Kelly and R. E. Walkup, Appl. Phys. Lett., **49**, 1478 (1986).
3. F. Alan McDonald, R. J. von Gutfeld and R. W. Dreyfus, Proc. IEEE 1986 Ultrasonics Sym. (1986), p. 403.
4. R. J. von Gutfeld, F. A. McDonald and R. W. Dreyfus, Appl. Phys. Lett. **49**, 1059 (1986).
5. R. W. Dreyfus, F. A. McDonald and R. J. von Gutfeld, Appl. Phys. Lett. **50**, 1491 (1987) and J. Vac. Sci. Tech. B, to be published Sept./Oct. (1987).

Adsorption-Desorption at a Solid Surface Detected by the Mirage Effect

F. Lepoutre[1], *G. Rousset*[2], *V. Plichon*[3], *and N. Rollat*[3]

[1]E.S.P.C.I., Laboratoire d'Optique, 10, rue Vauquelin, F-75231 Paris Cedex 05, France
[2]O.N.E.R.A., 20, avenue de la Division Leclerc, F-92320 Châtillon, France
[3]E.S.P.C.I., Laboratoire de Chimie et Electrochimie des Materiaux Moléculaires, 10, rue Vauquelin, F-75231 Paris Cedex 05, France

It has been known for several years that the photoacoustic effect is able to detect evaporation and sorption phenomena induced by a periodic heating. The first goal of such studies was to artificially enhance small photoacoustic signals obtained with weak excitations /1/. Indeed, in the generation of the photoacoustic pressure an "acoustic piston" must be added to the "thermal piston" (heat diffusion from sample to gas) to describe the oscillating thickness of the thin layer adsorbed on the sample (adsorption-desorption). Two theoretical models /2/ have been proposed to interpret these photoacoustic experiments. Nevertheless the photoacoustic detection is not directly sensitive to the main feature of the phenomenon, i.e. the mass transfer between the solid sample and the gas. In this paper we will show that the mirage effect may be a useful tool for that kind of investigation.

1. THEORETICAL MODEL

We consider the three media of the figure 1 : a solid sample (s), a liquid film (l) and a gas (g) in which the mirage detection will be achieved. We suppose that a volatile component B at saturation is introduced in air. The molecules of air are called A. At the steady state, the number of molecules of A, B per unit volume are called C_A, C_B and we note $C = C_A + C_B$.

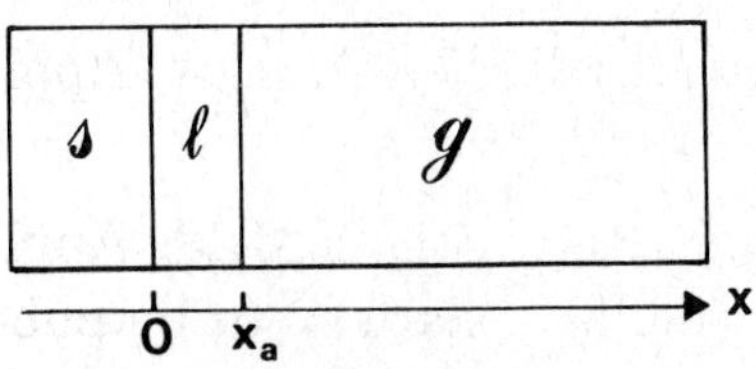

Figure 1 - Theoretical model : s : Sample ; l : liquid film ; g : gas.

If the steady temperature of s is not higher than the vaporization temperature of B a liquid film is formed on s. We irradiate periodically and assume that the radiative flux per unit surface, I_0, is absorbed at the surface of s (l is assumed transparent).

It is well known that the mirage effect is only sensitive to the thermal modes of the periodic temperature so that it is only necessary to solve ordinary thermal diffusion equations in the three media /3/. The complex amplitude, $\widetilde{T_i}$, of the periodic temperatures is then given by

$$\tilde{T}_s = A_s\, e^{(1+j)x/\mu_s}, \tag{1}$$

$$\tilde{T}_1 = A_1\, e^{(1+j)x/\mu_1} + B_1 e^{-(1+j)x/\mu_1}, \tag{2}$$

$$\tilde{T}_g = A_g\, e^{-(1+j)(x-x_a)/\mu_g} . \tag{3}$$

A_i, B_i are determined by the boundary conditions

$$A_s = A_1 + B_1, \tag{4}$$

$$\lambda_s \mu_s^{-1} = \lambda_1 \mu_1^{-1} (A_1 - B_1) + \frac{I_o}{1+j}, \tag{5}$$

$$A_g = A_1 d^+ + B_1 d^-, \tag{6}$$

$$\lambda_1 \mu_1^{-1} (A_1 d^+ - B_1 d^-) + \lambda_g \mu_g^{-1} A_g + \frac{\tilde{n}(x = x_a)L}{1 + j} = 0 . \tag{7}$$

μ_i is the thermal diffusion length of medium i ($\mu_i = (\alpha_i/\pi f)^{1/2}$, α_i thermal diffusivity), $d^{\pm} = \exp \pm (1+j)x_a/\mu_1$ and $\tilde{n}(x = a)L$ is the heat flux at $x = x_a$ produced by the flux of B at $x = x_a$ ($\tilde{n}(x = x_a)$) due to the vaporization of 1 (L heat of vaporization per mole). This flux $\tilde{n}(x = x_a)$ is related to the diffusion constant, D, of B through A by /4/

$$\tilde{n}(x = x_a) = -D/(1 - C_B/C)\partial\tilde{C}_B/\partial x(x = x_a). \tag{8}$$

The periodic gradient of B, $\partial\tilde{C}_B/\partial x$, in the gas g has the same form as the periodic temperature gradient since it is governed by the same equation of diffusion (α_g replaced by D) :

$$\frac{\partial\tilde{C}_B}{\partial x} = -(1+j)/\mu_D\, \tilde{C}_B(x=x_a)\quad e^{-(1+j)(x-x_a)/\mu_D} \tag{9}$$

with $\mu_D = (D/\pi f)^{1/2}$. (10)

The number of molecules B at $x = x_a$ is deduced from the ideal gas law and the Clapeyron law :

$$\tilde{C}_B(x = x_a) = (A_g C_B/T)\ (L/RT - 1), \tag{11}$$

so that equation (7) can be written

$$\lambda_1\, \mu_1^{-1}(A_1 d^+ - B_1 d^-) + (\lambda_g \mu_g^{-1} + \lambda_D \mu_D^{-1})\, A_g = 0. \tag{12}$$

We have called λ_D the quantity

$$\lambda_D = D(C_B/(1 - C_B/C))\ (L/T)\ (L/RT - 1). \tag{13}$$

A_g (the unknown of interest for the mirage detection) is then deduced from the system (4), (5), (6), (12) :

$$A_g = \frac{I_o}{1+j}\left[d^+(\lambda_s\mu_s^{-1}+\lambda_1\mu_1^{-1})+d^-(\lambda_s\mu_s^{-1}-\lambda_1\mu_1^{-1})+\left(d^+\left(1+\frac{\lambda_s\mu_s^{-1}}{\lambda_1\mu_1^{-1}}\right)+d^-\left(1-\frac{\lambda_s\mu_s^{-1}}{\lambda_1\mu_1^{-1}}\right)\right)(\lambda_g\mu_g^{-1}+\lambda_D\mu_D^{-1})\right]^{-1}. \tag{14}$$

If $x_A \ll \mu_1$, d^+ and d^- are almost equal to 1, (13) becomes

$$A_g = I_o\left[2(1+j)\lambda_s\mu_s^{-1}\left(1 + \frac{\lambda_g\mu_g^{-1}}{\lambda_s\mu_s^{-1}} + \frac{\lambda_D\mu_D^{-1}}{\lambda_s\mu_s^{-1}}\right)\right]^{-1} . \qquad (15)$$

(15) or (14) clearly show that the periodic temperature of g depends on the diffusion of B through A, and on the heat of vaporization L.

The mirage deflection, for a length l of interaction, is given by

$$\emptyset_n = \left[\left(\frac{1}{n}\frac{\partial n}{\partial T}\right) \frac{\partial \tilde{T}_g}{\partial x} + \left(\frac{1}{n}\frac{\partial n}{\partial C_B}\right) \frac{\partial \tilde{C}_B}{\partial x} \right] l , \qquad (16)$$

which can be written with the help of (3) and (9), (11) as

$$\emptyset_n = -(1 + j)\, A_g \left[\left(\frac{1}{n}\frac{\partial n}{\partial T}\right) \mu_g^{-1}\, e^{-(1+j)(x-x_a)/\mu_g} + \ldots\right.$$

$$\left.\ldots + \left(\frac{1}{n}\frac{\partial n}{\partial C_B}\right) \mu_D^{-1}\, \frac{C_B}{T}\left(\frac{L}{RT} - 1\right) e^{-(1+j)(x-x_a)/\mu_D} \right] . \qquad (17)$$

The constants $\left(\frac{1}{n}\frac{\partial n}{\partial T}\right)$ and $\left(\frac{1}{n}\frac{\partial n}{\partial C_B}\right)$ can be calculated from Gladstone's law and the additivity of refractivities in a mixture /5/ :

$$\frac{1}{n}\frac{\partial n}{\partial T} = -\frac{n-1}{T} \quad ; \quad \frac{1}{n}\frac{\partial n}{\partial C_B} = \frac{n_B - 1}{C_B} , \qquad (18)$$

where n is the refractive index of the mixture, and $\frac{n_B - 1}{C_B}$ is the refractivity of gas B (assuming $n_B \approx 1$).

The opposite signs of $\frac{1}{n}\frac{\partial n}{\partial T}$ and $\frac{1}{n}\frac{\partial n}{\partial C_B}$ lead to a phase shift of π between the two terms of (17) when the exponentials are close to 1.

2. EXPERIMENT

The figure 2 shows the experimental set-up. Two kinds of pump beams were used : a lamp and an Ar^+ laser. For the same powers the results obtained are similar.

Before the introduction of the molecules B, measurements of $\emptyset_n$ (amplitude and phase) are performed to determine the relative position of the probe beam and the sample surface. These measurements provide also the value of the thermal diffusivity of pure air ($2\ 10^{-5}\ m^2\ s^{-1}$). At time t = 0, a sufficient quantity of B is introduced in the closed cell to maintain at room temperature T a liquid-vapour equilibrium and $\emptyset_n$ (amplitude and phase) is recorded by a microcomputer versus time (see fig. 3). A decrease of $\emptyset_n$ is first observed (it corresponds to the cooling of g produced by the vaporization of B), then a strong but slow increase of the mirage deflection occurs. The signal goes through a maximum before reaching an asymptotic value. The enhancement E (ratio of amplitudes at $t\ \infty$ and $t < 0$) is depending upon the nature of B, the distance z of the probe beam from the sample surface, the modulation frequency and the power of the pump beam. A much more rapid variation of the phase of $\emptyset_n$ ($\Delta\Psi = \Psi(t\ \infty) - \Psi(t < 0)$) of approximately 180° appears. The asymptotic value of $\Delta\Psi$ depends on the same parameters as E. We limit the following discussion to the asymptotic behaviours ($t\ \infty$) observed with B = CH_2Cl_2 and a TiO_2 solid sample.

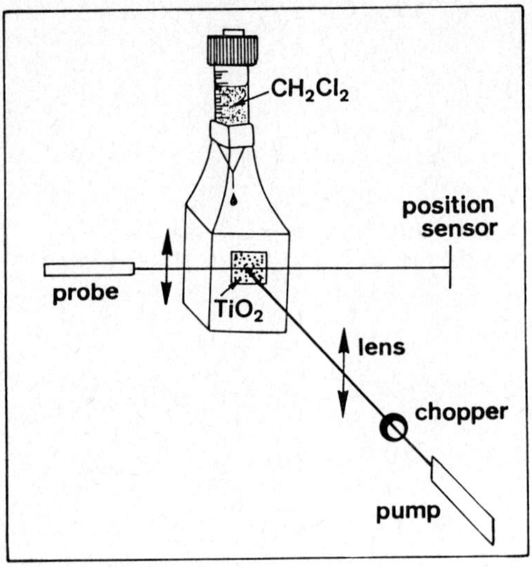

Figure 2 - Schematic experimental setup.

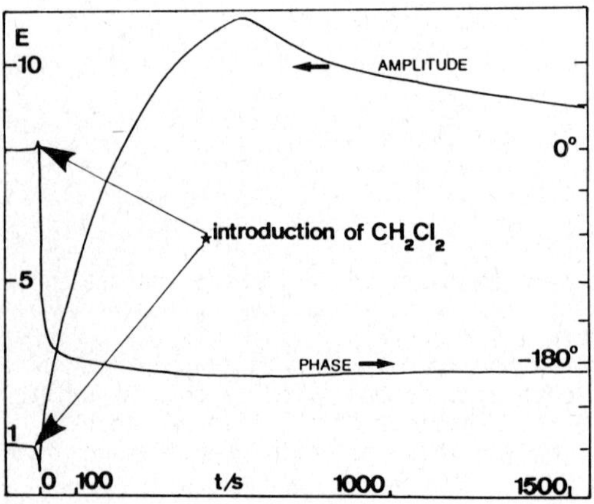

Figure 3 - Amplitude and phase of $\emptyset_n$ vs time t.

i) At power of the pump beam (I_o < $1 W/cm^2$), at z = 150 μm and f = 236 Hz we have obtained E ≈ 8 and $\Delta\Psi$ = 162°. This last value close to π means that in equation (17) the second term dominates. To know the value of D we have recorded the variations of the phase shift of $\emptyset_n$ vs $\sqrt{f}$. Figure 4 shows a linear dependence, the slope of the straight line gives D = 7 10^{-6} m^2 s^{-1}.

ii) At high power of the pump ($100 W/cm^2$) and in the same experimental conditions as i, no increase of the signal is observed and a very small phase shift $\Delta\Psi$ < 10° occurs). The same results are obtained when a continuous beam of 100 W/cm^2 unmodulated) is superimposed on the low power modulated pump beam (1 W/cm^2). That means that under experimental conditions (ii) we have increased the steady temperature of s enough to prevent the condensation of B on s. Once again, we have recorded the phase shift of $\emptyset_n$ vs $\sqrt{f}$. The straight line observed corresponds to the first term of (17) and gives α_g = 1.3 10^{-5} m^2/s (see Fig. 4).

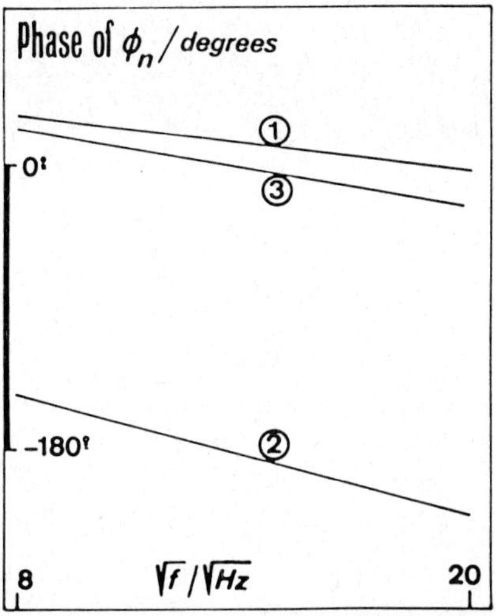

Figure 4 - Phase of $\emptyset_n$ vs $\sqrt{f}$. ① Pure air ; ② Air + CH_2Cl_2, pump beam power I_o = 1 W/cm^2 ; ③ Air + CH_2Cl_2, pump beam power I_o = 100 W/cm^2.

These numerical values are in rather good agreement with D, α_g, n, n_B and the thermal properties of B and s calculated or measured /6/ : we have deduced from the data E $\approx$ 10, α_g $\approx 10^{-5}$ m^2 s^{-1}, and D $\approx 10^{-5}$ m^2 s^{-1}. We want to emphasize that these calculations are only estimates since, to our knowledge, no precise measurements of such transport properties have been published. Nevertheless, the good orders of magnitude obtained, the phase shift $\Delta\Psi$ close to π, at low values of I_o, the disappearance of the enhancement E as the steady temperature of s increases prove that the phenomenon is a periodic vaporization of l as previously described by Korpiun and Srinivasan /2/. The mirage effect is able to detect directly this vaporization and, as shown by equation/17/, the behaviour of the film. We think that the study of the variations of $\emptyset_n$ vs t have potential applications too.

REFERENCES

1. T. Somasundaram, P. Ganguly, J. Phys. (Paris) 44, C6-239 (1983).
2. P. Korpiun, Appl. Phys. Lett. 44, 675 (1984).
   J. Srinivasan, R. Kumar and K.S. Gandhi, Applied Physics B 43, 35 (1987).
3. J.P. Roger and F. Lepoutre, J. Phys. (Paris) 44, C6-35 (1983).
4. R.B. Bird, W.E. Stewart and E.N. Lightfoot, "Transport Phenomena", p. 522, Wiley International Edition (New York, 1960).
5. M. Born and E. Wolf "Principles of Optics", p. 86, Pergamon Press (London, 1959).
6. See ref. 4, chapter 8.3, Lange's Handbook of Chemistry, McGraw-Hill, New York 1956, and CRC Handbook of Chemistry and Physics, CRC Press West Palm Beach (1978).

Detection of Adsorbed Gases by Mass Diffusion Waves

P. Korpiun, R. Osiander, B. Büchner, and W. Herrmann

Physik Department E13, TU München, James-Franck-Straße, D-8046 Garching, Fed. Rep. of Germany

The photoacoustic signal of porous samples in gas microphone cells may be relatively large. To a certain extent this phenomenon can be explained by light scattering /1/, the thermal properties of the sample /2/, and the intensive thermal coupling of the sample to the interstitial gas /3/. To explain measurements on charcoal /4/ with various transducer gases, temperature-dependent adsorption has to be taken into account.

We consider the simultaneous transport of heat and mass from a microporous sample ($z \leq 0$) to the gas ($0 \leq z \leq l_g$) and vice versa in z-direction. Sample (s) and gas (g) are assumed to be thermally thick. The light is absorbed at the boundary $z = 0$ of the opaque sample. The periodically varying transport of heat between sample and gas is described by the heat waves

$$\Phi_s(z,t) = \Theta \exp[i(k_s z+\omega t)] \text{ and } \Phi_g(z,t) = \Theta \exp[-i(k_g z-\omega t)]. \tag{1}$$

Here $\omega/2\pi$ is the modulation frequency, $k_n=(1-i)a_n$ the complex thermal wave number of medium labelled by n with $a_n = \sqrt{(\omega/2\alpha_n)}$, where $\alpha_n = \lambda_n/(\rho_n c_{pn})$ is the thermal diffusivity, λ_n the thermal conductivity, ρ_n the density and c_{pn} the isobaric specific heat capacity. The temperature-induced transport of mass in the sample and gas can be described by the concentration waves

$$\Psi_s(z,t) = C_s \exp[i(\xi_s z+\omega t)], \; \Psi_g(z,t) = C_g \exp[-i(\xi_g z-\omega t)]. \tag{2}$$

They are solutions of the diffusion equation in the medium n with the complex concentration wave number $\xi_n = (1-i)\sqrt{(\omega/2D_n)}$. D_s is the mass diffusion coefficient of the sample. $D_g = D/(1-P_s/P_t)$ is the effective diffusion coefficient /5/ that takes into account diffusive and convective flow of mass in the gas if D is the mutual diffusion coefficient, P_s the partial pressure of the gas component that is adsorbed, and P_t is the total pressure. At the sample to gas boundary $z = 0$ there hold following boundary conditions:

1. Temperature: $\Phi_s(0,t) = \Phi_g(0,t)$
2. Heat flow: $\lambda_s \mathrm{grad}\Phi_s(0,t) - \lambda_g \mathrm{grad}\Phi_g(0,t) - hD_g \mathrm{grad}\Psi_g = (Q_o/2)\exp(i\omega t)$ with the differential heat of adsorption h, and the heat Q_o supplied by light absorption to the sample
3. Concentration: $\Psi_s(0,t) = \frac{\rho_s}{M}[(\frac{\partial X}{\partial T})_P \Phi_g(0,t) + RT(\frac{\partial X}{\partial P})_T \Psi_g(0,t)]$, where M is

the molecular weight of the adsorbate and X the adsorbed mass at unit mass adsorber

4. Mass flow: $-j = D_s \mathrm{grad}\Psi_s(0,t) = D_g \mathrm{grad}\Psi_g(0,t)$.

Using the Clausius-Clapeyron relation for adsorption, $(\frac{\partial X}{\partial T})_P = -(\frac{\partial X}{\partial P})_T P_s h/RT^2$, the temperature and concentration amplitude is for $\lambda_s a_s \gg \lambda_g a_g$ and $D_s \ll D_g$

$$\Theta = \frac{Q_o}{2\sqrt{2}\lambda_s a_s} [1 + \frac{\sqrt{\alpha_s D_s} h^2 P_s \rho_s}{\lambda_s RT^2 M} (\frac{\partial X}{\partial P})_T]^{-1}, \tag{3}$$

$$C_g = \sqrt{D_s/D_g} \frac{P_s}{RT^2} \frac{\rho_s}{M} (\frac{\partial X}{\partial P})_T \Theta . \tag{4}$$

With the temperature and concentration waves (1) and (2) averaged along the gas length l_g, the relative pressure variation becomes

$$\frac{p(t)}{P_t} = \frac{\Theta \sqrt{\alpha_g}}{\sqrt{\omega} l_g T} [1 + \sqrt{D_s/\alpha_g} h \frac{\rho_s}{M} (\frac{\partial X}{\partial P})_T] . \tag{5}$$

Experimental data measured by WONG /4/ for He, H_2, Ar, N_2, and CO_2, on activated charcoal, referred to N_2, are listed in table 1. The results of our measurements on Degu Sorb (Degussa) are added. All measurements were performed at T = 300 K and 50 Hz. The results of both measurements agree very well and are within experimental errors consistent with amplitudes calculated with (5), using data for h and $(\partial X/\partial P)_T$ from literature /6/ and a Knudsen diffusion coefficient $D_s = 7.2\times10^{-3}/\sqrt{M}\ cm^2/s$ /7/.

Table 1: PA-amplitude of charcoal for several gases normalized to N_2

	He	H_2	Ar	N_2	CO_2	$CClF_3$
calcul. (5)	0.73±0.2	0.64	(0.51)	1	1.8	
exp. Wong /4/	0.81	0.71	1.12	1	1.96	
exp. this work	0.65±0.05		0.98	1	2.0	2.0

1. P. Helander, J. Lundström, D. McQueen: J. Appl. Phys. 51, 3841 (1980)
2. J. P. Monchalin, L. Bertrand, G. Rousset, F. Lepoutre: J. Appl. Phys. 56, 190 (1984)
3. S. J. McGovern, B. S. H. Royce, J. B. Benziger: J. Appl. Phys. 57, 1710 (1985)
4. K. Y. Wong: J. Appl. Phys. 47, 64 (1976)
5. P. Korpiun, W. Herrmann, R. Osiander: this issue
6. Landolt-Börnstein: Eigenschaften der Materie in ihren Aggregatszuständen, Vol.II, part 5a, ed. by K. Schäfer and E. Lax (Springer, Berlin, 1969)
7. A. L. Hines, R. N. Maddox: Mass Transfer (Prentice Hall, Englewood Cliffs, N. J. 1985)

Photoacoustic Analysis of Liquid Surface Films on Liquids

M.W. Sigrist and D. Scherrer

Institute of Quantum Electronics, ETH, CH-8093 Zürich, Switzerland

1. Introduction

In a recent review we have discussed the photoacoustic (PA) generation of plane acoustic waves in strongly absorbing or opaque liquids by pulsed laser radiation [1]. Good agreement with respect to acoustic waveforms and amplitudes has been found between experimental data and an analytical model for the two regimes of a free and a confined surface of the liquid. For the first time, we have performed laser-PA spectroscopy on opaque liquids as well as on absorbing liquid surface films spread on liquids, i.e. at liquid/liquid interfaces [2]. New spectroscopic results on solidified compared to liquid films of $\geq 0.5\ \mu m$ thickness present on the surface of water have been discussed recently [3]. Here we present a preliminary analytical model for PA studies on liquid surface films. Its predictions are compared with experimental data.

2. Theoretical Model

Strongly absorbing or opaque liquids permit the study of PA wave generation for different boundary conditions. As illustrated in Fig. 1 we have investigated the regimes of a free and a confined surface (realized by placing a transparent plate on the surface) as well as that of a liquid surface layer floating on the bulk liquid.

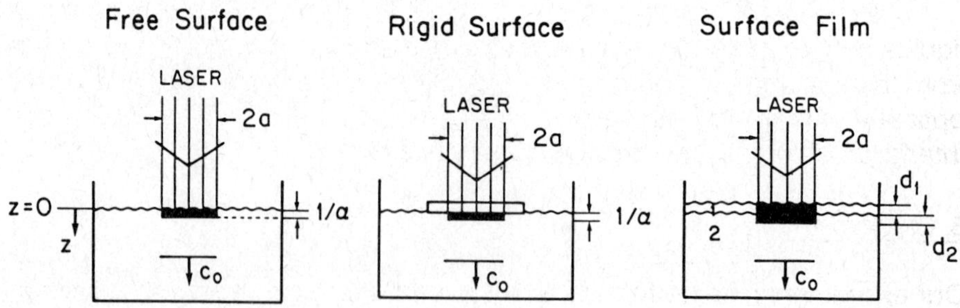

Fig. 1 Three different boundary conditions for the laser-liquid interaction

As indicated in Fig. 1 the absorption depth $1/\alpha$, where α represents the absorption coefficient of the liquid at the laser wavelength, is assumed to be considerably smaller than the diameter 2a of the unfocussed laser beam. For IR wavelengths, e.g. for CO_2 laser emission, this condition is fulfilled for many liquids because α usually greatly exceeds 10 cm^{-1}, i.e. $1/\alpha << 1$ mm $< 2a \cong 6$ mm. Thus, plane acoustic waves are generated which propagate through the liquid at sound speed c_0. With respect to the absorbing liquid surface layer with thickness d_1 and absorption coefficient α_1 only the case $d_1 < 1/\alpha_1$ is considered, i.e. the laser beam penetrates into the bulk liquid characterized by an absorption coefficient α_2 or absorption depth $d_2 = 1/\alpha_2$.

On the basis of the linearized hydrodynamical and heat-balance equations for a homogeneous isotropic medium we were able to derive an analytical solution for the acoustic signals which takes the actual temporal shape of the laser pulses into account. For laser pulse durations $\tau_L > 100$ ns, as used in our experiments, and liquids with $\alpha > 100$ cm^{-1}, the absorption length $1/\alpha$ is much shorter than the sound propagation distance during τ_L, i.e. $1/\alpha << c_0\tau_L$. In this case the peak pressures p^{max} for the three regimes of Fig. 1 can be approximated as follows:

$$p_f^{max} \propto \frac{\beta \cdot I_0}{\alpha \cdot \tau_R \cdot C_p} \qquad \text{for the free surface,} \qquad (1)$$

$$p_r^{max} = I_0 \cdot c_0 / C_p \qquad \text{for the rigid surface,} \qquad (2)$$

$$p^{max} = a \frac{\beta_2 I_0 e^{-\alpha_1 d_1}}{\alpha_2 C_{p_2} \tau_R} + b \frac{c_{01}^2 c_{02}^2 I_0 d_1 e^{-\alpha_1 d_1} \tau_R}{d_1 c_{01}^2 \rho_1 + d_2 c_{02}^2 \rho_2} \left[\frac{\rho_2 \beta_1 \alpha_1}{C_{p_1}} - \frac{\rho_1 \beta_2 \alpha_2}{C_{p_2}} \right] \qquad (3)$$

for the liquid layer (index 1) spread on the surface of a bulk liquid (index 2).

In (1-3) β represents the thermal expansion coefficient, C_p the specific heat and ρ the density of the liquid. I_0 is the laser intensity and τ_R the risetime of the laserpulse; a and b are constants.

Equation (3) for the regime of a liquid surface layer implies that both liquids contribute to the PA signal which is detected in the bulk liquid. Apart from thermal and mechanical parameters the PA signal also depends on the optical properties of both liquids. This result demonstrates that liquid/liquid interfaces can easily be investigated by pulsed PAS.

3. Experimental

Our experiments are performed with a hybrid CO_2 laser operating in a single longitudinal and TEM_{00} mode, step-tunable between 9.2 μm and 10.8 μm. The laser pulses of typically 140 ns halfwidth and 0.1 J energy are directed unfocussed onto the surface of the liquid in order to avoid its vaporization. The acoustic signals are detected in the bulk liquid by a home-made piezoelectric transducer of ≅ 5 ns risetime. The laser pulses and the corresponding acoustic transients are recorded by a LeCroy 9400 digital oscilloscope.

4. Results and Discussion

Excellent agreement between theory and experiment is obtained for the regimes of a rigid and free surface [1], whereas the preliminary calculation for the case of a liquid surface film only yields qualitative agreement as shown in Fig. 2.

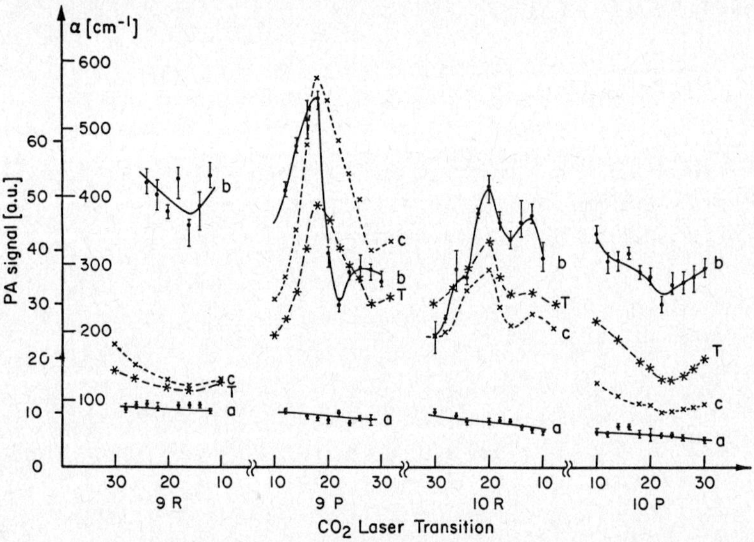

Fig. 2 Experimental (a,b), theoretical (T) and IR absorption (c) spectra for the 9R, 9P, 10R and 10P CO_2 laser branches: a -- PA spectra of H_2O with free surface, b -- PA spectra of H_2O with 8 μm thick surface film of medicated oil [2], T -- theoretical PA spectra obtained on the basis of (3) for case b, c -- IR absorption spectra of pure medicated oil (see scale for α)

The experimental PA (a,b) and IR spectra (c) of Fig. 2 have recently been discussed [2]. The enhancement of the PA amplitudes by the absorbing oil layer and the similarity between spectra b and c have been emphasized. Here we present in addition the predictions of our preliminary model for the regime of H_2O carrying a liquid surface film of medicated oil of a given thickness $d_1 = 8\ \mu m < 1/\alpha_1$, (spectra T). It is seen that the main features of the experimental PA spectra (b) are in agreement with the theoretical spectra yet the predicted absolute amplitudes are in general smaller than the measured ones. This discrepancy is not yet fully understood. It should be mentioned though that the condition $1/\alpha << c_0\tau_L$ is not fulfilled for the whole spectral range. Furthermore, the additional effect of thermal insulation at the liquid/liquid compared to a liquid/air interface has not yet been considered in this preliminary model.

In conclusion it is emphasized that the theoretically predicted spectral features of absorbing liquid films present on the surface of a bulk liquid are in agreement with experimental PA data. Thus, the noncontact pulsed PA technique permits an easy analysis of thin surface layers with thicknesses of $\geq 0.5\ \mu$m.

References

1. M.W. Sigrist: J.Appl. Phys. 60, R 83 (1986)
2. M.W. Sigrist, Z.H.Chen: Appl. Phys. B 43, 1 (1986)
3. M.W. Sigrist, Z.H. Chen, D. Scherrer: Proc. Eighth Int. Conf. on Laser Spectroscopy (EICOLS '87), Åre (S), June 22-26 (1987), to be published in Springer Ser. Opt. Sci. (Springer, Berlin, Heidelberg)

Transient Thermoreflectance of Thin Metal Films in the Picosecond Regime

A. Miklós and A. Lörincz

The Hungarian Academy of Sciences,
Department of Physics, Institute of Isotopes,
Budapest P.O. Box 77, Hungary H-1525

Picosecond transient thermoreflectance (PTTR) is a most appropriate tool for the studying of thermal transport properties of thin metal films [1]. Such investigations are needed since it is only in this form that many exciting materials like surface glazed metals or compositionally modulated structures can be produced. The advantage of the method is that in the picosecond regime the range of the heat diffusion process may be small in relation to film thicknesses, thereby allowing the experimenter to avoid problems due to the influence of the substrate on the heat transport.

PTTR is a pump-and-probe technique of high dynamic range [2]. It uses two picosecond pulses; the first to heat the sample, and the second, delayed one to detect reflection changes induced by the first pulse as a function of delay. The heat deposition is distributed, because the penetration depth for the light is about 20 nm. If this heat distribution is considered as the source term of the heat diffusion equation, the time dependence of the temperature at the surface and below may be calculated [3], see Fig.1.

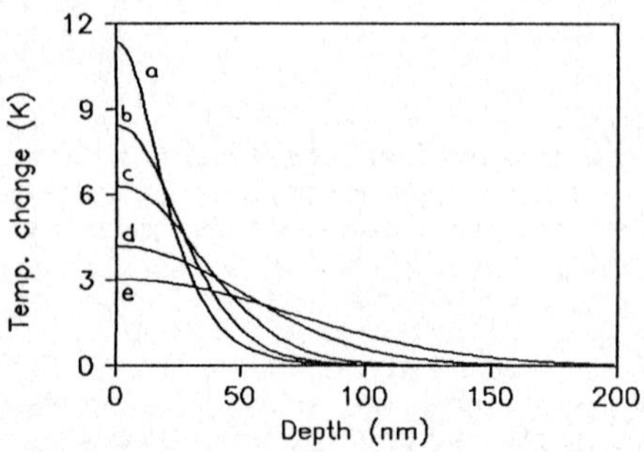

Fig. 1. Calculated temperature distributions for nickel sample at different delays after the arrival of the 5 ps heating pulse. Delays are 5, 10, 20, 50 and 100 ps for curves a,b,c,d and e, respectively.

Taking into account the temperature coefficient of the reflectivity and the temperature of the surface only, the transient reflectance is not given, as the distributed nature of the reflection process itself should be taken into account as well. According to the Maxwell theory of electrodynamics the reflected wave is generated in the inhomogeneous transition region between two media, where the gradients of the material parameters differ from zero. If the transition region is narrow compared with the wavelength, it may be considered as a discontinuity, and the solutions of the Maxwell equations may be matched with the help of boundary conditions.

However, in fast PTTR measurements the transient heating alters the material parameters in the region where the light is absorbed. The temperature gradient may be quite large, e.g. an increase of 10 K and penetration depth of 20 nm results in 5×10^8 K/m. The region where this gradient exists cannot be considered narrow since it is in the order of the wavelength in the material.

The reflection coefficient of a metal surface was calculated for normal incidence. The Maxwell equations of the problem may be reduced to a first-order 2x2 matrix differential equation [4]. This equation was solved for the homogeneous case, then the perturbation caused by the temperature dependence of the complex dielectric constant was taken into account. The details of the calculations are given elsewhere [5]. The results show that the temporal profile of the PTTR signal is strongly dependent on the temperature coefficients of the complex dielectric constant (Fig.2.).

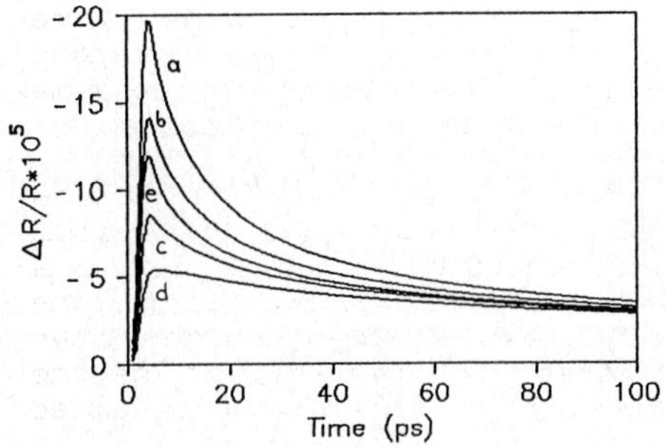

Fig. 2. Calculated transient thermoreflectance of nickel for different temperature coefficients

The values of the two unknown temperature coefficients were chosen to ensure a given value (-1×10^{-5}/K) for the temperature coefficient of the reflectivity. This value is characteristic for single-crystal nickel at the wavelength of 600 nm [6]. The complex dielectric constant assumes the form of $\varepsilon=\varepsilon_1(1+uT)+i\varepsilon_2(1+vT)$. (u,v) parameters in 10^5/K units were taken as (0.0;16.1), (-1.8;8.0), (-3.6;0.0), (-4.6;-4.6) for curves a,b,c,d respectively. Curve e shows the results of the calculation when the distributed nature of the reflectivity is neglected. (The FWHM of the heating pulse is 2.5 ps.) The difference between the reflectivity curves decreases as the temperature gradient disappears due to thermal diffusion. After 100-200 ps the curves are almost identical.

The distributed nature of the reflection should be considered for PTTR measurement of thermal diffusivity of thin films. The thermal diffusivity may be determined from the full curve taking into account the temperature coefficients of the optical constants, or the fitting should be restricted to times larger than 100 ps, thus spoiling the ability of this versatile method.

The actual values of parameters u and v may be determined by measuring the transient thermoreflectance as a function of the angle of incidence. Strong dependence was found for TM waves near the Brewster angle. The $\Delta R/R$ curves as a function of the angle of incidence are given in Fig.3 for different (u,v) pairs. The parameters for curves a,b,c,d are the same as for Fig.2. Curve e shows the results for the surface reflection model.

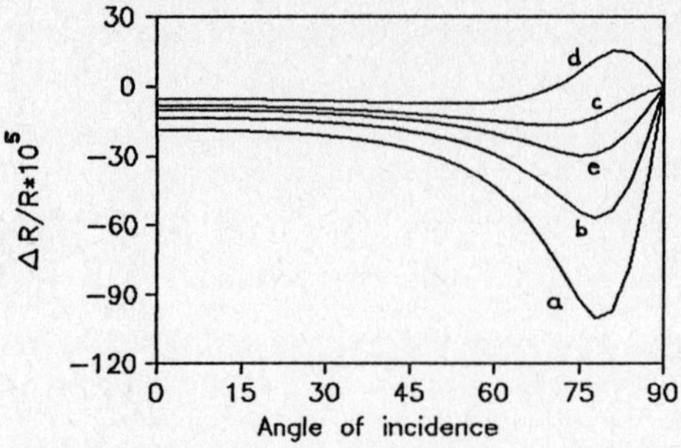

Fig.3. Calculated dependence of ΔR/R on the angle of incidence for different temperature coefficients at 5 ps delay.

References

1. C.A. Paddock and G.L. Eesley; J. Appl. Phys. 60 (1), 285-290 (1986)
2. L. Andor, A. Lörincz, J. Siemion, D.D. Smith and S.A. Rice; Rev. Sci. Instr. 55, 64 (1984)
3. J.H. Bechtel; J. Appl. Phys. 46, 1585 (1974)
4. M. Born and E. Wolf; Principles of Optics (Pergamon Press, Oxford 1975)
5. A. Miklós and A. Lörincz; submitted to J. Appl. Phys.
6. American Institute of Physics Handbook (McGraw-Hill, New York, 1972)

Energy Transfer from Energetic Ions to Thin Films Using a Pyroelectric Calorimeter

H. Coufal, H.F. Winters, and F. Sequeda

IBM Almaden Research Center, 650 Harry Road, San Jose, CA 95120, USA

The energy deposited by ion beams in metal films was measured with a pyroelectric calorimeter as a function of incident energy and the mass of ion and metal. The energy that an ion or neutral particle transfers to a lattice during a collision is important from the perspective of basic science and has significant impact on such diverse fields as thermonuclear fusion and deposition of thin films by physical sputtering. There has been no systematic experimental investigation of this phenomenon in the range of ion energies below a few keV. This is largely due to the difficulty that conventional ion scattering techniques encounter in making an accurate measurement of the kinetic energy of all the scattered ions (neutrals), a task that is extremely difficult for low energies. Photoacoustic or photothermal techniques that determine the energy deposited into the sample rather than the reflected energy of the particle should be able to overcome this problem.

Only a few photothermal techniques are compatible with the ultrahigh vacuum (UHV) requirements of an ion beam scattering experiment. Among these is a recently developed pyroelectric calorimeter which has been shown to combine UHV compatibility with utmost sensitivity [1]. It is particularly suitable for thin film samples and is the only one capable of coping with a surface roughness that increases during ion beam bombardment. Details of the calorimeter, the signal generation and detection process are described in full detail elsewhere [2,3].

The thin film material under study was deposited on top of the calorimeter which was subsequently mounted in a UHV system with a base pressure in the low 10^{-10} Torr range. The samples were cleaned by ion bombardment and the absence of impurities was demonstrated using Auger spectroscopy. Calibration of the calorimeter was accomplished by using a He-Ne laser to transfer a known amount of energy to the surface. Since the sensitivity of a pyroelectric calorimeter is a function of modulation frequency and the thermal and electronic time constants [4], it is imperative that the calibration and the measurements be accomplished under identical conditions. Therefore, the calibration was performed with calorimeter and sample in the UHV system using identical electronics by substituting the ion beam of interest with a modulated laser beam. The resulting photothermal signal from the pyroelectric calorimeter is amplified with a high input impedance, low noise preamplifier (Ithaco 1201) and detected with a two-phase lock-in-amplifier (PAR 5208). The incident laser power is determined with a power meter (Scientec 360203 head with 365 read out unit) which has been calibrated by electrical substitution heating using a power source (DataPrecision 8200) traceable to NBS standards. The transmission of the window and the reflectivity of the sample surface was determined in separate experiments to obtain a calibration with an overall accuracy of better than 5%. Measurements of the thermal diffusivity of the samples using a photothermal technique [4] showed that thermal wave attenuation in the samples is negligible considering the above quoted accuracy.

After the optical calibration, the calorimeter was exposed to an ion beam with the same modulation frequency. Due to the pronounced frequency dependence of the radiation induced thermal signal, it is crucial that the modulation frequencies during the calibration procedure and ion beam experiments are identical. A beat frequency technique was utilized to ensure that both frequencies agree within 0.02 Hz. The ion beam was not mass analyzed but the energy of the ionizing electrons was adjusted experimentally so that doubly charged ions were not generated. The ion beam with energies between 40 and 4000 eV had an energy spread of approximately 5 eV. The fact that the energy transfer from the ion beam to the calorimeter was found to be independent of pressure indicates that charge exchange collisions did not produce a significant number of energetic neutrals which subsequently collided with and transferred energy to the calorimeter. Ion bombardment of the sample did produce some changes in film properties during the course of the experiments. X-ray diffraction studies showed that the initial (111) preferential orientation was substantially reduced by ion bombardment. In addition, an increasing surface roughness was observed during the experiment slightly affecting the photothermal and the ion beam induced signals. At the conclusion of the experiment, to ensure that the calibration of the calorimeter is not affected by the sample changes, material loss or radiation damage, the complete optical calibration was repeated. No changes in the calibration were observed. The energy transfer per ion was calculated from the beam current, the ion energy and the previously determined calibration factor. It was found that the energy transfer is independent of the probe current in the range of 1 nA-100 nA utilized in our study.

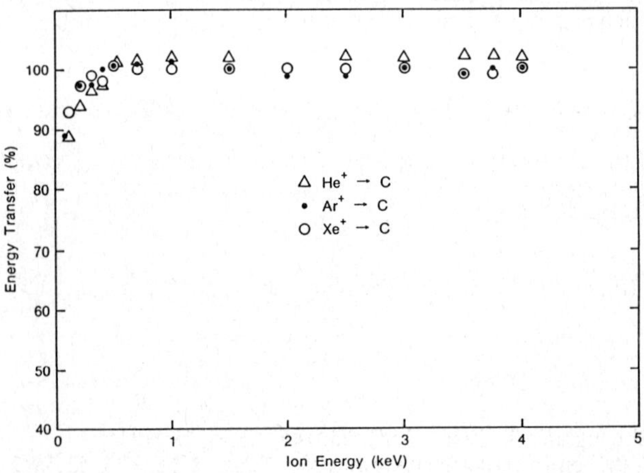

Fig. 1. Energy transfer from He^+, Ar^+ and Xe^+ ions to amorphous carbon as a function of ion energy.

Experimental results have been obtained for the transfer of energy from Xe^+, Ar^+ and He^+ to amorphous carbon and gold for ion energies between 40 eV and 4000 eV. Typical data for an amorphous carbon film with an estimated systematic error of less than 5% are shown in Fig. 1. Striking is the fact that the energy transfer is independent of the mass of the ion. This is, however, to be expected for a light target atom. For energies above a few hundred eV, essentially all of the ion energy is transferred to the lattice. This is expected theoretically based upon the assumption that energy is lost in a series of binary collisions with lattice atoms. However, for ion energies below 100 eV, a significant fraction of the energy is not transferred to the carbon lattice but is carried away by the reflected neutrals. The energy range between 100 eV and thermal energies is a transition region. Above approximately 100 eV, all of the incident ions penetrate into the sample and are thermalized; therefore, all of the energy is transferred. In contrast, atoms of Xe, Ar and He with thermal energies will all be reflected from the lattice with no energy being transferred. Figure 2 shows representative data for the energy transfer from He^+ to gold. Due to the larger mass of the target atoms, the scattering is partially elastic and hence no complete energy transfer is achieved. The above mentioned transition region extends in addition to higher ion energies.

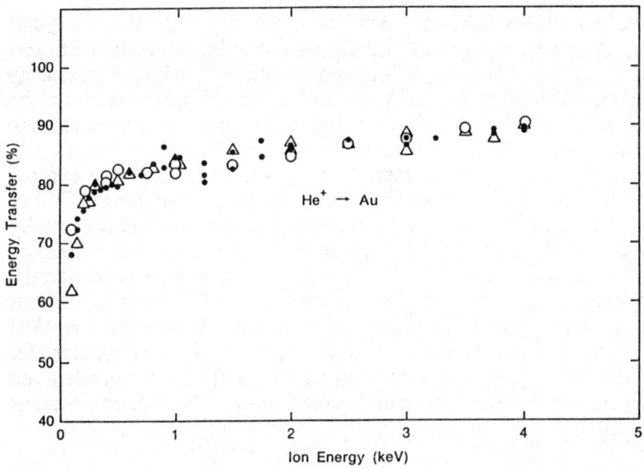

Fig. 2. Energy transfer from He$^+$ to a polycrystalline gold film as a function of ion energy. Shown are experiments with different samples and calorimeters.

The data shown were recorded with different calorimeters. The scattering of the data is within the above claimed accuracy showing that the underlying assumptions are actually justified. The absolute error of the measurements is estimated to be less than 10%. Further calibration procedures, which are expected to reduce the uncertainty in these results, and a more detailed analysis of the data are in progress.

Support by R. Grygier, T. Huang, G. Lim, D. Pearson and J. Salem is gratefully acknowledged.

References

1. H. Coufal, F. Träger, T. Chuang and A. Tam: Surf. Sci. 145, L504-508 (1984)
2. H. Coufal: IEEE Trans. Ultrason., Ferroelectrics, and Freq. Contr. UFFC-33, 507-512 (1986)
3. H. Coufal, R. Grygier, D. Horne and J. Fromm: J. Vac. Sci. Technol., in press
4. H. Coufal and P. Hefferle: Appl. Phys. A38, 213 (1985)

Simultaneous Photoacoustic and Photoelectrochemical Study of a TiO_2 Thin-Film Semiconductor Electrode

J. Rappich and J.K. Dohrmann

Institut für Physikalische und Theoretische Chemie der Freien Universität Berlin, Takustraße 3, D-1000 Berlin 33, Germany

We have extended our studies of electrode processes by in situ PAS /1/ to the photoanodic oxygen evolution from water at an n-semiconducting TiO_2 passive film. Similar to the photothermal technique /2/, the intrinsic quantum yield of the photocurrent can be obtained from the PA measurement without a calibration of the light source.

1. Experimental

PA signal and photocurrent at a passivated 0.1 mm Ti foil electrode (0.3 µm TiO_2 formed at 100 V) were measured during a voltage scan in 0.5 M Na_2SO_4 at pH 7 using the electrode/microphone assembly described previously /1/.

2. Results and Discussion

Figure 1 shows the potential dependence of the difference PA signal

$$L = \{P(U) - P(U_{fb})\}/P(U_{fb}) \tag{1}$$

of the electrode. $P(U)$ and $P(U_{fb})$ are the PA signals at potentials U and U_{fb}, respectively, where U_{fb} is the flatband potential (- 0.41 V vs. NHE). Also shown is the photocurrent-potential curve in terms of the quantum yield, η, referred to incident photons. $L(U)$ resembles the normalized temperature change of some semiconductor electrodes from photothermal measurements /2/.

For analysis we use the energy diagram of the metal/n-semiconductor/solution junction, Fig. 2, which is an extension of the diagram previously applied to the photothermal experiment /2b/. We assume proportionality between the PA

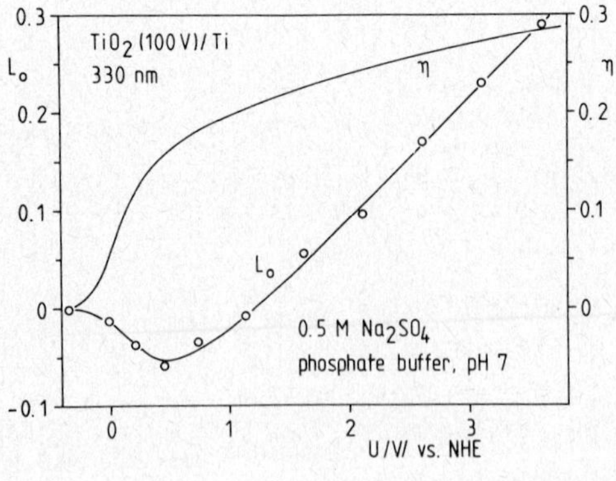

Fig. 1. Difference PA signal, L_o, extrapolated to zero modulation frequency, and photocurrent quantum yield, η, as a function of electrode potential for H_2O oxidation at the TiO_2 passive film ($2\ H_2O + 4\ h^+ \rightarrow O_2 + 4\ H^+$)

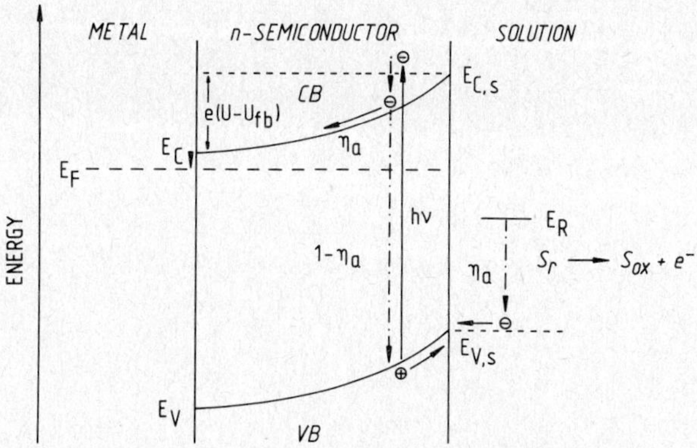

Fig. 2. Energy diagram and heat producing processes for the electrode/solution junction at $U > U_{fb}$. E_F: Fermi level; E_C, E_V, E_R: energy of the conduction band, valence band, and redox couple, S_r/S_{ox}, respectively; η_a: intrinsic quantum yield

signal and the heat produced upon relaxation of the electrons (holes) generated by absorption of photons and exclude luminescence. Considering the various heat sources at $U = U_{fb}$ (zero photocurrent, no band bending) and at $U > U_{fb}$ we obtain

$$L = \frac{\eta_a}{E}\{|e_o|(U-U_R) - T\Delta S_R\}, \tag{2}$$

where η_a: intrinsic quantum yield of the photocurrent as referred to absorbed photons, E: photon energy, U_R: redox potential, and $T\Delta S_R$: Peltier heat (ΔS_R: entropy change of the electrode reaction).

Application of (2) to the linear portion of L(U) in Fig. 1 gives the intrinsic quantum yield, $\eta_{a,s} = 0.46$ at 330 nm, in the saturation region of the photocurrent. The difference between η_a and η (Fig. 1) is accounted for by the reflectivity of the electrode. Since $U_R = 0.82$ V vs. NHE for water oxidation to give oxygen, we obtain 0.4 eV for the Peltier heat per electron transferred. Determination of $T\Delta S_R$ by use of (2) and the O_2/H_2O standard potential requires that the oxidation of any intermediate, e.g. H_2O_2, is fast compared with the modulation frequency.

3. Conclusion

The present work represents the first determination of the intrinsic quantum yield and of the Peltier heat for a photoelectrochemical process by in situ PA measurements at a semiconductor electrode. Since the photocurrent quantum yield has been measured simultaneously, we can also estimate the reflectivity, R, of the electrode/solution interface

$$R = 1 - (\eta/\eta_a), \tag{3}$$

where R = 0.4 at 330 nm in the present case.

In previous work the related PDS technique has been applied to the characterization of $CuInSe_2$ thin film electrodes under non-electrochemical conditions /3/ as well as to the study of the potential dependence at 505 nm of the PDS signal from a CdS single crystal electrode in alkaline polysulfide solution /4/. Although a curve similar to that of Figure 1 has been reported in the latter work, no analysis has been made for the intrinsic quantum yield and Peltier heat /4/.

4. Literature

1. (a) U. Sander, H.-H. Strehblow, J.K. Dohrmann: J. Phys. Chem. 85, 447 (1981); (b) J.K. Dohrmann, U. Sander: Ber. Bunsenges. Phys. Chem. 90, 605 (1986)
2. (a) A. Fujishima, Y. Maeda, K. Honda, G.H. Brilmyer, A.J. Bard: J. Electrochem. Soc. 127, 840 (1980); (b) Y. Maeda, A. Fujishima, K. Honda: Bull. Chem. Soc. Jpn. 55, 3373 (1982)
3. J.P. Roger, D. Fournier, A.C. Boccara, R. Noufi, D. Cahen: Thin Solid Films 128, 11 (1985)
4. R.E. Wagner, V.K.T. Wong, A. Mandelis: Analyst 111, 299 (1986)

Weak Absorbance Measurement of Optical Thin Films Using the Laser Photothermal Deflection Technique

Shi Bai-xuan, Hu Kai, and Chen Wen-bin

Optical Engineering Department, Zheliang University,
Hangzhou, People's Republic of China

It is important to measure the weak absorbance of optical thin films in order to improve the technique of coating optical thin films on the cavity mirrors of high-power lasers, and this has been investigated by *Boccara* and *Fournier* [1]. In this paper, the matrix transmission model of the photothermal effect in optical thin films with an arbitrary number of layers is established and applied to measure the absorbance of optical thin films with a single layer and multilayers. The sensitivity and the error in measurement are discussed, and the effectiveness of the application of the photothermal deflection technique to the weak absorbance measurement of optical thin films is proved.

When a thin film with an arbitrary number of layers is irradiated by a pump laser beam, the refractive index of the film and substrate will change due to the modulated temperature caused by the absorption of light. Passing through the sample, a probe beam will be deflected [2]. By the thin film approximation treatment and the photothermal matrix transmission model, the deflection angle in the substrate is given by

$$N = \frac{1}{n}\frac{dn}{dT} H_T(\sigma_0, \sigma_n, \kappa_0, \boldsymbol{S}, a) Q_\mathrm{s} \quad ,$$
$$H_T = \int_{\boldsymbol{S}} \nabla_\perp \iint f(\sigma_0, \sigma_n, \kappa_0, \kappa_n) \exp(-\sigma_n)\delta h(\delta) J_0(\delta r) d\delta \quad , \tag{1}$$

where $\sigma_i = \delta^2 + \mathrm{j}\omega/\alpha_i$, and $i = 0, 1, 2, \ldots, n-1$, n is the ordinal number of the coupling medium before the sample, each layer of the thin film and the substrate; α_i is the thermal diffusivity; ω is the modulating frequency; $h(\delta)$ is the spatial spectrum of the pump laser; δ is the spatial frequency; κ_i is thermal conductivity of medium i; $\boldsymbol{S}$ is the path of the probe beam; a is the radius of the pump beam at the waist; $Q_\mathrm{s} = \sum_{i=1}^{n-1} \beta_i l_i I_i$ is the absorption sum of each thin film; β_i, l_i are the optical asorption coefficient and the thickness of the ith layer film, respectively; and I_s is the intensity of the laser. Equation (1) demonstrates that the deflection angle of the probe beam is directly proportional to Q_s and the ratio depends only on the coupling medium and the thermal parameters of the substrate.

Using a two quadrant sensor, the differential output ΔV, which is proportional to the probe beam deflection angle, can be detected as

$$\Delta V_s = V_s \frac{1}{n}\frac{dn}{dT} H_T(\sigma_0, \sigma_n, \kappa_0, \kappa_n, S, a, \omega_0) Q_s \quad , \tag{2}$$

where V_s is the dc voltage sum of the two quadrants; ω_0 is the waist of the probe beam. If the sample is replaced by a vacuum-evaporated carbon film with the same substrate, the measured voltage is

$$\Delta V_c = V_c \frac{1}{n}\frac{dn}{dT} H_T(\sigma_0, \sigma_n, \kappa_0, \kappa_n, S, a, \omega_0) Q_c \quad , \tag{3}$$

where $Q_c = \beta_c l_c T_c$, so

$$Q_s = \left(\frac{Q_c}{\Delta V_c / V_c}\right) \cdot \frac{\Delta V_s}{V_s} \quad .$$

If both sides are divided by the incident laser intensity, the equation of absorbance measurement can be obtained:

$$A_s = \frac{A_c}{\Delta V_c / V_c} \cdot \frac{\Delta V_s}{V_s} \quad . \tag{4}$$

The absorbance of vacuum-evaporated carbon film, A_c, is high and can be detected using ordinary photometric methods. So the A_s of an optical thin film with multilayers could also be measured. With the collinear photothermal deflection arrangement, the absorbance of a series of single-layer and multilayer thin films, such as MgF_2, ZrO_2, TiO_2:MgO, ZiO_2:Y_2O_3 and interference filters, have been measured.

It is found from experiments that the sensitivity is mainly limited by noise caused by the drift of the laser beam. The observed effective noise value of the measurement system is $\Delta V = 0.015 \pm 0.006 \mu$V (V = 200 mV). If the pump beam is the 100 mW He-Ne laser, the sensitivity is $A_s = 2 \times 10^{-7}$. Analysis indicates that the error in measurement for most single-layer films is smaller than 10%, and that mainly the thermal conductivity of the film medium affects the error. For optical thin films with large thermal conductivity, such as MgO, the error could in principle reach 40%. Using the photoacoustic method, the error can be controlled to within 18%. So, the photoacoustic and photothermal methods have the same accuracy as the thermometric method [3], but the sensitivity is high and the measurement is much more convenient. The photothermal method provides an effective method of absorbance measurement for the study of thin film optics.

The authors wish to thank A.C. Tam for providing us with enlightening ideas and suggestions in an earlier phase of this work.

1. A.C. Boccara, D. Fournier: Opt. Lett. **5**, 377 (1980)
2. A.C. Boccara, D. Fournier, N.M. Amer, W.B. Jackson: Appl. Opt. **20**, 1333 (1981)
3. Jin Shang-zhong: *Laser and Infrared* (in Chinese), No. 6,7 (1985)

Standing Light Waves Investigated by Photothermal Spectroscopy

W. Knoll[1] *and H. Coufal*[2]

[1]Max-Planck-Institut für Polymerforschung, Jakob-Welder-Weg 11, D-6500 Mainz, Fed. Rep. of Germany
[2]IBM Almaden Research Center, 650 Harry Road, San Jose, CA 95120, USA

Photothermal (PT) Spectroscopy is based on the pyroelectric detection of heat that is generated by optical absorption and subsequent thermalization of the excitation energy by radiationless decay channels. This new method thus complements absorption and fluorescence spectroscopies /1/.

In the study presented here a cyanine dye (1-methyl-1'-octadecyl-2,2'-cyanine-iodide, S120 for short) monolayer organized in J-aggregates /2/ was utilized as an absorber to probe the light intensity in front of a flat silver mirror. Samples with different distances between the dye monolayer and the metal surface were prepared by intercalating the dye molecules into Langmuir-Blodgett (LB) assemblies composed of a varying number of transparent fatty acid bilayers. With a thickness of 5 nm per bilayer a spatial resolution sufficient for monitoring the local field intensity of the standing electromagnetic wave in front of the Ag mirror was obtained. The complete LB assembly of spacer layers plus dye layer was deposited directly onto one of the Ag-electrodes of a recently developed thin film calorimeter /3/. The sensitivity of this new detector is sufficient to determine the heat being generated by absorption and subsequent (partial) thermalization of photons in the dye monolayer. A schematic drawing of the sample detector configuration is given in Fig. 1.

Photothermal spectra were recorded from λ = 400 nm to λ = 700 nm using a conventional scanning spectrometer with modulated excitation and lock-in amplifier detection of the photothermal signal. To allow, after the conclusion of the photothermal experiment, for a characterization of the J-band monolayer by conventional absorption spectroscopy, the whole multilayer assembly was peeled off the calorimeter by means of a thin polyvinyl-

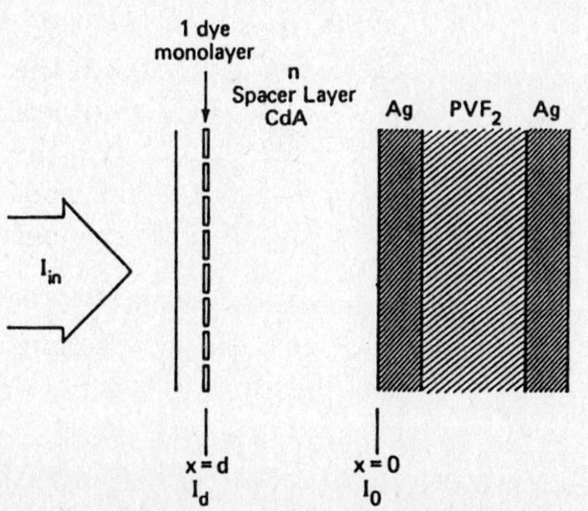

Fig. 1: Sample-detector configuration (schematic) for photothermal spectroscopy

alcohol film which was cast and dried on top of the LB sandwich. In both spectra the strongly absorbing J-band peak at λ = 578 nm is evident, though with different vibronic sideband and monomeric dye contributions at shorter wavelengths. The latter have a higher yield for radiationless deexcitation (and hence higher PT signal generation) as can also be deduced from their lower fluorescence intensity /2/.

The dye-Ag distance dependence of the photothermal signal is given by

$$\left\{ \frac{S\,(\lambda,d)}{A_J(\lambda)} + 1 \right\} \frac{A_{Ag}\,(\lambda)}{\zeta(\lambda,d)} = \frac{I(\lambda,d)}{I_o(\lambda)} \quad . \tag{1}$$

The bracket of the left side of eqn. 1 contains all the experimental data ($S(\lambda,d)$ = PT intensity, A_J = J-band absorption). The right side describes the variation of the local field intensity with distance d from the silver surface - the standing light wave first observed in 1890 in Wieners classical experiment /4/.

A comparison between the experimental data obtained by photothermal spectroscopy and the theoretically expected dye-Ag distance dependence of the field intensity is shown in Fig. 2 for λ = 578 nm. With the known absorbance of the silver electrode (A_{Ag} = 0.05 for λ = 578 nm) determined in a separate experiment the efficiency for radiationless deexcitation ζ = 0.33 ± 0.07 can be derived /5/.

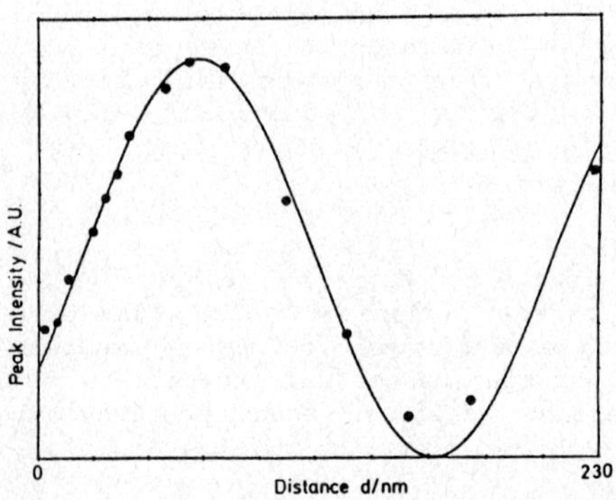

Fig. 2: Photothermal peak intensity of the dye J-band (full circles) as a function of the monolayer-silver surface separation. The full line corresponds to the electromagnetic field distribution of the standing light wave in front of the silver mirror.

References:
1. H. Coufal: Appl. Phys. Lett. 44, 59 (1984)
2. C. Duschl, M. Lösche, A. Miller, A. Fischer, H. Möhwald and W. Knoll: Thin Solid Films 133, 65 (1985)
3. H. J. Coufal, R. K. Grygier, D. E. Horn and J. E. Fromm:
4. A. Wiener: Wied. Ann. 40, 203 (1890)
5. W. Knoll and H. J. Coufal, Appl. Phys. Lett., in press

The Reverse Mirage Effect: Catching the Thermal Wave at the Solid-Liquid Interface

M.J. Smith and R.A. Palmer

Department of Chemistry, Duke University, Durham, NC 27706, USA

One very promising technique for avoiding the liquid opacity problem in FTIR mirage detection at the solid-liquid interface is the application of an experimental geometry (Fig. 1) different from the conventional front surface illumination mode.[1] In this variant of the method, which we will refer to as the "reverse" mirage effect, the IR beam passes through a transparent solid first and then is absorbed by the liquid deflection medium or by other species at the solid-liquid interface.[2] The probe beam passes through the liquid and grazes the lower surface of the solid (presuming upper surface illumination). The downward deflection of the probe beam resulting from the absorption of modulated infrared radiation transmitted by the solid and absorbed by the liquid or other chromophoric species at or near the solid-liquid interface is the photothermal signal (detected, as in the conventional experiment, by a silicon diode position detector). Thus, the liquid need only be transparent to the probe beam for a photothermal signal to be observed. This technique seems ideally suited to infrared detection and characterization of species which may be electrogenerated or otherwise located or generated at a solid-liquid interface, especially if the solid phase is a semiconductor such as silicon, germanium or gallium arsenide.

In this paper we present results obtained using the reverse mirage technique with single crystal silicon as the transparent solid, and pure acetonitrile as both the absorbing sample and liquid deflecting medium. Studies of the position of the probe laser beam center with respect to the Si/CH_3CN interface reveal some interesting qualities about photothermal detection within the absorbing medium. The resulting spectra are explained using the Rosencwaig-Gersho model.[3]

The mirage effect accessory, which fits into the standard sample compartment of the IBM Instruments IR/95, is mounted on a 1 inch thick aluminum plate supported on vibrational damping material. On this plate are mounted the detector optics consisting of 1) an IR pump beam focusing mirror, which turns the beam 90^0 downward and focuses it on the horizontal sample, 2) a sample positioning stage (x,y,z,rx), 3) a focused probe laser (0.8 mW HeNe, beam waist, 40 µm diameter), and 4) a bicell position sensor detector. Spectra were obtained using the slowest standard instrument mirror velocity (VEL = 0; 0.059 cm/sec). This yields an optical velocity of 0.235 cm/sec, resulting in photothermal modulation (Fourier) frequencies in the range from 141 Hz at 600 cm^{-1} to 940 Hz at 4000 cm^{-1}.

The solid-liquid interface to be observed is created by attaching a disk of IR transparent solid (e.g., silicon) to the bottom of a conical well using cyanoacrylate adhesive and then suspending this "cup" from the top of a 1x1x1 cm pyrex cuvette so that the bottom surface of the solid is 4-5 mm below the top of the cuvette. The cuvette is then filled with the deflecting liquid to a level 2-3 mm above the lower surface of the solid. The upper surface of the solid, inside the "cup", remains exposed to the N_2 atmosphere of the instrument (Fig. 1a). For the experiments reported here the solid was a 3x3x0.5 mm disk of single crystal silicon (110) etched clean with concentrated HF, rinsed with deionized H_2O and air dried. The liquid absorbing and deflecting phase was dry spectrograde CH_3CN. To measure the reverse mirage spectrum the solid-liquid interface is positioned at the intersection of the pump and probe beams. The sample's position is iteratively adjusted using the x,y,z,rx stage until the laser just grazes the bottom of the solid, and the interferogram $((I_A-I_B)/(I_A+I_B))$ from the position detector differential amplifier) as observed on the oscilloscope is optimized. The interferograms (typically $\leq$ 128) are then collected, coadded and transformed by the standard instrument software.

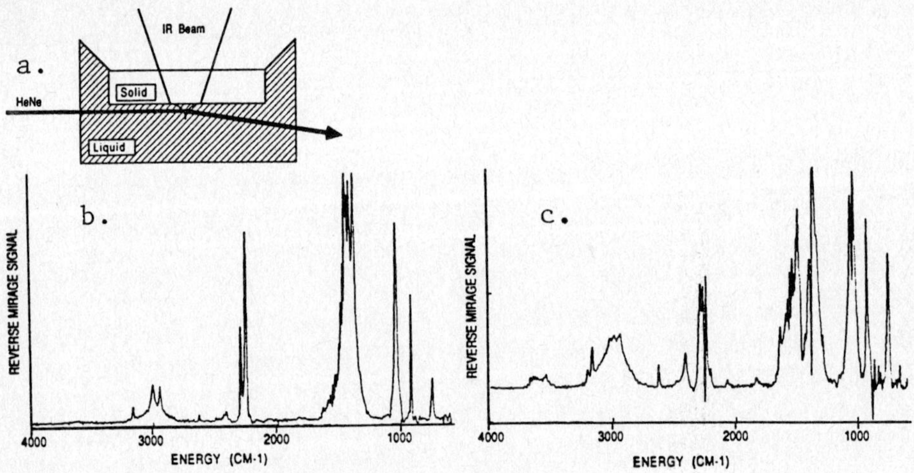

Fig 1. a. Reverse mirage effect. b. Si/CH_3CN, x = 50μm. c. Si/CH_3CN, x = 135 μm.

Figures 1b and 1c illustrate the reverse mirage detected spectrum of CH_3CN measured with two different probe-to-sample distances x. Figure 1b (x = 50 μm) resembles closely the transmission spectrum of CH_3CN but in 1c, where x = 135 μm, severe distortion of the spectrum is obvious. Specifically, the band intensity ratios are changed, and each of the stronger bands is inverted on itself, i.e., a minimum appears at the position of the true band maximum. By consideration of the Rosencwaig-Gercho model[3] it can be shown that this behavior is expected when the deflection medium is also the absorbing medium (here, the liquid). In this case the depth of initial heat distribution within the liquid varies inversely as its optical absorption coefficient α. This, in effect, compresses the thermal wave (and thus the refractive index gradient) closer to the surface for stronger bands than for weaker ones. Using the appropriate optical and thermal parameters for CH_3CN, the theory predicts that for a given value of x the deflection $\emptyset$ will first linearly increase with α and then begin to decrease. Thus, stronger bands will be inverted while weaker ones are undistorted. Furthermore, the range of linear response is predicted to be shorter for larger values of x and for shorter thermal diffusion lengths μ_f. These theoretical predictions have been quantitatively confirmed by extensive experiments.

These results clearly show that "catching the thermal wave" at the solid liquid interface using the reverse mirage effect places a high priority on optimum probe beam focusing and distance from the surface and on low modulation frequency. Failure to appreciate these factors could lead to severe misinterpretation of the spectral data.[4]

References

1. W. B. Jackson, N. M. Amer, A. C. Boccara and D. Fournier, Appl. Opt., 20, 1333 (1981).
2. R. A. Palmer and M. J. Smith, Can. J. Phys., 64, 1086 (1986).
3. A. Rosencwaig and A. Gersho, J. Appl. Phys., 47, 64 (1976).
4. M. J. Smith and R. A. Palmer, Appl. Spectros., 41, No.7 (1987).

Photothermal Investigations on Small Silver Particles

M. Bauer and H.D. Breuer

Institut für Physikalische Chemie, Universität des Saarlandes,
D-6600 Saarbrücken, Fed. Rep. of Germany

1. Introduction

In contrast to an infinitely large metal with defined optical constants small metal particles exhibit unique optical properties which strongly depend on particle size /1,2,3/ and the dielectric surroundings /4,5,6/. Spectra and size distributions of small silver particles are investigated by Photothermal Deflection Spectroscopy and Transmission Electron Microscopy.

2. Sample preparation

Unlike common preparatory methods like vacuum deposition etc. we photochemically reduced AgBr-crystals on filters, made of cellulose nitrate. Our method is based on aqueous solutions of $AgNO_3$ and KBr. These solutions were passed through the filters from both sides. The size of the AgBr-particles was thus limited by the defined diameter of the filter pores. The use of different concentrations allowed an additional control of the particle size. Using starting concentrations from 0.1 mol/l up to 10^{-6} mol/l we prepared silver particles of different size. The filters had pore diameters of 25 nm and 100 nm. The photochemical reduction to elementary silver was achieved by irradiating with monochromatic light of 2.91 eV (425 nm), from a 300 W CERMAX-lamp (ILC-Technology type LX 300 F) and a 0.25m Ebert-Monochromator (Jarrell-Ash 82-410).

3. Spectroscopy

Photothermal Deflection Spectroscopy (PDS) is a very powerful tool for investigating the absorption properties of small metal particles. Pure absorption spectra without scattering are obtained. The spectra were taken in the UV/VIS range from 350 nm to 750 nm. With decreasing concentration the maximum of absorption shifts towards shorter wavelengths (see Fig. 1) reaching a limit at 400 nm. This shift is due to a decrease of particle diameter. Reducing the concentration the halfwidth also decreases. This can be explained by the statistical distribution of the particle diameters. The wavelength of the absorption maximum and the halfwidth are functions of the particle size.

The filters can be made transparent by treating them with methanol and the absorption spectra can be measured in transmission. The maxima of absorption of these samples differ from those taken with PDS due to the different dielectric constant of the environment. All samples with concentrations between 10^{-6} mol/l and 0.005 mol/l show only one peak at 428 ± 8 nm.

In the range between 0.005 mol/l and 0.05 mol/l there are two maxima, one at 398 ± 3 nm and one at $500 + 10$ nm both resulting from an additional irra-

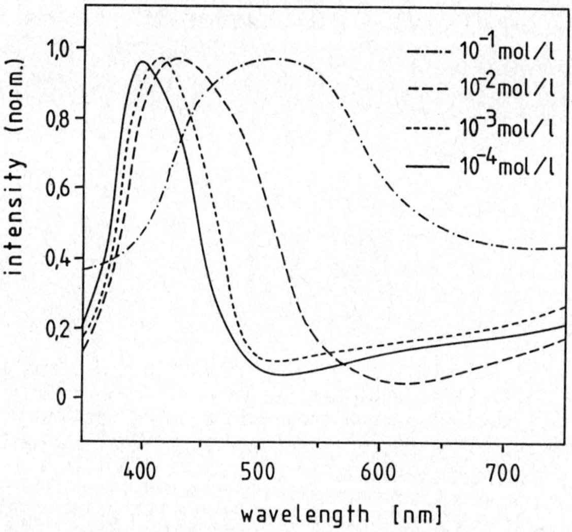

Fig. 1 PDS spectra

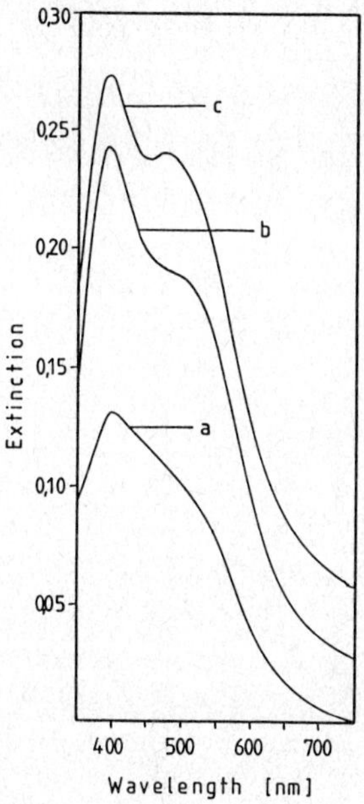

Fig. 2 UV/VIS-spectra of silver particles in a transparent filter (c_{KBr} = 0.01 ml/l) a) before additional irradiation; b) after 2 hours of irradiation; c) after 4 hours of irradiation

diation (see Fig. 2). The first band results from the plasma resonance of the free electrons of silver and the other one from the scattering of the particles. When the concentration exceeded 0.05 mol/l, the scattering prevailed and no pure absorption spectra could be recorded.

4. Electron microscopic studies

Electron microscopic investigations were made in order to relate the concentration to the actual particle size. For a concentration of 0.01 mol/l particles of a mean diameter of 35 nm were obtained (see Fig. 3). They appeared to be spherical and evenly distributed over the filter. At a concentration of 10^{-4} mol/l the mean diameter of 14 nm predominated (see Fig. 4).

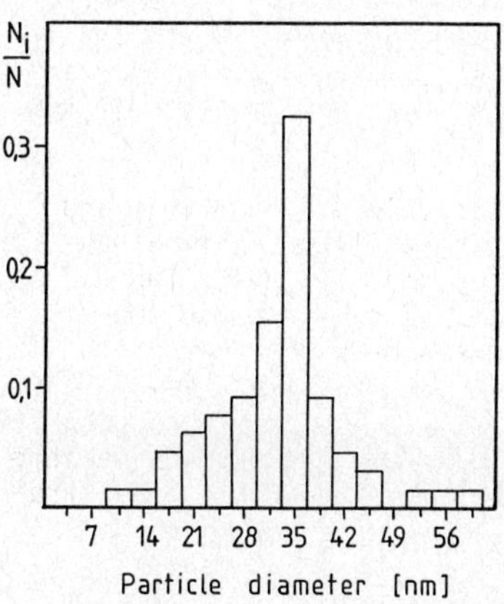

Fig. 3 Size distribution of silver particles produced at a KBr-concentration of 10^{-2} mol/l

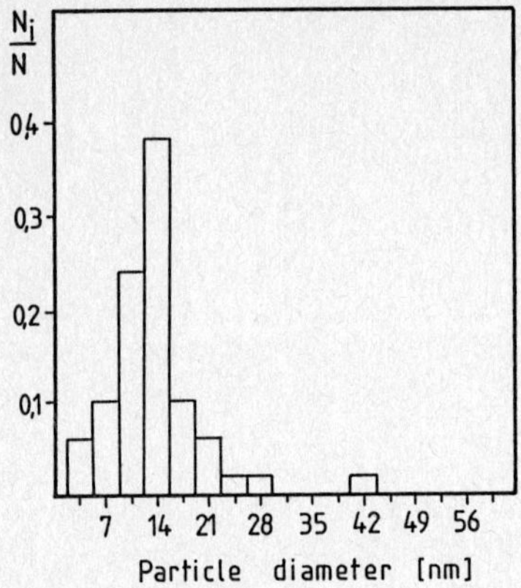

Fig. 4 Size distribution of silver particles produced at a KBr-concentration of 10^{-4} mol/l

1. J. C. Maxwell Garnett: Phil. Trans. Roy. Soc., London 203A, 385 (1904)
2. G. Mie: Annalen der Physik 25, 377 (1908)
3. E. David: Z. Physik 114, 389 (1939)
4. H. Wolter: Z. Physik 113, 547 (1939)
5. W. Hampe: Z. Physik 152, 476 (1958)
6. U. Kreibig, Z. Physik B21, 339 (1975)

Separation of Plasmon Polariton Modes in Small Metal Particles

U. Kreibig[1], *B. Schmitz*[2], *and H.D. Breuer*[2]

[1]FB11, Physik, Universität des Saarlandes, D-6600 Saarbrücken, Fed. Rep. of Germany

[2]FB13, Physikalische Chemie, Universität des Saarlandes, D-6600 Saarbrücken, Fed. Rep. of Germany

The plasmon polaritons excited optically in systems of small metal particles give rise to extinction which consists of both absorption (production of Joule heat) and elastic scattering (changes of propagation direction).

Since Mie (1) and Debye, the optical extinction is classified using multipole expansions of the electromagnetic fields, i.e. by introducing multipolar excitation "modes". The total extinction is then given by

$$E_{total}(\omega,R) = \sum_{\nu} A_{\nu}(\omega,R) + A'_{\nu}(\omega,R) + S_{\nu}(\omega,R) + S'_{\nu}(\omega,R),$$

with ω and R the 2π-frequency and the particle radius, respectively, and A_{ν} the absorption of the ν-th "electric", A'_{ν} the absorption of the ν-th "magnetic", S_{ν} the scattering loss of the ν-th "electric" and S'_{ν} the scattering of the ν-th "magnetic" mode.

Formally, the index ν of the orthogonal system of polariton modes goes to ∞; for $R \ll \lambda$ (λ:wavelength), however, only modes with small ν are excited to measurable amounts. E.g., for Ag-particles with 2R=19 nm it is predicted that only the dipolar plasmon absorption A_1 is important, for Ag-particles with 2R=63 nm, A_1, A_2, S_1 and S_2 , i.e. dipolar and quadrupolar absorption and scattering, should contribute to E_{total} with different peak heights (Fig.1), positions and widths.

To separate these contributions experimentally, both conventional extinction measurements and photothermal deflection (PT) measurements were performed

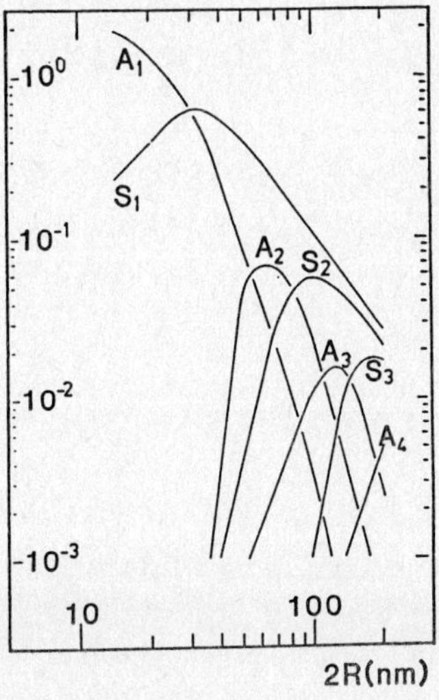

Fig.1. Silver particles: Calculated maximum values of the absorption (A_{ν}) and scattering (S_{ν}) modes versus particle diameter 2R (ν: mode index)

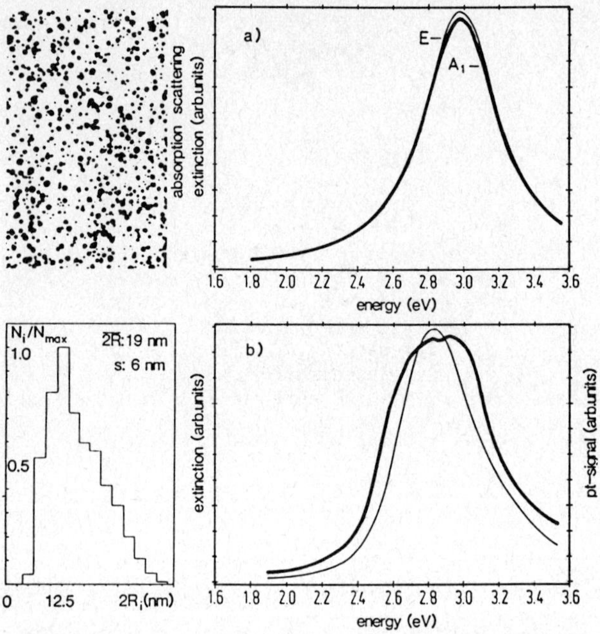

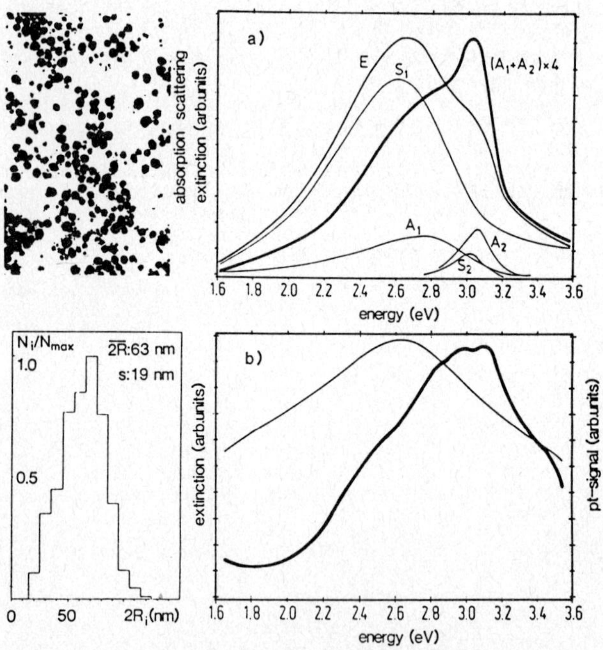

Fig.2 (top), Fig.3 (bottom). Ag-particles: TE-micrographs, size distributions, theoretical spectra (a) and measured spectra (b).(Spectra b: extinction (thin lines), PT-signal (bold lines))

on the same samples which, in addition, were analyzed by TEM [2]. The samples consisted of 2-dimensional arrangements of Ag-particles embedded in a dielectric matrix with low packing density to ensure that merely single particle excitations occurred.

Experimental results of two samples with different sizes, 2R=19 nm and 63 nm, are presented. It is confirmed that in the 19-nm particles mainly A_1 contributes to both kinds of spectra (Fig.2). In contrast to this, light excites the A_1, A_2 and S_1 mode in the 63 nm-particle sample which all contribute to the extinction. The PT signal, however, only registers the pure absorption A_1 and A_2. Comparing both kinds of spectra, the A and S contributions which have different peak positions and widths, can thus be separated. Particularly the quadrupolar plasmon absorption A_2 could clearly by resolved (Fig.3). Results are also presented for a densely packed Ag-particle system (2R=1o nm; filling factor ~ o.4) where collective interaction effects markedly influence the extinction. By comparing with the corresponding PT-spectrum, we find, especially in the IR region, large amounts of scattering (Fig.4). Since the single particles are too small to scatter, these losses are attributed to collective effects. Hence, the extinction of such samples cannot be described in the quasistatic approximation.
In Fig. 5 it is demonstrated that small Au-particles (2R=1o nm) show analogue effects to the Ag-particles of Fig.2.

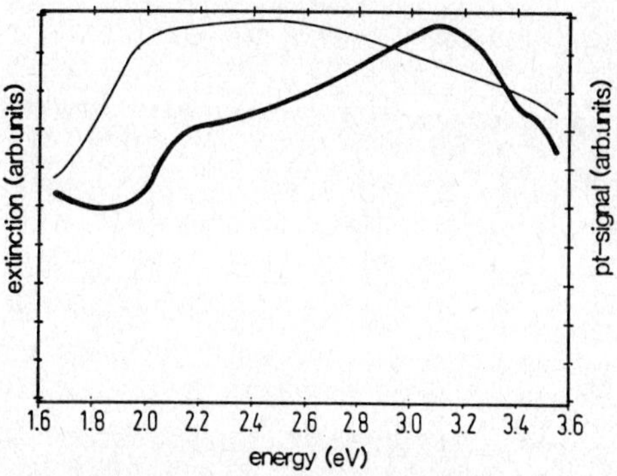

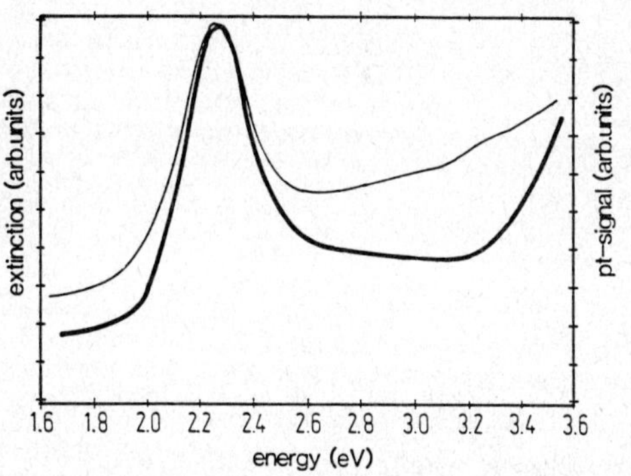

Fig.4 (left): Ag-particles, high filling factor. Fig.5 (right): Gold particles (extinction (thin lines); PT-signal (bold lines))

Literature: 1.) G.Mie: Ann.Physik 25, 377 (19o8)
2.) U.Kreibig, B.Schmitz, H.D.Breuer;Phys.Rev.B (1986)(in press)

Probe-Beam Deflection Measurement of Photothermal Desorption of Moisture Adsorbed on Gold or Silicon in Atmospheric Conditions

A.C. Tam and H. Schroeder

IBM Almaden Research Center, 650 Harry Road, San Jose, CA 95120, USA

Previous work done on laser induced thermal desorption of molecules on a solid surface [1], was mainly performed in vacuum, using particle detectors (e.g. mass spectrometer). The present work examines photo-thermal desorption from a surface in atmospheric conditions utilizing a probe-beam deflection [2] for adsorbed water in ordinary atmospheric conditions.

The experimental arrangement is shown in Fig. 1 [2]. A nitrogen laser pulse, of duration 8 nsec and energy adjustable (by attenuation) up to 1mJ, is directed onto a polished Si or Au surface. The UV laser beam is slightly focussed onto a spot size of $0.5 \times 4 mm^2$ and the absorbed maximum power density of 5 MW/cm^2 causes a surface temperature rise of up to 150K [3]. The pulse energy is kept well below the damage threshold. The temperature-controlled sample is situated in a small chamber, and dry or moist nitrogen at room temperature can be passed into the chamber to produce a "dry" surface, or a "moist" surface with adsorbed water respectively. A probe beam focussed (f=25mm) vertically above the irradiated surface (with typically 3 mm separation) is used to detect the shape of the photoacoustic pressure pulse.

The beam deflection is detected by a fast avalanche photodetector (risetime < 5ns) and is captured by a transient digitizer with a personal computer (IBM PC/AT).

The observed beam deflection $\phi(x,t)$ at distance x above the surface is related to the pressure $p(x, t)$ [2]:

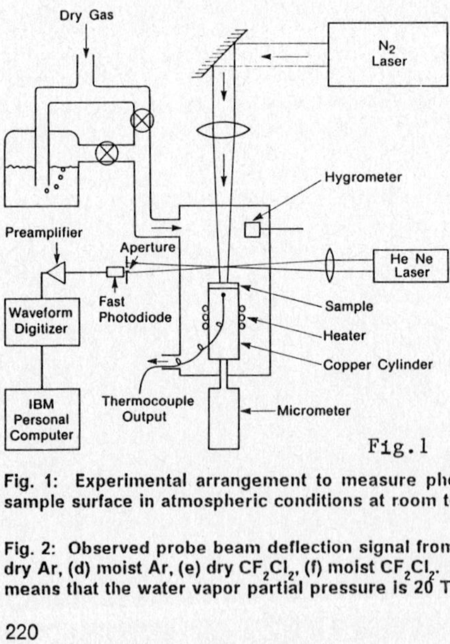

Fig.1

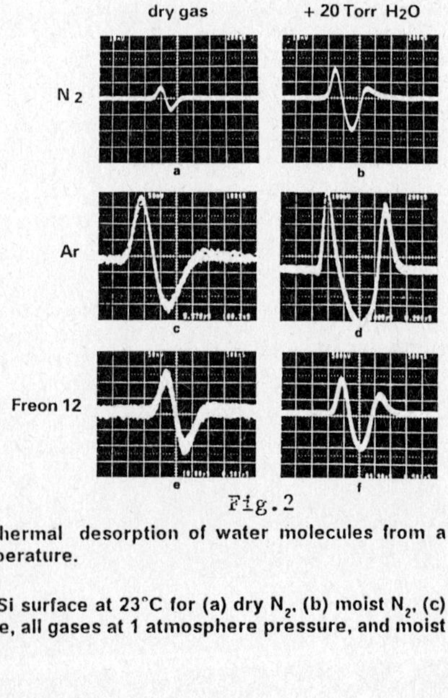

Fig.2

Fig. 1: Experimental arrangement to measure photothermal desorption of water molecules from a sample surface in atmospheric conditions at room temperature.

Fig. 2: Observed probe beam deflection signal from a Si surface at 23°C for (a) dry N_2, (b) moist N_2, (c) dry Ar, (d) moist Ar, (e) dry CF_2Cl_2, (f) moist CF_2Cl_2. Here, all gases at 1 atmosphere pressure, and moist means that the water vapor partial pressure is 20 Torr.

$$\phi(x,t) = \frac{1}{n}\frac{\partial n(x,t)}{\partial r}\, l = \frac{1}{n}\frac{2\pi a}{m}\, l\, \frac{M}{RT}\frac{1}{v}\frac{\partial p(x,t)}{\partial r}\,, \tag{1}$$

where n, a, and l are the refractive index, polarizability and interaction length, m is the mass of the molecule, R is the gas constant, T is the absolute temperature, and v is the acoustic velocity. The detector signal is proportional [4] to the beam deflection $\phi(x,t)$. Some observed deflection signals are shown in Fig. 2; the signal for the "dry" case is observed to be simply bipolar, since this is the derivative of the pressure pulse, which should thus be simply a compression pulse due to the "thermal piston" effect at the Si surface. However, the signal for the "moist" case in Fig. 2 is more complicated, broader, and much larger in amplitude; this derivative signal suggests that the pressure pulse produced at the Si surface is composed of three components: (1) "thermal piston" of the nitrogen gas adjacent to the Si surface, (2) desorption of the water molecules originally on the Si surface, and (3) re-adsorption of water molecules that is expected to cause the delayed positive peak in Fig. 2(b), whose magnitude is a measure of the "re-adsorption time" of water molecules onto the Si surface in atmospheric conditions. Similar observations have been made for a gold sample. We can define a "net moisture signal" $S_m(x, t)$ as the difference between the signal observed in moist nitrogen (e.g. Fig. 2b) and that observed in dry nitrogen (e.g. Fig. 2a). $S_m(x, t)$ is found to have two positive peaks, to be called "moist component first pulse" and "moist component second pulse." These pulse heights for various gold sample temperature

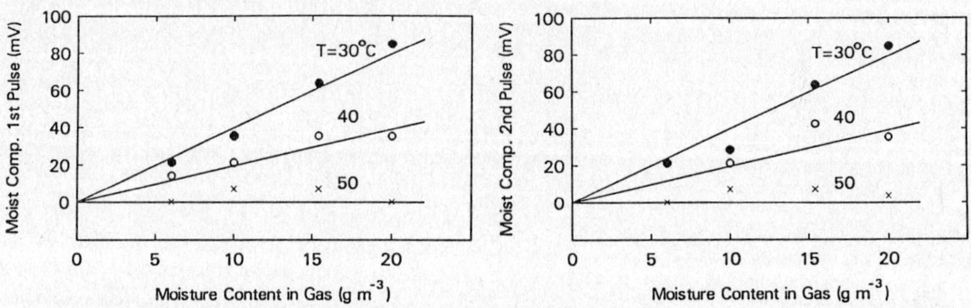

Fig. 3: Variation of the "moist" component of the first (left) and second (right) deflection peak on the excitation laser pulse for a gold surface in N_2 with different H_2O partial pressures as indicated: (a) is for a sample temperature being 30°C, (b) is for 40°C and (c) is for 50°C.

and moisture content in the nitrogen gas are shown in Fig. 3; we see that both components behave similarly, i.e., at a lower temperature, the net moisture signal increases rapidly with increased absolute moisture content in the gas due to the increased adsorbed water molecules in sample surface. Using ray optics and the refractive index of nitrogen and water, we estimate that the delayed peak corresponds to the adsorption of approximately a monolayer of water molecules on the Si surface. The delayed time interval between the two peaks suggests an "adsorption time" of several hundred nsec.

This work is supported in part by the U. S. Office of Naval Research.

References

1. T. J. Chuang: J. Vac. Sci. Technol. B3 , 1408 (1985).
2. A. C. Tam and W. P. Leung: Phys. Rev. Lett. 53 , 560 (1984).
3. D. Burgess, Jr., P. C. Stairs, and E. Weitz: J. Vac. Sci. Technol. A4 ,13 (1986).
4. H. Sontag and A. C. Tam: IEEE Trans. Ultrason. Ferroelectrics, Freq. Contr. UFFC-33 ,500 (1986).

Part V

Applications to Semiconductors

Theory of Bulk and Near Surface Effects on the Modulated Optical Reflectance in Silicon

J. Opsal and A. Rosencwaig

Therma-Wave, Inc., Fremont, CA 94539, USA

When a material is excited with an intensity-modulated energy source, or pump, its optical properties can be altered by the absorption of the incident energy. This results in the sample's complex refractive index undergoing periodic variations at the modulation frequency of the pump intensity excitation. These periodic variations in the sample's optical refractive index can be detected by measuring the modulated reflectance, scattering, transmittance or deflection of an optical probe that is reflected from or transmitted through the sample. Since the complex refractive index of most materials depends on temperature, a modulated temperature will induce a corresponding modulation in the refractive index and, consequently, a modulation in the optical reflectivity. In a semiconductor, the refractive index is also affected by the density of free carriers so that a modulated electron-hole plasma density will provide an additional modulation in the optical reflectivity.

To begin, let's consider qualitatively what happens when a laser beam is incident on a semiconductor. If the energy per photon E exceeds the band gap energy E_g then electrons will be excited from the valence band to an energy $E-E_g$ above the conduction band edge. These photoexcited free carriers will, within a fraction of a picosecond, give this excess energy to the lattice through nonradiative transitions to the unoccupied states near the bottom of the conduction band. After a much longer time, typically on the order of microseconds, these photoexcited carriers recombine with holes in the valence band giving up their remaining energy E_g to the lattice. Prior to this recombination there thus exists a plasma of electrons and holes whose density is governed by diffusion in a manner analogous to the flow of heat from a thermal source. Thus, if an incident laser beam is intensity modulated, we would expect to observe, in addition to the thermal wave, a modulated plasma density whose spatial profile is that of a critically damped wave, i.e., a plasma wave [1].

In order to obtain a quantitative description of these critical wave phenomena, we consider the wave equations for the plasma density, N(z), and temperature, T(z),

$$\nabla^2 N + p^2 N = -f(r,z)Q/DE, \tag{1}$$

$$\nabla^2 T + q^2 T = -f(r,z)Q(E-E_g)/\kappa E - NE_g/\kappa\tau - sNE_g\delta(z)/\kappa, \tag{2}$$

where p is the plasma wave vector defined by

$$p^2 = i(\omega\tau + i)/D\tau = 2i/\mu_p^2 - 1/l_p^2 \tag{3}$$

and q is the thermal wave vector defined by

$$q^2 = i\omega\rho C/\kappa = 2i/\mu_T^2 \,. \tag{4}$$

In the above equations D is the ambipolar diffusion coefficient, τ is the electron-hole recombination time, s is the surface recombination velocity, κ is the thermal conductivity, ρ is the density, and C is the specific heat. Also, in Eqs. (1) and (2), f(r,z)Q is the spatially dependent source term for which we've assumed a sinusoidal time dependence, $\exp(-i\omega t)$. In Eq. (3) we note that there are two characteristic lengths, the plasma wave diffusion length, $\mu_p = (2D/\omega)^{1/2}$, and the more familiar frequency independent plasma diffusion length, $l_p = (D\tau)^{1/2}$. Herein lies the essential difference between the plasma and thermal waves which is most pronounced at low modulation frequencies, i.e., $\omega\tau << 1$, where the plasma loses its wavelike properties, becoming purely diffusive and independent of the modulation frequency with a static diffusion length, l_p. However, in a semiconductor such as silicon and for modulation frequencies in the MHz regime, one can easily have $\omega\tau >> 1$ and the plasma density is then a critically damped wave analogous to the thermal wave. In fact, the plasma wave diffusion length, μ_p, is formally identical to the thermal wave diffusion length, $\mu_T = (2\kappa/\omega\rho C)^{1/2}$, if we simply replace the thermal diffusivity, $\kappa/\rho C$, by the plasma diffusion coefficient, D.

One of the complicating features of any wave equation is the source distribution, in this case f(r,z), which must be specified prior to obtaining a complete solution. One can, however, derive a formal solution without having to explicitly specify the source distribution. Since the wave equations are linear, the general solution for an arbitrary source can be formulated as a linear superposition of basis functions obtained from known sources. A particularly useful representation comes from the 1-dimensional δ-function distribution, $\delta(z-z')$, which leads to the 1-dimensional green's functions, $g_N(z,z')$ and $g_T(z,z')$. In terms of these green's functions, the complete solution in 1-dimension for an arbitrary source distribution, f(z'), is given by the integrals

$$N(z) = \frac{Q}{DE} \int dz'\; g_N(z,z') f(z'), \tag{5}$$

$$T(z) = \frac{Q}{\kappa}\left(\frac{E-E_g}{E}\right)\int dz'\; g_T(z,z') f(z')$$

$$+ \frac{E_g}{\kappa\tau}\int dz'\; g_T(z,z')N(z') + \frac{sE_g}{\kappa}\int dz'\; g_T(z,z')N(z')\delta(z') . \tag{6}$$

Thus, the thermal and plasma wave problem, like any other wave problem, is reduced to finding the solutions of a wave equation with a singular (but known) source. Assuming for the boundary conditions at z = 0, $-D(dN/dz) = -sN$ and $-\kappa(dT/dz) = sNE_g$, we have for the green's functions

$$g_N(z,z') = \frac{i}{2p}\left[e^{ip|z-z'|} + \left(\frac{1-is/pD}{1+is/pd}\right)e^{ip(z+z')}\right] \tag{7}$$

$$g_T(z,z') = \frac{i}{2q}\left[e^{iq|z-z'|} + e^{iq(z+z')}\right] \tag{8}$$

which can be thought of as arising from two localized sources; the first at z' and the second an image source at -z' introduced to satisfy the boundary conditions. When z' = 0, the two sources, of course, produce responses having the same spatial dependence but as z' moves away from the surface, the first term becomes dominant with g_N and g_T each approaching the appropriate infinite medium green's function. It is interesting to consider the behavior of the plasma wave response at points between the surface and the source, that is, for $0 < z < z'$. Obviously, $g_N(z,z')$ increases with increasing z reaching a maximum at $z = z'$. However, the manner in which this occurs is significant. For small z (i.e., $|p|z \ll 1$) the response is linear in z with a slope proportional to s/D. This means that for any distribution of sources and within a region where the surface recombination is dominant, the plasma density will always increase with increasing distance from the surface.

In order to appreciate the role of the source distribution on the plasma and thermal wave response, let's assume the exponential form, as often occurs in practice, $f(z') = \alpha\exp(-\alpha z')$, where α is the optical absorption coefficient. Performing the integrations indicated in Eqs. (5) and (6), we find for N(z) and T(z)

$$N(z) = \frac{iQ}{pDE}\left(\frac{\alpha}{\alpha-ip}\right)\left[\frac{\alpha e^{ipz} + ipe^{-\alpha z}}{\alpha + ip} - \frac{is/pD}{1+is/pD}e^{ipz}\right], \tag{9}$$

$$\begin{aligned} T(z) = {} & \frac{iQ}{q\kappa}\left(\frac{E-E_g}{E}\right)\left(\frac{\alpha}{\alpha-iq}\right)\left[\frac{\alpha e^{iqz} + iqe^{-\alpha z}}{\alpha + iq}\right] \\ & + \frac{iQ}{q\kappa}\left(\frac{E_g}{E}\right)\left(\frac{-1/p^2}{D\tau}\right)\left(\frac{p}{\alpha-ip}\right)\left(\frac{p}{\alpha+ip}\right)\left(\frac{\alpha}{\alpha-iq}\right)\left[\frac{\alpha e^{iqz} + iqe^{-\alpha z}}{\alpha + iq}\right] \\ & + \frac{iQ}{q\kappa}\left(\frac{E_g}{E}\right)\left(\frac{-1/p^2}{D\tau}\right)\left(\frac{\alpha}{\alpha-ip}\right)\left(\frac{\alpha}{\alpha+ip} - \frac{is/pD}{1+is/pD}\right)\left(\frac{p}{p+q}\right)\left[\frac{pe^{iqz} - qe^{ipz}}{p - q}\right] \\ & + \frac{iQ}{q\kappa}\left(\frac{E_g}{E}\right)\left(\frac{\alpha}{\alpha-ip}\right)\left(\frac{is/pD}{1+is/pD}\right)\left[\frac{e^{iq|z|} + e^{iqz}}{2}\right], \end{aligned} \tag{10}$$

and near the surface, i.e., for $|pz|$, $|qz|$, $\alpha z \ll 1$,

$$N(z) = \frac{iQ}{pDE}\left(\frac{\alpha}{\alpha-ip}\right)\left(\frac{1}{1+is/pD}\right)\left[1 + sz/D\right], \tag{11}$$

$$\begin{aligned} T(z) = {} & \frac{iQ}{q\kappa}\left(\frac{E-E_g}{E}\right)\left(\frac{\alpha}{\alpha-iq}\right) + \frac{iQ}{q\kappa}\left(\frac{E_g}{E}\right)\left(\frac{-1/p^2}{D\tau}\right)\left(\frac{\alpha p(\alpha-iq-ip+s/D)}{(\alpha-iq)(\alpha-ip)(p+q)(1+is/pD)}\right) \\ & + \frac{iQ}{q\kappa}\left(\frac{E_g}{E}\right)\left(\frac{\alpha}{\alpha-ip}\right)\left(\frac{is/pD}{1+is/pD}\right)\left[1 + iqz\right]. \end{aligned} \tag{12}$$

As noted earlier, the plasma density near the surface is a linearly increasing function of z whenever surface recombination is present. In a reciprocal manner we see from Eq. (12) that the temperature, because of the surface recombination heating, has a term that initially decreases linearly with distance from the surface. Of course, for distances far from the surface the behavior is determined by whichever exponentional terms in Eqs. (9) and (10) are most significant. In the limit of weak absorption, $\alpha \ll |p|$ and $\alpha \ll |q|$, the absorption is essentially uniform on the scale of a plasma or thermal diffusion length. The resulting plasma density profile is therefore the same as the source and independent of the plasma wave vector. In the absence of any surface recombination the temperature profile is similarly independent of the thermal wave vector. However, when surface recombination is present, the resulting temperature profile depends on the thermal and plasma wave vectors as well as the absorption coefficient. One way to appreciate the significance of this is to consider the dependence of the temperature on recombination effects. For distances far from the surface where $|pz|$ and $|qz| \gg 1$, only bulk recombination is significant. That is, $T(z)$ will increase by a factor $E/(E-E_g)$ as the bulk recombination lifetime varies from $\tau = \infty$ to $\tau = 0$. At the surface, however, the dependence of $T(0)$ on τ is affected by the amount of surface recombination that is present. With zero surface recombination the dependence on τ is the same at the surface as in the bulk. But as surface recombination increases, the surface temperature can <u>decrease</u> with decreasing τ. In the limit of strong surface recombination and with $|q| > |p|$, $T(0)$ will in fact decrease by a factor of $[E + (q/p-1)E_g]/E$ as τ decreases from ∞ to 0. In silicon, for example, $q/p \sim 4$ which means that this effect can be quite significant.

In the other extreme, $\alpha\mu \gg 1$, the spatial dependence is now set by the plasma and thermal wave vectors and Eqs. (9) and (10) reduce to

$$N(z) = \frac{iQ}{pDE}\left(\frac{1}{1+is/pD}\right)e^{ipz}, \tag{13}$$

$$T(z) = \frac{iQ}{q\kappa}\left(\frac{E-E_g}{E}\right)e^{iqz} + \frac{iQ}{q\kappa}\left(\frac{E_g}{E}\right)\left(\frac{-1/p^2}{D\tau}\right)\left(\frac{1}{1+is/pD}\right)\left(\frac{p}{p+q}\right)\left[\frac{pe^{iqz} - qe^{ipz}}{p-q}\right]$$

$$+ \frac{iQ}{q\kappa}\left(\frac{E_g}{E}\right)\left(\frac{is/pD}{1+is/pD}\right)e^{iqz} \quad . \tag{14}$$

To see how the thermal wave and plasma wave affect an optical probe beam, we consider the reflectance from a half-space. Letting R denote the unperturbed reflectance, then in terms of the complex index of refraction, $n + ik$, we have

$$R = \left|\frac{n + ik - 1}{n + ik + 1}\right|^2, \tag{15}$$

and applying a small perturbation, $\Delta n + i\Delta k$, we find for the modulated reflectance,

$$\Delta R/R = \mathrm{Re}\left(\frac{4}{(n+ik-1)(n+ik+1)}(\Delta n + i\Delta k)\right) \quad . \tag{16}$$

In deriving this expression we have tacitly assumed that the refractive index modulation is slowly varying with respect to the optical wavelength in the material. This is generally valid, since thermal and plasma wavelengths are typically much longer than the optical probe wavelength and, therefore, the modulated reflectance essentially measures the modulation of the refractive index at the surface. In semiconductors, however, there are electronic effects that can complicate the interpretation of a modulated reflectance measurement. As pointed out in Eq. (11) above, surface recombination gives rise to a plasma density that is an increasing function of z within an optical absorption length of the surface with a slope $\sim s/D$. For an optical probe wavelength, $\lambda = 633$ nm, and in Si with $n \sim 4$, this surface recombination-induced plasma density variation becomes significant on an optical probe scale if $s/D > 10/\mu m$. This is possible for damaged surfaces with $s > 10^5$ cm/sec and $D < 10$ cm^2/sec. Another effect to consider is carrier trapping at the surface that does not lead to recombination. In this case, as discussed in [2], one can have a very large density of electrons pinned at the surface if the trapped carriers are holes, and conversely, a large near-surface hole density when electrons are trapped. In any case, to properly understand the modulated reflectance in semiconductors we need to consider the effects of a spatial variation in the refractive index modulation that is not negligible on an optical scale. The simplest approach to understanding this problem is with a thin layer on a substrate for which the only differences between the layer and the substrate are their respective refractive index modulations. For a layer of thickness $d < 50$ Å and an index modulation Δn_s on a substrate with an index modulation Δn_b we find for the modulated reflectance

$$\Delta R/R = \frac{4n}{n^2 - 1}\left[\frac{\Delta n_b}{n} + \frac{4\pi nd}{\lambda}\left(\frac{2\pi nd}{\lambda} - \frac{2nk}{n^2 - 1} - \frac{\omega_{sc}}{\omega_{opt}}\right)\frac{\Delta n_s}{n}\right] \quad (17)$$

where ω_{sc} is the scattering rate governing the transverse mobility of carriers near the surface and ω_{opt} is the optical frequency of the probe laser. In obtaining this last result we've assumed the Drude effect

$$dn/dN = -\lambda^2 e^2/(2\pi nmc^2) \quad (18)$$

$$dk/dN = -(k/n + \omega_{sc}/\omega_{opt})(dn/dN) \quad (19)$$

which is valid for silicon where $k \ll n$. Equation (17) predicts a strong dependence of $\Delta R/R$ on the conditions of the surface of a semiconductor. In particular, for a processed Si wafer the measured $\Delta R/R$ signal can be expected to increase or decrease depending on the near-surface electronic, optical and carrier transport properties, as well as on the effective thickness of the near surface damaged region. These effects are, of course, in addition to the bulk transport and optical effects contained in Δn_b. However, being associated with the physical surface and consequently, most likely the electronic surface states, they are also subject to modification through the application of energy as, for example, is introduced by the pump laser. That is, if surface states are modified during a modulated reflectance measurement, then one might expect to observe a change in the $\Delta R/R$ signal with time. All of our experimental data on processed Si wafers exhibit a temporal behavior [2,3] that can be attributed to the surface of the sample. Thus we believe that a model of the kind presented here provides some of the explanation for the origin of a surface sensitive component in the modulated reflectance.

References

1. J. Opsal and A. Rosencwaig, Appl. Phys. Lett. 47, 498 (1985).
2. J. Opsal,M.W. Taylor,W.L. Smith and A. Rosencwaig, J. Appl. Pys. 61, 240 (1987).
3. A. Rosencwaig, J. Opsal, D.L. Willenborg, P. Geraghty and W.L. Smith, this issue.

Detection of Stable and Metastable Interface States in Silicon with Modulated Optical Reflectance

A. Rosencwaig, J. Opsal, D.L. Willenborg, P. Geraghty, and W.L. Smith

Therma-Wave, Inc., Fremont, CA 94539, USA

In modulated reflectance experiments on silicon wafers the signal arises primarily from the Drude modulation of the silicon refractive index from the laser generated electron-hole plasma density [1,2]. We have recently found that both the magnitude and phase of the modulated reflectance signal, $\Delta R/R$, are usually dependent on the duration of continuous pump and probe laser irradiation [1]. We attribute this temporal behavior of the $\Delta R/R$ signal to a laser induced modification of the defect-related electronic interface states at the Si/SiO_2 interface which act as trapping sites for the photogenerated carriers. Furthermore, there appear to be two broad categories of such interface states; stable and metastable. The stable states undergo temporal modification only under continuous high intensity laser irradiation, while the metastable states also exhibit spontaneous temporal evolution.

As discussed in the companion paper [2], the effect of the high intensity laser irradiation is to convert active (ie, detectable) interface states X^0 to inactive (ie, nondetectable) states X^1 by some energetic process. This energetic process may involve direct photon absorption or it may involve interactions between X^0 and local hot phonons generated by the nonradiative deexcitation of a photoelectron near the X^0 site or by an electron-hole recombination event at or near the X^0 site. In any case the conversion of X^0 to X^1 removes the effects of these electronic states on the $\Delta R/R$ signal.

Stable interface states do not spontaneously undergo the $X^0 \rightarrow X^1$ conversion process but rather need energy from the irradiating lasers to force this conversion. Thus X^1 will generally be at a higher energy level and upon cessation of laser irradiation the reverse $X^1 \rightarrow X^0$ process will usually occur in a spontaneous fashion over time.

Most Si wafers exhibit this reversible $X^0 \rightarrow X^1$ behavior. A notable exception are wafers with a high quality thermally grown SiO_2 oxide layer. These wafers exhibit no temporal dependence. Such wafers have very good Si/SiO_2 interfaces with most defect states compensated and with only intrinsic dangling bond interface states acting as amphoteric trapping sites. The fact that wafers that do not have such high quality Si/SiO_2 interfaces appear to exhibit a $\Delta R/R$ temporal behavior indicates that we are especially sensitive to the extrinsic defect-related states rather than to the intrinsic dangling bond states.

In Figs. 1 and 2 we show examples of the stable interface state behavior for a number of Si wafers. In the case of wafers that have undergone dry etching or ion implantation, the data was taken many days after the process so as to ensure that only the stable interface states were present. Of considerable interest is the situation in Fig. 1, where we present data on a wafer that has been ion implanted through a screen oxide. Before implantation this wafer exhibits no temporal laser irradiation behavior indicative of a high-quality Si/SiO_2 interface. After implantation the wafer

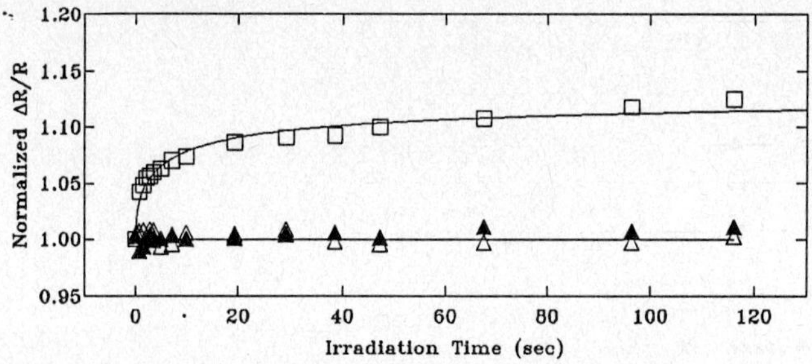

Fig.1 Temporal behavior of ΔR/R under continuous laser irradiation for a 10^{15} As implant through a 300 Å oxide layer: △- before implant; □- after implant; ▲- after 1000 °C anneal

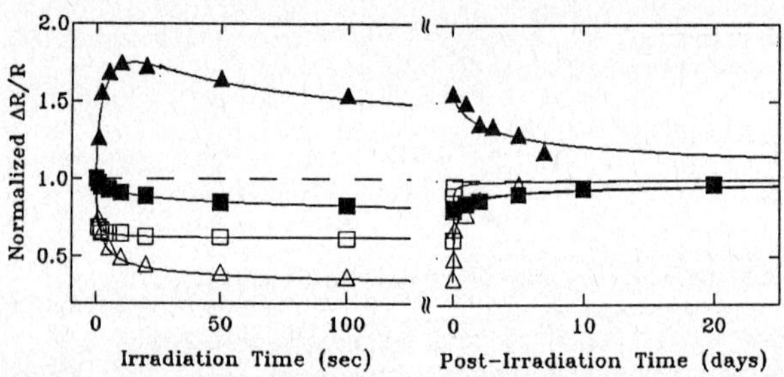

Fig.2 Temporal behavior of normalized ΔR/R under continuous laser irradiation for Si wafers with stable interface states: □- 0.21 Ω cm p-type epitaxial Si; ■- $1x10^{13}$ As implant; △- NF_3 plasma etch; ▲- Ar plasma etch

exhibits temporal behavior arising from the presence of defect-related interface states from both lattice damage and implanted ions. After a subsequent high temperature anneal there is no evidence of a temporal dependence indicating that the lattice damage has been removed and that states associated with the implanted ions at the Si/SiO_2 interface have been deactivated or made nondetectable. In Fig. 2 we show the laser induced temporal behavior for an epitaxial Si sample, two plasma etched samples and an ion implanted sample. We also show the reversible behavior after the laser irradiation ceases. Note that the reverse $X^1 \rightarrow X^0$ process is much slower at room temperature than the $X^0 \rightarrow X^1$ process.

When a silicon surface or Si/SiO_2 interface is subjected to an energetic physical or chemical process a variety of new electronic states may be generated and existing states altered. The new states that are active (detectable) trapping sites will affect the ΔR/R signal. Some of these new states are in energetically stable configurations and will behave as stable interface states exhibiting temporal behavior only under intense laser irradiation. Others are in energetically metastable configurations or are associated with transient defect species. Metastable states are thus trapping states that deactivate spontaneously with elapsed time and require no laser irradiation.

Figure 3 illustrates this spontaneous temporal behavior for a variety of Si wafers that have undergone physical and/or chemical processing such as ion implantation, dry etching or wet chemical dipping. In Fig. 3(a) we show the actual increases in ΔR/R signal due to the various processes as well as the

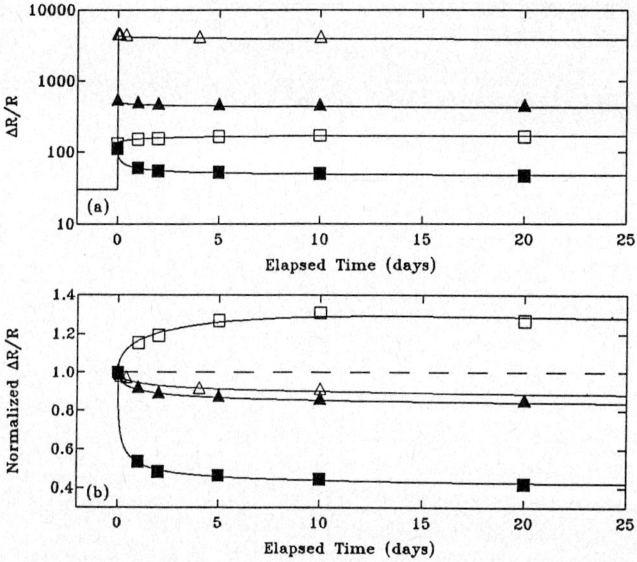

Fig.3 (a) ΔR/R for Si wafers before and after a physical or chemical process. (b) Temporal behavior of normalized ΔR/R with elapsed time after the physical or chemical process: △- $1x10^{13}$ As implant; ■- chemical clean with RCA dip (NH_3 + H_2O_2 + H_2O); ▲- NF_3 plasma etch; □- Ar plasma etch

subsequent temporal behavior. In Fig. 3(b) we show this post-processing temporal behavior more clearly by normalizing the ΔR/R signals to their initial post-process values. In these experiments the ΔR/R data is obtained while the wafer is moving rapidly under the laser beams so that any local region is exposed to the beams for only a few milliseconds. The observed temporal changes in the ΔR/R signal are thus due solely to the spontaneous deactivation of the metastable interface states with elapsed time after the physical or chemical process. After several days, the only interface states left are the stable states. If wafers with metastable states are subjected to continuous high intensity laser irradiation, the deactivation of the metastable states is greatly accelerated with complete metastable deexcitation usually occuring in seconds. Upon cessation of the laser irradiation, the metastable states do not reactivate. Thus, unlike the stable state situation, laser or temporal deactivation of metastable states is an irreversible process.

The ability to detect and quantify both stable and metastable interface states in Si provides a valuable tool for process development and control.

References

1. J. Opsal, M.W. Taylor, W.L. Smith and A. Rosencwaig, J. Appl. Phys. 61, 240 (1987).
2. J. Opsal and A. Rosencwaig, this issue.

Study of Photoacoustic Spectra of Amorphous Semiconductor Superlattice Films

Zhi-chao Wang[1], *Shu-ye Qiu*[2;3], *and Shu-yi Zhang*[2]

[1]Department of Physics, Nanjing University, Nanjing, People's Republic of China
[2]Institute of Acoustics, Nanjing University, Nanjing, People's Republic of China
[3]Present Address: Department of Physics, Shantou University, Guangdong Province, People's Republic of China

Some a-Si:H/a-SiN$_x$:H superlattice films have been studied by photoacoustic spectroscopy. The "blue shift" phenomenon induced by quantum size effects has been observed. Comparing the photoacoustic spectra and the optical absorption spectra of the same materials, it is shown that the nonradiative recombination of the photogenerated carriers (electrons and holes) in a-Si:H/a-SiN$_x$:H superlattices is much larger than that of intrinsic silicon and can be easily investigated by photoacoustic spectroscopy.

1. Introduction

Amorphous semiconductor superlattices, which are new materials, were first produced in 1983 [1]. Due to their special properties, they immediately attracted great interest. Studies of their electrical and optical properties were made and achieved great success. In this paper, we present a study of the photoacoustic spectroscopy (PAS) of a-Si:H/a-SiN$_x$:H superlattice films. The transport and the photoacoustic properties associated with quantum size effects and nonradiative recombination of carriers have been observed and discussed.

2. Sample Preparation

A series of samples of a-Si:H/a-SiN$_x$:H superlattice films were prepared by a plasma-assisted chemical vapor deposition (PCVD) device [2]. The reactive gases, SiH_4 for the a-Si:H layers and $SiH_4 + NH_3$ for the a-SiN$_x$:H layers, were changed periodically in the reaction chamber of the device. The sample series of five samples was deposited on glass substrates with a fixed silicon nitride (a-SiN$_x$:H) thickness (80 Å) and a range of amorphous silicon (a-Si:H) thicknesses varying from 300 Å to 16 Å. The a-Si:H and a-SiN$_x$:H monolayers were also prepared under the same conditions. The total thickness of each sample was about 1 μm. In addition, an intrinsic crystal silicon (c-Si) wafer with thickness 0.2 mm was prepared for comparison.

3. Experimental Results

3.1 PAS Spectra

The PAS spectra for the above eight samples were obtained at room temperature by a specially constructed single-beam photoacoustic spectrometer with a resolution of 8 nm [3]. The results of PAS are shown in Fig. 1a.

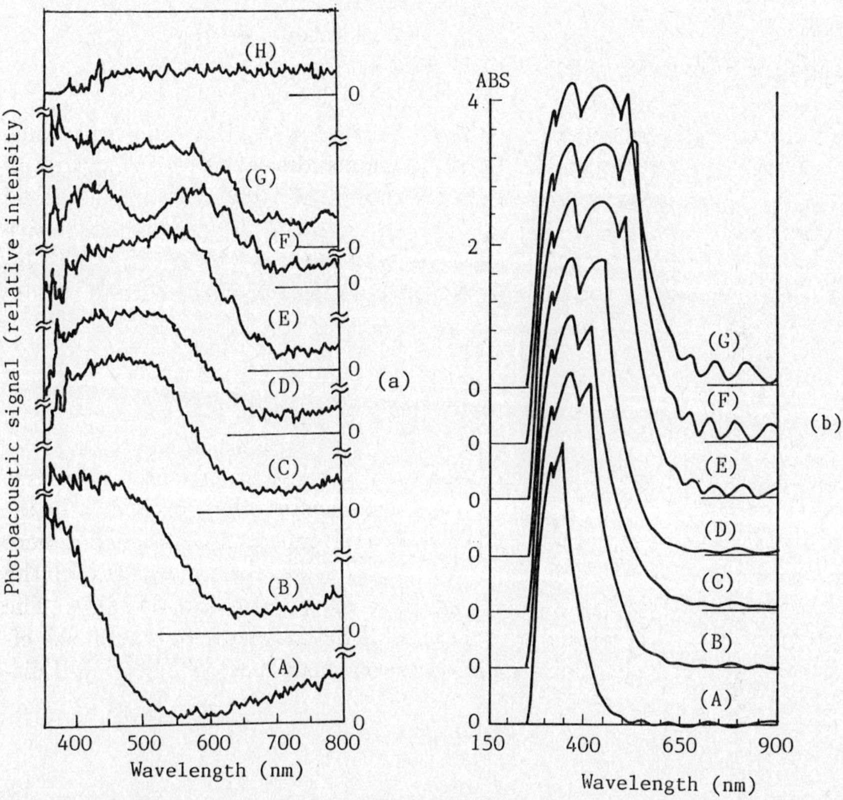

Fig. 1. (**a**) Photoacoustic spectra; (**b**) optical absorption spectra. (*A*) a-SiN$_x$:H; (*B*) a-Si:H(16 Å)/a-SiN$_x$:H(80 Å); (*C*) a-Si:H(32 Å)/a-SiN$_x$:H(80 Å); (*D*) a-Si:H(48 Å)/a-SiN$_x$:H(80 Å); (*E*) a-Si:H(100 Å)/a-SiN$_x$:H(80 Å); (*F*) a-Si:H(300 Å)/a-SiN$_x$:H(80 Å); (*G*) a-Si:H; (*H*) c-Si

3.2 Optical Absorption Spectra

To explain the physical mechanism of the PAS spectra, the optical absorption spectra of the same samples have also been measured by a Beckmann DU-8B spectrophotometer. The results are shown in Fig. 1b. Using Tauc's equation

$$(\alpha h\nu)^{1/2} = B(h\nu - E_g) \quad ,$$

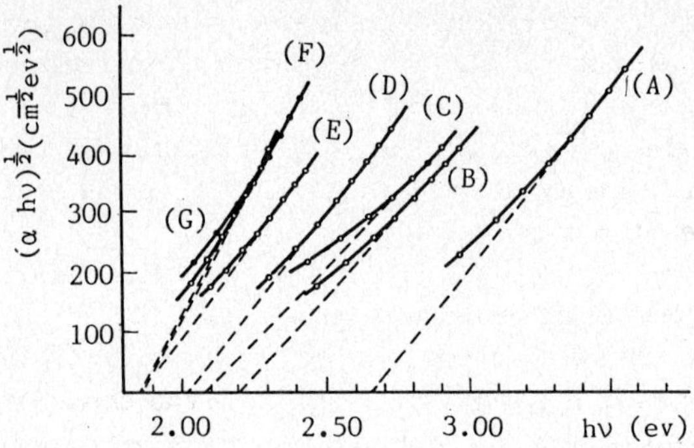

Fig. 2. $(\alpha h\nu)^{1/2}$ vs $h\nu$ curves.
(*A*) a-SiN$_x$:H
(*B*) a-Si:H(16 Å)/a-SiN$_x$:H(80 Å)
(*C*) a-Si:H(32 Å)/a-SiN$_x$:H(80 Å)
(*D*) a-Si:H(38 Å)/a-SiN$_x$:H(80 Å)
(*E*) a-Si:H(100 Å)/a-SiN$_x$:H(80 Å)
(*F*) a-Si:H(300 Å)/a-SiN$_x$:H(80 Å)
(*G*) a-Si:H

where α is the absorption coefficient, $h\nu$ is the incident photon energy, B is a constant and E_g is the optical energy gap, we can easily obtain the curves of $(\alpha h\nu)^{1/2}$ vs $h\nu$ from Fig. 1 as shown in Fig. 2. In Fig. 2, the intersections of the dotted lines and the horizontal axis represent the optical gap E_g of the corresponding samples.

4. Discussion and Conclusions

Our results lead to the following conclusions:

(1) In our case, the energy band structure of a-Si:H/a-SiN$_x$:H multilayer films is shown in Fig. 3; a-Si:H and a-SiN$_x$:H sublayers are potential wells and barriers, respectively. According to quantum theory, in multilayer films the energy states of the carriers (electrons and holes) located in a well should separate into many levels when the width of the well is small. Thereby, the gap E_g of a-Si:H/a-SiN$_x$:H superlattices should become larger. This is the so-called "quantum size effect". It can be verified by optical experiments.

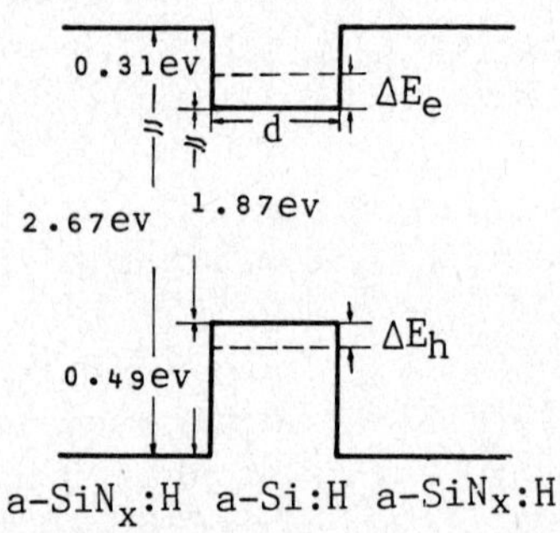

Fig. 3. The energy band structure of a a–Si : H/a–SiN$_x$: H superlattice

From Figs. 1a and b, we can see that the absorption edges appear in the photoacoustic spectra as well as the optical absorption spectra. It is very clear that the absorption edge shifts to a shorter wavelength of the incident light with decreasing a-Si:H thickness (potential well width) in the multilayer films;

this is called the "blue shift" of the absorption edge. It means that the gap becomes larger with decreasing thickness of the a-Si:H sublayer, because the intrinsic absorption is only produced for an incident photon with $h\nu > E_g$.

In accordance with Fig. 2, the resulting optical gap is a function of the thickness of the a-Si:H sublayer. This result agrees with that of *Abeles* and *Tiedje* [1] and has been interpreted as a quantum size effect.

In addition, for both samples a-Si:H(100 Å)/a-SiN$_x$:H(80 Å) and a-Si:H (300 Å)/a-SiN$_x$:H(80 Å), the locations of the absorption edges are almost the same as that of a-Si:H monolayer film, as shown by curves (E), (F) and (G) in Fig. 1, and their gap is almost the same, as shown in Fig. 2. We can conclude that these two multilayer samples do not exhibit apparent quantum size effects because they have much larger potential well widths.

(2) In amorphous semiconductors, the PAS spectra exhibit clear absorption edges, as shown in Fig. 1a. The PAS signal of a c-Si wafer of 0.2 mm thickness is very small even in the intrinsic absorption region. This implies that there are a great number of nonradiative recombination centers in the films.

In the a-Si:H film, there is a large density of defects ($10^{15} - 10^{16}$ cm^{-3}), some of them are the nonradiative recombination centers of the carriers and the others are the radiative recombination centers. It has been discovered experimentally that the nonradiative recombination of photogenerated carriers (electrons and holes) increases with the dangling-bond density [4, 5]. This means that the nonradiative recombination is related to the defects or the dangling-bonds (such as D^0 and $T_3^+ - T_3^-$ pairs) in an a-Si:H film.

In a-SiN$_x$:H films, except for the above conditions, the existence of N must be considered. From Fig. 1a, we can see that the PAS signal in a-SiN$_x$:H is much larger than that of a-Si:H, although they have the same thickness. So, the PAS signal of the a-SiN$_x$:H film is related to the number of Si-N bonds.

Of course, it is much more complicated for a-Si:H/a-SiN$_x$:H samples because the multilayer samples not only have more interfaces but also more components of a-SiN$_x$:H with decreasing thickness of a-Si:H sublayers. Thus, their PAS signals increase correspondingly.

(3) Comparison between Figs. 1a and b shows that both spectra of each sample have different shapes, although the location of the absorption edge for each sample in Fig. 1a agrees with the corresponding one in Fig. 1b.

In Fig. 1b, the absorption edges are considerably sharper and the differences of the absorption between the strong and the weak absorption regions are very large. In Fig. 1a, compared with Fig. 1b, the absorption edges slope more gently and the differences between the strong absorption and the weak absorption regions are smaller. Since there are many nonradiative recombination centers in the gap of the amorphous semiconductors, the PAS signals produced by nonradiative recombination are much larger, whether in the exponential or the weak absorption regions. However, in the intrinsic absorption (strong absorption) case, due to the face that most of the photogenerated carriers undergo radiative recombination, and only a small fraction of the carriers undergo nonradiative recombination, the photoacoustic signals do not increase very much, although a great number of photogenerated carriers are produced.

Thus, it can be shown that, under illumination with light, the carrier transitions between the band tail states of amorphous semiconductors occur mainly by nonradiative recombination, but the carrier transitions from band to band are mainly radiative.

Because the physical mechanisms of the photoacoustic effect in superlattices are very complicated, more detailed theoretical and experimental studies must be carried out.

5. References

1. B. Abeles, T. Tiedje: Phys. Rev. Lett. **51**, 2003 (1983)
2. Z.C. Wang, X.N. Liu, X.M. Feng, X.S. Gun: Acta Phys. Sin., to be published (in Chinese)
3. S.Y. Que, S.Y. Zhang, C.N. Hu, L.H. Wei: J. Appl. Sci. **4**, 207 (1986) (in Chinese)
4. K. Morigaki: Solid State Phys. **20**, 528 (1985) (in Japanese)
5. R. Street: In *Semiconductors and Semimetals*, Vol. 21, Part B, ed. by J.I. Pankove (Academic, New York 1984) Chap. 7

Transducer-Less SEAM of III–V Compound Semiconductors

N. Kultscher[1], *K. Steiner*[2], *and L.J. Balk*[1]

[1]Fachgebiet Werkstoffe der Elektrotechnik, Kommandantenstr. 60, D-4100 Duisburg 1, Fed. Rep. of Germany
[2]Fachgebiet Halbleitertechnik/-technologie, Kommandantenstr. 60, D-4100 Duisburg 1, Fed. Rep. of Germany

1. Introduction

Scanning electron acoustic microscopy (SEAM) gives information on thermoelastic and electronic specimen features /1/. For semiconducting material and devices good resolutions and high contrasts can be attained due to the simultaneous appearence of various sound generation mechanisms, the most important being coupling via excess carriers /2/. However, there are disadvantages in usual SEAM caused by the adaption of a detection transducer /3/. For piezoelectric materials (like III-V semiconductors) an additional transducer can be avoided as the sample itself can serve as detector. For the example of InP, requirements for the use of this technique are discussed.

2. Usual SEAM Set-up

In SEAM usually a separate piezoelectric transducer detects the acoustic waves /3/. The specimen has to be in acoustic contact with this detector to minimize reflection and absorption losses. To achieve such an acoustic contact, coupling fluids or glue can be used. However, this results in low signals. Higher signals can be obtained by pressing the sample onto the transducer, which can lead to artefacts in the images (e.g. contrasts caused by stress /4/) or even to specimen damage. Especially for measurements accompanying technology it is important to leave the sample untreated to enable its use in further processes. Moreover, it is important to quantify SEAM contrasts. Applying a transducer this is complicated, as its transfer function depends on frequency and on the transducer assembly geometry.

3. Transducer-less SEAM Set-up

Due to their crystal lattice III-V-semiconductors are piezoelectric /5/. In these materials a chopped electron beam generates acoustic signals directly via the ac electric field caused by excited excess carriers without the need for thermal waves /2/. The acoustic waves are always accompanied by an electric field caused by the displacement of the charged lattice sites /6/. This charge displacement is a measure for the magnitude of the acoustic wave and can be measured by electrodes connected to the specimen. Thus, it is not necessary to apply an extra transducer with all its disadvantages, and solely specimen features are responsible for the SEAM signal. Contacting the specimen is easy with micromanipulators or by bonding gold wires to existing electrodes. With such localized detectors it is possible to investigate the directivity of the sound generation /7/ or the contrast dependence on electrode positions. With shaped electrodes (interdigital transducers /8/) extremely high sensitivities can be attained. Despite these advantages some problems are important with this transducer-less technique. Illumination of a semiconductor with an electron beam can lead to competetive signals like absorbed electron current (AE) or electron beam induced current (EBIC), which have to be seperated from SEAM signals. It is necessary to use ohmic measuring contacts and to locate them several minority carrier diffusion lengths away from the investigated sample location. The top sample surface has to be connected to ground either directly or via a resistor.

4. Specimen Preparation

An overall ohmic contact has been applied to the bottom of an InP sample with a background doping of $n=1.4x10^{16}cm^{-3}$. The upper surface is selectively Zn diffused via a mask with $p=2x10^{18}cm^{-3}$ to a depth of ~4µm. Ohmic contacts (thickness 0.1µm) have been alloyed onto the surface by a second mask so that some contacts are located on top of a diffused zone and others directly on the substrate. Two contacts are connected by gold wire bonds.

5. Experimental Results

5.1 Electron Acoustic Micrographs

All images were taken at 30keV beam energy. Fig. 1a shows the secondary and backscattered electron (SE/BE) image of the specimen, in which the ohmic contacts and defects (A, B, and C) are to be seen. The SEAM magnitude image of the same area, employing detection at the bottom contact, gives more information (Fig. 1b). It shows dark regions where the Zn diffusion has taken place, like at site D, where a cross-like doping structure used to adjust the masks is shown, which in the SE/BE image, is hidden by the metal contact. The SEAM contrast is not due to the metal layer, as can be seen at another contact located directly on the substrate (E). The bright rim around this contact is caused by the process of alloying which is necessary to produce ohmic contacts in this material causing metal atoms to diffuse into the substrate. Other white rims of about diffusion length width are located around any structure where a pn-junction exists. This gives further information on the sample. One defect (A) disturbed the diffusion process, visible by the fact that the white rim does not reach area F. Similar effects show up beneath contacts B and C, as here the white zone does not surround the whole contact area thought to be the diffused region, because these defects occured already during the diffusion process. Besides these contrasts there are bright and dark spots and areas correlated to inhomogenities of the substrate itself. It is also possible to detect SEAM signals at the top surface of the specimen (Fig. 2, different sample area). The contrast in this image is restricted to small regions beside the pn-junction. Exactly at the location of the junction the SEAM signal is diminished due to the separation of excess carriers /2/.

5.2 Contrasts in Transducer-less SEAM

When comparing SEAM images by this transducer-less technique with conventional ones, contrasts seem to be inverted. This cannot be due to possible phase shifts as the images are recorded in the magnitude mode of the lock-in. A likely reason for the changed contrast is given by the form of the piezoelectric stress matrix of InP (and of most other III-V compounds/5/) which creates an electric field perpendicular to an applied stress. Usual transducers measure the longitudinal strain reaching the bottom specimen surface. The bottom sample contact for SEAM detection, however, measures those strain magnitudes being parallel to the surface. Use of surface SEAM detection gives a measure of strain magnitudes perpendicular to the surface. Therefore the appearance of these images should be comparable with conventional images. This can be recognized for the detection of the pn-junctions as black areas in Fig. 2. The contrasts obtained in the various detection

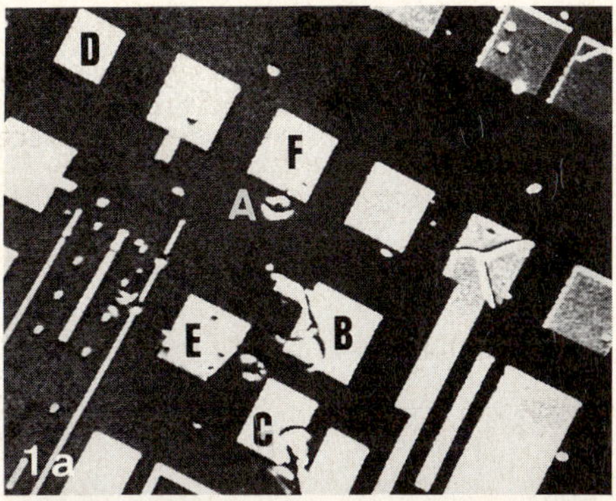

Fig. 1: Transducer-less SEAM image a) SE/BE, b) SEAM magnitude, transducer at bottom surface ⊢—⊣ 100µm

Fig. 2: Transducer-less SEAM image with surface detection ⊢—⊣ 100 µm

Fig. 3: On possible contributions to SEAM
a) SE/BE, b) AE, c) EBIC, d) SEAM magnitude, transducer at bottom surface ⊢—⊣ 400 µm

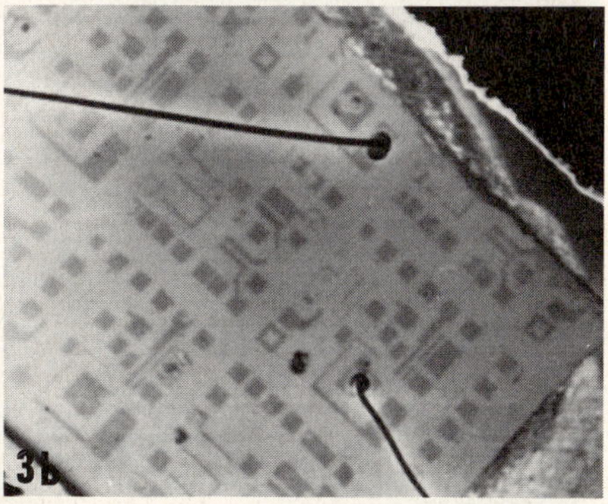

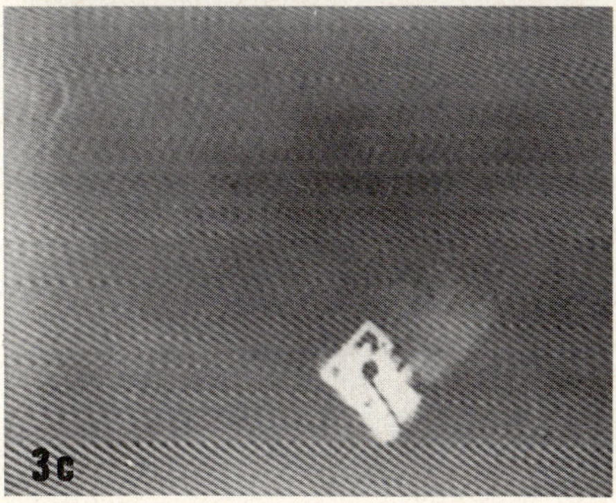

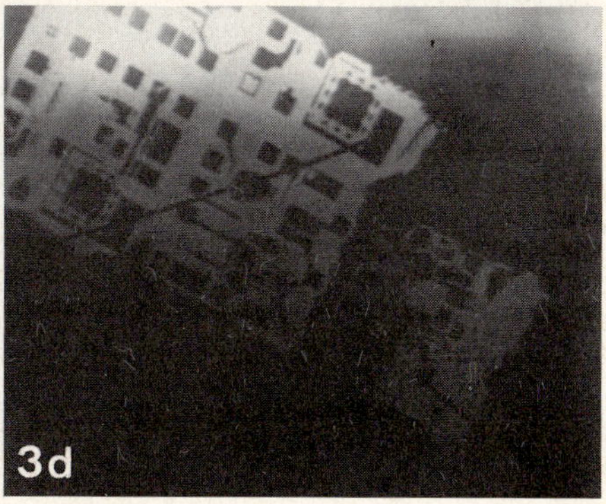

modes lead to the demand to treat the influence of electric fields on SEAM three-dimensionally. Fields being either parallel or perpendicular to the direction of sound detection show inverted contrasts, a fact due to the mentioned matrix properties.

5.3 On the Influence of Competitive Signals

SEAM signals in the transducer-less set-up may be falsified by other signals, the most important being AE and EBIC. AE is the difference between primary electrons and those leaving the sample. Thus, an AE image shows an inverted contrast with respect to the SE/BE image (Fig. 3a). Without shunting these charges to ground an AE image of the surface is measured by the SEAM equipment. The only features to be seen then are of the same contrast origin as in Fig. 3a. But, even if these AE are collected by surface electrodes and shunted to earth, it is possible to measure a non-acoustic signal between the top and the bottom electrodes due to a charge separation current (EBIC) at an electric barrier. This current decreases exponentially with the distance between the electron beam and the contacted area /9/. Thus, a reasonable EBIC signal can only be detected in the vicinity of such contacts (Fig. 3c). The corresponding SEAM image (Fig. 3d) is restricted to two areas which are separated by a dark region where no signal is measurable. This region is marked by a surface scratch (also visible in the SE/BE image) causing electrical insulation between the various surface areas. As only those contacted by a gold wire will be maintained on ground potential, only these will be imaged with a sufficient SEAM signal.

6. Conclusions

Transducer-less SEAM is able to give significant information about structure and defects of piezoelectric semiconductors without the risk of artefacts or specimen damage. Moreover, the restriction of the signal generation exclusively to the interesting material parameters enhances the chance of quantitative measurements. Provided the necessary care is taken to avoid competitive signals, this technique may become an excellent tool for the investigation of piezoelectric semiconductors.

7. Acknowledgements

The authors wish to thank Prof. E.Kubalek and Prof. K.Heime for helpful discussions of the experimental results. The work was funded by the Deutsche Forschungsgemeinschaft.

1. L.J. Balk and N. Kultscher: Beitr. elektronenmikroskop. Direktabb. Oberfl., BEDO-16, p. 107 (1983)
2. N. Kultscher and L.J. Balk: J. Scanning Electron Microscopy, Part I, p. 33 (1986)
3. G.S. Cargill III: In Scanned Image Microscopy, ed. by E.A. Ash, (Academic Press, London 1980) p. 319
4. L.J. Balk and N. Kultscher: J. de Physique, 45 (2), p. C2-869 (1984)
5. B.A. Auld: Acoustic Fields and Waves in Solids, Vol. 1, John Wiley&Sons, New York 1973
6. Yu.V. Gulyaev, I.I. Chusov, T.V. Bugaeva: Sov. Phys. Semicond. 19, p. 343 (1985)
7. R.P. Huebner and W. Metzger: J. Scanning Electron Microscopy, Part II, p. 617 (1985)
8. K. Yamanouchi, S. Kudo, Y. Wagatsuma: Japan J. Appl. Phys. 23, Suppl. 23-1, p. 191 (1984)
9. L.J. Balk, E. Menzel, E. Kubalek: 8th Int. Congr. X-ray Optics and Microanalysis, ed. by D.R. Beaman et al., (Pendell Publ. Co., Midland 1977), p.613

Time and Spatially Resolved Investigation of Transport in Semiconductors by the Mirage Effect

J. Pelzl[1], D. Fournier[2], and A.C. Boccara[2]

[1]Institut für Experimentalphysik, Ruhr Universität, D-4630 Bochum 1, Fed. Rep. of Germany

[2]Laboratoire d'Optique Physique, ESPCI, 10, rue Vauquelin, F-75231 Paris Cedex 05, France

1. Introduction

The mirage effect relies on the deflection of a light beam due to a gradient of the refractive index and therefore offers an ideal tool to study local changes of the optical properties as a function of time /1/. Time-dependent processes which are very efficient in producing a spatially varying refractive index comprise the diffusion of heat, of mass and of charge carriers. Recently, the mirage effect has been applied to silicon to study the transport of heat and of photocarriers /2,3/. In these experiments heat and plasma waves were generated in the course of the absorption of an intensity modulated argon ion laser beam. The deflection of a probe beam passing through the silicon specimen was measured as a function of the modulation frequency at different probe beam positions. An alternative method to investigate heat and plasma waves in regions near the surface is provided by the modulated reflectance, which has been successfully applied to the inspection of silicon wafers /4/. A common feature of both experiments is the use of the frequency domain as the dynamical scale. However, the different time scales involved in the heat and the carrier diffusion processes suggest time domain measurements to separate the two mechanisms. In this work we therefore set up an experiment where we measure the time dependence of the deflection of a probe laser beam inside the crystal in the course of photocarrier production in the surface region due to the absorption of a short laser pulse. The present studies are conducted on the same silicon specimen used in the frequency domain measurements of /2/ and /3/. This circumstance admits a quantitative comparison of the merits of both experimental procedures.

2. Experimental

The experimental set-up is basically composed of three units. The pulse laser system for the generation of the photocarriers in the surface region is a Nd-YAG laser followed by a frequency doubler. Attenuators are inserted in the light path to ensure a low power level of the absorbed pulse. The arrangement for the deflection measurement consists of a He-Ne laser running on the 1150 nm line focussed into the crystal and finally the signal processing unit with a transient digitizer having a time resolution of 100 ns per channel. The time variation of the deflection signal at different probe beam positions a distance x from the surface was accumulated up to 1000 times depending on the signal-to-noise ratio. Typical repetition rates were 1 - 10 Hz.

3. Results and Discussion

The vertical deflection of the probe beam is composed of two contributions having opposite directions (Fig.1). The negative deflection in the fast

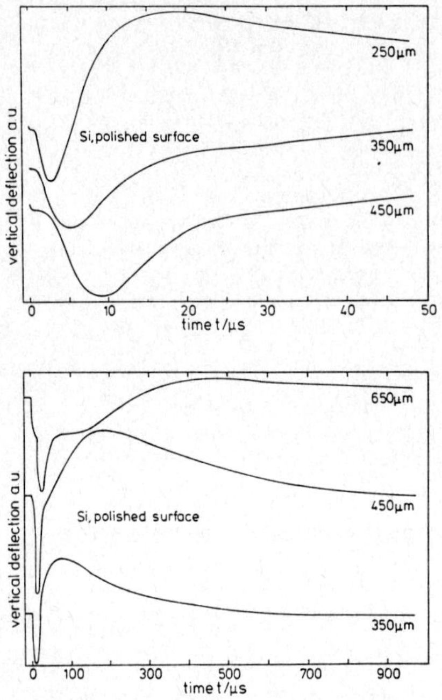

Fig.1 Long time (lower part) and short time (upper part) behaviour of the vertical deflection signal in a Si single crystal measured at different beam positions x from the illuminated surface

time regime results from the refractive index change due to the diffusing photo carriers, the positive signal that develops at longer time scales is caused by the temperature change associated with the diffusing heat pulse. The lower part of Fig.1 shows the evolution of the signals for different probe beam positions with respect to the surface where the laser pulse is absorbed ccmprising both the carrier peak and the thermal deflection signal. Additionally the carrier peaks occurring in the range below 0.1 ms have been measured with higher time resolutions. Three selected recordings are shown in the upper part of Fig.1.

To analyze the measured time dependence, the coupled carrier and heat diffusion equations already published in /3/ have been solved in the time domain representation. The source terms incorporated in the boundary conditions (eqs. 3 and 4 of /3/) which take into consideration the heat production and the photo carrier generation due to the absorption of light are approximated by a delta function. The carrier problem can be solved rigorously:

$$N(x,t) = n_{Ph} \cdot e^{-t/\tau} \left\{ \frac{e^{-\frac{x^2}{4Dt}}}{\sqrt{\pi D t}} - \frac{s}{D} e^{\left(\frac{s}{D} x + \frac{s^2}{D} t\right)} \cdot \mathrm{erf}\left[\frac{x}{\sqrt{4 D t}} + s \sqrt{\frac{t}{D}}\right] \right\};(1)$$

N is the minor-carrier density, D, τ and s are the diffusion coefficient, bulk recombination time and surface recombination velocity of these carriers. n_{ph} is the number of photons absorbed per cm^2. The thermal problem has been solved by neglecting the heat production due to the recombination processes in the bulk. In addition, the time evolution of the carrier re-

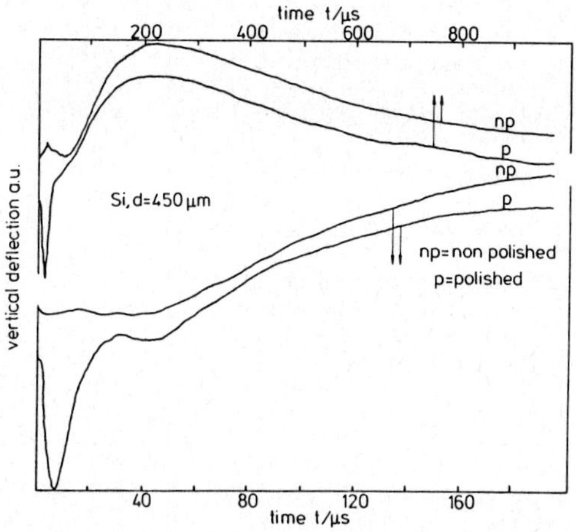

Fig.2 Comparison of the deflection signals obtained from a single Si crystal with a polished (p) and unpolished (np) surface illuminated by laser pulse with the probe beam passing at the same distance x

combination process at the surface has been approximated by a delta function. These assumptions, which do not severely affect the long time behaviour, lead to a simple relation for the heat flux f(x,t):

$$f(x,t) = C \cdot \left[x \,/\, \sqrt{4\pi D_{th} \cdot t^3} \right] \cdot e^{-\frac{x^2}{4D_{th} \cdot t}}, \tag{2}$$

where D_{th} is the thermal diffusivity of the solid. An experimental proof of the relative importance of the bulk recombination and surface recombination processes for the thermal response is provided by the comparative study shown in Fig.2. Here the vertical deflection signal has been measured under the same conditions except that in one experiment a polished surface was illuminated by the laser light whereas in the complementary measurement the surface was roughened. In the latter case a great number of traps in the surface region lead to a high surface recombination velocity, therefore hardly any minority-carriers will diffuse into the bulk. This is confirmed by the experimental observation. The recording obtained from the roughened Si surface does not further display the carrier peak in the deflection signal. In addition, the peak behaviour of the thermal response does not seem to be influenced significantly.

In order to deduce quantitative numbers on transport properties from the measured time dependences, we have tried to avoid complex and time-consuming fitting procedures. We only consider the maximum of the carrier and of thermal response and analyze their positions as a function of the probe beam distance x from the surface. The influence of the finite probe beam diameter has been taken into consideration by a convolution with a rectangular intensity profile. From the slopes of the asymptotic behaviours at large (Fig.3, lower part) and small (Fig.3, upper part) probe beam positions the following values are obtained: $D = 43\ cm^2/s$ and $\tau = 35$ µs. Within this procedure, the surface recombination velocity s is accessible only through the intersection of the linear behaviour where the separation of s and of the beam width becomes difficult. A fit to the time dependence of the carrier signal yields s = 1500 cm/s, a velocity that is much higher than the value reported in /3/. Analyzing the peak position of the thermal response in a similar treatment as already discussed, we obtain from the slope $D_{th} = 0.75\ cm^2/s$.

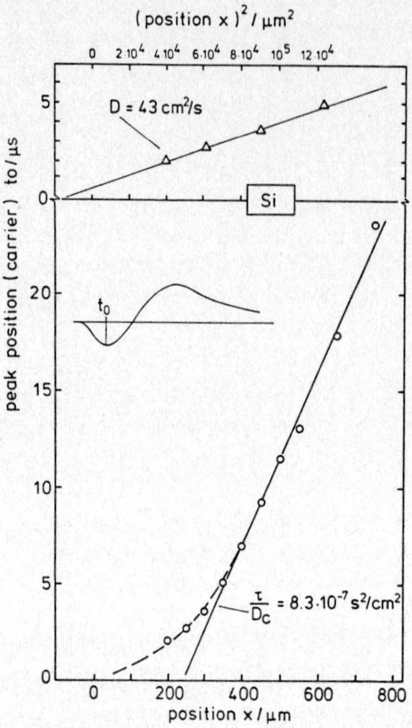

Fig.3 Peak position t_o of the carrier signal as a function of the probe beam distance. The theoretical asymptotic behaviours for short times (upper part) and for longer time scales (lower part) are adjusted to experimental data indicated by the open symbols.

Work supported by the Commission of the European Communities

References

1. W.B. Jackson, N.M. Amer, A.C. Boccara and D. Fournier: Appl.Optics 20, 1333(1981)
2. A. Skumanich, D. Fournier, A.C. Boccara and N.M. Amer: Appl.Phys.Lett. 47, 402(1985)
3. D. Fournier, A.C. Boccara, A. Skumanich and N.M. Amer: J.Appl.Phys. 59, 787(1986)
4. J. Opsal and A. Rosencwaig: Appl.Phys.Lett. 47, 498(1985)

Thermal and Optical Characterization of Thin Films by Mirage Detection

J.P. Roger, F. Lepoutre, D. Fournier, and A.C. Boccara

Ecole Supérieure de Physique et Chimie, Laboratoire d'Optique Physique, ER 5, CNRS, 10, rue Vauquelin, F-75231 Paris Cedex 05, France

Abstract : High frequency photothermal deflection measurements on semiconductor films lead to the determination of the films'thermal diffusivity. Spatial location of subbandgap absorption is deduced from signal phase spectral dependence.

The location of absorbing centers in thin films like stack of coatings (laser mirrors - antireflective coatings ...) or in semiconducting thin films (for instance dangling bonds in amorphous silicon ...) requires the determination of both spectral and thermal sample parameters.

Up to now, modulated photothermal deflection has been successfully applied to the thermal diffusivity determination of samples which exhibit a low characteristic frequency (f_c = diffusivity/(sample thickness)^2) /1/, or thermally thick samples /2/ by using various experimental and theoretical (1 D or 3 D) approaches. In this paper, we show that a photothermal deflection experiment can also be used, in a one-dimensional configuration, for the thermal characterization of thin films deposited on a substrate. Since thermal diffusivity is deduced from analysis of the photothermal signal beyond f_c, such a determination requires high-frequency measurements. In this case, when the thermal diffusion length of the deflecting medium becomes of the order of the probe beam diameter, the convolution between the probe beam and the thermal gradient strongly affects the frequency dependence of the photothermal signal. This problem, and also the critical determination of the distance between the sample and the probe beam, can be overcome, in a simple manner, by subtracting two phase measurements in which the thermal diffusion in the deflecting medium is introduced in exactly the same way. When the substrate is transparent, such a differential method may consist in the measurement of the relative phase lag $\Delta\varphi$ between the signals obtained for the rear surface and front surface excitations as defined in the inset of Fig. 1. We have applied this procedure to a 3.5 μm thick polycrystalline $CuInSe_2$ film deposited on glass. The sample is illuminated in air, by an unfocused Ar^+ laser so that the heat source is created at the film surface. The results reported in Fig. 1 lead to f_c = 54.7 kHz and to a value of the substrate to film effusivities ratio close to 1. The film thermal diffusivity value is found to be five times smaller than the single crystal one obtained by the 3 D method proposed by Kuo et al. /2/.

When the heating is only possible at one sample surface (rear or front) a differential method requires a calibration measurement with a second (thermally thick) sample. The distance probe beam to sample is carefully kept constant in the two phase measurements by positioning the sample surfaces on a reference plane. This second procedure has been applied to diffusivity measurements of polymer semiconductor thin films for which electrical heating is used.

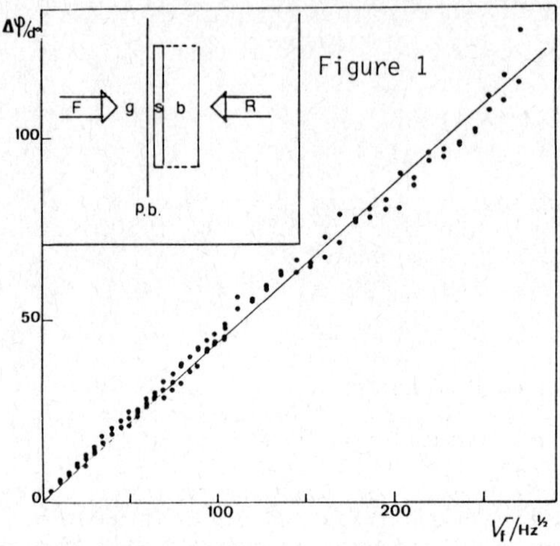

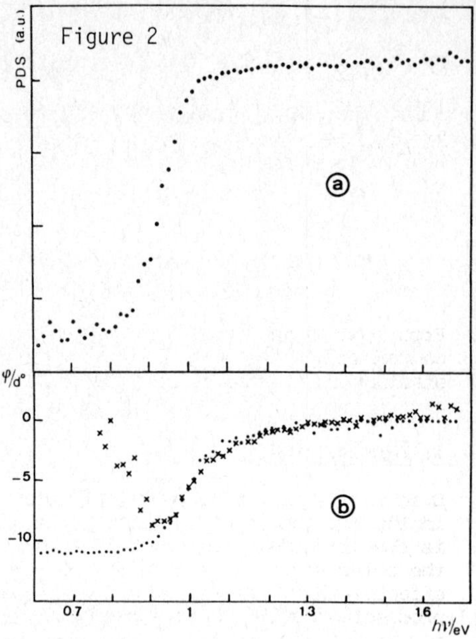

Figure 1 - Phase lag between rear surface and front surface excitations as a function of $\sqrt{f}$ for the 3.5 μm thick $CuInSe_2$ film.
Inset : principle of the two excitations. F : front surface ; R : rear surface ; g : deflecting medium ; s : sample ; b : substrate ; p.b. : probe beam.

Figure 2 - Photothermal deflection spectrum of 3.5 μm thick $CuInSe_2$ film : a) amplitude ; b) phase : × × × experimental, ... calculated from a) and the thermal parameters of the film by using the theoretical expression for the surface temperature.

In the case of the $CuInSe_2$ film, the previous thermal characterization allows us to perform a quantitative analysis of the spectroscopic properties. We present in Fig. 2 the phase and amplitude spectra obtained for this film in air at 3 kHz, a frequency at which the film is no longer thermally thin. We note that, in the subbandgap region, where the film is optically thin ($\alpha l \sim 0.1$), the phase of the signal turns up to the value obtained for surface absorption, i.e. for the photothermal saturation. This result, confirmed by the rear surface excitation phase spectrum, clearly shows that the absorbing centers giving rise to the subbandgap signal are located in the vicinity of the surface. In addition, the phase behavior in the bandgap region is well fitted by introducing the actual thermal parameters, previously determined, into the theoretical expressions for the surface temperature for both front and rear surface excitations.

References

1. G. Rousset and F. Lepoutre, Revue Phys. Appl. 17 (1982) 201.
2. P.K. Kuo, E.D. Sendler, L.D. Favro and R.L. Thomas, Can. J. Phys. 64 (1986) 1168.

Power Dissipation Mechanisms in Photovoltaic Cells

D. Cahen[1], *H. Flaisher*[1], *and M. Wolf*[2]

[1]The Weizmann Institute of Science, Rehovot, Israel, 76100
[2]University of Pennsylvania, Philadelphia, PA, USA

From (photo)acoustic and related measurements we can infer the relative contributions of several of the mechanisms that contribute to the total heat dissipation in a solar cell, including local cooling effects.

1. INTRODUCTION

Even in the most efficient of today's photovoltaic solar cells, over 70% of the incident photon power is dissipated, primarily as heat. While much of this loss is due to factors inherent to coupling the photovoltaic conversion process to the solar spectrum, a significant part of it results from carrier recombination effects that, in principle, are controllable via device materials and processing parameters. To be able to achieve such control, one needs to be able to assess the recombination rates that are effective in the device under operating conditions. The direct measurement of heat fluxes, caused by such recombination processes, is thus of interest for device optimization, as well as for improving our understanding of photovoltaic devices, esp. novel ones.

2. APPROACH

In order to be able to interpret (photo)calorimetric measurements on active electronic devices we need to start by considering the basic energy flux relationships, operative in a p-n junction. We can distinguish between fluxes between the device and its surroundings, which, in principle, are measurable, and internal fluxes, which often can be gotten at only by indirect methods. Naturally both types of fluxes need to be consistent.

2.1 External Energy Fluxes in a Photovoltaic Device

These are:

- Net electrical power input from an external electrical source:

$$Q(el) = J.V \ , \tag{1}$$

If this product is negative, we have energy output. In Figs.1 and 2 this flux is proportional to $q.V_B$.

- Net photon energy input, which needs to consider reflectance, transmittance and the spectral distribution of the source (cf. step 1 in Fig.2).
- Photon output, due to radiative recombination.
- Net heat flux to the surroundings (e.g. an acoustic chamber). Under some conditions influx may occur as well.
- Mechanical energy flow to the surroundings (generally negligible).

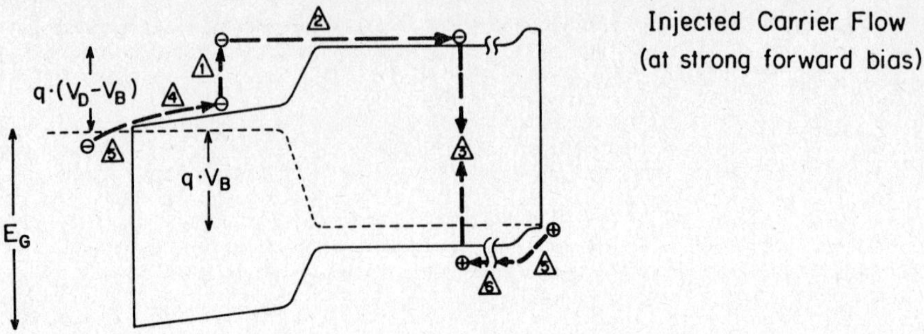

Figure 1
Schematic representation of injected carrier flow, shown under strong forward bias, where such flow becomes significant. The various steps are discussed in the text.

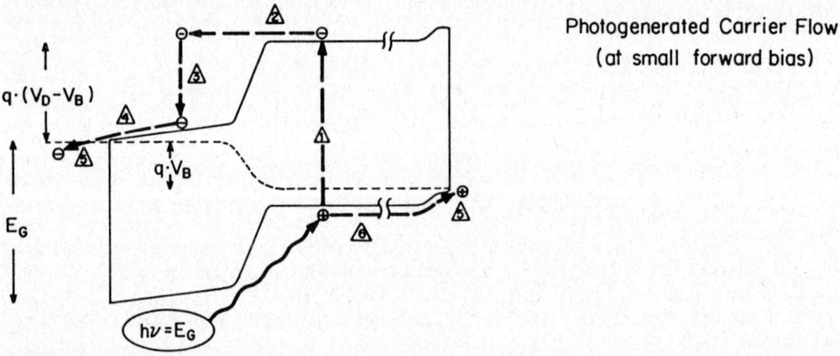

Figure 2
As fig. 1, for photogenerated carriers, under small forward bias conditions where this flow is dominant in a photodiode.

2.2 Internal Energy Fluxes in a Photovoltaic Device

(a) Heat, generated by non-radiative recombination of injected minority carriers. This is the flux of injected carriers, J_d, times the sum of steps 1 (negative), 3, 4 (negative) and 5 (negative) in Fig.1, which gives

$$Q(nrr,inj)=J_d.V_B \ . \tag{2}$$

We note that step 1 (and, to a minor extent, steps 4 and 5) implies that local cooling of the device occurs. When the heat flux to the surroundings is negative, cooling of the complete device will take place. Such will be the case when the quantum efficiency for radiative cooling is high enough. The occurrence of such a process has been deduced from luminescence data on GaAs diodes [1].

(b) Heat flux due to the absorbed incident photon flux, J_L, will occur as a result of thermalization of collected carriers down a potential gradient (cf. step 3 in Fig.2). This flux can be written as

$$Q(photon)=J_L.(E_G/q-V_B) \ . \tag{3}$$

In addition photons with energy $h\nu < E_G$ will contribute. Also photons with energy $h\nu > E_G$ will add $(h\nu - E_G)$. Other contributing factors are heat generation from absorbed photons that do not create free carriers, and recombination of the photogenerated free carrers.

(c) Joule losses: These refer to phenomena where majority carriers follow a potential gradient, and release the energy gained in this way, as heat. Thus the process described by (3) is strictly speaking also a Joule loss. As Joule losses refer to dissipation of electrical power, i.e. a J.V product, they can be represented by a resistance (if this resistance is constant, the losses are termed "ohmic"). In a modern photovoltaic device the losses will result from gradients in doping and the potential steps at the contacts (together expressed as series resistance). They lead to a junction voltage V(j) less than the terminal voltage, V(term). The series resistance is then given by

$$R_S = \{V(\text{term}) - V(j)\}/J. \tag{4}$$

Steps 4 and 5 in Fig.1 show these losses. In photovoltaic operation the current direction is reversed and V(j) becomes larger than V(term).

Leakage currents across the potential barrier resulting from space charge recombination and surface channeling are often expressed as shunt resistance. The external heat flux mentioned in section 2.1 above, will be the sum of all the internal ones discussed here.

3. RESULTS AND DISCUSSION

Previously we showed that the photoacoustic signal from a photovoltaic cell, completely enclosed in the acoustic chamber, is minimal at the maximum power voltage (V_{mp}) of the cell (optimum load resistance) [2]. This result can be understood as follows:

$$d(J.V)/dV_{mp} = 0\,. \tag{5}$$

Assuming a constant R_S and $J_o \ll J_d$ (mp), where J_o is the exchange current, we can derive the maximum power voltage as

$$V_{mp} = 1/\beta\ \ell n\ \{{}^{J}L/J_o e^{-\beta.JR}s\ (1+\beta.V_{mp})\};\quad (\beta = {}^{q}/_{nkT}). \tag{6}$$

The photoacoustic signal is given by

$$PA = S_1.J_L.(E_G/_q - V) + V_B.S.J_d = Q(\text{photon}) + Q(\text{nrr,inj}), \tag{7}$$

where the first type of term on the left hand side results from photogenerated carriers (cf. (3) and Fig.2) and the second type of term from injected ones (cf. (2) and Fig.1). S is the fraction of the area of the photovoltaic cell inside the acoustic chamber, and S_1 the ratio of the illuminated area of the cell inside the chamber to the total illumination area (if this area extends outside the acoustic chamber). It are these two factors that lead to differences between most photoacoustic [2] and photothermal radiometry [3], photopyroelectric [4] or thermistor measurements [5]. They express the different spatial characteristics of the internal fluxes given by (2) and (3). The photoacoustic signal will be minimal at

$$d(PA)/dV_{mp} = 0 \qquad (\text{PA min})\,. \tag{8}$$

Using (7) we find for the voltage at which the PA signal is minimal:

$$V_{PA\ min} = {}^{1}/\beta\ \ell n\ \{{}^{J}L.{}^{S}1/J_o e^{-\beta.JR}s\ (1+\beta.V_{PA\ min})S)\}. \tag{9}$$

Thus, for $S=S_1=1$, this voltage equals V_{mp} given in (6).

There is another acoustic way of obtaining V_{mp}, as described by Flaisher et al., [6]. In this voltacoustic method a small modulated voltage is superimposed on a ramp voltage and the resulting acoustic signal is measured when the photovoltaic cell is illuminated, as a function of applied voltage. This method allows a clear separation of J_L and J_d via the phase of the signal, which, over a certain voltage range changes by 180°. The voltage where the signal is zero is essentially the same as that given by (9).

Table 1 compares experimental values for the three voltages described above, for a few solar cells measured over load resistances (as contrasted to measurements with an external power supply, where (6) and (9) need to be modified [7]).

Table 1: Dark- & Photoacoustic vs. Electrical Data

PV Cell	Freq. [Hz]	S	Max. Power @ [mV]	PA min. @ [mV]	Zero Volt-Acoust. @ [mV]
Si	35	1	435	435	435
Si	70	0.22	493	546	545
Si	70	0.11	460	522	535
$CISe_2$	70	1	280	290(20)	280

$S_1=1$; ~ 2.5 AM1

It is important to note that the fact that these results fit the above given derivations provides strong (indirect) evidence for the cooling mechanism (esp. step 1 in Fig.1) mentioned above. Finally we point out that (7) provides a simple graphical way to determine separately the contributions of photogenerated- and injected carrier-generated heat fluxes ((2) and (3)) [7].

4. CONCLUSIONS

We have shown how calorimetric measurements can provide insight into internal energy fluxes in active diode devices and can be used to demonstrate the occurrence of different fluxes and quantitate them. In addition we have developed simple ways to detect local cooling in such devices. These developments may well be able to transform presently used qualitative radiometric investigations of active circuits into quantitative ones, and because they are based on detailed energy balance, provide the understanding needed to try to design devices that can take advantage of minimal heating or, even, cooling.

ACKNOWLEDGEMENTS

We thank the U.S.-Israel Binational Science Foundation, Jerusalem, Israel, for partial support. Some of our recent work has profited from support by the Belfer Foundation for Energy Research.

REFERENCES

1. G.C. Dousmanis, C.W. Mueller, H. Nelson and K.G. Petzinger, Phys. Rev. 133, A316 (1964).
2. D. Cahen, Appl. Phys. Lett., 33, 810 (1978).
3. D. Cahen, P.-E. Nordal and S.O. Kanstad, Appl. Phys. Lett., 49, 1351 (1986).
4. U.F. Faria Jr., C.C. Ghizoni, L.C.M. Miranda, H. Vargas. J. Appl. Phys. 59, 3394 (1986).
5. H. Koinuma, K. Hashimoto, K. Fueki, In "Conf. Rec. 18th IEEE PVSC, 1985" IEEE, NY (1985); pp.683-1688.
6. H. Flaisher, D. Cahen and M. Wolf, IEEE Trans. Electron. Electron Dev., ED-34, 457 (1987).
7. H. Flaisher, Ph.D. thesis, Feinberg Graduate School, The Weizmann Institute of Science, Rehovot, (1987), and to be published.

Characterization of the Photoacoustic Optical and Thermal Properties of Ion-Implanted Si and GaAs Layers

U. Zammit, M. Marinelli, F. Scudieri, and S. Martellucci

Dipartimento Ingegneria Meccanica, 2° Università di Roma "Tor Vergata", Via Orazio Raimondo, I-00173 Rome, Italy

The photoacoustic technique in the gas microphone configuration has been used to determine the thermal conductivity and optical absorption coefficient in ion-implanted semiconductor layers as a function of the implantation dose. The values for the thermal conductivity and for the absorption coefficient were found to progressively decrease and increase respectively as the implantation dose was increased.

In the present experiments the photoacoustic technique in the gas microphone configuration has been applied to the evaluation of the thermal conductivity k and optical absorption coefficient α of ion-implanted semiconductor layers. The experiment was carried out on a Si sample implanted on part of its surface with a dose of $5x10^{15} cm^{-2}$ 80 keV Si ions and on different GaAs samples implanted respectively with doses of $5x10^{13}$, 10^{14}, 10^{15}, cm^{-2} 100 keV Si ions. The depth of the implanted layers was 156 nm in the case if Si and 130 nm for GaAs. The photoacoustic signal amplitude and phase arising from both the crystalline and implanted parts of each sample were recorded for several values of the modulation frequency of the cw Ar ion laser (λ = 514.5 nm) which was used as the heating source. Phase data only have been considered since they turned out to be much more sensitive than the amplitude ones to the presence of the implanted layer on top of the crystalline bulk. Phase data were also independent of the laser beam intensity. The difference between the phases $\Delta\phi$ of the signals from the two parts of each samples was determined in order to avoid the influence of the instrumental phase offset. The model /1/ for the determination of the photoacoustic signal for a two layer sample was used to analyze the experimental data. The values of k and α for the implanted layers were then determined by finding those values which would yield the best fit of the experimental values of $\Delta\phi$ vs modulation frequency with their theoretical estimates, the values for the crystalline material being known. The experimental results of the phase difference vs frequency together with their corresponding fits are reported in reference /2/ while the values of k and α are shown in table I.

Values of k and α for amorphous Si are reported in the literature and therefore may be compared with the ones found in the present experiments. The value of α for glow discharge amorphous silicon is reported to be 10^5 cm^{-1} /3/, and compares very well with the one obtained in the present work. The value of k, $7x10^{-3}$ W $cm^{-1}K^{-1}$ reported in literature /4/, is quoted as approximate, stating that the true value should not differ by more than a factor of two. This again compares favourable with the result obtained in our experiment.

Table I. Thermal conductivity k and optical absorption coefficient for Si and GaAs ion implanted layers.

Sample	Si^+ dose/cm^{-2}	k/ W cm^{-1} K^{-1}	α /cm^{-1}
Si	$5x10^{15}$	(3.5 ± 0.2) $x10^{-3}$	$(9.9 \pm 0.4)x10^{4}$
GaAs	$5x10^{13}$	$(3.27 \pm 0.25)x10^{-3}$	$(1.17 \pm 0.4)x10^{5}$
	10^{4}	$(2.94 \pm 0.25)x10^{-3}$	$(1.23 \pm 0.4)x10^{5}$
	10^{15}	$(2.18 \pm 0.25)x10^{-3}$	$(1.42 \pm 0.4)x10^{5}$

The values of k for ion implanted semiconductors are more than two orders of magnitude lower than the values for the corresponding crystalline material. They progressively decrease as the ion implantation dose increases. The opposite trend is found in the case of the absorption coefficients.

Our results show that the photoacoustic technique in the gas cell microphone configuration is sensitive to the difference in physical parameters of ion-implanted semiconductor layers produced at different ion implantation doses, and has been able to discriminate between materials whose implantation doses differed by as little as a factor of 2. This suggests that the technique is very sensitive to the materials' structure and could be used as a useful diagnostic tool to monitor also the gradual recovery of ion implantation damage occurring during thermal or other kinds of treatments.

In conclusion, the photoacoustic technique in the gas microphone configuration has been used to investigate thermal and optical parameters in ion-implanted semiconductor layers, as a function of the implanted ion dose. Comparison of the results with existing data confirmed the reliability of the technique.

References

1. A. Mandelis, Y.C. Teng, B.S.H. Royce, J. Appl. Phys. 50 , 64 (1979)
2. U. Zammit, M. Marinelli, F. Scudieri, S. Martellucci, Appl. Phys. Lett., 50, 830 (1987)
3. S.T. Moss, "Handbook of Semiconductors", (North-Holland, Amsterdam, New York, Oxford, 1980) Vol. 3, p. 771
4. H.C. Webber, A.G. Cullis, N.G. Chew, Appl. Phys. Lett., 43 , 669 (1980)

Influence of Surface Photovoltage on PZT-PAS Detection with Semiconductor Samples

M. Watanabe

Production Engineering Research Laboratory, Hitachi Ltd.,
292 Yoshidamachi, Totsukaku, Yokohama 244, Japan

1 Introduction

Photoacoustic detection by a piezoelectric sensor often gives results different from those obtained by other methods such as gas-microphone detection. Elastic waves induced by sample bending[1],or the existence of defects or some other phase in the sample, which influences the propagation of the acoustic wave, etc., are all possible explanations. Here I report another cause, the surface photovoltage, that gives contributions to piezoelectric signal.

2 Experiment

P-type 4″ Si wafers(20~50 Ω·cm),the surfaces of which were covered with plasma-deposited Si_3N_4,were prepared. A Kr+ ion laser (647nm) was used as a light source with varying power 0~500mW. The laser beam was intensity modulated from 100Hz to 20kHz. The piezoelectric sensor used was PZT-5.

3 Results and Discussion

The PA signal amplitude from the sample was proportional to the laser power in gas-microphone detection,but,in piezoelectric detection,two electrical arrangements of the PZT device shown in Fig.1a and Fig.1b gave quite different results, as indicated in Fig.2a and Fig.2b. If the piezoelectric device had detected mechanical stress,the two arrangements should have given the same results. Therefore, the difference between Fig.2a and Fig.2b was attributed to electrical causes and not to those mentioned above.

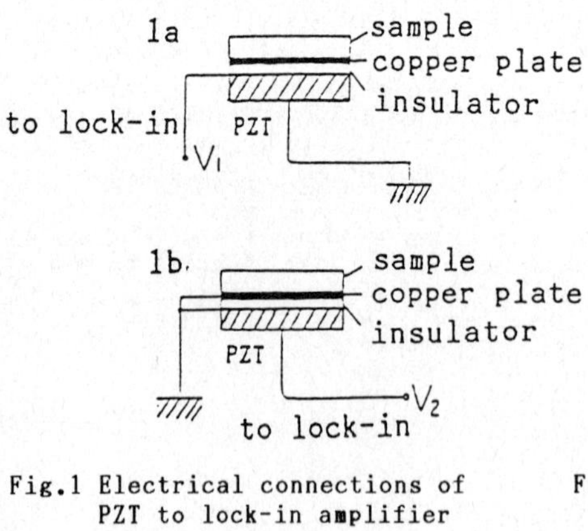

Fig.1 Electrical connections of PZT to lock-in amplifier

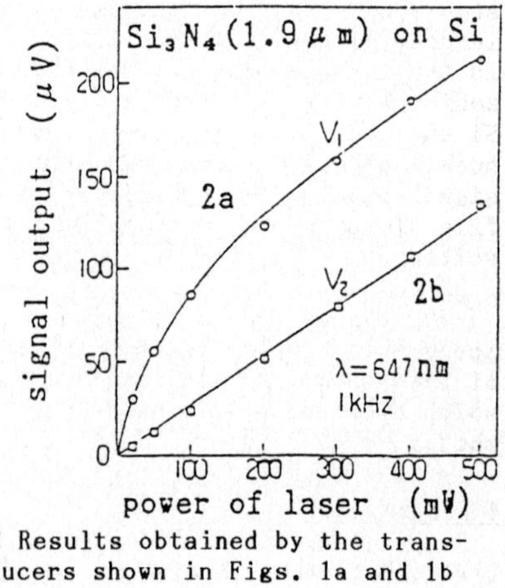

Fig.2 Results obtained by the transducers shown in Figs. 1a and 1b

When the sample side electrode was earthed and well shielded as Fig.1b, its electrical potential would not change even if the electrical field surroundings were changed by the light irradiation. However, such a change would result in a change of the electrical potential of the sample side electrode in Fig.1a. If the signal detected were of this kind, only an electrode such as a metal plate adherent to the sample as shown in Fig.3a would be able to sense it. Figure 3b is the result obtained by this arrangement, which shows the potential change of the plate when the samples were irradiated. The relation between signal amplitude and laser power was almost the same as that in Fig.2a.

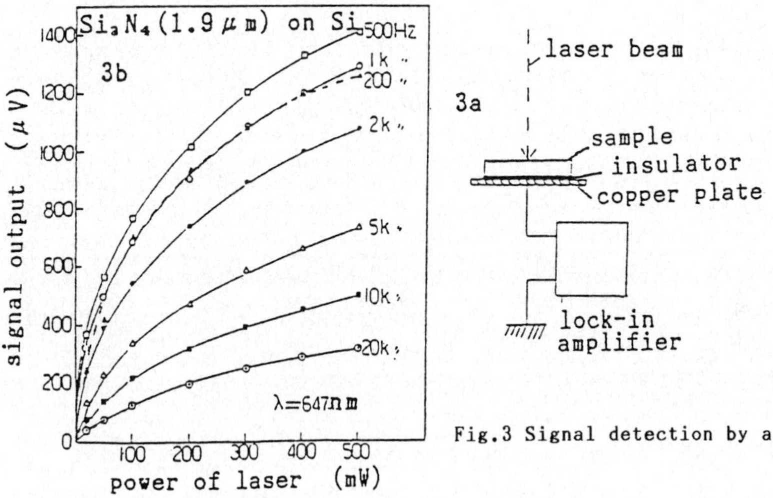

Fig.3 Signal detection by a metal plate

The photon energy in this experiment was less than the ionization energy of the sample material. So, the electrical potential change might be caused by the photoinduced positive and negative charge carriers which were distributed differently in the sample and attract one kind of charge to the plate surface. There are two possible explanations for the cause of such a distribution. One is the Dember effect, which is the photovoltage induced by different mobilities of positive and negative carieers. Another is the surface photovoltage induced by a space charge region which is associated with the potential difference at a junction or interface between Si and the film. Then other samples were measured with the plate. Although such a signal was observed when the film was a thermal oxide,there was no signal when Si had no film on it, nor for an n-type Si wafer with a Si_3N_4 film. These results indicate that the signal was due to a surface photovoltage.

There have been a lot of photoacoustic and thermal wave studies on semiconductor samples and many of them were by piezoelectric detection. Some of them reported anomalous frequency dependence of PN junction depth[2], which might well be explained by a surface photovoltage. Thus such results should be reconsidered in view of this effect.

4 References

1)W.Jackson and N.M.Amer, J. Appl. Phys. 51(6), 3343 June (1980)
2)T.Sawada and M Kasai, J. Spec. Soc. Jpn. 33, 307 (1984)

The Influence of Illumination on Photoacoustic Signals in Semiconductors

B.L. Yordanov, D.V. Ivanov, and J.I. Burov

Department of Solid State Physics, Faculty of Physics, Sofia University, Sofia-1126, Bulgaria

Recently, a number of reports on photoacoustics have appeared using photogeneration of excess carriers in order to measure some physical parameters of semiconductors /1-4/.

In the present paper a photoacoustic technique for semiconductor measurement is discussed. It uses a 1-D model based on the influence of the illumination, by means of the heat conductivity change on the surface thermoelastic deformation that takes place at constant heating.

If we have an infinite plate of a semiconductor material with thickness L that is heated constantly at a point of its surface and is cooled because of the heat transfer through the opposite surface, then a stationary temperature distribution will be established. At a point x away from the surface:

$$T^{r}(x)=(T_{o}+Q/h) + Q/k_{o}(L-x), \tag{1}$$

where k_o is the thermal conductivity; h is the heat transfer coefficient; Q is the incident energy flux and T_o is the temperature of the surroundings.

If the same plate is illuminated at the same point by a light source with intensity I, a stationary photocarrier distribution will be established:

$$n(x) = I\tau \exp\{-[R(\lambda) + (D\tau)^{-1/2}]x\}, \tag{2}$$

where τ is the mean lifetime of the photocarriers, $R(\lambda)$ is the absorption coefficient corresponding to light of wavelength λ, and D is the photocarrier diffusion coefficient.

If this illumination is modulated with low modulation frequencies (ω < 10 kHz) the stationary photocarrier concentration distribution in (2) will be still valid, because $\omega\tau \ll 1$ for most normal semiconductor materials.

It is known that in semiconductors there are two components of the thermal conductivity - the phonon one and the electronic carrier one. For carrier thermal conductivity there is a relation /5/

$$k_{n}(x) = AnT, \tag{3}$$

where A is a coefficient depending on the carrier mobility, energy gap and scattering mechanism of the photocarriers in the semiconductor. Here it is assumed that the bipolar mechanism of carrier heat conductivity is negligible and also $n_i << n$ in most cases (n_i is the intrinsic carrier concentration).

Hence, at a certain photocarrier concentration distribution (2) there will be a corresponding carrier heat conductivity distribution. The latter changes the stationary temperature distribution in (1) in this way:

$$T^n(x) = (T_o + QS_1/hS_2) + Q/S_2(L-x), \tag{4}$$

where

$$S_1 = k_o + AI\tau(T_o + Q/h)\exp[-L(D\tau)^{-1/2}] \tag{5}$$

and

$$S_2 = k_o + AI\tau(T_o + Q/h + QL/k_o). \tag{6}$$

Here it is assumed that photocarrier thermal conductivity does not depend on the temperature distribution $T^n(x)$, but on the temperature distribution $T^r(x)$, since this temperature change is negligible in value.

Thus, for the surface thermoelastic deformation change after illumination we have

$$\Delta U_x(0) = \alpha\int_0^L [T^r(x) - T^n(x)]dx, \tag{7}$$

where α is the thermal expansion coefficient.

This surface thermoelastic change was measured using "the one arm optical bridge" technique /6/, giving an accuracy of 10^{-12}m. When light fluxes with different absorption coefficients are applied, there will be different photocarrier distributions (2) and this will also change $T^n(x)$ and $\Delta U_x(0)$, which we measure. In this way one can obtain a relation between the semiconductor parameters and calculate τ or D numerically. Our evaluations for the accuracy of the presented technique show that τ or D can be measured with an accuracy of 5% in Si and CdS.

1. D. Fournier, C. Boccara, A. Skumanich, and N. Amer: J. Appl. Phys. 59, 787 (1986)
2. R. Stearns, B. Khuri-Yakub, and G. Kino: Appl. Phys. Lett. 45, 1181 (1984)
3. R. Stearns, and G. Kino: Appl. Phys. Lett. 47, 1048 (1985)
4. W. Lee Smith, A. Rosencwaig, and D. Willenborg: Appl. Phys. Lett. 47, 584 (1985)
5. A. Anselm: In: *Introduction to Semiconductor Theory*, 2nd. ed. (Nauka, Moscow 1982) p. 496 (in Russian)
6. M. Borissov, K. Bransalov, and J. Burov: Jap. J. Appl. Phys 15, 797 (1976)

Evolution of the Electron Acoustic Signal as a Function of Doping Level in III–V Compound Semiconductors

J.F. Bresse and A.C. Papadopoulo

C.N.E.T., Division PMM, Département MPD,
196 avenue H. Ravera, F-92220 Bagneux, France

In the case of GaAs silicon doped epitaxial layers, as an example, the evolution of the electron acoustic signal as a function of the doping level is given. The results are explained by the role played by defects connected with silicon complexes. They give rise to deep level emission and to a thermal conductivity decrease, as the electronic contribution is shown to play a minor role.

The electron acoustic signal (EAI) has been measured for : MBE-grown epitaxial layers (1 μm thick) of GaAs, silicon doped, on a semi-insulating substrate with different doping levels (measured by Hall effect). For the measurements, the operating frequency is chosen in order to obtain the maximum signal near the resonance frequency of the PZT transducer (250 kHz). At this frequency the thermal diffusion length is 5.6 μm for undoped GaAs [1]. The measurements are made at 20 keV beam energy, 2 μA beam current.

Low temperature cathodoluminescence experiments have been performed on the same samples. Integrated intensities for the Near Band Edge (NBE) emission (≃ 1.53 eV) and Deep Level (DL) emission (≃ 1.20 eV) associated with silicon complexes have been measured under the same experimental conditions (20 keV beam energy). The three variations of signal as a function of the doping level are reported in Fig. 1.

The NBE signal increases to a maximum value for a threshold value of the doping level where the deep level emission is observed. As the doping level increases the minority carrier lifetime, τ, and the diffusion constant, D, are reduced. This explains the evolution of the NBE signal but cannot explain the variation of the electron acoustic signal.

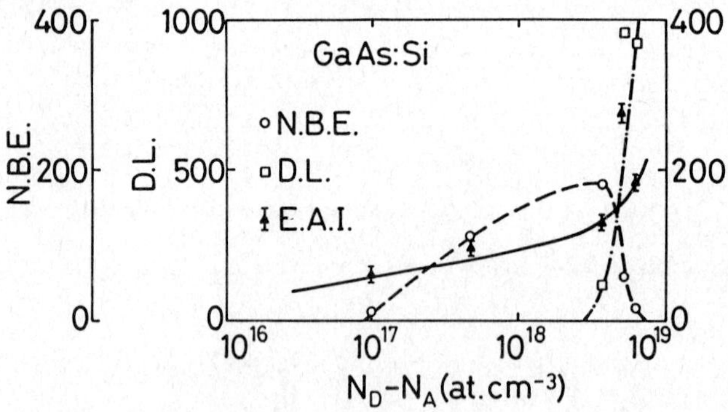

Fig. 1 : Evolution of electron acoustic signal (EAI) and cathodoluminescence emission (NBE), (DL).

In order to explain the electron acoustic signal variation, the evolution of the total thermal conductivity K is studied :

$K = K_L + K_e$,

where K_L is the lattice thermal conductivity, mainly given by phonon transport, and K_e is the electronic contribution to the thermal conductivity.

In semiconductors [2]

$$K_e = \frac{(5/2 - S)}{q^2} k^2 \sigma T + \frac{k^2 \sigma T}{q^2} \frac{(5 - 2S + E_G/kT)^2 n_i^2 \mu_n \mu_p}{(n \mu_n + p \mu_p)^2},$$

where k is the Boltzmann constant , σ the conductivity, E_G the semiconductor gap, q the electron charge, n (p) the electron (hole) concentration, n_i the intrinsic concentration, μ_n (μ_p) the electron (hole) mobility, and $S = -3/2$ for conductivity by ionized impurities.

In GaAs, $n_i \ll n$ or p, so the second term is negligible, and the calculated value of K_e is, for a doping level of 10^{19} atoms cm^{-3}, 1.7×10^{-2} $Wcm^{-1}K^{-1}$ for n type and 1.1×10^{-3} $Wcm^{-1}K^{-1}$ for p-type.

The evolution of the thermal conductivity cannot be explained by the electronic contribution to the thermal conductivity.

We may estimate the lattice thermal conductivity [3] as

$K_L = 1/3\, C\, v\, \ell$,

where C is the heat capacity, v the average particle velocity, and ℓ the mean free path.

The estimated mean free path in GaAs goes from 287 Å for a doping level of 10^{17} at.cm^{-3} to 28.7 Å for 10^{20} at.cm^{-3}. The doping level must take into account either electrically or non-electrically active sites.

The evolution of the photoacoustic signal as a function of the implanted dose in Si and GaAs has been reported [4]. The results are explained by the evolution of the thermal conductivity.

In our case, the variation of the thermal conductivity may explain the evolution of the electron acoustic signal by the reduction of the mean free path between collisions as the dopant incorporation in the lattice increases creating defects and silicon complexes revealed by deep level emission in cathodoluminescence.

ACKNOWLEDGEMENTS : The authors wish to thank F. Alexandre for furnishing silicon doped GaAs epitaxial layers and F. Auzel for helpful discussions.

REFERENCES

[1] A. Rosencwaig, Science, 218, 4569, 223 (1982)

[2] S.M. Sze : Physics of Semiconductor Devices, 2nd ed. (Wiley, New York 1981)

[3] C. Kittel : Introduction to Solid State Physics, 3rd ed. (Wiley, New York 1968)

[4] U. Zammit, M. Marinelli, F. Scudieri, S. Martelluci, Appl. Phys. Lett. 50, 830 (1987)

Fourier Transform Photothermal Deflection Spectroscopy of Impurity Centres in CdS Films: Doping and Annealing Effects

M. Fathallah[1], *B. Rezig*[2], *M. Zouaghi*[1], *N.M. Amer*[3], *J.P. Roger*[4], *A.C. Boccara*[4], *and D. Fournier*[4]

[1]Département de Physique, Faculté des Sciences de Tunis, Tunis Belvédère, Tunisia
[2]I.N.R.S.T., BP 95, Hammam-Lif, Tunisia
[3]IBM, T.J. Watson Research Center, Yorktown Heights, NY 10598, USA
[4]Laboratoire d'Optique, E.S.P.C.I., 10, rue Vauquelin, F-75231 Paris Cedex 05, France

Abstract : The subgap optical absorption of polycrystalline thin films was measured by Fourier Transform Photothermal Deflection Spectroscopy. Some impurity levels were identified : copper doping and annealing effects were studied.

CdS polycrystalline films are known to be very sensitive to their fabrication conditions as well as to their thermal history. This sensitivity is mainly related to some localized deep centres present in the layer. We used Fourier Transform Photothermal Deflection Spectroscopy (F.T.P.D.S.) /1/ to investigate the subgap optical absorption behaviour of this material and to identify some of these deep levels.

The studied layers were grown by the chemical spray pyrolysis method on Pyrex (or glass) kept at 430°-450°C. The aqueous solutions react at this temperature to give the CdS layers. The intentional doping was made by introducing the foreign species in the starting materials in their salt form. The optical density αl spectra were determined in the case of thermally thin samples from the PDS signal according to the relation /2/

$$\ell\alpha = \mathrm{Ln}\left(1 - \frac{S}{S_\infty}\right),$$

where α is the optical absorption coefficient and ℓ the thickness of the sample. S is the PDS signal and S_∞ the corresponding saturation value occurring when $\alpha\ell \gg 1$. Figure 1a shows the subgap optical absorption of undoped and copper-doped CdS films of the same thickness. Three bands are observed around 2.1, 1.4 and 0.9 eV. The 2.1 eV band (and to a lesser extent the 1.4 eV one) is significantly magnified by copper doping. This induces an overlapping between these two bands. However, the resolution was improved by using very pure (deionized) water in the spray solution (Fig. 1b). This confirms the presence of some undesired impurities even in the "undoped" layers.

Annealing of the two types of films in a N_2 atmosphere at T=200°-350°C improved the crystallinity of the samples as confirmed by X-ray diffraction, and decreased the resistivity by two orders of magnitude. The first effect is attributed to the increase of the grain size and the latter to the desorption of oxygen present at the surface and in the grain boundaries. Furthermore, the two major absorption bands were enhanced after

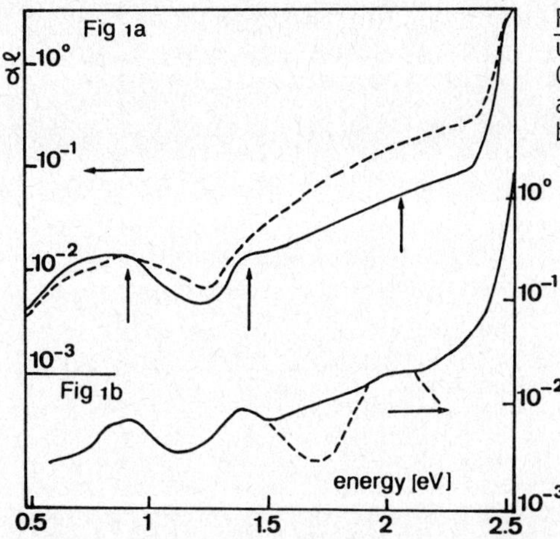

Fig. 1 Subgap absorption spectra of undoped (——) and Cu doped (---) CdS films.
a - distilled water
b - deionized water

annealing while the third one (0.9 eV) was reduced (Fig. 2). The first two bands could be related to the presence of copper in the films. This copper may exist in its two ionized states : the doubly ionized (Cu^{++}) located at 2.1 eV and a deeper one (Cu^{+}) around 1.4 eV below the conduction band. It is shown that copper doping induces a contraction in the lattice unit cell /3/. This favors the predominance of Cu^{++} ions known for their smaller volume and higher stability than Cu^{+} ions.

On the other hand, the primary effect of annealing in a N_2 atmosphere is to desorb oxygen. The observed decrease of the 0.9 eV band may be related to this mechanism in accordance with /4/. The enhancement of the 2.1 and 1.4 eV bands after annealing could be due to the increase respectively of Cu^{++} and Cu^{+} active ions. Indeed, it is thought that some weak Cu-O bonds in the grain boundaries were broken after annealing.

These results are in agreement with those obtained by photocopacitance measurements on similar samples /5/. Three recombination deep levels at the energies of 2.1, 1.5 and 0.93 eV below the conduction band were identified by this latter technique. Due to the contactless character of the PDS method, our results seem to be more consistent.

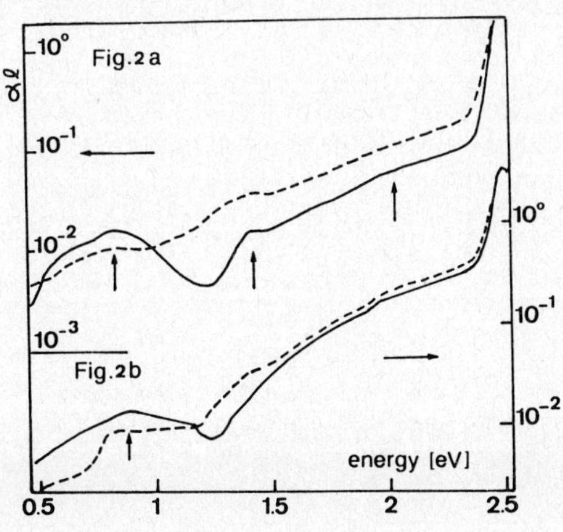

Fig. 2 Subgap absorption of as-made (——) and annealed (350°C, N_2) (---) CdS films
a - undoped layer
b - copper doped layer

References

1. D. Débarre, A.C. Boccara and D. Fournier, Appl. Opt. 20 (1981) 4281.
2. W.B. Jackson, N.M. Amer, A.C. Boccara and D. Fournier, Appl. Opt. 20 (1981) 1333.
3. M. Dachraoui, Thèse d'état (1986) Faculté des Sciences de Tunis.
4. I. Martil, G. Gonzalez-Diaz and F. Sanchez-Quesada, Solar Energy Materials 12 (1985) 345.
5. R. Rezig, H. Choukri, J. Bougnot, C. Llinares and M. Savelli, Solar Cells 14 (1985) 201.

Part VI

Ultrasonic Detection and Characterization

High Frequency Phonons on Surfaces and in Periodic Structures

K. Dransfeld, G.v. Eynatten, Z.M. Sun, and V. Uhlendorf

Fakultät für Physik, Universität Konstanz,
D-7750 Konstanz, Fed. Rep. of Germany

This paper is concerned with three areas of ultrasonics:

1. Thermoacoustic measurements of longitudinal and transverse acoustic vibrations.
2. Precision measurements of acoustic amplitudes by the Mössbauer effect (CEMS) and by tunnel microscopy (STM).
3. Acoustic waves in periodic submicron structures.

The methods described here, although useful also at low frequencies, are of particular interest for phonons having a wavelength in the submicron range where standard optical techniques, such as Brillouin scattering, can no longer be used.

1. Thermoacoustic Measurements of Longitudinal and Transverse Acoustic Amplitudes

In photoacoustic spectroscopy, usually, amplitude-modulated optical radiation incident on an absorbing sample leads to a time-dependent variation of the temperature of the solid surface and this in turn - due to thermal expansion - leads to a time-dependent acoustic signal both in the solid and in ambient gas. Here we are concerned with the fact that also acoustic vibrations of a solid can be detected in a gas cell by a microphone, although the surface of the vibrating sample does not necessarily show any temperature rise. How can this be understood ?

Let us for simplicity consider a vibrating quartz plate which is placed inside a photoacoustic gas cell and excited piezoelectrically by an electric field at 20 MHz in its acoustic resonant mode. If the exciting electric field is amplitude-modulated at a much lower frequency (for example at 80 Hz), the gas pressure in the cell varies accordingly. The resulting microphone signal is plotted in Fig. 1 as a function of the carrier frequency and shows a clear resonance response /1/ although the quartz sample is nearly at constant

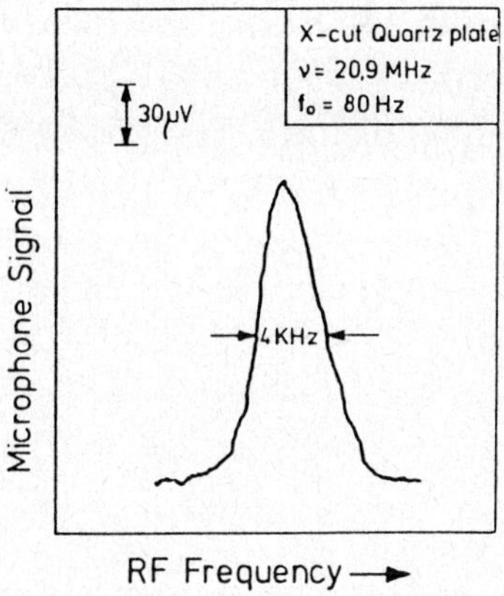

Fig. 1: Thermoacoustic recording of the resonance oscillation of an X-cut quartz plate around the resonance frequency of 20.90 MHz.

temperature. Clearly, the vibrating (but isothermal) quartz plate builds up the pressure in the gas cell.

If we are using – as shown in Fig. 1 – an X-cut quartz plate and excite its thickness vibration, all surface atoms move perpendicular to the surface at a time-dependent velocity

$$v_p = v_{po} \cdot \sin \omega t \qquad (1)$$

with $(\omega/2\pi) \simeq 20$ MHz and neglecting the modulation. Acoustic energy is radiated directly from the cool vibrating surface into the gas, finally heating it up at some distance from the solid surface. The acoustic power radiated into the gas and there converted into heat is

$$P_L = \rho_{gas} \cdot \frac{v_{po}^2}{2} C_{gas} \qquad (2)$$

with ρ_{gas} being the gas density and C_{gas} the acoustic velocity in the gas.

Of special interest is the case of a pure shear vibration (for example, if an AC-cut quartz plate is placed in the gas cell). Now the particle velocity v_p according to (1) is directed parallel to the surface. Therefore no longitudinal acoustic waves are radiated into the gas. If furthermore the surface is very smooth and all incident gas atoms are reflected specularly without changing their transverse momentum, no energy transfer occurs between the vibrating quartz and the surrounding gas. In consequence the vibration of the quartz is not damped by the gas and the gas is not heated up. There is no microphone signal.

If there is, however, full accommodation and the incident gas molecules lose their transverse momentum, each incident gas atom of mass m acquires – after re-emission – the additional kinetic energy $(m/2)v_p^2$ and, consequently, energy is transported from the vibrating quartz into the gas at a rate of $P_T = P_L/6$ (see (2) for the definition of P_L). In addition, the same power is now converted into heat at the sample surface. Therefore, for full accommodation the total rate of heating is $P_L/3$. As mentioned above: For accommodation zero and mirror-like reflections of all incident gas molecules there is no heating and no microphone signal at all.

In summary, the thermoacoustic method /1/ is very useful for measuring the acoustic surface velocity of a vibrating solid (including surface acoustic waves) up to very high frequencies, where the acoustic wavelength becomes comparable to the de Broglie wavelength of the impinging gas atoms, i.e. well into the THz-frequency regime.

For a surface undergoing a pure shear motion, thermoacoustic measurements can be used for the determination of the accommodation coefficient of the gas relative to the vibrating surface. For example, it is possible to change the gas (atomic or molecular) in the cell and to vary temperature or gas pressure. The variation of the accommodation coefficient with these parameters can thus be determined. For calibration purposes it is always possible to use an X-cut quartz.

In this context it is interesting to note that according to new molecular beam experiments /2/ the rotational degrees of freedom of molecules hitting a hot surface hardly participate in the energy exchange (by comparison with the translational degrees of freedom which alone carry nearly all the energy away from a hot surface).

The methods described above do not give information about the phonon <u>frequency</u>. For the cases of piezoelectric excitation discussed here this is no serious disadvantage since the phonon frequency is known from the excitation process.

2. Precision Measurement of Acoustic Amplitudes by the Mössbauer Effect (CEMS) and by Tunnel Microscopy (STM)

It is well known that the Mössbauer effect can be used for the detection of acoustic vibrations. For example, in the MHz-frequency range one can directly measure the intensity of the phonon-induced sidebands and derive from their intensity the amplitude of the acoustic vibration with very high accuracy. Furthermore, by using conversion electron Mössbauer spectroscopy (CEMS) instead of the more standard Mössbauer absorption spectroscopy one needs only very thin Fe films (100 Å and less).

We have recently used this method to measure the acoustic amplitude of an piezoelectrically driven PVDF film with an accuracy of 10^{-2} Å /3/. We thereby found that the piezoelectric constant of PVDF (polyvinylidenefluoride) in the MHz-frequency range is about two orders of magnitude weaker than at zero frequency. Moreover we could see that - without any applied electric field - the thermal acoustic vibration of a similar P(VDF/TrFE) film carrying two metal electrodes increased by 0.01 Å as soon as the elctrodes were electrically shorted /3/. CEMS is a very useful method for the detection of phonons down to acoustic wavelengths of the order of 100 Å. At higher frequencies a measurement of the Debye-Waller factor gives clear information about the acoustic amplitude relative to the X-ray wavelength.

The CEMS method seems to be very promising also for observing the thermal dynamic motion of a solid polymer at its surface: If one plates a 50 - 100 Å Fe film onto a polymer the Debye-Waller factor shows the onset of the polymeric surface mobility under the influence of heating or illumination when no such motion occurs inside the polymeric sample.

While the CEMS method of the Mössbauer effect gives very accurate information about the amplitude averaged over a large surface area <u>tunnel microscopy</u> gives information about the acoustic amplitude in the immediate neighbourhood of a single atom. If, for example, a surface acoustic wave travelling across the surface is being investigated by STM, the dc tunnel current adopts a higher value. From this current increase one can by calibration deduce the amplitude of the surface vibration, but not its frequency. Very recently, however, it has become possible also to measure directly the ac component of the tunnel current at about 100 MHz /4/.

3. Acoustic Waves in Submicron Periodic Structures

Let us now turn to <u>periodic structures</u>, to superlattices of submicron periodicity. They have interesting optical and electronic properties, but we will focus our attention on the excitation and propagation of <u>phonons</u> in these media. Structures of nanometer periodicities are becoming more and more important for data storage and part of their interest arises from possible applications of such small structures. They are, however, also of fundamental interest.

One of the best known applications of phonons in periodic media is in surface-acoustic-wave filters /5/, which are nowadays incorporated in almost all television sets. Metallic interdigital transducers form periodic structures on top of $LiNbO_3$ crystals for the excitation of SAW up to a few GHz. It is difficult to extend this technology to higher frequencies.

In the field of semiconductors, doped nipi- and hetero-structures are interesting not only from an electronic and optical but also from an acoustic point of view: If they are exposed to high frequency electric fields, intense acoustic waves can probably be excited /6/.

Supramolecular lattices in 2 or 3 dimensions with lattice constants well below 1 μ have been successfully produced by the condensation of block copolymers containing one hard and one soft component, for example, butadiene-styrene /7/. Another example of well reproducible multilayers of organic material are the Langmuir-Blodgett films /8/ which are also attracting more and more interest from industrial laboratories. Finally one should also mention liquid crystals, in particular the lyotropic liquid crystals which are made up of charged macromolecular particles, for example virus particles, which are dissolved in water together with their mobile counterions. They form a superlattice on account of their mutual repulsion, in a similar way to how colloidal lattices are formed. As an example we like to mention the lyotropic crystals of the tobacco mosaic virus /9/, with lattice constants of the order of 100 - 400 Å, corresponding to acoustic resonance frequencies of the order of 50 GHz.

These resonances can, in principle, be excited by ac electric fields since they represent infrared-active phonon modes of the lattice.

Let us look at a lyotropic superlattice, for example a virus crystal, a bit closer. For applications the absorption of ultrasonic waves in these structures must be kept low. As with any viscous liquid the acoustic absorption first rises with frequency until a critical frequency is reached for which the acoustic absorption coefficient times the acoustic wavelength is about unity. At still higher frequencies the liquid starts to become elastically a solid, i.e. it supports also shear waves and the absorption per wavelength decreases again. For applications of lyotropic liquid crystals at high acoustic frequencies it may, therefore, be useful to increase the viscosity of the solvent in order to reach a lower acoustic absorption per wavelength at the highest frequencies.

The first measurement of the ultrasonic absorption in a lyotropic crystal is shown in Fig. 2 /10/. This virus crystal can be rotated in a magnetic field, and, as can be seen from Fig. 2, the ultrasonic absorption in the MHz-frequency regime is highly anisotropic. At higher ultrasonic frequencies lyotropic liquid crystals have not yet been investigated. Clearly, the research on the behaviour of acoustic waves in these lyotropic periodic structures has hardly begun.

Let us finally consider a lyotropic fd-virus crystal at very high frequencies. Figure 3 shows the expected folded phonon branches for this liquid crystal. If the periodicity constant d of the superlattice is 300 Å the

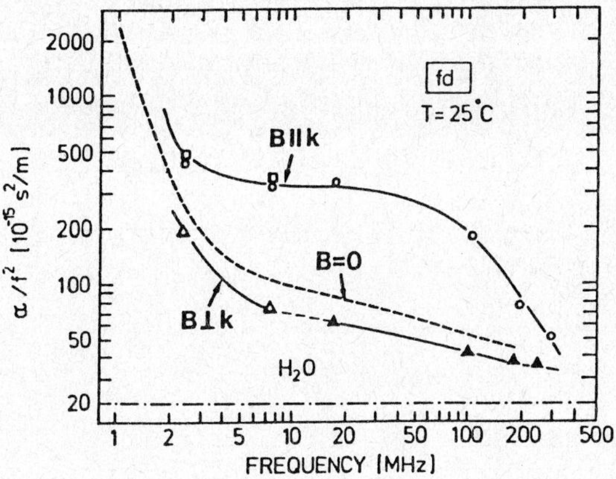

Fig. 2: The ultrasonic absorption in an orientated fd-virus crystal. (For further details see /10/.)

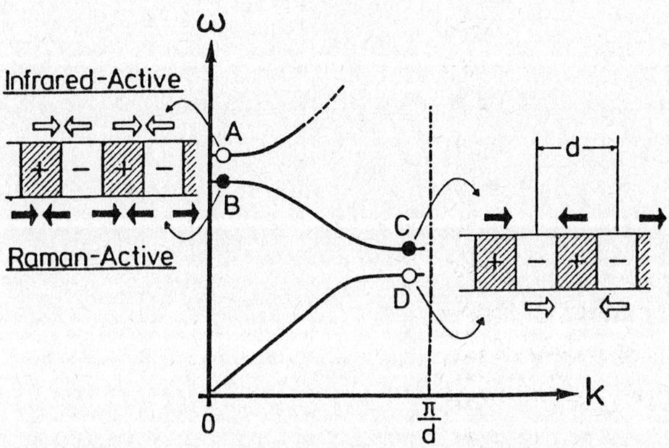

Fig. 3: Electrical excitation and optical detection of phonons in periodic structures.

infrared-active (A) and Raman-active (B) phonons would be located around 50 GHz. In general, one can
detect by light scattering only the Raman-active mode (B), which cannot, however, be excited by electric fields. The phonons at the zone boundary, (C) and (D), do not interact with macroscopic electromagnetic fields, but the infrared-active phonons (A) can be excited directly by macroscopic rf electric fields: Therefore one might expect a dielectric resonance absorption at 50 MHz, if the dielectric and acoustic background absorption is not too high. This resonance response may lead to interesting filter applications but it has not been observed yet.

We are grateful to the Deutsche Forschungsgemeinschaft for financial support (SFB 306).

1. F. Tsuruoka and K. Dransfeld, The vibration and rotation of piezoelectric particles in high frequency electric fields observed by a thermoacoustic method, Applied Physics A, 36, 125 (1985)
2 H. Walther, Laser studies of molecular dynamics at surfaces, Conf.on "Laser-Induced Processes in Matter", Helsinki; Aug. (1987)
3. G. von Eynatten, E. Fukada and K. Dransfeld, The dynamics of piezoelectric polymer materials at high frequencies measured with CEMS, Hyperfine Interaction (1987)
4. H. Walther, private communication
5. K. Dransfeld und E. Salzmann, Excitation, detection and attenuation of high frequency elastic surface waves (Review), in Physical Acoustics, Eds.: W.P. Mason and Anderson, Acad.Press, New York 1970, p. 219 - 272
6. K. Dransfeld, Quantum aspects of Rayleigh-waves: Surface and interface-waves at very high frequencies, in: Rayleigh-Wave Theory and Application, Vol. 2 of Springer Series on Waves-Phenomena, Eds. E.A. Ash and E.G.S. Paige, Springer (1985).
7. G. Kämpf, H. Krömer and M. Hoffmann, Long-range order of supramolecular structures in amorphous butadiene-styrene block copolymers, J. Macromol.Sci.-Phys. B6, 167 (1972)
8. See for example: P. Christie, G.G. Roberts and M.C. Petty, Spontaneous polarisation in organic superlattices, Appl.Phys.Lett. 48, 1101, (1986) or: G.W. Smith, M.F. Daniel, J.W. Barton and N. Ratcliffe, Pyroelectric activity in non-centrosymmetric Langmuir-Blodgett multilayer films, Thin Solid Films 132, 125 (1985)
or: L.M. Blinov, N.V. Dubinin, L.V. Mikhnev and S.G. Yudin, Polar Langmuir-Blodgett films, Thin Solid Films 120, 193 (1984)
9. U. Kreibig and C. Wetter, Light diffraction in vitro crystals of six Tobacco-Mosaic Viruses, Z.Naturforsch. 35c, 750 (1980)
10. K. Dransfeld, Z.M. Sun and V. Uhlendorf, Ultrasonic waves in magnetically oriented lyotropic liquid crystals, Proceed.Int.Conf.Biomolecules in High Magnetic Fields, Les Houches, 1986, Springer, Berlin

Acoustic Microscopy – State of the Art and Recent Developments

M. Hoppe[1] *and A. Atalar*[2]

[1]Ernst Leitz Wetzlar GmbH, P.O. Box 2020, D-6330 Wetzlar, Fed. Rep. of Germany
[2]Electrical and Electronics Engineering Department, Middle East Technical University, 06531 Ankara, Turkey

Technical aspects of a scanning acoustic microscope with broad frequency coverage (50 ... 2000 MHz) and some recent developments are described.

1. Introduction

The scanning acoustic microscope is finding applications as a powerful scientific instrument for imaging and characterization of materials. In this paper, we decribe the scanning acoustic microscope (ELSAM)™ developed at Ernst Leitz Wetzlar GmbH, West Germany, following the guidelines of the Stanford microscope /1/.

ELSAM is an acoustic microscope combined with an optical microscope capable of generating visual or hard-copy acoustic images. Use of microprocessors releases the user from routine adjustments of the complicated system, and also the hardware and wiring is simplified to result in a reliable system.

2. General Design Features and Mechanics

The wavelength of the acoustic waves used in imaging is determined by the frequency of operation. To serve both high resolution and high penetration depth applications, an acoustic microscope with broad frequency coverage is necessary. The 50 - 2000 MHz frequency band is selected to be the operation range. For ease of operation and for compatibility with many objects the coupling liquid is selected to be water. To get both optical and acoustical information from the same area of the object, a reflected-light microscope is combined with the acoustic microscope. The mechanical accuracy of conversion between the two microscopes is such that the obtained images are centered to within a few micrometers of each other.

The main sections of the acoustic microscope are its mechanical parts, its electrical parts and its acoustic part. The critical mechanical parts include the X-Y scan mechanism, the Z adjustment mechanics, and the object leveling apparatus. The X-Y scan mechanism is able to generate a 1 mm by 0.8 mm raster scan with less than 0.3 micrometer deviation in the Z direction. It is a electromechanical scan utilizing electromagnets and long leaf springs. Z adjustment can be done either manually or remotely by an electrical motor. The object leveling apparatus is coupled to the object stage to adjust the object surface parallel to the X-Y scan plane.

3. High Frequency Electronics

The heart of the electronics of the microscope is the high frequency part. The high frequency electronics operates in the pulse echo mode. It receives commands from a user terminal and sets its operating point accordingly. The 50 - 2000 MHz frequency band is divided between two units. The high frequency electronics operating at 0.8 GHz to 2.0 GHz is basically composed of transmitter oscillators, pulse generating circuits and a superheterodyne receiver. The transmitted signal is generated by varactor tuned oscillators whose frequency can be controlled by a voltage applied from a D/A converter driven by a computer. This signal is pulsed by a solidstate switch to generate 10 ns pulses. The switch is driven by a pulse drive electronics again controlled by the computer. The pulsed rf signal is amplified to a level of 1 W. The signal is fed to one of the arms of a double throw switch. The common arm of the switch is connected to the acoustic lens element. The second arm of the switch goes to a wide-band preamplifier. The pulse drive interface driving this switch is adjusted such that the first arm is connected to the common arm only while the first switch is on and the second arm is connected to common arm at a time slightly after the transmitter pulse is applied to the lens element. The output of the preamplifier is fed to a wide band mixer. The local oscillator is also pulsed to keep the IF amplifiers away from saturation caused by spurious reflections. The local oscillator is turned on only during the time the object pulse may appear. The necessary pulse is created by the pulse drive electronics using computer controlled digital delay techniques. The frequency of the local oscillator is corrected by the microprocessor to maximize the output voltage each time a new frequency is selected. The two intermediate frequency (IF) amplifiers in cascade provide the necessary gain. There is another switch between the two IF amplifiers to further reduce the undesired pulses. The gain of the IF amplifier is adjustable by the computer to the desired level. Finally, the output of the IF amplifier is detected by a detector diode to generate the video signal which is proportional to the amplitude of the received acoustic signal. The computer responsible for the whole high frequency electronics is built around a 8-bit single-chip microprocessor.

The high frequency electronics suitable for the 50 - 800 MHz range is based on a different principle. A 1 ns duration base-band pulse is used to excite the transducer. This pulse is fed to the transducer through a power divider. The output of the power divider is connected to a limiter to protect the input of the receiver amplifier. After amplification a mixer is used as a switch to gate out the unwanted parts of the incoming pulse train. After the time-gating operation further amplification is performed. High pass filters are placed in the receiving chain to get rid of switching spikes caused by the time gating operation. The amplitude detection of the amplifier output provides the necessary video signal for further processing.

4. Acoustic Objective

The heart of the acoustic part is the acoustic objective. It converts the electrical signals fed to it into ultrasonic signals, focusses them to a diffraction-limited spot and converts the reflected acoustic signals back to the electrical form. The lenses are manufactured from single-crystal sapphire material. A spherical lens cavity is ground on one side of the sapphire by mechanical means. A ZnO thin film tranducer is deposited on the flat side of the crystal. The tranducer generates planar acoustic wavefronts when used as a transmitter, and as a receiver it is sensitive to the shape of the wavefronts impinging on it. The lens cavity is coated with a quarter-wavelength-thick glass antireflection layer to reduce the reflection loss. The two-way conversion loss of tranducers is typically 10 DB. The lens units are housed in small metallic tubes with the high frequency connector on one end and the lens on the other. Matching networks are included within the lens housing.

Due to bandwidth limitations the whole frequency range cannot be taken care of with a single acoustic objective. Instead, the 0.8 to 2.0 GHz range is divided between two objectives: One centered at 1 GHz and the other at 1.7 GHz. The lower frequency 50 to 800 MHz range is covered by objectives operating at the following center frequencies: 100 MHz, 200 MHz and 400 MHz. A 60 MHz objective is in preparation. All the objectives have the associated matching networks to give the necessary bandwidth. The acoustic objectives differ not only in frequency but also in radius of lens cavity. At high frequencies the loss in the liquid medium is very high. In this case lenses with a cavity radius of 40 micrometers are used. On the other hand, at low frequencies the liquid losses drop very rapidly making large radius lenses feasible. At the low frequency end, 2000 μm radius lenses are used to increase the working distance and to make higher imaging depth possible. Working with large radius lenses is easy from both mechanical and electrical points of view. Mechanically there is a large clearance and electrically the various internal reflection pulses are well separated from each other. On the other hand, for a 40 μm radius lens there is only 60 ns separation between the first internal reflection pulse and the object pulse. Additionally, the size of the object pulse is 60 dB below the spurious pulse. Any reflections in electronics or in cables will manifest themselves as interference of the object pulse with a reference. The high frequency electronics described above is capable of separating the small object pulse from its large and close neighboring spurious pulses to generate interference free images.

The objectives of varying lens radius have varying time delays. The object pulse will not appear at the same place for the different objectives. Hence, the delays of time gating pulses as generated by the pulse drive circuitry should be different. This time delay adjustment is made automatically for every acoustic objective by the microprocessor.

5. Recent Developments

One of the important aspects of recent developments is the use of special lens configurations in conjunction with the acoustic microscope.

Regular acoustic lenses are spherical, thus providing no directional sensitivity. To overcome this, the following approach has been proposed:

Using cylindrical instead of spherical lenses results in high directional sensitivity due to highly directed surface wave excitation. This configuration is important for exact measurements of the anisotropy of surface wave velocity via V(z)-curves /2/. For imaging, the cylindrical lens suffers from bad resolution in the direction of the cylinder axis.

On the electronics side, highly sensitive phase measurement systems together with short pulse excitation of the lens transducer have been proposed /3/. These systems can be used to extract quantitative data with respect to surface wave velocity or surface topography with very high accuracy.

6. Outlook

The scanning acoustic microscope is a microscope capable of subsurface imaging with a resolution equalling a good optical microscope. Optically opaque materials or layers, which are unsuitable for the optical microscope, become the objects of the acoustic microscope. The acoustic microscope can be used with almost all objects and nondestructively. It is sensitive to a change in density or stiffness of the material. Voids within the body of the materials or delaminations in thin film structures are easily detected due to very high acoustic impedance change at the interface. The scanning acoustic microscope has found applications in such diverse fields as materials science, thin film technology, geology and biology, and new fields are emerging as the application research continues.

7. References

1. R. Lemons, C. F. Quate: Appl. Phys. Lett., 24:163. (1974)
2. R. Kushibiki, N. Chubachi: IEEE Sonics and Ultrasonics, SU-32, 2, p. 189 (1985)
3. K. Liang, S. D. Bennett, B. T. Khuri-Yakub, G. S.Kino: IEEE Sonics and Ultrasonics, SU-32, 2, p. 266 (1985)

Interferometric Detection of Ultrasound at Rough Surfaces Using Optical Phase Conjugation

M. Paul, B. Betz, and W. Arnold

Fraunhofer-Institute for Non-Destructive Testing, Bldg. 37, University of Saarbrücken, D-6600 Saarbrücken, Fed. Rep. of Germany

For the application of ultrasonic signals generated by short laser pulses via thermoelasticity and by ablation of surface material at solid surfaces, it is desirable to detect these signals in a contact-free manner at rough surfaces. If one could use optical interferometers as detectors this would constitute a new tool for the remote nondestructive testing of materials and components. Velocity interferometers are suitable instruments for this purpose /1/, however, their bandwidth is limited, in practice to about 10 MHz. We have developed a technique which allows one to combine the larger bandwidth attainable with heterodyne-interferometry with a large light-gathering power necessary for optical detection of ultrasonic signals from rough surfaces.

Our interferometric set-up can be separated into two basic optical systems. The interferometer itself is based on the heterodyne principle, ensuring large bandwidth, whereas the distorted wavefront of the signal beam after reflection from an optically rough surface is reconstructed by using optical phase-conjugation resulting in a large light-gathering power, at present $\sim 10^{-2}$ mm² sr. Without using phase-conjugation, the light-gathering power would be limited to λ^2 as given by the antenna theorem /2/.

The principle of operation of the interferometer is shown in Fig. 1. All optical beams are derived from an argon-ion laser used in the single-line mode at the wavelength λ = 514 nm. After the beam-splitter BS1 a small part of the beam is passed through a Bragg-cell which upshifts the light-frequency f_L to $f_L + f_B$ (here f_B = 160 MHz). A second beam splitter BS2 provides the signal beam (5 mW) which is focused onto the surface of the sample by the lens L1. A lens L2 gathers as much as possible scattered radiation from the rough surface and transmits it to the lens L3 which in turn focuses the distorted wavefront into the phase-conjugating mirror (PCM), in our case a $BaTiO_3$-crystal. The pump beams P1 and P2 carry the main power of the laser beam (175 mW) and are obtained via the beam

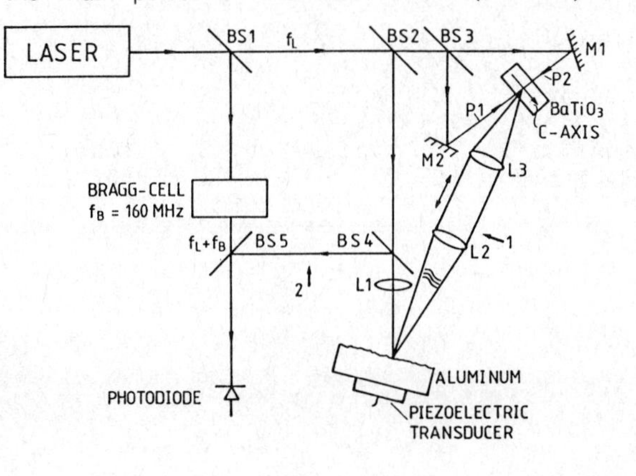

Figure 1: Principle of interferometry assembly.

splitter BS3 and the mirrors M1 and M2. The phase-conjugated replica of the distorted wavefront, as can be viewed by the speckle pattern at the location indicated by arrow 1 in Fig. 1, travels backward onto the same spot of the specimen and aberrations and phase distortions caused by the rough surface and the other parts of the optical system are removed and a well-defined beam is obtained (as can be viewed at the location indicated by arrow 2 in Fig. 1). The possibility to remove distortions in optical wavefronts by using phase-conjugation has been exploited and discussed previously /3/. The reconstructed beam is then combined with the reference beam passed through the Bragg-cell and both are received by a fast avalanche photodiode. From thereon the demodulation of the ultrasonic signals follows a well-known procedure /4,5/.

A backwall-echo sequence of longitudinal waves (P1, P2, P3) received by the present set-up is shown in Fig. 2. The ultrasonic pulses were generated on the opposite face of the sample by a piezoelectric transducer excited by a step voltage. The sample was an aluminium block whose surface exhibited a rugosity with a rms-value of ~ 4 μm. In spite of the rough surface the S/N ratio was ~30 dB. Further details of our scheme are published elsewhere /6/.

In summary, we have demonstrated the principle of an optical interferometer assembly enabling us to detect ultrasonic displacements at rough surfaces. The signal beam scattered from the surface is reconstructed by using phase-conjugation leading to an increase of the light-gathering power of the interferometer.

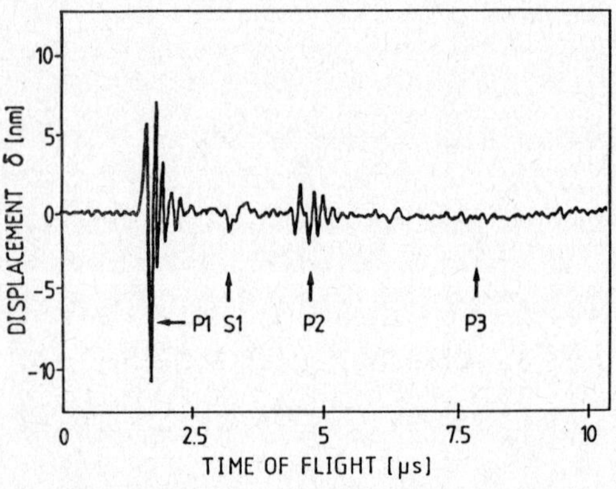

Figure 2: Received ultrasonic signal from a rough surface after demodulation

References

1. J.P. Monchalin: IEEE Trans. UFFC-33, 485 (1986), and ref. contained therein
2. A.E. Siegmann: Appl. Opt. 5, 1588 (1966)
3. D.M. Pepper: Opt. Eng., 21, 155 (1982) and ref. contained therein
4. Y. Martin and E.A. Ash: Phil. Trans. Roy. Soc. (London) A320, 257 (1986)
5. B. Cretin and P. Hauder: Proc. IEEE Ultrason. Symp., Ed. B.R. McAroy (1984), p. 656
6. M. Paul, B. Betz, and W. Arnold: Appl. Phys. Lett., 50, 14 (1987)

Interaction of a Photoacoustic SAW Pulse with Grain Structure in Solids

Zhang Xiaorong[1], *Gan Changming*[1], *and Zheng Legi*[2]

[1]Institute of Acoustics, Nanjing University, Nanjing, People's Republic of China
[2]Hehai University, Nanjing, People's Republic of China

A photoacoustic (PA) surface acoustic wave (SAW) pulse spectrometer consisting of an all-optic system is demonstrated. Some spectra recording the interaction of a PA SAW pulse with grain structure in a solid are presented.

The physical basis of pulsed PA generation and some of its applications have been described in [1-4]. The fundamental concept of an ultrasonic spectrometer has been discussed in [5]. Here we demonstrate a method of spectrum analysis of the profile of laser-generated SAW pulses. We call the apparatus a PA SAW spectrometer (LG-PASS for short).

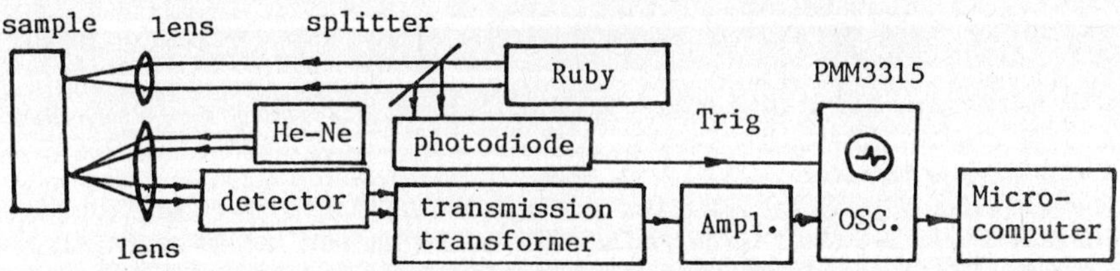

Fig.1 Scheme of a LG-PASS

A schematic of an all-optic LG-PASS is shown in Fig.1. A focused ruby laser was used as a pump beam to excite the SAW pulse. A focused He-Ne laser was used as a probe beam to detect the SAW pulse. Details are given in [6].The signal of the SAW pulse brought by the probe beam was received by position detector, recorded by a digital storage oscilloscope (PM3315) and analysed or processed by means of the fast Fourier transform (FFT). The bandwidth of the spectrum of ruby is about 11 MHz [6]. Five kinds of polished carbon steels with 3 cm diameter and 0.67 cm thickness were used as samples. The distance between the pump beam and probe beam at the sample surface is 0.9 cm. Five spectra of laser-generated SAW pulses at the surface of various samples are shown in Fig.2 and 3. In each figure, valley corresponds to the absorption peak of the SAW wave. Five photographs taken with a metallo-microscope (PMM) for these samples are shown in Figs. 4 and 5.From these figures it can be seen that the larger the grain structure, the higher the frequency at which the absorption peak appears.

According to ultrasonic theory [7,8], it is known that the ultrasonic attenuation depends on the grain structure of the sample. As the wavelength of an ultrasonic signal is the same order of magnitude as a grain in its path, the sound is strongly scattered or resonated,greatly attenuating the signal. It has been found that for approximately $\lambda < 3D$ (where λ is the wavelength and D the average grain diameter),the sound was com-

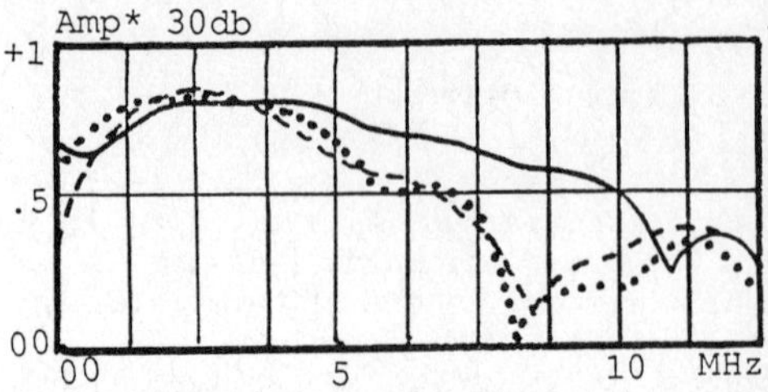

Fig.2 Spectra for samples 3(-), 4(---) and 5(....)

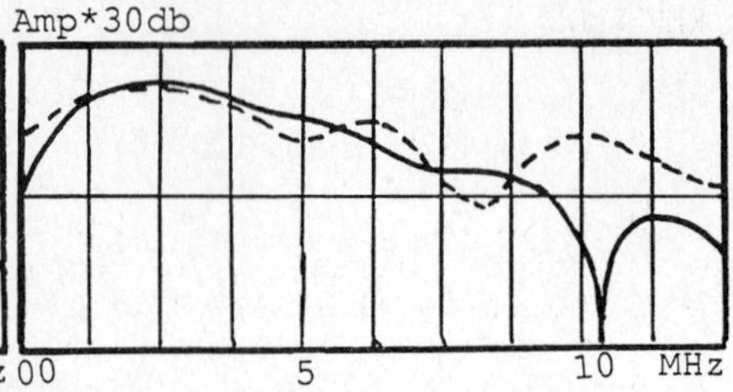

Fig.3 Spectra for samples 1(-), and 2(---)

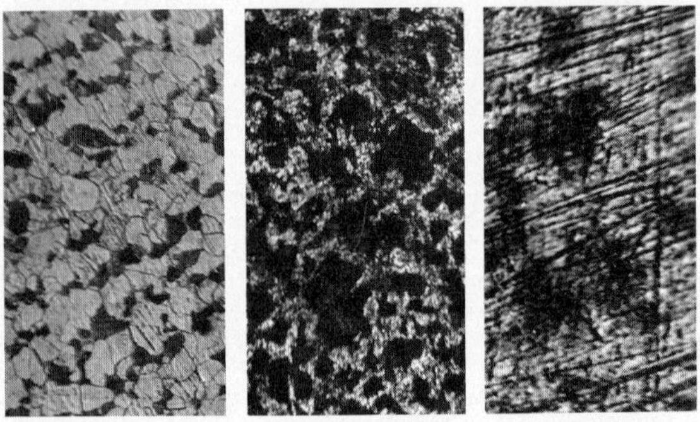

3# 4# 5#

Fig.4 Photographs taken with a metallo-microscope (PMM) with magnification 100X.

pletely attenuated by scattering in the material. These phenomena can be observed in Figs.2 and 3. The magnitude of $\lambda/3$ for samples 1, 2,3, 4 and 5, at the frequency where the SAW absorption peak appeared, are 123, 117, 84, 113, and 161 µm, respectively. However, from Figs.4 and 5 it can be seen that grains with such sizes exist in all the samples. The average grain size for samples 1, 2, 3, 4 and 5 are 116, 86, 78, 77, and 150 µm, respectively, which in each case is less than $\lambda/3$. This difference is due to the fact that the smaller grains within the sample, which cannot influence the ultrasonic attenuation at frequencies < 11 MHz, are included. So this method can be used to evaluate the grain size with the magnitude of the order of > 100 µm for solids. If the bandwdith can be increased, the estimation of grain size can be done precisely.

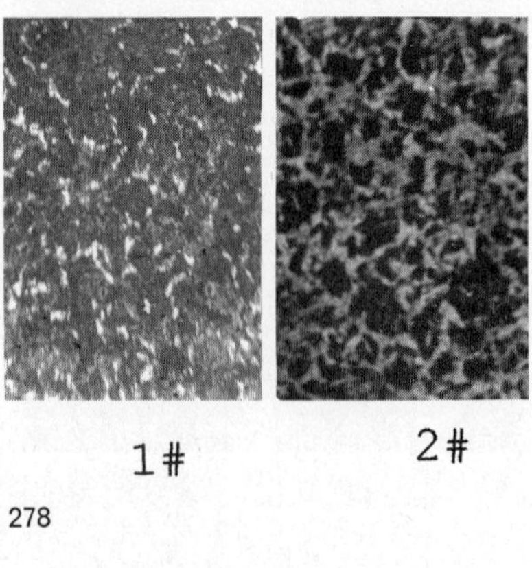

1# 2#

Fig.5 Photographs taken with a metallo-microscope (PMM) with magnification 80X.

REFERENCES

1. D. A. Hutchins: Can. J. Phys. 64, 1274 (1986)
2. A. C. Tam: Rev. Mod. Phys. 58, 412 (1986)
3. X.R. Zhang, C.M. Gan: Chinese Physics-Lasers,(to be published 1987)
4. X. R. Zhang and C. M. Gan et al,: Proceedings of the China-Japan joint conference on ultrasonics 455 (Nanjing Univ. 1987)
5. O. R. Gericke: in Ultrasonic Spectroscopy in NDE of Materials, ed. by J. J. Burke et al. (1979)
6. C.M. Gan, X.R. Zhang: Experimental Mechanics (to be published 1987)
7. E. P. Papadakis: Methods Exp. Phys. 19, 237 (1981)
8. W. P. Morse: Proc. Inter. School of Phys. Enrico Fermi Cours LXIII, ed. by D. Sett, 205 (1976)

Anomalous Photoacoustic Behavior of Semiconductors: Evidence for Thermally Generated Surface Deformations

H. Flaisher and D. Cahen

Department of Structural Chemistry, Weizmann Institute of Science, Rehovot, Israel 76100

I. ABSTRACT

The photoacoustic signal measured for a solid sample is generally thought to originate from plane thermal waves, created by exposing the sample to modulated irradiation. For illumination with energy near or below the bandgap energy of a semiconductor, photoacoustic response was observed which could not be explained with conventional, thermal wave theory. This behavior can be explained if it is assumed that the signal is dominated by thermal deformations, generated at the semiconductor surface.

II. INTRODUCTION

In general, the pressure waves detected in gas-microphone, photoacoustic (PA) spectroscopy originate from thermal waves in the sample /1/. One major feature of thermal wave response is that its phase should decrease with increasing optical absorption coefficient. After performing numerous PA experiments on various semiconductor wafers and Si photovoltaic cells, a phase **increase** was almost invariably measured with an increase in photon energy from below to above the bandgap. The main purpose of this work is to show that semiconductors do not behave according to thermal wave theory for irradiation energies near and/or below the bandgap energy. Here, we present both our experimental results and those in the literature which demonstrate this fact. With the aid of experiments in which the PA signal was detected from the unirradiated side of the semiconductor, we present evidence which suggests that thermally generated surface deformations may be responsible for this anomalous phase behavior, or APB.

III. RESULTS

For a thermally thick sample, one expects to observe a PA phase decrease of 45° when wavelength is varied from a region where the sample is optically transparent, to that where the sample is optically opaque. Figure 1 shows the wavelength dependence of both the normalized PA amplitude and phase recorded for a 1000 nm cut-off filter; it clearly behaves according to thermal wave theory.

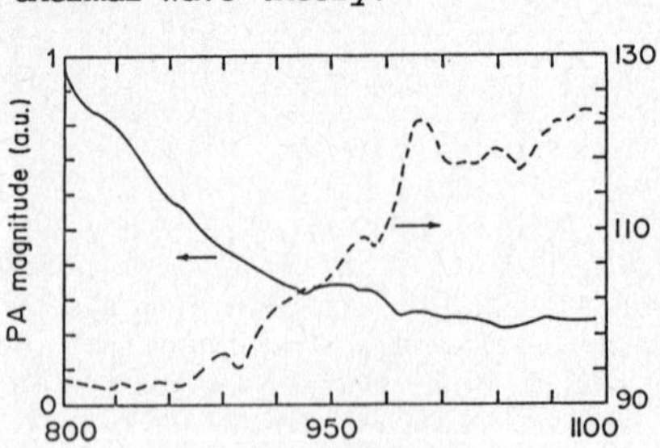

Figure 1 - Normalized PA magnitude (——) and phase (- - -) for a 1000 nm Schott cut-off filter as a function of irradiation wavelength (240 Hz)

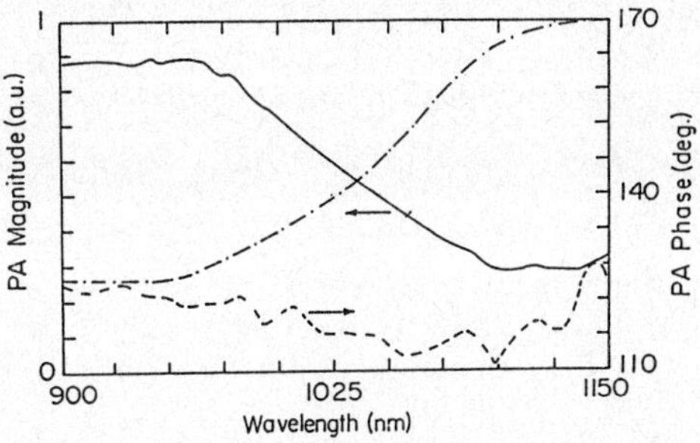

Figure 2 - Normalized PA magnitude (———) and phase (- - -) for a polished Si wafer as a function of irradiation wavelength (240 Hz)

Figure 2 shows the analogous scan for a polished Si wafer (bandgap at 1120 nm) carried out in the same PA chamber under the same experimental conditions. The PA phase clearly disobeys the approximate behavior expected from thermal wave theory, depicted by the dash-dot line in Fig. 2. The placement of this line implicitly assumes that thermal waves dominate the signal for irradiation energies greater than the bandgap.

This behavior typifies our measurements for both semiconductor wafers (Si, GaAs, GaP, CdSe, CdTe and WSe_2) at various modulation frequencies and Si solar cells under various electrical loading conditions.

IV. DISCUSSION

PA behavior consistent with thermal wave theory has been observed for $CdIn_2S_4$ /2/, GaAs /3/ and II-VI semiconductor powders /4/, while "two-layer" behavior /5,6/ was reported for $CdIn_2S_4$ /2/, GaAs /3/, II-VI semiconductor powders /4/ and CdS /7,8/. However, APB has been observed for CdS /7-9/ and GaAs /10/. While the major features of /7/ suggest two-layer behavior (although explained differently therein), one can discern the sub-bandgap PA phase to decrease to about 15° less than the supra-bandgap phase. No explanation was offered for APB in /8/ and in /9/, the phenomenon was explained on the basis of two-layer behavior. However, this explanation cannot be fully correct, since two-layer theory cannot predict the sub-bandgap PA phase to be less than the supra-bandgap phase, as was experimentally observed.

A viable explanation for APB was found in results of an experiment in which the PA response was detected from the **unilluminated**, back side of the sample (b-PA). Here, the side of the sample to be illuminated was coated with carbon black and then stuck with silicone grease to a clear microscope slide. The unilluminated side was then used as a window for the PA cell. Figure 3 shows the PA magnitude and phase for coated Al, glass and Si samples. The fact that glass has a response similar to that of Al proves that the PA signal here cannot be due to thermal waves, since the thermal diffusivity and conductivity of glass are greatly inferior to those of Al. On the other hand, since the thermal expansion coefficient of glass is on the same order of magnitude as that of Al, one concludes that **the b-PA signal for Al and glass results from bulk (probably flexural) vibrations.**

Next, observe that at low frequencies, the carbon-coated Si sample yields a signal of the same order of magnitude as that of Al. Since Si has thermal and elastic properties similar to those of Al, the low-frequency signal in Si is again attributed to sample vibrations. However, the phase of the signal in Si is about 90° less than that for Al and glass, suggesting that

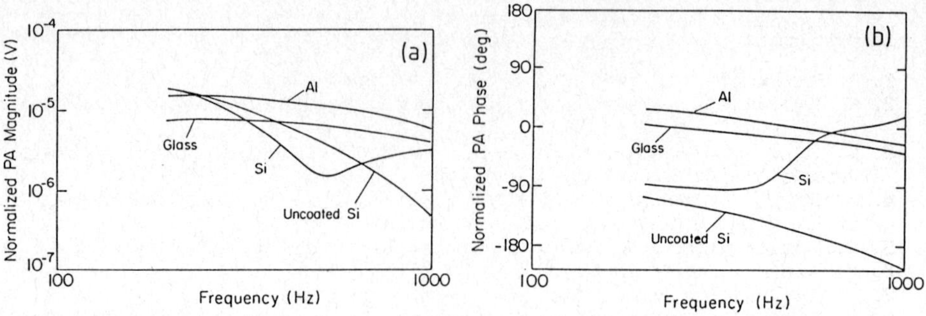

Figure 3 - b-PA magnitude (a) and phase (b) dependence on frequency for glass, Al and Si

the signal is not due to bulk flexural vibrations. Therefore, **the b-PA signal for Si appears to be related to sample vibrations of a nature different than bulk flexural vibrations.** In light of these and other /11/ considerations, the b-PA response for the coated semiconductor samples may be explained by the occurrence of thermally generated surface deformations (TGSD). At higher frequencies, both the signal and phase of the Si sample tend toward those for Al and glass, signifying that the b-PA response in this regime is dominated by bulk flexural vibrations.

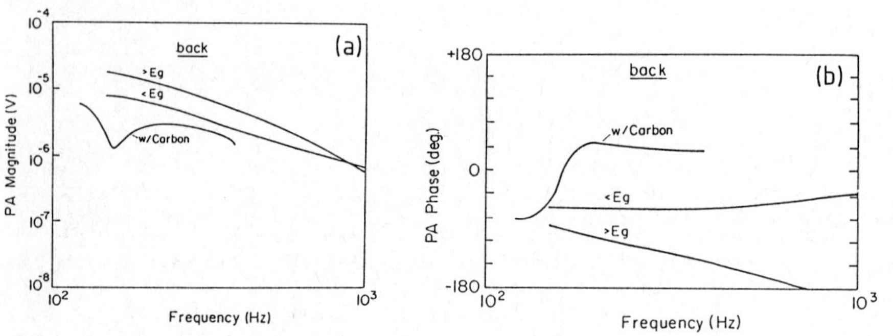

Figure 4 - b-PA magnitude (a) and phase (b) dependence on frequency for both coated and uncoated GaAs

Figure 4 shows curves for GaAs analogous to those for Si (Fig. 3) and all of the major features of Fig. 3 appear here too. Figure 4 also displays the b-PA frequency response obtained using sub-bandgap irradiation. The low frequency magnitude and phase approach the same values as those for the coated sample. Since the low-frequency response for coated Si and GaAs were attributed to TGSD, **the sub-bandgap, b-PA response for GaAs is also attributed to TGSD.** Finally, the b-PA and conventional PA response for GaAs obtained using sub-bandgap irradiation were measured and found to have the same magnitude and phase in the frequency range tested (150 Hz to 1 kHz). When taken into consideration with the conclusion above, this leads one to the final deduction that **APB results from TGSD.**

In summary, the PA response of semiconductors disobeys thermal wave theory for irradiation energy near or below the bandgap. Our observations and those in the literature suggest that TGSD may be responsible for this phenomenon. TGSD may also be responsible for two unexplained observations reported recently in the literature /12,13/.

V. REFERENCES

1. A. Rosencwaig, J. Appl. Phys. 49, 2905 (1978).
2. K. Yamashita, H. Kasahara, K. Yamamoto and K. Abe, Jpn. J. Appl. Phys. 21, Suppl. 21-3, 107 (1982).
3. Y. Fujii, H. Deguchi, H. Takeyama, A. Moritani and J. Nakai, Technology Reports of the Osaka University (Osaka, Japan), 31, 277 (1981).
4. Y. Fujii, A. Moritani, J. Nakai, H. Tai and S. Hori, Technology Reports of the Osaka University (Osaka, Japan), 31, 283 (1981).
5. H.S. Bennett and R.A. Forman, J. Appl. Phys. 48, 1432 (1977).
6. M. Morita, Jpn. J. Appl. Phys. 20, 835 (1981).
7. A. Mandelis and E.K.M. Siu, Phys. Rev. B 34, 7209 (1986).
8. R. Takaue, M. Matsunaga and K. Hosokawa, J. Appl. Phys. 56, 1543 (1984).
9. T. Dioszeghy and A. Mandelis, J. Phys. Chem. Sol. 47, 1115 (1986).
10. H. Tokumoto, Bul. Electrotech. Lab., 47, 97 (1983).
11. H. Flaisher and D. Cahen, to be published.
12. H. Sontag and A.C. Tam, Can. J. Phys., 64, 1330 (1986).
13. O. Goede, W. Heimbrodt and F. Sittel, Phys. Stat. Sol. (A), 93, 277 (1986).

Theoretical and Experimental Investigations of Broadband Thermoelastically Generated Ultrasonic Pulses

U. Schleichert[1], *M. Paul*[2], *B. Hoffmann*[2], *K.J. Langenberg*[2], *and W. Arnold*[2]

[1]Department of Electrical Engineering, University of Kassel, Wilhelmshöher Allee 71, D-3500 Kassel, Fed. Rep. of Germany
[2]Fraunhofer-Institute for Non-Destructive Testing, Bldg. 37, University of Saarbrücken, D-6600 Saarbrücken, Fed. Rep. of Germany

1. Introduction

Broadband ultrasonic pulses can be remotely generated by short laser pulses at solid surfaces via the thermoelastic effect and by ablation of surface material. Such a technique might be a viable tool in non-destructive evaluation/1/. This contribution firstly outlines the elements of a hitherto lacking three-dimensional theory of thermoelastic sound generation and secondly a comparison with experimental results is given.

2. Theoretical Model

In Fig.1 the steps of our theory are displayed schematically. Firstly, the laser beam is modelled as a rotationally symmetric beam of gaussian radial intensity variation. Its temporal dependence can be arbitrary. This beam impinges on the surface of the sample (z = 0), and part of it is absorbed within the optical skin depth $1/\alpha$ which is, in metals, of the order of 100 Å. The absorbed laser energy diffuses into the solid as determined by the heat-diffusion equation

$$-\kappa\Delta\Theta + \rho c\dot{\Theta} = p(R,t) = \alpha J_o \exp(-\alpha z)\exp(-r^2/w^2)f(t) \qquad (1)$$

Here, κ is the thermal conductivity, c is the specific heat, ρ is the density, w is the lateral width of the laser beam-profile, J_o is the inten-

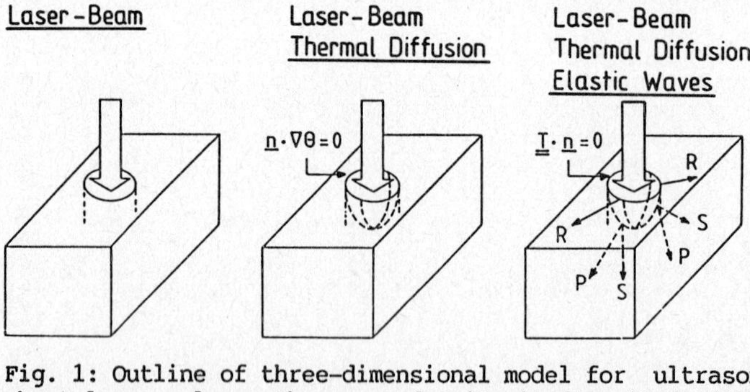

Fig. 1: Outline of three-dimensional model for ultrasonic generation by short laser pulses. The power density is deposited within the optical skin depth $1/\alpha$, and it diffuses into the solid as governed by the heat diffusion equation. The various ultrasonic modes are generated by thermal expansion taking into account the appropriate boundary condition.

sity of the laser pulse absorbed by the material at $r = z = 0$, $f(t)$ is its temporal shape, and $p(R,t)$ is the power density deposited. R is the position vector in cylindrical coordinates (r,θ,z), t is the time, the dot denotes the temporal derivative and Δ the Laplace operator. Because the temperature rise Θ entails a volume change due to thermal expansion (β is the linear expansion coefficient)

$$\Delta V/V = 3\beta\Theta(R,t) \tag{2}$$

a displacement field results which can be calculated from the wave equation for the potentials

$$\Delta\phi - \ddot{\phi}/c_P = (3\lambda + 2\mu)\beta\Theta(R,t)/(\lambda + 2\mu) \tag{3a}$$

$$\Delta\underline{\psi} - \ddot{\underline{\psi}}/c_S = 0 \tag{3b}$$

in the usual way. This displacement field contains one component, u_1, in radial direction and one, u_θ, in tangential direction (θ is the azimuth angle). In Eqs. (3a) and (3b) λ and μ are the Lame's constants, and c_P and c_S are the wave speeds for compressional and shear waves, respectively. Because of Eq.(2) (isotropic thermal expansion) the source term for ultrasonic generation appears only in Eq.(3a) describing compressional waves (P-). Shear waves (S-) are only generated via mode conversion due to the boundary conditions $\underline{\underline{T}}\cdot\underline{n} = 0$ for $z = 0$. Here, $\underline{\underline{T}}$ is the stress tensor and $\underline{n}$ is the unit vector perpendicular to the surface.

3. Results

The solutions of Eqs.(1) and (3) can be obtained with the aid of Green's functions. Their integration taking into account the explicit source terms can be carried out analytically for Eq.(1) and only numerically for Eq. (3). However, it is appropriate to consider some simplifications first. If we let $\kappa \to 0$ and assume a $\delta(t)$-like laser impulse, both Eqs. (1) and (3) can be solved analytically. In this case the temperature rise $\Theta(t)$ is a step-function $H(t)$:

$$\Theta(r,z,t) = (\alpha J_o/\rho c)\exp(-\alpha z)\exp(-r^2/w^2)H(t) \tag{4}$$

We further simplify by letting $w \to 0$. This amounts to a radial $\delta(r)/2\pi r$-dependence of the laser impulse (the factor $2\pi r$ appears because of the cylindrical coordinate system). The displacement field in both components u_1 and u_θ can now be calculated exactly and the resulting u_1-component is displayed in Fig. 2a-2c for different angles θ. It should be mentioned that in the calculations we made the transition $1/\alpha \to 0$ for simplicity. It is appropriate to call such a source Surface Center of Expansion (SCOE-source)/2/ because it can thought of as a small "breathing sphere" just below the surface.

We have also measured the waveforms experimentally. A half-cylinder of aluminum (radius l = 50 mm) was fabricated whose "curved" surface was facetted. The laser beam was of gaussian shape both in time (pulsewidth 2τ = 50 ns at 1/e points of the intensity) and lateral extent (width $2w$ = 4 mm). The u_1- component of the displacement field was recorded as a function of angle θ by means of a capacitance transducer whose ground plate was one of the facets. The bandwidth of the recording system was 1 kHz - 20 MHz. The laser pulses were directed onto the center of the plane surface of the half-cylinder.

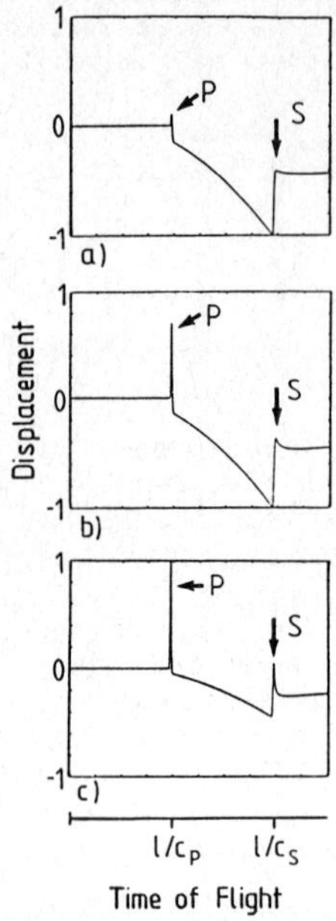

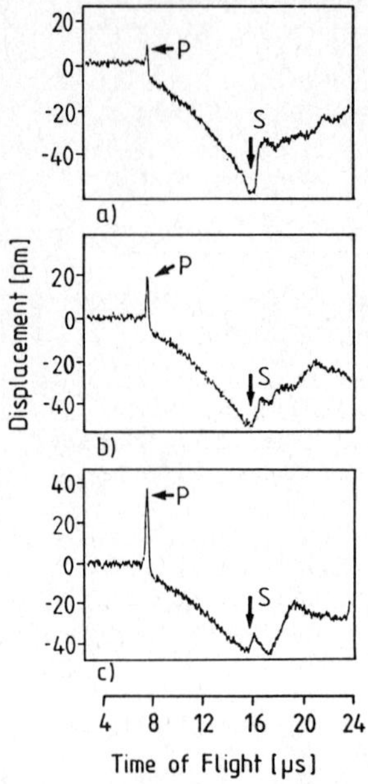

Figs.2a–c: Theoretical waveforms obtained at three different angles a) $\Theta = 10^{\circ}$, b) $\Theta = 20^{\circ}$, c) $\Theta = 30^{\circ}$ l is the pathlength.

Figs.3a–c: Experimental waveforms. The signals were measured with a capacitance transducer sensitive to normal displacements (Here w = 2 mm and J_o = 5 MW/cm^2). a) $\Theta = 10^{\circ}$, b) $\Theta = 20^{\circ}$, c) $\Theta = 30^{\circ}$

Figs. 3a – 3c show the experimentally determined u_1 – component at three different angles Θ. The overall agreement between experiment and theory is quite reasonable. This holds also for larger angles Θ (not shown here). First, a positive P-pulse (meaning excursion of the surface) is observed which arrives with the P-wave velocity followed by a negative signal (meaning retraction of the surface) which eventually has passed through with S-wave velocity. In this respect the displacement field resembles the one obtained from a radial point force acting at the surface of the sample/3,4/. This has been discussed previously/1,5/.

The u_1-component of the displacement field for $\Theta = 0^{\circ}$ as calculated with the same approximations as outlined above is displayed in Fig. 4 (solid line). Here, the positive P-pulse is absent, whereas experimentally it can be well observed (Fig. 5). This calls for a modification of the approximations put forward above. By taking into account the finite penetration depth of the temperature distribution as well as the finite extent of the laser pulse, one obtains in fact a displacement field similar to the experimentally observed one (dashed line in Fig. 4). Physically, this corresponds to ring sources below the surface weighted with the temperature distribution $\Theta(r,z,t)$ according to Eq.(4) whose depth is governed by the thermal diffusion length and by the optical penetration depth $1/\alpha$. This clearly indicates that for a quantitative description of the thermoelastic ultrasonic source the explicit solution of Eq.(1) has to be taken into account. At present, we are carrying out the corresponding

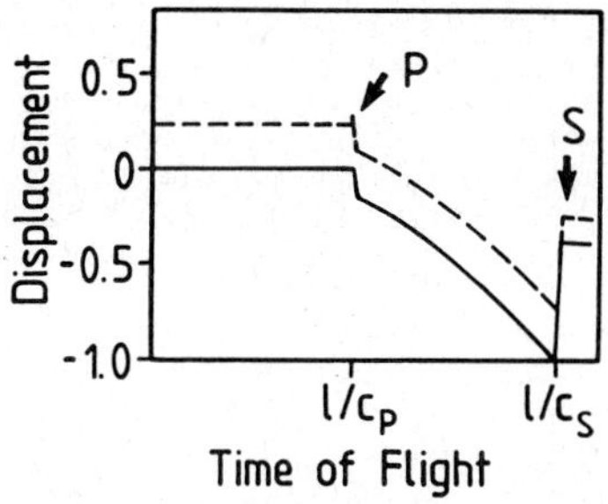

Fig. 4: Theoretical waveforms for for $\theta = 0^{o}$, solid line: SCOE-source, dashed line: ring source For clarity the dashed line is displaced.

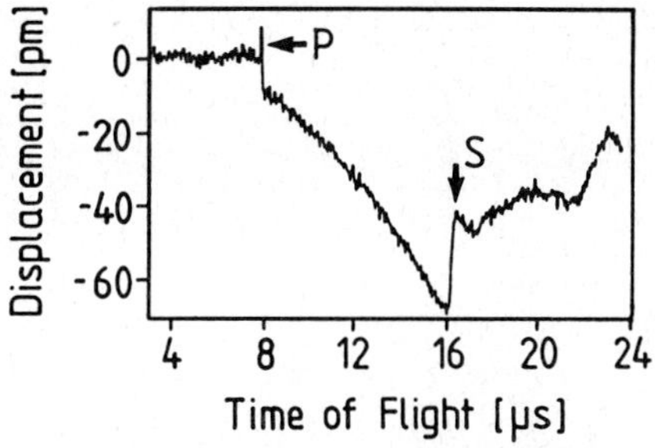

Fig. 5: Experimental waveform observed at $\theta = 0^{o}$. A positive P-pulse can be observed which cannot be explained by the SCOE-model.

numerical calculations which include not only the P- and S-wavefronts but also the generation of Rayleigh (R-)-waves/6,7/, and we will also make a quantitative comparison with the experimentally determined amplitudes for a given J_o-value.

4. Conclusion

In summary we have presented the outline of a three-dimensional theory describing the generation of ultrasound by short laser pulses in the thermoelastic regime. Whereas the simplifications underlying a SCOE source lead to a qualitative agreement with experiments, a quantitative agreement can only be achieved by taking into account the full three-dimensionality of the problem.

Acknowledgement

This contribution is based on work performed with the support of the Deutsche Forschungsgemeinschaft.

References

1. D.A. Hutchins: Can.J.Phys.,64, 1247 (1986) and references contained therein
2. L.R.F. Rose, J.Acoust.Soc.Am., 75, 723 (1984)
3. C.L. Pekeris and H. Lipson: J. Acoust.Soc.Am., 29, 1233 (1957)
4. Y.H. Pao, P.R. Gajewski, and A.N. Ceranglu: J.Acoust.Soc.Am., 65, 96 (1979)
5. R.J. Dewhurst, D.A. Hutchins, and S.B. Palmer: J.Appl.Phys.,53, 4064 (1982)
6. V.V. Krylov and V.I. Pavlov: Sov.Phys.Acoust.,28, 493 (1982)
7. W. Arnold, B. Betz, and B. Hoffmann: Appl.Phys.Lett.,47, 672 (1985)

Light Enhanced Ion-Acoustic Signal Generation in Si and GaAs

F.G. Satkiewicz, J.C. Murphy, and L.C. Aamodt

The Johns Hopkins University, Applied Physics Laboratory, Laurel, MD 20707, USA

1. Introduction

Modulated beams of Ar^+ ions have been used as acoustic sources for studies of ion-specimen interactions [1-3]. The technique is analogous to Scanning Electron Acoustic Microscopy (SEAM) [4] in using electrostatic beam blanking and piezoelectric detection. However, the ion-beam-specimen interactions differ from electron beams in causing surface erosion, ion implantation and, for reactive ion species, chemical reactions. This paper presents some preliminary information on the role of surface oxygen in elastic generation by rare gas ions on silicon and GaAs. Illumination of the specimens with unmodulated light above the semiconductor band gap changes the acoustic signal. The dependence of the acoustic signal on light wavelength, modulation frequency of the source, energy of the primary ion beam, and the electrical characteristics of the specimen are presented. A tentative model of the interactions which produce these effects in silicon is suggested. The model may be applicable to GaAs in revised form.

2. Experimental

In the experimental configuration, the ion generation/detection system is a SIMS (Secondary Ion Mass Spectrometry) spectrometer using a duoplasmatron ion source and a double focusing energy/mass analyzer [5]. The spectrometer was modified to incorporate amplitude modulation of the ion beam and a piezoelectric transducer to detect the modulated stress produced by the ions. The range of ion energies E for the Ar^+ ions used in this work was $2 < E < 10$ (keV). The instrumental configuration included a method of monitoring the specimen current produced by the beam and a low spectral resolution (10 nm) illumination system used to study the dependence of the ion-acoustic signal on light wavelength. Partial pressures of O_2 could be used in the target chamber to investigate the role of oxygen in the acoustic generation process.

Silicon and GaAs samples were examined. Most of the silicon measurements were made on [100] oriented wafers of n and p silicon, 533 µm and 355 µm thick respectively, both of resistivity $\rho \approx 5\ \Omega$ -cm. Some [111] specimens were also examined. For GaAs, [100] oriented n, p, and Cr-doped semi-insulating specimens were examined. These specimens were 15 mils thick. A second specimen of Cr-doped GaAs, 1 mm thick was also studied.

3. Presentation of Data

Figure 1 shows the time variation of the specimen current for Ar^+ ions on [100] n- and p-silicon samples with and without illumination in the presence of O_2. In the presence of O_2 the specimen current varied with time, increasing for the n-Si and showing little change for the p-Si samples. However, when simultaneously illuminated with light, the p-Si specimen current increased to a value near that of the n-Si sample with O_2 only. There was

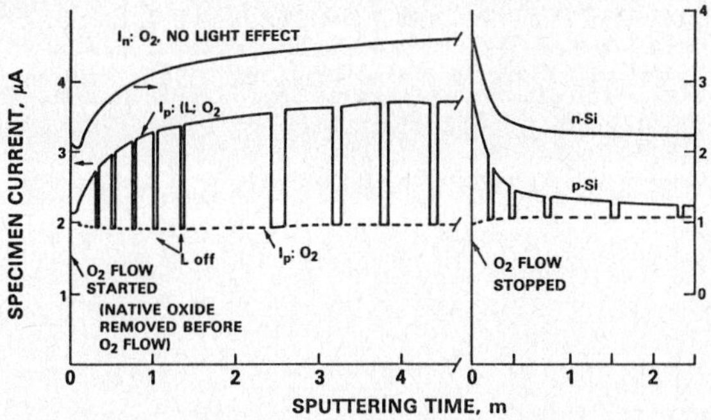

Fig. 1 Specimen current vs t; n, p silicon 5.5 keV Ar^+

no significant change in the current for the n-Si on illumination. The time constants for the two cases are approximately equal, suggesting that the accommodation of O_2 on silicon is similar for n- and p-doping. The spectral dependence observed for p-Si in Fig. 2 suggests the effect depends on carriers generated by interband optical transitions.

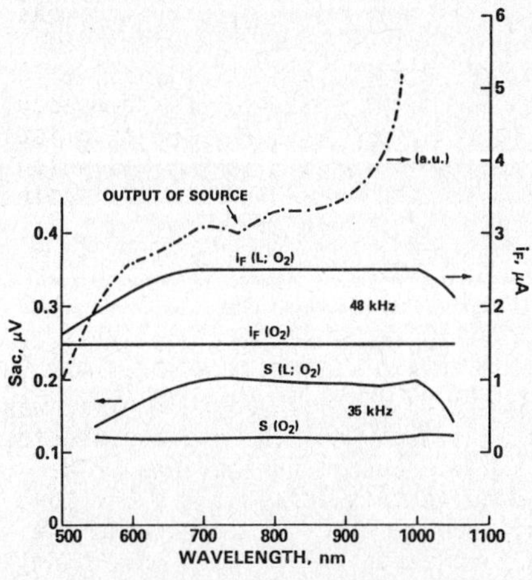

Fig. 2 Specimen current and acoustic response p-silicon vs λ 5.5 keV Ar^+

SIMS studies of the temporal changes in the Si^+ peak for a sample initially covered with a native oxide had showed an initial plateau in the peak intensity. (The Si^+ peak correlates with oxygen coverage.) Under the same experimental conditions the specimen currents for both n- and p-material showed no plateau suggesting that the oxide layer did not control the current generation or the acoustic generation discussed next.

The acoustic signal produced by the modulated ion beam also depended on the presence of O_2 and light. Figure 3 shows the change in acoustic signal in the presence of O_2 for n-Si and in the presence of both O_2 and light for p-Si. The maximum change in acoustic signal for n-Si with O_2 and for p-Si with O_2 and light is approximately equal. The spectral dependence of the acoustic response is similar to that of Fig. 2 for specimen current. Also, an acoustic signal is present without either perturbing agent.

Ion-acoustic signals in Cr-doped semi-insulating GaAs also depend on the wavelength of light illuminating the surface (Fig. 4). No effect of O_2 is

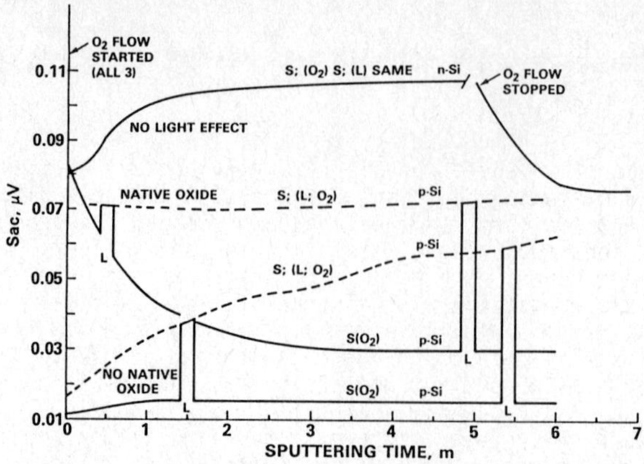

Fig. 3 Acoustic response vs t; n, p silicon 5.5 keV Ar^+

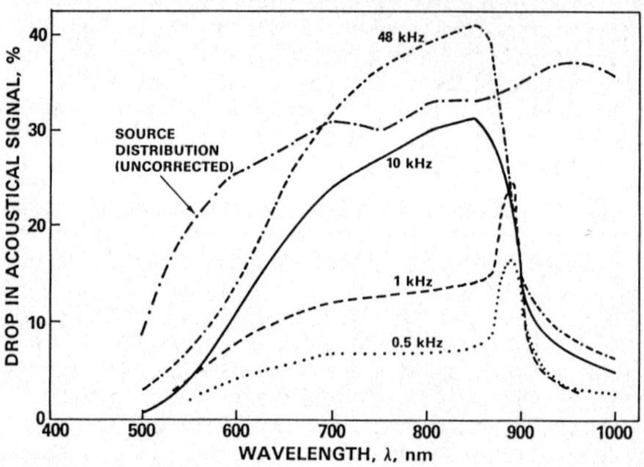

Fig. 4 GaAs acoustic response vs λ 5.5 keV Ar^+

seen. The peak in response at 880 nm at low frequencies may be associated with structure in the absorbance spectrum near 875 nm.

4. Discussion

The oxygen dependence of the ion-acoustic signal in silicon differs from that of metals [6]. For silicon, comparison of SIMS, specimen current and acoustic signal amplitudes suggest that the presence of a distinct physical oxide layer is unimportant. The enhancement observed in n-Si exists in p-Si only in combination with above bandgap light. For GaAs no dependence on oxygen has been observed.

A possible model of these observations presumes that the specimen current results from charge exchange between the incident ion beam and the specimen modified by currents due to transport of other ionic or electronic species in the target chamber. In our experiments the specimen current is positive having the same sign as the ion current. This requires either hole or electron transport in the silicon with the electron flow directed toward the silicon surface. Some possible charge exchange processes are:

$$Ar^+ + Si(sc) \rightarrow \{Ar^+ + e^-\} + \{Si + h^+\} \tag{1}$$

$$Ar^+ + Si(sc) \rightarrow \{Ar^+ + e^-\} + \{Si - e^-\} \tag{2}$$

$$Ar^+ + Si(sc) + X^{n-} \rightarrow Ar^+ + \{X^{n-} + e^-\} + \{Si + h^+\} \quad (3)$$

$$Ar^+ + Si(sc) + X^{n-} \rightarrow Ar^+ + \{X^{n-} + e^-\} + \{Si - e^-\} \quad (4)$$

where X denotes a species such as molecular or atomic oxygen present on the surface in the presence of argon ions. Equations (1) and (2) represent direct charge exchange between the Ar^+ ion beam and the valence or conduction band in silicon yielding neutralized ions and holes or electrons in the silicon. Equations (3) and (4) represent charge exchange via another species designated as X. Our specimen results are consistent with indirect charge transfer involving oxygen via (3), (4) being competitive with the direct transfer for the conditions of our experiment. However, the process depends on the particle densities and energies of available electrons and holes. For p-Si in the absence of light, the oxygen exchange pathway is inefficient, presumably because of a poor match between the populated energy levels in silicon and the energy level differences of the oxygen species on the surface. With light, photogenerated conduction band electrons are present and provide energy levels for charge exchange. For n-Si, conduction band electrons are present even without light and facilitate charge exchange via oxygen. No light effect is expected or observed in the n-Si case. In either n- or p-Si, the maximum specimen current should be equal, being limited by the ion current in the primary beam. This is the experimental result.

This model is consistent with the acoustic results. Acoustic generation requires energy deposition in the silicon to produce an elastic signal detectable by the transducer. In the suggested model, the acoustic signal generated on p-Si alone is relatively small and increases only slightly on exposure to oxygen. When oxygen plus light is used, the acoustic amplitude increased (Fig. 3) to a level approximately equal to that of the n-Si sample. For n-Si, an ion-acoustic signal is present without oxygen. In the presence of oxygen, the acoustic amplitude increased by 40% but no light effect was observed. This suggests that energy is deposited in the specimen by allowed charge exchange processes at the silicon surface via electrons in the silicon conduction band.

The change due to light in the ion-acoustic signal in semi-insulating GaAs also appears to involve electrons in the GaAs conduction band. A charge exchange mechanism could be involved. However, the absence of light enhancement for either n- or p-samples is unexplained. Also, while absorption spectra for the sample used to obtain Fig. 4 shows a transition near 880 nm, the reason the additional peak is seen only at frequencies near 1 kHz is unknown.

An unresolved question under active study is why acoustic generation using energetic ion beams (5 keV) is relatively small. This small signal allows the mechanism to be affected by optical transitions of low energy (1eV).

1. J.C. Murphy, F.G. Satkiewicz, and L.C. Aamodt, Bull. Am. Phys. Soc. 30, 474 (1985)
2. D.N. Rose, H.R. Turner, and K.O. Legg, Can. J. Phys. 64, 1284 (1986).
3. K. Kimura, K. Nakanishi, A. Nishimura, and M. Mannami, Jap. J. Appl. Phys. 24, L449-L450 (1985)
4. G.S. Cargill: In Scanned Image Microscopy, ed. by E.A. Ash, (Academic Press, London 1980) p. 319
5. J.C. Murphy, J.W. Maclachlan, R.B. Givens, F.G. Satkiewicz, and L.C. Aamodt: In Proc. Ultrasonics International 1985 (Butterworth, London 1986) p. 30.
6. F.G. Satkiewicz, J.C. Murphy, J.W. Maclachlan, and L.C. Aamodt: In Rev. Prog. Quantitative NDE, ed. D.O. Thompson and D.E. Chimenti (Plenum Press, New York 1987) p. 759

Supported by Department of the Navy under Contract No. N00039-87-C-5301.

Detector Strategy for Highly Versatile Scanning Electron Acoustic Microscopy (SEAM)

M. Domnik, M. Schöttler, and L.J. Balk

Universität Duisburg, Sonderforschungsbereich 254,
Fachgebiet Werkstoffe der Elektrotechnik,
Kommandantenstr. 60, D-4100 Duisburg 1, Fed. Rep. of Germany

1. Introduction

SEAM has become a convenient tool for imaging material properties with high spatial resolution and with the ability of subsurface non-destructive evaluation. Nearly any material can be analyzed by means of SEAM /1/. However, its versatility suffers from the fact that little attention has been paid to optimization of the detector necessary to monitor the acoustic wave signal as generated by the primary electron beam within the sample.

2. SEAM Detectors

The detector should

(i) deliver a high signal level for sound waves generated at typical frequencies of SEAM operation, usually covering the kHz regime;
(ii) enable quantitative measurements;
(iii) allow detection of the SEAM signal without need for mechanical contact between sample and detector(important for medical research);
(iv) permit extension of SEAM experiments up to very high frequencies, preferably into the GHz regime.

As these requirements cannot be met by one detector, various transducers are used in this work. Requirement (i) is met by the usual SEAM detector consisting of a lead zirconate titanate (PZT) ceramic as transducing material /2/. The ceramic is usually placed at the epicenter of the experiment and pressed to the sample bottom. Requirements (ii) and (iii) are achieved by a capacitive transducer/3,4/ (Fig. 1). This detector is a condensor arrangement with one plate the bottom sample surface and the other electrode in close distance ($\sim$10μm) to it. Requirement (iv) can be met using a high bandwidth material like polyvinylidene difluoride (PVDF) /5/.

3. Experimental Results

The frequency dependence of the transducer response was determined by various set-ups. In the kHz range SEAM magnitudes were measured via a lock-in amplifier, the samples being a brass alloy and zinc. The results were widely independent of the material. PZT showed highest signal levels, whereas the other transducers still deliver enough signal for recording micrographs as can be seen by the example of Fig. 2 for the capacitive transducer. In the MHz range SEAM measurements were carried out using a high frequency lock-in and an additional wideband preamplifier. PVDF showed best results here, though PZT still reveals enough sensitivity (Fig. 3). The GHz responses were determined by electrical determination of the transducers' transfer functions using subnanosecond primary beam pulses and a digital boxcar integrator. In principle PVDF and the capacitive transducer showed here sufficient signal levels (Fig. 4,for PVDF), however, up to now it was only possible to use PVDF successfully in this range. By time-resolved SEAM experiments 1μm-surface steps can be resolved due to an achieved 200ps resolution.

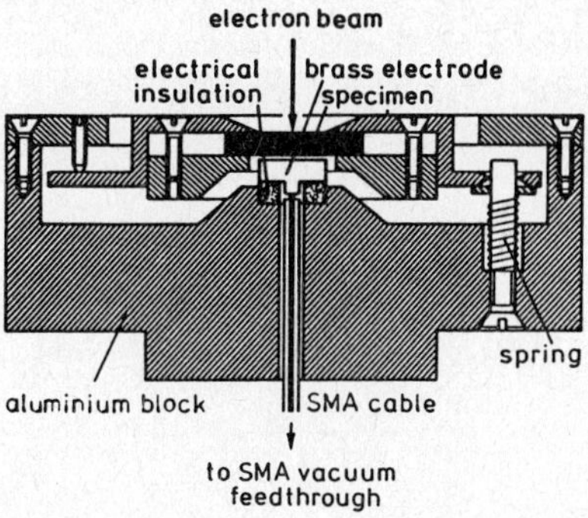

Fig. 1: Capacitive transducer

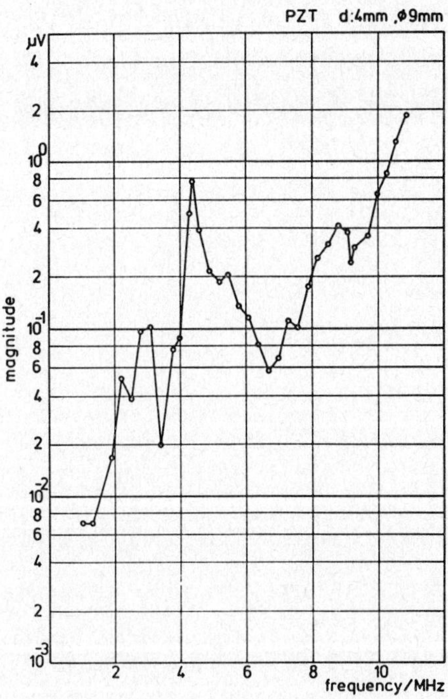

Fig. 3: MHz response of PZT

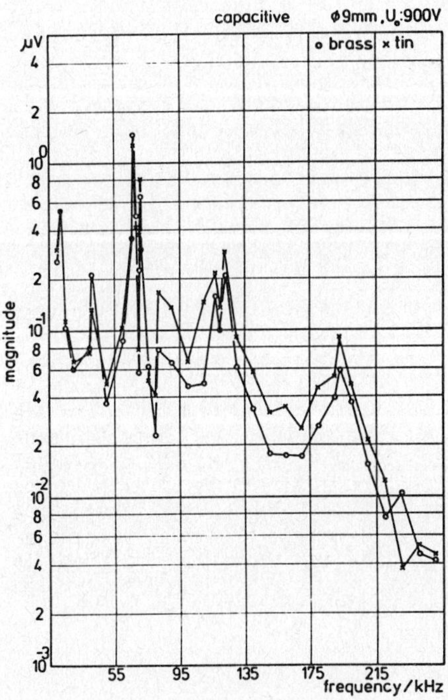

Fig. 2: kHz response of capacitive transducer

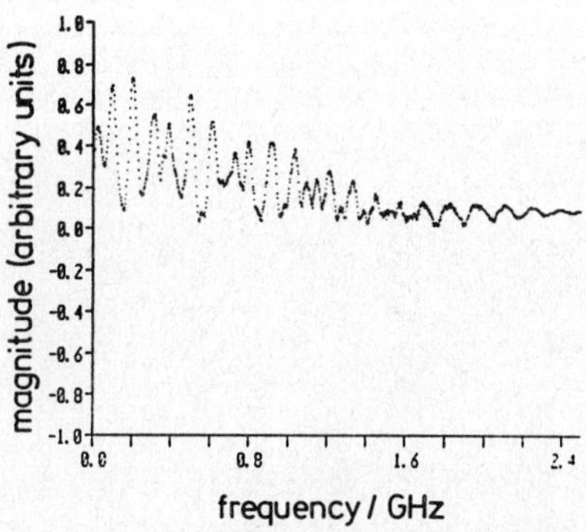

Fig. 4: GHz response of PVDF

4. References

1. L.J. Balk: in Advances in Electronics and Electron Physics, Academic Press (in press) (1987)
2. T.M. Proctor Jr.: J. Acoust. Soc. Am. 71, 1163 (1982)
3. W.B. Gauster and M.A. Breazeale: Rev. Sci. Instrum. 37, 1544 (1966)
4. L.J. Balk, M. Domnik, E. Böhm: at 4th Int.Symp. Optical & Optoelectronic Appl. Science & Eng., March 30-April 3, The Hague, Netherlands
5. H. Coufal: Appl. Phys. Lett. 44, 99 (1984)

Anisotropy in Thermoacoustic Imaging of Single Crystals

J.W. Maclachlan and J.C. Murphy

Center for Nondestructive Evaluation and Applied Physics Laboratory, The Johns Hopkins University, Laurel, MD 20707, USA

1. Introduction

The origins of the contrast observed in thermoacoustic imaging of material microstructure has received attention since the first such images were presented by CARGILL [1]. Two types of contrast are typically observed in images showing grain structure in metals - the contrast at the grain boundaries and the contrast between the grain interiors. Other possible sources of contrast in thermoacoustic images of microstructure include second phase structures such as martensite [2] and the presence of plastic deformation [3]. The contrast at the grain boundaries has been shown to be dependent on the thermal diffusion length in the material [4], but there has been much discussion as to whether the origin of the contrast between grain interiors has a thermal or an elastic origin. Observations by DAVIES and HOWIE [5] showed that bright grains (large thermoacoustic signal) in an electron-acoustic image of a Cu-Zn-Al specimen tended to have orientations close to [110] and [111] in the standard stereographic triangle while the dark grains (smaller signal) tended towards a [100] orientation. ROSENCWAIG and OPSAL [6] developed an expression for the ratio of the thermoelastic displacements in the [100] and [110] directions in a cubic single crystal and find that this ratio is dependent on the degree of elastic anisotropy and predict a greater displacement in the [100] direction.

The elastic moduli of a material form a fourth rank tensor and it is not possible to completely describe the elastic behavior of a crystal by a surface such as the representation quadric used to describe thermal conductivity in an anisotropic medium. A surface which has been used to give a sense of the variation in the elastic properties with direction is the Young's modulus surface. For cubic crystals, the reciprocal of the effective Young's modulus, E^*, is given by the expression [7]

$$\frac{1}{E^*} = S_{11} - 2(S_{11} - S_{12} - \tfrac{1}{2}S_{44})(\gamma_1^2\gamma_2^2 + \gamma_2^2\gamma_3^2 + \gamma_3^2\gamma_1^2), \qquad (1)$$

where S_{ij} are the elastic compliances and γ_i are the direction cosines. The effective Young's modulus as a function of orientation in the (100) plane is calculated from (1) and Fig. 1 shows plots for aluminum, silver, and copper.

2. Experimental

The experimental setup used for the thermoacoustic and optical beam deflection imaging is shown in Fig. 2. The pump beam from an argon ion laser was focused to a spot size of about 5 microns and amplitude-modulated using an acousto-optic modulator. The sample was mounted on the piezoelectric sensor and placed on X-Y stages which were scanned under computer control. Optical beam deflection images were produced using a HeNe probe beam and a position sensitive detector as shown in the figure. The signal collected from the detectors

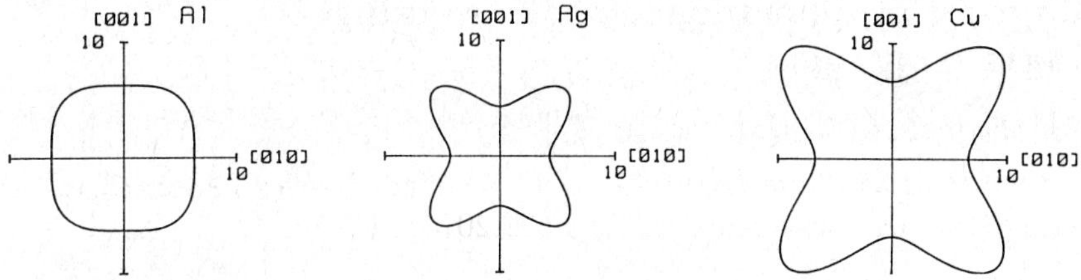

Fig. 1 Effective Young's modulus as a function of direction in the (100) plane for aluminum, silver, and copper single crystals. The axes are labelled in units of $10^{10}N/m^2$

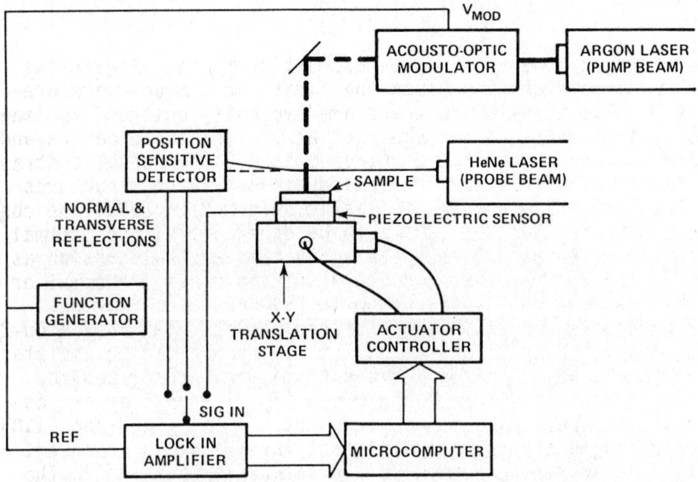

Fig. 2 Block diagram of experimental setup used for obtaining thermal wave images and linescans with laser beam excitation and detection by attached transducer and optical beam deflection methods

was analyzed with a lock-in amplifier, stored in the computer and linescan data and grey-scale and contour images were then produced with the aid of an image processor.

Thermoacoustic images were made of single crystals of aluminum, silver and copper which were oriented with the (100) plane as the plane of the sample. These are all cubic materials and are expected to have isotropic thermal conductivities and thermal expansion coefficients but anisotropic elastic properties (see Fig. 1). The [100] direction in the (100) plane for each sample was determined by Laue back-reflection techniques. It is important to note that the active area of the transducer used (NBS Conical Transducer) is only 1.8 mm and that the transducer was held fixed at the center of the sample. As the pump beam was scanned over the sample surface, the relative crystallographic orientation of the excitation source and the detector then varies for different locations in the area scan. Images were also performed using a transducer with a larger active area of 2 cm diameter for comparison with images collected with the point detector. Optical beam deflection images were made of the single crystals to determine whether there was any "thermal" structure present in the samples.

3. Results and Discussion

Contour plots of laser-acoustic magnitude images are shown in Fig. 3 for the aluminum, silver, and copper crystals where the modulation frequencies were 24.53, 13.9, and 14.12 kHz, respectively. The contour lines in each image connect points of equal intensity and the area of highest intensity is in the center of the sample, directly over the transducer. This is in agreement with earlier thermoacoustic imaging studies of buried defects which suggested the presence of a forward-directed radiation pattern for the thermoacoustic source [4]. The thermoacoustic signal level decreases as the distance away from the transducer increases, but the rate of this decrease shows a dependence on crystallographic orientation and on the type of material. This anisotropic behavior is far more pronounced in the case of the copper and silver crystals than for the aluminum sample. It was found that the images collected with the area transducer showed a uniform structure with no evidence of the anisotropy. The optical beam deflection images of these samples showed that there was no thermal structure present in these samples. The Laue back-reflection analysis showed that the directions in the crystal exhibiting the greatest thermoacoustic signal are the [100] directions while the signal is smallest in the [110] directions.

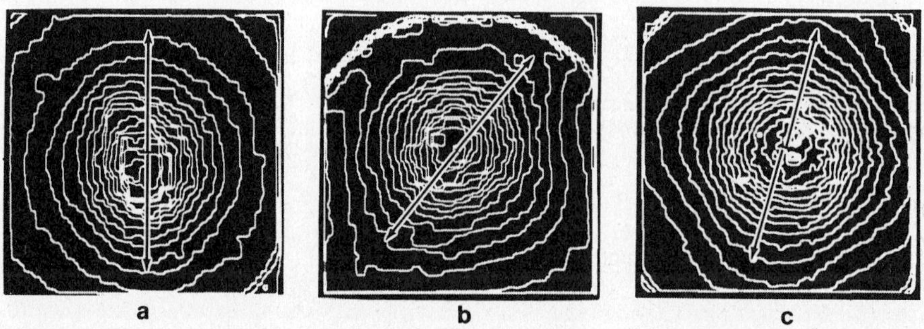

Fig. 3 Laser-acoustic contour images of single crystals, 10mmx10mm scans for (a) aluminum, (b) silver, (c) copper. The arrow on each image indicates the [100] direction as determined by Laue back-reflection techniques

The thermoacoustic contour image of one of the aluminum single crystals studied shown in Fig. 4(a) did not exhibit the symmetry observed in the image of the first aluminum crystal in Fig. 3(a), but showed enhanced thermoacoustic signal in the lower left corner of the image. Optical beam deflection images for the normal component and the transverse component of the probe beam are shown in Figs. 4(b) and (c) and again there is structure evident in the lower left corner of each image. An X-ray topograph of this crystal revealed that there was plastic deformation present in the lower left corner of the crystal.

These experimental measurements of elastic anisotropy in thermoacoustic images of single crystals indicate the role of elastic parameters in determining contrast between grains of different orientations in thermoacoustic images of microstructure. It is shown that grains in a [100] orientation are expected to show a stronger thermoacoustic signal than grains in a [110] direction in agreement with the analysis in [6], but in disagreement with the observations in [5]. Also, the degree of anisotropy in the thermoacoustic images correlates well with the degree of elastic anisotropy in the crystal as illustrated by the effective Young's modulus surfaces. Plastic deformation is

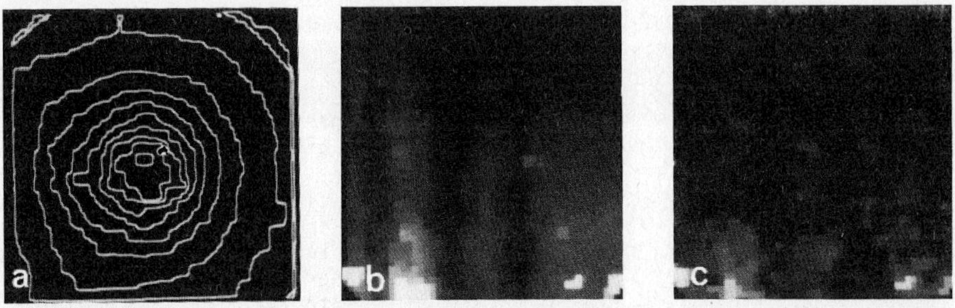

Fig. 4 Comparison of magnitude images of deformed aluminum single crystal, 10mm x 10mm scans. (a) Laser-acoustic, f=25.8 kHz, (b) Normal OBD, f=500 Hz, (c) Transverse OBD, f=500 Hz

also shown to be a source of contrast in images of microstructure, but since the deformation is imaged in purely thermal OBD images as well as in thermo-acoustic images, this contrast mechanism must have a thermal component in addition to any elastic contributions which may exist.

The authors gratefully acknowledge the assistance of Maureen Madey in these experiments. Funding was provided by the Center for Nondestructive Evaluation, The Johns Hopkins University, and by an AAUW International Fellowship for one of the authors (JWM).

References

1. G.S. Cargill: In Scanned Image Microscopy, ed. by E.A. Ash (Academic Press, London 1980) p. 319
2. L.J. Balk, D.G. Davies, N. Kultscher: Phys. Stat. Sol. (a) 82, 23 (1984)
3. M. Luukala, S.G. Askerov: Elect. Lett. 16, 84 (1980)
4. J.C. Murphy, J.W. Maclachlan, L.C. Aamodt: IEEE Trans. Ultrason., Ferroelectrics, Freq. Contr. UFFC-33, 529 (1986)
5. D.G. Davies, A. Howie: Inst. Phys. Conf. Ser. No. 68, Ch. 12, 467 (1984)
6. A. Rosencwaig, J. Opsal: IEEE Trans. Ultrason., Ferroelectrics, Freq. Contr. UFFC-33, 516 (1986)
7. E. Schmid, W. Boas: Plasticity of Crystals (Chapman and Hall, London 1950)

Sound Velocity Measurement of Chemical Solutions Using the Pulsed Photoacoustic Deflection Technique

Qi Yong, Shi Bai-xuan, and Chen Wen-bin

Optical Engineering Department, Zhejiang University, Hangzhou, People's Republic of China

It is well known that a sound pulse will be produced via the photoacoustic effect when a liquid is irradiated by a pulsed laser beam. The efficiency of sound generation depends on the photon energy density and on how the laser radiation is released. There are four mechanisms responsible for sound production in liquids by laser impact [1]: dielectric breakdown, vaporization, electrostriction and the thermoelastic process. To detect the sound pulse, stress transducers and high-speed cameras are generally used, or the optical density is measured by a probe beam. The latter method is called the pulsed photoacoustic deflection technique [2]. This method can be used to study the spectroscopic, thermal and acoustic properties of liquids [3].

In this paper, the photoacoustic effect via dielectric breakdown in liquids and the probe beam deflection technique are used to measure the sound velocity of chemical solutions. Because there are close relations between the equation of state and the sound velocity in a liquid [4], the probe deflection technique, which is a noncontact, fast and accurate method for ultrasonic measurement in liquids, can be used to study the thermodynamic properties of chemical solutions and to analyse the kinetics of physical and chemical processes in liquids.

In our experiment, a Q-switched YAG laser produces laser pulses of 10 ns duration and 50 mJ energy at 1.06 μm. The pulsed beam is focused into the photoacoustic cell which contains the sample liquid. The cell is designed to restrain the reflected sound. The temperature and pressure of the liquid can be changed. Two He-Ne probe beams, which are parallel to the pump beam and have 45 mm separation, pass through the cell and are detected by a position-sensitive sensor. The sound pulse causes a transient probe beam deflection and the sound velocity can be obtained by measuring the time difference between the deflection signals of two probe beams. The power density of the pump laser we used was always well above the breakdown threshold of the solutions studied. Therefore, the deflection signals were large and did not depend on the absorbance of the liquid. This made it easy to detect and analyse the signals.

Initial experiments show that the sound signals via the dielectric breakdown increase nonlinearly with the energy of the pump laser. The signal shape, which mainly depends on the temporal and spatial profile of the pump beam, is similar to that of the signal produced by the thermoelastic process. Remem-

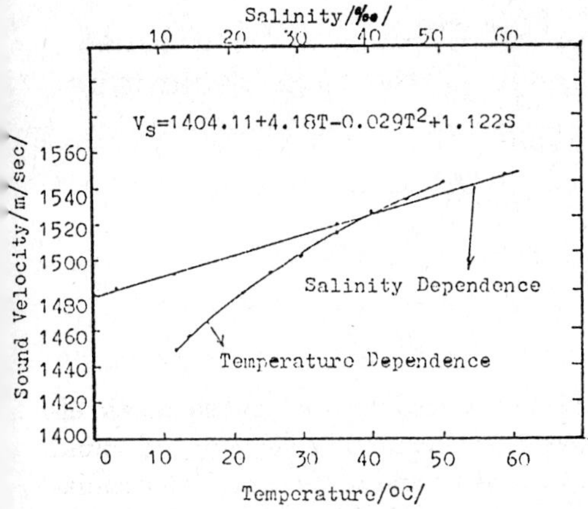

Fig. 1. Sound velocity in sea water

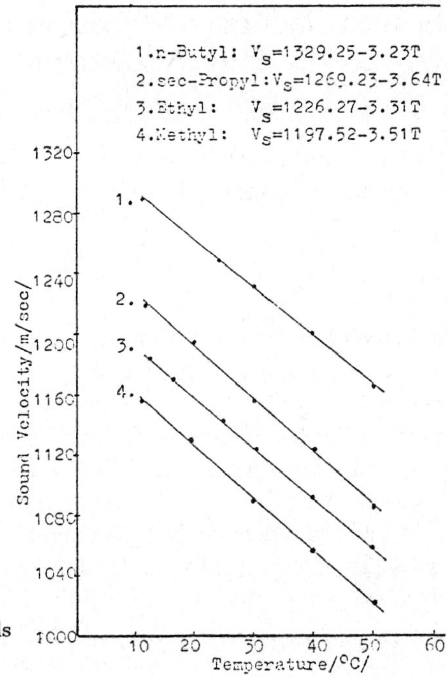

Fig. 2. Sound velocity in four primary alcohols

bering the initial nonlinear propagation of sound produced by breakdown, we investigated the linearity of the velocity. We found that when the sound pulse is more than 7 mm from the pump beam, its velocity is constant, so the sound velocity we measured was not affected by the nonlinearity because in the experiments the distance between the pump and probe beams was 25 mm. Figure 1 shows the temperature and salinity dependence of sound velocity data for sea water at a pressure of 1 atm. Figure 2 shows the sound velocity in methyl, ethyl, *sec*-propyl and *n*-butyl alcohols as a function of temperature. The experimental data indicate that in primary alcohols, the sound velocity depends linearly on temperature. Our results are in good agreement with those of [5]. The accuracy of data for *n*-butyl is 0.4%.

Acknowledgement. The authors wish to thank A.C. Tam for providing us with helpful ideas and suggestions in an earlier phase of this work.

References

1. M.W. Sigrist, F.K. Kneubuhl: J. Acoust. Soc. Am. **64**, 1652 (1978)
2. B. Sullivan, A.C. Tam: J. Acoust. Soc. Am. **75**, 437 (1984)
3. W. Zapka, A.C. Tam: Appl. Phys. Lett. **40**, 370 (1982)
4. B. Vodar, B. Le Neidre (Eds.): *Experimental Thermodynamics*, Vol. 2 (Pergamon, Oxford 1968) p. 527
5. W. Wilson, D. Bradley: J. Acoust. Soc. Am. **36**, 333 (1964)

Part VII

Mass and Heat Transfer

Heat Diffusion and Fractals: Heterogeneous Media and Rough Surfaces

A.C. Boccara and D. Fournier

Laboratoire d'Optique Physique, E.S.P.C.I,
10, rue Vauquelin, F-75231 Paris Cedex 05, France

After an introduction to geometrical and dynamical properties of fractal systems we investigate heat diffusion processes in disordered materials which can be mapped on fractal structures by heating the surface of a sample with a pulsed laser and by determining the time dependence of the surface temperature. Much care will be taken to describe the properties of the surface temperature when this surface is either tortuous or fractal. New theoretical results are given.

I - INTRODUCTION

Heat sources and samples with usual geometrical shapes such as planes, points, cylinders ..., have been mostly considered through the literature /1/. When highly complex systems have to be studied, it is difficult to account for the individual properties of each component to get the global properties of the system. For instance when one deals with heat diffusion one can account for a perturbated sample surface by introducing an "equivalent layer" whose thermal properties are different from the bulk ones /2/. Or one can try to modelize the complex sample by an assembly of elements of simple geometrical shapes (e.g., spheres, ... /3/).

Recently macroscopic self-similar objects have been considered in the framework of fractal theory developed by Mandelbrot /4/. We would like to demonstrate how such geometrical approach can be a great help in understanding the thermal behavior of rough surfaces and heterogeneous samples. These problems are of great interest both from a fundamental point of view (physics of disordered system) and for practical applications to characterize divided materials of high technological interest (ceramics, sintered materials ...).

Let us recall that a structure is self-similar if we cannot tell the difference in the structure as we change the scale. Among the fractal objects one can distinguish two kinds of self-similar structures, the geometrical ones which we will use as an example to define the fractal dimension $\bar{d}$ and the random self-similar structures which are of actual importance to describe physical systems. It is obvious that in real physical systems the scale range at which this self-similarity occurs is limited both in the small dimension limit (e.g., grains or atoms ...) and in the large dimension one where the objects behave like Euclidean entities.

Fig. 1 shows an example of self-similar geometrical construction called the Sierpinski gasket. We can see that if we change the unit length by 1/2 we have 3 equal pieces. Let us recall that for a straight line interval, a square or a cube if we change the scale by 1/2 we have $(1/2)^{-d}$ equal pieces of the initial object, d being equal to 1, 2 or 3 respectively. Here d is the usual Euclidean dimension. We can use this definition to get the dimension of the Sierpinski gasket:

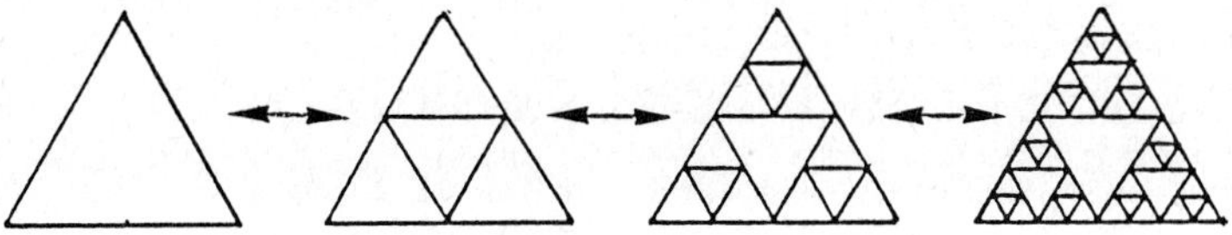

Fig. 1 : The Sierpinski gasket

$$(1/2)^{-\bar{d}} = 3 \qquad \text{thus} \qquad \bar{d} = 1.58 \;;$$

$\bar{d}$ is called the fractal dimension.

Now for application to heat diffusion it is worth focussing our attention on random self-similar structures. For such structures the above properties must be considered as an average. As an example Fig. 2 shows a so-called Brownian motion curve whose fractal dimension is 1.5 /1/.

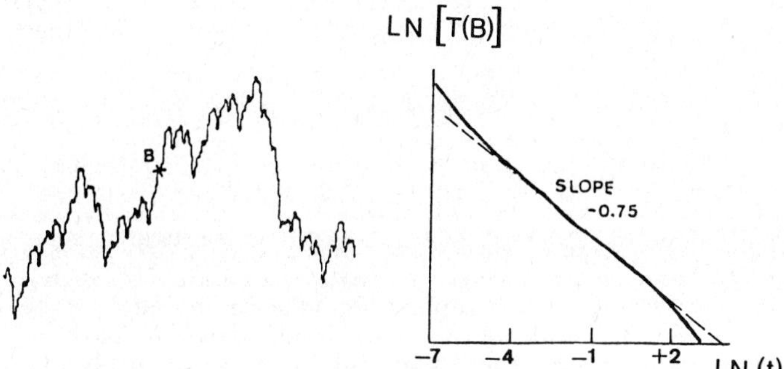

Fig. 2

Brownian scalar motion curve ($\bar{d}$ = 1.5) and the temperature at B as a function of time after a pulsed excitation.

We will label the dimensionality of the Euclidean space in which the structure is immersed d (d = 3 in the general case) ; $\bar{d}$ will be the notation for the fractal dimension and $\bar{\bar{d}}$, which will be introduced later, the spectral one. The diffusion coefficient will be labelled D.

We will first consider the case of a fractal heat source embedded in an Euclidean space of finite diffusion coefficient. Then we will discuss heat diffusion within the bulk of a fractal structure whose behavior involves two parameters ($\bar{d}$ and $\bar{\bar{d}}$). However, in some cases of random media, one can only heat a small part of the surface of these media. In this case the part of the structure which is really heated cannot reflect the self-similar properties of the whole structure ; in order to take into account this problem, we will consider in this second part the case of a fractal source of dimensionality $\bar{d}_S$, embedded in a fractal conducting structure of dimensionality $\bar{d}$.

II - EUCLIDEAN AND FRACTAL HEAT SOURCE.

In order to get a first insight into the behavior of fractals, let us consider the simple following situation : when a point is heated at time t = 0 by a fast heat pulse, the temperature in the surrounding three dimensional medium whose diffusivity is D is given by

$$T(r, t) = \frac{e^{-r^2/4Dt}}{(4Dt)^{3/2}}$$ (geometrical dimension of the source : d = 0, unit source excitation).

Integrating this contribution over a line source will lead to

$$T\ (r = 0,\ t) \sim (Dt)^{-1} \quad ; \quad (d = 1)$$

for a point located on the line source.

And over a plane source to

$$T(r = 0,\ t) \sim (Dt)^{-1/2} \quad ; \quad (d = 2)$$

for a point located on the plane source.

More generally assuming a source at time t = 0 with a fractal dimension $\bar{d}_s$ and a diffusion in a 3-dimensional space one finds that the temperature of the source decreases as

$$T(t) \sim (Dt)^{(\bar{d}_s-3)/2}.$$

The generalization of this equation to a $\bar{d}_s$ fractal source embedded in d-dimensional Euclidean space leads to

$$T(t) \sim (Dt)^{(\bar{d}_s-d)/2}. \qquad (1)$$

As an illustration of this result we have used the curve of Fig. 2 as a unit source at time t = 0 and have computed the temperature of the central point of this source. (In fact we have averaged the results over 20 curves). The source being limited in space (the curve is included in a 10 x 10 cm^2 square and the diffusivity of the surroundings is $D = 1\ cm^2/s$) we expect a crossover at long times, and because of quantification (2000 points to build the curve) crossover at short times ... In between, the slope is close to the expected value $-(3/2 - \bar{d}/2) = -0.75$.

III - DIFFUSION WITHIN RANDOM FRACTAL STRUCTURES.

Let us first recall that heat diffusion, like other diffusion processes, is related to the mean square displacement $\langle r^2 \rangle$ of the diffusers. For instance, due to phonon-phonon collisions or phonon-electron collisions the heat diffusers (phonons or electrons) exhibit a "random walk" and the mean square displacement is usually : $\langle r^2 \rangle \sim t$ in an Euclidean space, thus the diffusion coefficient $D = d/dt\ (\langle r^2 \rangle)$ is a constant. If the diffusing medium is not homogeneous but random (De Gennes' image of the drunk-man in a labyrinth) : $\langle r^2 \rangle \sim t^{2/dw} \sim$ with dw > 1, dw being the dimension of the random walk.

It has been demonstrated that in fractal space $\langle r^2 \rangle \sim t^{\bar{\bar{d}}/\bar{d}}$ and the diffusion coefficient D is time dependent, thus the usual heat diffusion solutions are no longer valid.

$\bar{\bar{d}}$, sometimes called the spectral dimension, is introduced in order to account for the peculiar diffusion processes /5/.

In some cases, the structure in which heat diffusion occurs exhibits a strong disorder and can be mapped on a fractal structure of fractal and fracton dimensions $\bar{d}$ and $\bar{\bar{d}}$ /5/. The heat source is obviously a part of this structure which will be assumed fractal of fractal dimensionality $\bar{d}_s$ ($\bar{d}_s < \bar{d}$). From Rammal and Toulouse /6/, one knows that $\bar{\bar{d}} < \bar{d}$.

Among the random fractal structures, particular attention has been devoted to the so-called "percolation network" /7/. To create a percolation network, each intersection of a d-dimensional grid is, e.g., occupied at random with probability p. A critical probability p_c is found such that for $p > p_c$ a connected cluster will cross the grid (infinite cluster in the case of an infinite grid). Such a structure, which can be easily generated by computer calculation /8/, is found to exhibit a fractal structure /9/ ($\bar{d} = 1.9$ for $d = 2$ and $\bar{d} = 2.6$ for $d = 3$). Moreover, Alexander and Orbach have conjectured that $\bar{\bar{d}} = 4/3$ /10/ for such a percolating network.

Let us go back to heat diffusion and suppose that at time $t = 0$ the sample surface is (uniformly) heated by a short pulse of heat, the heat source dimensionality being $\bar{d}_s$ (e.g. $\bar{d}_s = 0, 1, 2$ for a point, line and a plane source respectively). After a time t, heat has diffused over a volume V defined by $\langle r^2\rangle \sim t^{\bar{\bar{d}}/\bar{d}}$, with $V \sim \langle r^2\rangle^{\bar{d}/2}$. As the energy deposited at time $t = 0$ in this volume is proportional to the surface volume, this energy is $E \sim \langle r^2\rangle^{\bar{d}_s/2}$ and the temperature T(r) is proportional to E/V :

$$T(t) = E/V \sim \langle r^2\rangle^{(\bar{d}_s-\bar{d})} = (t^{\bar{\bar{d}}/\bar{d}})^{(\bar{d}_s-\bar{d})/2},$$

$$T(t) \sim t^{-\bar{\bar{d}}/2 + \bar{d}_s\,\bar{\bar{d}}/(2\bar{d})}. \qquad (2)$$

One can verify that for an Euclidean sample of dimension 3 ($d = \bar{d} = \bar{\bar{d}} = 3$) excited by a point, a line or a plane ($\bar{d}_s = 0, 1,$ or 2) one finds as in Section II

$$T \sim t^{(\bar{d}_s-d)/2}. \qquad (3)$$

IV - EXPERIMENTAL RESULTS.

We have checked the time behavior of the surface temperature of various optically opaque samples (both in the visible and in the IR). The sample surface is heated by a short ($\sim$ 10 ns) light pulse (0.53 nm) and its average temperature is monitored by a fast IR detector and averaged with a digital oscilloscope Lecroy 7600.

For an Euclidean sample and a fractal source such as a rough surface $3 > \bar{d}_s > 2$, formula (3) leads to

$$T \sim t^{-3/2 + \bar{d}_s/2} \sim t^{-\alpha} \qquad \text{with } 0 < \alpha < 1/2.$$

Thus the slope in log-log scales is smaller than the usual 1/2 for a plane excitation.

In a real physical situation we expect a crossover corresponding to a diffusion over a distance equal to the deepest structures of the surface. Indeed such behavior has been observed by us on opaque rough surfaces of carbon samples.

The crossover between the two lines moves towards the short time scales as the polishing is improved (smaller grain size of the polishing paper).

For a fractal sample, compact enough to assimilate its surface to a plane ($\bar{d}_s = 2$), one gets from formula (2)

$$T \sim T^{-(\bar{\bar{d}}/2) + (\bar{\bar{d}}/\bar{d})}. \qquad (4)$$

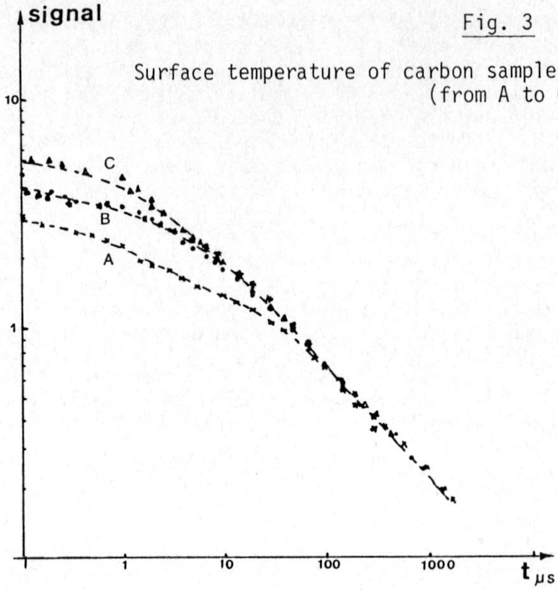

Fig. 3

Surface temperature of carbon samples with decreasing roughnesses (from A to C)

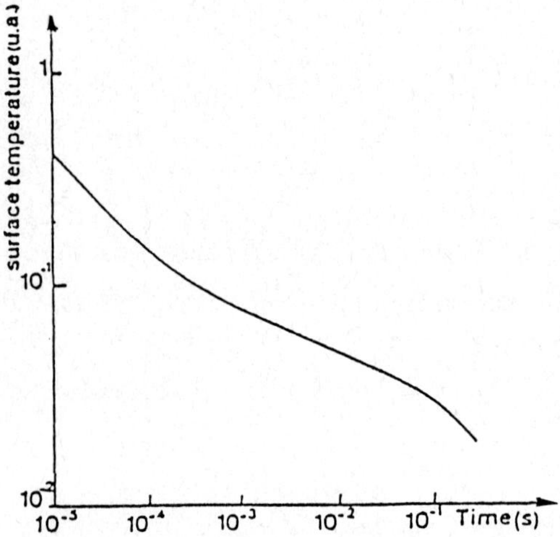

Fig. 4

Surface temperature of an assembly of copper spheres (diameter 100 μ)

To give an order of magnitude of $\bar{\bar{d}}/2 - \bar{\bar{d}}/\bar{d}$ let us consider the case of the 3-d percolating network ($\bar{d}$ =2.6 ; $\bar{\bar{d}}$ = 4/3) which has been used many times /9/ as a model for disordered systems. For such system

$$\bar{\bar{d}}/2 - \bar{\bar{d}}/\bar{d} = 0.154.$$

The exponent is thus much lower than 0.5 which is expected in the Euclidean case. We have looked at a larger variety of random disordered structures and have found in many occasions a power law over a large time scale with an exponent typically in the range 0.15 - 0.25. As an example Fig. 5 shows the result obtained for a weakly bonded assembly of copper spheres. Below a few hundred microseconds, the diffusion occurs within the spheres, then between spheres with a slope of about 0.2, finally, for times larger than 0.15 s the diffusion is Euclidean (slope $\sim$ 0.5).

In conclusion, we do not claim that the structures that we have studied are fractal systems ; but now that it is well established that nature very often exhibits a fractal geometry /4/, it appears interesting to examine heat diffusion in the framework of such theory.

The important point which has to be underlined is that, when diffusion processes takes place in random media (random structures, random bonds ...), the average square length covered by the diffuser during its random walk $\langle r^2 \rangle$, is a power function of the time whose exponent accounts for the dimensionality of the walk.

Acknowledgements : We would like to thank E. GUYON and P. EVESQUE for many stimulating and fruitful discussions.

REFERENCES

1. H.S. Carlsraw, J.C. Jaeger : "Conduction of heat in solids", Clarendon, Oxford (1959).
2. B.K. Bein, S. Krieger and J. Pelzl : Can. J. Phys. 64, 1208 (1986).
3. M. Hlavacek : Arch. Mech. (Varszawa), 32, 491 (1980).
4. B. Mandelbrot : "The fractal geometry of nature", Freeman, New-York (1983) and "Les objets fractals", Flammarion, Paris (1975).
5. R. Orbach : Science, 231, 814 (1986).
6. R. Rammal, G. Toulouse : J. de Physique (Paris) Lettres, 44, L-13 (1983).
7. R. Pynn, A. Skjeltorp, "Scaling phenomena in disordered systems", edit. NATO, ASI series Plenum (1985).
8. H.E. Stanley, N. Ostrowsky : "On growth and form" edit. NATO ASI series, Nijhoff Amsterdam (1985).
9. R. Zallen : "The Physics of amorphous materials", John Wiley (1983).
10. S. Alexander, R. Orbach, J. de Physique (Paris) Lettres, 43, 1-265 (1982).

Thermal Properties and Photoacoustic Signals of Rough and Porous Opaque Solids – Statistical Geometrical Model Considerations

B.K. Bein

Ruhr-Universität Bochum, Institut für Experimentalphysik VI, P.O. Box 102148, D-4630 Bochum, Fed. Rep. of Germany

Starting from basic assumptions on surface heating of solids, the effect of surface roughness, open porosity or vertical surface cracks on the effusivity $\sqrt{(k\rho c)}$ is analyzed. This quantity is the relevant thermophysical parameter for time-dependent surface heating/ cooling and heat transfer across material boundaries.

1. Introduction

When frequency-dependent photoacoustics are applied to powder samples or to rough, porous solid samples, the signal amplitude or the phase lag show frequency dependences which significantly deviate from the behaviour of compact homogeneous samples /1/. The optical, mechanical and thermophysical properties are affected by the random geometry of such samples and the measured photoacoustic signal in general is affected through various channels, e.g. micro- or macroscopic texture effects on the thermal transport /2/, interstitial gas effects /3/, or scattering of light inside the solid /4/. Here, in a simple, idealized model only the effect of surface roughness or open pores on the heat transport is studied theoretically. This may help to identify such a contribution among the other effects on the photoacoustic signal.

By considering a time-dependent surface heating process of opaque solids both from a macroscopic and a microscopic point of view, an equation is derived which relates the depth profile of the effusivity to the geometrical-structural features of the solid surface and to the thermal parameters of the respective compact material. In a second step, this equation is applied to a model of cylindrical open pores where the surface top and the bottom areas of the pores are heated by a radiation of normal incidence whereas the vertical pore walls are not directly heated. Two effects can be identified: One is related to the specific internal surface of the open porosity and the other is a spherical diffusion front effect. In a third step, an ensemble of pores of approximately equal width and of given depth distribution is considered. It is shown that the corresponding photoacoustic amplitude normalized against the amplitude of the respective compact material and recorded as a function of the modulation frequency can be considered as a measure for the cumulative pore depth distribution.

2. Comparison of the Surface Heating Process of a Rough and a Smooth Solid

On a macroscopic level, the surface temperatures of a smooth and a rough solid, heated respectively by the same absorbed radiation heat flux $F_s(t)$, increase as

$$T_{s,r}(t) = \frac{1}{\sqrt{\pi (k\rho c)_{s,r}}} \int_0^t dt' \, F_s(t-t') / \sqrt{t'} \quad , \tag{1}$$

and the ratio of the two surface temperatures is given by

$$T_s(t) \,/\, T_r(t) = \sqrt{(k\rho c)_r} \,/\, \sqrt{(k\rho c)_s} \quad . \tag{2}$$

Here, $T_s(t)$ and $\sqrt{(k\rho c)_s}$ are the temperature and the effusivity of the smooth surface, $T_r(t)$ is the locally averaged temperature and $\sqrt{(k\rho c)_r}$ the effective effusivity of the rough solid surface, respectively. In first approximation, conductional heat losses to the gas in front of the solid are negligible here due to the low effusivity ratio of gas-solid contacts; radiational heat losses can be excluded, if low surface temperatures are considered.

If only a short time interval δt of heating is considered, Eq.(1) can be solved for the smooth solid and can be transformed to give

$$T_s(\delta t) = \frac{2\,F_{so}\sqrt{\delta t}}{\sqrt{\pi (k\rho c)_s}} = \frac{2\,F_{so}\,\sqrt{\delta t}}{\sqrt{\pi}\,\sqrt{(k\rho c)_s}}\,\frac{\sqrt{\delta t}\,A_\perp\,\sqrt{(\rho c)_s}}{\sqrt{(\rho c)_s}\,\sqrt{\delta t}\,A_\perp} = \frac{2(F_{so}\,\delta t\,A_\perp)}{\sqrt{\pi}\,(\rho c)_s}\,\frac{\sqrt{(\rho c)_s}}{\sqrt{k_s\,\delta t}\,A_\perp}$$

$$= 2\,\delta Q\,/\,[\sqrt{\pi}\,(\rho c)_s\,\mu\,A_\perp] = 2\,\delta Q\,/\,[\sqrt{\pi}\,(\rho c)_s\,V_s(\delta t)] \tag{3}$$

Here, $A_\perp$ is the heated surface area (Fig. 1.) perpendicular to the incident heating radiation, $\delta Q = F_{so}\delta t\,A_\perp$ is the absorbed heat pulse, $\mu = \sqrt{k_s\delta t/(\rho c)_s}$ the thermal diffusion length and $V_s(\delta t) = \mu\,A_\perp$ the volume of the solid involved in the heating process. On the short time scale, Eq.(3) also applies to the heating process of the rough solid surface,

$$T_r(\delta t) = 2\,\delta Q\,/\,[\sqrt{\pi}\,(\rho c)_s\,V_r(\delta t)] \quad , \tag{4}$$

where $V_r(\delta t)$ represents the volume actually involved with its heat capacity in the heat absorption process. If the two solids consist of the same material, differing only by their surface geometries, the temperatures (3) and (4) are related by

$$T_s(\delta t)/T_r(\delta t) = V_r(\delta t)/V_s(\delta t) \quad . \tag{5}$$

By equivalence between the macroscopic description, Eq.(2), and the microscopic description, Eq.(5), the effusivity of the rough solid can be defined as

$$\sqrt{(k\rho c)_r} = \sqrt{(k\rho c)_s} * V_r(\delta t)/V_s(\delta t) \quad . \tag{6}$$

3. Model of Cylindrical Open Pores

Comparing a model with cylindrical open pores (Fig. 1.) and a smooth solid, the volumes involved in the heating process are

$$V_r(\delta t) = V_s(\delta t) + \delta V_r(\delta t) \;, \qquad V_s(\delta t) = A_\perp\,\mu \quad . \tag{7a, 7b}$$

The additional volume $\delta V_r(\delta t)$ accessible to the thermal diffusion length μ along the not directly heated vertical pore walls is calculated as

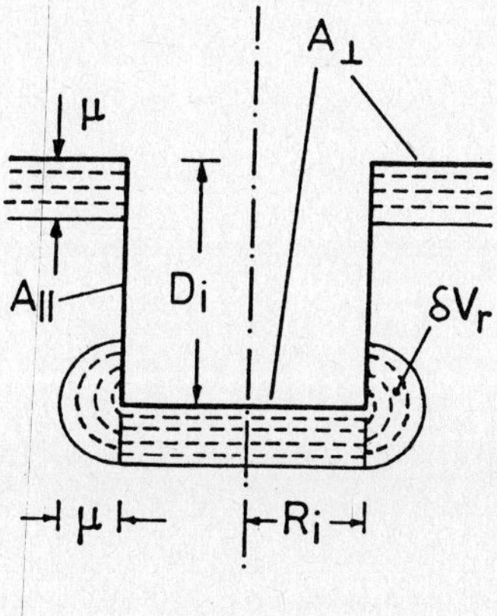

Figure 1.

Schematic of an open cylindrical pore with the diffusion fronts starting from the top of the surface and the bottom of the pore

$$\delta V_r(\delta t) = \sum_{i=1}^{N} (2\pi R_i + 8\mu/3)\,\mu^2\pi/2 \quad . \tag{8}$$

By inserting Eqs.(7a), (7b) and (8) into Eq.(6), a general expression for the effusivity depth profile is obtained

$$\sqrt{(k\rho c)_r} = \sqrt{(k\rho c)_s}\left[1 + \frac{\pi}{2A_\perp}\sum_{i=1}^{N} A_{\parallel i}\frac{\mu}{D_i} + \frac{1}{3A_\perp}\sum_{i=1}^{N} 4\pi\mu^2\right] \quad , \tag{9}$$

where $A_{\parallel i} = 2\pi R_i D_i$ is the cylindrical vertical pore wall area. For large open pores and small thermal diffusion lengths, the dominant first sum of (9) gives an effusivity profile increasing linearly with penetration depth μ. The quantity

$$S_r \sim \frac{1}{A_\perp}\sum_{i=1}^{N} A_{\parallel i} / D_i \tag{10}$$

of dimension $[\text{length}]^{-1}$ can here be interpreted as the specific internal surface of the open pores or of the surface roughness. For narrow deep pores, the second sum of Eq.(9) is dominant with its quadratic increase at larger thermal diffusion lengths. As long as the thermal diffusion length μ is smaller than half of the pore depth D_i, which means the thermal diffusion fronts from the bottom of the pore and from the top have not yet met,

$$\mu / D_i \leq 0.5 \quad , \tag{11}$$

each pore contributes with its not directly heated vertical wall $A_{\parallel i}$ to the specific internal surface in the linear term and with a spherical diffusion front $4\pi\mu^2$ to the quadratic term.

4. Analysis of Special Pore Ensembles

If an ensemble of open pores of approximately equal pore radius, $R_i = R_r$, and with a given pore depth distribution $dn = f(D)\,dD$ is considered, Eq.(9) can be written as

$$\sqrt{(k\rho c)_r} = \sqrt{(k\rho c)_s}\left[1 + \frac{\pi^2 R_r}{A_\perp}\mu\int_{D=2\mu}^{\infty} f(D)\,dD + \frac{4\pi}{3A_\perp}\mu^2\int_{D=2\mu}^{\infty} f(D)\,dD\right] . \tag{12}$$

Here, the lower integration limit of the cumulative pore depth distribution is given by condition (11) of not yet exhausted pore depth, $D \geq 2\mu$.

If for a number of N wide open pores only the term of linear increase of Eq.(12) is considered and if additionally a constant pore depth distribution

$$f(D) = N/D_m = \text{const.} \quad , \quad 0 < D < D_m \quad , \tag{13}$$

between D = 0 and the maximal pore depth D_m is assumed, an effusivity depth profile

$$\sqrt{(k\rho c)_r} = \sqrt{(k\rho c)_s}\left[1 + \frac{\pi A_\parallel}{A_\perp D_m}\mu\left(1 - \frac{2\mu}{D_m}\right)\right] \quad , \tag{14}$$

is obtained which, after a linear increase at low thermal diffusion lengths, reaches a relative maximum at $\mu = D_m/4$. For larger values μ the profile goes down again (Fig. 2). $A_\parallel = N\,2\pi R_r D_m/2$ is the internal surface in (14). In the interior, the model considered here is no longer valid: The effusivity has to approach the bulk value $\sqrt{(k\rho c)_s}$, if the solid is compact, or a value $\sqrt{(k\rho c)_p}$ which is reduced by the effect of the volume porosity p /5/. If a

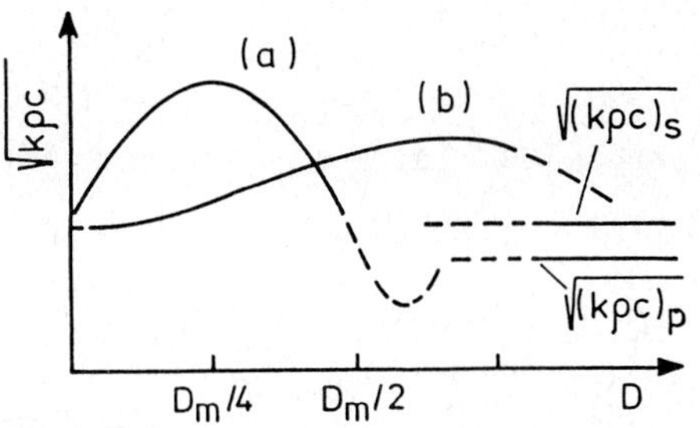

Fig. 2.

Effusivity depth profile for an ensemble of short and wide open pores of constant depth distribution (a), and for deep and narrow pores (b).

composite pore depth distribution is considered of different characteristic lengths, e.g. apart from short and wide open pores a second ensemble of N' deep and narrow pores such as vertical surface cracks in graphite, the non-linear term of Eq.(12) could give a contribution to the effusivity depth profile,

$$\sqrt{(k\rho c)_r} = \sqrt{(k\rho c)_s} \left[1 + \frac{4\pi N'}{3 A_\perp} \mu^2 \left(1 - \frac{2\mu}{D'_m} \right) \right] , \qquad (15)$$

which increases with the square of diffusion length μ and has its maximum at a deeper penetration depth (curve (b) in Fig. 2). Thus, a second relative maximum could be possible. Iterative surface roughness structures of multiple characteristic lengths, small pores inside larger ones with a fractal aspect within a limited range of length scales, also could give such depth profiles with more than one relative maximum.

If the photoacoustic amplitudes can be measured both for a smooth and a rough porous surface of the same material and if the heat conversion factors are known from additional reflectivity measurements /6/, the normalized amplitude

$$S_n = S_s / S_r \approx \sqrt{(k\rho c)_r} \; / \; \sqrt{(k\rho c)_s} \qquad (16)$$

can be compared to Eqs.(12), (14), or (15). Based on such statistical models, the normalized signal S_n can be considered as a measure of the cumulative pore depth distribution, and photoacoustics thus might supply the experimental data for a statistical description of roughness and open porosity.

Acknowledgement

In part, this work has been supported by Max-Planck-Institut für Plasma-Physik, Garching (bei München).

1. B.K. Bein, S. Krueger, and J. Pelzl: J. Nucl. Mater. 141-143, 119 (1986)
2. L.C. Aamodt and J.C. Murphy: Applied Optics 21, 111 (1982)
3. J.P. Monchalin, L. Bertrand, G. Rousset, and F. Lepoutre: J. Appl. Phys. 56, 190 (1984)
4. P. Helander and I. Lundstroem: J. Appl. Phys. 51, 3841 (1980)
5. B.K. Bein, S. Krueger, and J. Pelzl: Can. J. Phys. 64, 1208 (1986)
6. M. Wojczak, J. Pelzl, and B.K. Bein: In Photoacoustic and Photothermal Phenomena, P. Hess and J. Pelzl, eds., Springer Series in Optical Sciences (Springer Berlin, Heidelberg, 1987)

The Effect of Grain Size on the Photoacoustic Signal from Small Particles

U. Netzelmann[1], *J. Pelzl*[1], *and D. Schmalbein*[2]

[1]Institut für Experimentalphysik VI, Ruhr-Universität, D-4630 Bochum 1, Fed. Rep. of Germany
[2]Bruker Analytische Messtechnik GmbH, D-7512 Rheinstetten, Fed. Rep. of Germany

1. Introduction

The photoacoustic (PA) signal of a powdered sample increases with decreasing grain size, in addition there occurs a certain change of the phase angle. Qualitatively, these observations are accounted for by an enlarged effective solid-gas interface and by size-dependent optical properties and thermal coupling parameters. But up to now, no rigorous experimental or theoretical treatment of this problem exists, except some efforts devoted to the influence of the light scattering on the PA signal [1,2] and a macroscopic approach based on the model of a heterogeneous compact solid [3]. Common drawbacks of the hitherto published studies are the lack of quantitative data on the heat generation process, particularly in optical experiments, and the complications due to the interference of the optical and thermal properties, which are both a function of the grain size.

In the present work we have performed PA measurements on particles of paramagnetic $CuSO_4 \cdot 5H_2O$ which have been heated by resonance absorption of microwaves (EPR) in an external magnetic field. With this experimental approach most of the problems mentioned above could be reduced significantly. The particles are nearly transparent in the microwave region and an ideal volume heating is realized. The scattering problem occurring with light absorption is eliminated, as the particles are small compared with the electromagnetic wavelength. A major advantage is the ability to measure simultaneously the thermal response of the sample and the conventional EPR signal. This allows us to relate the PA signal to the actual number of absorbing paramagnetic centres in the PA cell and to deduce the PA signal per sample volume. Using a special technique, powdered samples have been prepared where the particles were virtually not in thermal contact with each other, so the complexity of the thermal problem is considerably reduced.

2. Theory

The theoretical description is based on the assumption of isolated particles. Neglecting any thermal interaction between the particles and between particles and the wall of the sample tube, the problem can be reduced to the treatment of a single spherical particle in a spherical cell, where the radius of the gas-filled cell is assumed to be much larger than the thermal diffusion length of the gas for all modulation frequencies. Further it is assumed that the heating by the microwave is homogeneous over the whole particle. Comparison of the theoretical results with the experimental data will show that the first and last requirements are met satisfactorily, whereas the heat contact to the wall cannot be neglected over the whole parameter range covered by the experiment.

The complex amplitude of the oscillating surface temperature of a particle with radius R is obtained by solving the diffusion equations in spherical coordinates. The acoustic pressure amplitude normalized to the sample volume is then determined by integration over the heat flow from the particle into the gas [4]. The magnitude and the phase angle of the normalized PA signal have been calculated as a function of the particle size and of modulation frequency. Figure 1 shows as a typical result the variation of the signal as a function of the sphere radius at the fixed modulation frequency of 300Hz.

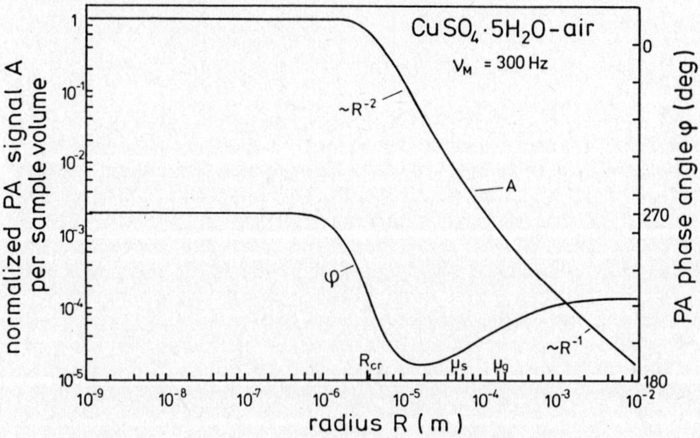

Fig. 1 Theoretical PA signal per sample volume for a homogeneously heated spherical particle of $CuSO_4 \cdot 5H_2O$ in a spherical gas cell as a function of the particle radius R at a modulation frequency of 300Hz.

The evolution of the PA signal as a function of the sphere radius covers three main regimes: $R < R_{cr}$, $R_{cr} < R < \mu_s$, $R > \mu_s$ and μ_g. The critical radius is defined by $R_{cr} = \mu_s[(3/2)\,\lambda_g/\lambda_s]^{1/2}$, where μ_s and λ_s are the thermal diffusion length and the thermal conductivity of the sample and μ_g and λ_g those of the gas. For very big spheres the frequency behaviour is the same as in the case of a semi-infinite solid with one-dimensional heat propagation. As a function of the grain size, the efficiency of signal generation varies as 1/R, as one would expect from the geometrical surface to volume ratio of the sphere. In the other limit for very small particles all power absorbed in the sample is transferred into the gas. The PA amplitude has a $1/\omega_M$ dependence. In this regime, both amplitude and phase angle are independent of the sphere radius.

3. Experimental

The experimental arrangement consists of an EPR spectrometer working with an X-band bridge and a conventional TE_{102} microwave resonator. For the PA measurements, the spectrometer has been modified [5]. A pin-diode modulator is inserted into the signal line from the klystron of the microwave bridge to the cavity admitting a 100% rectangular intensity modulation. The modulator is followed by a travelling-wave-tube (TWT) amplifier, which supplies a microwave power of up to 1.6W at 9.1GHz. Two directional couplers built in between the TWT and cavity allow incident and reflected microwave powers to be monitored.

The samples are contained in a standard quartz tube connected acoustically to a microphone by an airtight assembly [6]. Single crystalline particles of $CuSO_4 \cdot 5H_2O$ with definite grain sizes ranging from 10 μm to a few millimetres in diameter are selected from sieved fractions of the ground polycrystalline powder. Only a small number of particles with the same diameter are deposited on the inner tube walls of the PA cell in order to ensure that these particles are not in thermal contact with each other via the air or the backing.

4. Results and Discussion

A typical experimental result is displayed in the upper part of Fig. 2. The photoacoustic response of the paramagnetic resonance has been recorded at modulation frequencies ranging from 6Hz to 5kHz. For each frequency, the measured photoacoustic signal amplitude obtained from the small particles is first calibrated with the conventional detected signal and then divided by the corresponding calibrated signal recorded from a single compact crystallite of 1mm diameter. By this normalization procedure, the frequency response of the PA cell and that of the detection channel is eliminated. The photoacoustic yield of the large sample is set to one. The latter represents within a reasonable approximation a thermally thick solid. The theoretical curves shown in the same figure are calculated in the context of the model discussed above considering a single spherical particle in a spherical PA cell.

Comparison of the experimental and theoretical results shows a surprisingly good agreement for the shapes of the frequency dependence of both the amplitude and the phase lag. The maximum in the amplitude responses

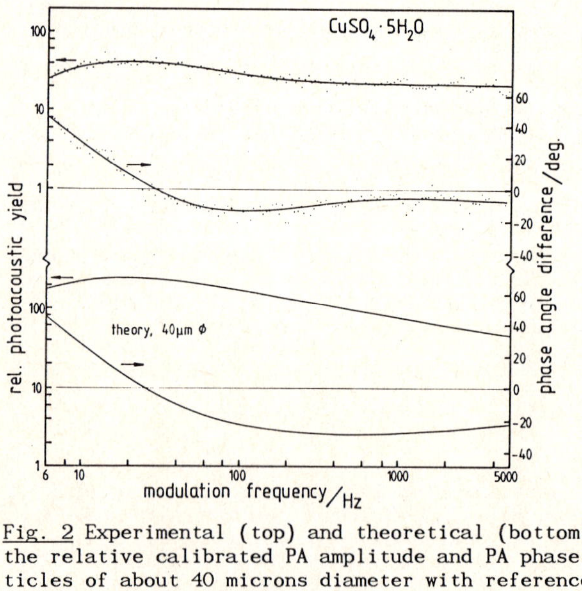

Fig. 2 Experimental (top) and theoretical (bottom) frequency dependence of the relative calibrated PA amplitude and PA phase angle increment from particles of about 40 microns diameter with reference to a thermally thick sample of about 1mm diameter.

occurs at frequencies where the grain size is equal to the critical radius R_{cr} and where the quasistatic limit is reached. In this region we also find the change of the sign of the phase angle. This frequency is much lower than the frequency where the small particle becomes thermally transparent. The measured value of the maximum photoacoustic yield is considerably larger than the value one would expect from the geometrical surface to volume ratio only. On the other hand, the simple model outlined above overestimates the PA response. This discrepancy is accounted for by the deviations of the reference sample from a spherical shape, by residual agglomeration of some of the small particles at the bottom of the quartz tube and by the thermal coupling of the particles with the wall. An estimate shows that the thermal relaxation into the walls which may be characterized by a relaxation time [7] reduces the heat flow into the gas and becomes more important for smaller particle sizes.

5. Conclusion

Detecting simultaneously conventional and photoacoustic EPR absorption in small, thermally isolated particles, we were able to study quantitatively the photoacoustic yield as a function of grain size and modulation frequency. The main features of the measured signals are explained by a single-particle model. The present results show clearly that small samples have extraordinarily increased signal generating efficiencies with a maximum value at a distinct frequency which increases with decreasing particle size.

Acknowledgements: This work is supported by the BMFT, project 13N 5379/1

References

1. P.Helander: J.Appl.Phys. 54,3410(1983)
2. Z.A.Yasa, W.B.Jackson and N.M.Amer: Appl.Optics 21,21(1982)
3. J.-P.Monchalin, L.Bertrand, G.Rousset, and F.Lepoutre: J.Appl.Phys. 56,190(1984)
4. J.Pelzl, B.K.Bein: Z.Angew.Chemie, Neue Folge 134,17(1983)
5. U. Netzelmann, J.Pelzl, D.Fournier and A.C.Boccara: Can.J.Phys. 64,1307(1986)
6. U.Netzelmann, E.v.Goldammer, J.Pelzl, H.Vargas: Appl.Optics 21,32(1982)
7. S.Utterback, F.Dacol, H.Ermert, R.Melcher: Appl.Phys.Lett. 46,1054(1985)

The Importance of an Adsorbed Liquid Layer for the Enhancement of Photoacoustic Signals

P. Ganguly and T. Somasundaram

Solid State and Structural Chemistry Unit,
Indian Institute of Science, Bangalore, India

1.Introduction

The enhancement (E) of photoacoustic signals from condensed phases due to the vapours of volatile liquids in the cell has been extensively studied by us [1,2]. The extent of enhancement (E) is quite different for porous and non-porous solids [3]. The exact mechanism of enhancement is not clear. In the theoretical models given so far, E is attributed to an oscillatory mass transfer to the gas phase [4,5]. TAM [6], in his review article, also believes that the enhancement is a generalisation of the gas evolution process. We have argued that enhancement could either be due to mass transfer or to a more efficient heat transfer to the gas phase [2]. We argue below that the enhancement is crucially dependent on the thickness of the adsorbed liquid layer on the surface of the solid.

2.Experimental

The experimental cell for studying the influence of the height of a sample above the liquid reservoir is similar to that shown in Fig.1 of Ref.3. In brief, the sample chamber contains an aluminum reservoir (~1cc capacity) and a sample holder (with 10mm dia. 2mm deep depression for the sample to sit). The reservoir can be filled with a liquid to the desired level. Two holes (~ 1.5mm dia) provided in the sample holder help to equilibrate the vapour with coupling gas. For the experiments with quartz crystal oscillator (QCO), the QCO has been coated with amorphous Se (a-Se) and introduced above the reservoir without making contact. The height of the liquid was controlled simply by injecting various amounts of liquid into the liquid reservoir. In the experiment designed to estimate the dependence of E on the thickness of the adsorbed layer, we introduced the QCO (coated with a thin film of Se) into the cell above the reservoir (Fig.1) and the frequency change associated with the amount of adsorbed substance

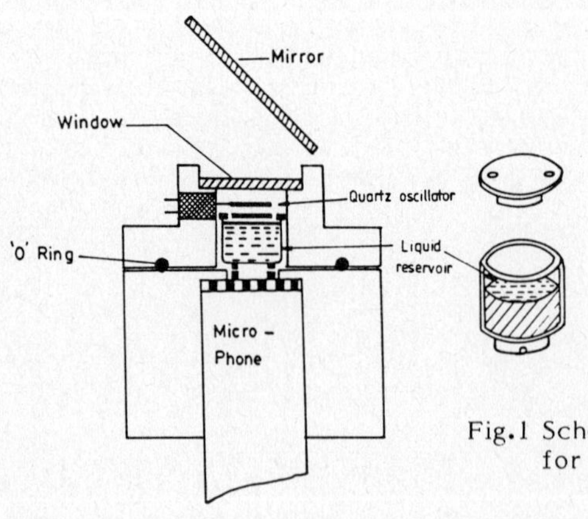

Fig.1 Schematic of the cell used for the enhancement studies

was calibrated by noting the change in frequency for known amounts of Se deposition. The oscillator diameter was close to that of the sample holder. The PA signal (I_{PA}) as a function of temperature (T) was followed for $CuSO_4.5H_2O$ mixed with carbon black powder. In another experiment I_{PA} from carbon black was followed as a function of T in the presence of $CuSO_4.5H_2O$. The cell used for this purpose is similar to that described by us earlier [7].

3. Results and Discussion

In Fig.2 we show I_{PA} as a function of time for various heights of the liquid. It is seen that as the height of the liquid in the reservoir is decreased, the rate of increase in the PA signal becomes less. This observation cannot be attributed to the mere increase in the volume of the gas because the signal does not change appreciably (when only air is the coupling gas in a related experiment) when the reservoir is completely filled or not with an inert material like alumina.

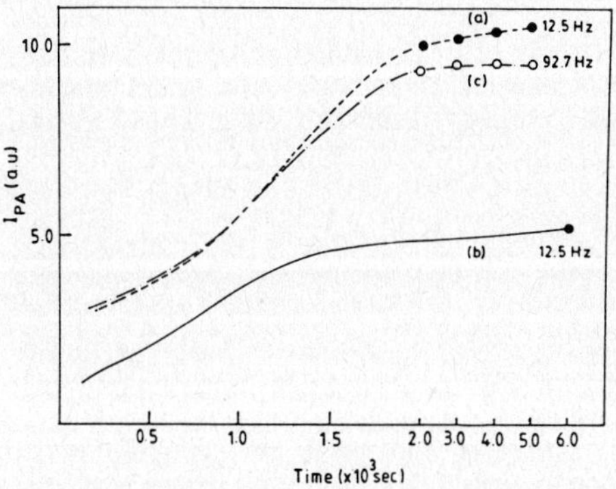

Fig.2 Plot of I_{PA} vs time for A.Se coating for various f. Amount of the liquid (acetone) in the reservoir a) & c) : 0.8cc and b) : 0.6cc. Note the change in time scale

The signal also saturates to a finite value for large values of time. At saturation the final value of E decreases as the height of the liquid in the reservoir decreases. When the liquid level in the reservoir is close to or in contact with the sample holder, the signal goes through a maximum as a function of time. It is known [8] that the thickness of a liquid layer on a plate kept horizontal above a liquid reservoir is dependent on the height of the plate above the reservoir. For a plate about a centimetre above the reservoir, the thickness is estimated [8] to be about 300Å, while it could it could be about 1μm thick or more when the height tends to zero. This result was not anticipated by us in the earlier experiments as we believed that the only critical factor was the vapour pressure.

In the experiment with the quartz oscillator we found that I_{PA} from a Se film coating on the oscillator is enhanced only 1.5 times in the presence of saturated vapours of ether at 296K compared to the more than 14-fold enhancement found when an identical sample was kept in the sample holder above the reservoir. From the frequency changes, the thickness of the adsorbed ether layer was estimated to be about 300Å.

PA signals from solids such as $CuSO_4.5H_2O$ which undergo dehydration in two stages at 130° and 250°C have been studied (after grinding $CuSO_4.5H_2O$ with carbon black) by following the signal from the mixture using white light. I_{PA} is found to be strongly enhanced in the first dehydration stage around 110°C but there is no effect in the second dehydration stage (Fig.3a). When $CuSO_4.5H_2O$ was kept separately in the cell and I_{PA} from carbon black alone as a function

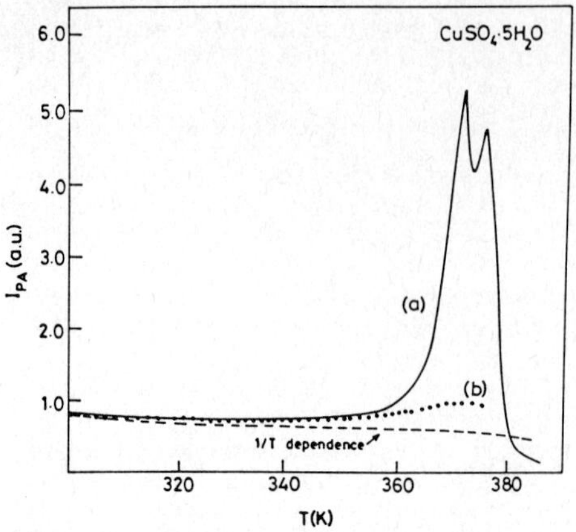

Fig.3 Plot of I_{PA} vs T for $CuSO_4 \cdot 5H_2O$ a) : with carbon black in the sample holder, b) : for carbon black in the sample holder and the sample inside the cell and c): 1/T dependence

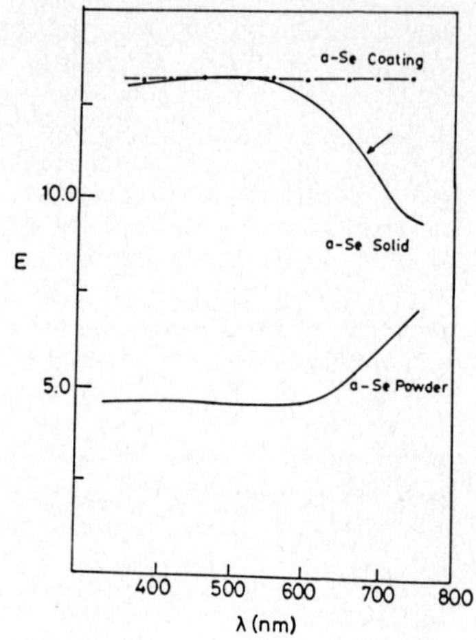

Fig.4 Enhancement for a-Se for the same amount of liquid (ether) in the reservoir. The arrow shows the point at which $\mu = 1/\beta$

of T was followed no marked anomaly or enhancement was seen even at 110°C (Fig.3b). These results again suggest that the vapour pressure of water is not really the most important factor but rather it is the presence of a liquid-like layer that is important.

In Fig.4 we show E of the PA signals as a function of wavelength for the same amount of liquid in the reservoir for powdered amorphous selenium (a-Se), and a 1 μm thick a-Se film on Teflon or aluminum, as well as for a (1.5 mm thick) disc of bulk a-Se at chopping frequency of 37Hz. At this frequency the thermal diffusion length (μ) of a-Se is $\sim 35 \mu$m. We see that for thin films ($\mu >$ thickness of film), E is independent of the thermal properties of the backing material (Teflon or aluminum) and is also independent of the wavelength of the light and hence the absorption coefficient (β) of the sample. In the other cases E shows marked changes when $\mu = 1/\beta$. None of the theoretical models proposed so far have been able to account for these observations.

A qualitative understanding of the above features may be obtained if we consider a model in which a thin optically absorbing layer (thickness $< \mu$) is sandwiched between two thermally thick optically non-absorbing layers A and B (thickness $\gg \mu$). The ratio of the heat conducted into the backing A and B is then given by $[(k_A \rho_A C_A)/(k_B \rho_B C_B)]^{1/2}$. It has been observed by us [9] that such an assumption explains quite satisfactorily the changes in I_{PA} as a function of the coupling gas. If A is a solid and B is a gas such as air, the ratio in most of the cases is nearly equal to 200. When B is a liquid the ratio becomes 3-5. Thus by having an adsorbed layer of a liquid of the order of one diffusion length thick, there may be a 40-70-fold increase in the amount of heat extracted from the solid by the liquid compared to a gas such as air. If the heat taken up by the liquid is now considered to cause an evaporation of the liquid molecules to the gas phase then it may be shown [10] that the maximum possible contribution by the mass transfer to the PA singal E_{max} is given by

$$E_{max} = [k_l \rho_l C_l)/(k_g \rho_g C_g)]^{1/2}(C_g^M T/L), \quad (1)$$

where the subscripts l and g refer to the liquid and gas, respectively; C_g^M is the specific heat of the gas per mole and L is the latent heat of evaporation of the liquid per mole. For ether and air this is roughly 25. We have observed that E as measured by the ratio of the PA signal in the presence of liquid to that of a coupling gas is independent of the thermal properties of the coupling gas [3]. It seems to us that these results cannot be explained on the basis of an oscillatory mass transfer model. Instead it seems to support that the presence of the liquid layer leads to a more efficient heat transfer in the sense that the evaporating liquid molecules transfer heat to the molecules of the coupling gas and condense back to a liquid. The temperature at the interface of the liquid and the gas therefore becomes important. For the more efficient heat transfer model the factor $C_g^M T/L$ in (1) is absent and we may write E'_{max} instead of E_{max}. E'_{max} is reduced by the following three factors : (i) when the thickness of the absorbed layer $\underline{l} < \mu_{\underline{l}}$, E'_{max} is reduced by a factor $1\text{-exp}\ (-\underline{l}\ /\mu_{\underline{l}})$; (ii) a finite thickness of the adsorbed layer would lead to a decrease in the temperature at $\underline{x}=\underline{l}$ where $\underline{x}$ is the distance from the surface of the solid. Since the evaporating molecules would acquire the temperature at the surface of the liquid layer, there could be a further decrease in E'_{max}. We assume that this is given by $\exp(-\underline{l}/\mu_{\underline{l}})$; and (iii) only a fraction of the heat taken up by the liquid layer would be transported to the gas phase. This we relate to the mole fraction of the saturated vapour in the gas phase ($\underline{y}$). The actual enhancement E" is then given by

$$E'' = E'_{max}\ (1\text{-exp}(-\underline{l}/\mu_{\underline{l}})\ \underline{y}\ \exp\ (-\underline{l}/\mu_{\underline{l}}). \tag{2}$$

This equation has a maximum at $\exp\ (-\underline{l}/\mu_{\underline{l}}) = 1/2$ so that at the maximum we obtain

$$E'' = 0.25\ \underline{y}\ E'_{max}. \tag{3}$$

For ether and air the enhancement is then given by 0.106-0.1415. Therefore when $\underline{l}$ = 300-400Å, the total enhancement (= 1+E") is nearly 1.106-1.141 which compares favourably with that found with the quartz oscillator.

The above discussion obviously requires further refinements. Nevertheless the results presented above show that the theories given earlier are quite inadequate in the sense that they have not taken into account the role played by the thickness of the adsorbed layer (Cf. Fig.2 and the experiment with QCO). Likewise the different enhancement seen for powders, single crystals and coatings (Fig.4) has also not been explained theoretically. The qualitative theory is also inadequate to explain the results of Fig.4 since the above model is only for the situation when the thickness of the absorbing layer $<\mu$. Further work is in progress in order to understand the pheonomenon further. The results however do indicate that with adequate precautions we would be able to obtain enhancements of the mirage effect [11] and that the technique of photoacoustic spectroscopy may be extended to the study of liquid films on surfaces [12].

References

*Contribution No. 473 from the Solid State and Structural Chemistry Unit

1. P. Ganguly, T. Somasundaram : Appl. Phys. Lett. **43**, 160 (1983)
2. T. Somasundaram, P. Ganguly : J. Appl. Phys. **57**, 5043 (1985)
3. P. Ganguly, T. Somasundaram : Appl. Phys. B **43**, 43 (1987)
4. P. Korpiun : Appl. Phys. Lett. **44**, 675 (1985)

5. J. Srinivasan, R. Kumar, K.S. Gandhi : Appl. Phys. B **43**, 35 (1987)
6. A.C. Tam : Rev. Mod. Phys. **58**, 381 (1986)
7. T. Somasundaram, P. Ganguly : J. Phys. (Paris) **44**, Suppl. C6-239 (1983)
8. P.G. de Gennes : Rev. Mod. Phys. **57**, 827 (1985)
9. T. Somasundaram : Ph.D. Thesis, Indian Institute of Science (1987)
10. P. Ganguly : Rev. Solid State Sci. **1**, 000 (1987).
11. A.C. Boccara, D. Fournier, J. Badoz : Appl. Phys. Lett. **36**, 160 (1980)
12. T. Somasundaram, P. Ganguly : (to be published)

Detection of Water Permeation with the Photoacoustic Effect

R. Osiander[1], *P. Korpiun*[1], *and W. Knoll*[2]

[1]Physik Department, TU München, James-Franck-Straße, D-8046 Garching, Fed. Rep. of Germany
[2]Max-Planck-Institut für Polymerforschung, Jakob-Welder-Weg 11, Postfach 3148, D-6500 Mainz, Fed. Rep. of Germany

The photoacoustic signal of liquids is quite well understood /1/. Periodic modulation of the surface temperature causes modulation of the equilibrium vapor pressure, which contributes to the pressure variations. Quantities as the diffusion coefficient in the sample and the surface resistance towards diffusion influence the signal in water-adsorbing substances. Periodic sorption, caused by low frequency pressure modulations, has been used already by EVNOCHIDES and HENLEY /2/ to measure diffusion coefficients and solubility constants.

1. Experimental Arrangement

The photoacoustic cell designed for measurements on polymer foils is shown in fig.1. The polymer foil is pressed on a porous microfilter, coated with a thin graphite layer for light absorption. At the back of the sample saturated water vapor atmosphere is kept in the water reservoir. In the cell gas the humidity falls from its equilibrium value at the sample surface to low humidities in a second reservoir filled with silica gel to maintain a constant vapor current through the sample.

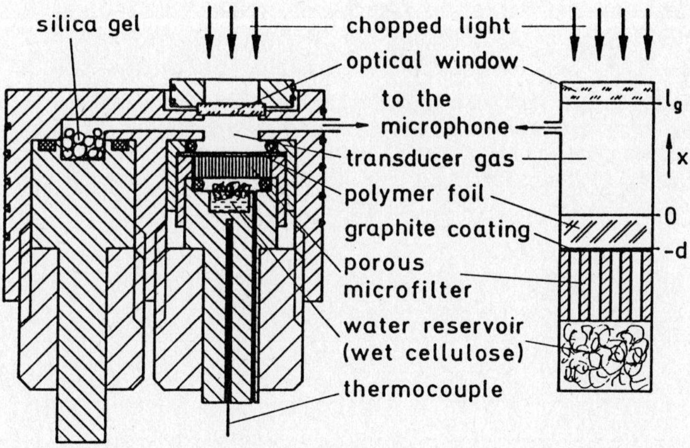

Fig.1: The photoacoustic cell and a scheme of the cell geometry

2. Theory

The scheme of the cell geometry is shown in fig.1. Light, modulated sinusoidally with frequency $\omega/2\pi$, is absorbed by the graphite coating at $x = -d$ and generates thermal waves in the sample and the gas. The temperature variation of the sample surface is $\phi(0,t) = T(0,t) - T_o = \phi_o e^{i\omega t}$ around an average temperature T_o. This temperature variation alters the equilibrium vapor pressure P_v and the equilibrium content of water per unit mass of the sample, $X(P_v,T)$, at the sample surface at $x = 0$. So the water concentration variation $C_s(0,t)$ in the sample at $x = 0$ is connected with the vapor concentration variation $C_g(o,t)$ in the gas at $x = 0$ by

$$C_s(0,t) = \rho_s \left(\frac{\partial X}{\partial T}\right)_{P_v} \phi(0,t) + \rho_s \left(\frac{\partial X}{\partial P_v}\right)_T C_g(0,t) RT . \tag{1}$$

Here R is the gas constant for water and ρ_s is the density of the sample. $\left(\frac{\partial X}{\partial T}\right)_{P_v}$ and $\left(\frac{\partial X}{\partial P_v}\right)_T$ are connected by the Clausius-Clapeyron relation

$$\left(\frac{\partial X}{\partial T}\right)_{P_v} = - \left(\frac{\partial X}{\partial P_v}\right)_T \frac{LP_v}{RT^2} . \tag{2}$$

The heat of sorption L for water is often very close to the heat of vaporization of water.

The water concentration variations in the sample and in the gas satisfy the one-dimensional diffusion equation. From the solutions we consider only the diffusion waves $C_s(x,t) = C_{so}\exp(\delta_s x+i\omega t)$ running to $x = -d$ in the sample and $C_g(x,t) = C_{go}\exp(-\delta_g x+i\omega t)$ running to l_g in the gas. Here $\delta_i = (1+i)\sqrt{\omega/2D_i}$ is the complex wave number with the diffusion coefficient D_s in the sample $(i = s)$ and an effective diffusion coefficient $D_g = D_{go}[P_o/(P_o-P_v)]$ in the gas $(i = g)$, taking into account convective flow /3/.

Now we assume a resistance r for the vapor diffusion at the sample surface, caused for example by a Langmuir-Blodgett layer. Then the continuity of mass flow at $x = 0$ demands

$$- D_s \frac{\partial C_s}{\partial x}(0,t) = \frac{1}{r} [C_g(0,t) - C_g^*(0,t)] = - D_g \frac{\partial C_g}{\partial x}(0,t) , \tag{3}$$

with the gas concentration $C_g^*(0,t) = C_{go}^* e^{i\omega t}$, that replaces C_g in (1) at the lower boundary of the resistance and $C_g(0,t)$ at the upper boundary. Calculating C_{go}^* and C_{so} from (3) and inserting in (1), one gets, together with (2), for the water concentration variation $C_g(x,t)$ in the gas

$$C_g(x,t) = \Phi_o \frac{LP_v}{R^2T^3} e^{-\delta_g x+i\omega t} \left[1 + \frac{\sqrt{D_g/D_s}}{\rho_s RT \left(\frac{\partial X}{\partial P_v}\right)_T} + rD_g\delta_g \right]^{-1}. \tag{4}$$

This leads to an enhancement of the pressure variation detected by the microphone. With the temperature and the concentration variation averaged over the gas volume /3/, the photoacoustic amplitude can be written as

$$A/A_o = 1 + \sqrt{D_g/\alpha_g}\,\frac{LP_v}{P_oRT}\,\Big[1 + \frac{\sqrt{D_g/D_s}}{\rho_s RT\big(\frac{\partial X}{\partial P_v}\big)_T} + (1+i)r\sqrt{\omega D/2}\,\Big]^{-1}, \qquad (5)$$

with A_o, the amplitude of the dry sample, the total pressure P_o and thermal diffusivity α_g of the gas. In the boundary condition for the flow of heat the contribution of the heat of sorption to the specific heat should be considered at higher temperatures /3/.

3. Experiments and Results

Measurements were done on polyethyleneterephthalate (PETP) foils (Hostaphan RE3 and RN12, Kalle-Höchst AG.) with 3 μm and 12 μm thickness. For further experiments the 3 μm foils were coated with 1 and 4 bilayers of cadmium arachidate, each of about 5 nm thickness, by the Langmuir-Blodgett dipping technique. The photoacoustic signal was measured as a function of temperature, because in that case the temperature dependence of the vapor pressure determines the signal of the wet samples. For a first approach the other quantities as D_s and $(\partial X/\partial P_v)_T$ are taken temperature independent. The experimental data are shown in fig.2. The data of the dry samples are smoothed and averaged over all measurements. Data taken with water in the reservoir are smoothed and related to the dry data. The solid lines are calculated with (5). For the PETP foils they show the best agreement with $(\partial X/\partial P_v)_T = 4.5\text{x}10^{-6}\ \text{Pa}^{-1}$, which is in the same order of magnitude as the values calculated from typical sorption isotherms for polymers /4/. The surface resistance r was assumed to vanish for the uncoated films. For the coated films the solid line was calculated with r = 0.35 s/cm.

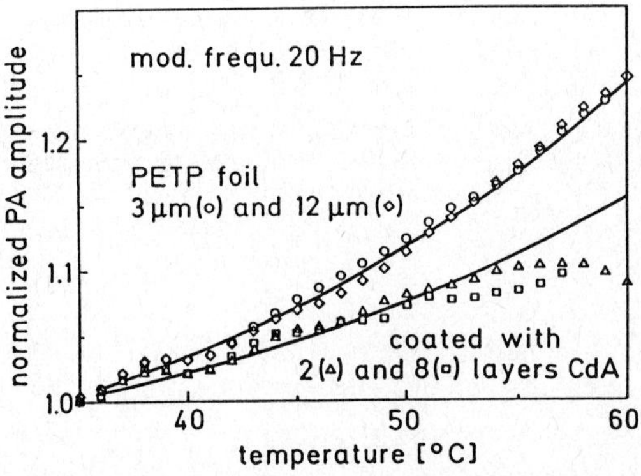

Fig.2: Photoacoustic amplitude for several PETP foils with water, related to the dry samples. The solid lines are calculated from (5)

4. Interpretation

The PA signal of the 3 μm and the 12 μm PETP foils shows the same temperature behaviour in the presence of water vapor. This agrees with the assumption that the vapor contributing to the signal comes from a layer of a thickness of one mass diffusion length $\mu_s = \sqrt{2D_s/\omega}$ in the sample. For PETP

D_s is about $3.9 \times 10^{-9} cm^2 s$ and for a modulation frequency of 20 Hz μ_s is smaller than 0.1 μm. The signal is governed by the diffusion in the sample and is smaller than that of liquid water.

The cadmium arachidate coating of the foils acts as a surface resistance to the diffusion. This reduces the signal, depending on the relation of r to the resistance $|D_g \delta_g|$ of a gas layer in the thickness of one mass diffusion length. From (5) the best agreement with the data measured on coated films of all thicknesses was found for r = 0.35 s/cm. Values measured for some other Langmuir-Blodgett monolayers on agar gel by BARNES and VANDERVEEN /5/ are of the same order of magnitude.

Acknowledgement

The authors gratefully acknowledge the help of Dr. S. Pauly of Kalle-Höchst AG., Wiesbaden, for providing the Hostaphan foils.

1. P. Korpiun: Appl. Phys. Lett. 44, 675 (1984)
2. S. K. Evnochides and E. J. Henley: J. Polym. Sci. A2 8, 1987, (1970)
3. P. Korpiun, W. Herrmann, R. Osiander: to be published in Photoacoustic and Photothermal Phenomena, P. Hess and J. Pelzl, eds., Springer Series in Optical Sciences; A. L. Schawlow, Series Editor, (Springer, Berlin, Heidelberg - 1987)
4. J. Crank, G. S. Park: in Diffusion in Polymers, Academic Press, London (1968)
5. R. J. Vanderveen and G. T. Barnes, Thin Solid Films 134, 227 (1985)

Effect of Diffusive and Convective Transport of Mass on the Photoacoustic Effect of Binary Mixtures of Liquids

P. Korpiun, W. Herrmann, and R. Osiander

Physik-Department E13, TU München, James-Franck-Straße, D-8046 Garching, Fed. Rep. of Germany

It is well-known that the photoacoustic amplitude of liquid samples in a gas-microphone cell increases strongly with temperature. This effect can be explained by an oscillating mass transfer from the sample into the gas, which is manifested by the vapour pressure of the liquid /1,2/. The description given earlier /1,2/ will now be completed and extended to binary mixtures of liquids.

1. Theory

The liquid sample is assumed to be a mixture of two components with the mole fractions x_1 and $x_2=1-x_1$, respectively. In the binary gas mixture above the sample (Fig. 1) that is in equilibrium with the liquid the mole fractions are $\xi_1=C_1/C_t=P_1/P_t$ for component 1 and $\xi_2=1-\xi_1=C_2/C_t=P_2/P_t$ for component 2. Here, C_1, C_2 are the mole concentrations and P_1, P_2 the partial equilibrium vapour pressures. At modulation frequencies $<10^4$ Hz the total molar concentration C_t and pressure P_t in the cell gas are

$$C_t(t) = C_1(z,t)+C_2(z,t), \quad P_t(t) = P_1(z,t)+P_2(z,t). \tag{1}$$

C_t and P_t only depend on time t and not on the position z, Fig. 2. If the liquid sample and gas are thermally thick and if the liquid is assumed to be optically opaque, temperature waves are generated in the form

$$\Phi_s = \Theta \exp[i(k_s z+\omega t)] \text{ and } \Phi_g = \Theta \exp[-i(k_g z-\omega t)] \tag{2}$$

in the sample (s) and the gas (g). $\omega/2\pi$ is the modulation frequency, $k_n=(1-i)a_n$ is the complex thermal wave number of the medium labelled by n with $a_n=\sqrt{(\omega/2\alpha_n)}$, where $\alpha_n=\lambda_n/(\rho_n c_{pn})$ is the thermal diffusivity, λ_n the thermal conductivity, ρ_n the density and c_{pn} the isobaric specific heat capacity. If the temperature T at the liquid/gas boundary at z=0 varies as $\Phi_s(0,t)=\Phi_g(0,t)$, the mole concentrations C_1 and C_2 are altered by evaporation or condensation. The vapour flows of both components are /3,4/

$$j_1 = w\, x_1 C_t = -D(\partial C_1/\partial z)+wC_1, \quad j_2 = w\, x_2 C_t = -D(\partial C_2/\partial z)+wC_2, \tag{3}$$

where D is the mutual diffusion coefficient in the gas and w at z=0 the total condensation velocity. The first term in each of the equations represents the diffusive flow, the second term the convective flow of mass in the gas. Because of (1), we have $\partial C_1/\partial z=-\partial_2 C/\partial z$. Therefore, eliminating w

(3), the flow j_i of component i is

$$j_i = -\frac{x_i C_t D}{(x_i C_t - C_t)} \frac{\partial C_i}{\partial z} . \tag{4}$$

Applying the equation of continuity $-\mathrm{div}\, j_i = \partial C_i/\partial t$ to (4) one obtains

$$\frac{x_i D}{x_i C_t - C_i} \left[\frac{\partial^2 C_i}{\partial z^2} + \frac{1}{x_i C_t - C_i} \left[\frac{\partial C_i}{\partial z}\right]^2\right] - \frac{1}{C_t} \frac{\partial C_1}{\partial t} = 0. \tag{5}$$

We express the concentration of both components in the gas by

$$C_i = C_{oi} + \Psi_i(z,t) \text{ with } \Psi_i = c_i \exp[-i(\varsigma z - \omega t)], \tag{6}$$

where C_{oi} is the average concentration of component i on which a concentration wave Ψ_i with the complex wave number ς_i is superposed. Since the condition $|\Psi_1| \ll C_{10}$ is fulfilled, $C_t - C_i$=const. The second term in (5) can be neglected if $|\Psi_i/C_i| \ll \xi_i - x_i$, that means, if the relative concentration variations are small compared to the difference between the mole fractions in the gas and the liquid. In this case (5) is an ordinary diffusion equation with the effective diffusion coefficient defined as

$$D_i^e = (x_i D)/(x_i - \xi_i) \tag{7}$$

for component i, and the wave number ς_i in (6) becomes

$$\varsigma_i = (1-i)[\omega(x_i - \xi_i)/(2x_i D)]^{1/2} .$$

The variation of the concentrations $\Psi_i(0,t)$, (6), with temperature at the liquid/gas boundary, z=0, is related to the temperature oscillation Φ_g by /1/ $\Psi_1(0,t)+\Psi_2(0,t) = c_1+c_2 = P_t/(RT)d(\ln P_t)/dT_i\ \Phi_g(0,t)$, where R is the gas constant. To simplify the calculation we assume in the following that the binary mixture is an ideal mixture. Then its vapour pressure P_t can be expressed in terms of the vapour pressure P_i^o of the pure components by Raoult's law: $P_t = x_1 P_1^o + (1-x_1)P_2^o$. Using further the Clausius-Clapeyron relation one obtains for the concentration amplitudes

$$c_1 = \frac{x_1 P_1^o L_1}{R^2 T^3} \Theta, \quad c_2 = \frac{(1-x_1)P_2^o L_2}{RT^3} \Theta, \tag{8}$$

where the L_i's are the heats of vaporization of the pure components. The boundary condition for the heat at z=0 is

$$Q + L_1 D_1^e \operatorname{grad} \Psi_1\Big|_{z=0} + L_2 D_2^e \operatorname{grad} \Psi_2\Big|_{z=0} = \lambda_s \frac{\partial \Phi_s}{\partial z}\Big|_{z=0} - \lambda_g \frac{\partial \Phi_g}{\partial z}\Big|_{z=0} \tag{9}$$

if the heat $Q = Q_o[1+\exp(i\omega t)]/2$ is supplied by light absorption to the sample. The temperature amplitude becomes for $\lambda_s a_s \gg \lambda_g a_g$ with $b_i = \xi_i/(1-i)$

$$\Theta = \frac{2^{-3/2}\, Q_o\, e^{-i\pi/4}}{\lambda_s a_s + [x_1 P_1 L_1^2 D_1^e b_1 + i(1-x_1) P_2 L_2^2 D_2^e b_2]/(R^2 T^3)} \, . \tag{10}$$

With the temperature (2) and concentration (7) variations averaged along the gas length l_g /2/ the relative pressure variation becomes /5/

$$\frac{p(t)}{P_o} = \gamma\left[\frac{\langle \Phi_g \rangle}{T} + \frac{\langle \Psi_1 + \Psi_2 \rangle}{C_o}\right] = \frac{\Theta e^{i(\omega t - \pi/4)}}{\sqrt{2}\, a_g l_g T}\left[1 + x_1 \frac{P_1 L_1}{P_o RT}\left[\frac{D_1^e}{\alpha_g}\right]^{\frac{1}{2}} - i x_2 \frac{L_2}{RT}\left[\frac{D_2^e}{\alpha_g}\right]^{\frac{1}{2}}\right], \tag{11}$$

where $\gamma = c_{pg}/c_{vg}$ is the ratio of the isobaric and isochoric specific heat capacity. Amplitude and phase angle of the pressure variation can be obtained from this complex expression in the usual way. Figure 1 shows a plot of these quantities calculated with (11) for the almost ideal solution of benzene and chloroform as a function of mole fraction x_1 of benzene at a temperature of 60 °C.

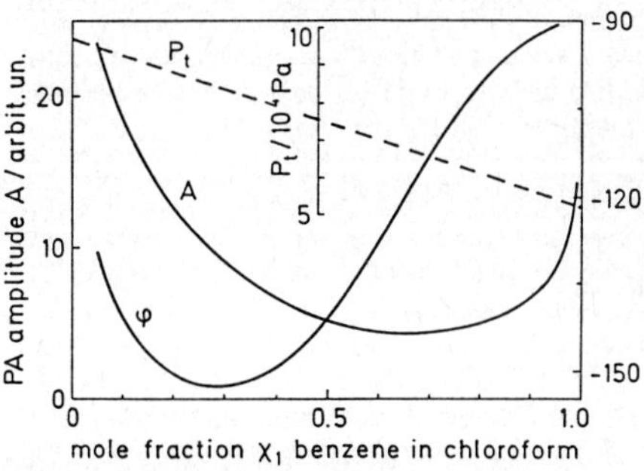

Fig.1: Calculated amplitude A ∝ p(t) and phase angle φ of a benzene and chloroform mixture vs. x_1 at T=60°C. Dashed line: total vapour pressure

2. One-Component Liquid in Gas Cell and Experimental Results

The experimental situation of a one-component liquid in a PA cell with transducer gas, e.g. air, can be considered as the limiting case for $x_1 \to 1$:

$$\frac{p(t)}{P_t} = \frac{Q_o \gamma \alpha_g \alpha_s [1 + (D^e/\alpha_g)^{1/2} L P_1/(P_t RT)]}{\omega \lambda_s l_g T [1 + (D^e/\alpha_s)^{1/2} P_1 L^2/(\rho_s c_{ps} R^2 T^3)]}\, e^{i(\omega t - \pi/2)} \tag{12}$$

with the effective diffusion coefficient (7)

$$D^e = D/(1 - P_1/P_t). \tag{13}$$

We measured the temperature and therefore vapour-pressure dependence of the PA signal of water and three organic liquids, which differ in their boiling points and heats of vaporization. We used a temperature-variable PA-cell with a condenser microphone serving as the detector and air as the transducer gas. In order to realize optically thick samples, we mixed the liquids with fine graphite powder. In Fig. 2 the amplitude A for the liquids water, methyl alcohol, isopropyl alcohol, and acetic acid, normalized to 30°C, is plotted against temperature together with the theoretical re-

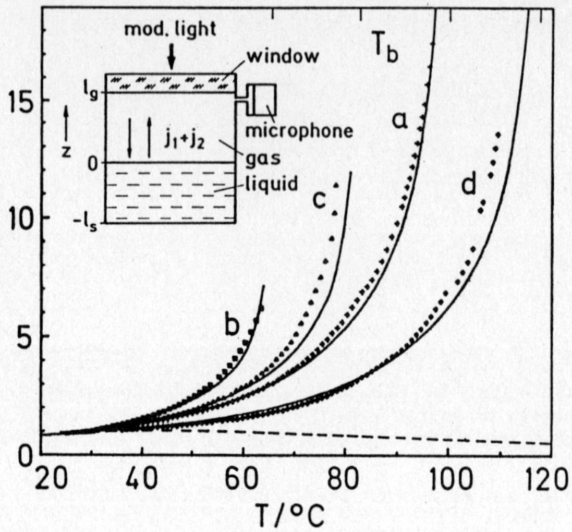

Fig.2: Normalized amplitude for liquids vs. temperature: (a) water, (b) methanol, (c) isopropanol, (d) acetic acid. Solid lines: theory (12); dashed line: RG-model; boiling points (T_b); insert: cell geometry

sults calculated with (12) for $x_1=1$. The temperature dependence of the amplitude expected from the RG-model without the effect of mass transfer is also shown. For the calculation we used values for the parameters involved and their temperature dependence as given in the literature /6/. Up to values near the boiling point, where experimental difficulties increase, the calculation agrees quite well with the measured curves. The relative magnitude of the different amplitudes is reproduced over the whole temperature range. In our measurements the phase angle did not vary significantly with temperature, as was expected from theory.

To sum up, the measurements show that at higher vapour pressures convective as well as diffusive transport significantly affects the photoacoustic amplitude.

1. P. Korpiun: Appl.Phys.Lett. 44, 675 (1984)
2. P. Korpiun, W. Herrmann, A. Kindermann, M. Rothmeyer, B. Büchner: Can.J.Phys. 64, 1042 (1986)
3. J.R. Welty, R.E. Wilson, C.E. Wicks: Fundamentals of Momentum, Heat and Mass Transfer, 2nd.ed. (John Wiley & Sons, New York 1976)
4. A.P. Colburn, T.B. Drew: Trans.Am.Inst.Chem.Egrs. 33, 197 (1937)
5. P. Korpiun, B. Büchner, R. Osiander, W. Herrmann: Appl.Phys.B (to be published)
6. N.B. Vargaftik: Handbook of Physical Properties of Liquids and Gases (Hemisphere, Washington 1983)

Diffusion Waves: Analogy to Thermal Waves

G. Busse, F. Twardon, and R. Müller*

Institut für Kunststoffprüfung und Kunststoffkunde,
Universität Stuttgart, D-7000 Stuttgart 80, Fed. Rep. of Germany
*Permanent address: Institut für Physik, FB ET, Universität der Bundeswehr München, D-8014 Neubiberg, Fed. Rep. of Germany

Processes described by the diffusion equation can be investigated by various methods. To give an example, thermal properties are analysed either in a steady state situation where the constant temperature gradient correlated with a constant heat flux is observed, or in a step response mode, or from the response to a modulated input. It has already been pointed out by ÅNGSTROM /1/ that the "thermal wave" - the way a temperature modulation propagates - provides information from the stationary phase shift and magnitude of the response.

The purpose of this paper is to demonstrate that the formalism used previously for material inspection with thermal wave transmission /2/ can be applied to monitor diffusion processes in polymers by concentration wave analysis. If the partial pressure is different on both sides, a "diffusion current" results where atoms or molecules are transported through the polymer material. Mass transport experiments with modulated input have been performed previously on metal /3/, electrolytes /4,5/, and, recently, on dialysis membranes /6/, while oscillatory sorption measurements have been performed earlier /7-9/.

1. METHOD OF ANALYSIS

For gas permeation the velocity of molecules is determined by the diffusion through the solid. The simplest situation (one dimensional diffusion in a homogeneous medium) is described by the linear equation correlating concentration c of molecules or atoms inside the solid sample to the coordinate x, time t, and diffusion coefficient D

$$\frac{\delta^2 c}{\delta x^2} - \frac{1}{D}\frac{\delta c}{\delta t} = 0. \qquad (1)$$

If the concentration at the surface x=0 of a semi-infinite sample is modulated in a sinusoidal way at angular frequency ω, then one finds in the stationary case (which is essentially achieved after less then 10 cycles /10/ in a near-surface region) a phase lag

$$\varphi = x\cdot(\omega/2D)^{0.5} \qquad (2)$$

between the concentration modulations at x=0 and distance x.

However, the plane concentration wave described here is highly damped since its amplitude decays to 1/e after the wave has travelled a distance $\sqrt{2D/\omega}$ (corresponding to thermal diffusion length μ in thermal wave propagation) which is the fraction $1/2\pi$ of a wavelength, while the phase shift increases to 1 rad. Phase velocity is given by $\sqrt{2\omega D}$. Non-harmonic waves will be distorted due to this strong dispersion. In a nonplanar geometry one has the same equ. (2) and hence the same phase velocity, however, modulation amplitude drops more rapidly.

As for reflection, interference, and stereoscopic effects,it is possible to transfer the thermal wave experience to permeation waves since the same equations have the same solutions. Therefore in the case of diffusion described by a constant value D (Fick's law) one finds that the phase shift between the concentration of two sides of a polymer plate (2) is proportional to the thickness of the plate (if the thickness is not small compared to $\sqrt{2D/\omega}$), thereby allowing for the dynamic determination of the diffusion constant and of parameters that may affect this value.

2. EXPERIMENTAL APPARATUS

We determined the diffusion coefficient D from the phase angle between the oscillating partial pressures on both sides of the sample (2). It is not possible to measure the partial pressure in the cell directly. However, if the gas fluxes are analysed after the cell an additional phase lag is possibly induced.

Two different methods have been used to solve this problem. The first is a vacuum method working with a low pressure on the permeent side of the sample while the other side is evacuated by the pump of a mass spectrometer (Det 2 in Fig. 1). The reference signal (needed for the correlation technique that provides the phase angle) is measured by a piezo-electric pressure gauge. The advantage is the fast response of the system. The second method is based on the carrier gas technique. Here the same detector systems (flame ionisation detectors of a gas chromatograph) and tube lengths are used on both sides, therefore the response time is the same for Det 1 and Det 2. Both arrangements use a computer controlled mass flow valve to produce the partial pressure modulation of the permeent. The two systems that we used were polycarbonate with carbon dioxide as permeent and silicone rubber with methane.

3. RESULTS

Our first experiments were made to test the performance of the experimental setup and to see how well the simple linear

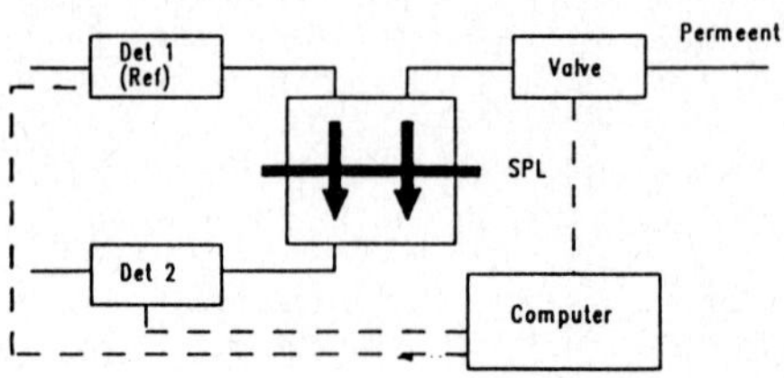

Fig. 1: Experimental arrangement for concentration wave analysis

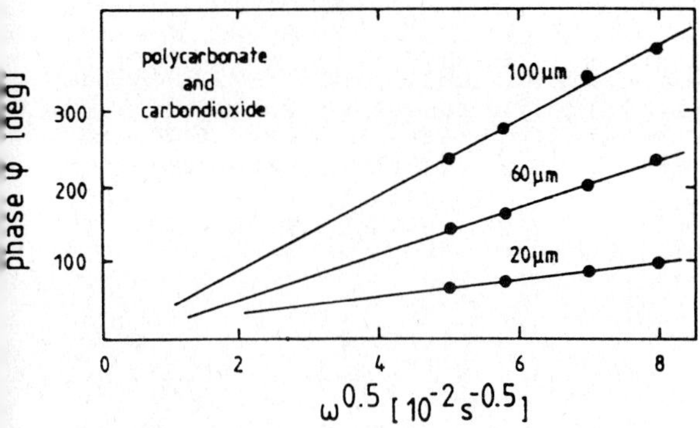

Fig. 2: Frequency-dependent concentration wave phase angle across polycarbonate samples of various thickness and the permeent carbon dioxide

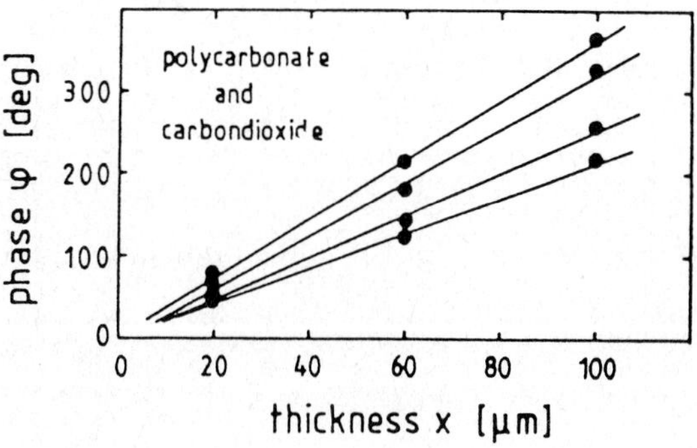

Fig. 3: Thickness-dependent concentration wave phase across polycarbonate samples at various frequencies and the permeent carbon dioxide

theory is applicable to a realistic system. The phase shift should be linear in sample thickness and in the square root of modulation frequency. The result that we found with the vacuum method (Figs. 2 and 3) confirms well the expected relation. Therefore we conclude that the model describes the system well enough within the range of frequencies that we used.

To eliminate the problem of zero phase lag calibration the slope of the lines in Figs. 2 and 3 was evaluated to give the diffusion constant D. This way we obtained for polycarbonate and carbon dioxide at 20°C

$$D = (8.4 \pm 1.4)\ 10^{-13}\ m^2\ s^{-1}.$$

With the step response measurement on the same sample we found

$$D = (7.0 \pm 1.5)\ 10^{-13}\ m^2\ s^{-1},$$

a value that is nearly consistent with the concentration wave result. In Fig. 4 the relation between phase shift and square root of modulation frequency measured with the carrier gas method for methane and silicone rubber can be seen.

One advantage of the permeation wave technique is its ability to monitor continously gradual changes of parameters. As an example we investigated the diffusion of methane through silicone rubber at various temperatures. According to the Arrhenius-law the influence of temperature T is described by

$$D(T) = D_0 \cdot \exp(-E/RT) \quad (3)$$

with activation energy E and gas constant R. Therefore the plot of log D versus 1/T gives a straight line with a slope depending on E. Our results for silicone rubber and methane are presented in Fig. 5. Curves of this kind have been published previously by BARRER and CHIO /11/. Therefore we conclude that the simple one-dimensional linear model for permeation through polymers is applicable. The technique of sinusoidal concentration modulation is obviously suited for non-destructive characterization of materials and of processes going on in them.

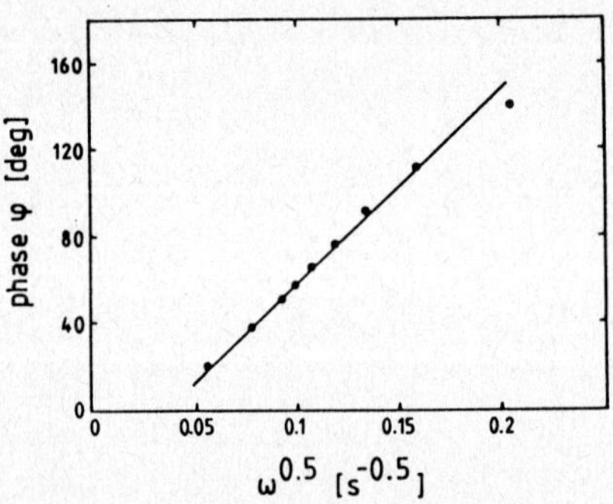

Fig. 4: Frequency-dependent phase angle of methane across silicone rubber at 20 °C

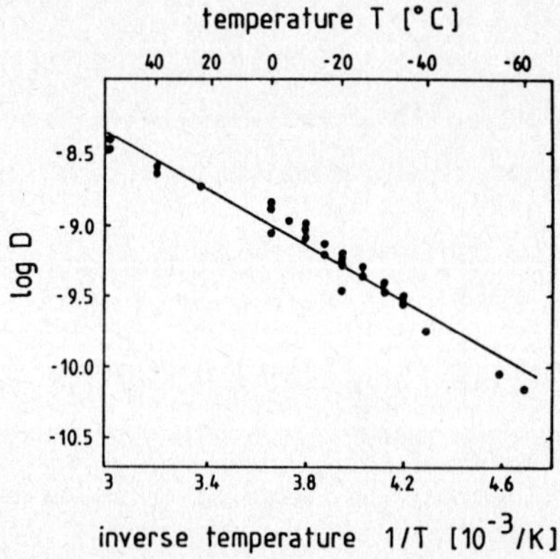

Fig. 5: Temperature dependency of methane diffusivity through silicone rubber

Acknowledgement: The authors are grateful to S. Weiss for her skillful assistance.

4. LITERATURE

1. M.A.J. Ångstrom: Philos. Mag. 25, 181 (1963)
2. G. Busse: Infrared Phys. 20, 419 (1980)
3. D.L. Cummings, R.L. Reuben, D.A. Blackburn: Metallurgical Transactions A, 15A, 639 (1984)
4. J.P. Roger, D. Fournier, A.C. Boccara: J. de Physique, Colloque C6, supplé. n° 10, 44, (1983), 313
5. B.S.H. Royce, D. Voss, A. Bocarsly: J. de Physique, Colloque C6, supplé. n° 10, 44, (1983), 325
6. R. Paterson, P. Doran: J. of Membr. Sci., 27, 105 (1986)
7. J.S. Vrentas, J.L. Duda, S.T. Ju, L.-W. Ni: J. of Membr. Sci., 18, 161 (1984)
8. S.K. Evnochides, E.J. Henley: J. of Poly. Sci., Part A-2, 8, 1987 (1970)
9. Ju Shiaw-Tzuu: Thesis, Pennsylvania State Uni. 1981
10. B. Rief: Can. J. Phys. 64, 1303 (1986)
11. R.M. Barrer, H.T. Chio: J. of Poly. Sci., Part C, 10, 111 (1965)

Influence of Flowing Gases on the Amplitude of Gas-Microphone-Detected Photoacoustic Signals from Porous and Non-Porous Solids

P. Ganguly and T. Somasundaram

Solid State and Structural Chemistry Unit, Indian Institute of Science, Bangalore 560012, India

The influence of flowing gases on the amplitude of gas-microphone detected photoacoustic signals from solids in an open photoacoustic cell is of interest not only because of its importance in the in situ study of catalysts during reaction conditions [1] but also because it serves as a test for theory for the PA effect [2-4]. Such experiments are different from those reported earlier [5] where the signal is generated by the gas molecules themselves. In the widely accepted theory for gas-microphone detected PA signals, the length of the gas (from the solid surface) that is heated during a chopping period is $\simeq 2\pi\mu_g$, where μ_g is the thermal diffusion length of the gas. The heating of the gas molecules in this region causes an increase in pressure which builds up to a maximum after every illumination period [6]. When a gas is flowing through the cell some of the heat is likely to be carried out of the cell, especially when the dimensions of the cell are comparable to $2\pi\mu_g$. Heat may be assumed to be taken up and lost by the flowing gas (thermalisation) within a distance $2\pi\mu_g$ from the beginning and end of the sample in the direction of flow. One should therefore expect a decrease in the amplitude of the PA signal as the flow-rate is increased, especially when the volume of the gas displaced during a chopping period is comparable to the volume of the cell or when the dimensions of the cell are comparable to the thermal diffusion length of the gas.

Two types of cells were used. Most of the experiments were carried out in a cell described earlier [1]. The cell used with amorphous selenium (a-Se) as sample is shown in Fig. 1. a-Se is coated on one side of the inner wall of the sample chamber by melting the powder under nitrogen. The illumination is from the opposite side and the entire sample chamber was illuminated. The distance between the sample of a-Se and the inlet and outlet walls was $<$ 1.0mm. In Fig. 2 we show a typical example of the results obtained from a coating of a-Se at 3.6Hz chopping frequency in the open PA cell for different flow-rates of nitrogen (normalized in terms of the ratio of gas displaced during a chopping period to the volume of the cell). The value of $2\pi\mu_g$ for nitrogen at 3.6Hz is 8.0mm which is very large compared to the dimensions of the cell. We see no dependence on the flow-rate even when ten times the volume of the cell is displaced during a chopping period. The slight increase at high flow-rates is found to be entirely accountable in terms of the increase in pressure in the sample chamber. Similar results were obtained for a number of solids taken as single crystals, thin films or powders or even some high-surface-area materials such as Cr_2O_3-Al_2O_3 or NiO-Al_2O_3 catalysts which have surface areas greater than 150 m^2/g. The only exception that is found is in case of carbon black powder (Fig.2) where the PA signal intensity is roughly inversely proportional to the flow-rate. In all the cases including carbon black the frequency dependence is not affected by the rate of gas flow. The results seem to suggest that the thermalisation of the flowing gas in the active region occurs at distances $<<2\pi\mu_g$. It could also suggest that the PA signal could be acoustic in nature (entirely mechanical piston effect) which is of course in contradiction

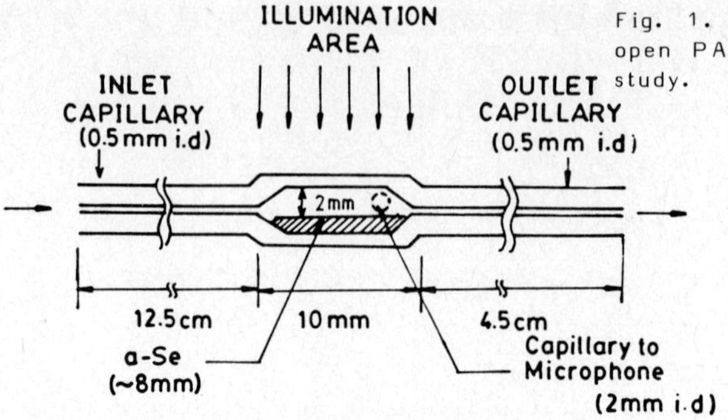

Fig. 1. Schematic diagram of the open PA cell used for the gas-flow study.

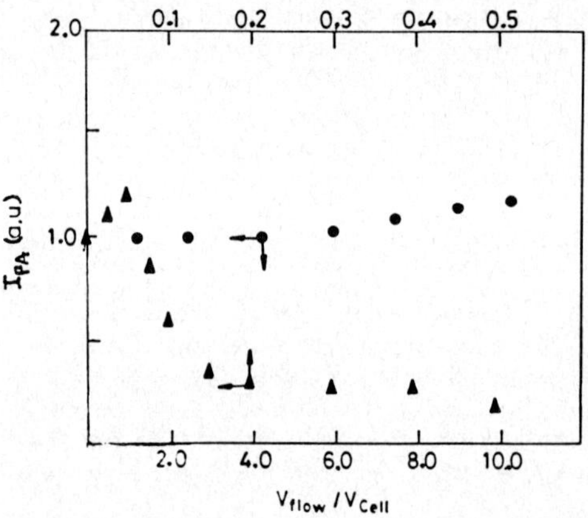

Fig. 2. Variation of PA amplitude as a function of ratio of the volume of the gas displaced during a chopping period to the volume of the cell. Carbon black (▲) and a-Se coating (●).

to all of the published work. The different behaviour seen with carbon black powder may be attributed to the highly porous nature of carbon black. Further studies are being carried out to understand better the generation of gas-microphone-detected PA signals, especially for powders.

References

*Contribution No. 474 from the Solid State and Structural Chemistry Unit

1. T. Somasundaram, P. Ganguly : Bull. Mater. Sci. **9**, 81 (1987)
2. A. Rosencwaig : In Photoacoustics and Photoacoustic Spectroscopy, ed. by P.J. Elving and J.D. Winefordner, Vol. 57, Chemical Analysis (John Wiley, New York 1980)
3. A.C. Tam : Rev. Mod. Phys. **58**, 381 (1986)
4. P. Ganguly, T. Somasundaram : Proc. Indian Acad. Sci. (Chem. Sci.) **98**, 305 (1987)
5. H. Sontag, A.C. Tam : Opt. Lett. **10**, 436 (1985)
6. L.C. Aamodt, J.C. Murphy : J. Appl. Phys. **49**, 3036 (1978)

Investigation of Anisotropic Thermal Diffusivity Using the "Mirage Effect"

Gan Changming and Zhang Xiaorong

Institute of Acoustics, Nanjing University,
Nanjing, People's Republic of China

I. INTRODUCTION

It is well known that the diffusivity of pure materials [1] and the changes of the thermal wavelength of Si wafers due to doping ions [2] can be investigated by the 'mirage effect' (i.e. OBD) [3-6] technique. It is also well known that many solids with anisotropic thermal diffusivities are of considerable importance in practice, e.g. crystals, sedimentary rocks, cold treated metals and thermal screen materisls used in spacecraft, etc. But, it is difficult to make accurate measurements of the conduction of heat in such solids, and in particular for crystals, few of the special problems have been solved [7]. It is thus very important to develop the theory and experimental techniques for anisotropic solids, because they would find a vast range of applications.

The OBD method has many advantages. In particular, the measurement is inherently local, so that measurements can be made on nonuniform or small samples. In principle, the diffusivity can be measured in different orientations of the anisotropic material by rotating the direction of the probe beam.

In this paper, experimental verification of the possibility of measuring an anisotropic thermal diffusivity using the 'mirage effect' technique is presented. The theoretical results are shown to be in agreement with experimental data obtained with a quartz plate.

II. THE ANISOTROPIC THERMAL CONDUCTIVITY OF QUARTZ

For an anisotropic solid, the thermal conductivity equation can be written as

$$k_{ij}\,(\delta^2 T/\delta x_i\,\delta x_j\,) + g = \rho C_p\,(\,\delta T/\delta t\,),\ i,j = 1,2,3, \qquad (1)$$

where g is the thermal source and T is the temperature at the position x and time t; ρ, C_p, and k_{ij} are respectively the density, specific heat and the thermal conductivity tensor of the solid. The number of the thermal conductivity components is diffrernt for various solids [7].For quartz, the number of independent thermal conductivity constants k_{ij} is 5 because it belongs to the trigonal crystal system.The matrix of constants k_{ij} for quartz can be given by

$$\begin{pmatrix} k_{11} & k_{12} & 0 \\ -k_{21} & k_{22} & 0 \\ 0 & 0 & k_{33} \end{pmatrix} . \qquad (2)$$

When we consider the conduction of semi-infinite quartz crystal along its surface, taking the X- and Z-axes in the surface, (1) can be rewritten as

$$k_{11} \; (\delta^2 T/\delta x^2) + k_{33} \; (\delta^2 T/\delta z^2) + g = \rho C_p \; (\delta T/\delta t), \quad (3)$$

from which it can be seen that the X- or Z- (i.e. C-) axis is the principal axis of thermal conductivity, and k_{11} and k_{33} are the principal thermal conductivities. So $D_{11} = k_{11} / \rho C_p$ and $D_{33} = k_{33} / \rho C_p$ are the principal thermal diffusivities.

If we make the additional tranformation

$$\xi = x \; (K / k_{11}), \quad \zeta = z \; (K / k_{33}),$$

where K may be chosen arbitrarily, then (3) becomes

$$K \; (\delta^2 T/\delta \xi^2 + \delta^2 T/\delta \zeta^2) + g = \rho C_p \; (\delta T/\delta t). \quad (4)$$

This equation has the same form as the equation for the isotropic solid in two dimensions. Thus this transformation reduces problems of the anitropic solid to the solution of corresponding problems of the isotropic solid when the solid is infinite. The isothermals are the family of ellipses

$$x^2 / k_{11} + z^2 / k_{33} = \text{const.}$$

i.e.

$$x^2 / D_{11} + z^2 / D_{33} = \text{const.} \quad (5)$$

In order to have the possibility to compare the theoretical values with the experimental results measured by means of the 'mirage effect', the k_{ij} in the above equations are replaced by the diffusivities D_{ij} . The shape of the thermal diffusivity ellipse is similar to that of the thermal conductivity. One diffusivity ellipse is shown in Fig.1. The directions of the thermal flux vectors are perpendicular to this curve. The diffusivities D_{33} along the Z-axis and D_{11} along the X-axis are calculated from the parameters listed in table 1, and are used as the long and the short principal axes, respectively. It can be seen that, usually, D is a function of φ ,the angle made with the Z-axis. The function $D(\varphi)$ can be obtained from the magnitudes of the half-widths of OBD profiles when the calculated values of D_{11} and D_{33} are used to calibrate the apparatus.

Table 1. Parameters for quartz [8]

SiO_2	k [W/cm deg]	C_p [cal/g deg]	ρ [g/cm^3]
//Z-	0.140	0.191	2.65
⊥ Z-	0.072	0.191	2.65

III. EXPERIMENT AND RESULTS

The experimental arrangement used for the OBD method is similar to that described in [2]. The CO_2 laser and He-Ne laser beams are focused carefully to make them as small as possible. The experimental principle is shown schematically in Fig.2. The sample used is a Y-cut polished quartz plate of 2cm length (parallel to the X-axis, i.e. perpendicular to the

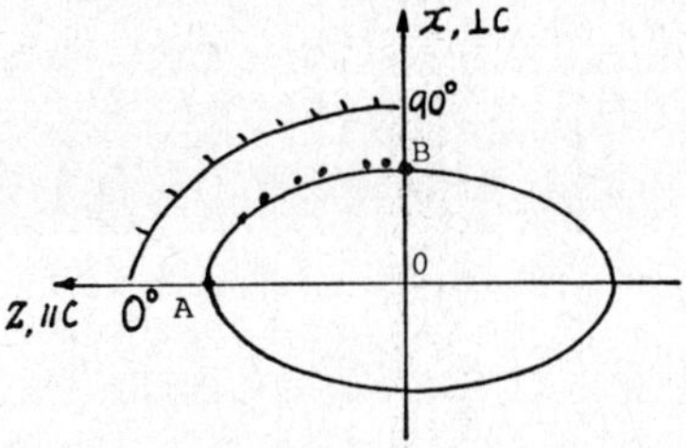

Fig.1. Isothermal curve or curve of D(φ)

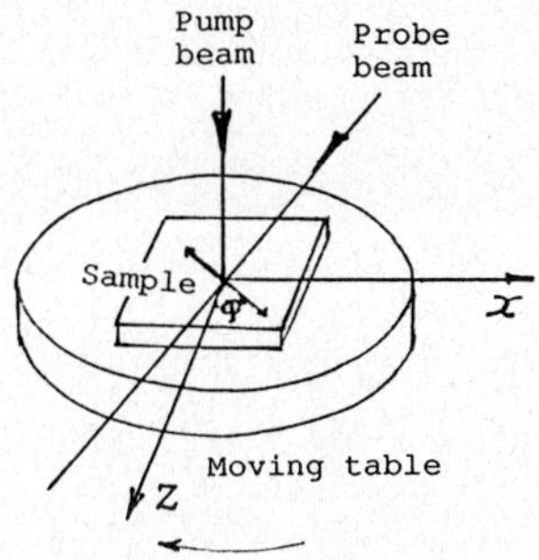

Fig.2. Principle of the experiment

Table 2. Normalized theoretical and experimental diffusivities of a quartz plate

[deg]	Difference of half-widths, Δ [cm]	Normalized diffusivities* Theoretical values	Experimental values
0	0.0450	1.00	1.00
10	0.0432	0.90	0.89
20	0.0409	0.78	0.77
40	0.0382	0.64	0.62
60	0.0368	0.56	0.56
70	0.0364	0.54	0.54
90	0.0359	0,52	0.52

* The values of normalized diffusivities 1 and 0.52 correspond to 0.066 cm^2/s and 0.034 cm^2/s, respectively.

Z-axis) and 1cm width (parallel to the Z-axis). The experimental system must be calibrated. Firstly the direction of probe beam was set parallel to the Z-axis, then the distance between the focused point of the laser beams and the sample surface was adjusted so that the measured D_{11} was close to the calculated length of the short major principal axis shown in Fig.1. Secondly, the sample was rotated to bring the direction of the probe beam perpendicular to the Z-axis, then the distance between the focused point and sample surface was adjused again,to make the measured D_{33} close to the calculatd length of the long principal axis are shown in Fig.1. The above two steps were repeated until both D_{11} and D_{33} were close to the calculated values. The system was then fixed so that only rotation of the sample within a horizontal plan was possible. Finally,the function D(φ) showing the dependence of the diffusivities on the angle can be measured. Here the values of measured D are obtained from the different half-widths Δ of two transverse OBD profiles at two frequncies. For comparison,theoretical values and experimental results (the solid points) shown in Fig.1 and Table 2. It can be seen that measured and theoretical results agree.

IV. SUMMARY

1. The thermal diffusivities D_{11} and D_{33} for a quartz plate along directions respectively perpendicular and parallel to the Z-axis can be obtained when the measurement system is adjusted carefully. The experimental values are adjusted to be close to the calculated values.

2. The dependence of the thermal diffusivity on the angle φ between the direction of the probe beam and the C-axis of the sample can be observed when the sample is rotated. The results measured are in good agreement with the calculated values.

3. It was confirmed that the anisotropic diffusivity of other anisotropic media can be studied using the OBD method, but it should noted that it is necessary to adjust the experimental apparatus very carefully.

REFERENCES

1. P. K. Kuo et al.: Can. J. Phys. 64, 1165,1168 (1986)
2. X. R. Zhang, C.M.Gan et al: Chin. Phys. Lett. 4,215 (1987)
3. A. C. Boccara, D. Fournier, J. Badoz: Appl. Phys. Lett.36,130(1980)
4. W. B. Jackson et al.: Appl. Opt. 20, 1333 (1981)
5. J. C. Murphy, L. C. Aamodt: J. Appl. Phys. 51, 4580(1980);ibid. 52, 4906 (1981)
6. L. C. Aamodt, J. C. Murphy: J. Appl. Phys. 54, 581 (1983)
7. M. N. Ozisik: Heat Conduction (Wiley, New York 1980)Chap. 15
8. Table of Physics Constants (in Chinese), (Chinese Science Press, Beijing 1980)

Fast Pulsed Photothermal Method Applied to Micrometer Coating Characterization

D.M. Boscher, D.L. Balageas, and A.A. Déom

O.N.E.R.A. 29, av. de la division Leclerc, F-92320 Châtillon, France

Pulsed back emission photothermal radiometry is now a well known method of detecting bounding defects in multilayer composites [1] or to measure thermal properties of coatings a few tenths of a micrometer thick [2]. The application proposed here is concerned with the measurement of the thermal diffusivity of 1 µm-thick amorphous carbon coatings which are used as non-reflective coatings for germanium infrared windows. Their hardnesses are very high so they are also called diamond-like coatings (DLC). For such coatings many problems in measuring thermal properties can arise, due to radiative effects and the necessary large bandwidth of the electronics.

The experimental set-up is drawn in Fig. 1. A ruby laser (JK Lasers 2000), working in a Q-switched mode, delivers a 1 J, 50 ns pulse at 0.695 µm wavelength. The laser beam is focused, after attenuation, on a hole in a hemispherical reflecting cavity in which the sample is positioned. The IR light emitted by the sample is collected by a mirror and focused on a HgCdTe detector. The purpose of the cavity is double [3] : it increases the absorptivity and the emissivity of the sample. The voltage signal given by the detector is amplified and then digitized. The total bandwidth of this system is about 70 MHz, which allows us to perform experiments with a time constant of 10 ns.

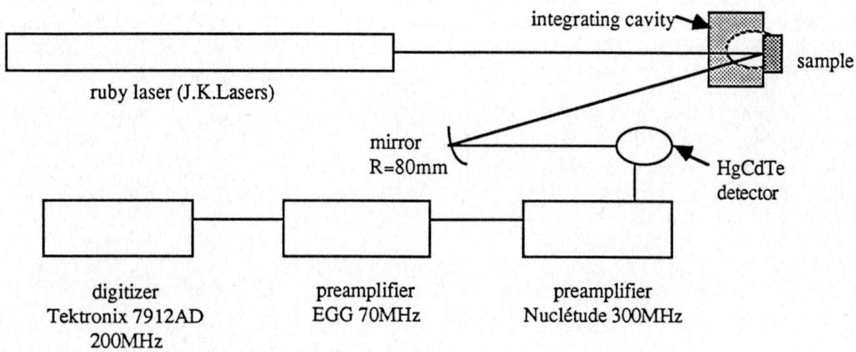

Fig. 1 : Experimental set-up for the 50 ns pulsed photothermal method

For a homogeneous and opaque thick sample, with constant thermal properties during the experiment, the temperature increase ΔT of the sample surface at a time t is given by

$$\Delta T = Q / b \sqrt{\pi} \sqrt{t} \quad , \tag{1}$$

where Q is the energy density absorbed by the sample, b the effusivity.

A first experiment was conducted to validate the set-up. An opaque and polished material was selected to avoid surface roughness and optical transmission problems. A copper sample

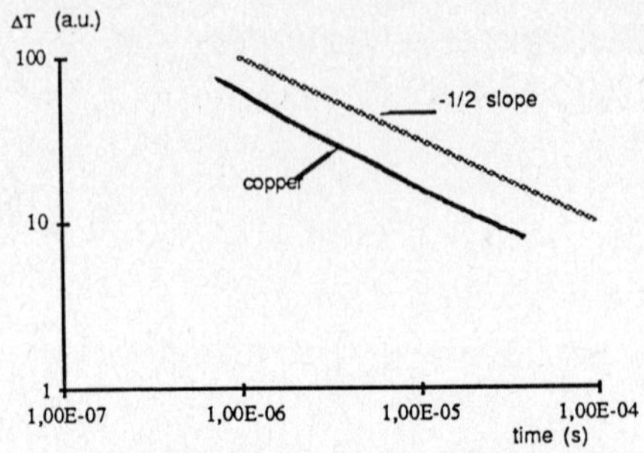

Fig. 2 : Temperature decrease of a polished copper sample. Comparison with a homogeneous surface absorbing and emitting sample (-1/2 slope)

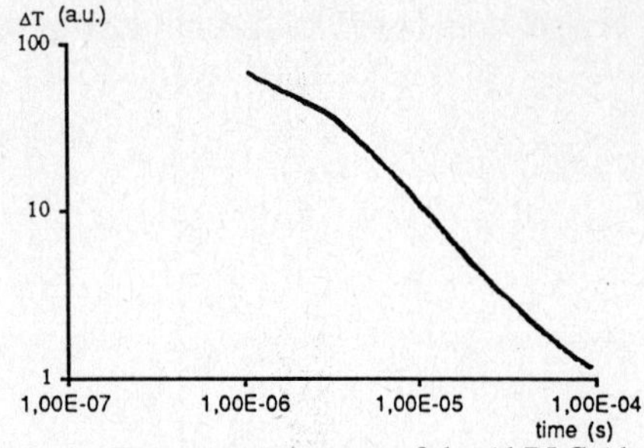

Fig. 3 : Temperature decrease of the Al-DLC-Al sample. The change in slope allows a measurement of DLC diffusivity

was used. The temperature increase for this sample is shown in Fig. 2 on a log-log scale. The -1/2 slope given by (1) is also represented. The agreement between the two slopes of the curves is good, even for the 1 μs time that corresponds to a depth of 10 μm from the front surface. For shorter times, a discrepancy appears which is due to surface, optical and pulse duration effects.

For a multilayer sample, the temperature-time curve is, on a log-log scale, a succession of straight lines with -1/2 slope, corresponding to equation (1) where b is the effusivity of the successive layers. Between two straight portions of the curve, a transition takes place. For a coating on a substrate, the occurrence time of the transition is of the order of the ratio L^2/a , where L is the thickness of the coating and a its diffusivity. The shape of the transition is related to the quality of the interface (its thermal resistance) [1]. So, by means of this method, it is possible to characterize the coating and its interface.

As the amorphous carbon is not opaque for infrared radiation, it is quite difficult to monitor its temperature evolution by radiometry. Another difficulty is due to the fact that the germanium substrate on which the carbon coating is normally deposited is highly transparent in the IR domain. For these reasons, a special sample was designed with three layers : a 1 mm aluminium substrate, a 1.1 μm carbon coating and an additional opaque layer (2.4 μm aluminium). The temperature decay obtained with this three-layer sample is shown on a log-log scale in Fig. 3. The part of the curve related to the first coating is not visible here, because it arises for times shorter than 1 μs. So the first part of the curve is due to the carbon coating, given a -1/2 slope straight line. Then, for times longer than 4 μs, the carbon-aluminium interface induces the change in slope. The comparison between the experimental curve and the results of an analytical model of the pulsed thermal response of layered materials [1] gives a good value of the DLC diffusivity : $a = 3 \times 10^{-6}\ m^2 s^{-1}$. With ρ = 1800 and C = 700 (SI), the corresponding thermal conductivity is $k = 4\ W\ m^{-1}\ K^{-1}$, which is extremely low compared to the diamond thermal conductivity.

This work was supported by the Direction des Recherches, Etudes et Techniques of the French Ministry of Defence.

1. D.L. Balageas, J. Krapez and P. Cielo ; J. Appl. Phys., 59 (1986) p.348
2. A.C. Tam, B. Sullivan ; Appl. Phys. Lett. 43 (1983) p.333
3. P. Cielo, S. Dallaire, G. Lamonde, S. Johar ; Can. J. Phys. 64 (1986) p.1217

Spectroscopic Determination of Thermal Diffusivity of Semiconductors by Photothermal Deflection Spectroscopy: Application to GaAs

N. Yacoubi and M. Fathallah

Département de Physique, Faculté des Sciences de Tunis,
1060 Tunis, Belvédère, Tunisia

Abstract : A method of determination of the thermal diffusivity of semiconductors is proposed where a one-dimensional analysis is adequate, using the relative influence of the thermal properties of the sample and backing on the phase of the photothermal deflection spectroscopy (PDS) signal.

Many methods have been proposed to measure the thermal properties of solids using the PDS signal /1-3/. They generally use an off-set distance between the pump and probe beams. A three-dimensional theory is needed to evaluate the PDS signal. The deflecting medium is chosen to have a small thermal coupling with the sample and the influence of its thermal parameters on heat diffusion is neglected /4/. In this paper, we propose a new method in which the probe beam is in the center of the pump beam. A one-dimensional analysis is sufficient and the difficult choice of the deflecting medium is avoided. This method is basically spectroscopic and is derived from the phase measurement of the fluid-sample interface temperature formulated by Rosencwaig and Gersho /5/.

The complex expression for the periodical temperature T_s at the fluid-sample interface is given by /5/

$$T_s = \frac{\alpha I_0}{2k_s(\alpha^2-\sigma_s^2)} \; \frac{(r-1)(b+1)\exp(\sigma_s\ell)-(r+1)(b-1)\exp(-\sigma_s\ell)-2(b-r)\exp(\alpha\ell)}{(f+1)(b+1)\exp(\sigma_s\ell)-(f-1)(b-1)\exp(-\sigma\ell_s)},$$

where α is the absorption coefficient, ℓ the thickness of the sample, I_0 the incident light intensity,

$$f = \frac{k_f}{k_s}\left(\frac{D_s}{D_f}\right)^{1/2}, \quad b = \frac{k_b}{k_s}\left(\frac{D_s}{D_b}\right)^{1/2}, \quad \sigma_s = (1+j)\left(\frac{\pi F}{D_s}\right)^{1/2} \text{ and } r = \frac{\alpha}{\alpha_s} \;;$$

k_i and D_i are respectively the thermal conductivity and diffusivity of the material i ; i takes the subscript f, s or b corresponding respectively to the fluid, the sample and the backing. F is the modulation frequency of the pump beam.

T_s may be written as

$T_s = |T_s|\exp(j\theta)$,

where $|T_S|$ is the amplitude and θ the phase. θ is independent of I_O and varies with the parameters α, D_S and b. We have theoretically studied (Fig. 1) the dependence of θ on the optical absorption coefficient α

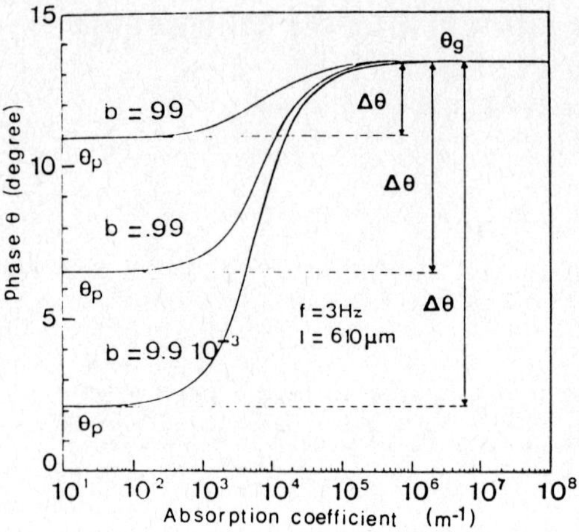

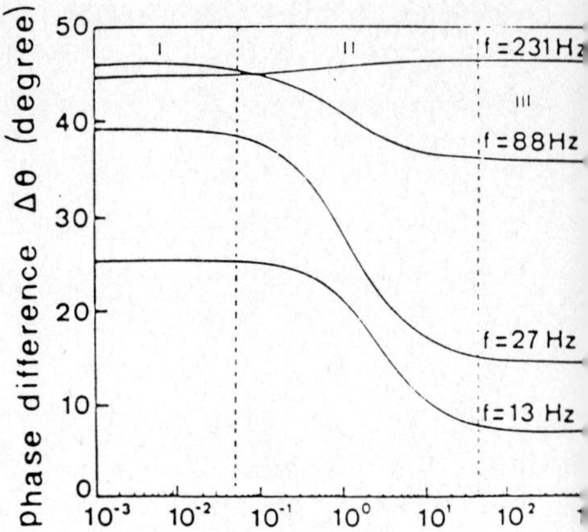

Figure 1. : Calculated PDS phase of the periodical temperature at the surface of a GaAs sample versus α for different b

Figure 2. Calculated phase difference Δθ in the surface of a GaAs sample versus b for different frequencies

for different values of b at 3 Hz modulation frequency. We find that θ has two limit values, $\theta = \theta_g$ for a high optical absorption coefficient and $\theta = \theta_p$ for the low α range. The phase difference $\Delta\theta = \theta_g - \theta_p$ between the two limits is a function of b. This difference is important for low b values (b<<1) and corresponds to a poor thermal backing. In this case, all the heat produced diffuses to the front of the sample. Meanwhile, Δθ is small for large b values (b>>1) which is the case for a good thermal backing. One part of the heat diffuses through the backing. In Fig. 2, we reported the theoretical variations of Δθ as a function of b for different modulation frequencies. It is shown that Δθ is independent of b at high frequencies (F > 230 Hz) for which the sample is thermally thick. However, for low modulation frequencies, Δθ has three domains of variation. Δθ is independent of b in the domains I and III which correspond respectively to the high and low values of b. Δθ is highly sensitive to intermediate b values. At low frequencies, the influence of the backing thermal properties may be ignored by the choice of regions I or III. In this case, a precise knowledge of the backing thermal parameters is not needed.

In Table 1 we list the different b values of the backings used and of various semiconductors.

Table 1. Values of b for various semiconductors and backings.

Semiconductors / Backing	GaAs	GaSb	Si	InAs	InP
Air	5×10^{-4}	9×10^{-4}	3.7×10^{-4}	9.4×10^{-4}	7×10^{-4}
Plexiglas	0.041	0.075	0.03	0.076	0.057

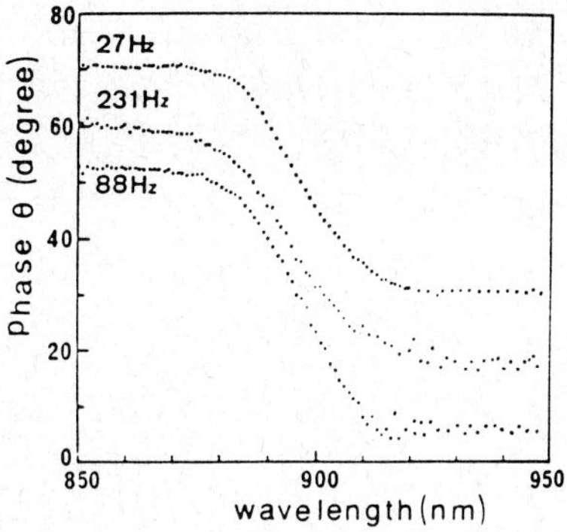

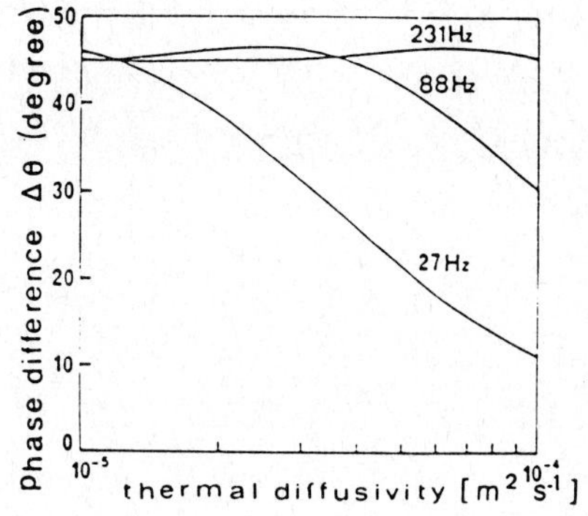

Figure 3. Experimental dependence of the PDS phase on wavelength for a GaAs sample

Figure 4. Theoretical dependence of $\Delta\theta$ at the surface of a 610 μm thick sample on thermal diffusivity.

We can conclude from this table that b is generally situated in domain I, where $\Delta\theta$ is independent of b. $\Delta\theta$ is indeed sensitive to the sample thermal diffusivity only at low frequencies. We may expect that a comparison of the experimental $\Delta\theta$ with the corresponding theoretical value will allow us to deduce the thermal diffusivity of the studied semiconductor. Our method has been applied to a GaAs sample of 610 μm thick with a Plexiglas backing. The sample was immersed in cedar oil used as a deflecting medium. The experimental setup utilized is described in /6/.

Figure 3 shows the experimental value of the photothermal signal phase of the GaAs sample as a function of wavelength. The measurement of the experimental variation $\Delta\theta$ of the phase between the high and low optical absorption and a comparison with the theoretical curve of $\Delta\theta$ as a function of the diffusivity of the sample (Fig. 4) allows us to deduce the thermal diffusivity of GaAs. Our result of 0.21 cm^2/s is in agreement with the literature values (0.21 - 0.26).

In conclusion, we have established a method of determination of thermal diffusivity of semiconductors based on the Rosencwaig and Gersho model. It has the advantage of using a one-dimensional treatment of the photothermal signal. Application of this method to thin films is of obvious interest.

References

1. G. Rousset and F. Lepoutre: Rev. Phys. Appl., 17 (1982).
2. P.K. Kuo, M.J. Lin, C.B. Reys, L.D. Favro, R.L. Thomas, S.Y. Zhang, L.J. Inglehart, D. Fournier, A.C. Boccara and N. Yacoubi : 4th Int. Topical Meeting on Photoacoustic, Thermal and Related Sciences, August 4-8, Québec (1985).
3. H. Dersh, A. Skumanich, M. Fathallah and N.M. Amer, J. Appl. Phys., in press.
4. L.C. Aamodt and J.C. Murphy: J. App. Phys. 52, 4903 (1981).
5. A. Rosencwaig and A. Gersho: J. Appl. Phys. 47, 1 (1976).
6. N. Yacoubi, B. Girault and J. Fesquet : Appl. Opt. 25, 24 (1986).

Thermal Diffusivity Measurements Using Piezoelectric Detection

M. Liezers[1] *and R.M. Miller*[2]

[1]Department of Instrumentation and Analytical Science, University of Manchester, Institute of Science and Technology, P.O. Box 88, Manchester, M60 1QD

[2]Unilever Research Port Sunlight Laboratory, Quarry Road East, Bebington, Wirral, L63 3JW, United Kingdom

1. Introduction

An accurate knowledge of the thermal properties of materials is of considerable importance in engineering, particularly where systems must operate in an environment which is not in thermal equilibrium. Although a large literature exists on the determination of thermal properties, existing techniques have limitations when dealing with composites and fragile materials. Thermal wave techniques have been successfully used for the determination of thermal properties, by measuring the propagation time, or phase lag, of the thermal wave passing through a sample of known geometry[1,2]. An advantage of this method is that the temperature gradient across the sample can be made extremely small, and the material will remain close to thermal equilibrium with its surroundings. Although a variety of detection schemes have been described, piezo-electric transducers appear not to have been used for this type of experiment[3]. Because of their convenience and sensitivity, we have investigated their use in the determination of thermal diffusivities.

2. Experimental Methods

Instrumentation used for these studies was similar to that previously reported for thermal wave imaging[4]. The light source was an amplitude modulated krypton ion laser which was focussed onto the surface of the sample. Samples were held in a Perspex C-clamp with the piezo-electric transducer pressed to the rear surface by a screw. The sample was illuminated on the front surface through a slot cut in the holder.

A variety of materials were studied, including pure metals, alloys and ceramics. Samples were in the form of small plates of varying sizes with thicknesses up to 5 mm. Samples of translucent or transparent materials were coated with a thin layer of copper on one face. This prevented direct laser radiation striking the transducer, and ensured that there would be a strong well-defined heat source at the surface of the sample. Experiments were carried out with a laser output power of 500 mW at 647 nm. The phase response of the samples was measured over a maximum frequency range of 0.5 Hz to 5 kHz.

3. Results and Discussion

Empirical observation of the frequency dependence of the phase lag for a wide variety of samples suggested that a clear relationship existed between the calculated thermal diffusion length in the sample (μ), and the observed phase lag. The pattern was:

1. μ = specimen thickness θ = 1 radian
2. μ = (specimen thickness/2) θ = 0 radians
3. μ = (specimen thickness/5) θ = -1 radian

By measuring the modulation frequency at which each of these key phase lags is observed, a relationship between the thermal diffusion length and frequency can be obtained, which leads directly to the thermal diffusivity. For example, for a sample of nickel alloy 1.28 mm thick, the 3 key frequencies were determined to be 0.16 Hz, 2.35 Hz and 15.1 Hz. These correspond to thermal diffusivities of 0.0295 cm^2s^{-1}, 0.0283 cm^2s^{-1} and 0.0291 cm^2s^{-1}. This compares with a literature value of 0.0303 cm^2s^{-1}[6].

Table 1 summarises the thermal diffusivity results obtained from 14 different materials using this approach.

Table 1: Comparison of Experimental and Literature Values for Thermal Diffusivity for Fourteen Materials

Material	Specimen Thickness (cm)	Mean Experimental $\alpha(cm^2s^{-1})$	Literature Value (cm^2s^{-1})
Copper	0.100-0.329	1.25	1.17
Aluminium	0.900-0.324	1.07	0.98
Molybdenum	0.034	0.48	0.53
Brass (α)	0.320	0.32	0.34
Nickel	0.048	0.21	0.22
Mild Steel	0.100-0.340	0.28	0.21
Titanium	0.158	0.08	0.09
Monel-K	0.264	0.06	0.05
18/8 Stainless Steel	0.112	0.038	0.04
Nimonic 75	0.128	0.029	0.030
Nimonic 90	0.124	0.032	0.031
Sapphire	0.093	0.10	0.13
Al_2O_3	0.101	0.071	0.068 -0.080
ZrO_2	0.222	0.0076	0.0068-0.0097

4. Conclusions

The measurement of thermal diffusivity using the thermal wave phase lag determined by a piezo-electric detector has been shown to be feasible. The technique is fast, accurate and easy to apply to materials of widely differing thermal properties. Good agreement is obtained between literature values and those determined experimentally in this study.

5. References

1. A Rosencwaig: In Photoacoustics and Photoacoustic Spectroscopy, John Wiley and Sons, New York (1980), Chap.20, p. 265
2. R Kordecki, B K Bein and J Pelzl, Can. J. Phys, 64, 1204 (1986)
3. A Biswas, T Ahmed and K W Johnson, Can. J. Phys., 64, 1984 (1986)
4. G F Kirkbright, M Liezers and R M Miller, Spectrochim. Acta B, 41, 741 (1986)
5. Y S Touloukian, R W Powell, C Y Ho and M C Nicolasen, Thermal Diffusivity, Plenum, New York (1973)

Analysis of Pulsed PA Signals Using the Finite Element Method

M. Kasai[1], *M. Ishioka*[2], *M. Kaihara*[2], *S. Fukushima*[1], *T. Sawada*[1], *and Y. Gohshi*[1]

[1]Department of Industrial Chemistry, Faculty of Engineering, The University of Tokyo, 7-3-1 Hongo, Bunkyo-ku, Tokyo 113, Japan
[2]Advanced Technology Research Center, Nippon Kohkan, K.K., 1-1 Minamiwatarida-cho, Kawasaki-ku, Kawasaki-shi, Kanagawa-ken 210, Japan

1. INTRODUCTION

Recently, PA (Photoacoustic) techniques have attracted remarkable attention due to their unique method. In particular, the PAM (Photoacoustic Microscope), a microscope using PA detection techniques, has been used in non-destructive evaluation of some semiconductor devices [1]-[4]. However, the mechanism of PA signal generation has not been fully elucidated.

The FEM (Finite Element Method), which is a series of computerized numerical calculations using the matrix method, is used in many fields, particularly mechanics and construction engineering, as a practical method for the simulation of thermal and/or elastic phenomena. In this paper, a simulation of the process by which PA signals are generated by a pulsed heat source is demonstrated using FEM. The distributions of thermal stress in real time are also discussed. Fourier transformed FEM data are compared with experimental results.

2. FEM CALCULATION

The FEM program used here was based on a commercial program (ADINAT) and improved by us. In this FEM calculation it is assumed that: 1) solutions of the temperature and strain are obtained independently of each other, i.e. cross terms concerned with thermal diffusion and thermoelastic deformation are not considered, 2) deformation is minute, 3) the sample is isotropic and elastic. Under these assumptions, the temperature at any time was calculated for each calculation point (node), and the thermal stress and deformation corresponding to the temperature solution at each node were then calculated using the principle of virtual work.

The model used in this calculation is shown in Fig.1. The sample assumed to be a brass cylinder 10 mm in diameter and 1 mm thick. It is also assumed that this sample is surrounded by air, and that the initial temperature is 20° C at all nodes in the model. A laser source of weak energy density has to be used to prevent the destruction of the brass sample. It is

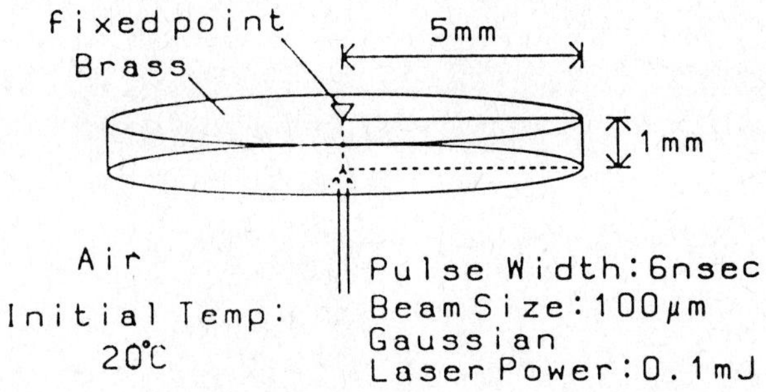

Fig.1 A model sample for the FEM calculation.

therefore assumed that a pulsed excitation laser with a 6 ns band width Gaussian beam, 100 μm in diameter and with 0.1 mJ light energy, is incident at the center of the base of the model and that 1 % of the energy is absorbed by this sample. There were 1702 nodes, which were situated densely (5 μm apart) near the heated area and more sparsely in other areas.

3. RESULTS AND DISCUSSION

3.1 Transient Temperature and Thermal Deformation

In Fig.2, the lower curve (left-hand axis) shows the transient temperature and the upper curve (right-hand axis) shows the transient thermal deformation at the center of the heated surface. The maximum temperature is about 300° C at 10 ns after the start and falls exponentially. The maximum shift for thermal deformation is about 100 Å about 50 ns after the start. The decay of this shift (in μs) is slower than the fall in temperature (in ns) at this point.

3.2 Distribution of thermal stress

The contour maps in Fig.3 show the distribution of thermal stress along the axis (normal stress from the heated surface

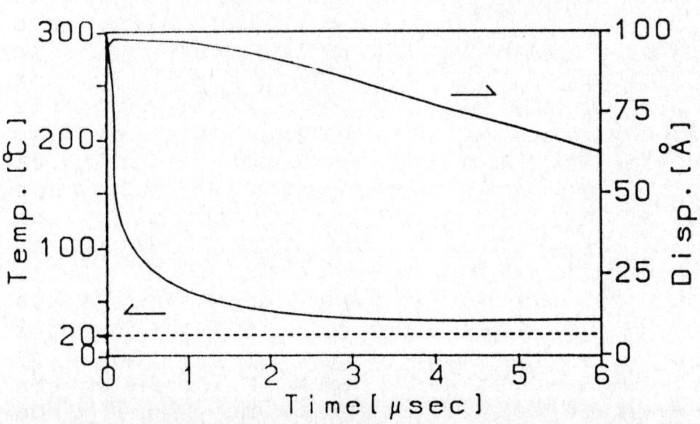

Fig.2 Transient temperature variation and thermal deformation at the center of the heated surface.

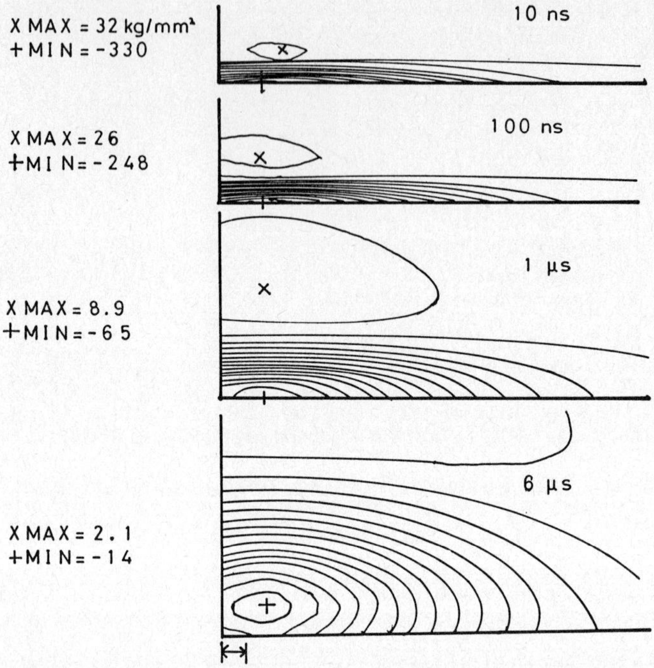

Fig.3 Distribution of thermal stress in the axis near the heated area as contour maps.

to subsurface) near the heated area at (a) 10 ns, (b) 100 ns, (c) 1 s and (d) 6 s after the start. "x" indicates the point where the stress to the front (from heated face to fixed face) has its maximum value, and "+" indicates the point where the stress to the back (from fixed face to heated face) has the maximum value. From these results, the change in the distribution of thermal stress can be seen as a kind of spatial wave spreading though the subsurface of the sample. It would appear that this is what is generally known as a thermoelastic wave or "thermal wave".

3.3 Comparison of FT FEM data with experimental results

If the results of FEM in Fig.2 are an impulse response, i.e. 6ns pulse excitation is approximately a delta function, FT (Fourier transformed) FEM results indicate the frequency response of the model used here. The solid line in Fig.4 (a) shows the FT results of FEM temperature data. Squares and circles in Fig.4 (a) indicate experimental results of PZT (piezoelectric transducer) and PTR (photothermal radiometry). Comparison of FT results of FEM temperature and experimental results of PZT and PTR gave good agreement for about f^{-1}, except the PZT results have a resonance peak at about 100 kHz (PZT) and 20 kHz (the scale of the measurement system). From these results, PA signals of PZT and PTR in homogeneous

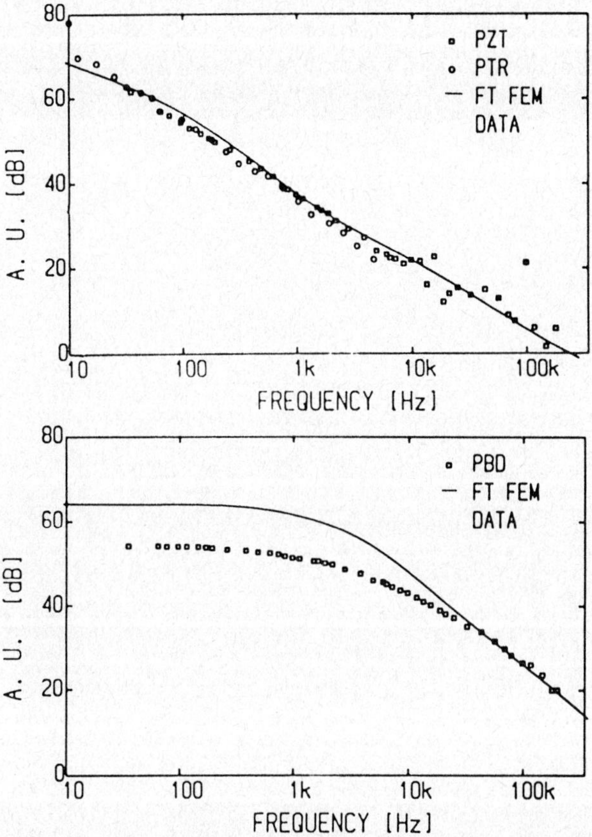

Fig.4 (a) FT FEM temperature data and experimental results (PZT & PTR), (b) FT FEM thermal deformation data and experimental results.

samples are proportional to the temperature at the heated surface of the sample.

The solid line in Fig.4(b) shows FT results of FEM thermal deformation data (squares indicate experimental results of thermal beam deflection), which detect the beam deflection reflected by the heated surface of the sample near the heated point. Although comparisons did not give good agreement, both results exhibited saturation at low frequency range. This is because a fixed point is the epicenter of the heated area here. One of the dominant factors in calculation of thermal deformation is the way in which the model is fixed. Therefore, if a model sample is fixed in the same way as the experimental condition, FT results of FEM thermal deformation data would be in good agreement with PBD results.

4.Conclusion

FEM may well prove to be a useful method for simulating various PA signals, because it gives a clear picture of temperature, thermal deformation and stress at any given time. Furthermore, it could be applied to a complex sample under complex conditions. For example, although it may be impossible to produce an analytical solution for a sample with thermal impurities or stressed regions under the surface, a solution can be found using FEM.

References

1. M.Kasai, H.Shimizu, T.Sawada, Y.Gohshi: Anl.Sci 1, 107 (1985)
2. M.Kasai, T.Sawada, Y.Gohshi: Jpn.J.Appl.Phys. s24, 220 (1985)
3. M.Kasai, T.Sawada, Y.Gohshi, T.Watanabe, K.Furuya: Jpn.J.Appl.Phys. s25, 229 (1986)
4. T.Sawada, Y.Gohshi, T.Watanabe, K.Furuya: Jpn.J.Appl.Phys. 24, L938 (1986)

Photoacoustic Characterization of CoO-Doped Soda-Lime Glasses: Thermal Diffusivity

M.L. Baesso[1], *Z.P. Arguello*[1], *A.C. Bento*[1], *H. Vargas*[1], *and L.C.M. Miranda*[2]

[1]Instituto de Fisica, Universidade Estadual de Campinas, 13081-Campinas, SP, Brazil

[2]Laboratório Associado de Sensores e Materiais, MCT/Instituto de Pesquisas Espaciais, 12201-São José dos Campos, SP, Brazil

The PA effect has been proved by several authors [1-5] to be a simple and reliable technique for the measurement of the thermal diffusivity. In particular, in Ref. [5] we have demonstrated the usefulness of a single modulation frequency method for measuring the thermal diffusivity of solid samples. The method consists in measuring the relative phase lag $\Delta\Phi = \Phi_F - \Phi_R$ at a single modulation frequency, between the rear surface illumination (R) and the front surface illumination (F) as follows. Using the Rosencwaig-Gersho theory [6] the phase lag for front- and rear-surface illumination, $\Delta\Phi$ is given by [5]

$$\tan(\Delta\Phi) = \tanh(\ell a)\cdot\tan(\ell a), \tag{1}$$

where ℓ is the sample thickness, $a = (\pi f/\alpha)^{1/2}$ is the sample thermal diffusion coefficient, f is the modulation frequency, and α is the sample thermal diffusivity. As evident from (1) the procedure to determine α is to substitute the experimental values of $\Delta\Phi$ in (1) and to solve it for $z = \ell a$. Knowing z, the thermal diffusivity is readily given by $\alpha = \pi f l^2/z^2$.

In this contribution we apply the method to the investigation of the effect of dopants on the thermal diffusivity. The samples used are soda-lime glass doped with varying concentrations (wt%) of cobalt oxide. The soda-lime glass was prepared from standard grade chemicals by usual methods. Its composition was 72 SiO_2, 10 CaO, 18 NaO_2 wt%. The experimental arrangement is basically the same as the one described in Ref. [4]. The samples were in the shape of 8mm diameter disks and the thickness varied from 325μm to 370μm. To ensure the optical opaqueness condition [5] implicit in (1), a thin circular Al foil 25μm thick and 3mm diameter was attached to each side of the sample using a thin layer of diffusion pump oil. In this way, we ensured the optical surface absorption condition, as discussed in Ref. [5]. In table 1 we summarise the results of our measurements of α at a modulation frequency of 12Hz. In Fig. 1 we show the dependence of the thermal diffusivity as a function of the CoO concentration. It follows from Fig. 1 that the method is capable of resolving the influence of at least 0.1% CoO concentration on the thermal diffusivity of soda-lime glass. The behaviour of α as a function of the dopant concentration shown in Fig. 1 is typical of what one would expect for a homogeneous binary mixture. According to theory of binary mixtures [7] the effective thermal diffusivity may be written as

$$\alpha_{eff} = \alpha_1 \{1+[(k_2/k_1)-1]x\} / \{1+[k_2\alpha_1/k_1\alpha_2-1]x\} . \tag{2}$$

Here, the subscript 1 (2) refers to the matrix (dopant) thermal properties. In the case of our soda-lime glass α_1=0.004cm^2/s, k_1=0.008W/cmK.

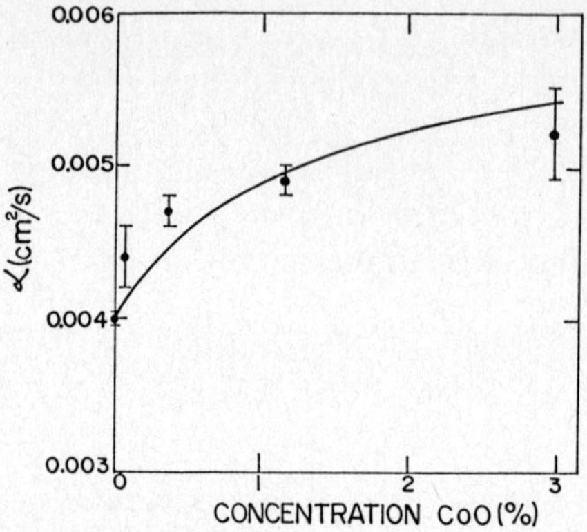

Fig. 1 Thermal diffusivity of soda-lime glass samples as a function of the CoO doping concentration

Table 1 Thermal diffusivity values obtained from the two-beam phase lag method for CoO-doped soda-lime glass samples

CoO doping (wt %)	sample thickness (μm)	thermal diffusivity (cm^2/s)
0.0	325	0.0040 ± 0.0001
0.1	335	0.0044 ± 0.0002
0.4	360	0.0047 ± 0.0001
1.2	370	0.0049 ± 0.0001
3.0	350	0.0052 ± 0.0003

Since the values of the thermal properties of CoO were not available to us, we have then used (2) to obtain the k_2 and α_2 from a data fitting procedure. The results we got are k_2=0.8W/cmK and α_2=0.0062cm^2/s. The value of k_2 is quite close to that of pure Co (≃1.0W/cmK). The solid curve in Fig. 1 corresponds to the plot of the effective thermal diffusivity of a binary mixture as given by (2) using the above values for α_2 and k_2. This result means that, at least qualitatively, the behaviour of α on the dopant concentration may be understood on the basis of binary mixture theory.

References

1. M.J.Adams and G.F.Kirkbright: Analyst 102,281(1977)
2. P.Charpentier, F.Lepoutre and L.Bertrand: J.Appl.Phys. 53,608(1982)
3. R.T.Swimm: Appl.Phys.Lett. 42,955(1983)
4. N.F.Leite, N.Cella, H.Vargas, L.C.M.Miranda: J.Appl.Phys. 61,3025(1987)
5. O.Pessoa Jr., C.L.Cesar, N.A.Patel, H.Vargas, C.C.Ghizoni and L.C.M.Miranda: J.Appl.Phys. 59,1316(1986)
6. A.Rosencwaig, A.Gersho: J.Appl.Phys. 47,64(1975)
7. A.W.Pratt: In Thermal Conductivity, ed. by R.P.Tye (Academic Press, New York 1969) p.301

Study of Complex Specific Heat in the Glass Transition Regime

B. Büchner and P. Korpiun

Physik Department E13, TU München, James-Franck-Straße, D-8046 Garching, Fed. Rep. of Germany

Glass has been used by mankind since ancient times and so the problem of the glass transition and the observed relaxational behaviour is very old. But a microscopic model of the glass transition based on first principles was developed only a few years ago by GÖTZE and coworkers /1-4/. According to this the freezing-in of the liquid amorphous structure is a dynamic transition that can be idealized as the extreme case of atomic motion. That means, the self-diffusion coefficient becomes zero and therefore the viscosity goes to infinity. Characteristics of the glass transition are its non-equilibrium nature and the associated jump in compressibility, thermal expansion coefficient and specific heat. The ideal glass transition occurs if the system cannot relax for an infinitely long time. For finite measuring times, the transition occurs if the system cannot reach equilibrium during the characteristic time scale. Therefore the transition depends on the measuring time or frequency, and the dispersion region, where the measured properties change from the equilibrium value of a liquid to that of a glass, is shifted to higher temperatures for higher frequencies.

All excitations contribute to the specific heat, which reflects the enthalpy of a system. But to extract parameters characteristic for enthalpy relaxation from common dc-measurement of the specific heat is very complicated /5/, because of the nonlinear nature of the perturbation /6/. Therefore, the ac-measurement of specific heat by BIRGE and NAGEL /7/ has aroused considerable interest. Recently we have shown that the PAE is suitable for the study of frequency-dependent specific heat in the linear response regime /8/.

1. Experimental

The photoacoustic measurements are performed on the ionic salt melt $0.6KNO_3-0.4Ca(NO_3)_2$, a system that is often used as a model substance for comparison with computer simulation /9/ because it has nearly argon structure. The sample with a thickness of ca. 2mm is coated with a graphite layer (<3 μm) to realize the case of a thermally thick and optically strong absorbing sample in a gas-microphone cell. The molten salt itself does not absorb the incident visible light. During the measurements the ambient temperature of the supercooled melt is cooled down with ca. 2.5 mK/s from 390K to about 320K. Four or five values of modulation frequency in the range of either 5 to 30, or 50 to 500, or 300 to 1500 Hz, are stepped each minute during one cooling. For further details of sample preparation and measurement performance, see /8/. Typical results for PA amplitude and phase angle versus

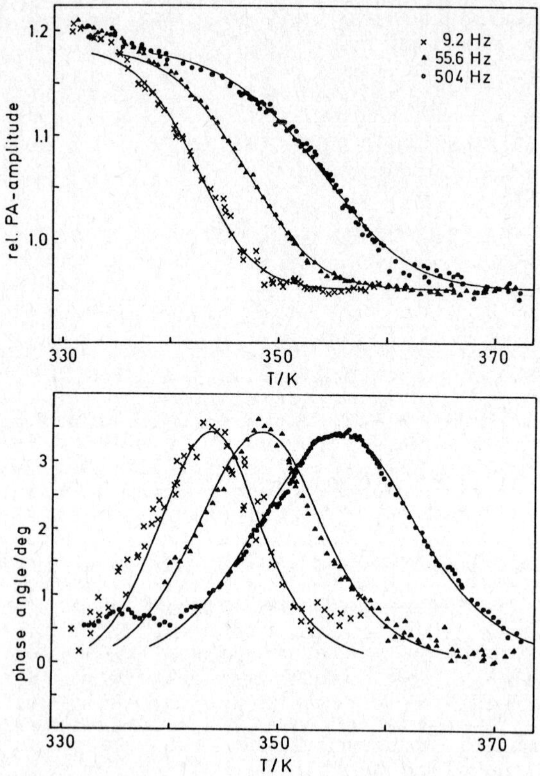

Fig.1: Relative PA amplitude and phase angle for various frequencies. Symbols: measured; lines: calculated with a temperature dependence (4), γ=0.661 and a KWW decay function, β=0.55.

temperature for different frequencies are shown in Fig.1. The temperature range where the PA amplitude changes from the approximately constant value in the liquid to a nearly constant value in the glass is defined as the "glass transition region" or "dispersion region". The curves are normalized to the constant high temperature values above this region.

2. PA Signal in the Glass Transition Region

For a thermally thick and optically strongly absorbing sample, the pressure oscillation in a gas-microphone cell is inversely proportional to the thermal effusivity of the sample, the square root of the conductivity λ and heat capacity C_p per unit volume. In the transition region, the frequency-dependent heat capacity can be written as a dynamical susceptibility $C_p(\omega)$, /10/, representing the linear response of enthalpy to variation in temperature. With a time-dependent correlation function $\Phi(t)$ that describes the decay of fluctuations according to the fluctuation-dissipation theorem, the complex heat capacity $C_p(\omega)$ is

$$C_p(\omega) = C_{p\infty} - (C_{po} - C_{p\infty}) \int_0^\infty e^{i\omega t} \frac{d\Phi(t)}{dt}\, dt \,, \tag{1}$$

where $C_{p\infty}$ and C_{po} are the constant values for high and low frequencies /11/. With such a $C_p(\omega)$ the PA amplitude $A(\omega)$ relative to the low-frequency limiting value A_o is

$$(\omega/\omega_o)A/A_o = C_p^{1/2}/(C_p'^2(\omega)+C_p''^2(\omega))^{1/4} = [(\gamma+\alpha N'(\omega))^2+\alpha^2 N''^2(\omega)]^{-1/4}, \quad (2)$$

and for the phase angle ϕ

$$\phi = \frac{1}{2}\arctan\frac{C_p''(\omega)}{C_p'(\omega)} = \frac{1}{2}\arctan\frac{\alpha N''(\omega)}{\gamma+\alpha N'(\omega)}, \quad (3)$$

with $\gamma=C_{p\infty}/C_{po}$, $\alpha=1-\gamma$. The normalized susceptibility N=N"+iN" is the Laplace transform of the time derivative of a decay function.

3. Results and Discussion

A transition temperature T_g is defined by the position of the maximum in the plot of phase angle shift versus temperature. For lower frequencies the dispersion region, and therefore T_g, is shifted to lower temperatures. A fit of the measuring frequencies ν_m versus transition temperature T_g to a scaling law

$$\nu_m = \nu_o\,[(T_g-T_o)/T_o]^{\mu} \quad (4)$$

is best for an ideal glass transition temperature T_o=323K and an exponent $\mu\approx 8.8$. As one can see from Fig.2, where $\log(T_g-T_o)$ is plotted versus $\log \nu_m$ for photoacoustic (X) and ultrasonic (◆) /12/ data, the exponent of the power law agrees very well, but the ideal glass transition temperature is slightly higher for the ultrasonic data (T_o=331K). The power-law dependence of the transition temperature is important, because the critical behaviour is a prediction of the mode-coupling theory /1-4/. But the exponent for the heat capacity of the real system $0.6KNO_3$-$0.4Ca(NO_3)_2$ is appreciably higher than the theoretically predicted μ~2. Considering a temperature dependence (4) of frequency or mean relaxation time, we can calculate temperature-dependent susceptibilities and can compare them with the measured data, according to (2,3). Many relaxation measurements can be interpreted by the Laplace transform of the time derivative of a "streched exponential" /13/, a decay function of the Kohlrausch-Williams-Watts (KWW) type /14,15/

$$\Phi(t) = \exp\{-t/\tau_o)^{\beta}\} . \quad (6)$$

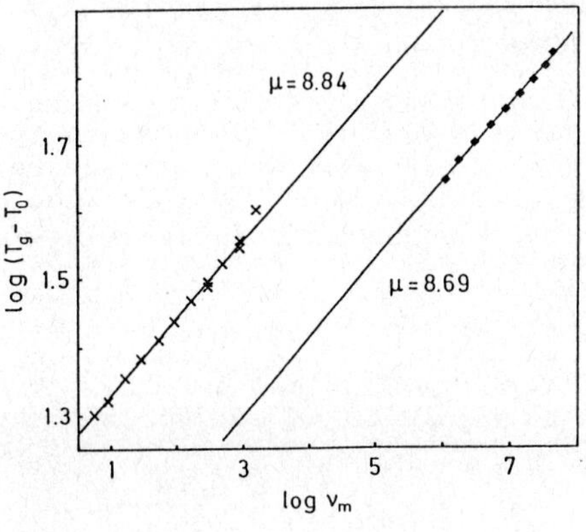

Fig.2: Log ν_m dependence of $\log(T_g-T_o)$ for heat capacity (PAE,X) and ultrasonic (/12/,◆) relaxation, (4)

The exponent β and a characteristic relaxation time τ_0 govern this correlation function. Assuming constant thermal conductivity, we can first determine the ratio γ, from the approximately constant high and low temperature values of PA amplitudes, and then β. The form for the susceptibility predicted by mode-coupling theory is about equivalent to that of a KWW-function /3/.

In conclusion, we get from our PA measurement of the enthalpy relaxation results that agree with ultrasonic and other measurements that are sensitive to density fluctuations and therefore to structural relaxation near the glass transition. Furthermore, the results are in qualitative agreement with the microscopic mode-coupling theory of the glass transition.

1. E.Leutheusser: Phys. Rev.A, 29, 2965 (1984)
2. U.Bengtzelius, W.Götze, A.Sjölander: J.Phys.C 17, 5915 (1984)
3. H.DeRaedt, W.Götze: J.Phys. C 19, 2607 (1986)
4. W.Götze: In Amorphous and Liquid Materials, ed.by E.Lüscher, G. Fritsch, G.Jacucci (M.Nijhoff Publishers, Dodrecht 1987) p.34
5. O.S.Narayanaswamy: J.Am.Ceram. Soc. 54, 491 (1971)
6. M.A.DeBolt, A.J.Easteal, P.B.Macedo, C.T.Moynihan: J.Am. Ceram.Soc. 59, 16 (1976)
7. N.O.Birge, S.R.Nagel: Phys.Rev.Lett. 54, 2674 (1985)
8. B.Büchner, P.Korpiun: Appl. Phys. B43, 29 (1987)
9. C.A.Angell, L.M.Torell: J.Chem.Phys. 78, 937 (1983)
10. R.Kubo: J.Phys.Soc.Japan, 12, 570 (1957)
11. N.O.Birge: Phys. Rev. B34, 1631 (1986)
12. R.Weiler, R.Bose, P.B.Macedo: J.Chem.Phys. 53, 1258 (1970)
13. J.Wong, C.A.Angell: In Glass structure by spectroscopy, (Dekker, New York 1976) chap. 11
14. R.Kohlrausch: Pogg.Ann.Phys.Chem. 72, 353 (1854)
15. G.Williams, D.C.Watts: Trans.Faraday Soc. 66, 80 (1970)

Photoacoustic Study of Critical Behaviour of Thermal Parameters in Liquid Crystal Phase Transitions

M. Marinelli, U. Zammit, F. Scudieri, R. Pizzoferrato, and S. Martellucci

Dip. Ingegneria Meccanica, 2° Università di Roma "Tor Vergata", Via Orazio Raimondo, I-00173 Roma, Italia

Application of the photoacoustic to the simultaneous characterization of the thermal diffusivity, specific heat capacity and thermal conductivity of a liquid crystal at a phase transition is presented. The first two show a critical decrease and increase, respectively, while the third one does not exhibit any critical behaviour.

A photoacoustic approach to the study of the behaviour of the thermal parameters near phase transitions of liquid crystals is presented. Such a technique is of great advantage since it enables a simultaneous determination of thermal diffusivity D, specific heat capacity c and, therefore, of the thermal conductivity k, versus temperature in all the liquid crystalline mesophases. Moreover in order to obtain a detectable photoacoustic signal, the rise of the sample surface temperature was estimated to be of the order 0.001 °C. This is very useful for the investigation of the behaviour of thermal parameters in the liquid crystal near the transition temperature, since a high temperature resolution is required for the determination of their critical behaviour.

Investigations have been carried out using a standard gas microphone lock detection arrangement. The heating source, which was modulated at 30 Hz, was a He-Ne laser operating at 3.39 μm, at which wavelength strong absorption of most of the liquid crystals occur due to the C-H bond. The experiment was performed for the smectic A - nematic phase transition for the 8CB liquid crystal, which is known to be near second order with a very small latent heat. The amplitude and phase of the photoacoustic signal were measured as a function of temperature at steps of 0.01 °C.

The sample thermal diffusivity, specific heat capacity and thermal conductivity have been calculated using the Rosencwaig and Gersho theory /1/ in the optically and thermally thick approximation, which is the case in our experimental conditions. In such an approximation the photoacoustic signal phase only depends /3/ on the parameter $p = \beta(\omega/2D)^{\frac{1}{2}}$ where ω and β are the sample chopping frequency and optical absorption coefficient respectively. The signal amplitude depends on both p and $q = (k_s c_s \rho_s / k_g c_g \rho_g)^{\frac{1}{2}}$ where ρ is the density; subscripts s and g refer to the sample and coupling gas respectively. The behaviour of each thermal parameter as a function of temperature can therefore be obtained from the photoacoustic signal amplitude and phase data.

Results are shown in Fig. 1. A sharp critical increase in the specific heat capacity and a critical decrease in the thermal diffusivity can be

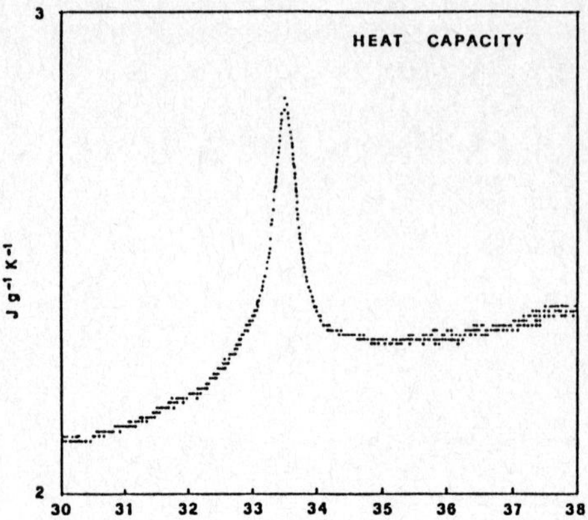

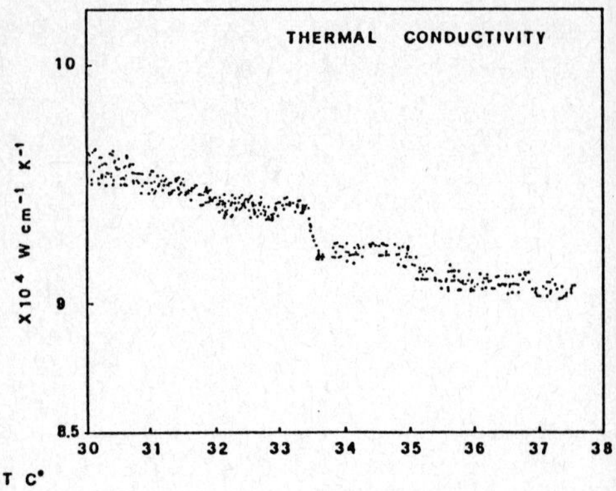

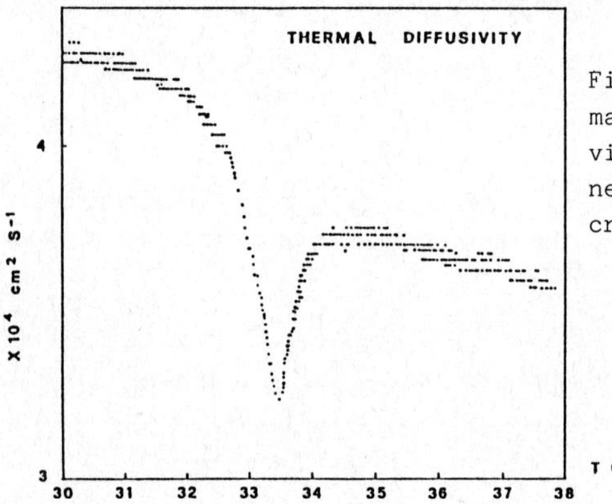

Fig. 1 Specific heat capacity, thermal conductivity and thermal diffusivity vs. temperature at the smectic A nematic phase transition of 8CB liquid crystal

seen, these two cancelling out in the thermal conductivity behaviour which only shows a discontinuity over the transition temperature. A peak value considerably lower than the one reported in the literature (4 - 4.5 J/g K) /2/ is evident for the specific heat capacity, together with a rounding-off effect for the points near the transition temperature. These two effects are due to the presence of a considerable amount of impurity in the material which, in this case, was used as delivered, with no further purifications. Critical exponents for the specific heat capacity α' ($T > T_{SN}$) and α ($T < T_{SN}$) were calculated not taking into consideration the points affected by the rounding-off, where T_{SN} is the smectic A nematic transition temperature. The results are $\alpha' = \alpha = 0.3$, which is in good agreement with the values reported in the literature. No data are available in the literature for comparison with the present ones for the thermal diffusivity and conductivity.
In conclusion, a new method for simultaneous measurements of thermal diffusivity, specific heat capacity and thermal conductivity at a liquid crystal phase transition is reported. Anomalies for thermal diffusivity and specific heat capacity have been found and critical exponents calculated, while the thermal conductivity shows only a discontinuity at the transition temperature.

REFERENCES

1. A. Rosencwaig, A. Gersho, J. Appl. Phys., 47 , 64 (1976)
2. J. Thoen, H. Marynissen, W. Van Dael, Phys. Rev. A, 5, 2886 (1982)
3. M. Marinelli, U. Zammit, F. Scudieri, S. Martellucci, J. Quartieri, F. Bloisi, L. Vicari, Nuovo Cimento D, in press

Study of Phase Transitions and Thermal Properties of PTS Single Crystals by the Photoacoustic Effect

Du Yinglei[1] *and Wu Baimei*[2]

[1]Fundamental Physics Centre, University of Science and Technology of China, Hefei, Anhui, People's Republic of China
[2]Department of Physics, University of Science and Technology of China, Hefei, Anhui, People's Republic of China

We report on the application of the photoacoustic effect to investigations of the phase transition in polybis-(p-toluene sulfonate) of 2,4-hexadiyne-1,6-diol(PTS)in a temperature range of 175—225K. The dependence of the PA signal amplitude and phase angle on the temperature was measured and is shown in Fig.1.

According to Korpiun and Tilgner's theory[1],one would expect the PA signal to become minimum and the phase angle to change from the maximum to the minimum at the first-order transition. This is because the heat produced by the adsorbed light is used as the required latent heat for the transition and therefore does not contribute to the PA signal[2,3]. The temperatures of the transition in heating and cooling have a shift about 1.4—9K [3]. In the case of a second-order transition, the PA signal and phase angle should exhibit a sudden jump at just the same temperature as the jump of the thermal parameters[4,5]. The shift of temperature introduced by the relaxation will not be obvious. From our experiments (see Fig.1) it is shown that the PA signal $Q_s(T)$ and shift of phase angle $\Delta\phi_s$ have a jump near 200K, and the shift of the temperature is not greater than 1K in heating and cooling. It seems that there exists a second-order phase transition near 200K[6].

From the R-G theory[7],in the case of an optically opaque and thermally thick sample the PA signal $Q_s(T)$ can be written as

$$Q_s(T)=AF(T)/[C_s(T)K_s(T)]^{1/2} , \tag{1}$$

where A is a coefficient independent of the temperature T; the function F(T) accounts for all the thermal properties of the cell and internal gas, and $C_s(T)$ and $K_s(T)$ are the specific heat

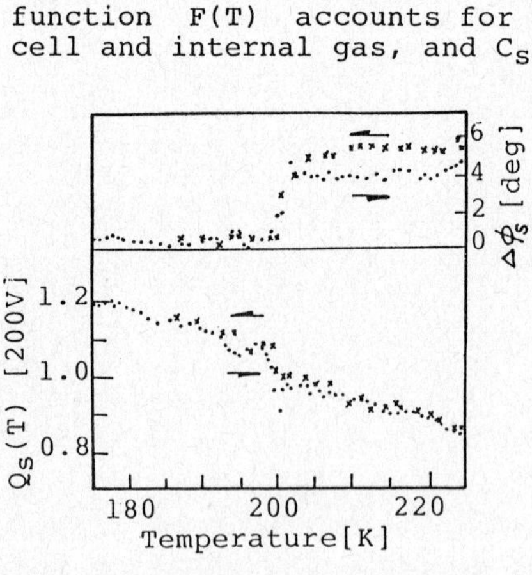

Fig.1 PA signal Q_s and shift of phase angle $\Delta\phi_s$ of PTS as a function of temperature T for He-Ne light illumination and chopping frequency of 30kHz....: heating rate 1K/min. ×××: cooling rate 2K/min.

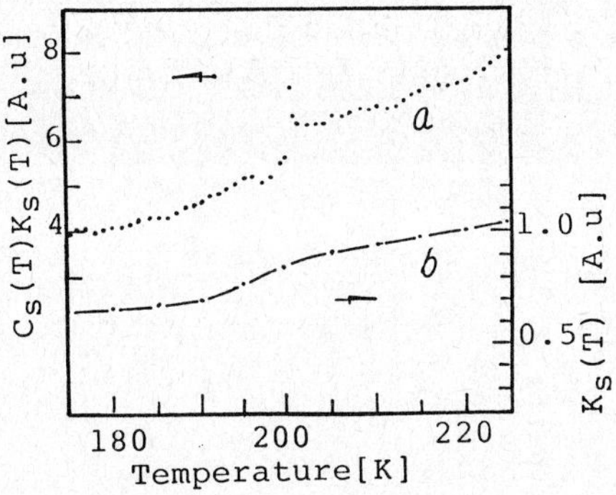

Fig.2 a: Temperature variation of $C_S(T)K_S(T)$ of PTS; b: Temperature variation of $K_S(T)$ of PTS found by dividing curve a by $C_S(T)$.

at constant pressure and the thermal conductivity of the sample respectively. The calibration of coefficient A and function F(T) was carried out with black carbon and quartz glass of known thermal properties. From (1), we get the quantity $C_S(T)K_S(T)$ as a function of temperature T (Fig.2 curve a). From the known specific heat $C_S(T)$ of PTS[8], the thermal conductivity $K_S(T)$ of PTS is deduced and shown in Fig.2, curve b.

The $C_S(T)K_S(T)$ of PTS has a sudden jump near 200K, and the increment of $C_S(T)K_S(T)$ in a temperature range of 5K around 200K is estimated to be about 5%. This is in agreement with DTA measurements[8]. We suggest that the jump of $C_S(T)K_S(T)$ probably comes from the contribution of the specific heat $C_S(T)$, which supports the existence of a second-order phase transition.

The thermal conductivity $K_S(T)$ of PTS is proportional to the temperature T (see Fig.2 curve b), which violates the 1/T law of nonmetallic crystal thermal conductivity at high temperature. This anomalous behavior is associated with the fact that the phonon mean free path becomes independent of the temperature with increasing degree of disorder of PTS[9].

References

1. P.Korpiun and R.Tilgner: J.Appl.Phys. 51, 6115(1980)
2. R.Florian, J.Pelzl, M.Rosenberg, H.Vargas and R.Wernhart: Phys. Stat. Sol. 48, K35(1978)
3. P.S.Bechthold, M.Campagna and T.Schober:Sol. Stat. Comm. 36, 225, (1980)
4. M.A.A.Siquira, C.C.Ghizoni, J.I.Vargas, E.A.Menezes, H.Vargas and L.C.M.Miranda: J.Appl. Phys. 51, 1403(1980)
5. J.Fernandez, J.Etrebarria, M.J.Tello and A.Lopez Echarri: J.Phys. D, Appl. Phys.16, 269(1983)
6. Du Yinglei and Wu Baimei: Acta Physica Sinica 36, 679(1987)
7. A.Rosencwaig and A.Gersho: J.Appl.Phys. 47, 64(1976)
8. I.Engeln and M.Meissner: J.Polym. Sci. 18, 2227(1980)
9. Du Yinglei and Wu Baimei: Acta Physica Sinica,To be published.

Part VIII

Thermal Wave Nondestructive Evaluation

Photothermal Imaging with Sub-100-nm Spatial Resolution

C.C. Williams and H.K. Wickramasinghe

IBM T.J. Watson Research Center, Yorktown Heights, NY 10598, USA

Abstract: A new high resolution thermal microscope has been demonstrated, capable of imaging thermal fields with sub-100-nm resolution. It is based upon a non-contacting near field thermal probe. The thermal probe consists of a thermocouple sensor on the end of a tip with 100 nm dimensions. The probe tip is scanned over a surface under servo control to maintain a constant gap between tip and surface. Simultaneously, the surface is periodically heated by an external source (current or laser) and the resultant ac temperature variation is measured. Two electronic images are produced, one of the surface topography and the other of the ac temperature distribution on the sample surface. The latest thermal images indicate a lateral resolution below 100 nm and a temperature sensitivity below 1 millidegree. These measurements demonstrate thermal microscopy with a spatial resolution which is an order of magnitude greater than has been achieved by other techniques to date. The basic theory and experimental work will be described. Thermal images of surfaces heated by electrical current and optically absorbed power will be presented.

1. Introduction

Over the past years, activity in the field of high resolution photoacoustic and photothermal imaging has depended on the use of focused acoustic detection [1] or focused optical beams [2-5] to achieve temperature mapping with lateral resolution on the micrometer scale. The use of a freely propagating focused optical beam (far field approach) in photothermal detection creates a limit to the spatial resolution of such an imaging system. This limit is on the order of the optical wavelength. More recently, the principles of near field imaging, originally demonstrated by Scanning Tunneling Microscopy (STM) [6], have been applied in the area of thermal imaging [7,8]. The resolution of a near field imaging system is not limited by the wavelength used, but rather the physical size of the probe used in the measurement. Since such probes can be made with dimensions on the order of 100 nm, resolution well below the optical wavelength limit can be achieved. In this paper, we report the demonstration of near field photothermal imaging with a spatial resolution below 100 nm.

2. Near Field Thermal Mapping

The near field mapping of surface temperature be achieved by scanning an ultra small temperature sensor across the surface of a heated sample. As the tip moves from a region of high temperature to a region of lower temperature, the temperature of the probe is modulated, and a signal is generated. This signal is then detected and simultaneously stored with the two dimensional coordinates of the probe location.

The spatial resolution of the resultant temperature map is limited by the greater of either the tip size or the gap between tip and surface. If the tip size and the gap are below 100 nm, the surface temperature can be mapped with comparable resolution.

The key element of this thermal microscope is the ultra small thermal probe which provides the sensitivity and the spatial resolution necessary to achieve high resolution temperature mapping. Such a probe was built. It consists of a conical tip with a thermocouple sensor at its end. See Figure 1. As shown schematically in the figure, a thermocouple sensor is produced at the tip by the junction of the dissimilar inner and outer conductors. An insulator separates these conductors in all areas remote from the tip. The thermocouple junction produces a temperature dependent voltage which can be sensed at the other end of the probe across the two conductors. This voltage provides the means for remotely sensing the local tip temperature as it is scanned laterally across the solid surface. The thermal probe tips can be made to have dimensions on the order of 100 nm. The minimum detectable change in tip temperature is less than 1 millidegree in a 100 Hz bandwidth.

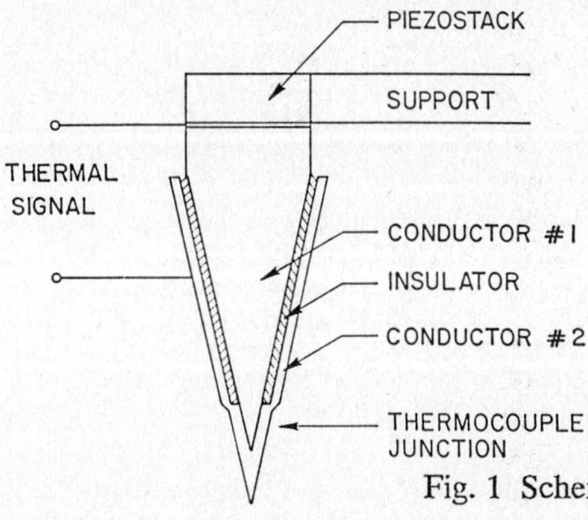

Fig. 1 Schematic diagram of the thermocouple probe.

To avoid the crashing of the thermal probe tip into the sample, a servo loop similar to that of the STM [6] is used. In contrast to the STM, however, this servo loop adjusts the vertical height of the probe to maintain constant the thermal flux between tip and sample, rather than electron tunneling current. To avoid problems with thermal variations in the ambient, the tip is vertically vibrated. The resultant ac temperature variation is maintained constant by the control loop. Under servo control, the system acts as a scanned surface profiler (STP) [7].

For temperature mapping, the sample is heated at a frequency outside the bandwidth of the profiler servo loop. If the sample temperature variation is small compared to the temperature difference between tip and sample, then simultaneous and independent images of surface topography and temperature can be obtained. This arrangement can be seen in Figure 2. The profiler feedback loop is locked to the gap modulation frequency f_1 and the sample temperature is modulated at a frequency f_2, where f_2 is far outside of the operating bandwidth of the servo loop, centered about f_1.

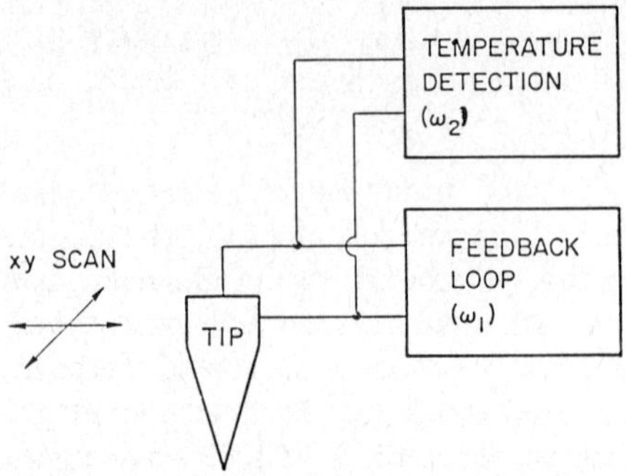

Fig. 2 Schematic diagram of the system used for simultaneous measurements of surface topography (servo control loop) and temperature mapping.

To facilitate imaging, the sample is mounted on a piezoelectric scanner which provides up to 100 micrometers of travel in the two lateral dimensions. The positioning and vibration of the tip are achieved by a single piezoelectric stack with 20 micrometers of travel and 3 kHz resonance frequency. The mechanical resonance of this vertical piezo limits the data aquisition under feedback control to 1 kHz.

3. Experimental Demonstration

To demonstrate the temperature mapping capabilities of the thermal probe under servo control, topographical and temperature images were obtained on an aluminum film heated via electrical current [8]. In one region on the film, a small particle was found. The topographical profile of a one micrometer square area of the film with the particle is seen in Figure 3a. The measurement indicates that the particle has a thickness of 30 nanometers. The film was heated with 20 mA of current at a 4 kHz frequency, creating a Joule heating at 8 kHz. A temperature map was made of the same area by measuring the voltage out of the thermal probe with a lock-in amplifier

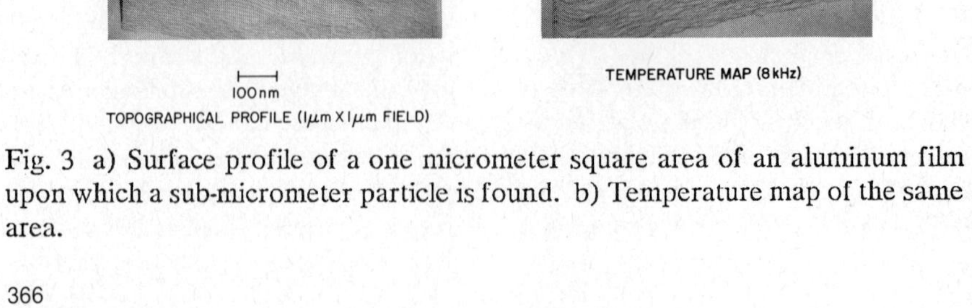

Fig. 3 a) Surface profile of a one micrometer square area of an aluminum film upon which a sub-micrometer particle is found. b) Temperature map of the same area.

at 8 kHz, with the profiler feedback loop closed. The temperature map can be seen in Figure 3b. Since the topographical and temperature maps were not taken simultaneously in this case, a slight drift is seen between the two images. The temperature change due to the presence of the particle is less than 1 millidegree centigrade. The lateral resolution in the temperature map appears to be less than 100 nm.

To demonstrate the principles of near field phtothermal imaging, a sample with high spatial frequencies was generated by electron beam lithography. The sample consisted of a 200 nm period electron resist grating atop a chromium film on a glass substrate (See Figure 4). A 1 mW He-Ne pump beam was focused through the glass substrate onto the back surface of the chromium film, where a portion of the beam is absorbed. The spot size was approximately 5 micrometers. The chromium layer thickness was 100 nm, with a negligibly small transmission at He-Ne wavelength. The laser was acousto-optically modulated at a frequency of 2.4 kHz. The thermal probe was brought close to the front surface of the sample, and under servo control was scanned laterally. The vertical vibration frequency of the tip was 1.1 kHz.

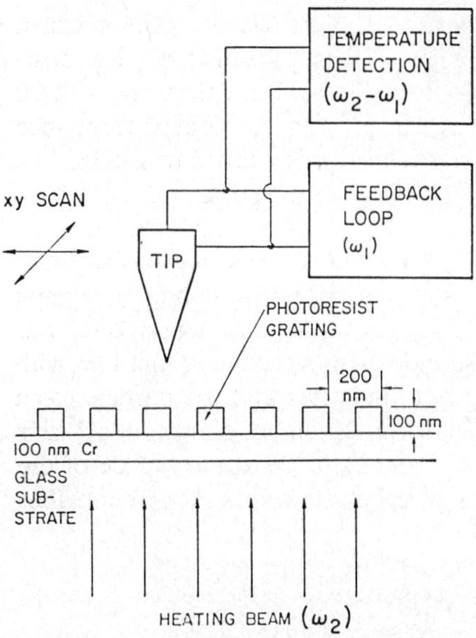

Fig. 4 System diagram describing the experimental configuration used for photothermal mapping of a 200 nm period electron resist grating. Optical power is modulated at a frequency f_2.

Figure 5a displays the topographical profile of the surface. Because the tip size is comparable to the space between the lines of the grating, only a 10 nm variation in height is seen in the topographical profile. The actual grating height was 100 nm. The simultaneous temperature map of the same area is shown in Figure 5b. The image displays the thermal signal at the difference frequency (1.3 kHz). The temperature variation over the field is on the order of 10^{-2} degrees centigrade. This

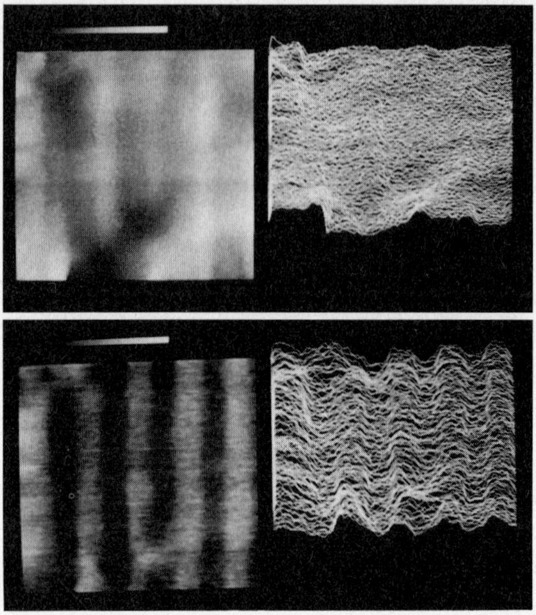

Fig. 5 a) Topographical profile of the resist grating as measured by the closed loop feedback signal. Area is 900 by 900 nm square.

b) Temperature map taken simultaneously of the same area.

variation is small compared to the constant background measured, which was more than an order of magnitude larger. As can be seen in figure 5b, the thermal variation at the grating period is easily resolved, and infact, the sharpness of the rising edges in the temperature map indicate a resolution well below 100 nm. It was observed that the presence of the heating beam did modify the appearance of the topographical map. This was most likely due to the fact that the ac temperature rise was not very small compared with the temperature difference between tip and sample.

4. Summary

The principles of near field imaging have been demonstrated in a new high resolution thermal microscope. The thermal interaction between an ultra small thermal sensor and a solid has been used to map the temperature distribution of surfaces with sub-100-nm lateral resolution and sub-1-millidegree temperature resolution. The temperature mapping and surface profiling has been demonstrated on samples heated by electric current and by laser absorption. We believe that high resolution thermal and photothermal microscopy will find application in biological spectroscopy with sub-100-nm resolution, thermal characterization of materials, and film thickness measurements.

5. Acknowledgements

We would like to thank Steve Rishton for the preparation of the electron resist grating sample, Bob Jackson for his assistance with the scan mechanics, and Russ Allen and Jim Doyle for their help in the production of the thermal probes.

6. References

[*1*] H.K. Wickramasinghe, R. Bray, V. Jipson, C. Quate, and J.R. Salceds, Appl. Phys. Lett. *33,* 923 (1978).

[*2*] S. Ameri, E.A. Ash, V. Neuman, and C.R. Petts, Electron. Lett. *17,* 337 (1981).

[*3*] M.A. Olmstead, N.M. Amer, S. Kohn, D. Fournier and A.C. Boccara, Appl. Phys. A *32,* 141 (1983).

[*4*] C.C. Williams, Appl. Phys. Lett. *44,* 1115 (1984).

[*5*] A. Rosencwaig, J. Opsal, W.L. Smith, and D.L. Willenborg, Appl. Phys. Lett. *46,* 1013 (1985).

[*6*] G. Binnig, H. Rohrer, Ch. Gerber, and E. Weibel, Phys. Rev. Lett. *49,* 57 (1982).

[*7*] C.C. Williams and H.K. Wickramasinghe, Appl. Phys. Lett. *49,* 1587 (1986).

[*8*] C.C. Williams and H.K. Wickramasinghe, 1986 Ultrasonics Symposium Proceedings, 393, ed. by B.R. McAvoy, (IEEE, New York 1986).

Mechanisms for the Generation and Scattering of Sound and Thermal Waves in Thermoacoustic Microscopes

L.D. Favro, S.M. Shepard, P.K. Kuo, and R.L. Thomas

Department of Physics, Wayne State University, Detroit, MI 48202, USA

1. Theory

Thermoacoustic microscopes [1] generate sound by depositing energy and momentum in a solid by means of an incident laser or particle beam. These beams may be either pulsed or periodically modulated and can generate sound either through thermal expansion or by direct momentum transfer to the lattice. In general, thermal waves and both longitudinal and transverse sound waves are generated. These various waves can then scatter from thermal and elastic inhomogeneities in the solid and, by several different mechanisms, produce an acoustic signal. This acoustic signal is detected by the microscope's acoustic transducer and used to produce an image. In this paper we describe a rigorous method for calculating the relative amplitudes of the various contributions to the acoustic signal and for assessing their importance as sources of contrast in images.

The elementary textbook picture of sound as consisting solely of a periodic propagating displacement of the solid, with no other associated phenomena, is adequate for most purposes, but it is too naive a picture when thermoacoustic effects are to be studied. This is because the temperature, T, and local displacement, $\boldsymbol{\xi}$, of the solid are coupled to each other through effects associated with thermal expansion. These effects are described by what might be called "the thermoacoustic wave equations" [2],

$$\kappa\nabla^2 T - \rho c(\partial T/\partial t) - T_0\alpha(3\lambda + 2\mu)(\partial/\partial t)\nabla\cdot\boldsymbol{\xi} = \nabla\cdot\mathbf{S}_e \quad \text{and} \tag{1}$$

$$\mu\nabla^2\boldsymbol{\xi} + (\lambda + \mu)\nabla\nabla\cdot\boldsymbol{\xi} - \rho(\partial^2\boldsymbol{\xi}/\partial t^2) - \alpha(3\lambda + 2\mu)\nabla T = \nabla\cdot\boldsymbol{\sigma}_e \ . \tag{2}$$

Here κ is the thermal conductivity, ρ the density, c the specific heat, T_0 the ambient absolute temperature, α the thermal expansion coefficient, λ the Lamé constant, and μ the shear modulus. The vector quantity $\mathbf{S}_e$ is the incoming beam's energy flux, and the tensor quantity $\boldsymbol{\sigma}_e$ is the beam's momentum flux. The divergences of these two fluxes respectively represent a heat source and a force density and are the sources of three kinds of waves (longitudinal acoustic, transverse acoustic, and thermal) in bulk material. Preliminary discussions of the origin and behavior of these waves have been presented elsewhere [3-5]. Here we present exact solutions to Eqs. (1) and (2) which describe the generation and scattering of all types of acoustic waves, both in the bulk and on the surface, for realistic sample geometries.

The thermoacoustic equations are solved by imagining a four-dimensional space in which the coordinates are the temperature and the three components of the displacement. It turns out to be convenient to scale the variables in such a way that the four-dimensional matrix-differential operator which represents these two equations is Hermitian. We have chosen to do this by defining a four-vector $\mathbf{V}(\mathbf{r})$ whose first coordinate is a quantity τ given by

$$\tau = \frac{\kappa T}{i\omega T_0(\lambda + 2\mu)} \tag{3}$$

(ω is the frequency of a periodic source or a frequency component of a pulsed source), and whose remaining three coordinates are just the components of $\boldsymbol{\xi}$ with no scale factors. The right hand sides of Eqs. (1) and (2), divided by κ and $(\lambda + 2\mu)$ respectively, are similarly taken to form a source four-vector $\mathbf{U}(\mathbf{r})$. The two equations can then be written symbolically as

$$\mathbf{O}\bullet\mathbf{V}(\mathbf{r}) = \mathbf{U}(\mathbf{r}) \,, \tag{4}$$

where $\mathbf{O}$ represents the matrix of differential operators on the left sides of (1) and (2). The operator $\mathbf{O}$ is Hermitian in the sense that it equals its transpose with the signs of all derivatives reversed. One then defines a Green's function $G_0(\mathbf{r})$ for a bulk material (with no boundaries) by the equation,

$$\mathbf{O}\bullet G_0(\mathbf{r}) = -\mathbf{1}\,\delta(\mathbf{r}) \,, \tag{5}$$

where $\mathbf{1}$ represents the unit four-matrix. So defined, $G_0(\mathbf{r})$ itself is a 4x4 matrix given by

$$G_0(r) = \frac{1}{4\pi(1 - R_1R_2)} \begin{bmatrix} \dfrac{-k_0^2 R_1}{R_2}\left(\dfrac{e^{iqr}}{r} - R_1R_2\dfrac{e^{ikr}}{r}\right) & -R_1\nabla\left(\dfrac{e^{iqr}}{r} - \dfrac{e^{ikr}}{r}\right) \\ R_1\nabla\left(\dfrac{e^{iqr}}{r} - \dfrac{e^{ikr}}{r}\right) & \dfrac{1}{k_0^2}\left\{\nabla\nabla\left[R_1R_2\dfrac{e^{iqr}}{r} - \dfrac{e^{ikr}}{r} + (1 - R_1R_2)\dfrac{e^{ik_sr}}{r}\right] + \right. \\ & \left.(1 - R_1R_2)k_s^2\dfrac{e^{ik_sr}}{r}\right\} \end{bmatrix} , \tag{6}$$

where

$$R_1 = \frac{\alpha(3\lambda + 2\mu)}{(\lambda + 2\mu)(q^2 - k_0^2)} \quad \text{and} \tag{7}$$

$$R_2 = \frac{-i\omega\alpha(3\lambda + 2\mu)T_0k_0^2}{\kappa(q^2 - k_0^2)} \,. \tag{8}$$

In these equations q represents the exact wave number for a thermal wave, k the exact wave number for the longitudinal sound wave, k_s the exact wave number for the transverse sound wave, and k_0 the longitudinal sound wave number with the thermal expansion coefficient, α, artificially set equal to zero. The derivations of the expressions for q and k have been presented elsewhere. [5] It should be noted that both the thermal wave (wave vector q) and the longitudinal wave (wave vector k) contribute both to the temperature (first row of $G_0(\mathbf{r})$), and to the displacement (rows two through four of $G_0(\mathbf{r})$). The shear wave contributes only to the displacement. The first column of the matrix, $G_0(\mathbf{r})$, represents the response, both thermal and acoustic, of a bulk medium to a heat source. It contains both longitudinal and thermal waves, but no shear waves because the heat source is not coupled to the shear-wave mode in Eq. (1). The next three columns of $G_0(\mathbf{r})$ represent the thermal and acoustic response of the medium to force densities in the x-, y-, and z-directions respectively. These columns contain all three types of waves, because a force density can couple to any of them. In the remainder of this paper we will, for mathematical simplicity, assume that the force density is only in the z-direction. This corresponds physically to the situation in which the incoming beam is in that direction.

To facilitate the treatment of sample boundaries which follows, we will replace the various spherical waves in the matrix, $G_0(\mathbf{r})$, by their cylindrical-wave expansions through the identity

$$\frac{e^{ikr}}{r} = i\int_0^\infty pdp\, J_0(pr)\, \frac{e^{ip_1|z|}}{p_1} , \tag{9}$$

where $J_0(pr)$ is a cylindrical Bessel function, and where $p_1 = (k^2 - p^2)^{1/2}$. The symbols p_2 and p_3 will be used in a similar fashion for shear and thermal waves respectively. Also, the symbol "r" is used here with two meanings: on the left it is the polar radial coordinate, and on the right it is the axial radial coordinate. A longitudinal acoustic wave with cylindrical symmetry can be written in the form

$$\xi = \int_0^\infty p\, dp\, a(p)\, [\hat{\mathbf{r}}\, p\, J_1(pr) - i\,\hat{\mathbf{z}}\, p_1\, J_0(pr)]\, e^{ip_1z} \quad \text{and} \quad T = R_2 \int_0^\infty p\, dp\, a(p)\, J_0(pr)\, e^{ip_1z} , \tag{10}$$

where a(p) is an arbitrary wave amplitude. There are two cylindrically polarized shear waves, one with radial polarization, and one with torsional polarization,

$$\xi = \int_0^\infty p\, dp\, b_r(p)\, [\, -i\,\hat{\mathbf{r}}\, p_2 J_1(pr) \ + \ \hat{\mathbf{z}}\, pJ_0(pr)]\, e^{ip_2z} , \tag{11}$$

and

$$\xi = \int_0^\infty p\, dp\, b_\theta(p)\, p\hat{\boldsymbol{\theta}}\, J_1(pr)\, e^{ip_2z} . \tag{12}$$

Finally, there is a cylindrically symmetric thermal wave,

$$T = \int_0^\infty p\, dp\, c(p)\, J_0(pr)\, e^{ip_3z} \quad \text{and} \quad \xi = R_1 \int_0^\infty pdp\, c(p)\, [\hat{\mathbf{r}}\, p\, J_1(pr) \ - \ i\,\hat{\mathbf{z}}\, p_3 J_0(pr)]\, e^{ip_3z} . \tag{13}$$

With the incoming momentum in the z-direction, as was assumed above, the torsional-wave terms in $G_0(\mathbf{r})$ will not contribute. One then has to deal with three types of waves, the longitudinal wave characterized by a(p), the radially polarized shear wave characterized by $b_r(p)$, and the thermal wave characterized by c(p). These can be imagined to form a three-dimensional vector defined by

$$\mathbf{x} = \begin{bmatrix} a(p) \\ b_r(p) \\ c(p) \end{bmatrix} , \tag{14}$$

whose components can be found by subsituting the identity (9) into Eq. (6). At this point we should make notice of the fact that Eq. (9), and hence the expression for $G_0(\mathbf{r})$, contains the absolute value of z while Eqs. (10-13) contain z itself. This difference arises from the fact that the Green's function represents the response to a point source, and therefore must produce waves propagating in opposite directions on opposite sides of the source. A corollary to this is that the relative phases of the different waves above and below the source depend on the number of times the exponentials in Eq. (6) are differentiated with respect to z. Since the response to a momentum source (columns two through four) contains one more gradient operator than the response to an energy source (column one), the phases of the waves produced by momentum sources are different from those produced by energy sources.

In either spherical or cylindrical coordinates, Eq. (6) represents an exact solution to the thermoacoustic equations in an infinite medium. A physical sample is, of course, not an infinite medium, so that $G_0(\mathbf{r})$ is not an appropriate Greens function for a real sample. The first step toward a more realistic model is to introduce the effect of the surface through which the incident beam has just passed. This effect is strongly dependent on the boundary conditions at that surface, but can be described in general by specifying how the waves represented symbolically by Eq. (14) scatter at the surface. This will be illustrated by using what is perhaps the simplest boundary, namely a thermally insulated and elastically unconstrained surface. We imagine the surface to be located at $z = 0$ and the point source in Eq. (5) to be moved to a point specified by $\mathbf{r}' = (0,0,z_0)$. The boundary conditions to be applied at $z = 0$ are then

$$\mathbf{n}\cdot\nabla T = 0 \qquad \text{and} \tag{15}$$

$$\mathbf{n}\cdot[-\lambda\mathbf{1}\nabla\cdot\boldsymbol{\xi} - \mu(\nabla\boldsymbol{\xi} + (\nabla\boldsymbol{\xi})^{\dagger}) + \alpha(3\lambda + 2\mu)T\mathbf{1}] = 0 \ . \tag{16}$$

These conditions are the appropriate ones for the vacuum conditions often found in thermoacoustic microscopes. One can satisfy these equations exactly by adding to the displaced Green's function, $G_0(\mathbf{r},z_0)$, an additional set of three waves (longitudinal, shear, and thermal) which propagate away from the surface and into the sample. These waves represent the waves which propagated from the source to the surface and scattered there. If we imagine the boundary conditions to be applied to the symbolic form of the waves given by Eq. (14), they will be satisfied exactly if

$$\mathbf{M}_- \mathbf{x}_{inc} + \mathbf{M}_+ \mathbf{x}_{scatt} = 0, \qquad \text{where} \tag{17}$$

$$\mathbf{M}_\pm = \begin{bmatrix} \mp p_1R_2 & 0 & \mp p_3 \\ (2p^2 - k_s^2) & \mp 2ipp_2 & R_1(2p^2 - k_s^2) \\ \pm 2ipp_1 & -(2p^2 - k_s^2) & \pm 2ipp_3R_1 \end{bmatrix} , \tag{18}$$

and where the incident wave, $\mathbf{x}_{inc}$, is obtained by combining Eqs. (6) and (9). An expression for the scattered wave, $\mathbf{x}_{scatt}$, can then be obtained by multiplying Eq. (17) by $\mathbf{M}_+^{-1}$. The result can be written in terms of a scattering matrix $\mathbf{S}$ $(= -\mathbf{M}_+^{-1}\mathbf{M}_-)$ as

$$\mathbf{x}_{scatt} = \mathbf{S}\ \mathbf{x}_{inc} \ . \tag{19}$$

In order to display an explicit expression for the matrix $\mathbf{S}$, we introduce the determinant, D, of $\mathbf{M}_\pm$,

$$D = (p_3 - p_1R_1R_2)(2p^2 - k_s^2)^2 + 4p^2p_1p_2p_3(1 - R_1R_2), \tag{20}$$

in terms of which $\mathbf{S}$ can be written as

$$\mathbf{S} = \frac{1}{D}\begin{bmatrix} \{4p^2p_1p_2p_3(1-R_1R_2) - (p_3+p_1R_1R_2)(2p^2-k_s^2)^2\} & -4ipp_2p_3(2p^2-k_s^2) & -2R_1p_3(2p^2-k_s^2)^2 \\ -4ipp_1p_3(2p^2-k_s^2)(1-R_1R_2) & \{4p^2p_1p_2p_3(1-R_1R_2) - (p_3-p_1R_1R_2)(2p^2-k_s^2)^2\} & -4iR_1pp_1p_3(2p^2-k_s^2)(1-R_1R_2) \\ 2R_2p_1(2p^2-k_s^2)^2 & 4iR_2pp_1p_2(2p^2-k_s^2) & \{4p^2p_1p_2p_3(1-R_1R_2) + (p_3+p_1R_1R_2)(2p^2-k_s^2)^2\} \end{bmatrix} . \tag{21}$$

We can now symbolically write the exact thermoacoustic Green's function for a semi-infinite sample as

$$G(\mathbf{r},z_0) = G_0(\mathbf{r},z_0) + \mathbf{S}\cdot G_0(\mathbf{r},z_0), \tag{22}$$

where the operator **S** is to be thought of as the application of the matrix of Eq. (21) to the cylindrical waves in $G_0(\mathbf{r},z_0)$.

The physical significance of the solution given by Eqs. (21) and (22) merits some attention. First, because it is appropriate to a semi-infinite solid, it may be used physically to describe situations in which the sample is so thick that reflections from the back side are negligble, or alternatively, for pulsed beams (where it represents just one Fourier component), in which the reflection from the back side has not yet returned. The term $G_0(\mathbf{r},z_0)$, of course, just represents the waves initially generated by the point source at z_0. The second term is somewhat more interesting and is perhaps best discussed in terms of the properties of the matrix **S**. The three diagonal elements of **S** just represent the reflection coefficients for longitudinal sound, transverse sound, and thermal waves at the sample surface. Depending on the depth of the source and on the relative phases of the waves on the two sides of the source, the reflected waves they represent may interfere either constructively of destructively with the initially generated waves. This interference is essentially different for waves which are generated by an energy source and waves which are generated by a momentum source. (See the discussion of phases on different sides of the source above.)

The off-diagonal elements of **S** represent mode-conversion coefficients, i.e. they describe the conversion of the energy from one kind of incident wave into energy in waves of the other two types as a result of their interaction with the surface. Since none of the off-diagonal terms vanish, it is clear that any one mode will scatter into three terms upon hitting the surface. The determinant, D, in the denominator of **S** is also of interest. The zeros of D correspond to singular points, or poles, of **S** in the complex p-plane. These poles represent the possibility of having surface wave modes. In fact, the pole of **S** which is closest to the real axis is responsible for Rayleigh waves. If the residue at that pole is non-zero, Rayleigh waves will be present in the sample. The remaining poles are far from the real axis and hence represent heavily damped surface waves. In fact, since Eq. (21) is an exact solution to the thermoacoustic equations, it describes <u>all</u> of the acoustic and thermal waves which can be generated by a "point source" of normally incident particles on a thick sample. The extension to a realistic source profile will be delayed until we have described the solution to the equations for a finite slab.

The symbolic description of the acoustic and thermal waves given in Eq. (14), together with the scattering matrix **S** which describes their interaction with the surface, provides a convenient method of constructing an exact solution to the thermoacoustic equations in a slab. We again place our point source at $\mathbf{r}' = (0,0,z_0)$ and place one surface at $z = 0$ and another at $z = w$. For purposes of definiteness we will imagine that the positive z-direction is down and that the particle or laser beam is incident on the surface at $z = 0$. We will also be a little more detailed about the handling of the waves propagating in the two directions from z_0. The quantity **x** in Eq. (14) will now be regarded as representing the waves described by Eqs. (10), (11), and (13) <u>without</u> the exponential dependence on z. That dependence will be described by multiplication of **x** by a propagation matrix $\phi(z)$ given by

$$\phi(z) = \begin{bmatrix} e^{ip_1 z} & 0 & 0 \\ 0 & e^{ip_2 z} & 0 \\ 0 & 0 & e^{ip_3 z} \end{bmatrix} . \tag{23}$$

(If one wishes, one can include frequency-dependent phenomenological attenuation factors for each of the three waves by changing the wave numbers in the exponentials in this matrix.) This will enable us to keep track of the waves propagating back and forth between the two surfaces of the sample. The wave initially propagating in the negative z-direction (i.e. "up") in our description of the directions above will be designated by $\mathbf{x}_{up}$ and the one initially propagating in the positive z-direction will be designated by $\mathbf{x}_{down}$. This distinction between up and down is important because the relative phases of the up and down components are different for waves originating in an energy source and those originating in a momentum source.

We now focus our attention on the wave specified by $\mathbf{x}_{up}$ as it leaves the source at z_0. As it propagates to the surface at $z = 0$, it picks up a factor of $\phi(z_0)$, and then, at the surface it gets multiplied by the scattering matrix $\mathbf{S}$. Propagation to the bottom surface at $z = w$ causes the introduction of a factor $\phi(w)$. At the bottom surface the directions of the incoming and outgoing waves are reversed relative to what they were at the top surface. The result of this is that the scattering at the bottom is described by $\mathbf{S}^{-1}$ rather than $\mathbf{S}$. The return trip to the top surface is again described by a factor of $\phi(w)$ and the process then repeats itself. Since the sequence of factors for each round trip from top, to bottom, to top again is always the same, the result is a geometric series whose sum, $[1 - \phi(w)\mathbf{S}^{-1}\phi(w)\mathbf{S}]^{-1}$, is easily obtained. A similar process can be used to handle $\mathbf{x}_{down}$ and the two contributions can be added to obtain total wave at any point in the sample. Again, it is necessary to make a distinction between the expression for the sum of the waves above the source and below it, so that we must give two expressions for the result. Above the source we obtain

$$\Sigma = [\phi(-z) + \phi(z)\mathbf{S}]\,[1 - \phi(w)\mathbf{S}^{-1}\phi(w)\mathbf{S}]^{-1}[\phi(z_0)\mathbf{x}_{up} + \phi(w)\mathbf{S}^{-1}\phi(w - z_0)\mathbf{x}_{down}] \tag{24}$$

and below it we find

$$\Sigma = [\phi(z - w) + \phi(w - z)\mathbf{S}^{-1}]\,[1 - \phi(w)\mathbf{S}\phi(w)\mathbf{S}^{-1}]^{-1}[\phi(w - z_0)\mathbf{x}_{down} + \phi(w)\mathbf{S}\phi(z_0)\mathbf{x}_{up}] \;. \tag{25}$$

The three component vector, Σ, is to be interpreted like $\mathbf{x}$ in Eq. (14) except that it now refers to the sum of all the waves in the sample and has the dependence on z given explicitly. The sign of z in the first factor of each of these expressions determines the direction of propagation of the waves it represents. In order to calculate the actual temperature and displacement in the sample, one must attach the appropriate combinations of Bessel functions (as given by Eqs. (10), (11), and (13)) to each of the three components Σ and integrate over p. So interpreted, Eqs. (24) and (25) represent an exact solution to the thermoacoustic equations in a slab of thickness w with a point source of either energy or momentum at a depth z_0. Being an exact solution they must contain all possible waves which can be generated by such a source. The presence of the bulk waves is obvious. Rayleigh waves, and other single-sided surface waves, again arise from the poles of the scattering matrix $\mathbf{S}$ (or equivalently $\mathbf{S}^{-1}$). Double sided surface waves, e.g. Lamb waves, arise from the poles of the factor representing the sum of the geometric series describing the multiple scattering between the two surfaces.

The point-source solution represented by Eqs. (24) and (25) can easily be generalized to realistic source geometries by integrating over the source point. The special case of a Gaussian transverse beam profile can be accommodated by use of the identity

$$\iint \frac{e^{\frac{-(x'^2+y'^2)}{R^2}}}{\pi R^2} J_0\{p[(x - x')^2 + (y - y')^2]\}\, dx'\, dy' = J_0[p(x^2 + y^2)^{1/2}]\, e^{\frac{-p^2R^2}{4}} \;. \tag{26}$$

Thus the generalization to a Gaussian transverse profile can be accomplished by simply inserting the exponential factor from the right hand side of Eq. (26) into each of the integrals in the expressions for the cylindrical waves. Its effect is to cut off the integrals for high radial wave numbers p. The depth profile of the source depends on the absorption profile of the incoming beam of particles, and is quite different for photons, electrons and ions. Since the absorption profile for photons is a simple exponential, and the dependence of the Green's function on the source depth z_0 is also exponential, the depth integration for photons is trivial. Electons and ions have more complicated energy and momentum absorption profiles which tend to peak near the ends of their ranges. For these particles one must use approximate analytic expressions and perform the integrations numerically.

In order to describe the scattering from defects in the sample, we will return to the point-source solution which we will now call $G(\mathbf{r},\mathbf{r}')$. It is evident that $G(\mathbf{r},\mathbf{r}')$ acts as a Green's function which satisfies the equation

$$\mathbf{O}\cdot G(\mathbf{r},\mathbf{r}') = -\mathbf{1}\ \delta(\mathbf{r} - \mathbf{r}') \ . \tag{27}$$

The strategy we have used for calculating the scattering from defects is the standard one which uses the Green's function for a perfect sample as the basis for an approximation scheme. Equations (4) and (27) are combined into an integral equation of the form

$$\mathbf{V}(\mathbf{r}) = -\int G(\mathbf{r},\mathbf{r}')\cdot\mathbf{U}(\mathbf{r}')\ d^3r' + \int [G(\mathbf{r},\mathbf{r}')\cdot\mathbf{O}'\cdot\mathbf{V}(\mathbf{r}') - \mathbf{V}(\mathbf{r}')\cdot\mathbf{O}'\cdot G^{\dagger}(\mathbf{r},\mathbf{r}')]\ d^3r \ , \tag{28}$$

where the prime on **O** indicates that the differential operators act on the primed variables and the dagger on G indicates the transpose of the matrix. The integration over the source term $\mathbf{U}(\mathbf{r}')$ on the right hand side has been described above. The second integral on the right hand side can be converted to a surface integral over the surface of the sample and over any defect, say a crack, which may be present. The integrals over the sample surface vanish (this is essentially a statement of the Hermiticity of **O**), and what remains is the integral over the defect. Since the resulting integral equation ordinarily does not lend itself to exact solutions, one usually substitutes some approximation to $\mathbf{V}(\mathbf{r}')$ into the surface integral term. For instance, the Born approximation consists of the substution of the first (source) term on the right hand side into the second term repetitively to generate an expansion in some small parameter. Since the Green's function used is exact for a perfect sample, the accuracy of the approximation to the scattered wave depends only on the degree of accuracy of the approximation to $\mathbf{V}(\mathbf{r}')$ on the right hand side, and the extent to which one is able to accurately model the boundary conditions on the defect.

2. Acknowledgements

This work was sponsored by the Army Research Office under contract No. DAAG29-84-K-0173, and by the Institute for Manufacturing Research, Wayne State University. One of us, S.M. Shepard, wishes to acknowledge fellowship support from the Allied Signal Engineered Materials Research Center.

3. References

1. J. Opsal and A. Rosencwaig: In J. Appl. Phys. **53**, 4240 (1982).

2. W. Nowacki: In Thermoelasticity (Pergamon Press, Oxford 1962).

3. L.D. Favro, P.K. Kuo, M.J. Lin, L.J. Inglehart, and R.L. Thomas: In Proc. 1984 IEEE Ultrasonics Symposium, 629 (1984).

4. L.D. Favro, P.K. Kuo, and R.L. Thomas: In Review of Progress in Quantatative NDE, Ed. D.O. Thompson and D. Chimenti, Vol. 5A, 439 (Plenum 1986).

5. L.D. Favro, P.K. Kuo, S.M. Shepard, and R.L. Thomas: In Proc. 1986 IEEE Ultrasonics Symposium, 399 (1986).

The Nature of Harmonic Heat Flow

L.C. Aamodt

The Johns Hopkins University, Applied Physics Laboratory,
Laurel, MD 20707, USA

The development of photothermal theory and methodology has produced an increased interest in harmonic heat flow. This interest has been enhanced through the application of these methods to the non-destructive evaluation and characterization of materials via imaging. This paper discusses various aspects of harmonic heat flow and their relationship to thermal parameters. Specific attention is paid to various regimes within which the harmonic temperature pattern is dominated by either a single thermal parameter or by a particular combination of thermal parameters.

The most commonly recognized characteristic of harmonic heating is its localized nature, which provides thermal-wave techniques with the spatial sensitivity needed for effective imaging. Another important characteristic is its narrow temporal frequency spectrum which provides the high sensitivity obtained in photothermal measurements. A third characteristic, of practical interest, is the ability to experimentally move between various operational regimes via changes in the modulation frequency.

If mechanical coupling is ignored, heat flow is governed by the three-dimensional heat equation,

$$\kappa\nabla^2 T(x,y,z,t) - C\, \partial T/\partial t(x,y,z,t) = -H(x,y,z,t), \qquad (1)$$

where κ is the thermal conductivity; C, the thermal capacity per unit volume; ω, the angular harmonic frequency; and H, the rate of heat generated per unit volume within the sample. κ and C are assumed to be independent of spatial coordinates and time. Furthermore, we assume that H can be factored into the product of a spatial and temporal term (i.e., $H(x,y,z,t) = H(x,y,z)\, H(t)$). This implies that all heat generation is in phase, or alternately, that the transit time of the heating agency used is instantaneous on the temporal scale we are concerned with.

Since our interest is only in harmonic heat flow created by heat generation within the sample (and not by harmonically varying temperature constraints at the system boundaries), we set $H(t) = \cos(\omega t)$ [or alternately $H(t) = \mathrm{Re}\, \exp(j\omega t)$, where the designation, Re (Real), will usually be suppressed]. Equation 1 now becomes

$$\nabla^2 T(x,y,z) - j\omega C/\kappa\, T(x,y,z) = -H(x,y,z)/\kappa, \qquad (2)$$

where T has also been factored into a product of spatial and temporal terms and the term $\exp(j\omega t)$ has been suppressed.

Supported by the U. S. Navy under contract N00039-87-C-5301.

Having suppressed the temporal terms, all remaining terms are purely spatial. We here consider ω to be a parameter, and not a variable, so we are effectively working at a particular point in the temporal frequency domain.

Several features of the harmonic heat equation, $(\nabla^2 - j\omega C/\kappa)T = -H/\kappa$ are of importance. First, we note that T is not a temperature in the usual sense. Rather, it is the spatially dependent amplitude of a temperature wave oscillating in time. As it oscillates, the "harmonic" temperature becomes negative as well as positive. Thus, it must always be "riding" on another temperature function. Specifically, T is the modulated "excess temperature" created by thermal modulation.

A second feature to be noted is that ω and C never appear separately in the heat equation, but only as the product, ωC. Because of this we can consider this product to be a single entity, which we call C_e, the "thermal capacity per unit time," or alternately, the "effective thermal capacity." This quantity is important, not only theoretically but also operationally, since it indicates that the relative importance of C and κ can be altered experimentally by changing the modulation frequency. As the ratio, $\omega C/\kappa$, changes, regimes appear in which capacitance or conductivity, or some specific combination of these parameters, becomes dominant.

An intuitive comparison of C and C_e can be made by considering a simple case of a sample element heated by a constant heating source and small enough so that heat conduction can be ignored. Solving the heat equation for the time-dependent case, $T = (H/C)t$. Thus $1/C$ indicates the rate of heating. The larger the C, the slower the rate. The equivalent result in the harmonic case is $T = (H/j\omega C)$ (where j indicates the phasing). The "heating rate" here is $1/\omega C$. The higher the frequency, the smaller the period of heating, and thus the smaller the amplitude of the thermal wave created by a given heat input with C fixed.

The heat equation contains two constants, κ and C. These are considered to be the basic thermal parameters. This selection is not fundamental, however, and any pair of independent combinations of κ and C could also be used, for example, the thermal diffusivity ($\alpha = \kappa/C$) and the thermal effusivity ($e = \sqrt{(\kappa C)}$), or alternately, the "absolute" thermal diffusion length [$\mu = \sqrt{(\kappa/\omega C)} = \sqrt{(\kappa/C_e)}$] and the effective thermal effusivity, [$\epsilon = \sqrt{(\omega C\kappa)} = \sqrt{(\kappa C_e)}$]. The latter two terms both involve the effective thermal capacity [$\kappa = e\sqrt{\alpha} = \epsilon\,\mu$; $C = e/\sqrt{\alpha} = (\epsilon/\mu)/\omega$].

For harmonic heat flow, the last set of thermal parameters frequently simplify notation, and provide intuitive insight. Using these parameters, the heat equation becomes

$$\nabla^2 T(x,y,z) - j/\mu^2\, T(x,y,z) = -(E/\mu)H(x,y,z), \qquad (3)$$

where $E = 1/\epsilon$.

The local, spatial dependence of the harmonic temperature on particular thermal parameters is directly apparent from (1), $(\nabla^2 - jC_e/\kappa)T = -H/\kappa$. At points where no heat is generated (H=0), the temperature depends exclusively on the thermal diffusivity (or thermal diffusion length) except for possible dependencies that might be introduced by boundary conditions. At other points in the sample where the temperature or flux are spatially uniform ($\nabla^2 T \rightarrow 0$), $\omega C/\kappa$ dominates ∇^2 and the temperature depends predominantly on the effective thermal capacity. Under the opposite condition, i.e., low-frequency modulation and rapid spatial variations in the temperature, ∇^2 can dominate $\omega C/\kappa$ and the temperature then depends predominantly on thermal conductivity. This dependence, however, does not necessarily result in the functional regimes described above, since these regimes are produced by the integrated effect of these local dependencies over sample surfaces or volumes.

When optical excitation is used, the situation is further complicated, and regimes can often be identified as being regimes with or without photothermal (or photoacoustic) saturation. Regimes can also be created when the spatial extent of the harmonic heating is comparable to the thermal diffusion length in the material being heated. As an illustration, when localized excitation is used, photothermal detection (which involves heat flow both in a sample material and in the gas layer above the sample) has five identifiable regimes, and photoacoustic detection has four.

The above generalizations can be made more quantitative in some thermal systems. For example, if a thermally thick planar sample interfacing with a highly insulating media is heated internally, the surface temperature can be written as the product $E\,\tau(x,y,z,\mu)$, where $\tau(x,y,z,\mu)$ depends upon thermal parameters only through the thermal diffusion length, μ. For this special case, and any other case where this type of factoring is possible, we can define a local thermal dependence (LTD) on the thermal parameters. This quantity is a measure of how sensitive the temperature is to small changes in ω, κ, or C at a particular value of ω, κ, and C, and is defined as $\mathrm{LTD} = (\partial \log_e(\tau)/\partial \log_e(\mu))_\epsilon$.

If $\log_e \tau(\mu)$ is expanded in a Taylor series around $\tau(\mu_0)$ and only the first two terms of the expansion are retained, $T(x,y,z) = \tau(x,y,z,\mu_0)/\epsilon(\mu/\mu_0)^{\mathrm{LTD}} = [\tau(x,y,z,\mu_0)/(\mu_0)^{\mathrm{LTD}}]\,[(\kappa)^{(\mathrm{LTD}-1)/2}/(\omega C)^{(\mathrm{LTD}+1)/2}]$. When LTD has the values, -1, 0 and 1, T has a special dependence on thermal properties. When LTD = -1, T depends exclusively on κ; when LTD = 0, it depends exclusively on $\sqrt{\kappa\omega C}$; and when LTD = 1, T depends exclusively on ωC. It is of experimental interest to note that each of these three regimes has a different frequency dependence, specifically ω^0, $\omega^{-1/2}$, and ω^{-1}. Intermediate frequency dependence indicates the proportional effect of κ and C.

Evaluating LTD for a planar sample interfacing with a good insulator and heated by a Gaussian optical excitation beam with radius, R,

$$\mathrm{LTD} = \frac{\mathrm{Re}\iint \exp(-\lambda^2 R^2/2)\ \mu(\partial D/\partial\mu)\ \exp(j\ \vec{\lambda}\cdot\vec{\rho})\ d\vec{\lambda}}{\mathrm{Re}\iint \exp(-\lambda^2 R^2/2)\ D\ \exp(j\ \vec{\lambda}\cdot\vec{\rho})\ d\vec{\lambda}} \tag{4}$$

for surface heating, and

$$\mathrm{LTD} = \frac{\mathrm{Re}\iint \exp(-\lambda^2 R^2/2)\ D\ \mu\ (\partial D/\partial\mu)(2+\beta D)/(1+\beta D)^2\ \exp(j\ \vec{\lambda}\cdot\vec{\rho})\,d\vec{\lambda}}{\mathrm{Re}\iint \exp(-\lambda^2 R^2/2)\ D^2/(1+\beta D)\ \exp(j\ \vec{\lambda}\cdot\vec{\rho})\ d\vec{\lambda}} \tag{5}$$

for exponential heating.

Here λ is the spatial frequency; D is the effective thermal diffusion length = $1/\sqrt{(\lambda^2 + j/\mu^2)}$; and β is the optical absorption coefficient.

For broadband heating, LTD = 1 for surface heating, indicating that the temperature is locally dependent on the effective thermal capacity only. For exponential heating, the local thermal dependence varies from one to zero, ranging from total dependence on ωC for small values of the product $\beta\mu$ to exclusive dependence of $\kappa\omega C$ for large values of $\beta\mu$. (See Fig. 1.)

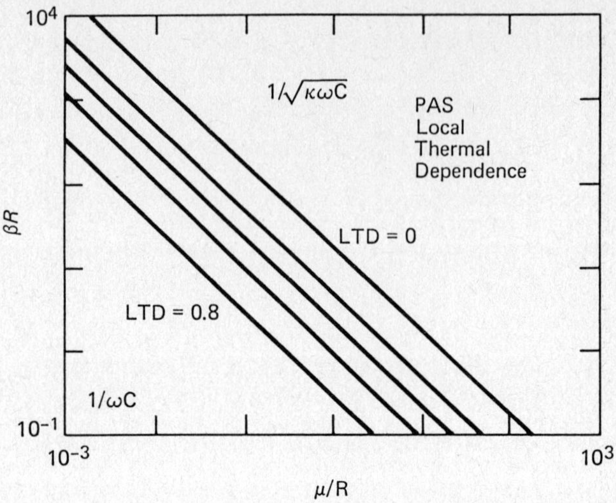

Fig. 1. Local thermal dependence for broad beam excitation.

New, Rigorous Inversion Imaging Technique for Thermal Waves

J. Burt and A. Gaudefroy

Physics Department, York University, 4700 Keele Street, Downsview, Ontario, M3J 1P3, Canada

1. Introduction

XY thermal wave scans of material samples can produce images directly if the energy source chopping frequency is high enough. Application is frequently found for this in integrated circuit examination [1]. If the chopping frequency is low on the other hand, blurred images are produced [2]. To deblur these images requires numerical processing of the data. Little success has been attained in this endeavour to date. Rather, modelling of the physical process and comparison of the modelled and measured scan signals have been the main emphasis.

The imaging methods reported previously for thermal wave measured data are what Sabatier [3] has called direct imaging methods. In such methods, a model is first postulated which will represent the underlying surface structure and then calculations are carried out and the results compared with the actual experimental data. If there is agreement, then it means we have made a good guess about the structure and proved that such a mathematically modelled structure can duplicate actual measurements. Although it sounds as if we must know the answer before we ask the question, in inversion and imaging theory such an a priori modelling is not necessarily a bad thing. Indeed, it merely makes use of our prior knowledge. In the case where the model does not reproduce the experimental results, the disagreement may suggest how to modify the model so that we get a better representation for the image. Such a procedure of repeated iterations over initial guesses is generally successful in tomography and is implicit in the ART algorithm technique [4] used in that field. In many cases the procedure succeeds even when the initial guess is a poor one.

One of us chose a model consisting of an aluminum sample with a hole [5] and used ray theory to see whether it could predict the received data [2]. The same approach was successful in modelling thermal wave data from a thin steel blade [6]. A second example is that given by Thomas et al. [7] who used a Green's function manipulation in modelling the photoacoustic cell in thermal wave scanning over samples with steps.

More satisfactory than the solution of the direct imaging problem is the solution of the inverse problem. This approach requires that we make no presupposition about the underlying structure in a sample but that we merely process the received experimental thermal wave scan data and produce an image of what lies under the surface. So far, no one has inverted thermal wave scan data. We wish to outline how a new technique may be applied to such data and how it should eventually produce good inverted images.

Before discussing the new technique, we should point out two important works on thermal inversion, though not specifically on the thermal wave technique. The first is the interesting article by Bouc et al. [8] where the general problem of inversion of thermal data from an elastic solid is discussed in detail and applied to imaging of a crack. The second article of particular importance is the mathematical analysis of Cannon and Esteva [9], where the condition of uniqueness in inversion is discussed for the heat equation.

2. Theory

The imaging theory we describe is an application of Gaudefroy's [10-19] theory of optimization, which is that we reconstruct the original wave field by means of point sources making use only of the data gathered by receivers. Clearly, the more points, the better the image.

Let n be the number of replacement sources and F the thermal wave field. F satisfies a Helmholtz equation:

$$\nabla^2 F - \frac{i\omega}{\chi} F = \Sigma \text{ sources}, \qquad (1)$$

where ω is the angular frequency of the chopping, χ is the thermal diffusivity, and Σ represents the net effect of both the sources and the boundary conditions. Equation (1) can be found in the photoacoustic literature and in Carslaw and Jaeger [20].

Next, we write the equation obeyed by the replacement sources in terms of the field R_i which each source produces:

$$\nabla^2 R_i - \frac{i\omega}{\chi} R_i = \delta(\underline{x} - \underline{x}_i), \qquad i = 1,..,n ; \qquad (2)$$

here δ is the Dirac delta function, $\underline{x}$ the vector space coordinates, and $\underline{x}_i$ the coordinates of the replacement source.

Both eqn.(1) and eqn. (2) can be generalized beyond the form shown. In fact, they are convolution equations in the sense of distribution theory, but, for simplicity we have not written them in that form. Because they are convolution equations, we can use for the R_i the Green's functions for an infinite domain. These are readily available and the boundary conditions will be simply handled by the convolution product. Otherwise, in complicated geometries it is not easy to find a Green's function.

The Huyghen's equations which determine the position in space of the optimal set of replacement sources and the source intensities have the form:

$$HN(R_1,...R_i,...R_n) = HN(R_1,...F,...R_n), \qquad i = 1,..n \quad . \qquad (3)$$

Each of the n equations has the same left-hand side while in the right-hand side the R_i functions are replaced by F. The HN themselves are ratios of Gram determinants. HN is evaluated as

$$HN = \frac{\det\left(\frac{\partial V_{R_i}}{\partial x} \; .... \; \frac{\partial V_{R_n}}{\partial x}\right)}{\det\left(V_{R_i} \; , ... \; V_{R_n}\right)} ,$$

where $V_{R_j} = \begin{bmatrix} (R_i, R_j) \\ \vdots \\ (R_n, R_j) \end{bmatrix}$, $\frac{\partial V_{R_j}}{\partial x} = \begin{bmatrix} \left(\frac{\partial R_i}{\partial x}, R_j\right) \\ \vdots \\ \left(\frac{\partial R_n}{\partial x}, R_j\right) \end{bmatrix}$.

The Huyghens equations (3) are nonlinear equations in the coordinates of the replacement sources. Solving them tells us where to position the sources to form an image. Furthermore, the equations can be rewritten in a form to show what the replacement source intensities must be. We do not give that development here, it will appear in a forthcoming publication.

The theory outlined here applies equally well to any physical system described by a linear partial differential equation. We have already applied the technique to 5 different physical systems, of which 3 are well known in imaging: (1) thermal waves, (2) acoustic waves, (3) electrostatic field. Thus, we are able to employ in this article on thermal waves, techniques which were confirmed by laboratory measurements using an electrostatic sheet model [10] or by the problem of acoustic waves reflecting from a uniform, surface layer [11].

3. Results and Conclusion

Two methods were used to solve the Huyghens equations: the direct method and the source-drive method. The latter method calculates the difference between the intensities of the replacement sources found by a set of linear equations related to the denominator terms of eqn.(3) and the intensities derived from the numerator terms. Curves of minimum intensity difference are plotted on the XY plane (we have studied only 2-D problems, so far). It can be shown that where the intensity difference is a minimum is a location of a replacement source. The XY plane represents a cross-sectional cut through the sample being studied. We used a 1 cm x 1 cm aluminum sample whose decay length is 1 mm for 20Hz chopping. For the case of $n = 2$, i.e., 2 equivalent sources, we get one curve for one set of 6 receivers. If we use another set of 6 receivers, we get another curve which intersects the first two points. The system we have modelled is the laser spot (1 point) and a point reflector (the second replacement point) and these two points are found by the curve intersections. Equivalently, we may regard the point reflector instead as the source image which represents the effect of a reflecting layer parallel to the sample surface at half the image depth. Since the actual coordinates depend on the spacing of the 20 x 20 grid, the location accuracy is only within 20%. We then feed these coordinates into a non-linear equation solving program (the direct method) which solves eqn.(3) directly, and thus obtain the location of the replacement sources with only a 3% error.

In spite of the success with 2 points, we have not been able to proceed immediately to add further points and obtain higher resolution in our image. For one thing, even with $n = 2$ we do not always obtain convergence using the direct method if the initial values are not properly chosen. Also the prefiltering accomplished by the source-drive method depends on the grid size used. Our goal is to re-develop eqn.(3) in a form which exploits their quasi-linearity so that we may proceed immediately with the direct method without having to pick good initial values. Further, as n becomes greater than 2 the computing time for the source-drive method becomes prohibitive.

We have found satisfactory resolution of eqn.(3) for an equation mimicking the acoustic Helmholtz equation [10] with a single replacement source. In seismic work this is interesting since it is the earthquake location problem. For this case, pairs, rather than sixes, of receivers are taken and for a uniform medium it turns out that there is a simple, even geometric construction, to locate the single replacement source. A given receiver pair produces a circle, so 3 pairs produce 3 circles which intersect in a unique point. When we proceed to the true acoustic Helmholtz equation, the circles become families of sinuous circles and the determination of a true minimum in the source-drive method is more difficult. In comparison, the thermal wave case is less complicated, probably because of the severe damping of thermal waves. However, we have now included the direct method, which is important since it calculates only the replacement source points and does not produce spurious minima.

The method is not confined to a single physical system so that imaging analogues may be used in the laboratory. For example, a new algorithm to solve the Huyghens equations could be tested in an acoustic tank and then applied to a thermal wave theoretical study by changing only the form of the Green's function in the computer program.

In conclusion, the Huyghens number equations present a new and rigorous solution to image inversion. It is a solution which produces images by successive, best approximations. At present the image quality is rudimentary but we expect to correct this in the near future.

References

1. A. Rosencwaig: J. de Phys. 44, C6-437--C6-452 (1983)
2. G. Busse: Appl. Phys. Lett. 35, 759-763 (1979)
3. P.C. Sabatier: Applied Inverse Problems (Springer, Berlin 1978)
4. M.B. Katz: Questions of Uniqueness and Resolution in Reconstruction from Projections (Springer, Berlin 1978)
5. J.A. Burt: Can. Assoc. Physicists Annual Congress (1982)
6. J.A. Burt: J. de Phys. 44, C6-453--C6-457 (1983)
7. R.L. Thomas, J.J. Pouch, Y.H. Wong, L.D. Favro, P.K. Kuo, A. Rosencwaig: J. Appl. Phys. 52, 1152-1158 (1980)
8. B. Nayroles, R. Bouc, H. Caumon, J.C. Chezeaux, E. Giacometti: Int. J. Engng. Sci. 19, 929-947 (1981)
9. J.R. Cannon, S.P. Esteva: Inverse Problems 2, 395-403 (1986)
10. A. Gaudefroy, J.A. Burt: Canadian Society of Exploration Geophysicists National Convention, Calgary (1987)
11. J.A. Burt, A. Gaudefroy: 16th International Symposium on Acoustical Imaging, Chicago, (1987)
12. A. Gaudefroy: C.R. Acad. Sc. Paris, 296B, 1139-42 (1983)
13. A. Gaudefroy: Acoustics Letters 4, 145-9 (1981)
14. A. Gaudefroy: Acoustics Letters 4 136-44 (1981)
15. A. Gaudefroy: C.R. Acad. Sc. Paris 290A, 67-9 (1980)
16. A. Gaudefroy: C.R. Acad. Sc. Paris, 290B, 405-406 (1980)
17. A. Gaudefroy: C.R. Acad. Sc. Paris 290B, 187-9 (1980)
18. A. Gaudefroy: C.R. Acad. Sc. Paris 288B, 379-82 (1979)
19. A. Gaudefroy: C.R. Acad. Paris, 288B, 405-6 (1979)
20. H.S. Carslaw, J.C. Jaeger: Conduction of Heat in Solids (Clarendon, Oxford 1959)

Image Distortion in Optical-Beam-Deflection Imaging

L.C. Aamodt, J.C. Murphy, and J.W. Maclachlan

The Johns Hopkins University, Applied Physics Laboratory,
Laurel, MD 20707, USA

Images obtained using mirage (or optical-beam-deflection (OBD)) detection are frequently interpreted using a simplified theory which assumes that the probe beam has an infinitesimal cross-section, and that its path just grazes the sample surface. Under these assumptions, the probe beam deflection component, S_n, normal to the sample surface is nearly proportional to the average temperature along the probe beam path (normal mode) and the deflection component, S_t, parallel to the sample surface is proportional to the average transverse temperature gradient along the probe beam path (transverse mode).

At times, the near proportionality of S_n is overlooked, and the photothermal signal is viewed as being proportional to the average temperature along the probe beam path.

This simple interpretation is attractive since it makes intuitive reasoning about heat flow easy, but the question arises as to its validity under laboratory, rather than ideal, conditions. In this paper we investigate the nature of the distortion introduced by actual laboratory conditions as an aid in determining the validity of using the simplified theory under actual working conditions.

Two types of images can be acquired using OBD detection. A "detailed thermal map" of the sample surface (in the vicinity of a particular point) can be obtained by locally heating that point and monitoring the photothermal signal as the probe beam is moved relative to the heated point. Alternately, a "standard image" of the sample surface can be obtained by fixing the relative orientation of the heating source and probe beam, and sweeping these over a matrix of points on the sample surface, heating each point in turn. Plotting the photothermal signal measured at each matrix point (using pseudo color or a suitable gray scale to indicate intensity) provides the thermal image.

Let $T(x,y,t)$ be the temperature on a sample surface and $T_a(x)$ its average value along the projection of the probe beam path on the sample surface. If the gas above the sample surface (air) is a good insulator relative to the sample, then, matching boundary conditions at the sample/gas interface, the two probe-beam-deflection components are

$$S_n = A/(2\pi)^2 \iint \tilde{T}_a(\lambda,\omega)\, \tilde{W}_n\, \tilde{R}_e \exp[j(\lambda x+\omega t)]\, d\lambda\, d\omega, \quad (1)$$

$$S_t = A/(2\pi)^2 \iint \widetilde{dT_a/dx}(\lambda,\omega)\, \tilde{W}_t\, \tilde{R}_e \exp[j(\lambda x+\omega t)]\, d\lambda\, d\omega, \quad (2)$$

where $\tilde{W}_n(\lambda,\omega,z_o) = \exp(-z_o/D)/D$ and $\tilde{W}_t(\lambda,\omega,z_o) = \exp(-z_o/D)$ are distortion terms introduced by elevating the probe beam above the sample surface and

Supported by the U. S. Navy under contract N00039-87-C-5301.

$\tilde{T}_a(\lambda,\omega)$ is the spatial-temporal frequency spectrum of $T_a(x,t)$. [$\tilde{R}_e(\lambda,\omega,z_0) = \exp(r_p^2/2d^2)\ \mathrm{erfc}[(r_p/D - z_0/r_p)/\sqrt{2}]$ is the distortion factor introduced by an extended probe beam. Its effects are not considered here, but will be the subject of a future publication.] For small variations in temperature, A is constant. (Integrals without limits are integrated between $-\infty$ and ∞.)

In (1) and (2), z_0 is the elevation of the probe beam above the sample surface; r_p, the probe beam radius; λ, the spatial frequency; ω, the angular temporal frequency; κ, the gas thermal conductivity; C, the gas thermal capacity (per unit volume); $\alpha=\kappa/C$, the gas thermal diffusivity; $d = \sqrt{(\alpha/j\omega)}$, the complex thermal diffusion length in the gas; $1/D = \sqrt{(\lambda^2 + 1/d^2)}$, the reciprocal effective complex thermal diffusion length in the gas; and x, the spatial coordinate on the sample surface transverse to the projection of the probe beam path.

When an infinitesimal probe beam just grazes the sample surface, $\tilde{W}_t=1$ and $\tilde{W}_n=1/D$. Integrating (1) and (2), $S_t = A\ dT_a/dx(x,t)$ and $S_n = T_{distorted}(x,t)$ where $\tilde{T}_{distorted} = \tilde{T}_a/D = \tilde{T}_a \sqrt{(\lambda^2 + 1/d^2.)}$ Here D^{-1} can be viewed as a distortion operator converting T_a into $T_{distorted}$. This distortion is inherent in mirage detection, and comes from the basic character of the thermal lens deflecting the probe beam. It does not disappear when $z_0=0$. (This deflection is proportional to the vertical slope of the gas temperature regardless of probe beam elevation.)

Spatial features in T_a that are broader than a gas thermal diffusion length are spatially unaltered since in the integration, $\tilde{T}_a$ cuts off at some value of λ small enough so that λ^2 is insignificant compared with $1/d^2$. Features narrower than a gas thermal diffusion length are distorted. (See Fig. 1.)

The distortion created in a detailed thermal mapping by non-idealized conditions can be studied by convoluting (1) and (2):

$$S_n = A \iint T_a(x-u,t-\tau)\ W_n(u,\tau)\ du\ d\tau,$$

$$S_t = A \iint dT_a/dx(x-u,t-\tau)\ W_t(u,\tau)\ du\ d\tau\ .$$

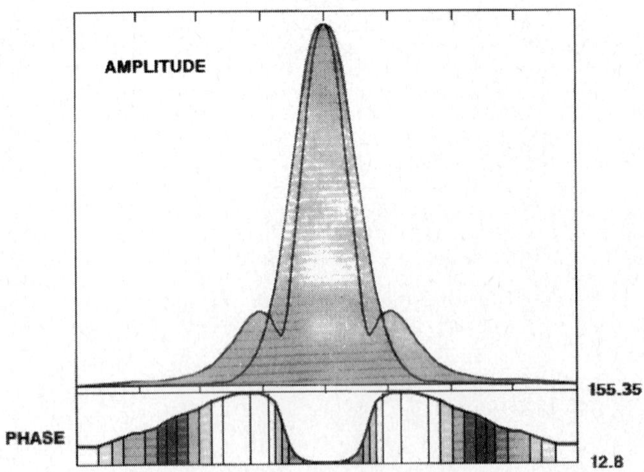

Fig. 1. Superposition of $T_{distorted}$ and T_a (assumed to be gaussian with radius d/2)

Under ideal conditions $W_n(u,\tau)$ and $W_t(u,\tau)$ would be delta functions in space and time. S_n would then map the temperature pattern, $T_a(x,t)$ and S_t, the gradient, $dT_a/dx\ (x,t)$. Under non-idealized conditions, $W_n(u,\tau)$ and $W_t(u,\tau)$ act as "smearing functions" distorting the idealized patterns. Evaluating these factors,

$$W_t\ (u,\tau) = (\pi\ z_0\ T_0)^{-1}\ (T_0/\tau)^2\ \exp[T_0(1+m^2)/\tau],$$

$$W_n\ (u,\tau) = W_t\ (2(T_0/\tau)-1)\ ,$$

where $T_0 = z_0{}^2/4\alpha$ and $m = x/z_0$. (See Fig. 2.) As z_0 decreases both smearing functions sharpen and in the limit approach delta functions. The peak of each function decreases rapidly with elevation, decreasing as $z_0{}^{-3}$ for the transverse-mode and $z_0{}^{-4}$ for the normal-mode. In general, temporal smearing is small, but spatial smearing can be significant. The peak delay is $z_0{}^{-2}/(q\alpha)$, where $q=7 + \sqrt{(33)}$ (normal-mode) and 8 (transverse-mode).

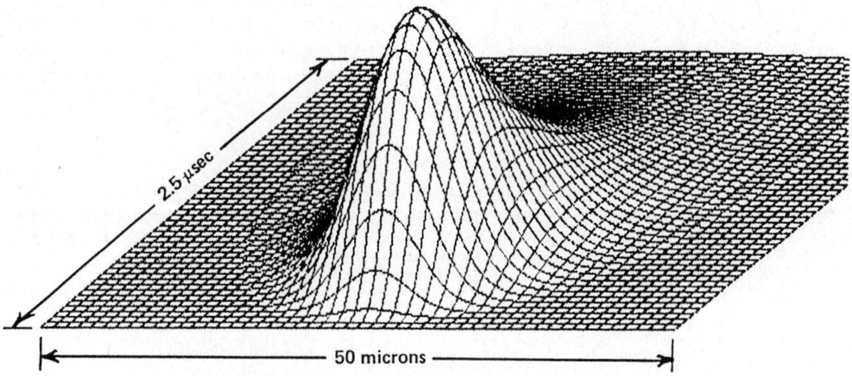

Fig. 2. Topographical plot of W_n; z_0 = 10 microns

In a "standard" thermal mapping each matrix point has a different associated temperature pattern (say $T_a{}^{(m)}(x)$ at the mth matrix point). We define the distortion at a particular point as the ratio of the actual image intensity, S_n, to the image intensity under idealized conditions, S'_n, at that point. To determine this we consider $T_a{}^{(m)}(x)$ [with spectra $\tilde{T}_a{}^{(m)}(\lambda)$] to be the "signal" processed by the OBD detector. By analogy with communication theory, distortion then arises when the response function, $\tilde{W}_n$, of the OBD detector is "faulty" (i.e., is not flat (broadband)).

The magnitude of the normal mode response at a probe beam elevation of 1 micron is shown in Fig. 3 for various modulation frequencies. The response is essentially a band pass filter peaking at 1 μm^{-1}. Lower spatial frequencies depend upon the temporal modulation frequency, while higher spatial frequencies are independent of ω. As a consequence, the distortion of broad spatial features of the temperature pattern vary with modulation, but the rapidly varying spatial features do not.

If the excitation and probe beam axes coincide,

$$S_n/S'_n = \int\tilde{T}_a(\lambda)\ \tilde{W}_n(\lambda)\ d\lambda\ /\ \int\tilde{T}_a(\lambda)\ d\lambda\ .$$

Since this ratio depends upon the spectral shape (radius) of the temperature

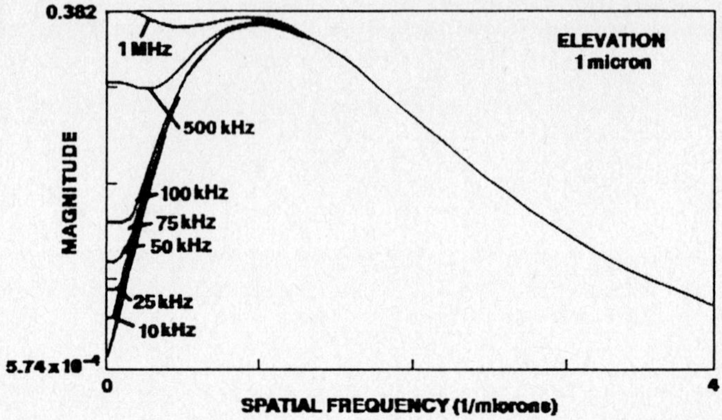

Fig. 3. Normal mode spatial frequency response function; z_0 = 1 micron

pattern, we call it the "shape factor." (See Fig. 4.) As shown in this figure, the intensity varies slowly as the radius decreases and then precipitously drops off for narrow patterns.

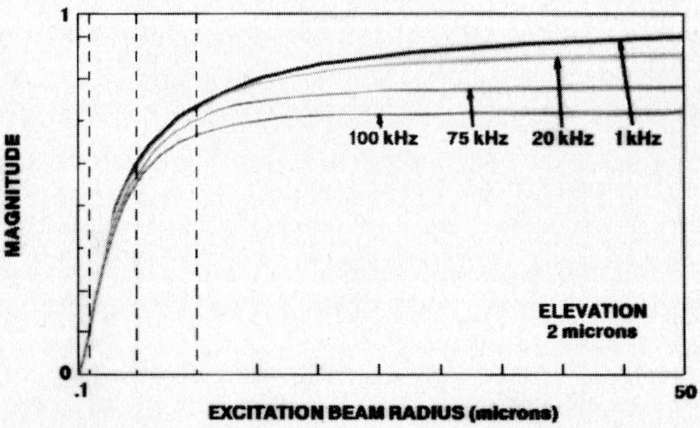

Fig. 4. Shape factor; normal mode; z_0 = 2 microns; excitation beam assumed to be gaussian

Profile of the Local AC Temperature Distribution in a Thermally Insulating Solid

L. J. Inglehart[1] *and E. Le Gal La Salle*[2]

[1]Center for Nondestructive Evaluation, Materials Science and Engineering, The Johns Hopkins University, Baltimore, MD 21218, USA
[2]Laboratoire d'Optique Physique, Ecole Superieure de Physique et de Chimie, 10, rue Vauquelin, F-75231 Paris, France

Many applications in thermal wave physics involve calculating the periodic temperature distribution $T(\vec{r},t)$, everywhere for a source of known geometry. For calculations involving more than one medium, such as a gas and a solid, or a layered structure, we assume particular boundary conditions for the continuity of temperature and the heat flux, which affect the final temperature distribution. KUO et al. [1] have calculated the periodic temperature distribution assuming a point heat source for a homogeneous, isotropic solid, when the thermal diffusivity of the solid is greater than, equal to, and less than that of the gas, predicting the distortion of the temperature profile near the gas-solid interface. We present a direct experimental measurement of the temperature profile in an insulating solid, which demonstrates the effect of the mismatch in thermal properties at the interface and provides a picture of the temperature distribution in the solid near the interface. We describe the experimental parameters, present our results, and briefly compare our results with the theoretical picture of [1].

1. THEORY

Mirage effect detection [2] is used to probe the temperature profile of the gas and solid. The deflection, $\vec{\varphi}$, of the probe beam is related to the temperature in the probed media by

$$\vec{\varphi} = \frac{1}{n}\frac{dn}{dT}\int \nabla T \times d\vec{l}, \tag{1}$$

where n is the index of refraction, T is the temperature, and l is the path of the probe beam. Since $(dn/dT)_s \gg (dn/dT)_g$, and we expect $\nabla T_s > \nabla T_g$, we expect that $|\vec{\varphi}_s| \gg |\vec{\varphi}_g|$, which is the case for both the normal, φ_n, and transverse, φ_t, deflection. In addition, we expect the direction of ∇T to change as we go from the gas into the solid. For the normal deflection, $\nabla T = dT/dz$, which contains a cosine term giving it symmetry about the source [see Fig. 1]. Whereas for the transverse deflection, we have $\nabla T = dT/dy$, which contains a sine term making it asymmetric about the source: φ_t goes to zero and changes phase by 180° at the origin [see Fig. 2].

2. EXPERIMENT

Lock-in detection was used to monitor the in-phase and quadrature signals of a quadrant detector configured to give both the normal and transverse deflections of the probe beam. The sample material, vitreous silica, absorbs at the source wavelength (CO_2, 9.2 µm, 100 µm dia.), and transmits at the probe wavelength (HeNe, 40 µm dia.). The data presented

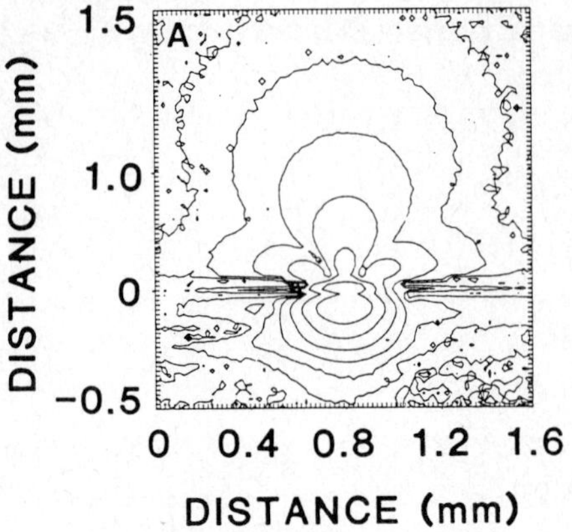

Fig. 1. Contour of the log of the magnitude of the normal deflection. Source is centered at z=0, y=0.8 mm.

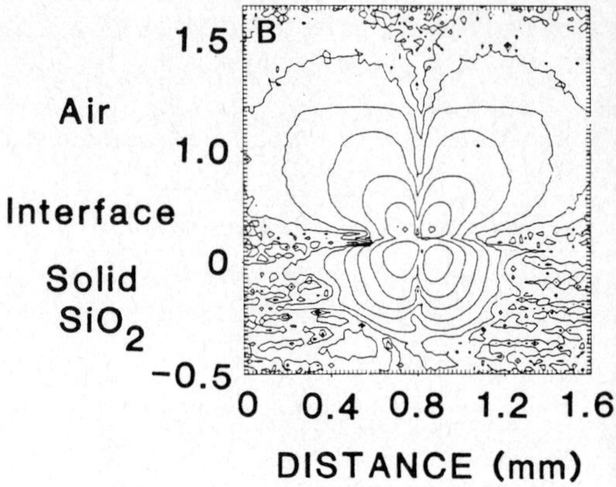

Fig. 2. Contour of the log of the magnitude of the transverse deflection. Source centered at 0.8 mm.

were recorded at a chopping frequency of 100 Hz, giving a thermal diffusion length of 310 μm in air, and 50 μm in the solid.

A line profile of the temperature distribution at each height of the probe was obtained by displacing the heat source transverse to the probe beam (transverse offset, 20 μm step size). The probe beam was maintained parallel to the sample surface and directed through the heated region to the quadrant detector. The height of the probe relative to the sample (z) was changed by stepping the sample in 20 μm steps. A complete profile was obtained by collecting the in-phase and quadrature signals of the normal and transverse deflection signals as a function of transverse offset for several heights of the probe beam.

From the in-phase and quadrature data we compute the magnitude and phase of φ_n and φ_t. The log of the magnitudes are presented as contour plots in Figs. 1 and 2, respectively. The interface is determined for both deflections. However, we note that the probe beam is strongly scattered at the surface and the data in a region of 50 μm above and below the interface should not be considered. The distribution decays more rapidly in the solid as predicted by [1], and is measurable to a distance of about six thermal diffusion lengths in both the solid and the gas.

In Fig. 1, there is a maximum in the solid at a distance of ~2 diffusion lengths below the surface, preceded by a local minima and a local maxima. In the gas near the interface, lobes which are symmetric about the source are present as predicted theoretically [3] when $\alpha_s < \alpha_g$. The lobes are observed in the solid also, but are decreased in size by 4-5 orders of magnitude. The behavior is similar for φ_t in Fig. 2, with the exception that the lobes are not observed for this large diameter heat source. We have observed from the line profiles that the relative amplitude of the normal component is larger when probing the solid than the transverse component, but that the relative amplitude of the transverse component is larger than the normal component when probing the gas.

Finally, a 2-D cross-sectional view of the total temperature distribution intersecting the gas/solid interface is presented in Fig. 3.

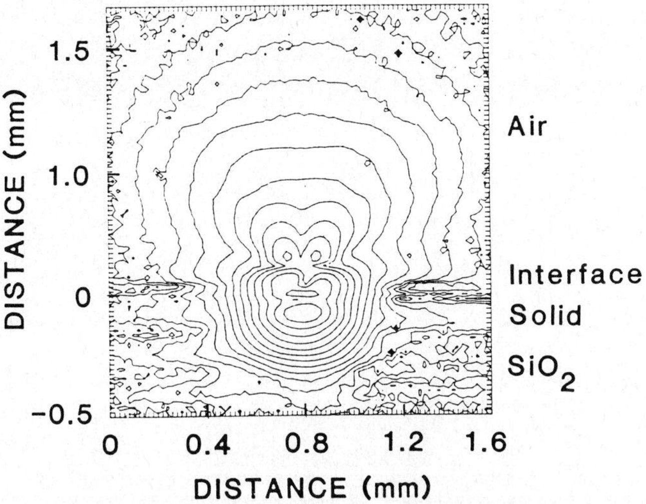

Fig. 3. Contour of the log of the magnitude of the total deflection, M. The heat source is centered at z=0, y=0.8 mm.

The contour is a plot of the log of the magnitude of the total deflection, M, where

$$M = [|\varphi_n|^2 + |\varphi_t|^2]^{\frac{1}{2}}. \tag{2}$$

This result may be interpreted and understood by comparison with the individual normal and transverse contour images of Figs. 1 and 2. The effect of the transverse component is found to dominate over the the normal component near the interface along the source, while the normal component dominates near the interface away from the source.

This work was sponsored by the National Bureau of Standards, Office of Nondestructive Evaluation.

3. REFERENCES

1. P.K. Kuo, L.J. Inglehart, E.D. Sendler, M.J. Lin, L.D. Favro, and R.L. Thomas in Review of Progress in Quantitative NDE, edited by D.O. Thompson and D.E. Chimenti (Plenum Press, New York, 1985) Vol. 4, p. 745.

2. A.C. Boccara, D. Fournier, and J. Badoz, Appl. Phys. Lett., 36 130 (1980).

3. L.J. Inglehart, J. Jaarinen, P.K. Kuo, E.H. Le Gal La Salle, in Review of Progress in Quantitative NDE, edited by D.O. Thompson and D.E. Chimenti (Plenum Press, New York, 1987) Vol. 6A, p. 263.

Photothermal Evaluation of Layered Samples with High Accuracy Based on 3-D Analysis of Thermal Waves

M. Beyfuss, R. Tilgner, and J. Baumann

Siemens AG, Corporate Research and Technology,
D-8000 München, Fed. Rep. of Germany

1. Introduction

For many practical applications, especially during semiconductor device processing, it is important to know the geometrical (thickness d) and other physical properties (density ρ, specific heat c, and thermal conductivity λ) of layered samples. As shown by several authors in recent years /1-6/, photothermal investigation is useful for evaluation of both surface layers /1,2,5,6/ and buried layers /3,4/, even at very small thicknesses compared with the thermal diffusion length μ.

To achieve a high precision in the determination of the geometrical and thermal data combined with high lateral resolution one has to take into account the 3-D propagation of thermal waves within the sample. In the following we show that the 3-D-behaviour can lead to strong shifts in the amplitude- and phase-response as compared with 1-D treatment.

2. Thermal waves in layered samples

3-D calculations of thermal diffusion, following a periodical heating on the surface, have been published by several authors so far /1,7-11/. Considering a two-layer sample, which is illuminated by a chopped laser beam of Gaussian profile (Fig. 1), we extended the one-layer approach of McDONALD /8/ to obtain the resulting complex surface temperature.

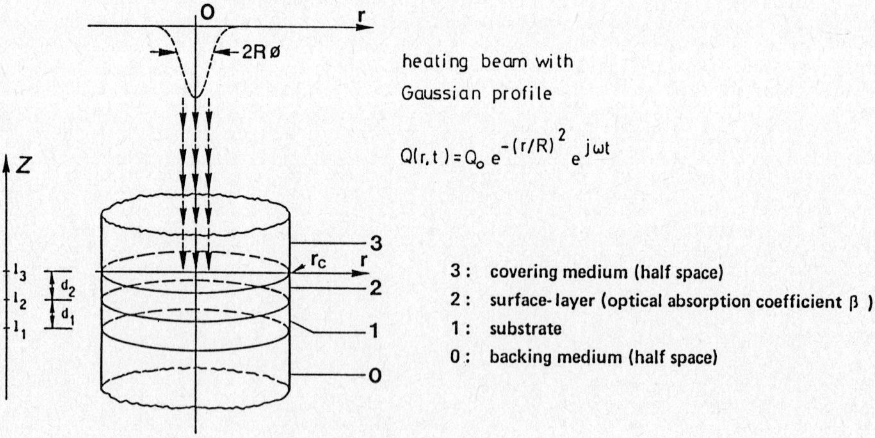

Fig. 1: 3-D model of a two-layer sample consisting of stratified discs of radius $r_c \gg R, \mu_i$

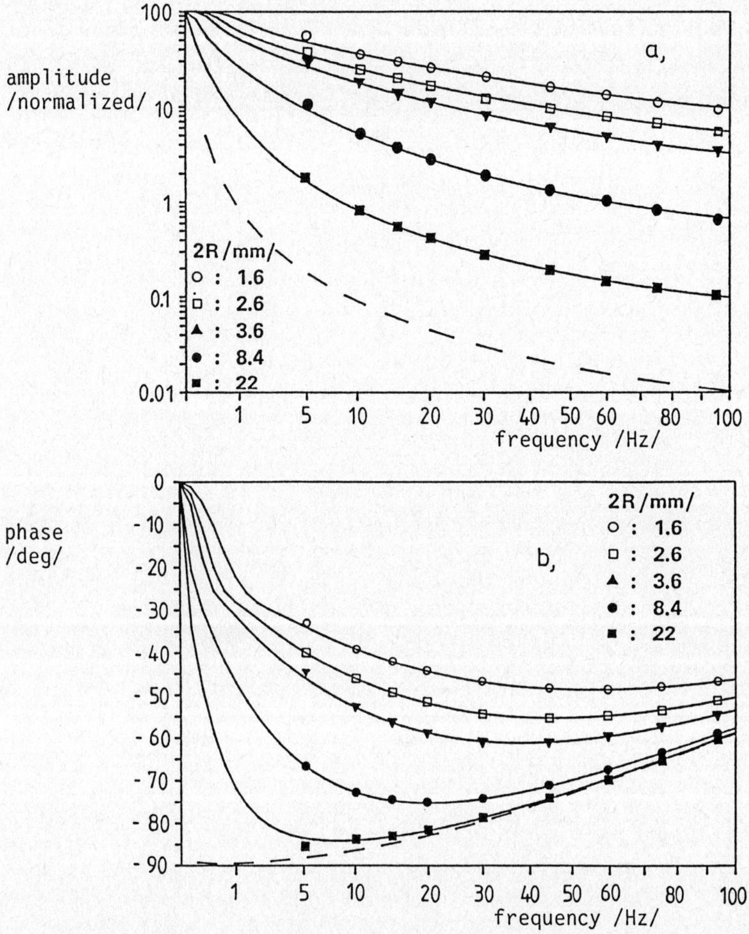

Fig. 2: Amplitude (a) and phase (b) of the surface temperature. Solid lines correspond to 3-D calculations, dashed lines correspond to 1-D calculations. Data: $(\rho c)_{Si}$ =1.71 J $cm^{-3} K^{-1}$, λ_{Si}=1.41 W $cm^{-1} K^{-1}$, d_{Si} =380 µm, $(\rho c)_{TiN}$ =3.5 J $cm^{-3} K^{-1}$, λ_{TiN}=0.019 W $cm^{-1} K^{-1}$, d_{TiN} =1.4 µm, $\beta d_{TiN} \gg 1$

Within a frequency range from 5 Hz to 100 Hz the thermal response of the sample was measured for 1/e-laser-beam-diameters varying thermally thick silicon (Fig. 2). Because the infrared-detector is sensitive for a finite area on the sample's surface, one has to integrate over this area to calculate the total temperature signal.

As one can see from Fig. 2, the experimental data fit very well the calculations. Obviously the thermal response of the sample depends very strongly upon the 1/e-diameter of the laser beam. Looking at the phase shift, which is usually more sensitive in photothermal investigations, we obtain an increasing deviation from the 1-D calculation /2/ with decreasing beam diameter.

Table 1: Calculations of $\Delta\varphi$ for a thermally thick silicon sample at r=0 (a), and for our actual sample integrating the surface temperature as mentioned above (b)

$2R/\mu_{si}$	1	2	5	10	20
a, $\Delta\varphi$ /deg/	31	22	7.7	2.2	0.6
b, $\Delta\varphi$/deg/ (5-100Hz)	50-16	37-15	15-9.1	4.4-2.9	1.1-0.8

Further calculations show that in the limit of a thermally thick silicon sample the difference in phase between 3-D and 1-D treatment is only a function of the quotient of the beam diameter and the thermal diffusion length in silicon (Table 1):

$$\Delta\varphi = \varphi_{3\text{-}D} - \varphi_{1\text{-}D} = F(2R/\mu_{si}).$$

In the case of our actual sample this simple relation is modified due to the additional absorber layer, the finite thickness of the substrate and the integrating sensing mode mentioned above. For $\Delta\varphi \lesssim 1°$ a value of $2R/\mu_{si} \gtrsim 20$ is necessary.

Taking $(\rho c)=3.5\ \mathrm{J\ cm^{-3}K^{-1}}$ from /14/ and d =1.4 μm (SEM-measurement) we determined $\lambda = (1.9\pm0.2)\times10^{-2}\ \mathrm{W\ cm^{-1}K^{-1}}$ for the TiN-layer. This accuracy can be obtained only by photoacoustic or photothermal methods.

Compared with 1-D calculations of layered samples we have an additional radial dependence of the surface temperature due to the intensity-distribution of the laser beam as well as the lateral diffusion of the thermal wave.

3. Experimental results

In a standard experimental set-up /12,13/ we used an 3W Ar^{+}-laser to illuminate the sample's surface periodically. The resulting temperature-oscillation was measured in a single-sided method focusing the surface infrared radiation with a concave mirror into a HgCdTe detector. A lock-in amplifier enables one to measure the absolute phase shift between the laser illumination and the temperature signal in this experimental scheme.

We examined a TiN-coated silicon wafer with different diameters of the laser beam. Using a 30μm-pinhole and a photodetector the actual profiles of the laser beam were measured.

4. Conclusion

To reach a high accuracy in the determination of the geometrical and thermal properties of thin layers on substrates, one has to examine carefully, whether an 1-D approach to thermal waves holds or not. The result depends strongly upon the ratio between the diameter of the heating beam and the thermal diffusion length. For a consistent interpretation of the experimental results by a 1-D calculation we found that $2R/\mu \gtrsim 20$ must be realized. Looking at photothermal mapping of layered structures by localized illumination (corresponding experiments are in progress), a consideration of the 3-D behaviour will also be very important in evaluating the thermal response quantitatively.

/1/ J.Opsal, A. Rosencwaig, D.C. Willenborg: Appl. Opt. 22, 3169 (1983)
/2/ R. Tilgner, J. Baumann, M. Beyfuß: Can. J. Phys. 64, 1287 (1986)
/3/ A.C. Tam, H. Sontag: Appl. Phys. Lett. 49, 1761 (1986)
/4/ J. Baumann, R. Tilgner: J. Appl. Phys. 58, 1982 (1985)
/5/ R.T. Swimm: Appl. Phys. Lett. 42, 955 (1983)
/6/ C.E. Yeack, R.L. Melcher, S.S Jha: J. Appl. Phys. 53, 3947 (1982)
/7/ F.A. McDonald: Appl. Phys. Lett. 36, 123 (1980)
/8/ F.A. McDonald: J. Appl. Phys. 52, 381 (1981)
/9/ H.C. Chow: J. Appl. Phys. 51, 4053 (1980)
/10/ L. C. Aamodt, J. C. Murphy: J. Appl. Phys. 52, 4903 (1981)
/11/ M. V. Iravani, H. K. Wickramasinghe: J. Appl. Phys. 58, 122 (1985)
/12/ P.-E. Nordal, S. O. Kanstad: Phys. Scr. 20, 659 (1979)
/13/ G. Busse: Infrared.Phys. 20, 419 (1980)
/14/ Y. Touloukian (Ed.): In Thermophys. prop. of matter. TPRC data series, Vol.2, (IFI/Plenum, New York, N.Y., 1970)

Note:
Similar work has been carried out independently by:
J. Jaarinen, C.B. Reyes, I.C. Oppenheim, L.D. Favro, P.K. Kuo, R.L. Thomas: Rev. Progress in Quantitative NDE, Vol 6B, 1347

The 3-D Character of Mirage Detection: Three Examples of Practical Importance

E. Le Gal La Salle[1], *J.P. Roger*[1], *F. Lepoutre*[1], *and L.J. Inglehart*[2]

[1]E.S.P.C.I., Laboratoire d'Optique, ER 5, CNRS,
10, rue Vauquelin, F-75231 Paris Cedex 06, France
[2]National Bureau of Standards, Bldg. 233/A331,
Gaithersburg, MD 20899, USA

Abstract : This paper presents three cases of usual experimental conditions in which a three-dimensional analysis is absolutely necessary. Applications of these effects to NDE and thermal characterization are discussed.

We define the 1-D model as a calculation assuming, for the temperature heat diffusion in one direction and for the mirage detection, an infinitely thin probe beam. In many experiments these conditions are rather well fulfilled and the 1-D results can be used for their interpretation. But sometimes important disagreements with the 1-D theory can be observed. We present here three experimental situations in which 3-D effects of particular importance occur. The 3-D calculations used in this paper result from a development in series of the normal deflection $\varnothing_n$. We do not give this long equation here, very similar expressions have been published previously /1/.

First example : a thick solid sample (in Fig. 1, stainless steel) is set in a vessel full of liquid paraffin, and excited by a modulated argon ion laser. The probe beam (radius 65 μm) is deflected in the liquid. Even at low modulation frequencies ($25 < f < 400$ Hz) the phase of $\varnothing_n$ does not vary linearly versus $\sqrt{f}$ as predicted by the 1-D model. This is not surprising since the thermal diffusion length in liquid paraffin is then always larger than the probe beam radius. But close to particular normal offset ($z = 150$ μm) and modulation frequency ($f = 80$ Hz) a catastrophic effect appears : the variations of the phase of $\varnothing_n$ become considerable and the amplitude of $\varnothing_n$ vanishes. This phenomenon is correctly analysed by the 3-D expression of $\varnothing_n$ (see continuous curves of Fig. 1) and can be understood as a destructive contribution between the different deflections undergone by the rays of the probe beam crossing opposite thermal gradients. This result points out some problems which may appear in photothermal spectroscopy when the samples are set in liquid to enhance the mirage signals. In usual Fourier Transform spectrometers the modulation frequency depends upon the wavenumber so that at some wavelengths the probe beam size effect can produce strong distortions of the reconstructed spectrum.

Second example : The second example deals with the study of low thermal diffusers. In Fig. 2, the sample is a 2 mm thick polymer. The pump beam (argon ion laser) is assumed to be absorbed at the surface of the polymer. At low modulation frequencies ($f < 40$ Hz) the phase of $\varnothing_n$ does not vary linearly versus $\sqrt{f}$. The 3-D calculated curves show that the thermal diffusivity is equal to some 10^{-7} m^2 s^{-1}, i.e. two orders of magnitude smaller than the one of air. This large difference is the reason for the nonlinearity of the phase versus $\sqrt{f}$. This result can be applied to measurement of low thermal diffusivities which are generally difficult to perform by the other photothermal techniques /2/.

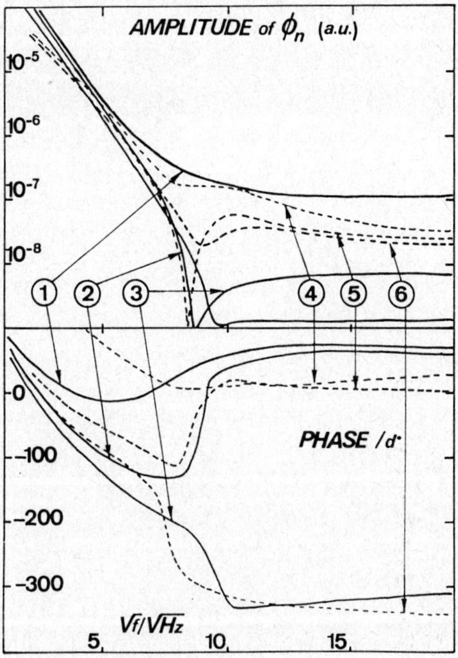

Figure 1 - The continuous curves were calculated from the 3-D expression of ϕ_n for three different normal offsets z : (1) z = 120 µm, (2) z = 150µm, (3) z = 160 µm. The dashed curves are fit of experimental results obtained at different normal offsets : (4) z = 130 ± 5 µm, (5) z = 150 ± 5 µm, (6) z = 160 ± 5 µm.

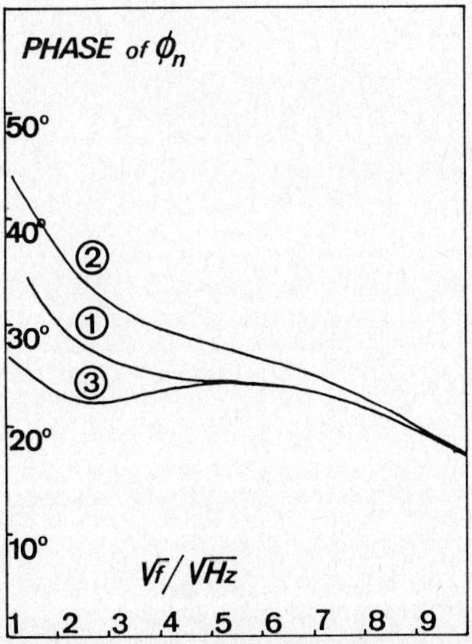

Figure 2 - Curve (1) : fit of experiments with a polymer sample. Curves (2) and (3) : calculation from the 3-D expression of ϕ_n with sample thermal diffusivities equal to 10^{-6} m^2 s^{-1} and 10^{-7} m^2 s^{-1} respectively.

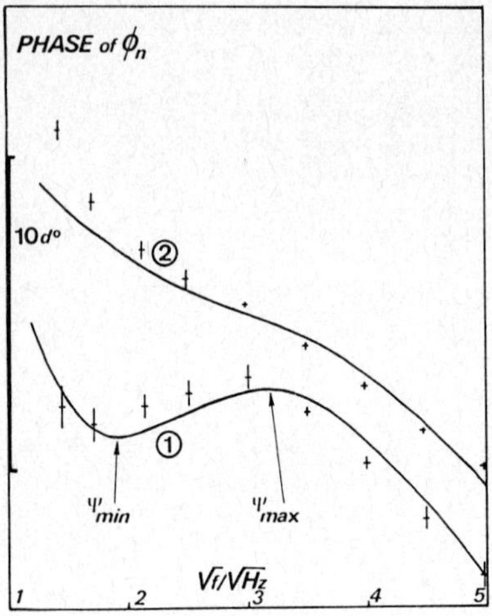

Figure 3 - The experimental points have been obtained with a pipe of titanium (thermal diffusivity : 10^{-5} m^2 s^{-1}) containing a 5 µm thick slab of air located at 900 µm below its surface. The upper points were obtained with a pump beam radius equal to 0.7 mm and the lower points with a pump beam radius equal to 2 mm. The continuous curves were calculated from the 3-D expression of $\emptyset_n$.

Third example : Figure 3 illustrates the importance of the 3-D character of N.D.E. by mirage effect. The sample is a pipe of titanium containing a 5 µm thick slab of air located at 900 µm under the surface. The pump beam comes from an argon ion laser and the deflecting medium is air. When the pump beam radius is decreased from 2 mm to 0.7 mm the contrast $(\Psi_{max} - \Psi_{min})/\Psi_{max}$ (see Fig. 3) decreases strongly and the detection of the defect becomes difficult. The continuous curves calculated from the 3-D model fit the experimental points rather well. The phenomenon can also be interpreted physically as follows : when the pump beam radius is of the same order of magnitude as the distance at which the defect is located under the surface, the thermal waves are almost plane when they reach the defect. The reflected part of these waves is then important in the area of the surface probed by the probe beam. On the other hand, when the pump beam radius is small, the thermal waves on the defect are quasi spherical and an important part of the reflection is lost for the probe beam. The defect "disappears" /3/. The consequence of this effect is very important in N.D.E. : to detect a defect at a depth d below the surface a pump beam radius larger than d is necessary.

References

1. W.B. Jackson, N.M. Amer, A.C. Boccara and D. Fournier, J. Appl. Phys., 52, 4903 (1981).
. L.C. Aamodt and J.C. Murphy, J. Appl. Phys., 54, 591 (1983).
. F.A. Mc Donald, G.C. Wetsel Jr. and G.E. Jamieson, Can. J. Phys., 64, 1265 (1986).
. P.K. Kuo, L.J. Inglehart, E.D. Sendler, M.J. Lin, L.D. Favro

and R.L. Thomas, Review of progress in quantitative non destructive evaluation, vol. 4B Edited by D.O. Thompson and D. Chimenti. Plenum Publishing Corporation, New York, 1985, p. 745.
2. F. Lepoutre, Techniques de l'Ingénieur, R 2959 (1986).
3. F. Lepoutre, D. Fournier and A.C. Boccara. Proceedings of the Second International Symposium on Nondestructive Material Property Characterization, Montreal (Canada) July 1986. To be published by Plenum Publishing Corporation, New York (1987).

Nondestructive Evaluation of Solids by Photothermal Interferometry on Nonreflective Surfaces

Z. Sodnik and H.J. Tiziani

Institut für Technische Optik, Universität Stuttgart, Pfaffenwaldring 9, D-7000 Stuttgart 80, Fed. Rep. of Germany

1. INTRODUCTION

For nondestructive material evaluation (NDE) ultrasonic and X-ray methods have been well established for many years. About two decades ago an old effect, discovered by BELL /1/, was theoretically examined by WHITE /2/. A comprehensive theory for NDE utilizing a laser heat source was given by ROSENCWAIG et al./3/. It is called the thermoacoustic effect and describes the phenomenon of sound generation, when periodically heating the surface of a body with an audio frequency. Since heat generation is usually done optically, using a modulated laser beam or a laser pulse /4/, the process is called photoacoustic or photothermal depending on whether the main interest is in the acoustic or the thermal aspect respectively. One should however keep in mind that all effects happen at the same time. At first, light is converted into heat, which heats the surrounding air and generates sound. Second, heat causes thermal expansion, leading to sound transport in the material. We decided to call our experiment a photothermal process, although both effects are involved.

In a photothermal process the absorption of modulated light modifies the surface temperature of a specimen and causes a heat diffusion, called a thermal wave, to propagate through the material. Inhomogeneities in the material can be detected by directly /5/ or indirectly monitoring the thermal wave. In our indirect measurements we detect the periodically changing thermal expansion - caused by a modulated laser source - interferometrically.

Changes in the amplitude or the phase of the expansion, with respect to the temperature modulation, lead to the detection of subsurface defects.

2. EXPERIMENTAL ARRANGEMENT

A chopperwheel (CP) modulates the intensity of an argon laserbeam (L), which is focussed onto a specimen´s surface, generating a temperature variation and a periodical spherical heat flow into the material as shown in Fig.1.

The specimen (SP) is mounted onto a stepper-motor-driven and computer-controlled table (T), which can be moved in X and Y directions with a position accuracy of one micrometer.

To avoid polishing of a surface we developed a tactile interferometrical sensor head (SH). After beam expansion (BE), spatial filtering (SF), and passing a beam splitter (BS), one beam of the interferometer, normally focussed onto the specimen´s surface, is now focussed onto a small steel pin, which is in mechanical contact with the specimen´s surface. In order not to damage the surface or to misalign the interferometer, the sensor is removed

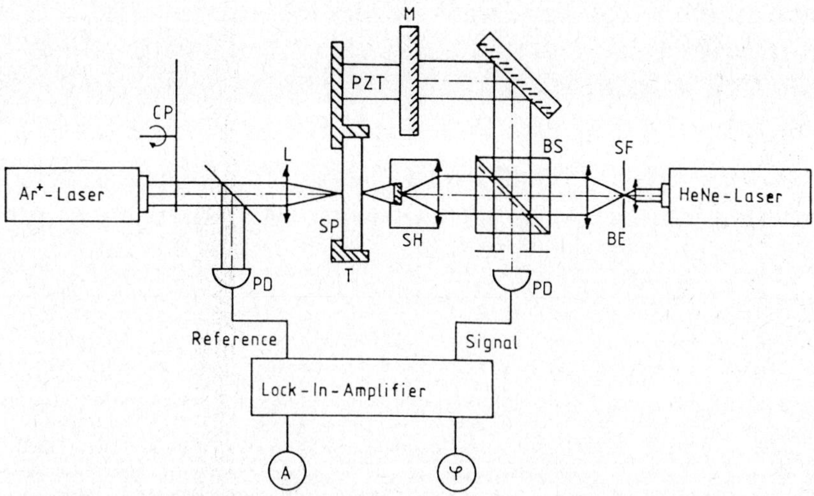

Figure 1: Experimental arrangement using a tactile sensor head

from the surface by an electromagnet, while moving the specimen to the next scanning position. The pin is attached to the focussing lens, its function is to transmit the specimen´s surface expansion to a small mirror, which is glued to its back.

For stability reasons it was found useful to have the reference beam reflected back from a mirror (M), which is attached close to the specimen. In this way vibrations have less effect on the optical path difference in the interferometer. A computer-controlled piezo element (PZT) moves the reference mirror in order to adjust the interferometer.

The detection takes place on the rear surface of the specimen, opposite to the argon laser focus. It is also possible to generate and detect thermal waves on the same side of the specimen /6,7/, using the effect of thermal waves being "reflected" at subsurface defects.

The zero order interference fringe is observed by a photodiode (PD) and its signal is fed into a lock-in amplifier. In amplitude and phase measurements subsurface defects can be detected.

3. THEORETICAL CONSIDERATIONS

Heat conduction in solids is a three-dimensional diffusion phenomenon /8/, which can be solved for a thermal point source in an infinite homogeneous body. Due to the fact that expansions are measured, rather than radiation, our theory had to rely on a quasi-static approximation in the periodic case, under consideration of boundary conditions. We ignored the size of the argon laser focus spot because it is much smaller than the thermal wavelength and found a one-dimensional model in good agreement with the results. Assuming T to be the amplitude of the temperature variation at the argon laser point of impact, C the thermal expansion coefficient, μ the thermal diffusion length and ignoring the time dependence, then the expansion δ in a distance l from the heat source was found to be

$$\delta(l) = \frac{C\,T}{l^2} \int_0^l x^2 \, \mathrm{EXP}\left[-\frac{x}{\mu}(1+i)\right] dx . \tag{1}$$

4. EXPERIMENTAL VERIFICATION OF THEORETICAL RESULTS

To verify the theory, an aluminum wedge of a thickness variation between 150 to 2550 micrometers was examined using two different modulation frequencies. The theory (dashed line) predicts an unexpected result, that the amplitude decreases at the thin end of the specimen, but it cannot properly explain the results at its very thin end (at 20 Hz). Equation (1) is basically developed from the theory of an infinite body, and a major difference occurs at the thin end of the wedge. With increasing thickness of the specimen or at a higher frequency, both phase and amplitude follow the theoretically predicted variations.

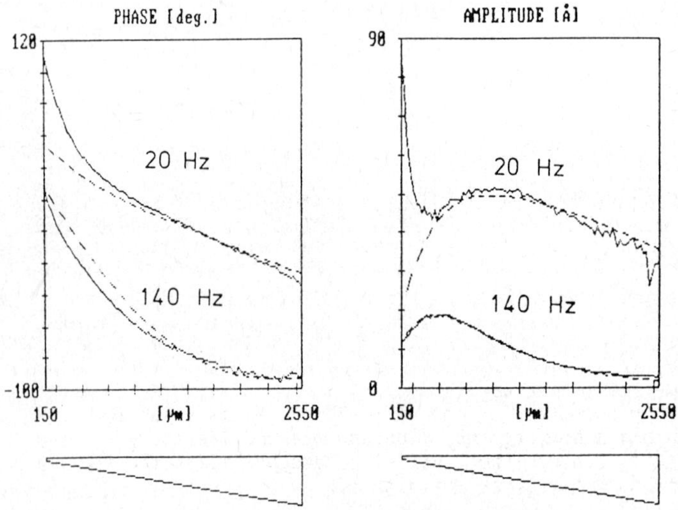

Figure 2: Experimental and theoretical (dashed) results obtained on a wedge

An example for subsurface defect detection is shown in Fig.3. The specimen is a block of aluminum of 2.5 mm thickness, with three drilled holes of 0.8 mm diameter. The gap between the first and the second hole is 0.8 mm wide and between the second and the third 0.4 mm wide. The upper graphs are results obtained at a modulation frequency of 20 and the lower of 140 Hz. Additionally a comparison is given between results obtained with an interferometer focussed onto the specimen´s surface (upper curve) and one using the tactile sensor head (lower curve). As one can see the sensor head causes a phase decay of about 5 degrees, an amplitude damping of about 10 % and a small increase in amplitude´s noise. The phase response however, which is most important for the detection of subsurface defects, because unlike the amplitude it is not dependent on varying argon laser absorptions, is in both cases the same. This indicates, that there is no obvious loss in resolution. The width of the photoacoustic signal, being larger than the width of the holes, has its origin in the detection of expansions, which are influenced by the presence of a defect in a bigger radius then given by the thermal diffusion length.

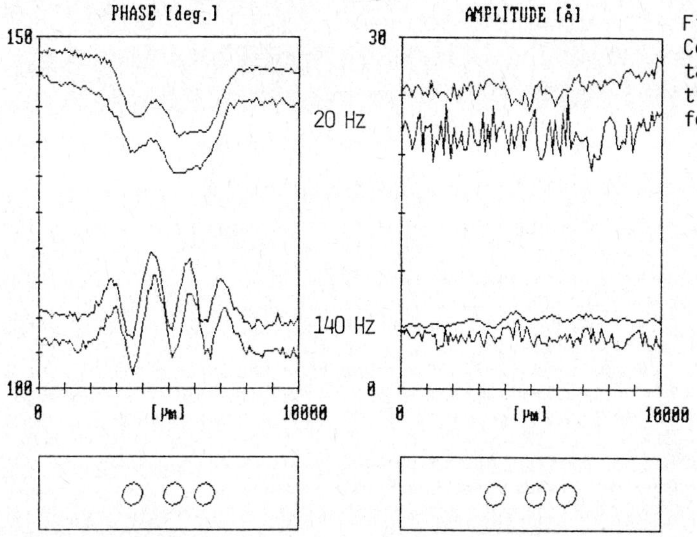

Figure 3: Comparison between tactile and non-tactile subsurface defect detection

5. CONCLUSION

Thermal expansion interferometry is a simple and very sensitive tool for nondestructive material analysis. In its basic form highly reflective surfaces are needed however. We developed a modified experimental set-up that can be used on any surface and found a theoretical description in good agreement with the results.

6. ACKNOWLEDGEMENT

This project was partly supported by the BMFT (Bundes Ministerium für Forschung und Technologie) Contract No. 1306 and the DFG (Deutsche Forschungs Gemeinschaft) Contract No. 2552. The responsibility for the contents of the paper is entirely by the authors.

The authors would like to thank Dr. W. Arnold, Prof. J. Burt and Dr. G. Busse for many helpful discussions and providing additional information.

7. REFERENCES

1. A.G. Bell: Am. J. Sci. 20 (1880), 305
2. R.M. White: J. Appl. Phys. 34 (1963) 3559
3. A. Rosencwaig and A. Gersho: J. Appl. Phys. 47 (1976) 64
4. W. Arnold: Acta Phys. Slov. 36 (1986), No.1
5. G. Busse: Appl. Opt. 21 (1982) 107
6. Z. Sodnik and H.J. Tiziani: Optics Comm. 58 (1986) 295
7. Y. Martin and E.A. Ash: Elect. Lett. 18 (1982) 763
8. H.S. Carslaw and J.C. Jaeger: In Conduction of heat in solids, 2nd.ed. (Clarendon Press, Oxford 1973)

Enhanced Thermographic Imaging for Subsurface Flaw Detection by Full-Field Heating of the Inspected Surface

P. Cielo, J.C. Krapez, and X. Maldague

Industrial Materials Research Institute, National Research Council Canada, 75 de Mortagne Blvd., Boucherville, Québec J4B 6Y4, Canada

1. Introduction

Subsurface imaging using the thermal wave generated by the absorption of a focused laser beam is an attractive approach for the nondestructive testing (NDT) of industrial materials and structures (see, e.g., ref [1] for a review). As this technology matures toward an industrial-floor procedure, practical considerations such as the time required to scan relatively large parts are receiving an increasing amount of attention.

A natural choice to improve the scanning speed is to resort to full-field heating and thermographic imaging of the whole heated surface in parallel. Qualitative applications of thermography to quality control in the industrial laboratory [2, 3] or in the production line [4] have been recently reported. The application of quantitative photothermal models to thermal NDE, coupled to the important advances in modern image-processing capabilities, should open new possibilities for industrial noncontact inspection.

This paper describes a number of possible approaches for subsurface defect detection by pulsed thermographic NDT. Either time pulsed 2-D heating or 1-D line scanning (space pulsed) are described, with a particular emphasis on the different signal processing techniques which are available for enhanced visualization of subsurface defects.

2. Surface Heating Configurations

Figure 1 shows two surface scanning configurations for thermographic NDT. In the diagram shown on the left the whole surface is pulse heated by a cluster of heating lamps, and the thermal distribution over the surface is observed during and after heating by an infrared (IR) camera. At our institute, up to 6 quartz-halogen lamps of 1000 W each, with paraboloidal back reflector are used to heat a typically 20 cm-diameter area of graphite-epoxy material with pulse durations from 0.1 to 10 seconds. Such an approach has been followed, e.g., in refs. [2] and [3]. Another approach, which is described, e.g., in ref. [5], produces the surface scan by keeping a line heater (tungsten-filament or arc-discharge tube of a few KW power, typically 10 to 50 cm-long, backed by a parabolic mirror) in close proximity to the surface and displacing the sample from right to left as in Fig. 1(b).

The substantial similarity of the two techniques for high field-of-view/sample-thickness ratios is apparent if we plot the temperature increase vs. time in the time-pulsed approach and vs. position in the space-pulsed method, as shown in Fig. 1. In a log-log scale, the temperature history curve shows the expected slope of -1/2 with a defect-free sample, if we neglect the initial transient pulse perturbation and the final finite-

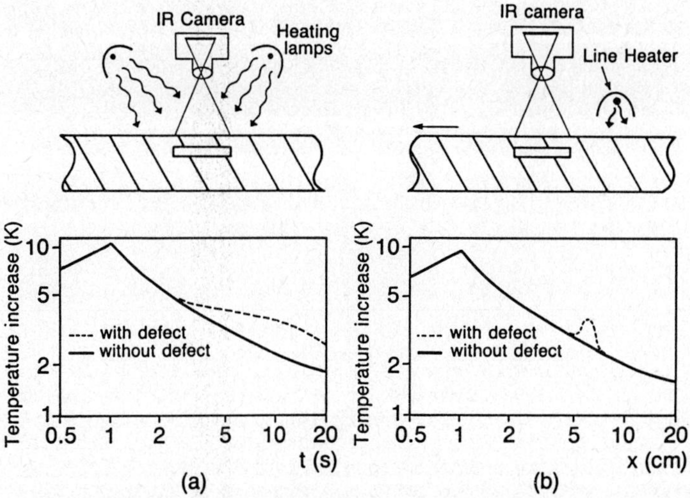

Fig. 1: Schematic diagrams and computed signals for the (a) time pulsed and (b) space pulsed thermographic NDT configurations.

thickness saturation. In both cases, a 5 mm-thick sheet with thermal conductivity K = 0.01 W/cm K, density ρ = 1 g/cm^3, specific heat C_p = 2 J/g K and thermal diffusivity $\alpha = K/\rho C_p$ = 0.5 mm^2/s, typical of a carbon-powder-filled polymer, was assumed in the 2-D finite-difference computations. The effect of a subsurface defect (air-filled slot whose cross section, shown in Fig. 1, was 0.25 mm x 10 mm at a depth of 2 mm below the surface) is also shown in Fig. 1. Equivalent results are obtained in the time or space pulsed cases in terms of defect visibility vs. the ratio between its lateral width and depth, as shown in Fig. 2. Larger ratios may be required with anisotropic materials.

3. Applications to Material Inspection

An application of time-pulsed thermography to the detection of subsurface damage produced by small weight impact on foam backed graphite-epoxy laminates is shown in Fig. 3. A drop-weight of 2.6 kg with an instrumented striking tup of 1.27 cm in diameter was used with impact energies of the

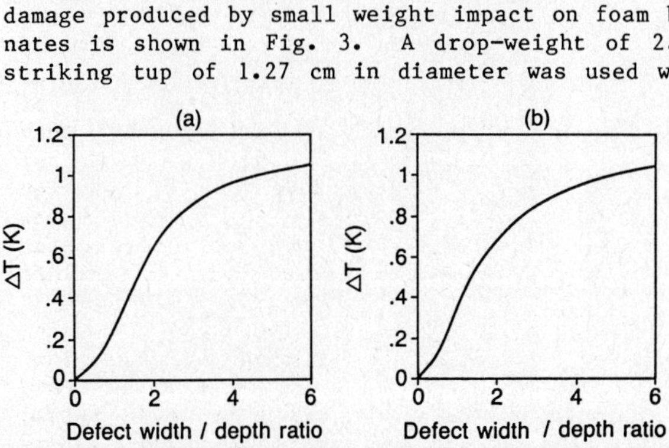

Fig. 2: Temperature differential between the defect and nondefect regions after 6 s from heating for the (a) time pulsed and (b) space pulsed configurations.

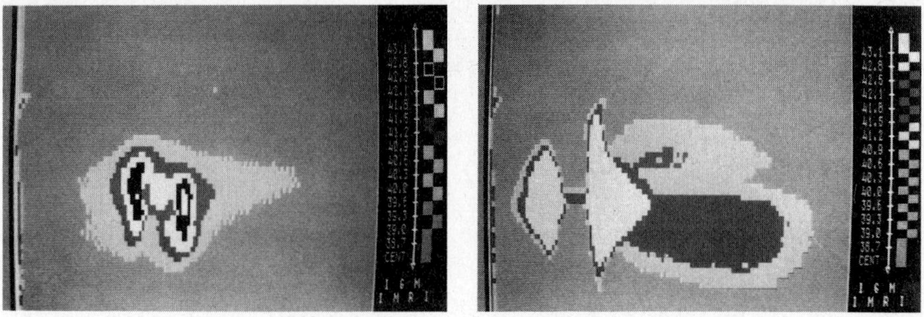

Fig. 3: Examples of time pulsed thermographs: these thermal images indicate the presence of subsurface impact damage in graphite-epoxy laminates.

order of 5 joules to produce delamination damage which was barely visible on the surface [6]. The thermal images obtained with an AGEMA SW-782 camera using the configuration shown in Fig. 1(a) are shown in Fig. 3 for an 8-ply, 1.14 mm-thick laminate with (a) $(0, \pm 45, 90)_s$ and (b) $(0, 90)_{2s}$ fiber orientation lay-up. The imaged area, nearly 5 cm x 5 cm in Fig. 3, shows the presence of subsurface delaminations which extend over several centimeters in the horizontal direction in the case of the more fragile 0°-90° fiber lay-up sample.

An application of the space-pulsed configuration (see Fig. 1(b)) as applied to the inspection of Aℓ-epoxy-foam sandwich laminates [7] is shown in Fig. 4. A quartz-halogen line heater, of 15 cm active length, 1500 W with parabolic back-reflector was used to scan the surface of the black-painted laminate which was moving at a speed of 11.5 cm/s. This results in a scanning rate of more than 100 cm^2/s with a pixel resolution of the order of 1 mm^2, nearly three orders of magnitude faster than the scanning rate achievable with an ultrasonic C-scanner [3].

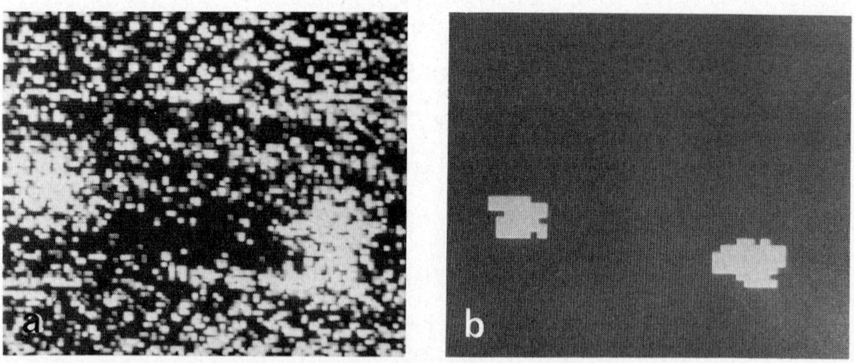

Fig. 4: An example of space pulsed thermographic defect detection in an aluminum-epoxy laminate: The two thermographs were obtained after normalization with a defect-free sample (a) before and (b) after time-resolved image processing. Two subsurface unbond defects are clearly visible in the processed thermograph.

Considerable enhancement of the defect visibility can be obtained by using suitable image processing techniques to reduce the amount of noise which arises under practical circumstances because of heating nonuniformities or variations of the surface emissivity and reflectivity. Heat nonuniformities can be compensated to a large extent by space normalization procedures (comparing the thermograph with the image obtained on a defect-free area) while variations of the surface emissivity can be compensated using time-domain normalization, by comparing subsequent images obtained on the same area at different times after heating to evaluate the rate of temperature decay across the object [3, 6]. Another example of time-resolved image processing procedure is shown in Fig. 4, where the picture on the left was obtained with the apparatus shown in Fig. 1(b) after normalization by comparison with a defect-free sample. The picture shown on the right of Fig. 4 was obtained after cross-correlation with a similar picture obtained at a slightly delayed time and after application of defect enhancement algorithms, and shows a substantial reduction of the uncorrelated noise.

4. Conclusion

Nondestructive testing by full-field heating and whole-thermal-image processing is a particularly effective procedure. As compared to more conventional NDT approaches, such as ultrasound and X-rays, this approach has the advantages of being very fast and of being applicable in situ with minimum sample preparation requirements. Either 2-D or line-scanned heating configurations can be effectively used, the latter being more appropriate for high-volume scanning of relatively large and flat panels.

References

1. G. Busse, IEEE Trans. Son. Ultrason., SU-32, 355 (1985).
2. W.N. Reynolds, Can. J. Phys. 64, 1150 (1986).
3. P. Cielo, R. Lewak and D.L. Balageas, in: Thermosense VIII, H. Kaplan ed., vol. 581, p. 47 (SPIE, Bellingham, WA, 1985).
4. D. Holmsten, in: Optical Techniques for Industrial Inspection, P. Cielo ed., vol. 665, p. 75 (SPIE, Bellingham, WA, 1986).
5. H. Tretout and J.Y. Marin, in: IR Technology and Applications, L.R. Baker et al. eds., vol. 590, p. 277 (SPIE, Bellingham, WA, 1985).
6. P. Cielo, X. Maldague, A.A. Deom and R. Lewak, Mater. Eval., 45, 452 (1987).
7. X. Maldague, P. Cielo, P.J. Ashley and B. Farahbakhsh, "Thermographic NDT of alumimum laminates", 5th Pan-Pacific Conf. NDT, Vancouver, Apr. 7-10, 1987. To be published in Can. Soc. NDT Journ.

Nondestructive Evaluation of Ceramic and Metallic Components by Photoacoustic Microscopy

B. Hoffmann, H. Peukert, and W. Arnold*

Fraunhofer-Institute for Non-Destructive Testing, University, Bldg. 37, D-6600 Saarbrücken 11, Fed. Rep. of Germany

We developed a photoacoustic microscope that allows rapid image build-up. Besides the well-known contrast mechanism due to changes of thermal properties, acoustic scattering can contribute considerably to the contrast of the images obtained.

1. Introduction

In recent years photoacoustic microscopy (PAM) has been developed as a laboratory instrument for the nondestructive examination of surface and subsurface features of opaque solids [1,2]. In order to test the viability of this technique for practical applications, we used this technique to investigate metals, technical ceramics, and the bonding of thin layers.

2. Experimental Apparatus and Results

Figure 1 shows schematically our experimental apparatus. An argon laser beam is modulated by an electro-optic cell and scanned over the surface of the sample by two mirrors mounted on microcomputer-controlled translation stages. The laser beam is focused down to a spotsize of approximately 10 μm at the surface of the sample.

The acoustic signals, generated by the modulated heat source via the photoacoustic effect, can be detected by various techniques. Using modulation frequencies in the range of 1 kHz - 100 kHz, the wavelength of the acoustic waves is in most cases larger than the dimension of the specimen to be tested so that one has to detect vibrations rather than propagating waves [3,4]. These vibrations can be detected with high sensitivity by accelerometers in conjunction with a charge amplifier.

The contrast of the signal detected contains contributions not only from changes of the thermal properties [5] of the specimen, but also from the scattering of the acoustic wave in its near-field at subsurface flaws. This scattering mechanism is analogous to superresolution in microwave imaging [6] and optics [7]. It can be clearly demonstrated by a simple experiment. Figure 2 shows the PAM-images of a 30 μm pin-hole in an aluminium sheet with a thickness of 100 μm together with the corresponding specimen geometry. The width, x, of the phase signal as well as of the amplitude signal is appreciably larger than the pin-hole diameter. This is due to the acoustic near-field effect. The width, x, is determined by the

*permanent address: Rheinisch-Westfälischer Technischer Überwachungsverein Essen, Langemarckstraße 20, D-4300 Essen, Fed. Rep. Germany

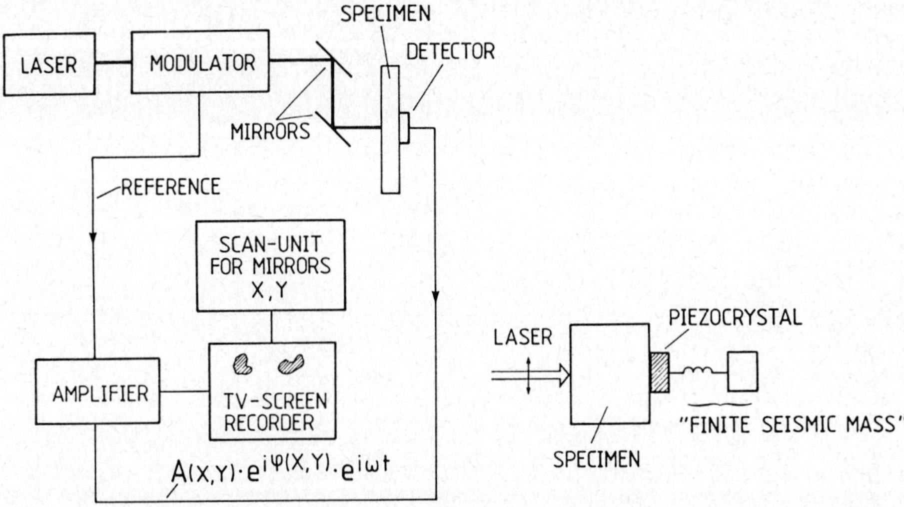

Fig. 1. Principle of apparatus for photoacoustic imaging. The vibrations of the specimen are detected by accelerometers driven close to their resonance (from [3]).

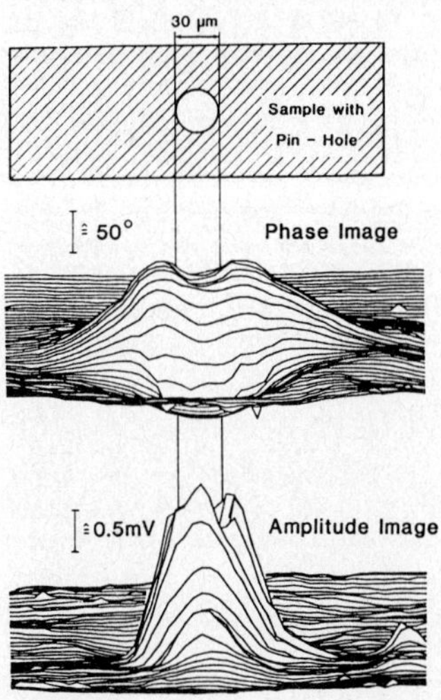

Fig. 2. PAM-images of a pin-hole (ø 30 μm) in an aluminium sheet (thickness 100 μm); scan-area 300 x 300 μm^2; scan-width 6 μm; modulation frequency 16.7 kHz

depth of the pin-hole rather than the width. The signal does not vanish when the laser spot passes through the center of the hole, because the beam rapidly broadens outside the focus along its axis. We have exploited near-field scattering previously to estimate the depth of cracks much larger than the thermal diffusion length [3]. Figure 3 shows the photoacoustic signal of a line-scan over a crack (Fig. 3a) together with the corresponding measuring geometry (Fig. 3b). The measured signal width, x, far exceeds the true geometrical width, w, of the crack [3]. Similar acoustic scattering mechanisms in PAM-images have also been observed [8,9] and discussed [10] by others recently.

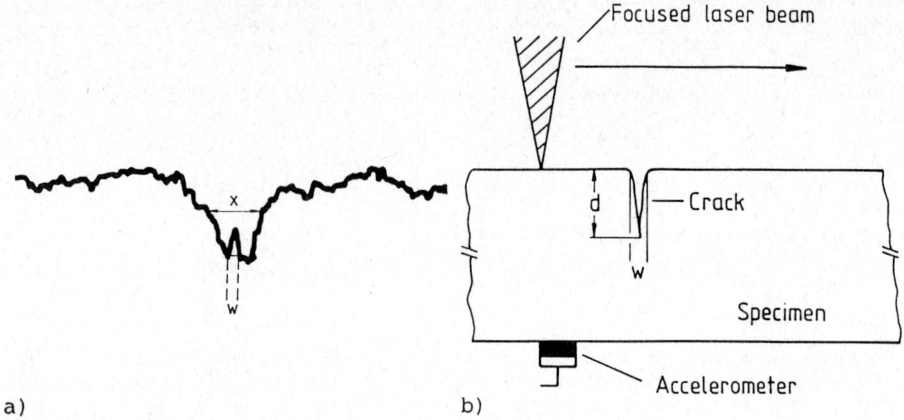

Fig. 3. Crack depth estimation by PAM; a) phase of acoustic signal (line-scan); b) measuring geometry (schematically)

By exploiting the increased sensitivity of accelerometers close to their resonance, we were able to decrease the integration time of the lock-in amplifier. The scanning speed is limited by the integration time of the lock-in amplifier and thus by the modulation frequency chosen and by the lateral resolution desired. With a lateral resolution of 10 μm a scanning speed of up to 10 mm/s is possible. At present the build-up time for an image (5 x 5 mm², distance between line-scans 100 μm) is less than 2 min. An increase of the lateral resolution will lead to a corresponding decrease of the scanning speed, if the time constant of the lock-in amplifier is fixed. Improving the resolution for example to 1 μm requires not only a decrease of the laser spot to 1 μm, but also a decrease of the thermal diffusion length to the same size, so that modulation frequencies higher than 1 MHz will be required. The use of higher frequencies allows in principle lower integration times, and thus higher scanning speeds are possible. In reality, however, the integration time must be increased again, as the PAM-signal decreases with the modulation frequency as $1/\omega$ [11].

Despite the difficulty in interpreting photoacoustic images, we made serial investigations on ceramic flexure bars made out of silicon nitride and silicon carbide [12]. Fig. 4 shows a typical phase-image of such a ce-

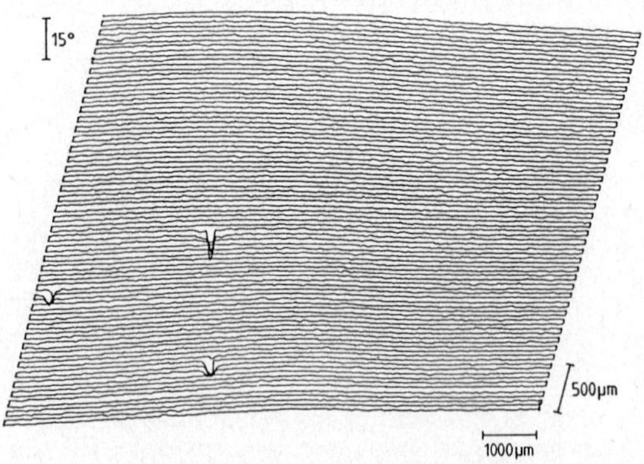

Fig. 4. Phase image of a ceramic flexure bar (10 x 4.2 mm²; distance between line-scans 10 μm; modulation frequency 28.7 kHz)

ramic specimen. It exhibits 3 signals from flaws not optically visible. We determined the tensile stress surface of the specimens for the following four-point crack-test as the one with the most indications of flaws. The results of these destructive bending tests will be compared with the results of the nondestructive PAM-investigations so that we will learn more how to interpret our photoacoustic images. Parts of the flexure bars were also investigated by acoustic surface waves of 10 MHz and a comparison of the results shows that most of the flaws are indicated by both techniques [13]. This corroborates our findings that acoustic scattering contributes considerably to the contrast in PAM-images.

3. Conclusion

We have shown that by using resonant acoustic detecting systems in PAM it is possible to increase the scanning speed to 10 mm/s, and that acoustic scattering contributes to the contrast in PAM-images.

References

1. see for example the collected papers in Proc. 4th Internat. Topical Meeting of Photoacoustic, Thermal and Related Sciences, Can. J. Phys. 64 (1986)
2. A.C. Tam: Rev. Mod. Phys. 58, 381 (1986)
3. W. Arnold, B. Hoffmann, and H. Willems: Z. Phys. B64, 31 (1986)
4. B. Hoffmann, H. Willems, H. Peukert, and W. Arnold: Proc. Jahrestagung der DGZfP (1987) in press
5. G. Busse and A. Rosencwaig: Appl. Phys. Lett. 36, 815 (1980)
6. E.A. Ash and G. Nicols: Nature 237, 510 (1972)
7. E. Betzig, A. Harootunian, A. Lewis, and M. Isaacson: Appl. Opt. 25, 1980 (1986)
8. H.I. Ringermacher and C.A. Kittredge: In Rev. Prog. Quant. Nondestr. Eval. 6B ed. by D.O. Thompson and D.E. Chimenti, (Plenum Press, New York 1987) p.1231
9. H.I. Ringermacher and L. Jackman: In Rev. Prog. Quant. Nondestr. Eval. 5A ed. by D.O. Thompson and D.E. Chimenti, (Plenum Press, New York 1986) p.567
10. L.D. Favro, P.K. Kuo and R.L. Thomas: In Rev. Prog. Quant. Nondestr. Eval. 5A ed. by D.O. Thompson and D.E. Chimenti, (Plenum Press, New York 1986) p.439
11. W. Jackson and N.A. Amer: J. Appl. Phys. 51, 3343 (1980)
12. H. Reiter, H. Peukert, and B. Hoffmann: IzfP-Report 870313-TW (Saarbrücken, 1987) unpublished
13. H. Reiter, B. Hoffmann, R. Karos, and W. Arnold: IzfP-Report 860103-TW (Saarbrücken, 1986) unpublished

Automatic Measurements of Thermal Resistances by the Mirage Effect

F. Lepoutre, F. Charbonnier, M. Le Liboux, R. Nahoum, D. Fournier, and A.C. Boccara

I.D.E.M. (Institut pour la Diffusion de l'Evaluation et de la Mesure), 10, rue Vauquelin, F-75005 Paris, France

Abstract - An industrial mirage effect bench monitored by a microcomputer is used to measure automatically thermal resistances between metallic samples.

It has been shown that photothermal methods are able to detect subsurface defects if they are located at depths smaller than a thermal diffusion length μ (in metals less than 1 mm typically). At these depths, it is possible to measure with a good sensitivity thicknesses (dimension perpendicular to the surface) of layers of air in metals between 0.1 μm and 10 μm.Defects thinner than 0.1 μm are almost invisible (they become thermally transparent) and above 10 μm they appear infinite (they are thermally opaque so that any increase above 10 μm does not affect the signals /1/).

In many industrial applications (boilers, heat exchangers, etc.) this sensitivity range (0.1 μm to 10 μm) is very interesting. Indeed the thermal contact between two metals can be described by layers of air. For instance, with a heat flow of 10 W/cm^2 a thermal resistance corresponding to a thickness of air of 1 μm introduces a loss of 4°C which corresponds to a prohibitive loss of efficiency. In other respects the commercially available industrial N.D.E. methods cannot easily detect such small defects.

These reasons are sufficient to build a photothermal apparatus. We have chosen the mirage method for three reasons :

- It is a non-contact method.
- It can be totally automated.
- Banks of three-dimensional data are available /2/.

The apparatus is shown diagrammatically in Fig. 1. The sample is a pipe (length 4 m, diameter 16 mm). The thermal resistance is located at 900 μm from the surface inside the pipe. The pipe is placed on a rail (2) insulated against the ground vibrations (1). It is then automatically moved (translation + rotation (6))towards the test area (4). The pump beam coming from a halogen lamp (250 W) is mechanically chopped (5).

The compact mirage bench /3/ is shown in Fig. 2. The noise at 10 Hz is less than 10^{-9} radians on the sensor position (9).The critical distance between the probe beam (2) and the sample surface (7) (top of the pipe) is monitored by a translation stage (4) and mechanical systems (3), (5), (6) with a precision better than 1 μm. The computer stores the phase of the normal deflection ϕ_n vs the square root of the modulation frequency ($\sqrt{f}$) via a lock-in amplifier. At the end of the measurement, the microcomputer calls files containing the theoretical data of ϕ_n vs $\sqrt{f}$ for diffe-

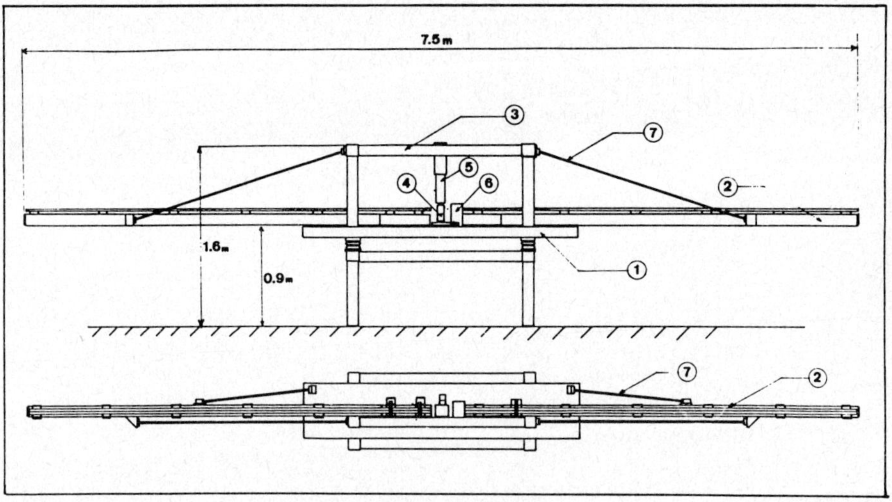

Figure 1 - Schematic of photothermal apparatus.
① : Marble insulated from ground vibrations ; ② : Rail supporting the pipe ; ③ : Mechanical holder of the halogen lamp ; ④ : Mirage bench ; ⑤ Light box (halogen lamp, chopper, optical comonents) ; ⑥ : translation + rotation stages ; ⑦ : Stay.

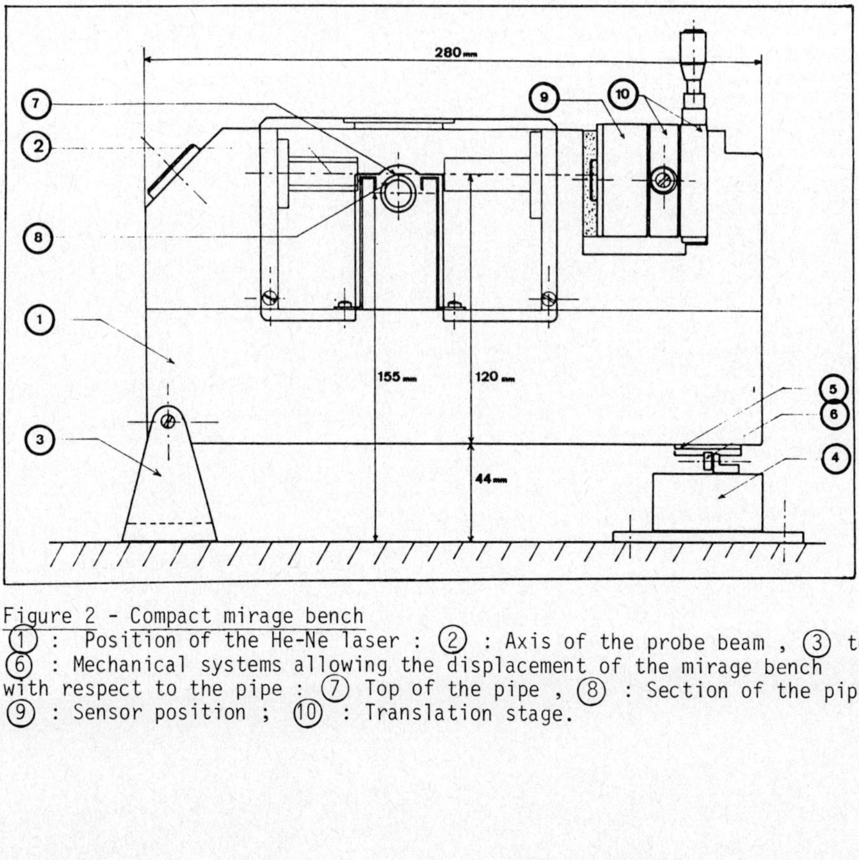

Figure 2 - Compact mirage bench
① : Position of the He-Ne laser : ② : Axis of the probe beam , ③ to ⑥ : Mechanical systems allowing the displacement of the mirage bench with respect to the pipe : ⑦ Top of the pipe , ⑧ : Section of the pipe ; ⑨ : Sensor position ; ⑩ : Translation stage.

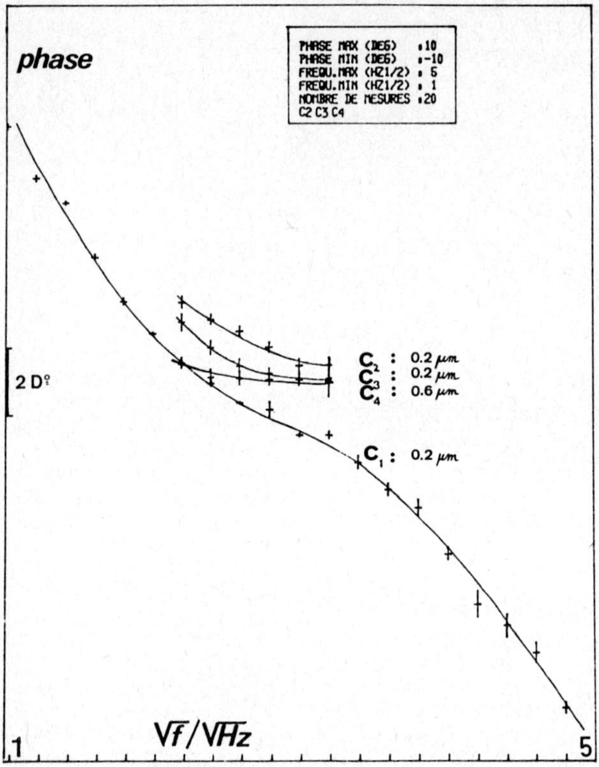

Figure 3 - Example of plotted results
C_1, C_2, C_3, C_4 correspond to four tested points on a same section of a pipe. For each point, the thickness of the detected slab of air is given by a comparison with the 3-D theoretical data.

rent values of the air layer thickness. The results of this comparison can be displayed immediately on the screen or with a short delay on a plotter (see Fig. 3). The complete check of a pipe (points separated by 10 cm) takes a few hours but requires only two manual operations, at the beginning and the end of operation.

References

1. F. Lepoutre, D. Fournier and A.C. Boccara, J. Appl. Phys. 57, 1009 (1985).
2. F. Lepoutre, B.K. Bein and L.J. Inglehart, Can. J. Phys. 64, 1037 (1986).
3. F. Charbonnier and D. Fournier, Rev. Sci. Instrum. 57, 1126 (1986).

Parallel Thermal Wave Imaging Using a Vector Lock-In Video Technique

P.K. Kuo[1], *Z.J. Feng*[1], *T. Ahmed*[1], *L.D. Favro*[1], *R.L. Thomas*[1], *and J. Hartikainen*[2]

[1]Department of Physics, Wayne State University, Detroit, MI 48202, USA
[2]Department of Physics, University of Helsinki, Siltavuorenpenger 20 D, SF-00170 Helsiniki, Finland

1.0 Introduction

The appearance of the IR video camera has extended the wavelength range of the visible video camera to the thermal IR range (3-12 μm), thus providing a powerful tool to researchers in thermal wave imaging. However, imaging in the thermal IR range has its special handicaps, not shared by its visible counterpart. Most objects in conventional photography reflect rather than emit light of their own. As a result one often has the freedom to choose the intensity, direction and color of illumination to accentuate the aspects of the object to be photographed. In thermal IR imaging the situation is very different in that nearly all objects emit thermal radiation of their own, in addition to reflecting radiation of other objects. What is recorded in a thermograph is always a mixture of emitted and reflected radiation, some of which even comes from components of the camera itself, including lenses and their supporting structures. This problem is particularly severe in the 8-12 μm range, because it corresponds to the peak of blackbody radiation at room temperature. It is this same range of wavelength that is most relevent in non-destructive evaluation. In conventional scanned thermal wave imaging applications this problem is overcome by the use of a lock-in analyzer synchronized to the source of the thermal wave. Without the lock-in technique, the IR video camera is capable of observing only very slow thermal phenomena[1], despite the fact that the intrinsic band width of the camera is very broad. This limitation offsets the main advantage of the IR video camera, namely its high data-acquisition rate. In this paper we report on instrumentation development which combines the lock-in technique with the IR video camera. With this technique the information of each pixel of an image is handled in the manner of a lock-in analyzer, while the object is illuminated (i.e., heated) or stimulated (e.g., joule heating) with a signal which is synchronous with the reference signal of the lock-in detection. This way the unsynchronous background radiation is rejected and the signal-to-noise ratio is enhanced.

2.0 Instrumentation

Figure 1 shows a block diagram of this system. The signal from a conventional IR video camera (Inframetrics IR-600) is processed by a video signal processor (manufactured by Datacube, Inc.). The processing consists of digitizing at 10 MHz and digitally merging the video signal with the sine and cosine functions of the phase of the reference signal and accumulating them in separate image buffers (each 512x512x16 bits). When a predetermined number of frames have been accumulated, the results are normalized and displayed as in-phase and quadrature images, in the same manner as the in-phase and quadrature signals are displayed by a vector lock-in analyzer. The postprocessor (a color work-station by Sun Microsystems 3/160C) handles the phase adjustment, pseudocolor and gray scale image display, as well as image storage and retrieval. The phase and timing control assures the synchronization between the heat source and the lock-in analysis. In this arrangement, the lock-in frequency can be chosen from sub-Hz frequencies up to the pixel rate of the IR camera, or the speed of the video signal processing system, whichever is less. For the present system, it is the IR camera speed of about 4 MHz which sets the intrinsic limit of lock-in frequency at about 2 MHz.

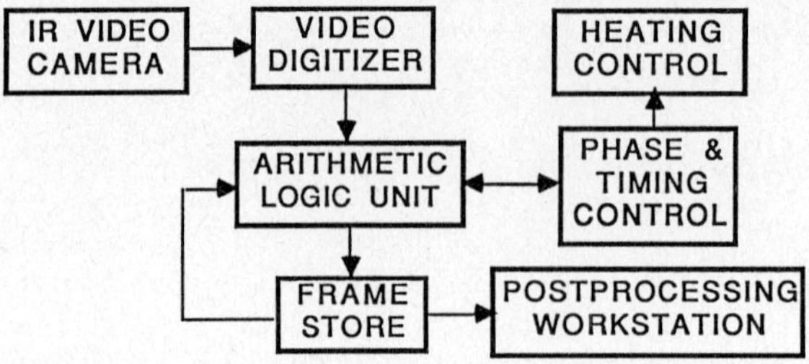

Fig. 1 Block diagram of the vector lock-in IR video imaging system

3.0 Results

To illustrate the principle of area-wide lock-in detection, in Fig. 2a we show a video picture of a circuit board containing an LED display lamp with all its segments turned on by a dc current. Just below it there is another LED, three of whose segments are energized by a weaker current modulated by the reference signal. Figure 2b shows the results of the in-phase image after just two cycles of accumulation. One sees that the only discernable features are the modulated LED segments, all other features having been eliminated by lock-in detection. There is, however, a considerable amount of noise in the background. Increasing the number of cycles of averaging decreases the background noise but keeps the magnitude of the modulated signal the same. Figure 3 plots the variance of the images versus the cycle number. This number plays the same role as the time constant in the conventional lock-in analyzer. Thus, the observed linear relation (see Fig. 3) is just the expected result. The simultaneous acquisition of both the in-phase and the quadrature images is of practical importance here because it makes possible the reconstruction of the image of any other required phase from stored image data. With lock-in analysis capability, video thermal wave imaging now enjoys the same advantage as conventional scanned thermal wave imaging, namely, the retention of the phase information contained in the images. Such information can be very useful, for example, in providing characteristic signatures of different types of subsurface defects. [2,3]

We have applied this video IR lock-in analyzing system to a sample of laminar structure. The sample consists of a copper film microbridge (a few microns thick) on a polymer substrate, covered by a thin (about 10 μm) film of Kapton [4] as shown in Fig. 4. The microbridge in the copper film provides a convenient means of localized joule heating. Figures 5a and 5b show the in-phase and quadrature images of this sample when it is being heated by a

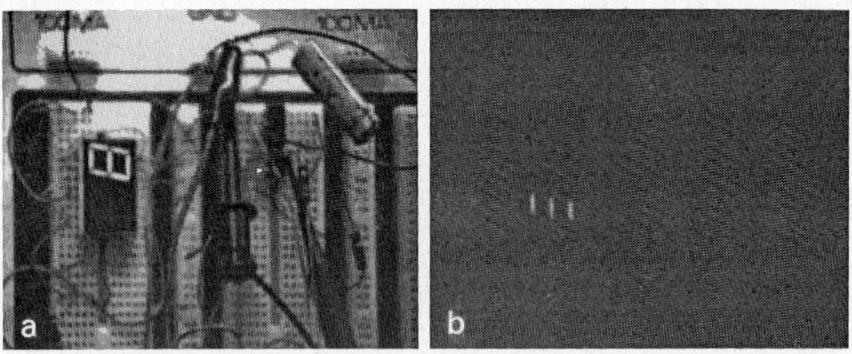

Fig. 2 A video image of a circuit board (a), together with its lock-in image after just two cycles of averaging (b)

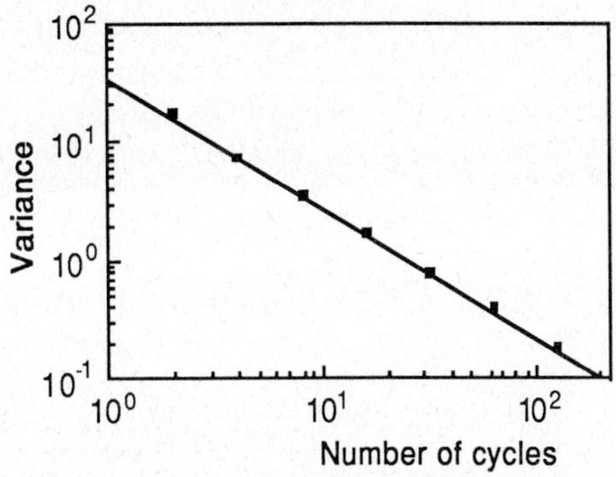

Fig. 3 Plot of the variance of the lock-in image of Fig. 2b as a function of the number of accumulating cycles

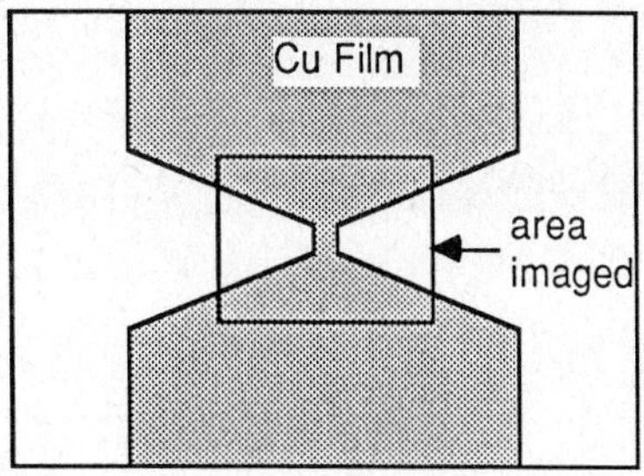

Fig. 4 Schematic diagram of the samples used in Fig. 5, showing the shape of the copper microbridge

square-wave current at 15 Hz. Figures 5c and 5d show the same images (taken with the same average input power) of another sample with a simulated debond built into it. The simulated debond was produced by inserting a film of Teflon [5] underneath the microbridge while it is being bonded to the substrate. It is evident that the increased thermal resistance due to the debond causes the ac temperature to increase sharply. It should be pointed out that since copper is a very poor emitter in this wavelength range, the radiation seen in the images comes mainly from the Kapton top layer which is heated by the copper film. The phase information

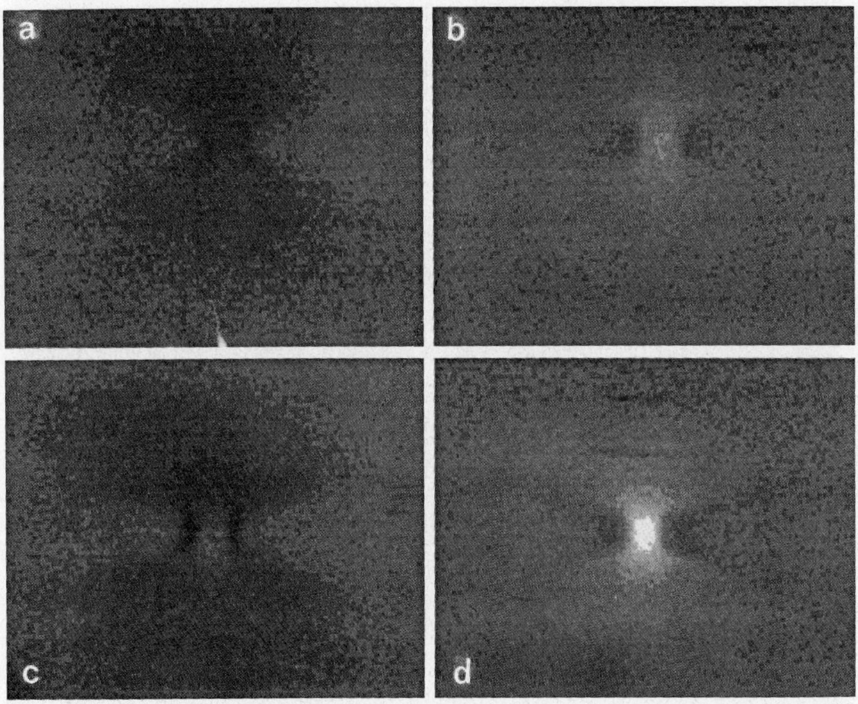

Fig. 5 In-phase (a) and quadrature (b) lock-in images of a Cu microbridge; and the same images (c,d) of a microbridge with a simulated debond. The ac current was adjusted to give the same average power input to both samples

contained in the in-phase and quadrature images corresponds to the time delay associated with the thermal diffusion through the "debond".

4.0 Summary

We have demonstrated the viability of the concept of area-wide lock-in detection and have applied it to thermal wave imaging. This technique, which combines the advantages of narrow band detection and high speed data acquisition, should be a useful tool in the area of non-destructive evaluation.

5.0 Acknowledgements

This work was sponsored by the Army Research Office under contract No. DAAG29-84-K-0173, and by the Institute for Manufacturing Research, Wayne State University.

6.0 References

1. W.N. Reynolds: In Can. J. Phys. **64**, 1150 (1986).

2. K.R. Grice, L.J. Inglehart, L.D. Favro, P.K. Kuo, and R.L. Thomas: In J. Appl. Phys. **54**, 6245 (1983).

3. P.K. Kuo, L.D. Favro, L.J. Inglehart, and R.L. Thomas: In J. Appl. Phys. **53**, 1258 (1983).

4. Polyimide film, reg. U.S. Patent Office, E.I. du Pont Nemours and Co., Inc.

5. Tetrafluoroethylene film, reg. U.S. Patent Office, E.I. du Pont Nemours and Co., Inc.

Pulsed Photothermal Radiometry for Measuring Subsurface Air Gap Thickness or Contact Thermal Resistance

A.C. Tam, H. Sontag, and W.P. Leung

IBM Almaden Research Center, 650 Harry Road, San Jose, CA 95120, USA

We apply the pulsed photothermal radiometry (PPTR) technique [1-5] to investigate a layered structure composed of an opaque film separated from a thick substrate by an air gap of thickness δ, or more generally, by a contact thermal resistance R. The PPTR signal shape at certain delay times from the excitation pulse is found to be very sensitive to the value of δ up to a few hundred micrometers in thickness. We show how the signal shape can be deconvoluted to give δ, thus providing a quantitative nondestructive tool for delamination mapping. The present technique is also used to more generally quantify the value of R between a coating and a substrate; if we expect that R is related to adhesion strength (with R decreasing for better adhesion), the measurement of R also provides a new nondestructive detection method for adhesion strengths.

The experimental arrangement is shown in Fig. 1. The sample film is polycarbonate (PC) of 20 μm thickness containing 27% carbon black particles. The film is mounted flat and one side is irradiated by an unfocused 8 ns pulse from a nitrogen laser (pulse energy 1 mJ). At the back, the air gap between the film and a given backing material (e.g., polished germanium) is controlled to 1 μm accuracy by a piezoelectric translator. The center of the uniformly heated

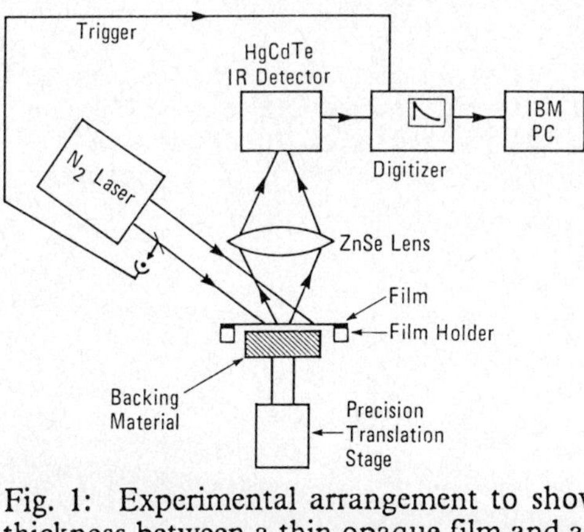

Fig. 1: Experimental arrangement to show PPTR measurements of air gap thickness between a thin opaque film and a thick backing material (substrate). The position of contact between the backing and the film is determined by measuring the electrical resistance between the conducting film and the backing.

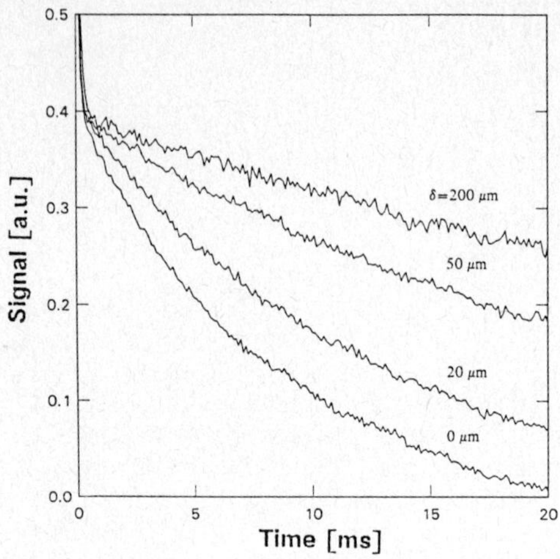

Fig. 2: Observed PPTR signal $S(t)$ for a carbon-loaded polycarbonate film at various separations from a germanium substrate.

region is imaged onto a DC-coupled HgCdTe detector (sensitive range being 7-12 μm) using an anti-reflection-coated ZnSe lens. The transient PPTR signal is recorded on a transient waveform recorder and subsequently processed on an IBM PC. Typical signals are shown in Fig. 2.

Maximum surface temperatures reached directly after absorption of the nitrogen laser pulse are < 20K above room temperature T_o. After 1 ms however, the temperature is almost constant across the film and less than 0.4K above T_o. Under these conditions convection is not expected to affect our measurements, which are taken on a time scale of 100 ms. On this time scale, lateral heat diffusion within the film can also be neglected, since the laser spot is large (1 cm² in area) at the sample. However, nonuniformity in the laser intensity profile can contribute to observable lateral diffusion effects.

The PPTR signal can be calculated by assuming that there is no heat loss at the front surface ($x = 0$) while the heat loss through the air gap at $x = \ell$ (ℓ being the film thickness) is represented by a thermal resistance R. Following our previous approach [1], we solve the heat diffusion equation for the temperature rise $\theta(x,t)$ in the film as a function of position x and time t.

$$\frac{\partial \theta}{\partial t} - D\frac{\partial^2 \theta}{\partial x^2} = 0\,, \qquad 0 \le x \le \ell\,, \tag{1}$$

subjected to the initial condition

$$\theta(x,0) = A\alpha e^{-\alpha x}\,, \qquad 0 \le x \le \ell\,, \tag{2}$$

and the following radiation boundary conditions:

$$\frac{\partial \theta}{\partial x} = 0\,, \qquad x = 0\,, \tag{3}$$

$$\frac{\partial\theta}{\partial x} = -h\theta\,, \qquad x = \ell\,. \tag{4}$$

Here $h = 1/RK$, and D and K are the thermal diffusivity and conductivity of the film, respectively; A is a constant depending on laser energy and heat capacity of the film; α is absorption coefficient of the film at the excitation wavelength. Note also that $1/h$ is the equivalent thickness of a material identical to the film that has the same thermal resistance as the air gap.

The condition set in (4) also implies that the backing material has a much larger thermal conductivity than that of the film so that its temperature does not rise during the heat conduction process. Equation (1) can be solved using the general solution suggested by Carslaw and Jaeger [6] and matching the initial and boundary conditions. The PPTR signal monitored at $x = 0$ is given by

$$S(t) = B\int_0^{\ell} \alpha'\theta(x,t)e^{-\alpha' x}dx\,, \tag{5}$$

where B is a constant factor depending only on the sensitivity of the infrared (IR) detecting system, the emissivity and the average temperature of the sample; and α' is the IR absorption coefficient of the film averaged over the detection spectrum.

The integral in (5) finally yields

$$S(t) = 2AB\sum_{n=1}^{\infty} \frac{G(\beta_n\ell\,,\alpha'\ell)G(\beta_n\ell\,,\alpha\ell)}{\ell + \dfrac{\sin^2\beta_n\ell}{h}}\, e^{-t/\tau_n}\,, \tag{6}$$

where β_n is the nth root of the transcendental equation

$$h = \beta_n \tan\beta_n\ell\,, \qquad n = 1,2,3,\ldots\,, \tag{7}$$

$$G(a,b) = \frac{b}{a^2+b^2}\,[(a\sin a - b\cos a)e^{-b} + b] \quad \text{and} \tag{8}$$

$$\tau_n = \frac{1}{D\beta_n^2}\,. \tag{9}$$

One can easily verify that when $h \to 0$, as in the case of an isolated film, (6) reduces to eq. (13) of [1]. It can also be shown that

$$\frac{\ell^2}{\left(n-\frac{1}{2}\right)^2\pi^2 D} \le \tau_n \le \frac{\ell^2}{(n-1)^2\pi^2 D} \qquad n = 1,2,3\ldots\,, \tag{10}$$

and τ_1 is usually much larger then τ_n ($n \neq 1$) for a given value of h. Hence for $t > \tau_1$, only the first term in (6) is significant. Furthermore if $\alpha\ell, \alpha'\ell$ are large enough (say greater than 20), as in our case, $G \to 1$ and (6) becomes

$$S(t > \tau_1) = \frac{2ABe^{-D\beta_1^2 t}}{\ell + \dfrac{\sin^2\beta_1\ell}{h}}\,. \tag{11}$$

Therefore, we can deduce β_1 and hence h (from (7)) as well as the thermal resistance R from the slope g of a log $S(t)$ vs. t plot; the effective gap width δ can also be found from $\delta = K'R$, where K' is the conductivity of the gas in the gap. In fact, if h is small (say $\leq \frac{0.1}{\ell}$) then, $\beta_1\ell << 1$, Eqs. (7) and (11) give the slope g to be

$$g \approx \frac{K'}{\rho C \ell \delta}, \tag{12}$$

where ρ and C are the density and specific heat of the film, in agreement with [2]. In practice, we should not use (11) for $t >> \tau_1$, especially when the heat loss is large and/or the thermal conductivity of the backing material is not high enough, since under these conditions, the boundary condition (4) may not be valid any more.

Figure 2 shows the experimental PPTR signal $S(t)$ of the PC film at various separations from a germanium substrate. The values of $1/h$, which should be proportional to the gap width, are plotted as a function of δ in Fig. 3. To avoid the effect of the rising temperature at the backing substrate, the slope g is determined using data at different time intervals according to the width of the air gap. The agreement between the data and theory is good for air gaps of width 10-200 μm. Below 10 μm, the error due to surface roughness and alignment dominate. We have also measured the thermal contact resistance of the PC film in direct contact with a piece of brass under different pressures. The results (Fig. 4) indicate that the thermal contact resistance at the interface is essentially

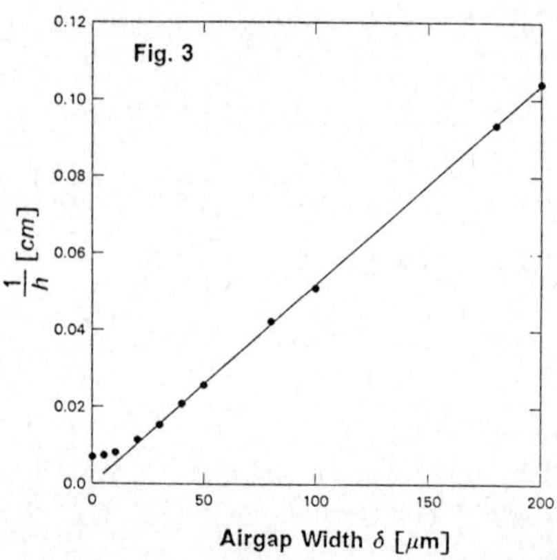

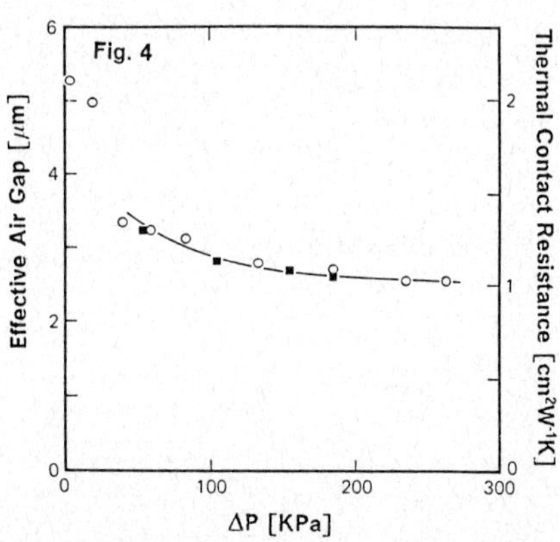

Fig. 3: Plot of $1/h$ vs δ; h is found from the average slope of log $S(t)$ against t between 0.8 and 20 ms for $\delta > 30$ μm, between 0.4 and 10 ms for $\delta = 20$ and 30 μm, and between 0.4 and 5.4 ms for $\delta < 20$ μm. Equation (12) applies when $1/h \geq 0.02$.

Fig. 4: Effective air gap width and thermal contact resistance of a carbon-loaded PC film in direct contact with a piece of thick brass. ΔP is the pressure difference between the front and back surfaces of the film. ○ is for a pressure of 84 kPa at the back surface; ▪ is for a pressure of 5 kPa at the back surface with respect to atmospheric pressure.

independent of pressure in the range of 40-250 kPa [6,7], and is equivalent to a 3 μm wide air gap. This value is close to the measured 2 μm peak-to-peak roughness on the film surface using a telestep machine (Dektak Model II A). The surface roughness of the brass substrate is found to be better than 1 μm. At low pressures, the width of the air gap increases due to the slight buckling of the film. We have also tried to deconvolute the contact surface area by repeating the experiment with gases of different thermal conductivities. Details on this and other related work will be published elsewhere [9].

In the calculation of the air gap width δ, we used the thermal conductivity of air found in [10], which is not correct if δ is less than a few mean free paths. The change in thermal conductivity of gases in these circumstances can have a significant effect on the thermal resistance between two contacting or nearly contacting surfaces. More details will be published.

This work is supported in part by the Office of Naval Research. The present address of H. Sontag is: Dornier-System GmbH, West Germany. The permanent address of W. P. Leung is: Physics Department, The Chinese University of Hong Kong.

References

1. W. P. Leung and A. C. Tam: J. Appl. Phys. 56, 153 (1984)
2. A. C. Tam and H. Sontag: Appl. Phys. Lett. 49, 1761 (1986)
3. D. L. Balageas, J. C. Krapez and P. Cielo: J. Appl. Phys. 59, 348 (1986)
4. F. Gitzhofer, L. Pawlowski, D. Lombard, C. Martin, R. Kaczmarck and M. Boulos: High Temp. - High Press. 17, 563 (1985)
5. R. E. Imhof, D. J. S. Birch, F. R. Thornley, J. R. Gilchrist, and T. A. Strivens: J. Phys. E. 17, 521 (1984)
6. H. S. Carslaw and J. C. Jaeger: *Conduction of Heat in Solids*, 2nd edition (Clarendon, Oxford, 1959), p. 114
7. A. M. Clausing and B. T. Chao: J. Heat Transfer 87, 243 (1965)
8. E. Fried and M. J. Kelley: In *Thermophysics and Temperature Control*, ed. by G. Heller (Academic Press, New York 1966), p. 697
9. W. P. Leung and A. C. Tam, submitted to *Appl. Phys. Lett.*
10. *Handbook of Chemistry and Physics, 60th Edition*, ed. by R. Weast (CRC Press Inc., Boca Raton, Florida 1979), p. E-2

Thermal Wave Characterization of Translucent Ceramic Coatings

J.D. Morris, D.P. Almond, P.M. Patel, and H. Reiter

School of Materials Science, Bath University, Claverton Down, Bath, Avon BA2 7AY, United Kingdom

1 Introduction

Thermal wave interferometry has been successfully used to measure the thickness of metallic plasma-sprayed coatings, and to detect the presence of defects such as debonded regions, inclusions or air gaps [1]. Such metallic coatings are opaque and relatively thermally conductive. In contrast to this, ceramic coatings such as Yttria-stabilised Zirconia (YSZ) or Alumina are translucent and relatively poor thermal conductors. The effect of translucency on the thermal wave response of ceramic coatings is described below.

2 Theory

Thermal waves can be generated in coatings by heating due to the absorption of periodic radiation such as that from a modulated laser. In an opaque coating, light absorption, and thus thermal wave generation, is confined to the surface. In a translucent medium, light absorption also occurs below the surface which gives rise to subsurface-generated thermal waves. These subsurface thermal waves interfere with each other and with surface generated thermal waves, causing a phase shift with respect to an otherwise identical opaque coating.

Using equations derived by BENNETT and PATTY [2], it can be shown that the normalised phase of the signal at very large coating thickness (i.e. thickness > $1 \cdot 5\mu$, the thermal diffusion length) is given by arc $\tan(-(\beta+\sigma)/\beta)$ where σ is the complex thermal wavenumber. Where the coating is opaque, $\beta \gg \sigma$ and the phase is $-45°$; where the coating is translucent, β approximates σ and the phase is less than $-45°$.

3 Results and Discussion

Two types of YSZ coating of various known thicknesses were measured using a range of frequencies, and the results were normalised to 1 Hz. Theoretical computer generated curves derived from equations in [2] were fitted to the data (Fig. 1). The physical values giving optimum fit are listed in Table 1.

A third type of Zirconia-coated sample was partially darkened with colloidal graphite to allow a comparison between

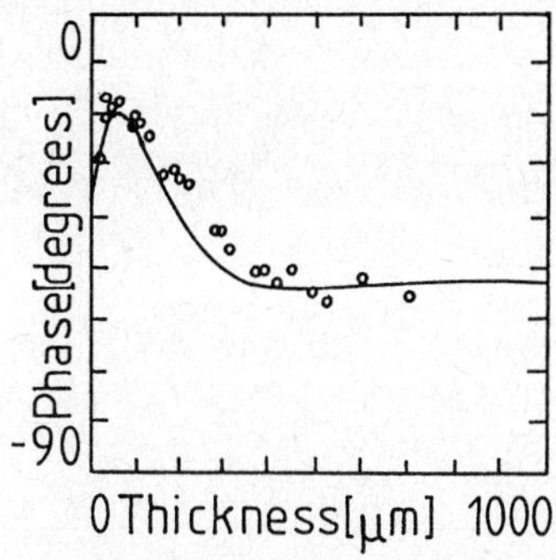

Figure 1 (a):
Theoretical and experimental data for Type I YSZ on stainless steel. Experimental data normalised to 1 Hz
ß=3.10^4/m
k=0·35 W/m/K

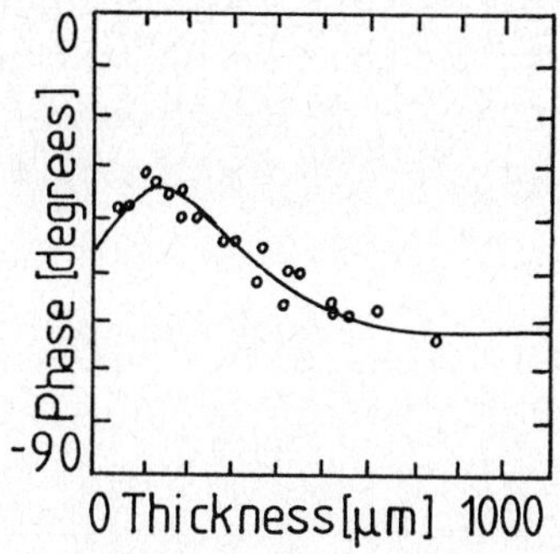

Figure 1 (b):
Theoretical and experimental data for Type II YSZ on stainless steel.
Experimental data normalised to 1 Hz
ß=1.10^3/m
k=1 W/m/K

opaque and translucent zirconias. The results were compared with theoretical models (Fig. 2). High modulation frequencies were used to eliminate effects from reflections at the coating-substrate boundary.

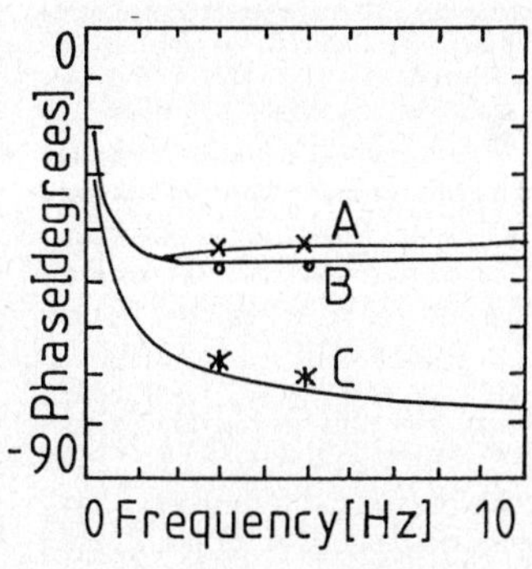

Figure 2:
Theoretical and experimental data for colloidal graphite on Type III YSZ on stainless steel

Experimental data
× 12 μm graphite
∘ 22 μm graphite
✳ 0 μm graphite

Theoretical data
A 12 & 22 μm graphite
B uncoated "opaque" YSZ
C uncoated translucent YSZ

Table 1: Properties of YSZ coatings

Coating	Density [kg/m^3]	Cp [J/kg/K]	k [W/m/K]	ß [/m]
Type I	5000	500	0·35	3.10^4
Type II	5000	500	1·0	1.10^4
Type III	5000	500	0·35	1.10^4

It can be seen from Fig. 1 that the Type-I YSZ curve tails off at about -56° indicating that ß is about 3.10^4/m. The curve for the Type-II YSZ tails off at a phase of about -62° indicating that ß is approximately 10^4/m. After allowing for

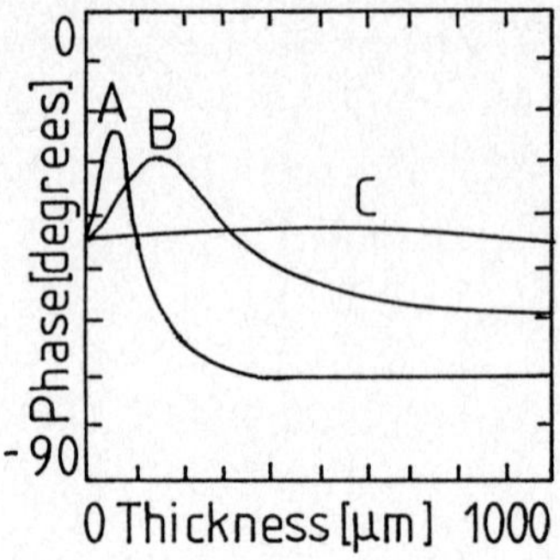

Figure 3:
The effect of thermal conductivity on the phase-thickness plot of YSZ on stainless steel

A k= 0·1 W/m/K
B k= 1·0 W/m/K
C k=10·0 W/m/K

the effects of the graphite layer, the results in Fig. 2 suggest a value of ß of between 7.10^3/m and 10^4/m for the undarkened coating.

Figure 3 shows modelled curves for YSZ on stainless steel with differing values of k, the thermal conductivity. It can be seen that there is a large difference within the range of values used. Literature values of k range between 0·1 W/m/K for sprayed coatings [3] and 1·9 W/m/K for bulk ceramic [4]. Curve fitting the data to the models suggested that k was 0·35 W/m/K for type I YSZ, and 1 W/m/K for the type II YSZ. These values correspond well with other values quoted in the literature [5,6,7].

4 Conclusions

It is possible to determine the thermal conductivity of the coating either by curve fitting or by noting the position of peak phase shift. The optical absorption coefficient ß can be determined by curve fitting, or by finding the large-thickness phase and using the relation phase=arc tan(-(ß+σ)/ß).

Artificially darkening the coating, e.g. with colloidal graphite, has more effect on the phase than would be expected if the coating could be simply made opaque. This can nonetheless be used as a method to determine ß provided the physical properties of the dark layer are known so that the effects of the darkening layer can be accounted for.

5 Acknowledgements

The authors gratefully acknowledge the support of the Central Electricity Generating Board, the Science and Engineering Research Council, and Ministry of Defence.

6 References

1. D.P.Almond et al, Materials Evaluation,Apr 1987,pp1986-1987
2. C.A.Bennett and R.R.Patty, Appl. Optics,V21(1),1 Jan 82, pp49-54
3. M.K.Hobbs, Surfacing Journal,V16(4), 1985,pp101-108
4. K.E.Wilkes and J.F.Lagedrost, NASA C-121144, Battelle Columbus Labs,Ohio
5. R.Brandt et al,High Temp High Press,V18,1986,pp65-77
6. R.McPherson,Thin Solid Films,V112,1984,pp89-95
7. P.Cielo and S.Dallaire,Presented at 1985 ASM Metals Congress, Toronto,Canada,12-17 Oct 1985

Photothermal Nondestructive Inspection of Paint and Coatings

G. Busse, D. Vergne, and B. Wetzel*

Institut für Kunststoffprüfung und Kunststoffkunde,
Universität Stuttgart, D-7000 Stuttgart 80, Fed. Rep. of Germany
*Permanent address: Institut für Physik, FB ET, Universität der Bundeswehr München, D-8014 Neubiberg, Fed. Rep. of Germany

The investigation of coatings and their defects has been of interest since the early applications of thermal waves for NDE purposes /1/ because paint and coatings are important materials to protect metals and polymers against environmental influences (e.g. aggressive media). To optimize the protection one needs to monitor the process conditions, coating thickness, and coating deterioration. In situ monitoring is difficult if the paint is still wet. Therefore we investigated which information is provided by remote photothermal inspection, both in thermal wave transmission (if sample thickness is not large compared to thermal diffusion length) and front surface arrangement (applicable also to thermally thick samples) at 18 Hz modulation frequency.

1. DRYING AND PRETREATMENT

The drying process of black paint on a metal substrate is shown in Fig. 1. This result has been obtained in a transmission arrangement. Therefore the phase angle is proportional to the ratio of thickness and thermal diffusion length /2/, both being dependent on time in this process. However, the curve in Fig. 1 could be used in a production process to determine when painted parts can be processed.

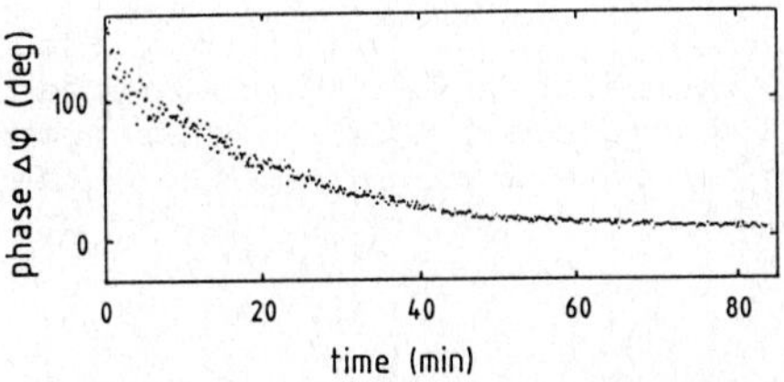

Fig. 1: Drying process of paint on metal observed with thermal wave transmission

No averaging of the data points has been performed to show the improvement of the signal to noise ratio as a function of time.

We found that the temperature history has an effect on the polymer coating. The reason may be different degrees of crosslinking, or, as well, changes that affect the boundary conditions. As an example, Fig. 2 shows results obtained on four coated polymer samples where two samples were cured for 30 minutes and two for 120 minutes (dashed curve). Only two data points are shown for each curve. However, the general shape is known from other measurements (see e.g. Figs. 3 and 4). We hope to use data of this kind for process optimization.

Also it should be mentioned that this result and those described later were obtained in a single-ended arrangement that is of more interest for applications.

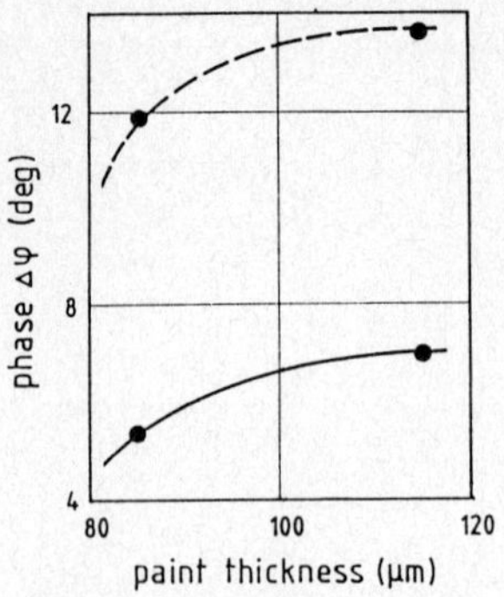

Fig. 2: Influence of heat treatment on initially identical samples (paint on polymer). Duration of elevated temperature 30 minutes (bottom) and 120 minutes (top)

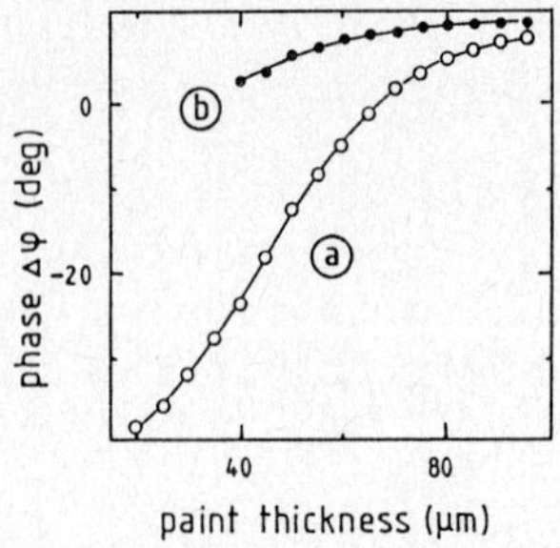

Fig. 3: Influence of substrate on thickness dependent phase:
metal (a), polymer (b).
the same paint has been used

2. THICKNESS MEASUREMENTS

Besides process monitoring, the determination of paint thickness is of interest, especially for polymer substrates where at present no nondestructive method is available.

The difficulty in thermal wave applications is the small reflection at the substrate/coating interface. In Fig. 3, curve "a" shows how signal phase depends on paint thickness for a metal substrate, while "b" shows the corresponding data for a polymer substrate. The slope is obviously reduced thereby making calibration more difficult.

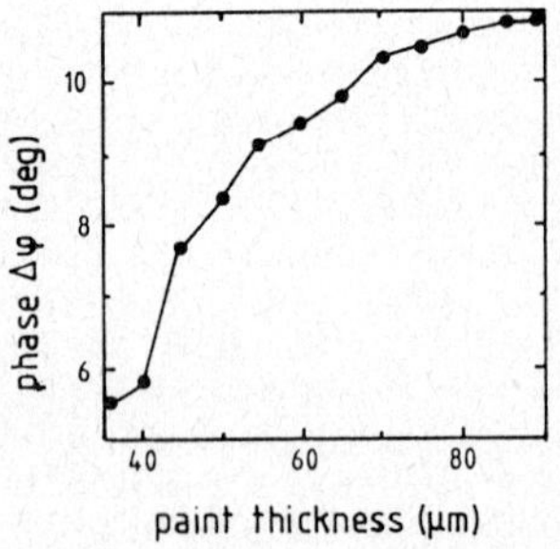

Fig. 4: Signal phase depending on paint thickness on polymer substrate.
Phase accuracy is better than ∓ 0.2 degrees

However, a detailed analysis of curve "b" shows (Fig. 4) that with a phase angle accuracy of ∓ 0.2 degrees a paint thickness accuracy of nearly ∓ 2 μm at 50 μm and ∓ 6 μm at 75 μm can be achieved that is good enough for most applications e.g. in car manufacturing industries. The flat part of the curve occurs when paint thickness is about twice the thermal diffusion length /3/, measurements of larger paint thickness are therefore possible at lower frequencies. The surface area that is needed for evaluation has a diameter of less than 0.3 mm.

3. INFLUENCE OF SUBSTRATE

As these front surface measurements depend on boundary reflection, it is obvious that the calibration curves need to be made for each polymer substrate. As an illustration, Fig. 5 shows the paint thickness dependent signal phase for 5 different polymers (and the same paint). For certain paint/polymer combinations the curve can be rather flat (e.g. for PA in Fig. 5).

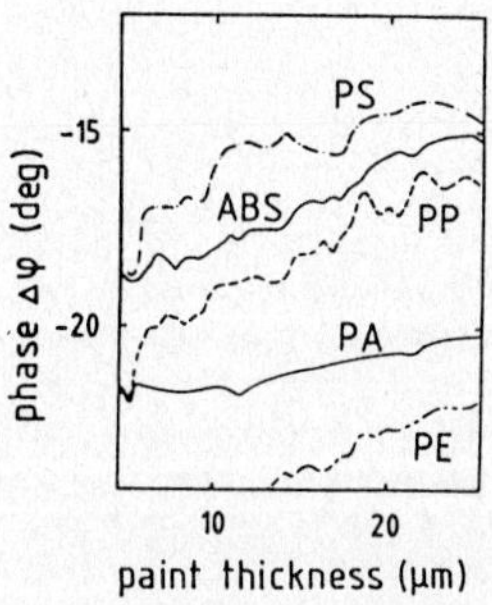

Fig. 5: Influence of paint thickness for various polymer substrates

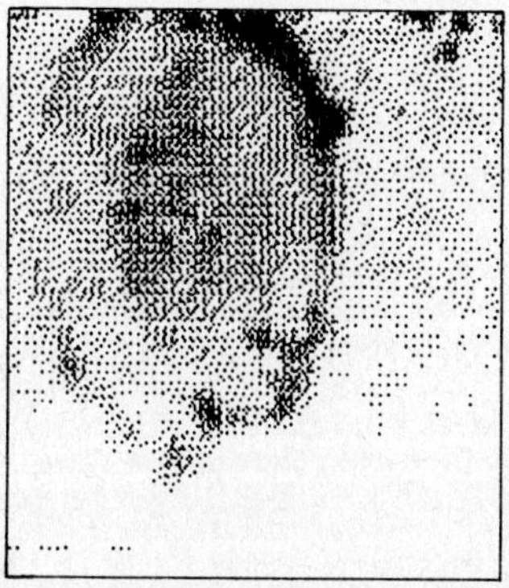

Fig. 6: Detection of finger print on polymer surface under layer of black paint.

These curves do not look as smooth as the one in Fig. 4 which was obtained at larger thicknesses. The structures in Fig. 5 are presumably related to local inhomogeneities which are difficult to avoid if the paint is very thin.

From the importance of the boundary it is obvious that the signal may be affected by surface contaminations. Figure 6 is the result of a raster scan across a sample where a finger print had been applied to the polymer surface before coating it with black paint.

The investigation of coatings and their defects is obviously more difficult if the substrate is not a metal but a polymer. However, our examples indicate that remote and nondestructive photothermal analysis is well applicable to monitor processes, thicknesses, and defects of painted polymer systems.

Acknowledgement: The authors gratefully acknowledge financial support by the German ministry of research and technology (BMFT) under Contract-No. 13 N 5318/1.

4. LITERATURE

1. G. Busse, A. Ograbek: J. Appl. Phys. 51, 3576 (1980)
2. G. Busse: Infrared Physics 20, 419 (1980)
3. G. Busse: Appl. Phys. Lett. 35, 759 (1979)

Analysis of Thermal Wave Reflectivity at Rough Surfaces in Contact

P.M. Patel, D.P. Almond, and H. Reiter

School of Materials Science, University of Bath, Claverton Down, Bath, Avon BA27AY, United Kingdom

INTRODUCTION

The detection of subsurface structure in opaque and translucent materials by thermal waves has been widely reported over the past decade for fatigue cracks, subsurface channels, drilled holes, spherical and disc-shaped air-filled defects and partially bonded surface coatings /1,2/. These investigations have led to the development of two models for the characterisation of subsurface "defects":-

i) the discrete air gap layer, and
ii) the thermal contact resistance.

Recent work on defect detection in coated structures has shown the inadequacy of the simple air gap model in explaining the thermal wave response of "real defects" /2/. The presence of contact points between delaminations in these types of bonding defects resulted in the departure of the thermal wave response from that of an air gap layer. This paper examines the usage of the thermal contact resistance model for coating delaminations modelled as rough surfaces in contact. A comparison of the detectability of contacting surfaces by thermal and ultrasonic waves is also discussed.

THEORY

The thermal contact resistance between rough surfaces in contact can be related to the true contact area by applying a statistical description to the surface topography /3,4/. In the following analysis the assumptions below are made:

i) The thermal contact resistance at a contacting surface may be divided into two components:

a) the thermal resistance due to contact between surface asperities (the constriction resistance /5/), R_m, and
b) the thermal resistance due to the trapped fluid (gas) layer between the mating surface, R_g.

ii) R_m and R_g are independent of each other and the total resistance of the interface is given by

$$\left(\frac{1}{R}\right)=\left(\frac{1}{R_m}\right)+\left(\frac{1}{R_g}\right). \tag{1}$$

iii) The contact topography of the rough surface follows the plastic contact theory. This theory assumes that when two rough surfaces (rms h_1 and h_2) are pressed into contact, the surface deformation can be modelled as being that of a rigid flat plane on a surface having a combined roughness $h_c=(h_1^2+h_2^2)^{1/2}$. Further, the asperities of the rough surface deform plastically with increasing load, and at any load, the true contact consists of a large number of equi-area circular contacts of radius r.

For a contact surface formed between a rigid flat plane and a rough surface having a normal distribution of conical asperities /6/, the thermal resistance R_m can be obtained by applying the relationships

$$\left(\frac{A_t}{A_o}\right)=\left(\frac{S}{P_m}\right)=N\pi r^2, \qquad (2)$$

$$r=\left(\frac{2}{\pi}\right)\left(\frac{h_c}{u}\right)(3.48h_c+4.69), \qquad (3)$$

$$R_m=\frac{1}{(2Nrk_m)}, \qquad (4)$$

where

- A_t is the real area of contact,
- A_o is the apparent area of contact,
- S is the applied stress,
- P_m is the hardness of the deforming material,
- N is the number of the contact spots per unit area,
- u is the distance between the mean plane of the rough surface and the perfectly flat surface,
- k_m is the harmonic mean thermal conductivity.

The thermal resistance, R_g, can as a first approximation be written as /5/

$$R_g=\frac{h_c}{k_g}, \qquad (5)$$

and the corresponding change in the magnitude of the thermal wave reflection coefficient , Γ, may be evaluated from the relation

$$\Gamma=\frac{(1-a)+Rk_2\sigma_2}{(1+a)+Rk_2\sigma_2}, \qquad a=\frac{k_2\sigma_2}{k_1\sigma_1}, \qquad (6)$$

assuming one-dimensional heat flow.
a is the thermal effusivity ratio; k_i and σ_i are the thermal conductivity and thermal wave vector of medium i (i=1 , 2).

DISCUSSION

In fig. 1a and b, the thermal contact resistance, for contact spots in vacuum and air, is plotted against the normalised contact area (A_t/A_o) for surface roughness

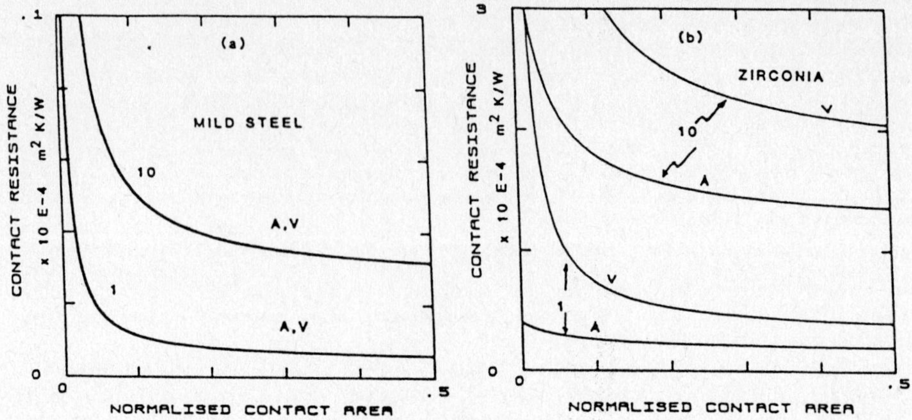

Figure 1a and 1b: The dependence of thermal contact on the fractional contact area for mild steel and zirconia samples for contact in air (A) and in vacuum (V) with interface roughnesses of 1 and 10 μm.

values of 1 and 10 μm rms and harmonic mean thermal conductivities of 64 and 1 W/m K. These figures show that thermal contact resistance decreases rapidly with increasing contact area ratio. The thermal contact resistance is lower for contact with smooth surfaces and materials with high thermal conductivity. Air present between the contacting surfaces also reduces the thermal resistance of the joint, the overall reduction being dependent on the spot thermal resistance ,R_m.

Figure 2a and b show the normalised phase angle dependence on contact area of thermal contact resistance in mild steel and

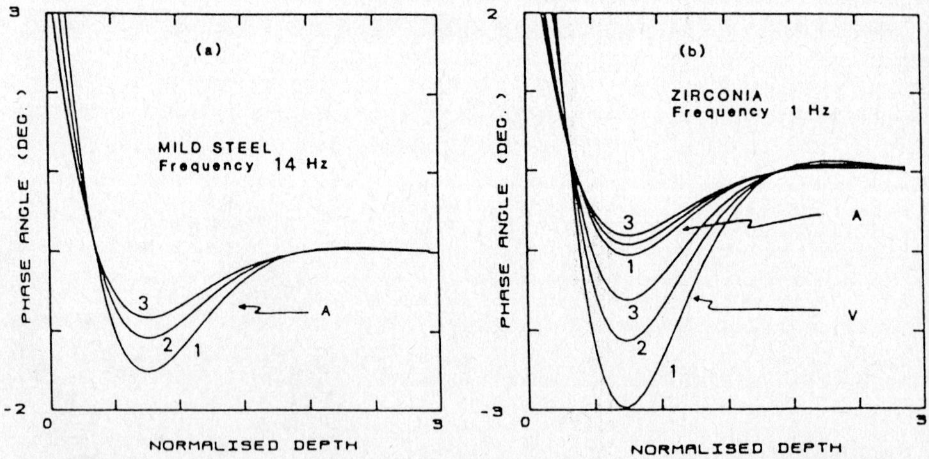

Figure 2a and b: Plots of the thermal wave phase angle variation with normalised depth and fractional contact area of 1) 0.1, 2) 0.2 and 3)0.5 of partially contacting interfaces in mild steel (Fig. 2a) and zirconia (Fig. 2b) with a roughness of 1 μm. (A= contact in air, V= contact in vacuum.)

opaque zirconia samples as a function of the normalised depth (depth /thermal diffusion length) beneath the surface - assuming the thermal waves are being generated and detected at the surface. The surface roughness of the contacting faces for these two plots is 1 μm rms. For an experimental phase angle noise of 1 degree, it is seen from the plots that fractional contact areas up to 0.3 can be readily detected by thermal waves for these samples. It is also seen from fig. 2b that the presence of air between the contact surface of low conductivity materials reduces the detectability of this interface.

To compare the sensitivity of thermal and ultrasonic waves to contacting surfaces we assume that, for comparable resolution between the two techniques, the thermal diffusion length equals the ultrasonic wavelength. This allows the determination of the appropriate thermal wave test frequency. Next it is assumed that a suitable noise level be set to the detected signal: an experimental phase angle noise of 1 degree for thermal waves and a 1dB signal change for ultrasonics. Figure 3 shows the variation of the ultrasonic reflection coefficient magnitude (computed from equation 19 in /6/) against the normalised contact area for a mild steel sample with a rms interface roughness 1 μm and 10 MHz. ultrasonic wave. In a through transmission ultrasonic test a 1 dB signal change corresponds to detecting a contact area ratio of about 0.1. This value is slightly less then the contact area ratio predicted for thermal waves of 0.3.

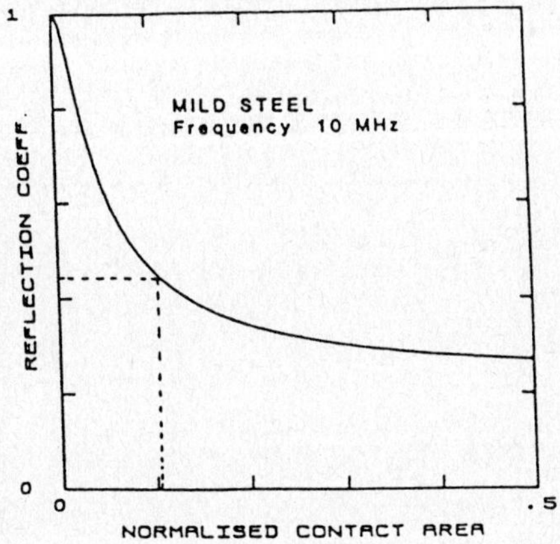

Figure 3. A plot of the change in the ultrasonic reflection coefficient magnitude with fractional contact area in a mild steel sample of roughness 1 μm and an ultrasonic frequency of 10 MHz.

CONCLUSION

A thermal resistance model using a statistical description of the interface roughness has been applied to determine the sensitivity of thermal wave methods in detecting contacting surfaces. The analysis predicts that contact between surfaces of high thermal conductivity and effusivity are more readily detectable and also thermal waves are more sensitive in detecting contacting surfaces than ultrasonics.

ACKNOWLEDGEMENTS

This work is supported by the Science and Engineering Research Council and the Central Electricity Generating Board, U.K.

REFERENCES

1. **G. Busse:** IEEE SU-32(2) p355-346 (1985)
2. **P.M. Patel, D.P. Almond and H. Reiter**
   Appl. Phys. B43 p9-15 (1987)
3. **C.V. Madhusudana and L.S. Fletcher**
   A.I.A.A., J. 24(3) p510-523 (1986)
4. **T. Tsukizoe and T. Hisakado**
   A.S.M.E., J. Lub. Technol. p81-88 (1968)
5. **Yu. P. Shlykov and Ye. A. Ganin**
   Int. J. Heat and Mass Transfer 7 p921-929 (1964)
6. **N.F. Haine:** CEGB Report RD/B/N 4744 (1980)

Photoacoustic Measurement of Subsurface Air Gaps

A. Lachaîne

Department of Physics, Royal Military College,
Kingston, Ontario, K7K 5LO, Canada

Various authors [1-3] have reported measurements on multilayer samples in an effort to detect and characterize the various layers. The complete analysis of the thermal wave reflections [4] between the various layers is in general quite complicated. We report on a simple method of interpreting photoacoustic "signatures" to obtain estimates of <u>both</u> the depth and the thickness of a buried layer, in this case, air gaps of various thicknesses sandwiched between two layers of Pb metal. The samples which were investigated consisted of an optically opaque first layer of Pb, thickness ℓ_1 = 140 µm, a second layer of air, thickness ℓ_2, and a third layer of Pb (thermally thick). Chopped white light was made incident on the first layer of the samples inside a closed photoacoustic cell and the amplitude and phase of the signal were measured as a function of chopper frequency, f. Details of the experimental set-up are published elsewhere [5]. Measurements were first done on a reference sample which was a thermally thick piece of Pb, thickness 2.0 mm (no air gap). Subsequent measurements on all other samples were normalized to this reference sample.

The analysis of the curves is done by using the concept of characteristic frequencies $f_i = \alpha_i/\ell_i^2 (i = 1,2)$ where α_i is the thermal diffusivity of the i-th layer and ℓ_i is its thickness.

The amplitude vs. chopper frequency is plotted on a log-log scale in Fig. 1. The bottom curve (D) is the reference sample (no air gap) while the top three curves are for air gaps of ℓ_2 = 2.0 mm (A), 370 µm (B) and 250 µm (C). One can clearly see the effect of the thermal wave interferences in the air gaps up until the characteristic frequency of the first layer of Pb is reached, $f_1 = \alpha_1/\ell_1^2$, where all four curves meet. Thus, if α_1 is known, the thickness of the first layer, or depth of the air gap, ℓ_1, can be estimated by reading off this point of intersection.

The phase vs chopper frequency is plotted in Fig. 2. The top curve (A) is for a thermally thick air gap (ℓ_2 = 2.0 mm) while the lower two curves are for air gaps of ℓ_2 = 370 µm (B) and 250 µm (C). The transition from thermally thin to thermally thick air gaps is spread out over a region delimited by a minimum and a maximum characteristic frequency, f_{2min} and f_{2max}, giving an estimate for the air gap thickness between ℓ_{2max} and ℓ_{2min} respectively. Table I summarizes the values for the depth ℓ_1 and the thickness ℓ_2 (average of ℓ_{2min} and ℓ_{2max}) of the air gaps compared with the nominal values as measured with a micrometer.

This research was supported by the Ministry of National Defence of Canada under CRAD/ARP #3610-811.

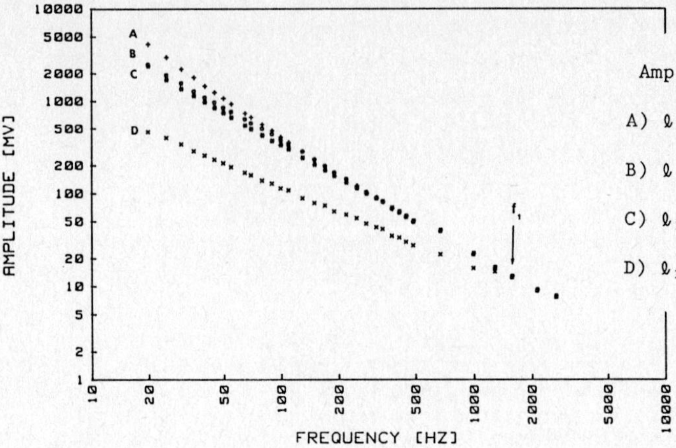

Figure 1

Amplitude vs. Frequency

A) ℓ_1 = 140μm; ℓ_2 = 2.0mm

B) ℓ_1 = 140μm; ℓ_2 = 370μm

C) ℓ_1 = 140μm; ℓ_2 = 250μm

D) ℓ_1 = 2.0mm(no air gap)

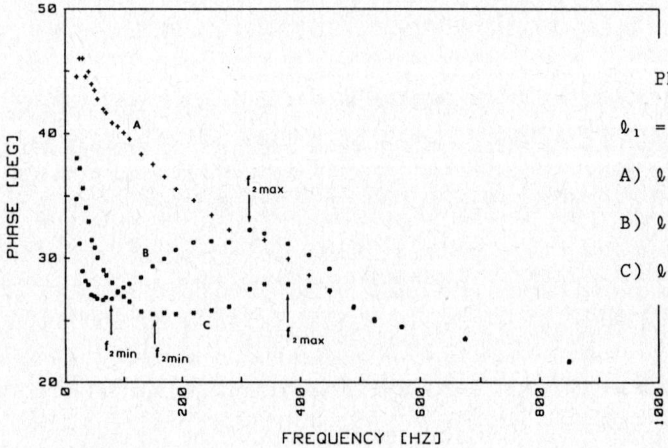

Figure 2

Phase vs. Frequency

ℓ_1 = 140μm

A) ℓ_2 = 2.0mm

B) ℓ_2 = 370μm

C) ℓ_2 = 250μm

Table I

Nominal values(μm)		Characteristic Frequencies(Hz)			Measured Values(μm)	
ℓ_1	ℓ_2	f_1	f_{2min}	f_{2max}	ℓ_1	ℓ_2(avge)
140	250	1600	150	380	130	290
140	370	1600	80	320	130	365

Values used for α_1 = 0.25 cm^2s^{-1}(Pb) and α_2 = 0.19 cm^2s^{-1}(air) [6]

1. J. Baumann, R. Tilgner: J. Appl. Phys. 58, 1982 (1985).
2. A.C. Tam, H. Sontag: Appl. Phys. Lett. 49, 1761 (1986).
3. D.P. Almond, P.M. Patel, H. Reiter: Mat. Eval. 45, 471 (1987).
4. J. Opsal, A. Rosencwaig: J. Appl. Phys. 53, 4240 (1982).
5. A. Lachaîne: J. Appl. Phys. 57, 5075 (1985).
6. Y.S. Touloukian: Thermophysical Properties of Matter (IFI/Plenum, New York, 1970).

Thermal Wave Imaging of Ceramic Thermal Barrier Coatings

M. Liezers[1] *and R.M. Miller*[2]

[1]Department of Instrumentation and Analytical Science, University of Manchester Institute of Science and Technology, P.O. Box 88, Manchester, M60 1QD, United Kingdom
[2]Unilever Research Port Sunlight Laboratory, Quarry Road East, Bebington, Wirral, L63 3JW, United Kingdom

1. Introduction

There is an ever increasing demand for materials capable of extended operation at very high temperatures. One approach to the problem is the use of refractory ceramic coatings on metal substrates. These materials offer high temperature stability coupled with corrosion resistance. To ensure good coating quality an intermediate bond coat is applied between the substrate and the ceramic thermal barrier coating. This acts as a matching layer between the coating and the substrate to ensure good adhesion. In evaluating possible types of coating and bond coat, it is necessary to assess the expected life time in use. It is possible, by alternate heating and cooling of test pieces, to simulate in the laboratory many thousands of cycles under operational conditions. Samples eventually fail catastrophically, and the number of cycles is a measure of the coating quality. Unfortunately, this method is extremely time consuming. With thermal wave imaging it might be possible to detect the onset of failure before visible evidence is available.

2. Experimental Method

The instrumentation used for this study is as previously described[1,2]. The light source was an amplitude modulated krypton ion laser operating at 647 nm at a power of 500 mW. The scan area was approximately 20 mm x 20 mm.

Samples were mounted as previously described with a piezo-electric transducer held in contact with the rear surface of the sample using a Perspex C-clamp[2].

Four samples were obtained consisting of steel plates 6 cm x 6 cm and 0.25 cm thick. These were coated on one surface with a test bond coating, and a zirconia thermal barrier coating approximately 0.3 mm thick[3]. The samples represented different stages of the coating's life. They ranged from a fresh sample, to a sample which had been cycled more than 1200 times. Damage ranged from no visible coating failure to extensive delamination and cracking of the coating. Samples were selected so as to include two examples either side of the anticipated point of initial coating failure.

3. Results and Discussion

Initial experiments were carried out at the sites of visible coating failure in the most highly stressed samples. Figure 1 shows the amplitude image and Fig. 2 the phase image of a site recorded at 10 Hz

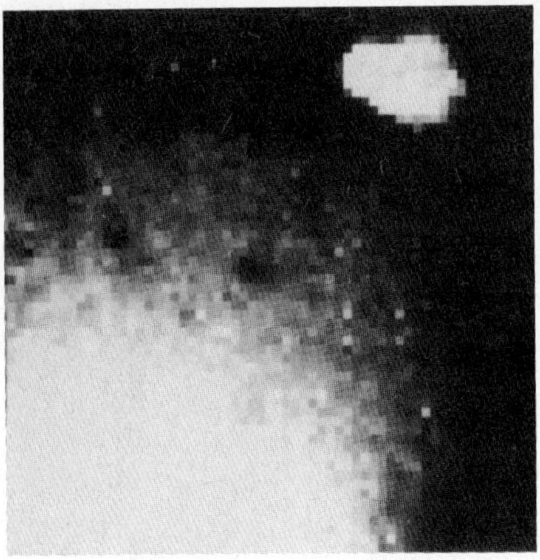

Figure 1. Thermal wave amplitude image of failed coating recorded at 10 Hz

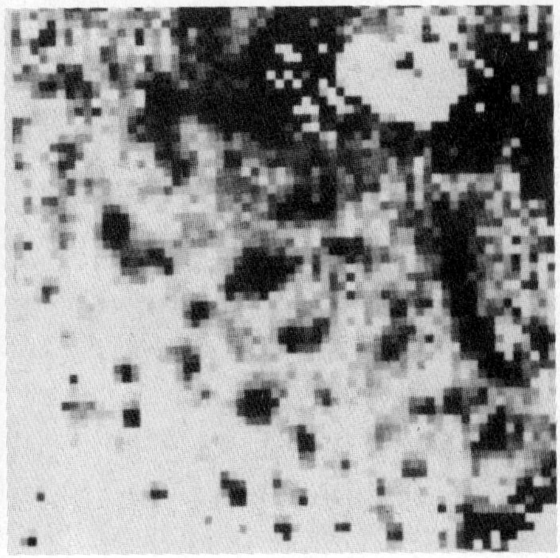

Figure 2. Thermal wave phase image of failed coating recorded at 10 Hz

modulation frequency. Sites of obvious coating failure are clearly visible in both amplitude and phase images. Where the substrate was exposed, the signal amplitude was at a maximum, and the phase lag at a minimum, recording as a distinct white patch. On either side of the exposed substrate area, two dark regions of different character were visible. These had a very low signal amplitude, a high signal phase lag, and had a high contrast with the surrounding intact coating regions. This suggested extremely poor coating substrate contact, and close optical examination revealed swollen regions of the coating suggesting delamination.

The phase and amplitude response of different regions of the sample were recorded from 5-500 Hz. Figure 3 shows the frequency response for

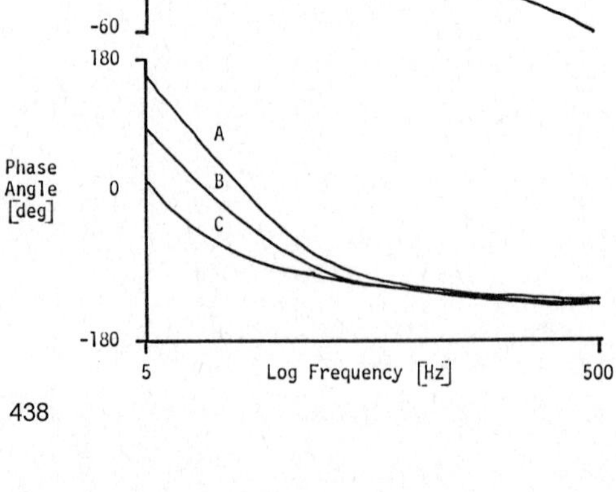

Figure 3. Frequency response A: exposed substrate, B: normal coating, C: delaminated coating

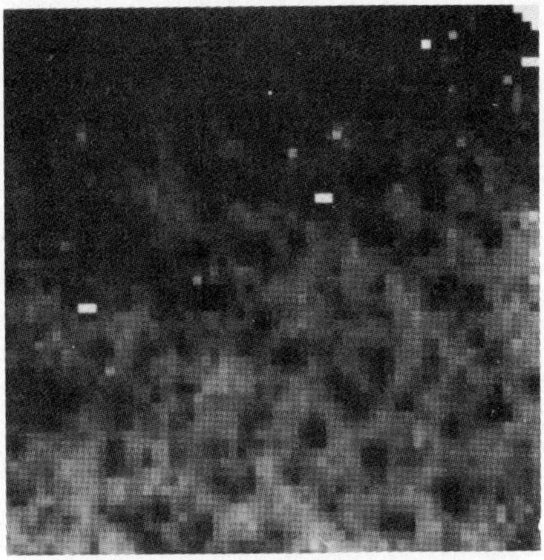

Figure 4: Thermal wave phase image of incipient coating failure recorded at 10 Hz

exposed oxidised substrate, normal coating and delaminated coating. The oxidised susbstrate gives the highest amplitude at all frequencies, and the delaminated coating the lowest, as would be predicted from simple considerations of source amplitude and thermal impedance. The phase response shows similar behaviour.

A further type of feature can be seen in both images, consisting of small dark regions 1-4 pixels across. These are more clearly seen in the phase image. Close optical examination of the samples reveals that these features are associated with small locally thickened regions of the coating which appear to contain embedded particles 50-200 micrometers in size. The frequency response of these features was slightly different to that observed for the main areas, suggesting a defect in the coating above the bond coat layer.

Having obtained some confidence in the ability to locate major defects, attention turned to the sample where no visible damage had occurred. Thermal wave imaging of this sample revealed a number of small features with phase and amplitude characteristics similar to the delamination regions and small defects in the failed sample. The phase image is shown in Fig. 4. These features were of varying size, and in the case of the larger features, some slight swelling of the coating could be detected optically. This suggests that the failure mechanism could be progressive oxidation of the bond coat starting with a number of sites across the coating. Initial failure points could develop at defects within the coating leading to bond coat oxidation. As these defect points grow and develop, they link up to form larger delamination regions leading to catastrophic failure of the coating.

These results indicate that thermal wave imaging can be used to detect bond coat failure at an earlier stage than conventional methods. This will provide further information on the mode of failure of the coatings, and will allow more rapid assessment of coating quality.

1. G F Kirkbright, M Liezers, R M Miller and Y Sugitani, Analyst, 109, 465 (1984)
2. G F Kirkbright, M Liezers and R M Miller, Spectrochim. Acta B, 41, 741 (1986)
3. Samples supplied by Dr R Taylor, Department of Metallurgy, UMIST, Manchester, UK

Laser-Generated Thermal Wave Interference for NDT of Hard Coatings on Boiler Tubes and Reactor Components

J. Corbett, M.B.C. Quigley, B. Hart, and B.L. Smith

Central Electricity Generating Board,
Marchwood Engineering Laboratories,
Marchwood, Southampton, SO44ZB, United Kingdom

Thick, hard plasma sprayed coatings of materials such as NiCr and Alumina are widely used by the Central Electricity Generating Board. These provide protection against corrosion and erosion in the boilers and superheaters of coal fired stations and for components in many other areas. Good quality control during manufacture is important and the wide variation in operating conditions necessitates regular inspection during their lifetime. All of the generally used techniques are destructive and/or contacting and photoacoustic inspection is not suitable because of the strong scattering of acoustic waves due to the coating structure. This is characterized by an irregular surface, variation in particle shape, porosity, oxide and grit inclusions and variable substrate contact. Photothermal interferometry provides a non-contacting and accurate technique for measuring coating thickness and locating and characterising many types of defect.

The technique is based on the periodic heating of a surface by a modulated light source - a 5 W Ar-ion laser. This provides a high power, controllable heat input. The theory of propagation of thermal waves is covered in (1,2,3) and has been applied to coatings in (4). The test equipment is based on that described in (4).

The results of coating thickness measurements performed on LC1B coated onto mild steel are shown in Figure 1. The coating thickness varies from

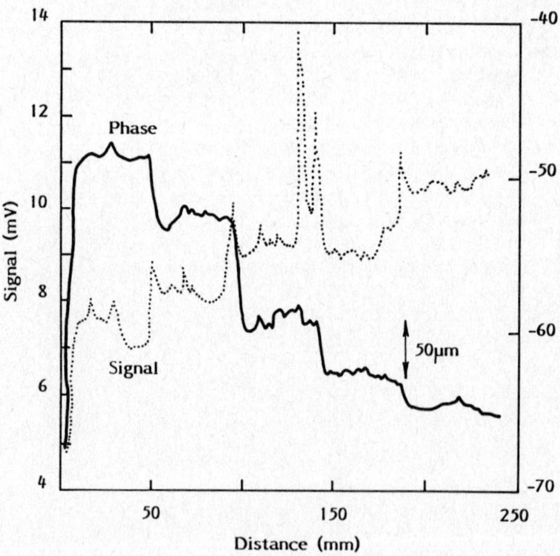

Fig. 1

50 to 250 μm in 50 μm steps. The sensitivity is optimised by choosing a thermal diffusion length, μ, which is comparable to the coating thickness. $\mu = (2K/\rho C\omega)^{\frac{1}{2}}$ where K = thermal conductivity, ρ = density, C = specific heat and ω = chopping angular frequency, and can thus be varied via the chopping frequency. At 10 Hz a phase difference of ∿ 10° is observed when the coating thickness increases from 150 to 200 μm, i.e. 0.1° per 0.5 μm. Since the lock-in amplifier can measure phase differences to 0.1° accuracy when the time constant is 3 s, sub-micron thickness resolution is, in principle, possible. However, the results of photothermal measurements will be affected by surface texture and coating porosity.

The most common types of defect in a plasma sprayed coating are areas of delamination or debond i.e. thin air layers between the coating and substrate parallel to the surface. To simulate these, thin disc-shaped defects of varying size were produced at depths of 0.2 to 1.4 mm below the surface of steel plates. Figure 2 shows the result of scanning across the surface 0.4 mm above a 0.5 mm defect. A laser beam focussed to ∿ 100 μm spot was used. While the technique is, in principle, capable of this order of resolution, the defect depth (required μ) and surface roughness will limit the resolution capability. Areas of variable thermal contact can be dealt with as described in (5). Other types of defects, e.g. vertical and angled cracks, have been detected. However, characterisation requires a full three-dimensional treatment. A 3-D heat flow model is also required when the defect size is comparable to the coating thickness or the laser spot size and is currently being developed.

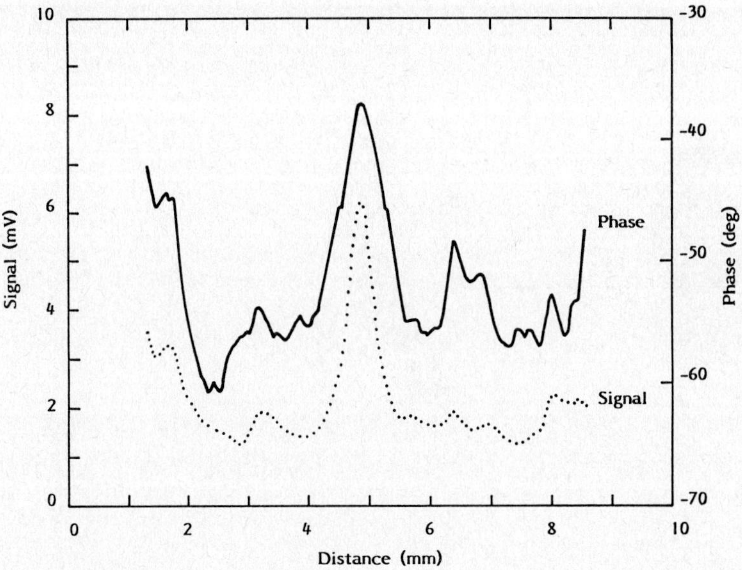

Fig. 2

While plasma sprayed components can now be tested for thickness or debonding in the laboratory, a portable system for quality control and in-situ testing is desirable. To this end, a portable diode laser based system is being developed.

ACKNOWLEDGEMENTS

The authors gratefully acknowledge the help of their colleagues at the Marchwood Engineering Laboratories and the Department of Materials Science, Bath University. This paper is published by permission of the Central Electricity Generating Board.

REFERENCES

1. H.S. Carslaw, J.C. Jaeger: Conduction of Heat in Solids (Oxford University press, Oxford, 1959).
2. A. Rosencwaig, A. Geersho: J. Appl. Phys 47, 1, 64 (1976).
3. G.A. Bennett, R.R. Patty: Appl. Opt., 46, 11, 1013, (1982).
4. P.M. Patel, D.P. Almond: J. Mat. Sci. 20, 955 (1985).
5. P.M. Patel, D.P. Almond, H. Reiter: Appl. Phys. B 43, 9, (1987)

A New PTR Imaging System for the Characterization of Layered Materials

E. Van Schel, E. Merienne, C. Menu, P. Poplimont, and M. Egée

Service Universitaire d'Energétique,
BP 347, F-51062 Reims Cedex, France

Based on a previous experimental and theoretical study /1, 2/, we describe here a new method for the characterization of the thermophysical and structural parameters of the upper layer and/or the interface of semitransparent bilayered materials (thickness l, thermal contact resistance R, thermal diffusivity α, absorption coefficient β).

The original character of this method lies in its dealing with a modelisation specific to semi-transparent bilayers. It is indeed the very existence of such a modelisation which has allowed us to result in a straightforward process: either with the measurements (only two frequencies are required) or with the comparison with the reference sample of known thickness.
We cannot here show the mathematical developments described thereafter /3/ for lack of space. This method allows us to measure- on a structure of the preceding type - one of the following parameters (each of the remaining parameters being considered as constant)

- the thickness of a coating (l') laid on a substrate
- the thermal contact resistance (R') between two layers
- the absorption coefficient (β) for a given wavelength and/or the thermal diffusivity (α)

In order to estimate the thickness (l') of a coating (its diffusivity and absorption coefficient being unknown), we make a series of measurements on the sample.

1) determination of α and β.
Their simultaneous determination lies in the measurement of the product $\beta(\alpha)^{1/2}$ and can be obtained through several stages:
- a phase and amplitude measurement at a high frequency (Fh = 10 Fb) on a reference sample of known thickness (l) and perfect adherence
- a phase (or amplitude) measurement at a low frequency (Fb= $\alpha/(\pi(l)^2)$ on the same sample
- a series of calculations resting on the theoretical model, which allows us to determine α and β.

2) determination of l':
A phase (or amplitude) measurement at the same frequency on the sample of unknown thickness (with a supposed perfect adherence) and the use of another series of calculations which thus allows us to determine l'.
The determination of the thermal resistance is obtained through the same stages
- a phase (or amplitude) measurement at a low frequency (Fb) on a sample of perfect adherence (R=0)
- a phase (or amplitude) measurement on a sample of unknown adherence (R'=/=0)
- a series of calculation allows us to determine R'

The experimental setup used in this work is composed of the following elements: a modulated source, a detection chain, a lock-in amplifier and an X-Y table. The choice of the different measurement procedures, the detection and treatment of the signals, the presentation of the results, and the XY-displacement of the sample are all controlled by a microcomputer.

In order to estimate the limits and the accuracy of the method, we have made a series of measurements on samples with different values of the thickness and diffusivity. We also changed the nature, colour and thickness of the substrate (steel or aluminium, thickness) and different pairs of frequencies were tried. The sensitivity of the method (usually between 5 and 10%) depends on three main factors: 1) the accuracy of the measurement of the geometrical thickness of the coating on the reference sample ; 2) the choice of the coating thickness of the sample: it must be chosen according to the range of the thicknesses to be to determined; 3) the selection of the low and high modulation frequencies.
The time needed to scan the whole sample depends on the coating thickness (a few seconds in the range 50-100 µm). The physical parameters can be calculated in a time shorter than the scan duration.

The system described here is mainly to be considered as a laboratory instrument used for many applications: adjustment, analysis, control of coating apposition processes, ageing studies of thin layered materials, study in time and space of physico-chemical changes (drying, sticking, polymerisation).

1. R. Dartois: Thesis, Université de Reims (1986)
2. M. Egée, R. Dartois, J. Marx, C. Bissieux: Can. J. Phys., 64, 1297 (1986)
3. Patent n° 8601613, Université de Reims Champagne Ardenne

Thermal Depth Profiling of a Plasma-Exposed Metallic Plate

M. Wojczak, J. Pelzl, and B.K. Bein

Ruhr-Universität Bochum, Institut für Experimentalphysik VI,
P.O. Box 102148, D-4630 Bochum, Fed. Rep. of Germany

To establish thermal wave analysis as a tool to characterize plasma-induced changes of metal surfaces, a neutralizer plate has been analysed with the help of frequency-dependent photoacoustic measurements. The plate consisting of the titanium alloy $TiAl_6V_4$ had been exposed to the plasma in the lower divertor chamber on the ion side of the tokamak ASDEX (Max-Planck-Institut für Plasmaphysik, Garching, FRG) during about 20000 discharges of about 2.5s duration.

In order to distinguish between the changes of the optical properties induced by plasma-surface interactions and the changes of the thermal properties, the photothermal conversion factor $\eta = 1 - r$ was determined from the reflectivity r measured for the wave length of the laser light used in the photoacoustic experiment.

Depending on the position of the measurement on the neutralizer plate relative to the intersection with the plasma separatrix, three different types of thermal depth profiles have been observed.

(1) In the region of the main plasma-solid contact [1], about 1.3 cm wide, where the peak of the power deposition profile was located and where the plate shows a bright metallic surface, an increase of the effusivity up to 20% above its original value could be

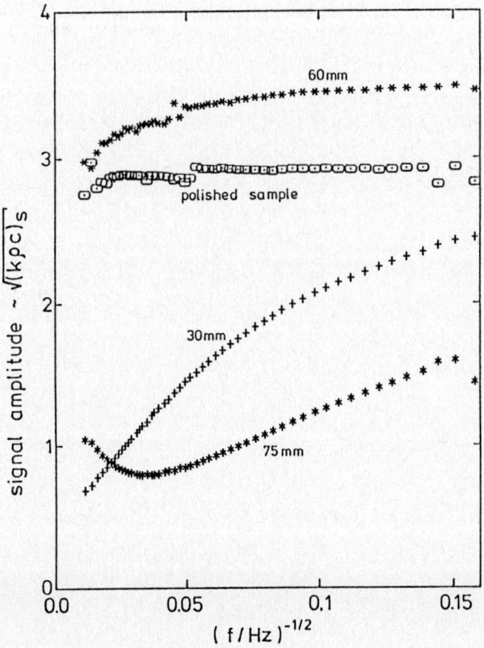

Fig. 1

Normalized amplitudes from the main plasma / neutralizer plate contact region (position 60 mm), from a polished material sample, from the region of plasma-deposited material (position 30 mm) and from a narrow region above the intersection plasma separatrix / neutralizer plate (position 75 mm) detected from the measured normalized photoacoustic amplitudes at larger thermal wave lengths (Fig.1, position 60 mm). These changes of the thermal properties can be related to annealing effects by the plasma heat. Additionally, near the surface, at small thermal diffusion lengths the effusivity profile shows a slight decrease which probably can be related to particle bombardment effects (lattice damage).

(2) Far away from the main plasma-solid contact (Fig.1, pos. 30 mm) where a gray or darker layer of deposited material (carbon or titanium) is observed, thermal depth profiles are found with an effusivity value below that of the original material (polished sample). The reduced thermal properties of the deposited material and the contact resistance between the deposition layer and the metal may be responsible. With increasing thermal diffusion length, the effusivity value increases again.

(3) In a limited region above the bright main plasma-solid region (position 75 mm in Fig.1 and in Fig.2), a change of sign is found for the normalized phase and the amplitude first decreases with increasing penetration depth of the thermal wave. This can in general be explained by a reduced effusivity value below the solid surface, as expected for subsurface defects or delaminated coatings [2,3]. Subsurface defects due to bombardment by accelerated particles [1] might be an explanation here, close to the intersection between separatrix and neutralizer plate.

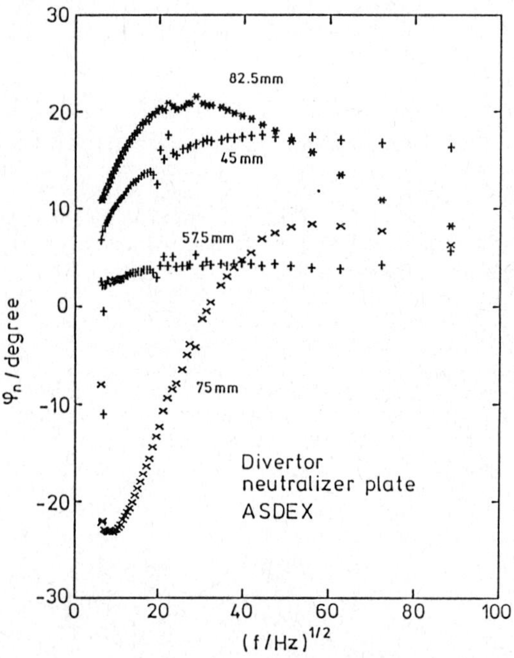

Fig. 2

Normalized phase lags from the region of plasma-deposited materials (position 82.5 mm and 45 mm), from the main plasma / neutralizer plate contact region (position 57.5 mm), from the narrow region above the intersection plasma separatrix / neutralizer plate (position 75 mm)

[1] H. Rapp, H. Niedermeyer, M. Kornherr and ASDEX team, Proc. 12th Symp. on Fusion Technology, Avignon, France, 1986

[2] G. Busse and A. Rosencwaig, Appl. Phys. Lett. 36, 815 (1980)

[3] P.M. Patel and D.P. Almond, J. of Materials Science 20, 955 (1985)

Nondestructive Testing of Carbon Fibre Reinforced Plastics (CFRP) by Thermal Wave Radiometry

B. Rief, G. Busse, and P. Eyerer*

Institut für Kunststoffprüfung und Kunststoffkunde, Universität Stuttgart, D-7000 Stuttgart 80, Fed. Rep. of Germany

Photothermal inspection /1/ of glass fibre reinforced plastics has been performed previously by BUSSE and EYERER /2/. INGLEHART et al. used mirage effect detection of thermal waves to investigate nondestructive testing of carbon fibre reinforced plastics (CFRP) /3/. TITTMANN et al. generated thermal waves by ultrasonics /4/. Preliminary results on photothermal inspection of CFRP have been published /5,6/. CFRP is well suited for thermal wave analysis since its black colour provides highly efficient deposition of optical energy, while the large thermal diffusivity allows for offsets that exceed significantly the sizes of the optical focus and of the infrared-detector spot.

1. EXPERIMENTAL ARRANGEMENT

The experiments were performed with the focused beam of an Argon ion laser and with spatially resolved detection of modulated thermal infrared emission.
Both the laser spot and the detector spot were on the front surface of the sample. Scan experiments could be performed either by moving the sample in a raster-like fashion with respect to the stationary spots (this arrangement is suited to reveal local imperfections), or, alternatively, the offset of the detector spot with respect to the stationary laser focus was scanned. Such an experiment provides the local thermal diffusivity, since the phase shift is linear in the distance travelled by the thermal wave. This arrangement was used for the first experiment described in the following.

2. CHARACTERISATION OF CFRP

Carbon fibres show graphite layers in their elementary structure due to the sp^2-hybridization of the electron orbitals. The mobility of the - electrons in this structure causes the high electrical and thermal conductivity along the graphite layers. Young's modulus and strength based on the σ - bonds are directly influenced by the orientation of this crystal structure with respect to the fibre axis.
The increase of the Young's modulus correlates with the increasing degree of this orientation. The structure of graphite layers can be analysed by high resolution transmission electron microscopy and by X-ray diffraction. For the photothermal wave analysis we used unidirectional prepregs with different fibres, however, with the same resin.

---

*) Permanent address: Institut für Physik, FB ET, Universität der Bundeswehr, D-8014 Neubiberg, Fed. Rep. Germany

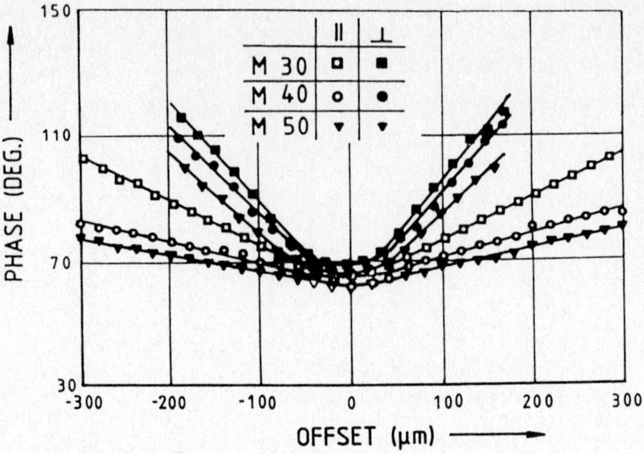

Fig. 1: Identification of various High Modulus Fibres: Phase-angle as a function of the distance z (offset)between the laser focus and the infrared-detector spot. Offset orientated along (II) or perpendicular (I) to the prepreg fibres

Figure 1 presents the results for high modulus prepregs. The orientation of the offset was along or perpendicular to the fibres. The interesting result is that for the offset along the fibre the slope of the profile decreases with the increase of Young's modulus (M 30, M 40, M 50). This is the consequence of the higher thermal diffusivity for fibres with increasing Young's modulus. The slope change of the profile for the orientation with the offset perpendicular to the fibres is partly caused by the increase of thermal diffusivity perpendicular to the fibre axis. In addition, with a finite size of the detector spot, the signal is a superposition of thermal wave paths which also have components along the fibres, though the detector is moved perpendicular to the fibre orientation.

Stored prepreg material may suffer from aging which may reduce the material quality. Therefore we performed experiments on epoxy based prepreg samples that had been exposed to elevated temperatures (120°C) for certain lengths of time ranging from 0 to 25 minutes. The curing process of these artificially aged samples was monitored with signal phase at a constant offset. To avoid an increase of temperature due to the laser beam, the sample was moved back and forth during this time.
The result (Fig. 2) shows how the temperature history affects the curing process.

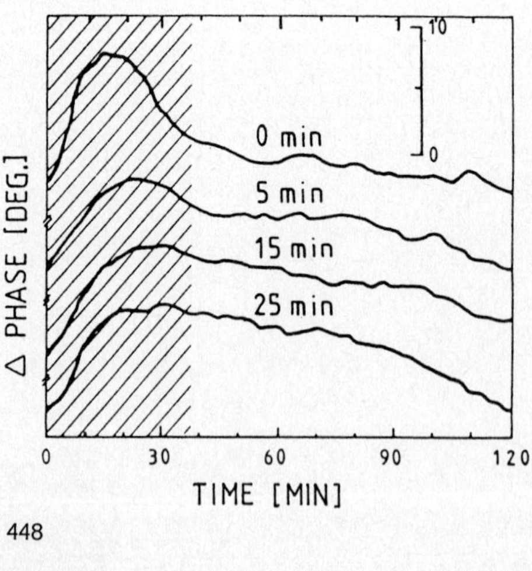

Fig. 2: Influence of artificial aging (various durations at 120°C) on time dependent phase angle Δ φ during curing at 170°C

Even the previous heating of only 5 minutes duration can be detected. From these results we suppose that photothermal inspection might be useful to monitor aging effects.

3. DEFECT DETECTION

The quality of CRFP may suffer also from imperfections induced in the manufacturing process. To investigate these effects we performed scans across samples provided with artificial faults.

Detection of material (PTFE or aluminium) imbedded between the first and the second laminate layer is presented in Fig. 3 a and b. The nature of the planar defect compared to CFRP is evidently of importance for the sign of the signal phase change. This difference can be interpreted in terms of thermal wave reflection at different boundary conditions.

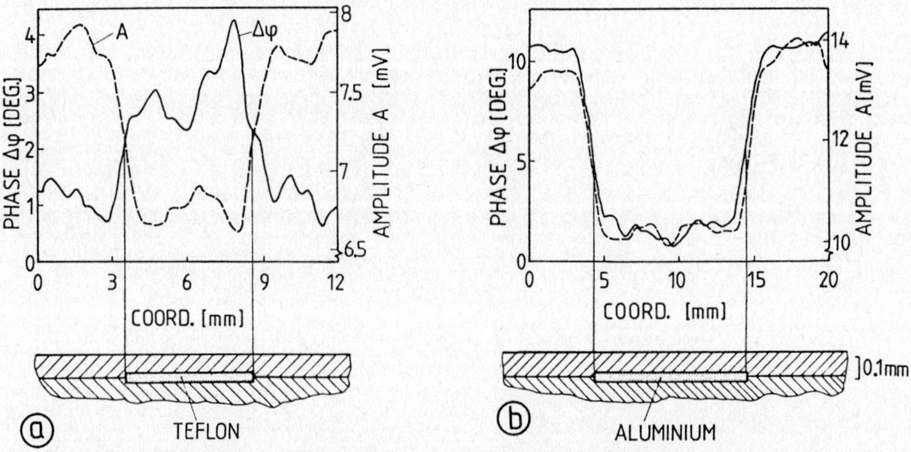

Fig. 3: Photothermal radiometry scan with signal magnitude A and phase Δφ across layer of Teflon (a) or aluminium (b) imbedded in a CRFP laminate sample

A result is shown in Fig. 4 for a matrix crack oriented perpendicular both to the scan direction and the offset between the laser spot and the detector spot D. Signal changes occur at the crack.

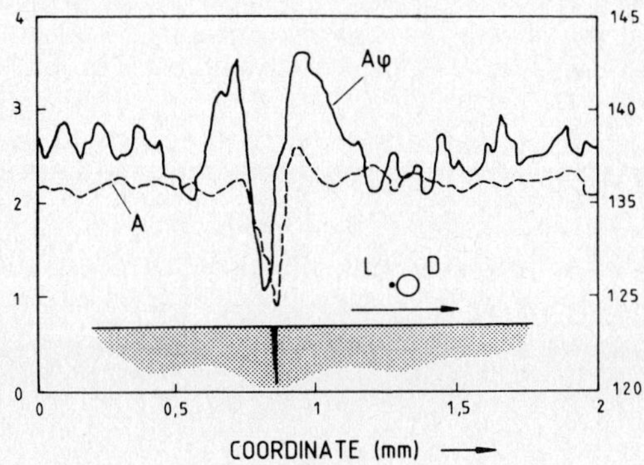

Fig. 4: Scan across CFRP with a matrix crack. L and D indicate laser and detector spot, respectively

4. CONCLUSION

On CFRP material, we could demonstrate that Young's modulus can be estimated from thermal properties observed in a remote and nondestructive way. Also one can monitor the stage of artificial aging. Fault detection is possible both for vertical and horizontal boundaries of the defect. From these observations we conclude that photothermal radiometry is a valuable tool for NDE of carbon fibre reinforced plastics.

Acknowledgement: The authors gratefully acknowledge financial support by the German Ministry of Research and Technology (BMFT) under Contract-No. 13 N 5318/1.

5. LITERATURE

1. P.-E. Nordal, S. O. Kanstad: Physica Scripta, 20, 659 (1979)
2. G. Busse, P. Eyerer: Appl.Phys.Lett., 42, 355 (1983)
3. L. J. Inglehart, F. Lepoutre, F. Charbonnier: J. Appl. Phys., 59, 234 (1986)
4. B. R. Tittmann, R. S. Lineberger: Proc. Ultrasonics symposium 1984, Vol. 2, Dallas
5. B. Rief, G. Busse: Gummi Faser Kunststoffe, 40, 188 (1987)
6. B. Rief, G. Busse, P. Eyerer: Proc. 6th International Conference on Composite Materials, Elsevier Sc. Pub. London, 1987, Vol. 1, 349

Inhomogeneous Ferromagnetic Resonance Absorption in Ferrites Detected by Photothermal Techniques

O. von Geisau, U. Netzelmann, and J. Pelzl

Institut für Experimentalphysik AG VI, Ruhr-Universität Bochum, D-4630 Bochum, Fed. Rep. of Germany

Photoacoustic (PA) detection and the photothermal deflection (PD) technique have proved to be valuable tools for magnetic depth profiling and lateral imaging of magnetically inhomogeneous materials [1]. In this work, these techniques have been applied to investigate the lateral and depth dependence of ferromagnetic resonance (FMR) microwave absorption in thin ferrite plates. Although this material possesses homogeneous magnetic and thermal properties, strong inhomogeneities in microwave absorption exist which are due to interfering electromagnetic waves.

1. Ferromagnetic Resonance Microwave Absorption in Thin Ferrite Plates

Microwave absorption in thin ferrite plates magnetized by an external magnetic field generally cannot be described by a Beer absorption law. Wave propagation inside the gyrotropic medium is predominantly determined by the real and imaginary part of the propagation constant, α and β, for transverse plane wave propagation given by [2]

$$\left.\begin{matrix}\beta\\ \alpha\end{matrix}\right\} = \omega\sqrt{\mu_0\epsilon/2}\sqrt{|\mu_r| \left\{\begin{matrix}+\\ -\end{matrix}\right\} \mu_r'}\ , \tag{1}$$

where $\mu_r=\mu_r'-i\mu_r''$ is the effective complex-valued relative scalar permeability. Close to the FMR, μ_r and the propagation constant are strongly dependent on the value of the external field. For the case of damped resonance the dependence can be approximated by [3]

$$\mu_r \approx \left[(B_0+M_0)^2-B^2+2i\Delta B(B+\frac{B}{B_0}M_0)\right]\cdot\left[B_0(B_0+M_0)-B^2+2i\Delta B(B+\frac{B}{2B_0}M_0)\right]^{-1}, \tag{2}$$

where B_0 is the external magnetic field, M_0 the saturation magnetization, ΔB the line width of the resonance curve and B the internal resonance field. The microwave absorption profile inside the sample with a thickness d is determined by the propagation constants (1) and the electromagnetic boundary conditions. For one-dimensional wave propagation a calculation of the absorbed power density P as a function of the depth z yields

$$P(z) \sim \mu_r''\, b_1^2\, (\cosh(2\alpha(d-z)) + \cos(2\beta(d-z))), \tag{3}$$

where b_1 is the high-frequency magnetic field. Depending on the values of α and β, the absorption profile inside the ferrite can be homogeneous, exponential or can show oscillatory behaviour known as body resonances [2].

2. Experimental and Results

The experimental setup is based on an X-band EPR spectrometer which is modified for PA and PD measurements. The microwave power of the klystron

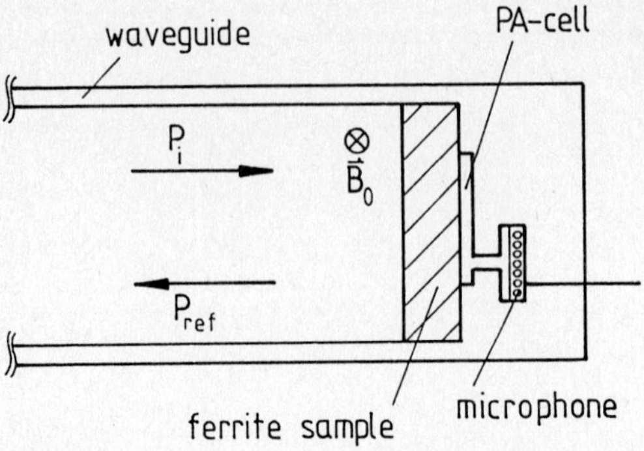

Fig. 1
The waveguide PA-cell

(200mW) is 100% square wave modulated by a PIN-diode modulator and - for PD measurements - fed into a traveling wave tube amplifier with a maximum output power of 20W. Conventional absorption measurements are effected by diode detection of the incident and the reflected microwave power (Fig. 1). The samples are thin ferrite plates (23mm x 10mm, thickness 0.3mm..3mm) fixed in front of a shortening wall of the waveguide. The external magnetic field is applied in the plane parallel to the air-ferrite interface, admitting predominantly transverse electromagnetic wave propagation. The PA-cell is constructed as a flat cavity in the waveguide short, whereas for the PD experiment a slitted waveguide configuration [1] is used.

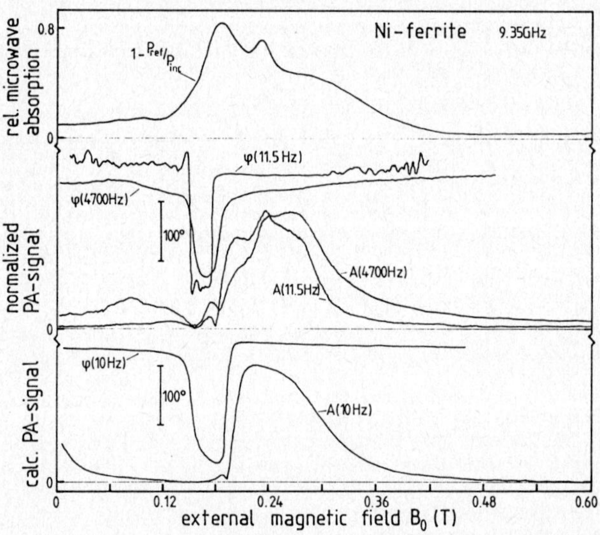

Fig. 2
The conventional signal (upper part) and the measured (centre part) and calculated (lower part) PA-signal from a Ni-ferrite sample

Fig. 2 shows experimental results obtained from a 1.15mm thick, sintered Ni-ferrite sample as used for microwave applications. In the upper trace, the relative integral absorption over the whole sample is displayed as a result of the conventional reflection measurement. A subsidiary absorption peak occurring at 0.24T is a typical feature of standing wave modes in ferrites. In the centre traces of Fig. 2, the simultaneously measured PA-signal in thermal transmission normalized to the incident power P_{inc} is shown for the two modulation frequencies 11.5Hz and 4700Hz. The PA-signals differ distinctly from the integral conventional signal. The dominant property of the PA-signals is the small amplitude around 0.15T, which corresponds to a minimum of the PA-signal phase. Both the amplitude signal shape and the depth of the phase minimum vary with the modulation frequency. The behaviour of the PA-signal measured in thermal transmission has a qualitative explanation in the magnetic field dependence of the absorption properties of the ferrite, changing from a transparent state at very low and very high fields to a nearly opaque state at resonance. In the latter case, damping of the thermal waves passing the sample causes the minima in amplitude and phase.

3. Theoretical Analysis

A theoretical model for the PA-signal from the ferrite has to allow all of the heat source distributions of (3). For this purpose, we derived a one-dimensional three-layer model of thermal wave propagation with arbitary distribution of heat sources P(z). In a first step, a Green's function type solution G(z,z') is constructed, which is a measure for the temperature response at the point z from a δ-shaped unit heat source in a fixed depth z' inside the sample (Fig. 3). In addition to the usual boundary conditions, the heat flux at the imaginary interface z' must be increased by the amount delivered by the heat source. For example, the solution at the solid-gas interface is given by:

$$G(0,z') = \frac{1}{\lambda_s \sigma_s} \frac{(1+b)\cdot e^{-\sigma_s(z'-d)} + (1-b)\cdot e^{\sigma_s(z'-d)}}{(1+b)(1+g)\cdot e^{\sigma_s d} - (1-b)(1-g)\cdot e^{-\sigma_s d}} , \qquad (4)$$

where the notations of Rosencwaig and Gersho [4] have been used. The temperature oscillation response at the point z caused by the source profile P(z') is then given by

$$\vartheta(z) = \int_0^d P(z')\, G(z,z')\, dz'. \qquad (5)$$

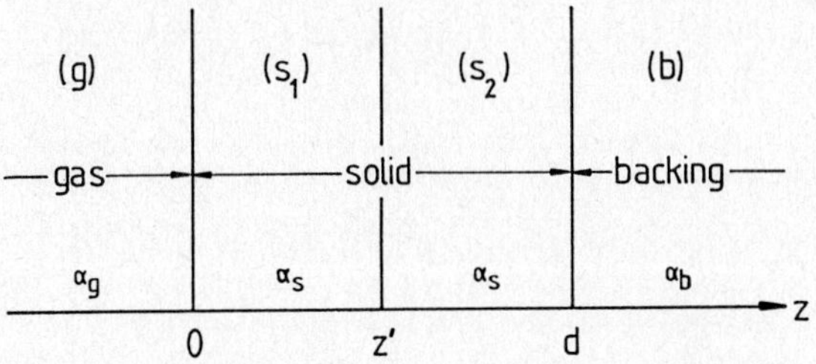

Fig. 3
Three layer model with δ-heat source at z'

To calculate the PA-signal of the ferrite in thermal transmission, we used G(d,z') with the backing material, gas and the source distribution P(z') given in (3). The calculated signal at the back side of the sample (Fig. 2, lower traces) reproduces characteristic features of the measured PA-signal, but it fails to describe details of the signal, esp. at high modulation frequencies. We attribute this mainly to the existence of higher electromagnetic wave modes as well as an inhomogeneous demagnetization field, which generate a three-dimensional absorption pattern in the sample.

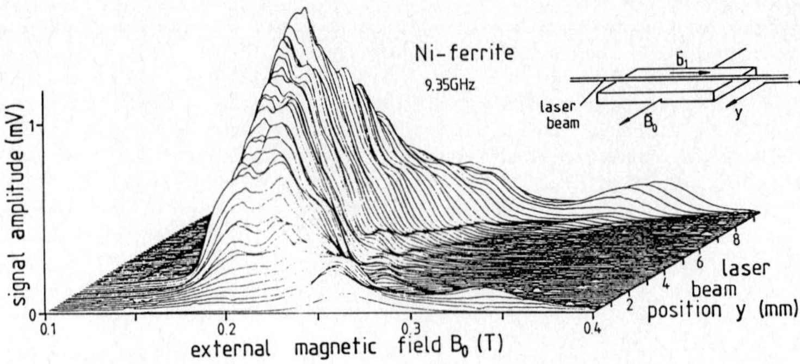

Fig. 4 The PD-FMR signal as a function of the laser beam position

4. Photothermal Deflection (PD) Measurements

Experimental evidence of the lateral inhomogeneity has been obtained by an additional experiment using the PD technique in thermal reflection and transmission. In Fig. 4 we have plotted the PD-signal in thermal reflection for a scan of the laser beam parallel to the long edge of the sample (see inset of Fig. 4). The modulation frequency was 1kHz. There is a clear variation of the FMR absorption symmetrical to the centre of the ferrite. The peak of the signal shifts to higher values of the external magnetic field B_0 for regions near the margins of the sample. This observation can be accounted for by influence of the inhomogeneous demagnetization field inside the ferrite, which reduces the effective internal field. A similar behaviour show the smaller subsidiary peaks, which are averaged out in the conventional signal shown in Fig. 2. Using varying modulation frequencies, one can also find a frequency dependence of the signal shape indicating a local depth dependence of the microwave absorption.

5. Conclusion

We have shown that in contrast to the spatially averaged information of the conventional technique, PA and PD detection give evidence of the depth dependence as well as the lateral dependence of ferromagnetic resonance absorption in thin ferrite plates. There is a marked deviation from one-dimensional behaviour generally used to describe absorption profiles in these materials. This information on lateral and depth resolved inhomogeneities, not accessible to simple models of wave propagation, could be useful for the development of components in microwave technology.

Acknowledgements: This work is supported by the BMFT, project 13N 5379/1

References

1. U.Netzelmann, J.Pelzl, D.Fournier, A.Boccara: Can.J.Phys. 64,1307(1986)
2. B.Lax: Microwave ferrites and ferrimagnetics, New York 1962
3. A.Gurevich: Ferrites at microwave frequencies, London 1963
4. A.Rosencwaig, A.Gersho: J.Appl.Phys. 47,64(1976)

Part IX

Experimental Techniques

Frequency Modulation Time Delay Photopyroelectric Spectrometry (FM-TD P^2ES)

A. Mandelis and J.F. Power

Photoacoustic and Photothermal Sciences Laboratory,
Department of Mechanical Engineering, University of Toronto,
Toronto, Ontario M5S 1A4, Canada

I. INTRODUCTION

Photopyroelectric spectrometry is a novel and powerful technique for the measurement of the thermal properties of materials [1-4]. The strategy of photopyroelectric spectrometry uses light absorption by a sample to generate thermal waves in the material, which are detected by means of a contact pyroelectric detector on the back surface of the sample. The use of a pulsed laser as the excitation source enables the recovery of transient information which is directly interpretable in terms of transit times for thermal signals through the material via a Green's function model. Pulsed laser excitation, however, requires large peak power, with the possibility of optical damage to the material under test. An alternative to the use of pulsed lasers is broadband modulated CW excitation with the recovery of the impulse response by FFT methods [5,6]. In this work, we have demonstrated the application of Frequency Modulation Time Delay Spectrometry (FM-TDS), a minimum phase, wideband, excitation method, with a superior dynamic range and coherence [6,7], to photopyroelectric measurements on thin solid samples. Photopyroelectric detection is best suited for these experiments, due to the wide flat frequency response of thin film PVDF pyroelectric transducers [8], which guarantee undistorted time and frequency-domain spectral feature retrieval. We have also made a comparative study of two wideband excitation strategies in the photopyroelectric spectrometry of solids: (a) Stochastic (white noise) excitation, and (b) frequency modulation time delay spectrometry (FM-TDS). Wideband noise excitation has been used in recent photopyroelectric work to generate fast frequency response profiles in a fraction of the time required by lock-in amplification [5]. Because of the Fourier Transform relationship existing between the frequency response (H(f)) and the impulse response, pulsed laser equivalent responses are available from the wideband methods by use of FFT analysis. Frequency modulation time delay spectrometry (FM-TDS) promises to yield further improvements over the random noise excitation strategy [6], because of its minimum phase character [7], which gives rise to extended dynamic range and noise rejection.

II. EXPERIMENTAL

The design of our Time Delay Domain Photopyroelectric detection system is summarized in Fig. 1. The excitation source was a Coherent Innova 90 CW Argon Ion laser operating single line at 488 nm. The excitation power used in these experiments ranged from 100-200 mW depending on the sample thickness. The central excitation/detection component in the system was an HP 3562 Fast Fourier Transform analyzer equipped with an internal frequency synthesizer. The synthesizer was capable of generating fixed sine and random noise waveforms, as well as linear frequency sweeps, with modulation bandwidths up to 100 kHz. The sweep source was used to drive an acousto-optic modulator (Isomet 1201E) to modulate the intensity of the Ar^+ beam using a knife edge. The output of the acousto-optic modulator was optimized for modulation depth and minimum distortion, by adjusting the peak to peak voltage applied to the driver circuit to a value of 400 mV. A reference beam was obtained by recording a fraction of the excitation signal with a beam splitter (BS) and a photodiode (PD). The photodiode input was used as a measure of the excitation waveform $x(t)$. In some experiments, the output of the HP

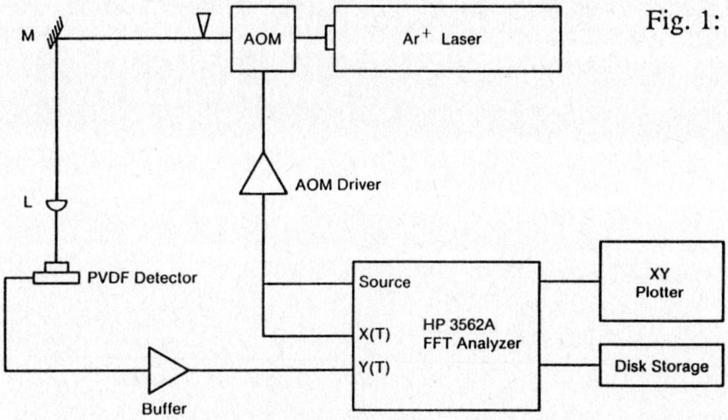

Fig. 1: FM-TD P^2E spectrometer

3562A internal synthesizer was used to supply the $x(t)$ waveform. The main excitation beam was directed through a microscope objective lens (magnification: 32×) (L1) to simulate a point source, and focussed onto the surface of a thin film pyroelectric detector (PED) or a sample. The detector consisted of a 28 μm film of polyvinylidene difluoride (PVDF) (Penwalt Corporation "KYNAR" Piezo Film) [9] supported by a stainless steel backing. The detector was enclosed in an assembly supplied by Inficon to provide electrical contact and to ensure shielding from electromagnetic RF interference. Thermal contact between the sample and the detector was optimized by using a thin layer of thermally conducting grease (No. 120-8, Wakefield). The output of the detector was fed to a high impedance buffer preamplifier (Comlinear, CLC-B-600, 3pF parallel 3MΩ input impedance; 600 MHz bandwidth), and applied to the $y(t)$ system input of the HP 3562A FFT analyzer. The recovery of all frequency and time delay domain functions was achieved via the analyzer computations.

III. NATURE OF RECOVERED FM-TDS P^2E SIGNALS

Any photothermal system may be modelled as a general linear system or "black box" (Fig. 2) with impulse response $h(\tau)$, input wavetrain $x(t)$ and output response $y(t)$. In experiments involving broadband excitation, the power spectrum of the input, $G_{xx}(f)$, is assumed to be flat, or "white", relative to the power spectrum of the response, $G_{yy}(f)$ [5,6,10]. In FM-TDS as in random noise excitation methods [6,11] the frequency response is recovered from a cross spectral measurement as

$$H(f) = \frac{G_{xy}(f)}{G_{xx}(f)} , \tag{1}$$

so that if $G_{xx}(f) \cong 1$, then $H(f) \cong G_{xy}(f)$. Equivalently, a flat input power spectrum yields an autocorrelation function, $R_{xx}(\tau)$, which is mathematically equivalent to a Dirac delta function, by inverse transformation to the time delay domain:

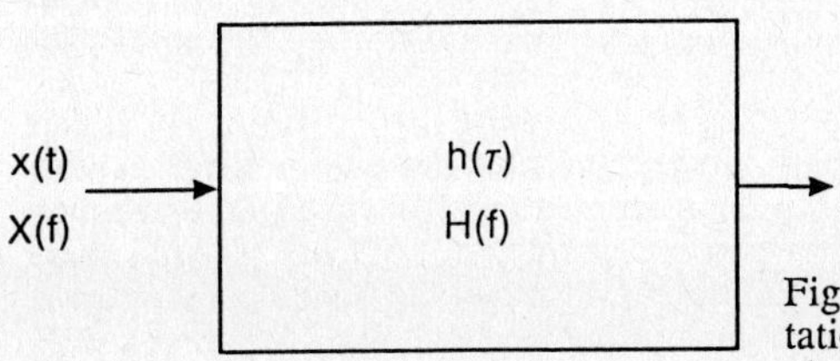

Fig. 2: Generalized "Black-Box" representation of frequency response, H(f), for a photothermal system

$$R_{xy}(\tau) = R_{xx}(\tau) * h(\tau)$$
$$= \delta(\tau) * h(\tau) \tag{2}$$
or
$$R_{xy}(\tau) = h(\tau) \ .$$

In the case of FM-TDS where the excitation wavetrain is a linear frequency sweep (Fig. 3), the criterion for flatness of the input autospectrum, $G_{xx}(f)$ is met when the sweep time, T, is much longer than the longest time delay component, τ, in the system response [6,7], and when the product of the modulation bandwidth, Δf, and T is very large, well above the limit of the instrumental uncertainty principle:

$$(\Delta f)T \gg 1 \ . \tag{3}$$

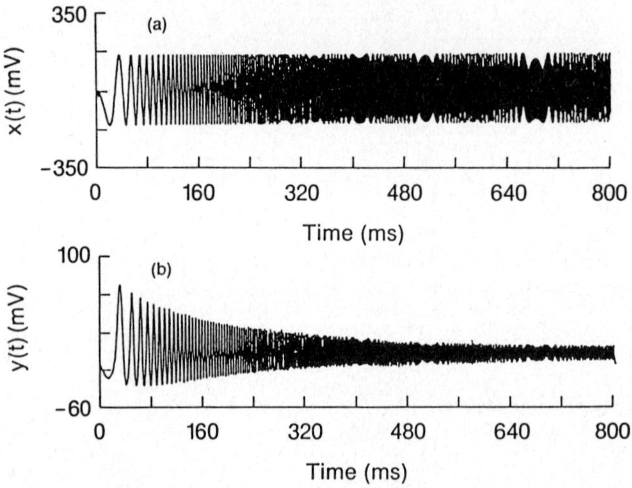

Fig. 3: FM Time delay swept wave excitation and response signals (a) input wavetrain, x(t); (b) PVDF thin film detector response, y(t). $\Delta f = 0\text{-}1$ kHz. Sweep rate: 1.25 kHz/s.

Typical excitation and response signals, for a 28 μm PVDF pyroelectric thin film detector, in an FM-TDS experiment, are shown in Fig. 3. Swept waves were generated in the HP 3562A internal source with frequency spans ranging from 0 to 25 Hz (for thermally thick samples with slow time delay domain response) to up to 0-25 kHz (for thermally thin samples, with risetimes approaching that of the pyroelectric film itself). The swept waves generated within the HP 3562A were used to drive the acousto-optic modulator. Measurements carried out with the HP 3562A FFT analyzer were made in the linear resolution mode. In that mode the sweep time was related to the frequency span by the relationship

$$T(sec) = \frac{800}{\Delta f(Hz)} \ , \tag{4}$$

where the frequency span Δf is the modulation bandwidth of the carrier ($\Delta f = f_2 - f_1$) with $f_1=0$, and f_2 the maximum frequency in the sweep. The product $(\Delta f)T$ approaches 10^3, which ensures flatness of the excitation autospectrum, $G_{xx}(f)$ [7,12]. In all cases, the sweep was unidirectional from f_1 to f_2, and the resolution of the input data was 2048 points/sweep, with the sampling rate set above the aliasing frequency at 2.56 f_2.

IV. THERMAL DIFFUSIVITY AND CONDUCTIVITY MEASUREMENTS ON THIN SOLID SAMPLES

In order to characterize the response of our instrument and demonstrate its capabilities for thermal parameter measurements, we have recorded and theoretically modelled the responses of a series of well-characterized samples. These samples consisted of a series of quartz microscope slides and cover slips of varying thickness. Each series of samples was prepared from a single slide, cut in many smaller pieces, to ensure material uniformity. The pieces were etched in 50% HF/50% H_2O to the desired thickness. The front surfaces of these samples were coated with a thin layer of water soluble ink, which served as a blackbody layer. The thinness of the ink layer and its large absorption coefficient ensured that the laser radiation absorbed at the sample surface acted as an ideal plane heat source. The impulse responses of these samples may be readily predicted from a Green's function model of transient heat conduction in a four layer system [13].

The response for the pyroelectric detector element under load has the form [9]

$$V(t) = K \frac{\partial \langle T_d(x,t) \rangle}{\partial t} , \tag{5}$$

where $\langle T_d(x,t) \rangle$ is the spatially averaged temperature profile in the film, $V(t)$ is the voltage recorded at the output of the buffer preamplifier and K is a constant which incorporates the electrical properties of the pyroelectric film.

The theory [13], interpreted for thermally thick samples, gives a pyroelectric response of the form

$$V(t) = \frac{KA}{t^{3/2}} \sum_{n=0}^{\infty} (-1)^n \gamma^n \left\{ \tau_{1n}^{1/2}\, e^{-\tau_{1n}/4t} - 2\tau_{2n}^{1/2} e^{-\tau_{2n}/4t} + \tau_{3n}^{1/2} e^{-\tau_{3n}/4t} \right\} , \tag{6}$$

where

$$\tau_{1n}^{1/2} = \frac{2nd}{\alpha_3^{1/2}} + \frac{l}{\alpha_2^{1/2}} ,$$

$$\tau_{2n}^{1/2} = \frac{(2n+1)d}{\alpha_3^{1/2}} + \frac{l}{\alpha_2^{1/2}} ,$$

$$\tau_{3n}^{1/2} = \frac{2(n+1)d}{\alpha_3^{1/2}} + \frac{l}{\alpha_2^{1/2}} .$$

The factor A is a constant which incorporates the static thermal properties of the sample/pyroelectric system:

$$A = -\frac{2(\alpha_3 \alpha_2)^{1/2}}{\alpha_2 (b_{32}+1) 4\sqrt{\pi}} . \tag{7}$$

The subscripts 3 and 2 refer to the pyroelectric film and sample, respectively. Here, α_i is the thermal diffusivity of material (i). The coupling coefficients b_{ij} are defined as $b_{ij} = k_i \alpha_j^{1/2} / k_j \alpha_i^{1/2}$, where k_i is thermal conductivity. The parameter $\gamma = (b_{32}-1)/(b_{32}+1)$, and l,d refer to the thickness of the sample and film, respectively. Other assumptions of Eq. (6) are that there are large thermal mismatches at the sample/gas and film/backing interfaces; that is, $b_{12} \ll 1$ and $b_{43} \gg 1$ [13].

In this work, we have compared the results predicted by this model with our experimental impulse responses. Because the absolute intensity of the recovered signals is a function of instrumental factors such as irradiation power, amplifier gain and excitation geometry, we have normalized the impulse responses to give $V(t) = 1$ at the peak of the time delay response.

Figure 4 shows the comparison between experiment and theory for a 500 μm sample of quartz. The impulse response profile was fitted assuming $\alpha_2 = 4.0 \times 10^{-7}$ m^2/s for quartz, which shows excellent agreement with literature values: $\alpha_2 = 4.0 \times 10^{-7}$ m^2/s [14] and $\alpha_2 = 4.4 \times 10^{-7}$ m^2/s [15]. The value of the parameter γ which gave the best fit to

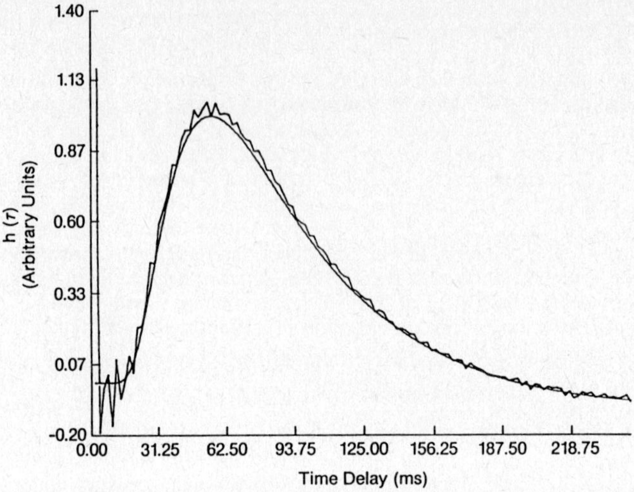

Fig. 4: Theoretical and experimental impulse response profiles obtained for a 500 μm quartz sample.

experiment was $\gamma = 0.7$. From this parameter, it was possible to calculate the thermal conductivity for the sample as $k_2 = 2.2$ W/m K. Agreement between theory and experiment is excellent in the time delay domain from 0 to 1 s.

We have also tested the methodology using metallic samples (aluminum, stainless steel) which were contacted to the pyroelectric (PVDF) detector using thermal compound. In the case of the metal samples, $b_{32} \ll 1$, so that $\gamma = -1$ in (6). The expression is valid provided $l^2/4\alpha_2\tau \gg d^2/4\alpha_3\tau$ [13]. Thermal diffusivity measurements on these samples are reported in Table 1, which also reports the sample thicknesses and time delay ranges for which the measurements were made. These conditions fall well within the approximations required by the model.

The results of Table 1 show a reasonable agreement between measured and literature values of α_2 for a wide range of different materials. The main factor controlling the agreement between theory and experiment for these results is the thermal contact resistance present at the sample/detector interface. Flatness of the sample surface promotes a

Table 1: Measured diffusivity values for selected materials

Material	α_2 (m²/s) Experimental, This Work	α_2 (m²/s) Literature Values	Measurement Conditions
Quartz, fused	4.0×10^{-7}	4.0×10^{-7} (ref. 14) 4.4×10^{-7} (ref. 15)	l_{range}: 300 μm-1 mm τ_{range}: < 1 s.
Stainless Steel	2.2×10^{-6}	3.5×10^{-6} (ref. 16)	l_{range}: 390 μm-1 mm τ_{range}: < 25 ms.
Aluminum	38×10^{-6}	97.1×10^{-6} (ref. 16) 68×10^{-6} (cast) (ref. 16)	l_{range}: 500 μm-2 mm τ_{range}: < 10 ms.

good thermal contact with the detector film while warpage or other surface deformities cause poor contact and an overestimated value of the measured thermal diffusivity. A residual contact resistance persisted in our sample/pyroelectric system, even when thermal coupling compound was used, and contributed appreciable errors in α_2 at time delays earlier than 5 ms.

V. INSTRUMENT PERFORMANCE AND SIGNAL RECOVERY TECHNIQUES

All wideband methods have as their basis the excitation of a linear, time invariant system (with frequency response H(f)) by a waveform having a flat power spectrum ($G_{xx}(f)$) over the response bandwidth of the system. The standard method of extracting frequency response data from systems excited by random noise is correlation and spectral analysis [11]. Frequency response information is normally recovered from a cross spectral measurement: $H(f) = G_{xy}(f)/G_{xx}(f)$ so that $H(f) \approx G_{xy}(f)$ if $G_{xx}(f)=1$. By inverse Fourier transformation, the equivalent condition in the time domain is $h(\tau) \approx R_{xy}(\tau)$ when $R_{xx}(\tau) \approx \delta(\tau)$. Here $h(\tau)$ is the impulse response.

In FM-TDS, excitation is with a linear cosine sweep, with time varying "instantaneous" frequency f_i [6,7],

$$x(t) = \cos(St^2) = \cos[2\pi f_i(t)t] \quad , \tag{8}$$

where $f_i(t)=St$ and S is the sweep rate $S=\Delta f/T$, $\Delta f=f_{\max}-f_{\min}$ and T is the sweep time. The initial phase and lowest frequency in the sweep are both assumed to be zero.

The FM-TDS strategy, like any wideband excitation technique requires that $R_{xx}(\tau) \approx \delta(\tau)$. This condition is met for linear sine sweeps with sufficiently slow sweep times and sufficiently wide modulation bandwidths, Δf, such that

$$\frac{1}{\sqrt{8(\Delta f)T}} \ll 1 \ . \tag{9}$$

Frequency response data were recovered from the cross-spectral density functions $H(f) = G_{xy}(f)/G_{xx}(f)$. Impulse response data were obtained by inverse Fourier transformation of H(f). Coherence data were calculated as $\gamma^2 = G_{xy}^2(f)/G_{xx}(f)G_{yy}(f)$.

In this work, we initially compared very slow sinusoidal sweeps (generated by the HP3562A) with data acquired by lock-in amplification and found that the frequency response data obtained by the two methods were congruent. The very slow sine sweeps were, thereafter, used as a standard for comparison with the frequency response data recovered by wideband excitation methods.

Frequency response information for samples of varying thickness of glass were recorded with (a) very slow sine sweeps; (b) random noise excitation; and (c) FM-TDS fast sine sweeps. The envelopes of H(f) and $\phi(f)$ agreed well with the slow sine standard for both methods. Inverse transformation of these responses to the time delay domain again showed excellent agreement between the methods and with the theory [13] indicating that the recorded impulse response profiles conformed to the "true" $h(\tau)$ for the sample/pyroelectric system (Fig. 5).

Figure 6 shows the equivalence between $h(\tau)$ and $R_{xy}(\tau)$ for a 400 µm glass sample using the FM-TDS method. The data indicate the condition $R_{xy}(\tau) \approx h(\tau)$, which demonstrates that $R_{xx}(\tau) = \delta(\tau)$, to a good approximation. This condition is verified in Fig. 6c (lower trace) which shows that $R_{xx}(\tau)$ is band-limited by the frequency span Δf, of the excitation.

While both methods gave essentially band-limited responses within a given frequency span setting, the coherence of the FM-TDS data was generally much higher than that of the random noise recording over much of the frequency span. The result is an improved signal-to-noise ratio and dynamic range of the FM-TDS method, as predicted

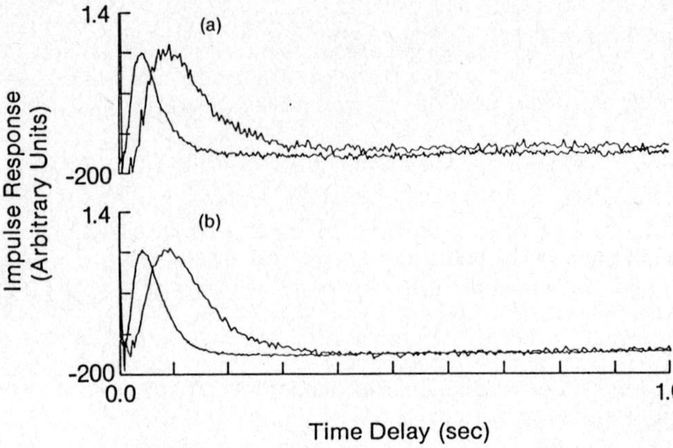

Fig. 5: Comparative impulse response profiles for samples of quartz: (a) random noise recordings (b) FM-TDS recordings. Sample thicknesses were 400 μm (early peak) and 600 μm (later peak). Time delay parameters: <u>random noise</u>: $\tau_d = 42.97$ ms (400 μm) and 85.94 ms (600 μm); Peak widths: $\Delta\tau_d = 35.2$ ms (400 μm) and 78.16 ms (600 μm); <u>FM-TDS results</u>: $\tau_d = 42.97$ ms (400 μ m) and 85.94 ms (600 μm); Peak widths: $\Delta\tau_d = 37.11$ ms (400 μm) and 78.16 ms (600 μm). $\Delta f = 200$ Hz. Both magnitudes were normalized to a maximum of 1 (arbitrary units).

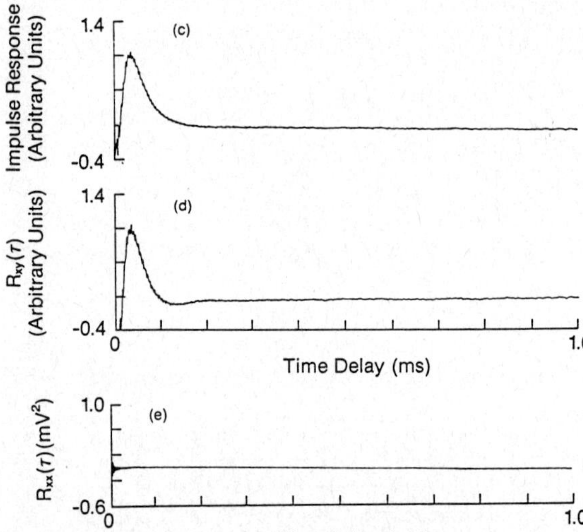

Fig. 6: Impulse response (upper curve), cross correlation function (middle curve) and excitation autocorrelation function (lower curve) for a 400 μm glass sample using FM-TDS excitation. Peak delay in both cases was 35.16 ms. Modulation bandwidth 0-200 Hz.

by the minimum phase character of the linear sine sweep [6,7]. This gives FM-TDS a major speed advantage over random noise, because data acquisition times are significantly reduced. The application potential of our technique to NDE of materials, especially those with low damage threshold to laser pulsing, appears very promising.

REFERENCES

1. A. Tam and H. Coufal: Appl. Phys. Lett. **42,** 33 (1983)
2. H. Coufal and P. Hefferle: Can. J. Phys., **64,** 1200 (1986)
3. A. Mandelis: Chem. Phys. Lett., **108,** 388 (1984)
4. C.E. Yeack, R.L. Melcher, and S.S. Jha: J. Appl. Phys., **53,** 3947 (1982)
5. H. Coufal: J. Photoacoust., **1,** 413 (1983-4)
6. A. Mandelis, L.M.L. Borm and J. Tiessinga: Rev. Sci. Instrum., **57,** 617-621, (Part I); 622-629, (Part II); 670, (Part III), (1986)
7. H. Biering and O.Z. Pedersen: Brüel & Kjaer Tech. Rev., **1,** 5-42, (Part I) and 43, (Part II), (1983)
8. L. Bui, H.J. Shaw and L.T. Zitelli: Electron. Lett., **12,** 393, (1976)
9. KYNAR^{TM} Piezo Film Applications Manual, Pennwalt Corp. (1983)
10. A. Mandelis: IEEE Trans. UFFC, **UFFC-33,** 590 (1986)
11. J.S. Bendat and A.G. Piersol: In *Engineering Applications of Correlation and Spectral Analysis* (Wiley, N.Y., 1980)
12. R.C. Heyser: J. Audio Eng. Soc. **15,** 370 (1967)
13. J.F. Power and A. Mandelis: Rev. Sci. Instrum. (in press)
14. C.R.C. Handbook of Chemistry and Physics, Chemical Rubber Company, Cleveland, OH, (1964)
15. A. Rosencwaig: In *Photoacoustics and Photoacoustic Spectroscopy* (Wiley, N.Y.,1980) Ch. 2
16. F.P. Incropero and D.P. de Witt: *Introduction to Heat Transfer* (Wiley, N.Y., 1985), Appendix A, pp. 667-696.

Numerical Simulation of the Signal Generation and Detection Process in a Pyroelectric Calorimeter

J. Fromm and H. Coufal

IBM Almaden Research Center, 650 Harry Road,
San Jose, CA 95120, USA

Pyroelectric thin film calorimeters [1] have proven to have unsurpassed time resolution and sensitivity as thermal wave detectors. Theoretical understanding of the complex signal generation and detection process, however, has been rather limited. Interpretation of experimental data was qualitative and based only on prominent features, such as time to maximum and amplitude of maximum in a pulsed experiment. To achieve a more full understanding required for quantitative analysis of experimental data, an extensive investigation of the signal generation process was carried out here.

The active element in the calorimeter is a 16 μm thick ferroelectric β-poly(vinylidene fluoride) (β-PVDF) foil coated with 135 nm thick silver electrodes on both sides. The 1/2" diameter element is supported around its perimeter by a metal ring that serves as mechanical support, heat sink and electrical contact. The signal, a charge that is proportional to the energy content of the calorimeter, is detected with a transient digitizer. Details of the calorimeter and the associated detection electronics will be described elsewhere in more detail [2].

A finite difference method was used to model the heat diffusion process within and between electrode, sample and the heat sink. The assumptions were that the calorimeter materials are homogeneous and their properties temperature independent. For excitation, a pulse from an XeCl Excimer laser with a pulse energy of 5 mJ and a constant intensity across an 8 mm diameter area was used. The heat diffusion problem was treated in two steps. During the first step (Fig. 1), experimental data for the temporal profile of the laser pulse and the optical absorption coefficient of silver were used to calculate the heating and the thermal diffusion within the metal electrode and an adjacent 0.675 μm thick sheet of the pyroelectric PVDF. This step involved a one-dimensional calculation with a grid size of 13.5 nm and a time step of 0.45 ps. This approach is justified by the negligible radial heat flow during the laser pulse. The second step (Fig. 2) treated the thermal diffusion in the pyroelectric material and the electrodes with a two-dimensional model assuming cylindrical symmetry. A grid size of 125.0 nm and a time step of 119.2 ns was required for this step. As compared to an earlier approach [3] that used a three-step simulation with variable grid size, the simulation presented here provides equivalent resolution with a minimum of computational effort and complexity. The numerical solution of the heat diffusion problem provides a set of time dependent isotherms. From these isotherms, the induced charge as a function of time was derived using literature values of the pyroelectric coefficient. Treating the calorimeter as a non-ideal current source, connected via a complex transmission line to a non-ideal digitizer, allowed an analytic treatment of the electrical properties of the detector and the associated electronics. Figures 1 and 2 show the calculated and the observed response of the calorimeter. The agreement is excellent considering that no adjustable parameters have been used in the simulation. We thus have developed a quantitative numerical model of the calorimeter. This model allows correlation of the observed signal with surface temperatures or thermal properties of a sample. This agreement not only justifies the assumptions made, but also validates data that are not readily accessible to an experiment but of considerable interest, such as the surface temperature. As an example, Fig. 3 shows the observed electrical signal as a function of surface temperature. During the pulse, they are approximately proportional, enabling use of the calorimeter as a thermometer. The same is true at late times when thermal equilibrium between metal

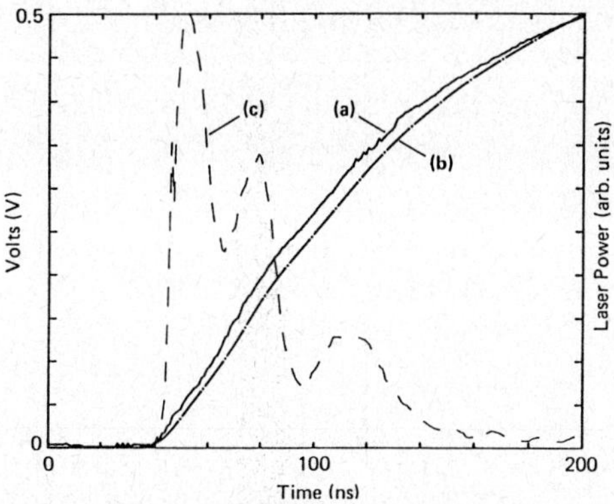

Fig. 1. Comparison of experimentally observed (a) and simulated signal (b) during the laser pulse (c).

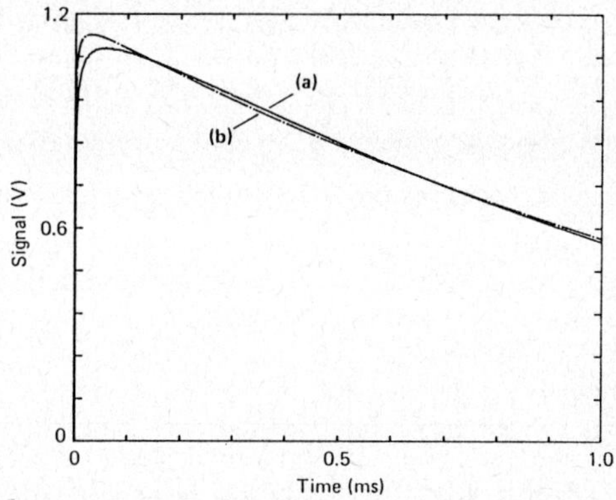

Fig. 2. Comparison of experimentally observed (a) and simulated signal (b) after the laser pulse.

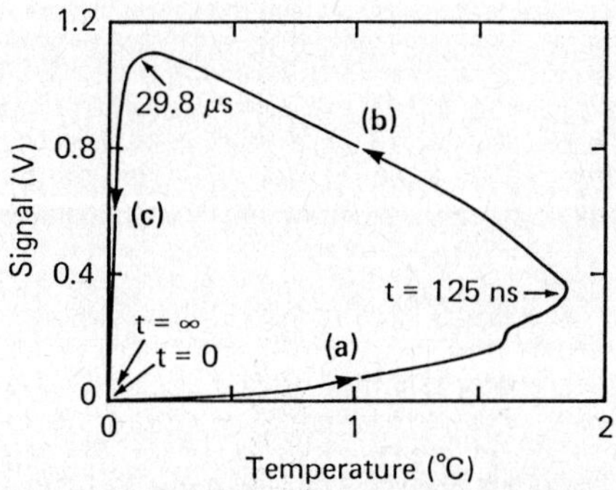

Fig. 3. Observed electrical signal versus surface temperature showing three domains with different functional behavior: (a) heating of the electrode during the pulse with small thermal leakage from the electrode into the pyroelectric material, (b) upon completion of the pulse diffusion from the electrode into the adjacent layer of the pyroelectric and (c) thermal and electrical losses lead back to the starting conditions.

electrode and the adjacent sheet of β-PVDF is achieved. In between is a range where the signal generated by the detector increases due to thermal diffusion from the electrode into the pyroelectric active material. This is accompanied by a substantial drop of surface temperature. It also provides insight into the signal generation and detection process that will allow further improvements of time resolution and sensitivity to be made. Advances of this model will allow us to explore temperature dependence of sample properties and calorimeter characteristics. First-order phase transitions in the sample or the temperature dependence of the pyroelectric coefficient are examples of questions that can now be addressed. Furthermore, the influence of the polarization profile in the pyroelectric and inhomogeneities of samples will be studied, enabling the quantitative analysis of experimental data for complex samples.

References

1. H. Coufal: IEEE Trans. UFFC-33, 507-512 (1986)
2. H. Coufal, R. Grygier, D. Horne and J. Fromm: J. Vac. Sci. Technol. A, in press
3. J. Fromm and H. Coufal: Technical Digest of this meeting, paper WB7

Optical Bistability by Photothermal Displacement

W. Lukosz, P. Pirani, and V. Briguet

Optics Laboratory, Swiss Federal Institute of Technology,
CH-8093 Zürich, Switzerland

1. Introduction

We have demonstrated that photothermal displacement is a new mechanism for optical bistability (OB) in integrated optics. The bistable elements investigated are prism couplers on planar waveguides. Input power P is the power of the laser beam incident on the prism coupler, output power P_R the reflected power (see Fig. 1). Bistability means that at constant input power P the output power P_R can take on either of two different constant values. The system can be switched from one state to the other one by a momentary increase or decrease, respectively, of the input power P. The origin of the OB is the following: a part $P_a=\eta_a P$ of the power coupled into the waveguide is absorbed in the coupling region, either in the waveguiding film F itself, if it is absorbing (as in the configuration shown in Fig. 1a), or in a metal film M (in the configurations shown in Figs. 1b and c). This heating causes a local temperature increase and, consequently, a thermal expansion of the prism and the substrate S or the substrate S_M. (Since the waveguiding film F and the metal film M are very thin, their thermal expansions have a negligible effect.) This expansion produces a photothermal displacement or 'buckling' of the surfaces of the waveguiding film F and of the prism or of the metal film M on the substrate S_M, respectively. This buckling leads to a local reduction in the width d of the air gap C in the coupling region. The incoupling efficiency η_a is a function of d. In the OB regime the two

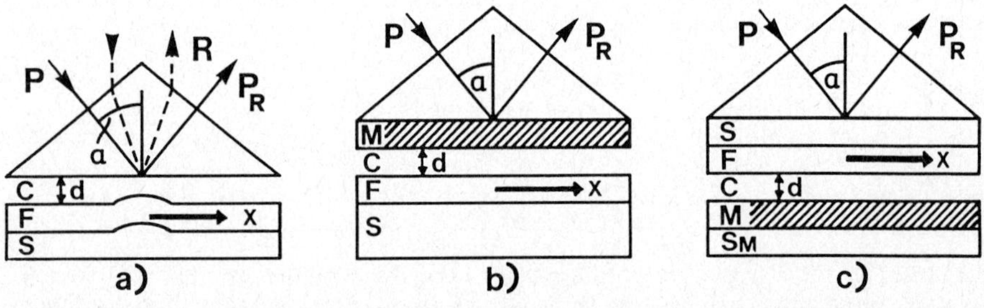

Fig. 1. Prism coupling into planar waveguides. P, input power; α, angle of incidence; P_R, output power; F, waveguiding film of refractive index n_F and thickness d_F; S, substrate; R, reflectance of He-Ne probe beam (- - -) used to monitor the width d of the air gap C. The buckling of the surface of the waveguide but not that of the prism is shown in a). The waveguiding film F is absorbing in configuration a) and non-absorbing in b) and c). M, thin metal film of thickness d_M. In configuration c), S is a thin film of thickness d_S and S_M is a (glass) substrate for M.

bistable states of the system at constant input power P differ in output power P_R and in the value of the width d of the air gap C. If the guided mode is strongly absorbed and does not propagate out of the coupling region, the output power is $P_R=P-P_a=(1-\eta_a)P$.

2. Theory

We calculated the incoupling efficiency η_a as a function of the width d of the air gap C in the prism coupler-waveguide configurations shown in Figs. 1a-c (cf. refs. [1], [2]). We assumed that the incident laser beam is focused to a spot of dimension L_x in the x direction. The results shown in Figs. 2a-c are for excitation of the TE_o mode at wavelength $\lambda=514.5$ nm, and $L_x=200$ µm.

We briefly indicate the underlying physical effects: a guided mode is optimally excited if the incoupling condition $N=n_p \sin\alpha$ is satisfied, where N is the effective index of the guided mode, n_p the refractive index of the prism, and α the angle of incidence. Because of the perturbation of the mode by the presence of the prism and/or the metal film M, the index N depends on the air gap width d; its value at $d\geq\lambda$ is denoted by $N_o=n_p \sin\alpha_o$. The index shift $\Delta N(d)\equiv N(d)-N_o$ is one effect which influences the form of the $\eta_a(d)$ curves. To describe the deviation of the chosen angle of incidence α from the angle α_o, we use the angular detuning variable $\bar{N}_o\equiv n_p(\sin\alpha_o-\sin\alpha)$.

The excited guided mode is strongly absorbed. The absorption coefficient γ_a of the guided mode is a constant in case a), but it increases with decreasing d in cases b) and c) because of the increasing strength of the guided mode's electric field in the metal film M. An important parameter in the incoupling theory is the coupling strength γ_c between the incident wave and the guided mode. While γ_c increases with decreasing d in cases a) and b), it is constant in case c). This dependence of γ_a and γ_c on d, and the effective index shift $\Delta N(d)$ determine the function $\eta_a(d)$. From incoupling theory we derived that $\eta_a=\{4\gamma_a\gamma_c/[\gamma^2+(2\Delta k_x)^2]\}f(u,v)$, where $\Delta k_x\equiv(2\pi/\lambda)[\bar{N}_o+\Delta N(d)]$, $\gamma\equiv\gamma_a+\gamma_c$, and f is a function of the dimensionless variables $u\equiv\gamma L_x$ and $v\equiv\Delta k_x L_x$. An effect not taken into account in this expression for η_a is the direct absorption of the incident beam in the metal film M in the cases b)

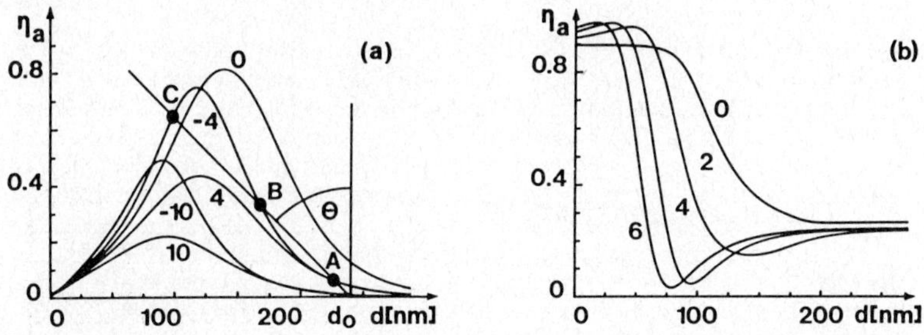

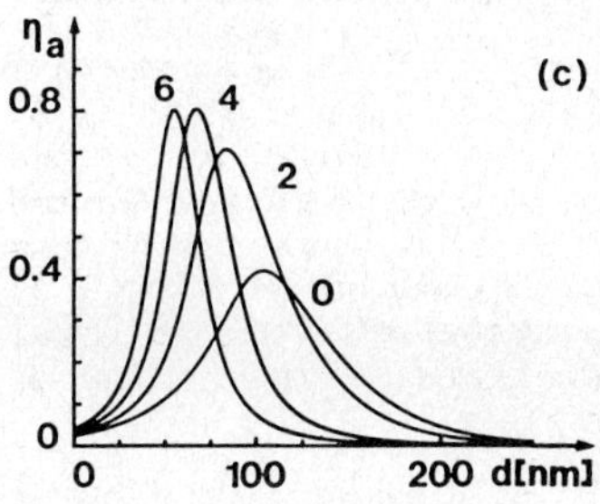

Figs. 2a-c. Calculated incoupling efficiency curves η_a versus gap width d for the configurations shown in Figs. 1a-c. Parameter is the angular detuning $\bar{N}_o$ in units 10^{-3}. Intersections A and C of the straight line with the curve $\eta_a(d)$ are the bistable states at constant input power P. a) indium-tin-oxide waveguiding film F, $n_F=2.0$, $d_F=92$ nm, N=1.61. b) and c) SiO_2-TiO_2 waveguiding film F, $n_F=1.8$, $d_F=185$ nm, N=1.62; gold film M, $d_M=25$ nm in case b) and $d_M=80$ nm in case c).

and c). We included this effect by considering these cases as attenuated total reflection (ATR) configurations and deriving the incoupling efficiency η_a from the calculated reflectances [2].

From the linearity of the laws of heat conduction and thermo-elasticity, it follows that in the steady state the reduction of the gap width in the coupling region is proportional to $P_a=\eta_a P$, i.e., we have

$$d = d_o - \tan\Theta \, \eta_a(d) \, , \qquad (1)$$

d_o being the initial gap width which is adjusted at low input power, and $\tan\Theta=g(\beta/\Lambda)P$, where β is the thermal expansion coefficient and Λ the heat conductivity. For the substrates (soda lime glass) and the prism couplers (Schott glass LaSF22) used in our experiments, the values taken from the literature are $\beta/\Lambda=9$ and 8.2 µm/W, respectively. The dimensionless factor g depends on the geometry of the problem, in particular on the beam width L_x and on the imposed thermal boundary conditions. To obtain a rough estimate for g we use the following oversimplified model: the gap width change Δd is caused alone by expansion of the substrate, a plane parallel plate of thickness d_S. We assume a cylindrical heat sink of radius R around the uniformly heated coupling region of diameter L_x and solve the two-dimensional heat conduction problem. The temperature increase in the centre of the coupling region is $\Delta T=gP_a/(d_S\Lambda)$, where $g=[1+2\ln(2R/L_x)]/(4\pi)$ is independent of d_S; the expansion is $\Delta d=d_S\beta\Delta T$. For $2R/L_x=50-100$ we obtain $g=0.7-0.8$.

Equation (1) is solved graphically in Fig.2. The intersections of the straight line with slope $-1/\tan\Theta$ with the $\eta_a(d)$ curve give the stationary states of the system. Points A and C are the two stable states, B is unstable. From this graphical solution it follows: to obtain OB the initial gap width d_o has to be chosen somewhat larger than the gap width for optimum low power incoupling. The gap width changes Δd between the two bistable states A and C are about 100 nm in case a), and about 40 nm in cases b) and c). The minimum input powers P at which OB is expected to occur are P=18, 5, and 5 mW, respectively, in cases a),b) and c).

3. Experiments

We observed OB coupling argon laser light into absorbing waveguides in the configuration shown in Fig. 1a. Our first results were obtained with evaporated ZnS and ZnSe waveguides [3], and the results reported below with indium-tin-oxide (ITO) waveguides (λ=514.5nm, TE_o mode). In Fig. 3 we show hysteresis curves of output power P_R versus input power P. Also shown in Fig. 3 is the reflectance R of a low power He-Ne laser probe beam with which we monitored the changes of the width d of the air gap (see also ref. [1]). Switching from one to the other of the bistable states can be effected not only by a momentary change in input power P, but also by a momentary change in the angle of incidence α. Hysteresis curves, output power P_R versus angle of incidence α, are shown in Fig. 4. The theoretical results were calculated with the incoupling efficiencies $\eta_a(d)$ shown in Fig. 2a and with the value $g(\beta/\Lambda)=5.25$ µm/W. Good qualitative agreement between theory and experiment was achieved also for the hysteresis curves P_R versus P.

We observed self-pulsing (SP) in waveguides consisting of an SiO_2-TiO_2 waveguiding film on an absorbing ITO layer on a glass substrate [4]. In SP the output power P_R varies periodically with time at constant input power P. The pulsations are caused by the interplay of two thermal effects, the photo-thermally induced changes in the air gap width d and desorption and re-adsorption, respectively, of the H_2O molecules from the SiO_2-TiO_2 film.

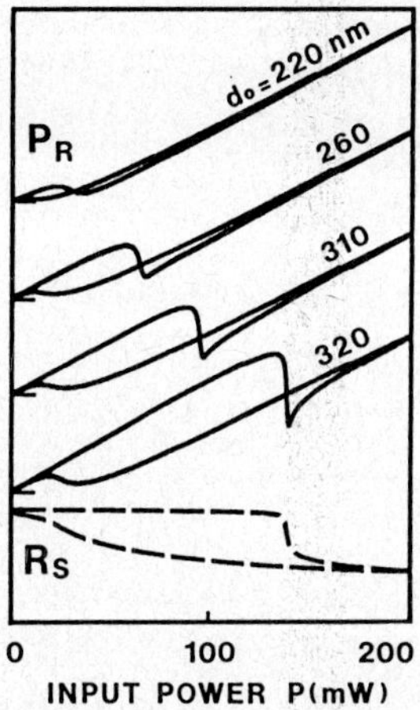

Fig. 3. Output power P_R versus input power P. Parameter is the initial gap width d_o in nm. ---- reflectance R_s of probe beam versus P for bottom hysteresis curve. Scan time 20 sec.

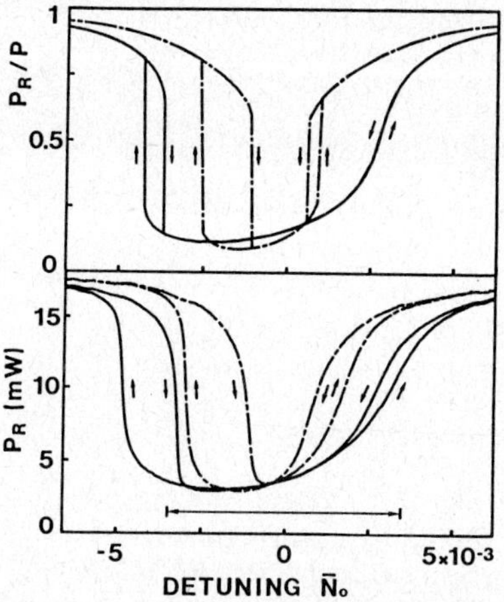

Fig. 4 Output power P_R versus angular detuning $\bar{N}_o$ at constant input power P=25 mW. Top: theoretical results. Bottom: experimental results for $d_o \approx 170$ nm (----) and $d_o \approx 210$ nm (-·-·-). Angular scan time 160 sec. <----> arrow indicates half-width of low power incoupling efficiency curve at large gap widths.

4. Conclusions and Outlook

We observed OB with input powers P≥20 mW coupling argon laser light into absorbing planar waveguides in the prism coupler configuration shown in Fig. 1a. Both the good agreement between theory and experiment and the direct measurement of the changes in the width d of the air gap C prove that the OB is caused by the photothermal effect. From the results of our theory we predict that OB will also occur in the configurations shown in Figs. 1b and c.

5. References

1. W. Lukosz, P. Pirani, and V. Briguet: Opt. Lett. 12, 263 (1987)
2. W. Lukosz and P. Pirani, to be published
3. W. Lukosz, P. Pirani, and V. Briguet: In Optical Bistability III, Springer Proc. in Physics, Vol. 8 (Springer, Berlin, Heidelberg 1986), p.109
4. V. Briguet, P. Pirani, and W. Lukosz: Opt. Lett. 12, 525 (1987)

Investigation of Material with a Single Laser Photo-Displacement Technique

K.H. Yang, S.Y. Zhang, and L. Chen

Institute of Acoustics, Nanjing University,
Nanjing, People's Republic of China

1. Introduction

A photothermal displacement technique has been developed as a sensitive method for determining the optical and thermal features in a variety of materials. To date, common methods have used two distinct lasers for pumping and probing, which made the alignment of the optical system difficult [1, 2]. Recently, we developed a new method using only one laser for both excitation and detection of the photothermal displacement [3, 4]. This technique avoids the difficulty of the superposition of the pumping and the probing beams, and thus it can measure the maximum displacement and improve the detection sensitivity. In this paper, the performance of this method in characterizing materials is described.

2. Experimental Setup

The new configuration is sketched in Fig. 1. The beamsplitter used in the interferometer has three reflection films on both its surfaces. The film M1 has reflectivity exceeding 90%, M2 is of total reflectivity, and M3 is a semi-reflecting film. With this interferometer two complementary interferences of the same intensity are produced. The differential output of both interfering signals at 2ω is

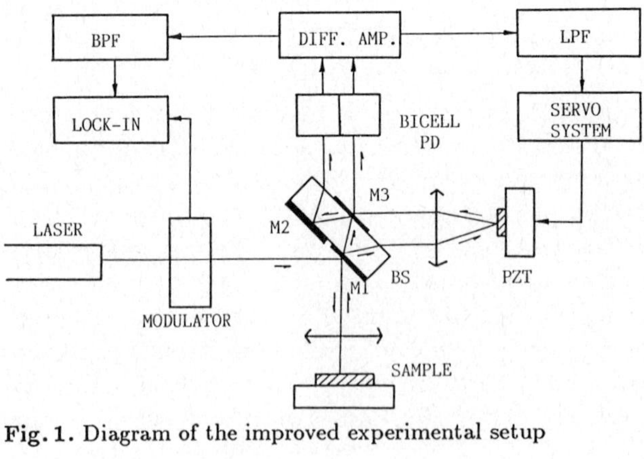

Fig. 1. Diagram of the improved experimental setup

$$S = S_0\sqrt{I_{\mathrm{s}} I_{\mathrm{r}}}(4\pi\eta/\lambda)\cos(2\omega + \theta) \quad , \tag{1}$$

where S_0 is the photodetector transduction factor, and I_{s} and I_{r} are the peak intensities of two interfering beams. The photothermal displacement amplitude ν and phase θ can be measured by a lock-in amplifier set at 2ω.

The laser beam is modulated by an AOM, and focused to 10 μm diameter on the sample surface. The detection sensitivity measured in the experiment is 10^{-3} Å/$\sqrt{\mathrm{Hz}}$.

3. Results

The theoretical analysis, based on a one-dimensional model, shows that the photothermal displacement signal has $1/\omega$-dependence and $-90°$ phase angle if the laser power is completely absorbed on the sample surface. However, in practice such behavior depends on the material properties, which is the baseline for the material investigation [5].

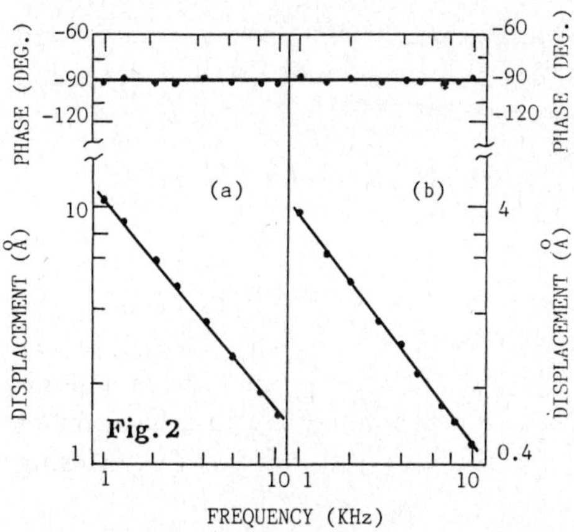

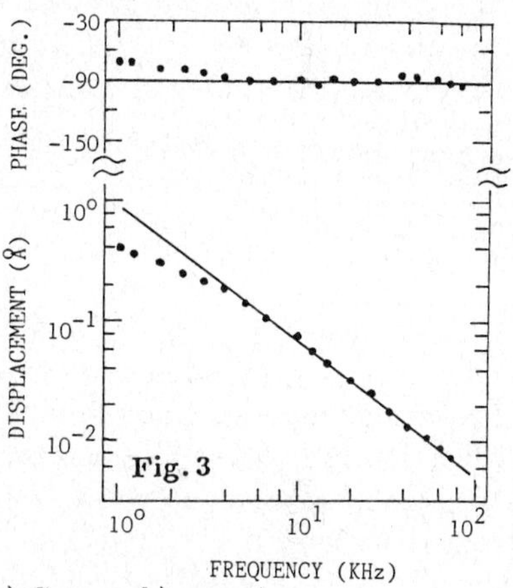

Fig. 2. Magnitude and phase of the displacements. a) Copper, b) ceramics

Fig. 3. Photodisplacement for an amorphous silicon sample

The frequency dependences of the displacement signals in several materials have been studied with our method. Figure 2 shows experimental results for polished copper and PZT ceramics samples. The linear relationships indicate that in both cases the signals approach the behavior of the $1/\omega$-dependence and the phase of $-90°$ in the measured frequency range. The absolute measured values agree with the theoretical values to within 5%. Figure 3 shows the results obtained on the amorphous silicon sample. The sample was prepared using a PVCD device. The frequency behavior given above is reached in the range 10–80 kHz, but is not found at low frequency. This result is similar to the previous work by *Dersch* and *Amer* [5]. Since amorphous silicon is not absolutely opaque, and non-radiative recombination of photogenerated carriers occurs, the frequency dependence of the signal is complicated. Further investigations on amorphous silicon are in progress.

In conclusion, the method using one laser for generating and detecting photothermal displacement instead of two lasers provides an effective and practical means for studying the thermal properties of materials.

4. References

1. Y. Martin, H.K. Wickramasinghe, E.A. Ash: Proc. 1982 IEEE Ultrason. Symp., 563
2. M.A. Olmstead, N.M. Amer, S. Kohn, D. Fournier, A.C. Boccara: Appl. Phys. A **32**, 141 (1983)
3. K.H. Yang, L. Chen, S.Y. Zhang, D.T. Gao: Proc. Int. Workshop on Acoustic NDE, T.1 (1985)
4. L. Chen, K.H. Yang, S.Y. Zhang: Appl. Phys. Lett. **50**, 1349 (1987)
5. H. Dersch, N.M. Amer: Appl. Phys. Lett. **47**, 292 (1985)

Displacement Sensor by Measuring the Phase of Thermal Waves in Air

M. Oksanen and M. Luukkala

University of Helsinki, Department of Physics,
Siltavuorenpenger 20 D, SF-00170 Helsinki, Finland

Thermal waves can be launched and detected in air by using a thin, periodically heated wire as a transmitter and a foil thermocouple as a receiver. The wavelength of the low frequency waves used is around 5 mm. A phase resolution of 1 degree gives a displacement resolution of about 14 μm.

The measurement geometry dictates the phase-displacement relationship of the received signal. The wire is assumed to be much longer and the radial dimension much smaller than the thermal wavelength, and its temperature is assumed to follow the current $I(t)=1/2*I_0\cos(\omega t)+I_0$. Under these circumstances, the wire can be treated as an infinitely long and thin periodical temperature boundary condition. Solving the differential equation for the conduction of heat CARSLAW,JAEGER [1] gives the temperature-distance relation, and the phase-distance relation can be evaluated. It can be given as

$$\phi(r) = \arctan\{\frac{\mathrm{kei}(\kappa r)}{\mathrm{ker}(\kappa r)}\} . \tag{1}$$

In (1) $\kappa = (\omega/\alpha)^{1/2}$, ω = angular frequency of the excitation, α = thermal diffusivity of the media, ker and kei are Kelvin functions SPIEGEL [2]. Figure 1 shows $\phi(r)$ for f=10Hz, $\alpha = 0.3\ \mathrm{s/cm^2}$ (air).

The sensor consists of a current-heated (diameter = 18μm) tungsten wire and a Rdf Corporation, Hudson,N.H.,USA, No. 20109 foil thermocouple. The dimensions of the sensitive area are small enough so that the measured phase-distance relationship approximates the calculated relation. The signal from the thermocouple is amplified with an Analog Devices AD325 and filtered with a narrow 4th degree,Q=20, National MF10 Switched Capacitor bandpass filter. The clock signal for the filters is used to generate the heating current waveform in order to keep the phase shift of the filters constant. The sensor was tested with a Princeton Applied Research 5204 lock-in analyzer. Phase vs. distance is shown in Figure 1. The frequency is 10Hz and the time constant is 1s.

The measured phase has arbitrary zero points for phase and distance. The discrepancy between theory and measurements can be explained by the uncertainty in the thermal diffusivity of air for thermal waves. The dc value given in CARSLAW,JAEGER [3],

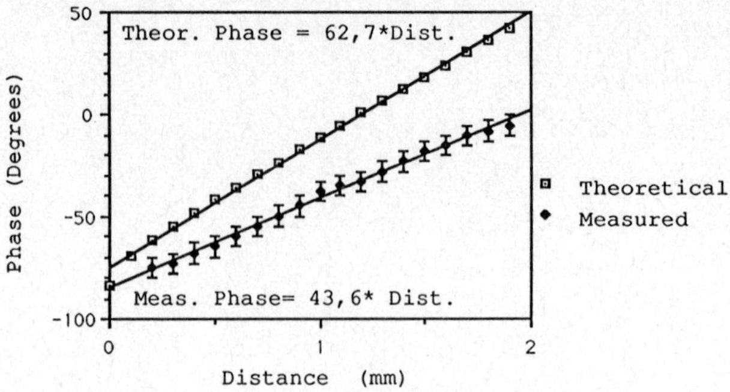

Figure 1 Phase vs. Distance f = 10 Hz

0.187 s/cm^2 seems too small, and 0.3 s/cm^2 used gives better, but not satisfactory results.

Measurement accuracy can be increased by changing the media to decrease the wave length or by just making the measurement progressively slower by increasing time constants in phase evaluation. The sensor is rather slow because of the need for filtering and the 1s time constant in the phase detector. Encapsulation has to be done carefully since any airflow is seen as phase noise. This idea would find use in any movement-based measurements or proximity sensors.

1. H.S.Carslaw, J.C. Jaeger: Conduction of Heat in Solids (Oxford University Press 1959) p.17
2. Murray R. Spiegel: Mathematical Handbook Of Formulas and Tables (McGraw-Hill Company 1968) p.140
3. H.S.Carslaw, J.C. Jaeger: Conduction of Heat in Solids (Oxford University Press 1959) appendix 4

Photothermal Imaging with a Spatially Multiplexed Excitation

D. Fournier and J. Badoz

Laboratoire d'Optique Physique, ER 5, CNRS,
Ecole Supérieure de Physique et Chimie,
10, rue Vauquelin, F-75231 Paris Cedex 05, France

ABSTRACT

A new set-up combining a F.T. spectrometer and a dispersive spectrograph is presented for a spatially multiplexed excitation photothermal imaging of opaque samples.

When running a photothermal imaging experiment of opaque samples, the irradiation is often obtained by using a focussed laser beam (0.1 to a few Watt with a spot size of 1 mm² to 1 μ²) scanned across the sample. The detection is achieved either with photoacoustic sensors (PZT or photoacoustic cell), photothermal radiometers or photothermal deflexion ones. The high density of energy impinging on the sample may produce irreversible damage and that is why a great deal of effort has been made in order to reduce the light density. COUFAL was the first to propose to use 1 D Hadamard or Fourier coding of the excitation light in order to reduce the energy density without decreasing the signal to noise ratio /1, 2/. More recently 2-D Hadamard transform masks have been successfully used by WEI et al. /3/. For our part, we have demonstrated the possibility of using a tomographic approach based on the linear geometry of the "Mirage" detection /4/.

In this work we demonstrate that it is possible to associate a Fourier transform spectrometer and a dispersive spectrograph to get a spatial coding of the excitation beam and consequently of the heated area. By using the linear geometry of the "Mirage" detection, it is possible to record a signal corresponding to a full line in one scan of the interferometer. This method avoids the using of masks which have to be generated, photographed, very carefully positioned and moved when a good resolution is needed.

The schematic diagram of the set-up is shown on Figure 1. The output light of the Michelson interferometer is focussed on the entrance slit of the dis-

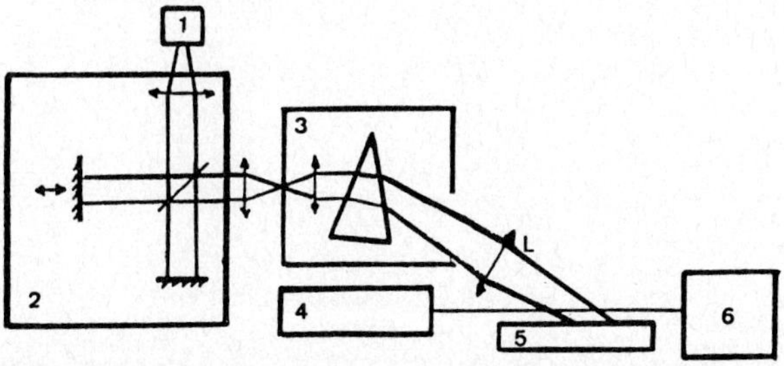

Fig. 1

Schematic experimental setup : 1 : quartz halogen source ; 2 : Fourier transform interferometer ; 3 : dispersive spectrograph , 4 : He-Ne laser ; 5 : opaque sample ; 6 : position sensor.

persive spectrograph. At the output of the spectrograph the spectrum is spread out in the focal plane of the objective L where each point is irradiated by a specific wavelength which is coded by the Michelson Interferometer. After a Fourier transform calculation of the signal issued from the Mirage detection, we get a signal versus wavelength and therefore versus position.

The detection of the He-Ne probe beam is the complex sum of all the elementary deflections induced by each wavelength of the spectrum. Let us point out that one can use either a continuous rapid scan F.T. Interferometer (like most of the medium IR FT Interferometers) or a step by step scanning one /5/ which allows each wavelength to be modulated at the same frequency. We have used a step by step scanning with which an easier analysis of the phase and the amplitude of the signals with conventional lock-in techniques can be obtained.

In our preliminary experiment the average energy density available on the sample is about 1 mW/cm² for a spatial resolution of 300 μ. Let us point out that despite the very low energy density used in this experiment the signal to noise ratio of the experiment is rather good due to Fourier transform spatial multiplexing of the signal. Indeed, Fig. 2 shows the amplitude and phase photothermal signal obtained with a carbon epoxy composite sample. The geometrical structure is compared to the photothermal information /6/.

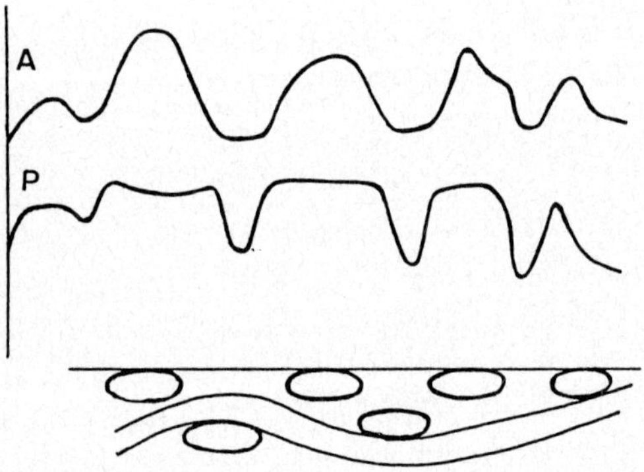

Fig. 2

Amplitude (A) and phase (P) signals for a carbon-epoxy composite sample (for more details see J. Appl. Phys. 59, 235 (1986)).

In conclusion, we have demonstrated the feasibility of a spatially multiplexed excitation photothermal imaging of opaque samples. The opaqueness can be overcome by a preliminary spectral analysis of the sample with the same set-up. The resolution can be highly improved as can the number of effective points which can be multiplexed by using a grating dispersive spectrometer.

At last a two-dimensional multiplexing can be achieved by coupling this set up with a multiplexed mirage probe beam /7/.

1. H. Coufal, U. Moller, S. Schneider : Appl. Opt. 21, 116(1982).
2. H. Coufal : In Proceedings of the Second Topical Meeting on Photoacoustic Spectroscopy, Berkeley, 22-25 June 1981.
3. L.W. Wei, S.Y. Zhang, J.S. Xu : In Proceedings of International Workshop on Acoustic Non Destructive Evaluation. Nanjing, Nov.4-10 (1985).

4. A.C. Boccara, D. Fournier, J. Badoz: Appl. Phys. Lett. 36, 130 (1980).
5. D. Débarre, A.C. Boccara, D. Fournier: App. Opt. 20, 4281 (1981).
6. L. Inglehart, F. Lepoutre, F. Charbonnier: J. Appl. Phys. 59, 235 (1986).
7. F. Charbonnier, D. Fournier, A.C. Boccara, : In Proceedings of 5th International Meeting on Photoacoustic and Photothermal Phenomena, Heidelberg, July 27-30 (1987).

Spatially Multiplexed Photothermal Detection Using a Charge Coupled Linear Photodiode Array

F. Charbonnier, D. Fournier, and A.C. Boccara

Laboratoire d'Optique Physique, ER 5, CNRS,
Ecole Supérieure de Physique et Chimie,
10, rue Vauquelin, F-75231 Paris Cedex 05, France

Abstract : An experimental setup using a CCD linear array of detectors and numerical signal processing is proposed for spatially multiplexed mirage detection.

It is well known that either because of the low frequency excitation which is used in thermal wave imaging or because of the time constant of the lock-in detection which is needed to obtain a good signal-to-noise ratio such experiments are time consuming /1/. We propose a new experimental scheme for mirage detection allowing us to measure simultaneously a large number of pixels which correspond to a linear geometry of the excitation pump.

Figure 1 shows the experimental scheme. The cylindrical lens (1) allows us to obtain a heated line on the sample under investigation. The probe beam produced, for example, by an He-Ne Laser (2) and expanded with an afocal system (3), is focused by a cylindrical lens (4) above the heated line. Each pixel of this line is then conjugated (5) with a diode in the CCD array (6). The angular deflection due to the modulated thermal gradient is converted into an intensity modulation by a razor blade (7). Figure 2 explains this conversion in more detail, and shows the cylindrical geometry of the probe beam. For this preliminary instrumental study, we have used a 256 photodiode linear array (HAMAMATSU) /2/ which can be read typically at a maximum rate of 100 kHz : this array is coupled with an A/D converter, which limits the actual acquisition speed of the signal.

First results lead us to believe that it is possible to read an (nxn) image in (n) measurements (here = n = 256), so the recording time is divided by n.

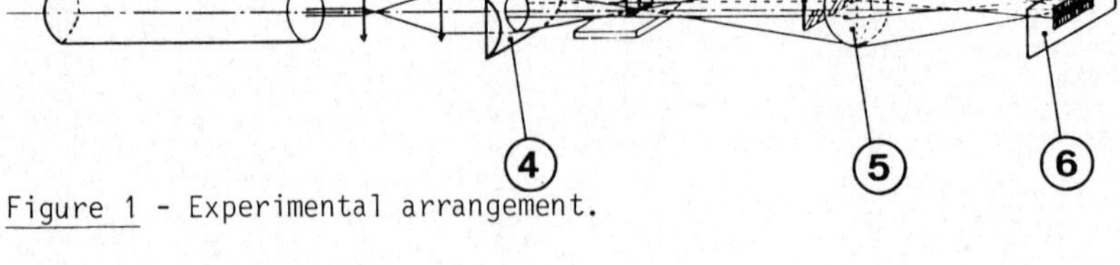

Figure 1 - Experimental arrangement.

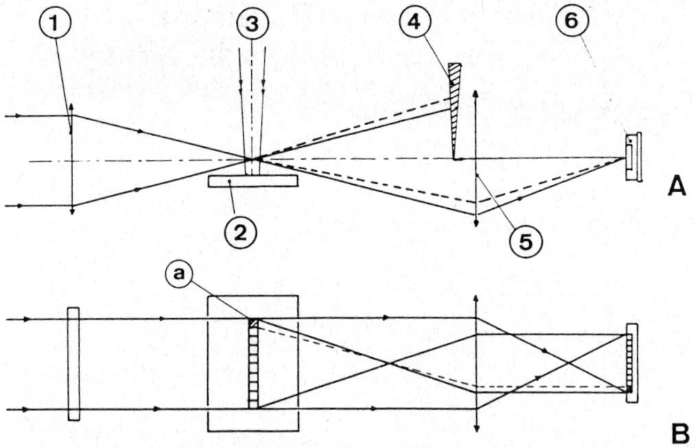

Figure 2 - Experimental arrangement conjugating the heated area and the multidetector.
a) Side view : ① Cylindrical lens ; ② Sample ; ③ Excitation beam ; ④ Razor blade ; ⑤ Spherical lens ; ⑥ Photodiode array.
b) Top view : The pixel ⓐ of the excitation line is conjugated with one photodiode of the CCD array.

The numerical processing of the signal is based on the intrinsic characteristic of the photodiode array used in this experiment. Indeed, the signal value is averaged during the time between two readings of the CCD register from the photodiode array . We have used a reading rate triggered on the modulated excitation, for instance four readings of each photodiode during one period.

It is then straightforward to get the phase and the amplitude information from the signals of each diode (Fig. 3).

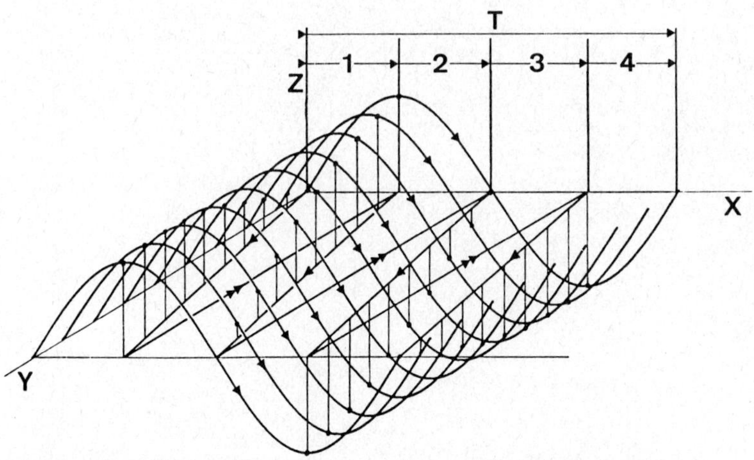

Figure 3 - The reading process of the CCD detector.
X-axis : time ; Y-axis : photodiode elements ; Z-axis : modulated intensity. The line OA shows one of the four successive linear readings 1 2 3 4 , versus time, done during a period (T) of the excitation modulation. For one reading, the difference of phase between the first and the last diodes is $\pi/2$.

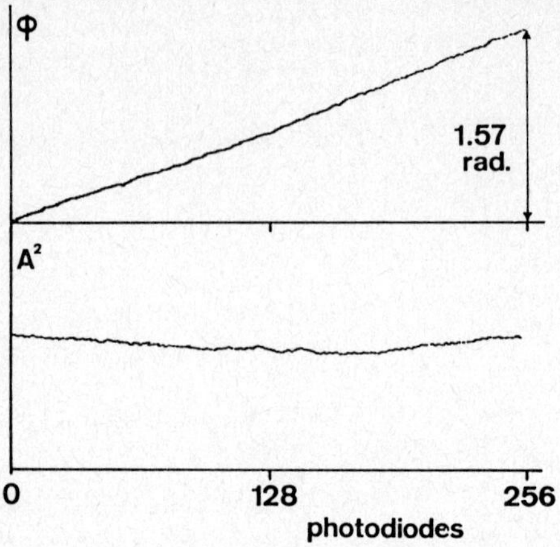

Figure 4 - Experimental results of uniform illumination of the 256 photodiodes. (Square of amplitude A^2, and phase Ø)

In order to evaluate the signal/noise ratio of this setup and to check the validity of the numerical processing, we used a white modulated source to illuminate the array. Figure 4 shows the amplitude and phase signals obtained with four readings per period.

In an actual mirage experiment we will have to correct for both the non-uniformity of the illumination and the linear phase shift associated with this technique.

We believe that such an approach can be a stimulating starting point for an efficient multiplexing of the photothermal signals both for spectroscopy and imaging.

References

1. A.C. Boccara, D. Fournier, J. Badoz : Appl. Phys. Lett. 36, 130 (1980).
2. J.B. BARTON, D.J. BURT, et al.

   Charge-Coupled Devices and Systems, ed. by M.J. Howes and D.V. Morgan (J. Wiley, New York, 1979).

Laser Diodes and Optical Fibers: Two New Approaches for Mirage Detection Sensors

F. Charbonnier, M. Nerozzi, M. Le Liboux, D. Fournier, and A.C. Boccara

Laboratoire d'Optique Physique, ER 5, CNRS,
Ecole Supérieure de Physique et Chimie,
10, rue Vauquelin, F-75231 Paris Cedex 05, France

Abstract : Two sensors using mirage detection are proposed: they use laser diodes or optical fibers in order to reduce the overall dimensions of the setup. We recently proposed /1/ that the traditional mirage setup, which is usually about 1 meter long, can be replaced by a more compact design (∿ 25 cm).

For numerous applications in NDE or spectroscopy it is important to carry out the detection close to the object under investigation rather than moving this object (i.e., in situ inspection in an industrial or hostile environment).

In this contribution we present two experimental realizations which fulfil these above requirements and which have now to be considered as "mirage sensors" rather than "mirage setups".

The first sensor is built on a single block of aluminium, machined in order to integrate all the optical, mechanical and electrical parts. The traditional He-Ne laser has been replaced by a laser diode (Sharp LT022, Hitachi 7801, ...) coupled to an optical system with a suitable numerical aperture collimating the output light cone from the source. This arrangement provides a 50 µm beam waist and a confocal length of 11 mm to probe the temperature gradient. Figure 1 shows the experimental realization.

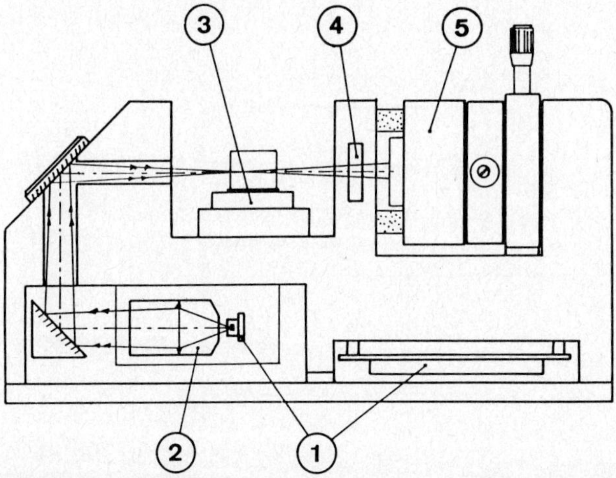

Figure 1 - Laser diode mirage sensor : ① Laser diode and its electronic driver ; ② Collimator ; ③ Sample holder ; ④ Filter ; ⑤ Position sensor.

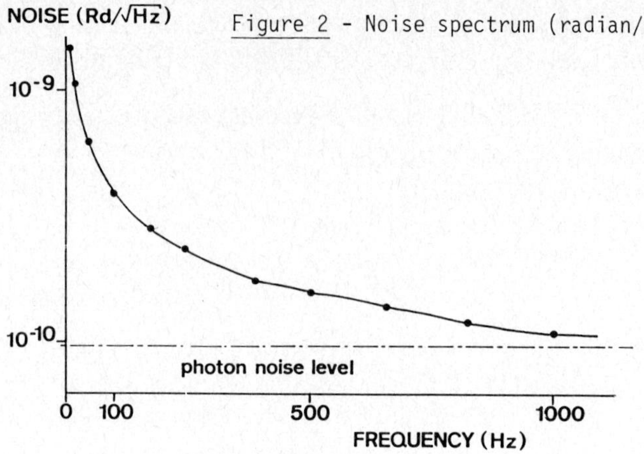

Figure 2 - Noise spectrum (radian/ $\sqrt{HZ}$), versus frequency.

The noise level of this kind of sensor was found to be very good because at frequencies higher than 1000 Hz one reaches the photon noise level, which is the ultimate limit for mirage detection ($\sim 10^{-10}$ rd/ $\sqrt{Hz}$) (Fig. 2). At last, this all-solid-state mirage detector presents a much higher immunity towards mechanical disturbances than the ones using He-Ne probe lasers.

The second sensor is smaller, lighter and more compact. It can be easily moved on the surface of a large sample under examination. Both probe and heating beams are brought into the sensor by two optical fibers. The probe beam is obtained by coupling an He-Ne laser with a 6 μ monomode optical fiber. A 70 μ beam waist and a 24 mm confocal length is obtained with a micro-lens suitably matched to the output of the optical fiber. The use of a monomode fiber was found to be necessary because of the high level of mode noise (speckle fluctuations) carried by a multimode optical fiber. The excitation beam is carried by a 400 μ multimode fiber positioned directly in front of the sample. It is obvious that the fluctuations of the heating spot will lead to minor

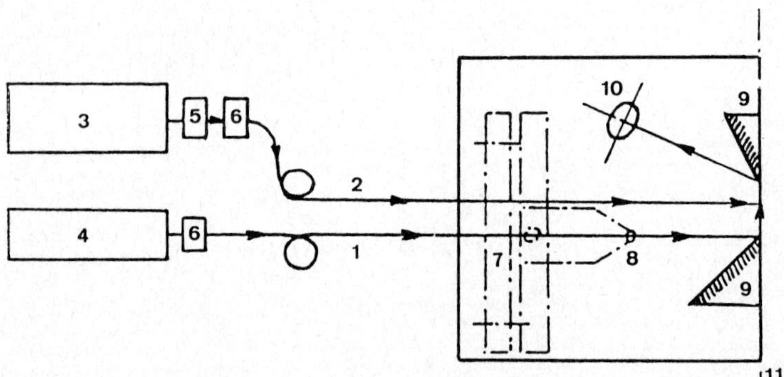

Figure 3 - Optical fiber sensor : ① Optical fiber probe ; ② Optical fiber pump ; ③ Excitation power laser ; ④ He-Ne laser ; ⑤ Optical modulator ; ⑥ Light injection lens ; ⑦ Micropositioners ; ⑧ Output fiber lens ; ⑨ Reflecting metalized prisms ; ⑩ Position sensor ; ⑪ Sample surface reference.

effects on the stability of the signal. Figure 3 illustrates the setup. Up to now the noise of this sensor is larger than the preceding one. This noise, which is about 10^{-8} rad/ Hz, at 30 Hz is namely due to the residual vibrations in the position of the monomode fiber.

These two sensors are only the beginning of the miniaturization of the "mirage" cell. A smaller setup can be imagined. Moreover, it is now possible to conceive of an all-optical sensor (without electrical connections) which would be very helpful in hostile electromagnetic environment.

Reference

1. F. Charbonnier and D. Fournier, Rev. Sci. Instrum. 57 (6) (1986).

Comparison of the Sensitivity and Time Resolution of the PVF_2 Foil and Beam Deflection Detection Methods for Pulsed Photoacoustic Signals in Liquids

S.J. Komorowski[1] *and E.M. Eyring*[2]

[1]Polish Academy of Sciences, Physical Chemistry Institute, PL-Warsaw, Poland

[2]Department of Chemistry, University of Utah, Salt Lake City, UT 84112, USA

In this communication two different approaches to the detection of short acoustic wave pulses generated in liquids by focused laser beams are compared (Fig. 1.). In PVF_2 PAS (photoacoustic spectroscopy) signals are detected with a home made poly(vinylidene difluoride) (PVF_2) foil transducer immersed in the sample liquid. The foil is 28 μm thick, is gold electrodized on both sides, and has a 6 mm^2 transducer active area [1].

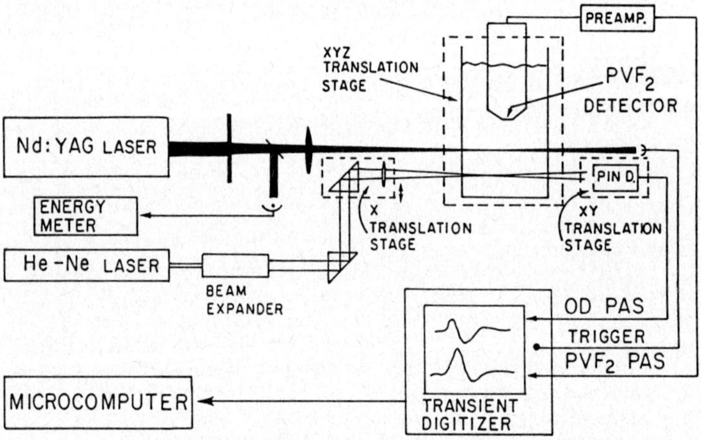

Fig. 1. Schematic of experimental set-up for comparison of the optical beam deflection and PVF_2 foil detection methods in pulsed PAS

In optical detection photoacoustic spectroscopy (OD PAS) the same photoacoustic signals are observed optically [2]. The deflection of the He-Ne laser beam, due to the change of the refractive index of the solution and caused by the acoustic wave crossing the He-Ne laser beam, is observed with a small surface fast PIN diode with a preamplifier of 1 kHz to 100 MHz bandwidth. The fastest PVF_2 PAS signal observed is 22 ns fwhm and the fastest corresponding OD PAS signal is 24 ns from maximum to minimum (Fig. 2.). The DC slow photodiode (SPD) preamplifier set-up. Thus, the probe beam is split in two at the beam splitter (BS) so that the slow TL signal can be observed simultaneously with the combined fast TL and OD PA signals. Thus, the entire nanosecond to hundreds of milliseconds time scale for heat releasing processes is monitored in a single experiment, Fig. 2. A key advantage is the constancy of laser beam geometries in all the measurements.

Usefulness of this new method was verified by measuring the dependence of the TL and OD PA signals on absorption of a $CoCl_2$ in methanol solution and

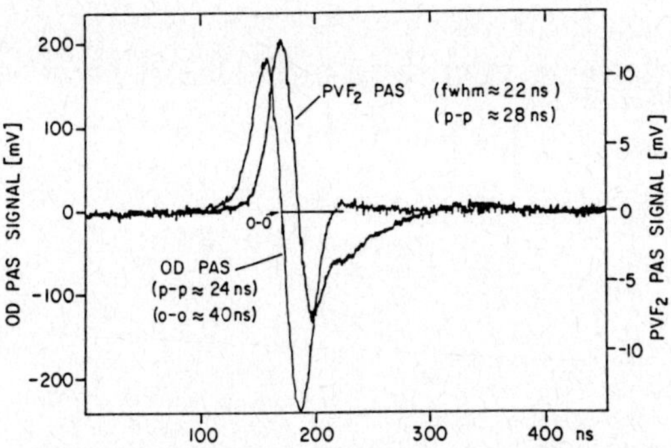

Fig. 2. Example of optical detection of a photoacoustic signal and the thermal lens signal in an experiment with $CoCl_2$ dissolved in methanol

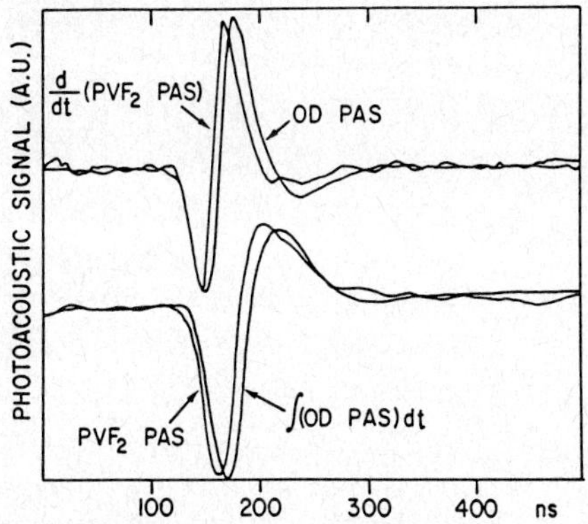

Fig. 3. Comparison of the fast photodiode signals for DODCI and $CoCl_2$ in methanol/30% water (v/v) solution at concentrations giving rise to a 0.01 cm^{-1} absorption in each case

on the focusing of the λ = 532 nm, 200 μJ/pulse excitation beam. For both TL and OD PA signals, the dependence on absorption (from 0.15 to 0.003 cm^{-1}) is linear. The ratio of the amplitude of the TL signal to that of the OD PA signal remains constant as the excitation laser beam is defocused from a diameter of 50 to 100 μm (and both signals diminish.) In a further test the laser dye 3,3'-diethyl-oxadicarbocyanine iodide (DODCI) was studied in water-methanol solution. A pulsed photoacoustic study of this system limited to time scales longer than a microsecond was reported recently [2]. In the present study signals from DODCI were compared with those from $CoCl_2$. For comparable TL signals from these two solutes there is a significant difference in the amplitude of the fast OD PA signals (Fig. 3). The existence of some slow (> 70 ns) heat releasing process is indicated. This can be in DODCI itself, in some impurity, or in oxygen. (The DODCI was used as received from Exciton, and the solution was not deoxygenated.) This conclusion is further supported by the slower increase of the TL signal during the first microsecond after excitation for DODCI compared to $CoCl_2$.

This research was funded by the Department of Energy, Office of Basic Energy Sciences.

1. S. J. Komorowski and E. M. Eyring: the present volume
2. G. M. Bilmes, J. O. Tocho and S. E. Braslavsky: Chem. Phys. Lett. 134, 335 (1987)

Single-Beam Thermal Wave Microscopes

A. Lörincz[1] *and L. Andor*[2]

[1]Department of Physics, Institute of Isotopes, The Hungarian Academy of Sciences, P.O. Box 77, H-1525 Budapest, Hungary

[2]Research Institute for Technical Physics, The Hungarian Academy of Sciences, P.O. Box 76, H-1440 Budapest 1, Hungary

Technological needs have led to a growing interest in non-touching, non-destructive material imaging techniques. One of the most recent of such techniques is the thermal wave method with unique capabilities such as depth profiling and subsurface imaging. Several methods have been developed to detect the thermal waves, including gas-microphone photoacoustic measurements, piezoelectric and infrared detection, interferometric, optical beam deflection and optical reflection techniques [1,2].

There is a simplifying possibility inherent in thermal wave methods, which involves detecting the thermal waves through the reflected part of the first chopped laser beam. Such a solution simultaneously solves certain drawbacks of the conventional approaches, in that it eliminates the problem of "making a second contact" for detection such as with a piezoelectric crystal or the gas environment for a microphone, or a second laser beam. This possibility has been already touched upon in the literature for the case of interferometric detection [3], but to date no detailed study has been done. In this paper two versions of the basic idea are investigated; the dark field method and the Zernike or phase contrast method.

The basic physical process may be originated by a focused, intensity modulated laser beam. The modulation leads to periodic local heating and launches the thermal waves. These waves are diffusive and their behaviour is controlled and influenced by the parameters of the light beam as well as by those of the material; these include absorption coefficient, focus diameter, modulation frequency, heat capacity, thermal diffusion etc. The local ac and dc deformation may be calculated in a rigorous manner [1]. It will be assumed that the modulation frequency is high enough to have a thermal wavelength comparable with the diameter of the focused light beam. Now the major part of the ac component of the photothermal deformation may be approximated as a Gaussian amplitude of constant phase.

The object we are dealing with may be considered as a phase object for a reflection mode scanning optical microscope. There are different methods to detect phase objects, e.g. the dark field method and the phase contrast technique [4]. Calculations done in Fresnel approximation show that the phase contrast method offers linear signal characteristics, while the dark field technique has a mixed linear and quadratic response depending on the chopping frequency [5].

It is easy to see what sort of electronics is required to extract the information from the reflected beam in single-beam detection schemes. For normal double modulated pump and probe experiments the electronics should be set to the sum or the difference frequency, whereas in single-beam schemes the modulation frequencies are identical, so one can detect at the second harmonics of the fundamental frequency. One may anticipate some difficulties due to the fact that the modulation may contain overtones of the fundamental frequency resulting in spurious signals, whereas nonlinearities in the detection scheme may create them. To reduce this problem, square modulation of 50% duty cycle, fast switches, fast detectors and fast electronics may be used since an ideal square wave of 50% duty cycle has no second harmonics content. Methods, where both direct and scattered parts of the heating beam reach the detector - such as the interferometric technique or defocussing arrangements - may be sensitive to this problem, especially at high chopping frequencies. In the phase contrast method the problem is already reduced, and special phase contrast increasing tools like the absorbing phase ring can further decrease it below the observation threshold. In the dark field method the direct passage of the light is obscured, and only scattered or diffracted light may reach the detector; that is, the dark field method is the least sensitive in this respect.

One can calculate the transmission of these microscopes by solving the three dimensional heat diffusion equation and the Navier-Stokes equation for the surface displacement [1], and plugging the results into the transmission integrals [4]. Assuming $\alpha_{th}/\kappa_{th}=2*10^{-5}$ cm/W and an incident power of 10 mW one has a transmission density of $2*10^{-3}$ cm^{-2} for the coherent and a transmission of 10^{-7} for the traditional dark field microscopes. In the case of the Zernike phase contrast method these numbers can be enhanced at the expense of higher spurious background signal levels. Details of the calculation are given elsewhere [5].

References

1. M.A. Olmstead, N.M. Amer, S. Kohn, D. Fournier and A.C. Boccara; Appl. Phys. 32, 141 (1983) and references therein
2. A. Rosencwaig, J. Opsal, W.L. Smith and D.L. Willenborg; Appl. Phys. Lett. 46, 1013 (1985)
3. S. Ameri, E.A. Ash, V. Neuman and C.R. Petts; Electron. Lett. 17, 337 (1981)
4. C.J.R. Sheppard and T. Wilson; Phil. Trans. Roy. Soc. 295, 513 (1980)
5. A. Lörincz; submitted to Appl. Phys. B

The Simultaneous Detection of Photoacoustic Waves by Two Techniques

M. Terzić[1], *J. Diaci*[2], *and J. Možina*

[1]University of Novi Sad, Institute of Physics, P.O. Box 224, YU-21001 Novi Sad, Yugoslavia

[2]University of Ljubljana, Faculty of Mechanical Engineering, P.O. Box 394, YU-61001 Ljubljana, Yugoslavia

Various methods are currently used for the detection of photoacoustic waves [1,2,3,4]. The simultaneous application of two different techniques for the detection of signals generated in the same sample and under the same conditions has not been frequently used [5]. Simultaneous detection of photoacoustic waves by a microphone and by the deflection of a transverse probe laser beam was used in this work to investigate the possibility of obtaining more information about a metal surface covered with a thin layer of black varnish and to test the feasibility of quantitative analytical determination in optically thick liquids. We present here the description of the apparatus containing a two-liquid photoacoustic cell.

The radiation of a pulsed Nd-YAG laser (1.06 μm) was used to generate the photoacoustic waves. It yields the energy of 30 mJ per pulse, A He-Ne laser beam perpendicularly oriented to the Nd-YAG laser beam direction was used as a probe (Fig. 1), The deflection of the probe beam due to the acoustic wave was sensed by a fast photodiode. The acoustic wave was detected simultaneously with a condenser microphone (B & K model 4138 and 4133 with the preamplifier model 2619). The second photodiode positioned to detect the part of the primary laser-light-pulse reflected from the absorbing surface was used to trigger the system. The signals obtained from the microphone and the deflection sensing diode were recorded with a Nicolet 4094A oscilloscope and stored on floppy disks.

The acoustic waves were measured with this combined technique on a metal disk sample (47 mm diameter, 0.8 mm thick) covered with black varnish of various origins. The second kind of samples were solutions of $CuSO_4 \cdot 5H_2O$ in

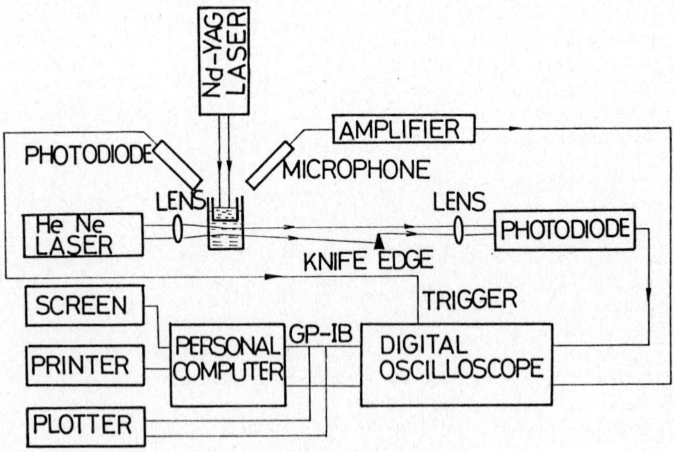

Fig. 1
Experimental setup

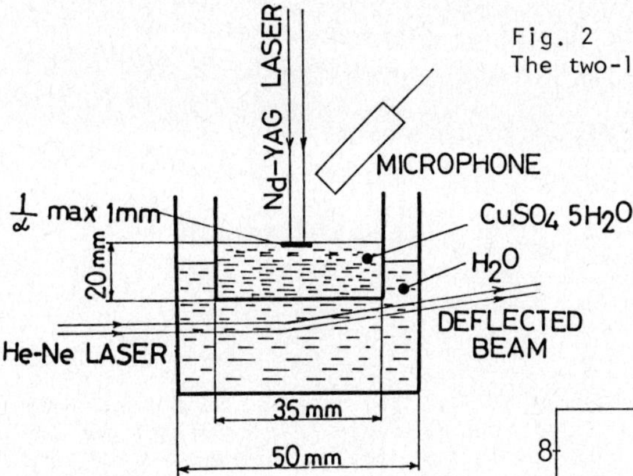

Fig. 2
The two-liquid photoacoustic cell

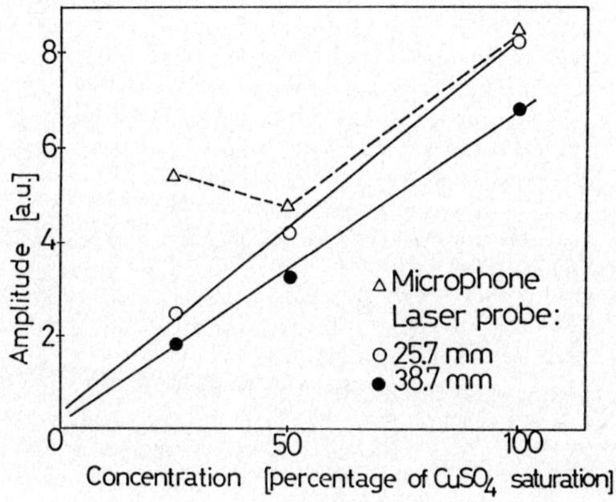

Fig. 3
The waveform amplitude dependence on the solution concentration for various depths of the probe beam

distilled water with various concentrations (100% = saturated at 20°C, 50%, 25%). Because these solutions are not transparent to the probe beam radiation, we constructed a photoacoustic cell which consists of an outer cell and of a smaller cell with a cylindric shape. The inner cell contains the solution and it is placed so that the Nd-YAG laser beam is oriented perpendicularly to the surface of the liquid. The photoacoustic waves propagate through the solution to the destilled water in the outer cell (Fig. 2). The preliminary results of these experiments are presented in Fig. 3 as the waveform amplitude dependence on the $CuSO_4 \cdot 5H_2O$ concentration.

These investigations are of considerable interest for liquids in which the detection is otherwise impossible or very difficult due to large probe beam absorption, in corrosive liquids, etc. The described apparatus is currently also used to investigate the behaviour of acoustic waves during degrading of the black varnish layer due to plasma generation above the metal surface.

References

1. M.W. Sigrist, J. Appl. Phys. 60, R83 (1986)
2. A.C. Tam, Rev. Mod. Phys. 58, 381 (1986)
3. W.B. Jackson, N.M. Amer, A.C. Boccara and D. Fournier, Appl. Opt. 20, 1333 (1981)
4. M.J. Smith and R.A. Palmer, Digest of the 5th International Topical Meeting on Photoacoustic and Photothermal Phenomena, Heidelberg, FRG, paper ThB7 (1987)
5. J. Možina and J, Diaci, J. Phys. Colloq. C6, 73 (1983)

Stray-Light Corrections in Frequency-Dependent Photoacoustics with Transparent Cells

R. Kordecki, J. Pelzl, and B.K. Bein

Institut für Experimentalphysik VI, Ruhr-Universität Bochum, P.O. Box 102148, D-4630 Bochum, Fed. Rep. of Germany

Photoacoustic measurements of the thermal diffusivity of amorphous metallic foils [1] and depth profiling studies on graphite [2] have shown that the signals from the PA cell walls generated either by the incident laser beam or by light diffusively reflected at the sample surface are not negligible. For the system "opaque cell / transparent sample" it was shown that stray-light corrections seem to be more important at higher modulation frequencies. This was explained by the different frequency dependence of the cell and sample signals [3]: The PA signal of a transparent solid varies as $f^{-3/2}$ and falls off more rapidly with increasing frequency than the signal of the opaque cell material which varies with the inverse of the modulation frequency, f^{-1}. Here we discuss the stray-light corrections for the system "transparent cell / opaque samples", both for the signal amplitude and phase shift.

The influence of the stray-light on the measured PA amplitudes for the system "transparent cell/ opaque sample" is shown in Fig. 1. In the limit of low frequencies the uncorrected signal amplitudes approach the correct values for $f \to 0$, but the slope of the measured curves is not correct. For higher frequencies, on the other hand, the slopes are correct, but an offset of the quantity $(A^2 \cdot f)$ is observed. This offset is constant because the signal amplitudes both from the opaque thin samples, illuminated symmetri-

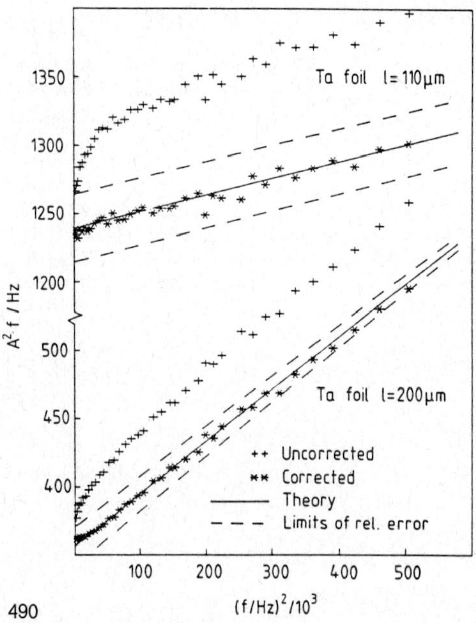

Fig. 1: For the corrected signals, the PA amplitudes A of two tantalum foils of thermal diffusivity α and different thickness ℓ agree with theory, $A^2 f = 2\alpha/\pi\ell^2 + \pi\ell^2/12\alpha\ f^2$, within the limits of 2% relative error, whereas the magnitude and slope of the uncorrected signals cannot be interpreted consistently

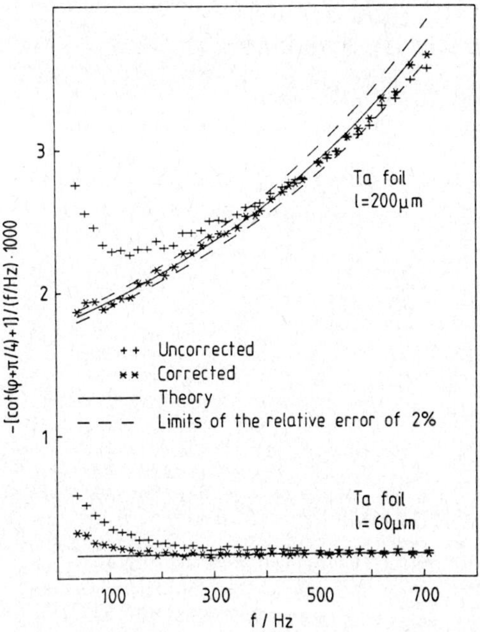

Fig. 2: The PA phase lags of two tantalum foils of different thickness, plotted as functions of the modulation frequency f, show a systematical phase offset at lower frequencies for uncorrected data, while the stray-light corrected signals are in agreement with the theoretical solution within the limits of 2% relative error

cally at the front and rear surface in the cell cavity /1/, and from the transparent cell walls vary with $f^{-3/2}$.

The uncorrected phase lags in Fig. 2 deviate from the theoretical curves at low frequencies. This is explained by the large phase difference existing between the contributions from the opaque samples and the transparent cell walls at low frequencies. Thus also the phase of the vector of the total signal is affected. At higher modulation frequencies, the phases of the different signal contributions become equal.

This example demonstrates the necessity and efficiency of stray-light corrections in reducing the systematic errors of PA measurements for both the amplitude and the phase. In contrast to the interpretation of the PA amplitude of the combination opaque cell / transparent sample [3], the understanding of the stray-light effect is not as easy here and the stray-light correction procedure is necessary for the whole frequency range.

1. R.Kordecki, B.K.Bein and J.Pelzl, Can. J. of Phys. 64, 1204 (1986)
2. B.K.Bein, S.Krüger and J.Pelzl, J. Nucl. Materials 141/143, 119 (1986)
3. S.Krüger, R.Kordecki, J.Pelzl and B.K.Bein, J. Appl. Phys. 62, 55 (1987)

A Magnetostrictively Linearised Gas Coupled Microphone Cell for Quantitative PAS

M.R. Brant[1] *and M.L. Waller*[2]

[1]Manchester Polytechnic, Chester Street, Manchester, M1 5GD, United Kingdom

[2]DIAS, UMIST, P.O. Box 88, Manchester, M60 1QD, United Kingdom

A linearised gas coupled microphone photoacoustic cell has been designed for quantitative photoacoustic spectroscopy. Design principles will be discussed.

1. Introduction

The magnitude of a photoacoustic signal depends upon a variety of factors including the thermal and absorptive properties of the sample involved and the characteristics of the spectrometer itself. MILLER [1] recognised that a limiting factor is the response of the microphone. COX and COLEMAN [2] have highlighted the non-linear transient response and resulting limitation on qualitative comparisons. A system has been designed which operates at almost constant capacitance by means of magnetrostrictive compensation within the sample cell. The magnetostrictive element operates in the inverse transducer mode thus making the response almost independent of the microphone response.

The magnetostrictive transducer comprises a small array of 1 cm long nickel tubes resonated at approximately 200 kHz. Coupling into the cell is via a stiff membrane which transmits amplitude variations of the order of 50 Angstroms. This is comparable to the deflection of a capacitative microphone for a pressure differential of 1 Pa. The amplitude of these variations is modulated in response to pressure variations within the cell detected by the microphone, which produces the error signal to drive the magnetostrictive transducer.

1.1 The Measurement System

A block diagram of the system is shown in Fig. 1. The cell can be considered as the "summing point" in operational amplifier terminology. A head amplifier, followed by a notch filter and active phase advance, provides the forward gain, and the feedback path comprises an amplitude modulator, a power stage and the magnetostrictive element. From feedback theory, the transfer function depends upon the feedback path, provided the loop gain is much greater than unity. Thus, the linearity of the system depends predominantly upon the magnetostrictive element and the modulator. Calculations show that 2% linearity between the cell input and the circuit output is attainable.

1.2 System dynamics

When considering system dynamics, it should be noted that the microphone is a second order element and consequently some phase lead compensation is required to ensure adequate stability. Residual carrier breakthrough, which has the effect of reducing loop gain due to carrier suppression at

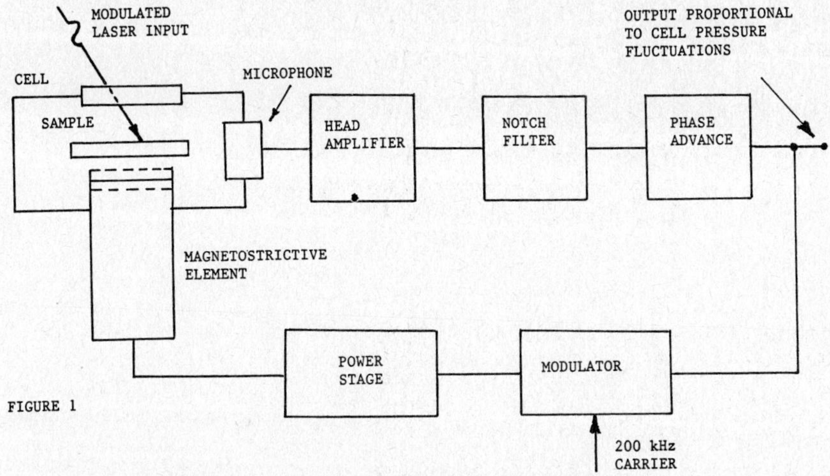

FIGURE 1

the modulator control input, can be removed by means of a notch filter. In the loop analysis, the notch filter can be represented by a low pass filter.

The magnetostrictive transducer has a high Q at resonance, thus phase lag at modulation frequencies (up to 20 kHz) is negligible. A greater delay occurs in the modulator and consequently there is a case for operating the transducer at the first overtone, albeit at reduced efficiency.

1.3 Transducer Design

An array of four tubular transducers (10mm x 3mm) is mounted in rosette manner into the side of the PAS cell. The number of tubes is limited by the winding requirements for each drive coil. Coupling into the cell is via a stiff membrane which is glued to the ends of the tubes (small screws give unwanted local resonances). The whole is encapsulated in a high thermal conductivity epoxy which serves the triple purposes of cooling, providing a high inertial mass, and sealing the transducer into the cell.

1.4 Literature

1. R. M. Miller: Digital Simulation of Photoacoustic Impulse Responses, Can. J. Phys., 1985.
2. M. F. Cox, G. N. Coleman: Measurement of Single-pulse Photoacoustic Signals, Anal. Chem., 1981, 53, 2034-2036.

Measurement of Optical Coating Absorptivity Using Optoacoustic Detection

J. Diaci and J. Možina

Faculty of Mechanical Engineering, University of Ljubljana, P.O. Box 394, YU-61000 Ljubljana, Yugoslavia

Optoacoustic measurements of low–level optical absorption have been reported by several authors using modulated cw excitation with microphone or piezoelectric detection /1,2/ and pulsed excitation with piezoelectric detection /3,4/. Contrary to commonly accepted view it is proposed in this article that gas–microphone detection can also be used with pulsed excitation to measure low–level absorptivity of optical materials.

To compare the sensitivity of gas–microphone and piezoelectric detection we used the experimental system for simultaneous detection of both signals shown in Fig. 1 /5/. Bruel & Kjaer microphone type 4138 and broad–band piezoelectric transducer were used as sensing elements for optoacoustic signals. The signals were digitized by Nicolet 4175 fast sampling unit and stored by personal computer, where averaging and amplitude analysis were carried out. The basis for the comparison of microphone and piezoelectric responses were peak–to–peak amplitudes vs. incident laser energy for the specimen of known absorptivity value (Al HR coating on BK–7 glass substrate – A = 0.13) Resulting linear dependencies, as evident from Fig. 2, were used as calibration curves for other specimens.

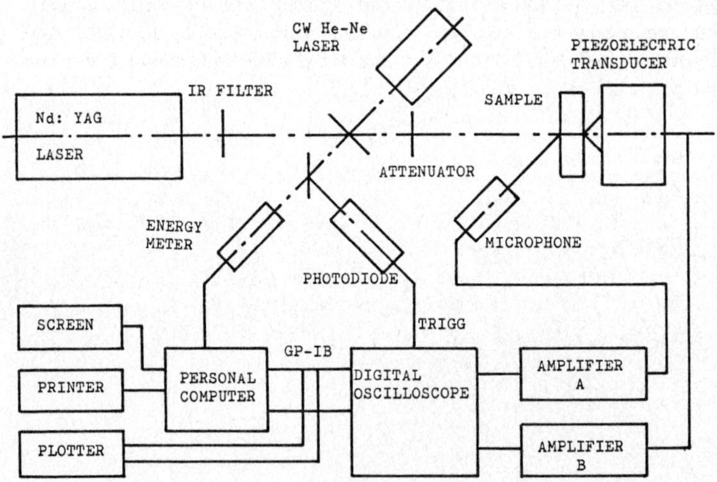

Fig. 1 Shematic of the experimental set–up for simultaneous detection of gas – microphone and piezoelectric optoacoustic responses of optical components

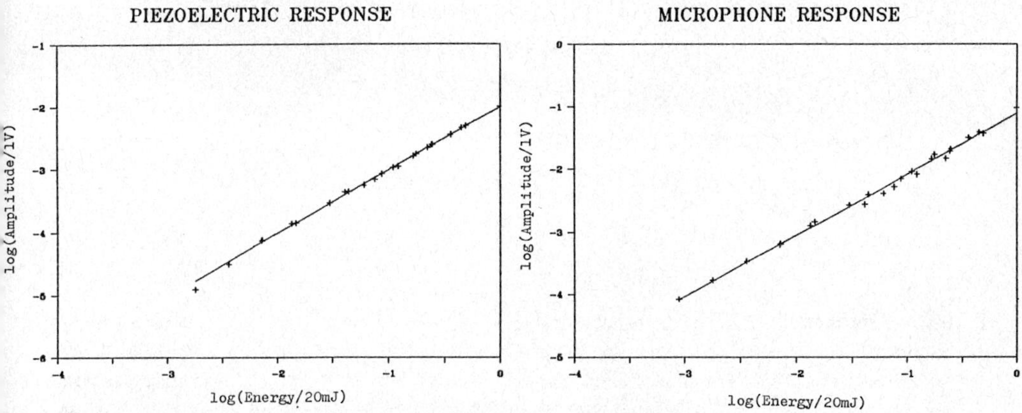

Fig. 2 Optoacoustic signal amplitudes vs. incident laser energy (Al high–reflective coating on BK–7 glass substrate)

Several other HR and AR coating OA responses were measured. By the comparison of their OA signal amplitudes to the OA amplitude of the calibration sample, their absorptivities were determined. The results obtained by this method were compared to the results of the reflection – transmission method and good agreement was found. Using microphone detection we could measure absorptions in the $2x10^{-4}$ range (SiO_2/TiO_2 HR coating), but we believe that the method can be further refined to enable absorptivity measurements in the 10^{-5} range.

Although higher sensitivities have been achieved using other optoacoustic techniques, it must be kept in mind that pulsed microphone detection is very simple and free from geometry and acoustic–coupling variations. In addition, due to much greater sound transit time delay, the scattered light disturbance can be effectively separated from the main optoacoustic signal.

REFERENCES

1. H.S. Bennett and R.A. Forman: Appl. Opt. 16, 2834 (1977)
2. A. Hordvik and L. Skolnik: Appl. Opt. 16, 2919 (1977)
3. A. Rosencwaig and T.W. Hindley: Appl. Opt. 20, 606 (1981)
4. Y. Bae, J.J. Song and Y.B. Kim: Appl. Opt. 21, 35 (1982)
5. J. Možina and J. Diaci: J. Physique, C6 suppl. 10, 44, C6–73 (1983)

Time-Resolved IR Video Imaging with Synchronized Scanned Laser Heating

P.K. Kuo[1], *I.C. Oppenheim*[1], *L.D. Favro*[1], *Z.J. Feng*[1], *R.L. Thomas*[1], *J. Hartikainen*[2], *and L.J. Inglehart*[3]

[1]Department of Physics, Wayne State University, Detroit, MI 48202, USA
[2]Department of Physics, University of Helsinki, Siltavuorenpenger 20 D, SF-00170 Helsinki, Finland
[3]Materials Science and Engineering, The Johns Hopkins University, Baltimore, MD 21218, USA

1.0 Description of the instrumentation

Recent advances in the speed and sensitivity of cooled HgCdTe detectors have made it possible to construct an infrared video camera with a single detector by using mirror scanning techniques. The commercial availability of this kind of camera has led to interesting techniques for active thermography measurements. For example, they have been used with flash lamps for imaging defects in materials [1], and with a scanned IR source for measuring thermal diffusivities [2,3]. With an infrared camera, vast amounts of thermal information from an entire area can be gathered in parallel. This method has obvious advantages in imaging as compared to the more traditional methods of acoustic detection, optical beam deflection or focused (unscanned) IR detection, each of which require a slow scan of the sample to obtain an image. The traditional methods, however, due to the continuous nature of their detectors, can gather fast temporal thermal information by using modulated heat sources [4]. The intrinsic bandwidth of detection is only limited by that of the detectors. The infrared video camera, due to its fixed rate of 50 or 60 fields per second, cannot catch fast thermal phenomena in the millisecond range when used with a fixed area source (flash lamp) or a scanned (unsynchronized) IR source. The method described here combines some of the advantages of both methods by using a focused and scanned optical laser beam, synchronized with the vertical scan of the video camera. It is capable of observing thermal phenomena on a sub-millisecond time scale and also can acquire data on an entire vertically scanned line at the video field rate.

A block diagram of the instrumentation is shown in Fig. 1. The instrument uses an Ar-ion laser, an A/O modulator for beam blanking during retrace, and a microcomputer-controlled

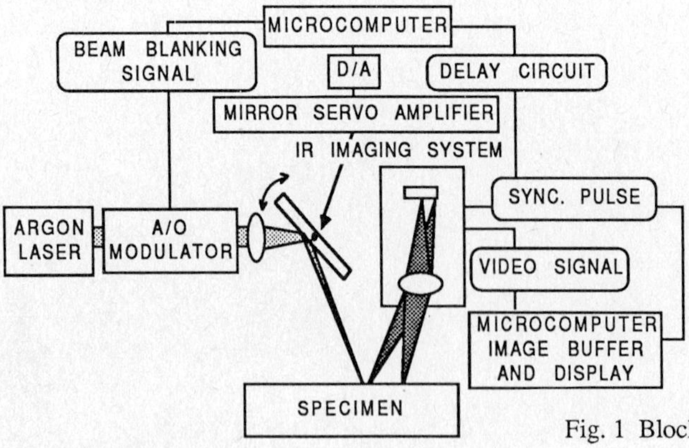

Fig. 1 Block diagram of imaging system

mirror scanning system which is synchronized with the vertical scanning of the video camera (Inframetrics Model IR-600 operated in the 8-12 μm band, either with a 1.8 inch focal length microscope objective or a 6 inch close-up lens). A second microcomputer is used for image processing. For time scales of 0.1 ms or longer, the video camera's horizontal scan time (64 μs) is fast enough to be considered as being instantaneous. Thus, the video camera can be considered as providing a moving line of point detectors, gathering data in parallel. The mirror scanning system is programmed to keep the focused laser beam a fixed temporal interval ahead of the moving line of detectors. The fixed temporal advance can be programmably controlled to better than 0.1 ms accuracy. This arrangement improves the effective detection bandwidth of the infrared camera to 16 kHz (the horizontal scan rate) by trading off some parallelism, i.e., thermal data are acquired only one line at a time. The spatial resolution is limited by the size of the focused laser beam and not by the camera's IR detector [5]. The focused source, coupled with spatially distributed point detectors, provides the opportunity to measure lateral heat flow. This capability turns out to be very advantageous when applied to thin (submicron) film structures. With an unfocused heat source, the pattern of heat flow is primarily one-dimensional. To be able to measure thermal properties of submicron films with such a system one would need response times in the nanosecond range [6]. The technique described here, shown schematically in Fig. 2, is capable of detecting the presence of very thin films. An example is given below in which a 0.1 μm boron film is detected underneath a 0.6 μm ZrN film on a silicon substrate.

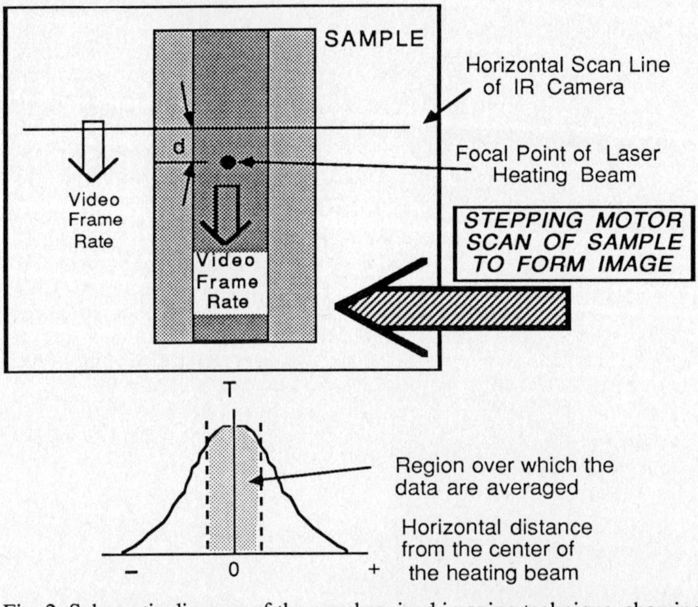

Fig. 2 Schematic diagram of the synchronized imaging technique, showing the data acquisition for one x-position of the stepping motor scan of the sample (top). A vertical strip of the data at that position is averaged over the region shown (bottom) and stored to form one vertical image line, with the full image composed digitally by stepping the x-position of the sample at right angles to the direction of the heating line

We have developed imaging algorithms which utilize the hardware of the IR-600, but which are optimized for use with the scanned heating spot. These algorithms enable averaging over multiple frames and are capable of suppressing the dc IR background signal relative to the signal in the immediate vicinity of the heating spot.

There are obviously many variations of this scanning technique which can be used for special purposes. The principal advantage of the method used here over the flash lamp technique for IR imaging is that it can be operated in a much shorter time regime (the horizontal scan time, 64 μs, as compared to the video frame time, 16.7 ms). The shorter time scale makes the instrument particularly suitable for studies of submicron films (it of course can also be used for longer time scale measurements for other kinds of samples). Another advantage is that it is sensitive to lateral heat flow because it has a point heat source and a point detector which can be programmably displaced relative to one another. The system can operate either in reflection, as shown in Fig. 1, or in transmission if the substrate is transparent in either the visible or the infrared. An advantage in the transmission mode is the ease of achieving short working distances and hence high spatial resolution.

2.0 Experimental results

Figure 3 shows a schematic diagram, and Fig. 4 an IR image of a 0.1 μm film of boron deposited underneath a 0.6 μm film of ZrN on a silicon substrate, obtained with this instrument. This image was taken in reflection, using the 6 inch IR close-up lens, and clearly indicates the presence of the subsurface boron film through its influence on the short-time-scale transverse heat flow. The boron film is undetectable optically, and also cannot be seen using conventional IR thermography.

The technique should be sensitive to defects in thin film structures. To illustrate, in Fig. 5 we show an image of a 0.1 μm film of boron on a glass substrate, obtained with this instrument. This image was taken in reflection, using the 6 inch IR close-up lens. The image covers a 1.5 mm x 6.5 mm region of the film which contains regions of delaminations.

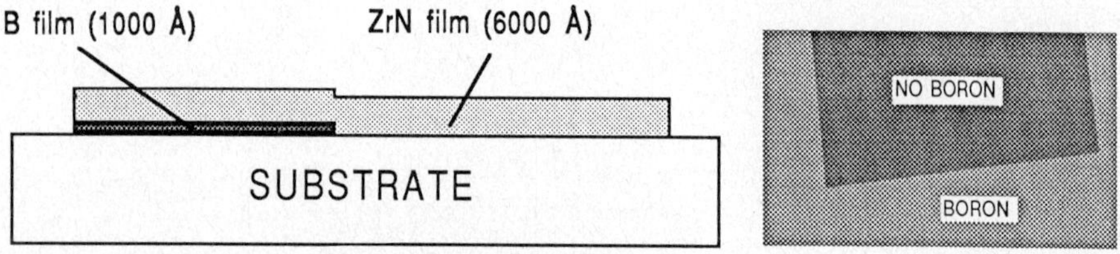

Fig 3 Schematic diagram (left) of ZrN/B/Silicon sample whose scanned IR thermal wave image is shown in Fig. 4. Schematic diagram of boron film location and shape (right)

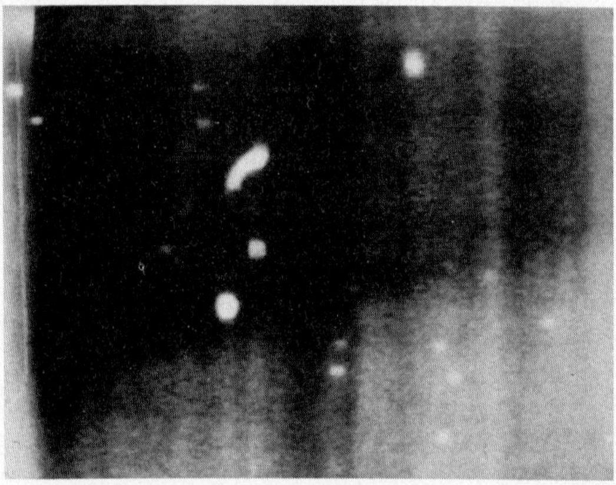

Fig. 4 Scanned IR thermal wave image of the ZrN/B/Silicon sample shown schematically in Fig. 3. The darker (cooler) region is a region for which the B film is absent (by masking)

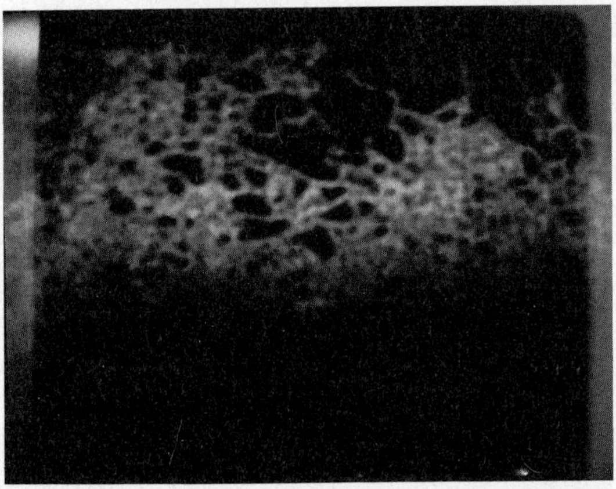

Fig. 5 IR image of a 0.1 μm film of boron on a glass substrate.

3.0 Acknowledgements

This work was sponsored by the Army Research Office under contract No. DAAG29-84-K-0173, and by the Institute for Manufacturing Research, Wayne State University.

4.0 References

1. W.N. Reynolds: In Can. J. Phys. **64**, 1150 (1986).

2. D.M.Heath, C.S. Welch, W.P. Winfree, J.S. Heyman, and W.E. Miller: In Review of Progress in Quantitative Nondestructive Evaluation, D.O. Thompson and D.E. Chimenti, Eds., Vol. **5B**, 1125 Plenum, New York (1986).

3. C.S. Welch, D.M.Heath, and W.P. Winfree: In Review of Progress in Quantitative Nondestructive Evaluation, D.O. Thompson and D.E. Chimenti, Eds., Vol. **5B**, 1133 Plenum, New York (1986).

4. See, for example, A. Rosencwaig: In Science **218**, 223 (1982).

5. L.J. Inglehart, K.R. Grice, L.D. Favro, P.K. Kuo, and R.L. Thomas: In Appl. Phys. Lett. **43**, 446 (1983).

6. G.L. Easley, B.M. Clemens, and C.A. Paddock: In Appl. Phys. Lett. **50**, 717 (1987).

Temporal Moment Method in Pulsed Photothermal Radiometry: Application to Carbon Epoxy NDT

D.L. Balageas, D.M. Boscher, and A.A. Déom

O.N.E.R.A., 29 av. de la division Leclerc, F-92320 Châtillon, France

1. THEORY

In pulsed back emission photothermal radiometry, a sample is heated for a short while by an IR or visible source. The surface temperature increase is observed by means of an IR camera [1]. The method of the zero-order temporal moment consists in integrating the temperature increase over the time. For a homogeneous medium, the temperature-time history being $\Delta T_{R=0}(t)$, this moment

$$M_o = \int_o^\infty \Delta T_{R=0}(t)\, dt$$

tends to infinity. When an internal defect is present, characterized by its thermal resistance R and its depth z from the front surface, the perturbation of the temperature-time history, $\Delta T(t)$, induces a variation in the moment, and it may be analytically demonstrated by the use of the Laplace transform or the Fourier transform that the difference between the two moments is

$$\Delta M_o = \int_o^\infty [\Delta T_R(t) - \Delta T_{R=0}(t)]\ dt = Q.R.(1-z/L)^2,$$

where Q is the energy density deposited in the sample and L its thickness. It is very important to point out that, if the sample is thick, M is equal to Q.R and is independent of the defect depth. Experimentally, it may be more interesting to use this parameter for identifying the thermal resistance of the defect than the correlation previously proposed [2]. As a matter of fact, obtaining ΔM_o corresponds to the integration of temperature, an operation enhancing the signal-to-noise ratio. In this way, it becomes possible to obtain sensitivity as good as those of modulated flux methods using coherent detection. It is then possible to observe defects far in depth from the surface. We also mention that the temporal moment method may be used for correcting the pulsed thermogram for loss effects [3].

2. EXPERIMENT

This method was applied to the localization of Teflon inclusions, simulating delaminations in a carbon epoxy sample manufactured by AMD.BA. The location depths and sizes are varied (see Fig. 1). The setup consists of a bank of IR quartz lamps and a drop shutter to heat a large

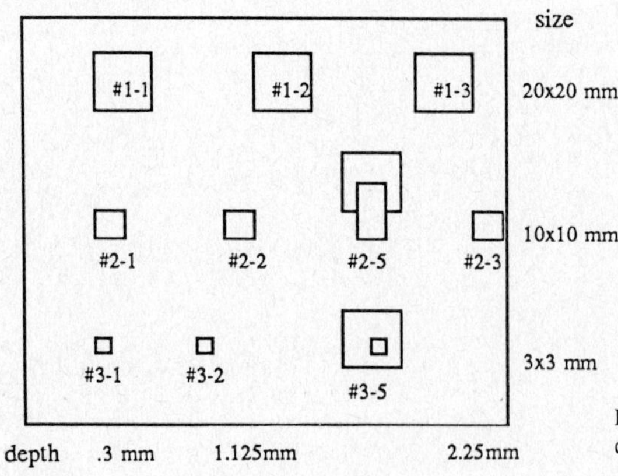

Fig. 1 : Artificial defect arrangement in the carbon epoxy sample

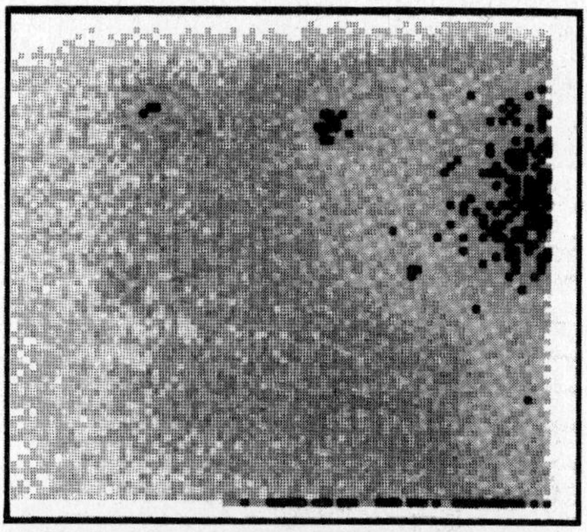

Fig. 2 : Temperature increase, in arbitrary units, at 4s after the pulse heating, of a carbon epoxy sample containing Teflon inclusions (image size 160x130 mm)

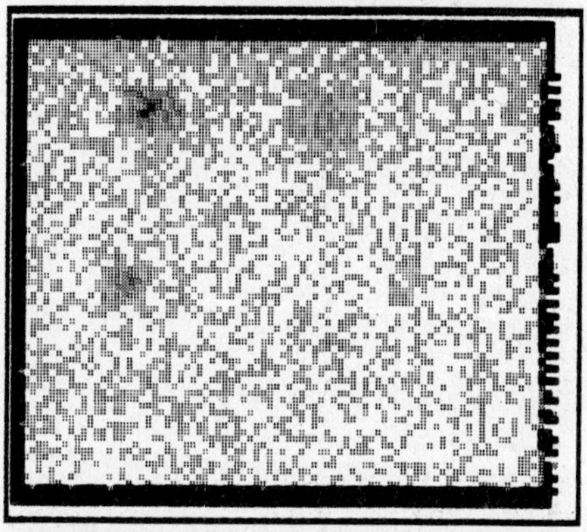

Fig. 3 : The same as Fig. 2, after energy normalization

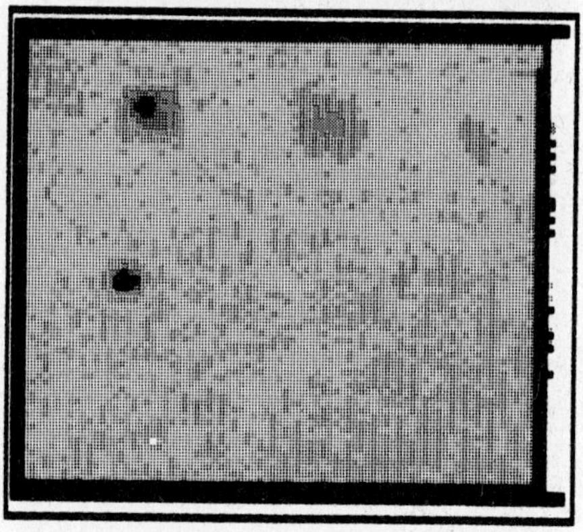

Fig. 4 : Temporal moment image calculated by 300 frames addition (arbitrary units)

surface (more than 200x200 mm) of the sample. The signal is observed using an AGA 782 IR camera and images may be recorded at a rate of 25 frames per second, for several minutes.

3. RESULTS

Figure 2 shows a typical image, recorded 4 s after the pulse heating. The background non-uniformity is due to an inhomogeneous heating flux and emissivity variations. In this picture, only defects #1-1, 1-2 and 2-1 are visible. It is possible to take into account the non-uniformity in deposited light and emissivity variations. In effect, the temperature decrease just after the pulsed heating is not perturbed by the presence of defects, and depends only on the effusivity of the sample, Q and the surface emissivity ε. Therefore, the later images may be divided by one of the first images, leading to normalized images independent of Q and ε. Figure 3 is an example of this treatment.

In a third step, integration is carried out over all the recorded frames, in order to obtain the zero-order moment. Figure 4 shows such an addition, calculated with 300 images. It can be seen clearly that the defects #1-1 and 1-2 then have nearly the same contrast, in spite of their difference in depth. The existing difference is due to the $(1-z/L)^2$ factor and possible thermal resistance variations from one inclusion to another. It is also possible to observe many other defects on this image, this being due to the decrease of noise (#1-3, 2-2, 2-5 and 3-1).

This work was supported by the Direction des Recherches Etudes et Techniques of the French Ministry of Defence.

1. W.N. Reynolds , Can. J. Phys. 64 (1986) p. 1150
2. D.L. Balageas, A.A. Deom, D.M. Boscher , Materials Evaluation 45,4 (1987) p.461
3. D. Balageas, D. Boscher , C.R.Acad.Sc. 305,II (1987) p.13

Double-Dispersive Opto-Thermal Transient Emission Radiometry

R.E. Imhof, C.J. Whitters, D.J.S. Birch, and F.R. Thornley

Photophysics Research, Strathclyde University, Glasgow G40NG, Scotland

We report the development of a new technique for non-destructive characterisation of opaque materials. Double-dispersive opto-thermal transient emission radiometry is based on the same measurement principle as conventional OTTER [1,2], but includes infrared dispersion as part of the signal detection system, to enable opto-thermal transients to be measured at selected infrared wavelengths.

A schematic diagram of the apparatus is shown in Fig 1. A Spectron Nd:YAG laser and harmonic generator is used to excite samples with 1064, 532, 355 or 266 nm radiation. The thermal infrared signal is passed through a grating monochromator and detected with a cooled Cadmium Mercury Telluride detector. Signals are generated at a rate of ~10 Hz and averaged with a transient digitiser interfaced to a PDP-11/2 computer [3].

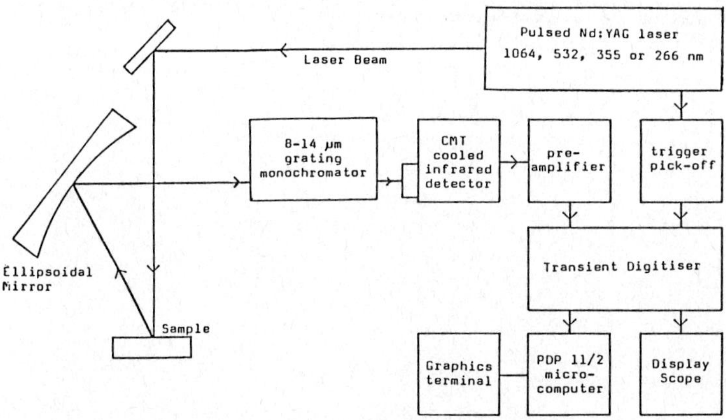

Figure 1. Schematic diagram of the apparatus

The samples used in these preliminary experiments were TiO_2 pigment, compacted to form tablets and excited with 355 nm radiation. Figure 2 shows the first observation of a dependence of opto-thermal decay profile on emission wavelength. Least-squares analysis can be used to calculate best-fit decay-times, which relate to the thermal and optical properties of the surface.

The technique offers the ability to separate the effects associated with the absorption of exciting light from those of the infrared emission. This selectivity makes possible the characterisation of

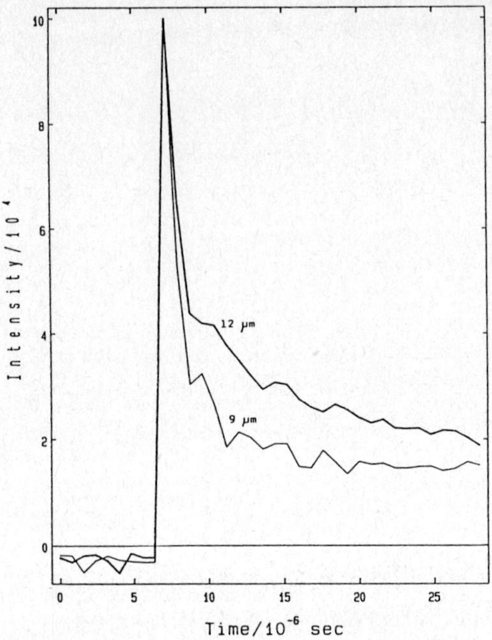

Figure 2. Opto-thermal decay signals from TiO_2 observed at 9 and 12 μm emission wavelength

sub-surface properties by correlating decay-time changes with infrared absorbance. It may also prove useful for calibrating the absorbance scale of diffuse reflectance infrared spectra.

We thank the Paul Instrument Fund and SERC for equipment grants, SERC and ICI Paints Division PLC for a co-operative research studentship held by one of us (CJW).

References

1. A.C. Tam, B. Sullivan: Appl. Phys. Lett. 43, 333-335 (1983)
2. R.E. Imhof, D.J.S. Birch, F.R. Thornley, J.R. Gilchrist, T.A. Strivens: J. Phys. E 17, 521-525 (1984)
3. J.R. Gilchrist, A. Hallam, R.E. Imhof, D.J.S. Birch: Comput. Enhanced Spectrosc. 3, 41-47 (1986)

Photothermally Induced Mode Interaction in Optical Fibers

N. Tankovsky, B.L. Yordanov, and A. Kebedjiev

Sofia University, Faculty of Physics, 5. A. Ivanov Blvd., and
Institute of Physics, Bulgarian Academy of Science, Sofia-1126, Bulgaria

1. Introduction

The effect of temperature on light propagation in optical fibers was studied in the work by Shatalov et al. /1/ recently. In their experiments the whole length of the fiber was heated and the induced phase changes in the light propagation were observed by means of interference pattern detection. Another experimental method was applied by Burt et al. /2/ who used the energy loss of the light propagating along the fiber axis as a heat source and measured the elastic response of the fiber with help of a piezo-electric transducer.

In this work an alternative experimental arrangement has been used. The fiber has been heated locally by an external source and the near field light distribution has been investigated.

2. Experimental

The local heating has been realized by light absorption from a 300 mW Ar-ion laser. The laser beam propagating in the direction perpendicular to the fiber axis has been focussed on the cladding of the fiber by means of a cylindrical lens. The fiber modes have been excited at the wavelength of 638 nm by a single-mode He-Ne laser of which the beam has been fed into the fiber by a focussing lens. The mode distribution at the output of the fiber has been observed with help of a linear photodiode array. The scanning system of the photodiode array has been controlled by a micro-computer. The interrupter of the heating laser beam has been synchronized with the starting command of the scanner. The mode pictures have been memory stored in real time. With this experimental procedure it was possible to study directly the time dependence of the mode interactions.

2. Results and Discussion

Efficient photothermally induced mode-coupling was observed both in graded-index multimode fibers and in step-index single mode fibers. In first approximation the coupling of the m-th mode amplitude to the n-th mode amplitude is given by the relation

$$A_m(z) = A_m(0) + A_n(0) \int_0^z g_{mn} \exp((q_m - q_n)x)\, dx,$$

where the coupling coefficient g_{mn} is given by

$$g_{mn} = k^2/2\, q_m \iint (dn)^2 E_m E_n r \, dr \, d\beta \, .$$

Here q is the propagation constant, dn is the perturbed refractive index, k is the wavenumber, E_m is the electric field distribution in the m-th mode of the fiber /3/.

Evidently, the light interactions and the elastic responses are much faster processes than the thermal flux propagation. Therefore the experimentally measured time dependence of the signal change due to the mode interactions in the course of the external local heating reflects the time behaviour of the thermally induced fiber perturbation.

In a multimode fiber an efficient energy transfer between adjacent modes has been observed. Two mechanisms of mode interaction could be distinguished: thermally induced mode coupling and mode interference caused by thermal changes of the propagation constant.

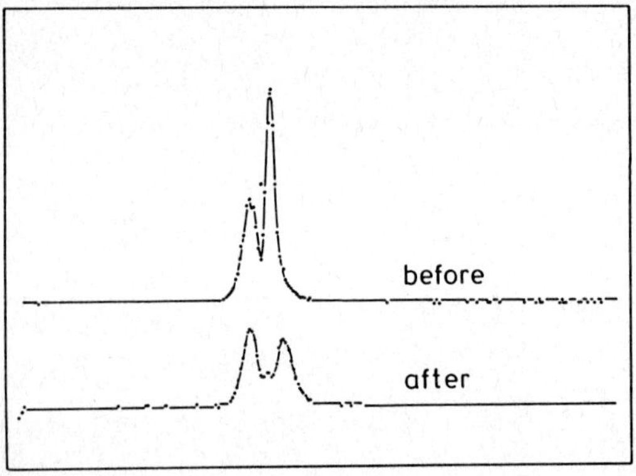

Fig.1 Light mode distribution before (upper trace) and 6 s after (lower trace) the heating was switched on.

In addition, a single-mode fiber with a critical wavelength of 1.15 to 1.22 μm has been explored. In this fiber mode losses up to 100% have been observed (Fig.1). The cladding propagation modes play an important role in the thermo-optical interactions. This has been proved by comparing results on mode interaction before and after the cladding had been locally destroyed.

After having interrupted the external heat source the mode structure was recovered exactly but the heating and cooling cycle showed a hysteresis loop. The time required to build up the the thermally induced local perturbations on the fiber depends on the fiber structure and size, on the thermoelastic parameters of the material and on the presence of defects. Therefore the described technique may serve as a tool for a local characterization of fibers.

References

1. F. Shatalov, G. Gukov: J. Techn. Phys. 55,4, 757(1981), (in Russian)
2. J.A. Burt, K.J. Eberling, D. Efthimiades: Opt. Comm. 32,1, 59(1986)
3. H.G. Unger: In "Planar Optical Waveguides and Fibers", Clarendon Press, Oxford 1977

Part X

Biological and Medical Applications

Pulsed-Laser Optoacoustic and Thermal-Lensing Studies of Relaxation Processes in Biological Systems and Molecules of Biological Interest

S.E. Braslavsky

Max-Planck-Institut für Strahlenchemie,
D-4330 Mülheim a.d. Ruhr, Fed. Rep. of Germany

1. Introduction

Electronically excited states of molecular species can relax through radiative (*e. g.*, fluorescence) and radiationless processes. The latter may be of physical or chemical (*e. g.* photoreactions) nature and, being exothermic, release heat to the medium. Classically, photophysicists and photochemists have measured the emission characteristics (yields and lifetimes) and the production of photoproducts directly, and they drew indirect conclusions about the magnitude of the radiationless processes from these observations. Flash photolytic studies monitoring various (optical, magnetic, electric) properties of the short-lived transients (*e. g.*, singlets, triplets, isomers, radicals) produced upon excitation have also served to follow their kinetic behaviour. These methods rely on different properties (*e. g.*, optical absorption) of the transients and the ground starting material, a prerequisite which is not always fulfilled. Additionally, it is not always possible to extract in this way information about the time-resolved thermodynamic behaviour of a system.

The direct measurement of the heat deposited in the medium upon absorption of a light pulse can provide kinetic and calorimetric information on any short-lived product generated directly or indirectly by absorption of electromagnetic radiation. In condensed matter the time evolution of this heat may be monitored, for instance, by recording the pressure wave generated by the heat pulse (laser-induced optoacoustic spectroscopy, LIOAS) with a piezoelectric element, or by measuring the transient refractive index changes with a CW laser beam (time-resolved thermal lensing, TRTL).

2. Time resolution of pulsed photothermal methods

The problem of the time resolution of photothermal methods in weakly absorbing condensed media has been treated by several authors and recently reviewed by TAM [1]. The case of strongly absorbing media has been treated by SIGRIST [2] and by HUTCHINS [3]. For excitation within the electronic absorption band of biological systems, weakly absorbing solutions are normally handled. In this case the heat source is cylindrical, and for systems delivering all the absorbed light energy as heat, the amplitude of the optoacoustic signal depends on the laser pulse duration and on the acoustic transit time τ_a. $\tau_a = R/v_a$ is the time required by the acoustic wave to traverse the laser beam radius R with the sound velocity v_a [4]. The width of the optoacoustic signal is also determined by τ_a [5]. Thus, for nanosecond laser pulses, in addition to the slow detector response, τ_a is the factor determining signal amplitude and width, especially when using relatively broad beams (R = 0.5-1 mm, corresponding to τ_a = 300-600 ns). This fact has not been sufficiently taken into account in the past [6]. In common solvents τ_a may be varied from *ca.* 30 ns to *ca.* 5 µs by varying the laser beam radius. It is clear that the response time of detector and amplifying system should be short relative to τ_a, particularly when using focused beams corresponding to τ_a in the ns range. In optoacoustics, the use of nonresonant piezoelectric detectors, *e. g.*, polyvinylidenefluoride (PVF_2) films, eliminates the additional restrictions imposed on the time resolution by the formerly used resonant, ceramic (PZT) detectors and permits us to follow the real-time course of the pressure wave [5,7]. This has been confirmed in our laboratory by varying τ_a in measurements of the LIOAS signal of copper chloride in dilute solutions as a reference sample (in a reference sample all absorbed energy is released as heat in times shorter than the acoustic transit time [8]). The full width of the LIOAS signal at 1/e of the signal maximum linearly correlated with the laser waist W (full width at $1/e^2$ of signal intensity maximum) through the slope $1.47/v_a$ [9]. This indicates that all heat released to the medium in times shorter than $1.47\ W/v_a$ is registered in the LIOAS signal height and does not enlarge the signal width. We have called this heat the prompt heat.

The prompt heat dissipation leads also to a rapid refractive index change which in TRTL produces a sharp voltage jump when following the irradiance changes of the probe beam with a diode [10-12].

When relaxation processes with time constants comparable to τ_a take place, the time resolution is determined by a convolution of τ_a and τ_{NR}, the relaxation time of the radiationless processes. For a particular τ_a, both amplitude and width of the LIOAS signal are affected by the intervention of energy storing species with lifetimes in the range of or longer than τ_a, but still shorter than the heat-diffusion rates in solution [9]. The latter upper time-limit has been observed to be around 10 µs for LIOAS [13].

In TRTL, relaxation processes with time constants $> \tau_a$ lead to a slow refractive index change. The thermal recovery of the lens limits the longest observable lifetimes to *ca.* 2 ms [10].

Simple energy balance considerations permit us to expand the expression for the light energy absorbed by the system, $N_A \cdot h \cdot \nu$, to a sum of three terms: i) the energy emitted radiatively (*e. g.*, fluorescence, $\Phi_f \cdot N_A \cdot h \cdot \nu_f$), ii) the prompt heat, $\alpha \cdot N_A \cdot h \cdot \nu$, and iii) the energy stored in transients and products formed, $\Phi_{st} \cdot E_{st}$. In these expressions N_A is the Avogadro number, h the Planck constant, ν the frequency of the exciting laser pulse, Φ_f the fluorescence quantum yield, ν_f the average frequency of fluorescence emission, α the fraction of heat dissipated promptly, Φ_{st} the quantum yield of production of the energy storing species, and E_{st} their molar energy content. From the energy balance equation (1)

$$N_A \cdot h \cdot \nu = \Phi_f \cdot N_A \cdot h \cdot \nu_f + \alpha \cdot N_A \cdot h \cdot \nu + \Phi_{st} \cdot E_{st} \qquad (1)$$

the total energy stored by species with lifetimes $> \tau_a$, represented by the third term, is derived from LIOAS measurements of α and the emission properties of the sample. α is readily obtained from the ratio of the energy normalized amplitudes of the optoacoustic signals from sample and reference [8].

From similar energy balance considerations the ratio between the slow (with respect to τ_a), ΔU, and the total, U, refractive index changes obtained by TRTL is related to the photophysical properties of the sample through the expression [11,12]

$$\Delta U/U = \Phi_{st} \cdot E_{st}/N_A \cdot h \cdot (\nu - \Phi_f \cdot \nu_f). \qquad (2)$$

Both LIOAS and TRTL yield the product $\Phi_{st} \cdot E_{st}$. Provided one of the terms is known from independent measurements, the other is easily calculated.

In order to avoid complications derived from saturation and/or multiphotonic processes, linearity of the amplitude of the LIOAS signal with the laser-pulse energy has to be thoroughly maintained. For the same reason, the TRTL signals proportional to the slow (ΔU) and total heat (U) should bear a linear relationship.

3. Transients lifetimes and calorimetric evaluations with LIOAS

For the well known excited-state system of benzophenone, the triplet lifetimes in acetonitrile at various I^- concentrations were determined by LIOAS, varying signal amplitudes and widths [14]. Using the amplitude, measurements of the prompt heat fraction, α, were performed employing various transit times. The heat stored, $1 - \alpha$, attributed to the benzophenone triplet, yielded a different slope for each I^- concentration when logarithmically plotted against τ_a. They corresponded to the benzophenone triplet lifetime under the particular conditions, as confirmed by comparison with parallel flash photolytic measurements. Similar results were obtained by signal deconvolution, using as reference the signal from benzophenone at an I^- concentration sufficient to quench all triplets (*i. e.*, all absorbed light energy is released as prompt heat) [13]. For the deconvolution the sum of one single exponential function and a time shift sufficed to obtain lifetimes in excellent agreement with those from flash photolysis. The time shift between the LIOAS signal maximum for sample and reference confirms the calculations by KUO *et al.*, [5]. The shift reflects the fact that a retarded heat pulse shows a maximum at a later time than a prompt pulse.

Similar methods (*i. e.*, the measurement of the heat fraction dissipated promptly at various transit times) served to determine a lifetime of 150 ± 30 ns for the photoisomer of phycocyanobilin, a tetrapyrrol related to biliverdin and derived from the chromophore of the algae pigment phycocyanine [15]. In our previous LIOAS studies of biliverdin dimethyl ester using a ceramic detector, we had established for the first time the formation of a photoisomer for this type of molecules without being able, however, to determine its lifetime [8].

Within the frame of calorimetric measurements, chlorophylls were studied as well, with the aim of assessing the general assumption that in cyclic tetrapyrrols the sum of the quantum yield of fluorescence plus that of chemical reaction from the excited singlet adds up to one. Our LIOAS determinations using a PZT detector showed that, at least in organic solvents and at room temperature, this assumption is not valid, since 10% of the light energy absorbed by chlorophyll-a is lost by internal conversion [16]. However, this is not the case for chlorophyll-b, which exhibits no internal conversion, within the experimental error of the LIOAS measurements [16].

In the case of the photochromic plant pigment, phytochrome, we measured the normalized optoacoustic signals, with a PVF_2 film at transit times ranging from 30 to 530 ns, in order to determine the heat dissipated promptly. These results together with the spectroscopic and kinetic data on phytochrome, allowed the estimation of an upper limit of 0.5 for the quantum yield of the photochemical primary process(es) yielding the two short-lived transients I_{700} [9]. The short lifetimes of the two I_{700} transients (3 and 20 μs at room temperature as determined by flash photolysis [17]) offered a unique possibility for the estimation by LIOAS of their quantum yield of formation.

The application of LIOAS to photosynthetic systems offers, beyond the conventional photoacoustic method [18], the advantage of a better time resolution and the exclusion of possible interference by photosynthetic oxygen evolution. We present in this conference [19] the results of measurements of the prompt heat release by intact cells of the cyanobacterium *Synechococcus sp* and by isolated photosystems I and II particles. Calorimetric evaluations can be made of the various electron transfer steps involved in the photosynthetic chain.

4. Absolute quantum yields and lifetimes determinations by TRTL

Absolute quantum yields for singlet molecular oxygen ($O_2,^1\Delta_g$) production by several sensitizers were measured by TRTL [11]. Contrary to LIOAS, the determination by TRTL quantum yields of production of transient species storing part of the absorbed energy longer than τ_a, does not require any reference, since the total and the slow heat are measured simultaneously in the same experiment [see eq. (2)] [12]. This fact made possible the measurement of quantum yields of $O_2,^1\Delta_g$ production by various dyes utilized in the photodynamic treatment of tumors, in media where no sensitization reference is available [20]. Equation (2) and the known value of the $O_2,^1\Delta_g$ molar energy content were used. The measurement of these quantum yields by any other method, including $O_2,^1\Delta_g$ infra-red emission, requires a reference with a known yield of $O_2,^1\Delta_g$ formation in the same medium, which is often difficult.

The lifetimes of the energy storing species are simply derived from the time profile of the slow heat emission in TRTL. The lifetime of $O_2,^1\Delta_g$ was measured in various solvents and the values obtained were in good agreement with those determined by $O_2,^1\Delta_g$ emission [11,20].

5. Comparison of LIOAS and TRTL

TRTL requires transparency of the sample for the monitoring laser beam. This is often not fulfilled in biological tissues or preparations. In these cases LIOAS has a better chance of application. However, the electronic amplification system necessary for time-resolved LIOAS requires careful control in order not to deform the signals. In addition, either programs for signal deconvolution, or determinations of the laser beam diameter are required for LIOAS in order to derive the lifetimes of the transients involved, while this information is directly obtained from the TRTL signals.

Acknowledgements: I am indebted to Professor Kurt Schaffner for his constant encouragement and support and to my collaborators for their dedication and creative imagination.

6. References

1. A. C. Tam: Rev. Mod. Phys. 58, 381 (1986)
2. M. W. Sigrist: J. Appl. Phys. 60, R83 (1986)
3. D. A. Hutchins: Can. J. Phys. 64, 1247 (1986)
4. H. M. Lai, K. Young: J. Acous. Soc. Am. 72, 2000 (1982)
5. C-Y. Kuo, M. F. Vieira, C. K. N. Patel: J. Appl. Phys. 55, 3333 (1984)
6. J. E. Rudzki, J. L. Goodman, K. S. Peters: J. Am. Chem. Soc. 107, 7849 (1985)

7. A. C. Tam, H. Coufal: Appl. Phy. Lett. 42, 33 (1984)
8. S. E. Braslavsky, R. M. Ellul, R. G. Weiss, H. Al-Ekabi, K. Schaffner: Tetrahedron 39, 1909 (1983)
9. K. Heihoff, S. E. Braslavsky, K. Schaffner: Biochem. 26, 1422 (1987)
10. A. J. Twarowski, D. S. Kliger: Chem. Phys. 20, 253 (1977)
11. G. Rossbroich, N. A. García, S. E. Braslavsky J. Photochem. 31, 37 (1985)
12. R. W. Redmond, S. E. Braslavsky: Oral presentation MA4, these Proceedings.
13. K. Heihoff, S. E. Braslavsky: Chem. Phys. Lett. 131, 183 (1987)
14. K. Heihoff, S. E. Braslavsky: Poster presentation MA9, these Proceedings.
15. K. Heihoff, D. Schneider, S. E. Braslavsky, K. Schaffner: in preparation.
16. M. Jabben, García, N. A., S. E. Braslavsky, K. Schaffner: Photochem. Photobiol. 43, 127 (1986)
17. B. P. Ruzsicska, S. E. Braslavsky, K. Schaffner: Photochem. Photobiol. 41, 681 (1985)
18. S. Malkin, D. Cahen: Photochem. Photobiol. 29, 803 (1979)
19. C. Nitsch, S. E. Braslavsky, G. H. Schatz: Oral presentation MAA3, these Proceedings.
20. R. W. Redmond, K. Heihoff, S. E. Braslavsky, T. G. Truscott: Photochem. Photobiol. 45, 209 (1987)

Photoacoustic Spectroscopy of Biological Photoreceptor Membranes

R.M. Leblanc and C.N. N'soukpoé-Kossi

Centre de Recherche en Photobiophysique, Université du Québec à Trois-Rivières, C.P. 500, Trois-Rivières, Québec, G9A 5H7, Canada

1. INTRODUCTION

The most important systems involved in photoreception are the visual and photosynthetic membranes. One of the most striking facts about these systems is the existence of ordered structures or pigments in the organelles. These pigments can be excited either directly by light absorption or by excitation energy transfer from other molecules absorbing in shorter wavelength regions. The efficiency of energy transfer strongly depends upon the acceptor transition moment and the mutual orientation of donor and acceptor. In order to properly study the properties of these organelles, the synthesis of aggregates of definite order is of prime importance. To achieve the necessary orientation and aggregate state, several sample preparation techniques have been employed, including polyvinylalcohol (PVA) films, Langmuir-Blodgett films, and sometimes vesicles and black lipid membranes. Besides, subcellular and intact tissue fractions have also been utilized. Visual as well as photosynthetic systems are concerned with energy conversion, storage and dissipation. Furthermore, photosynthesis is concerned with energy transfer (ET). All these phenomena can be investigated by photoacoustic spectroscopy (PAS). It is the purpose of this paper to report the most important developments in this special area of photobiophysics.

2. PHOTOSYNTHESIS

2.1 Model Systems

Model systems usually used comprise amorphous and solid films, mono-and multilayers (Langmuir-Blodgett films) and PVA matrix. It is well known that chlorophyll _a_ (Chl _a_) molecules can associate with each other or with other molecules to form aggregated structures, the main sites of association being magnesium atoms, functional groups and conjugated double bonds.

It has been stated that the aggregation state of Chl _a_ in monolayers is different from that in multilayers [1-3]. In a comparative study [4], absorption and photoacoustic (PA) spectra have been recorded for Chl _a_ in the amorphous solid state, in mono- and bilayers in order to gain more insight about the difference in aggregation pattern of Chl _a_ in the different states. Figure 1 shows the absorption and PA spectra of Chl _a_ in amorphous solid state (Fig. 1a) and in a lecithin matrix (Fig. 1b). One can see in the PA spectrum the appearance of a shoulder beside the 440-nm and 678-nm bands (Fig. 1a) suggesting the existence of a different aggregate of Chl _a_ in the amorphous film compared with that in the lecithin matrix (Fig. 1b). It was also observed that humidity level broadened the PA half-bandwidth, probably as a result of strong interactions with water. In Chl _a_-lecithin films, sharpening of the absorption and PA bands was observed (Fig. 1b) associated

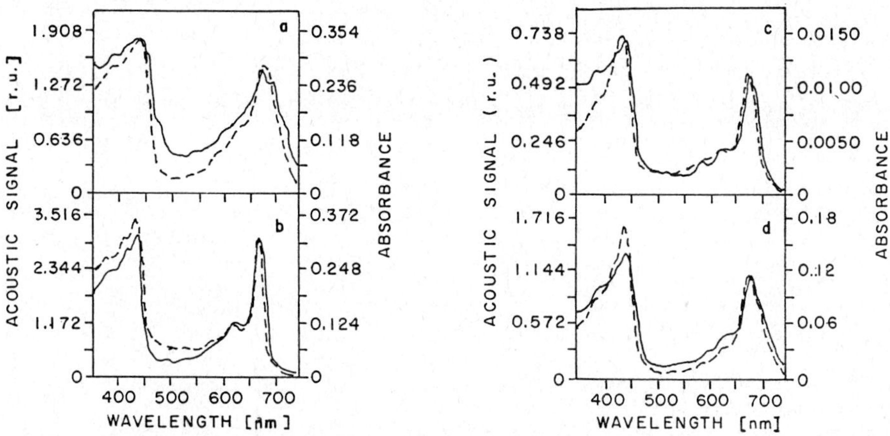

Fig. 1: Absorption (---) and PA (—) spectra of Chl a in (a) amorphous solid film, (b) lecithin matrix, (c) monolayer, and (d) multilayer (12 monolayers)

with a blue-shift compared with amorphous solid films of Chl a with maxima at 435 nm and 667 nm. It has been suggested that predominance of monomeric Chl a in lecithin solid films is the cause of the results obtained in the Chl a-lecithin films [4].

In monolayer and multilayer states (Fig. 1, c and d), general agreement was observed between absorption and PA spectra of Chl a, with maxima located at 440 nm and 678 nm for one layer and at 436-439 nm and 678-680 nm for multilayers. However, from a quantitative viewpoint, the amplitude of absorption and PA signals does not proportionally follow the increasing number of Chl layers. This non-linear response may be explained on the basis of Lambert-Beer's law deviation (for absorption) and optical saturation (for PAS).

The study of many aspects of photosynthesis has been carried out with various photosynthetic pigments incorporated in PVA films, in stretched and unstretched conditions for different orientations of the chromophores. PAS of biliproteins, namely R-phycoerythrin (PE) and C-phycocyanin (PC) and other mixtures in anisotropic and isotropic PVA films were measured under illumination with polarized and natural light [5]. Also deuterated PVA samples were investigated. The yields of fluorescence of various chromophores were obtained from PA, absorption, fluorescence and fluorescence lifetime measurements. The results are summarized in tables 1 and 2. We can see that,

Table 1. Yield of fluorescence obtained from PAS

Sample	λ Absorbed (nm)	(ϕ)
PC	620	0.67
PC (deut.)	620	0.37
PE (PUB)	500	0.67
PE (LWPE)	540	0.67
PE (LWPE)	560	0.66
PE (deut.)(PUB)	500	0.42
PE (deut.)(LWPE)	540	0.52
PE (deut.)(LWPE)	560	0.51
M (PE + PC)(PUB)	500	0.52
M (PE + PC)(LWPE)	540	0.56
M (PE + PC)(LWPE)	560	0.61

Table 2. PAS to absorption ratio (PAS/α) of PE for two polarized light components

λ (nm)	500	540	560
$(\frac{PAS}{\alpha})_{\parallel}$	0.37	0.38	0.38
$(\frac{PAS}{\alpha})_{\perp}$	0.43	0.40	0.44

Note: PUB = phycourobilin, LWPE = phycoerythrobilin chromophores of PE, M = mixture changing the chromophore environment influences the yield of excitation energy path between biliproteins, changing their thermal dissipation of excitation and excitation energy transfer. The various chromophores clearly show their different abilities for thermal deactivation and for transfer of the excitation to other chromophores. In PVA, the excitation energy transfer was found to be lower than in *in vivo* structures. For R-PE in buffer solution, the quantum yield of fluorescence $\eta_1 = 0.56$ [6]. Fluorescence quantum yield for Chl a in ether solution is 0.22. The value η_2 for chlorophyllin (Chllin) is lower than that of Chl a. Comparing fluorescence intensities of PE and Chllin in PVA, excited in the region of similar absorption, the ratio $\eta_2/\eta_1 = 0.21$ [7]. Therefore, the quantum yield of Chllin fluorescence (η_2) in PVA is about 0.12 assuming that η_1 in PVA is similar to that in buffer solution. Evaluated fluorescence yield of excitation energy transfer, ϕ_{ET}, for the two absorption bands ϕ_{ET} (500 nm) = 0.14 and ϕ_{ET} (560 nm) = 0.33 was obtained [7]. Analysis of spectra (not shown) suggested that part of the energy absorbed by Chl a at the Soret band migrates to Chl b with a yield of 0.16.

In order to establish the influence of PVA anisotropy on PA amplitude, the stretched and unstretched PVA films coloured by Congo red were also examined [8]. We observed that films of PE and Chllin mixture, when illuminated in the region of PE absorption showed higher PA signal than the films with PE alone. This was explained by the excitation energy transfer from efficiently fluorescent PE to Chllin, which has much lower yield of fluorescence. The stretched films always showed a higher PA signal than isotropic (unstretched) films, probably due to increase in the thermal conductivity of PVA. But normalized PA signals of Congo red and Chllin in stretched and unstretched PVA were similar.

Polarized PA, absorption and fluorescence spectra of chloroplasts and thylakoids (isolated from pea leaves) in stretched and unstretched PVA films have also been recorded [9]. The intensity ratio of fluorescence bands at 674, 700 and 750 nm, and the polarized fluorescence excitation spectra were strongly dependent on light polarization and film stretching with thylakoids in stretched films exhibiting predominantly 674-nm emission. The polarized absorption and PA spectra are depicted in Figs.2 and 3, respectively. The changes in polarized PA spectra caused by film stretching are accompanied by

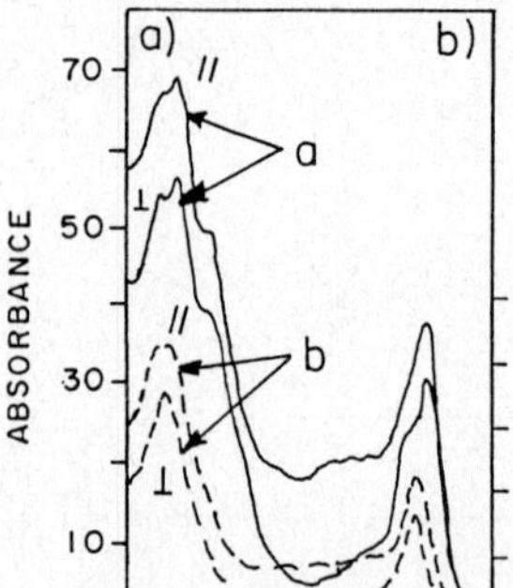

Fig. 2 Polarized absorption spectra of chloroplats (a) and thylakoids (b) in stretched PVA

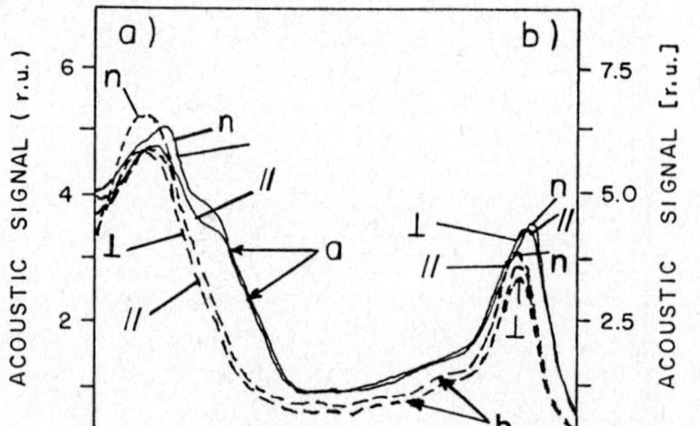

Fig. 3. PA spectra of chloroplasts (a) and thylakoids (b) in stretched PVA (n, natural; ∥, parallel and ⊥, perpendicular)

changes in the fluorescence data. The observed spectral changes were explained by reorientation of pigment molecules influencing the yield of excitation transfer between different pigments.

2.2 Cellular and Subcellular Fractions

2.2.1 Bundle Sheath Cells

In C_4-photosynthesis plants, two types of cells are involved in the photosynthetic activity: mesophyll cells located in the stroma of the leaf, and bundle sheath (BS) cells (also called Kranz cells) organized around the vascular tissue. In these plants, carbon assimilation begins in the mesophyll chloroplasts where CO_2 is fixed to C_3-compounds (e.g. pyruvate, alanine, 3-phosphoglycerate or PGA) by a carboxylation process, leading to the formation of malate and aspartate, the principal metabolites which diffuse into bundle sheath chloroplasts. There decarboxylation will take place to produce pyruvate, alanine and PEP which return to the mesophyll cells. The CO_2 formed by decarboxylation is metabolized in the bundle cells. The energy storage and participation of the photosystems (PS) have been monitored by PAS in *Zea mays* [10-12]. It has been observed that PSI activity in BS cell chloroplasts is partially supported by PSII activity, probably by preventing overoxidation of electron carrier associated with PSII activity. Very recently [13], we have observed that PSI may be reduced by some electron donor, other than water and that the role of PSII would be the regulation of PSI electron transport and prevention of its overoxidation.

2.2.2 PSII Submembrane Fractions

Regarding the role played by pheophytin (Pheo) in photosynthesis, it was believed that Pheo is an intermediate electron acceptor in PSII prior to a quinone species Q which is now known to be Q_A [14,15]. PA characteristics (e.g. energy storage and heat dissipation) measurements of PSII core-enriched particles from barley have been undertaken [16]. A well defined dip at about 540 nm in the ρ/α vs λ spectrum has been observed (Fig. 4) which did not disappear when the samples were irradiated in the presence of FeCN and DPC (cf. ρ_{ni}/α vs λ spectrum), i.e., under conditions of electron flow (ρ is relative PA response; α, the fraction of light absorbed).

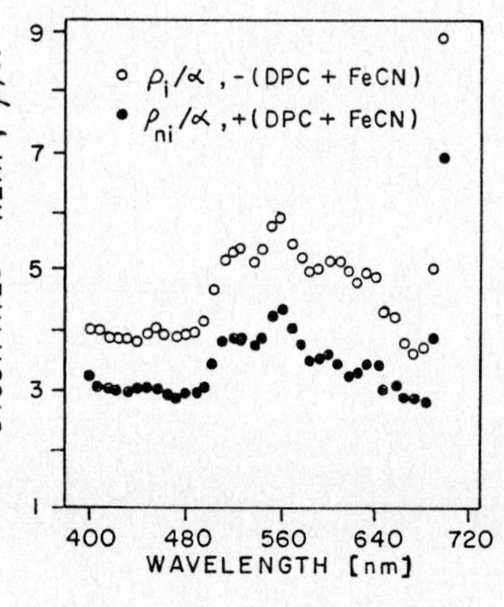

Fig. 4. Heat dissipation spectra of PSII core-enriched particles from barley obtained under conditions of electron flow (ρ_{ni}/α) and when no electron flow was observed (ρ_i/α)

$$\rho = \alpha \left(1 - \frac{\phi \, \Delta E \, \lambda}{Nhc}\right)$$

$$\text{or } \rho/\alpha = 1 - \frac{\phi \, \Delta E \, \lambda}{Nhc}, \qquad (1)$$

where ϕ is quantum yield; ΔE, internal energy change; λ, wavelength; N, Avogadro number; h, Planck constant; c, speed of light.

The decrease of ρ/α near 540 nm should be due at least partly to the function of Pheo in energy storage in PSII particles. This finding could open a new avenue for the study of PSII photochemistry with PAS.

2.2.3 Photosynthetic Bacteria

Antenna carotenoids have been investigated by PAS in *Rhodospirillum rubrum* and *Rhodopseudomonas sphaeroides*. For *Rhp. sphaeroides*, absorption and PA spectra have been found to match, an expected result since the carotenoid of that bacterium is an antenna pigment which is as efficient as bacteriochlorophyll itself. In *Rhs. rubrum*, spirilloxanthin has a low antenna efficiency. Thus, if the energy absorbed by bacteriochlorophyll is efficiently stored in the photosynthetic process, it should result in PA loss, and there should be a smaller loss when energy is absorbed by spirilloxanthin [17]. In whole bacteria and chromatophores, the effectiveness of spirilloxanthin to generate PA signals is greater than that of bacteriochlorophyll. This indicates a more efficient storage of the energy absorbed by bacteriochlorophyll.

2.2.4 *Anacystis nidulans*

The PA spectrum of *A. nidulans* recorded at room temperature is qualitatively similar to the low-temperature absorption or fluorescence excitation spectra [18]. The bands of holochromes are well defined compared with absorption spectra recorded at room temperature (Table 3). Photoacoustic data from *A. nidulans* were used to obtain the action spectrum for photochemical energy storage. It was found that at least 60% of the 630-nm photons absorbed perform photochemical work, and about half of the useful energy is stored as stable products. Although it cannot be separated from the purely

Table 3. Photoacoustic and absorption peaks in spectra of *Anacystis nidulans*

Pigment	Absorption spectrum 298-305 K	Absorption spectrum 77 K	Photoacoustic spectrum (298 K)
Chlorophyll a (red region)		670 (S)	670 (S)
	690 (P)	679 (P)	678 (P)
		686 (S)	690 (S;weak)
		705 (S)	
Carotenoids	460-465 (S)	465-470 (S)	455 (P/S)
			480 (P)
			490 (P)
	490-505 (S)	502 (P)	500 (P)
Phycocyanin	580 (S)	580 (S)	550-610 (Several shoulders)
	625 (P)	625 (P)	625 (P)
	635 (S;weak)	634 (P)	635 (P)

Note. S = shoulder, P = peak. Values in nm.

thermal effect, the contribution of modulated oxygen evolution to the PA signal of algae was estimated to be relatively small, as deduced from DCMU and actinic light-inhibition experiments [19].

2.2.5 Algae, Chloroplasts and Whole Leaves

Usually, in oxygen evolving systems such as algae, chloroplasts and leaves, three important factors contributing to the PA signal should be considered: the thermal component(ρ_{th}), the photochemical loss (ρ_L) and the oxygen evolution (ρ_{O2}). In dim modulated light, the measured PA signal includes two components, ρ_{th} and ρ_{O2}, acting in the same direction with increasing pressure in the PA cell. Under photosynthesis-saturation conditions (i.e. using saturating actinic light), ρ_{O2} is suppressed, resulting in a decrease of total PA signal (ρ_T) while photochemically stored energy is released and increases ρ_T. Commonly, ρ_{O2} is predominant at low modulation frequencies (< 100 Hz), giving net decrease of ρ_T (negative effect) whereas ρ_L is predominant at high frequencies ($\geq$ 250 Hz) where net increase of ρ_T is observed (positive effect). This is especially the case for tobacco whole leaves [20]. But with *Impatiens petersiana* chloroplasts and *A. nidulans*, positive effect was observed (Fig. 5) from 25 Hz to 250 Hz [19] while we obtained a low-frequency negative effect and high-frequency positive effect in whole leaves of *I. petersiana* and mays (unpublished data). The absence of negative effect in *I. petersiana* chloroplasts and *A. nidulans* was explained by the weakness and efficient damping of oxygen signal compared with thermal component. In the case of maple leaf, a positive effect was observed upon actinic light irradiation (20-500 Hz) in fresh-cut leaves followed by negative effect (after) some 75% drying of the leaves at room temperature) in the same frequency range (unpublished data). This phenomenon was found to be reversible upon rehydration. It appears quite clear that water content is playing an important role in the sign of the effect of actinic light, at least in maple leaf. Probably, for this C_3-plant for which photorespiration is considerable, the hydration state of leaves would have some regulatory effect

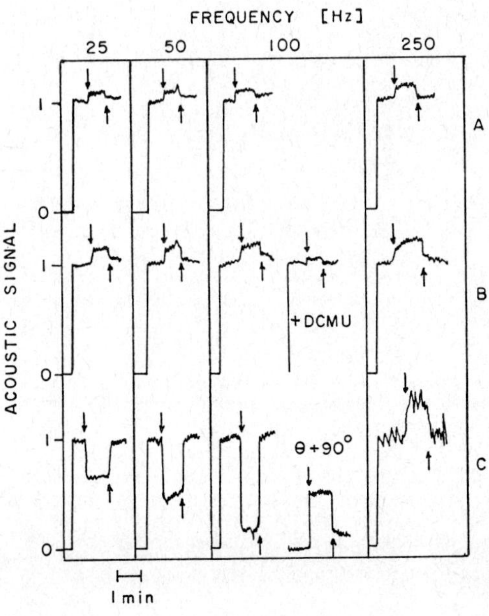

Fig. 5. Effect of actinic beam on the PA signal of (A) chloroplasts from *I. petersiana*, (B) *A. nidulans*, and (C) leaf from *I. petersiana*. (↓) and (↑) indicate that actinic light is switched on and off, respectively

on oxygen evolution and diffusion into internal gaseous network and on photorespiration equilibrium. Also, anatomical characteristics of the maple leaf could be of concern. Deeper investigation is required to gain better insight into this mechanism.

Before closing this section, let us say a word about photosynthetic responses of leaves to water stress. The effects of rapid dehydration on the photochemical apparatus under *in vivo* conditions have been examined [21]. The Emerson enhancement, as measured by PAS represents a convenient and direct monitor for the changes in the relative activities of the photosystems. Typical changes in Emerson enhancement were observed during state 1 - state 2 transitions, which were used to quantitate these states and the transitions between them. The magnitude of the enhancement is defined as the ratio of the O_2 signal obtained with far-red background light to the O_2 signal without background light. A rather surprising result was the finding that the enhancement phenomenon still persisted in dessicated leaves. The most prominent effect of water stress was in state 2. It was concluded from this experiment that in presence of water stress, there is still a tendency to achieve state 2 (equal activities in PSI and PSII), although photochemical activities remain in favour of PSII.

3. VISION

PAS could be conveniently used to monitor energy storage in the visual process. Equation 1 could be rewritten for this purpose as

$$\rho/\alpha = k\,[1 - \frac{\sum_i \phi_i\,\Delta E_i\,\lambda}{Nhc}]\,, \tag{2}$$

where k is an instrumental constant.

For a single process having constant values of ϕ and ΔE, ρ/α shows a linear dependence on λ, and once extrapolated to $\rho/\alpha = 0$, one obtains

$$\phi\,\Delta E = \frac{Nhc}{\lambda}\,. \tag{3}$$

Extrapolation of this dissipation spectrum to λ= 1207 nm and using ϕ= 0.67 [22] for the formation of bathorhodopsin from rhodopsin, a value of 145 kJ mole^{-1} was obtained for ΔE, energy stored [23], in agreement with Cooper [24].

Besides energy conversion determinations, PAS could also be utilized for the analysis of cellular morphology and its modifications for instance in diseased tissue. In connection with these aspects, posterior ocular tissues have been examined [25] at low temperature (77 K) by the depth profile methodology. The low-temperature PA cell previously described [26] allows spectroscopic measurements without bleaching of the visual pigment, rhodopsin. The results are shown in Fig. 6. At frequencies between 50 and 600 Hz, the photoreceptor being on top of the sample, rhodopsin-bathorhodopsin photostationary mixture appears to be the major absorbing species (Fig. 6a). Besides the 340 and 515-nm bands of rhodopsin, at high frequencies a small band around 410 nm was also observed which has been ascribed to blood hemoglobin. At lower frequencies, deeper layers were probed giving rise to increased PA amplitude at 412 nm. These PA data indicate that, in the 25-μm layer which was probed between 600 and 50 Hz, the first half (0-13 μm) contains essentially rhodopsin (rod outer segments). As the probed layer extends to the inner segments the hemoglobin and neural material become

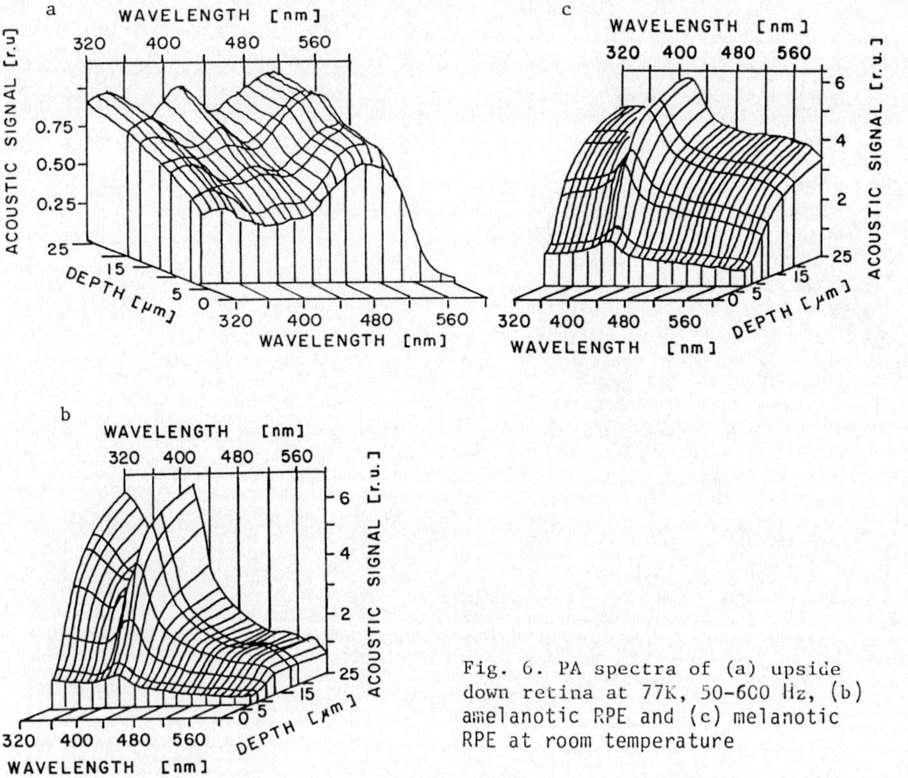

Fig. 6. PA spectra of (a) upside down retina at 77K, 50-600 Hz, (b) amelanotic RPE and (c) melanotic RPE at room temperature

more and more important. Using the retinal pigment epithelium (RPE), we were able to detect the amelanotic (*tapetum lucidum*) and melanotic (*tapetum nigris*) RPE shown in Fig. 6,b and c, respectively, at room temperature. From these PA data and others not shown here, it could be possible to sketch a complete chromophoric map of the posterior ocular tissue by juxtaposition of PA spectra, a result which is difficult to achieve by other methods.

As far as the visual system is concerned, the depth profile methodology is expected to become very useful in detecting and evaluating some visual diseases such as photodynamic damage, retinal distrophy or *retinitis pigmentosa*. Furthermore, PAS proved very useful for analysis of energy storage and energy transfer in visual and photosynthetic systems using model as well as *in vivo* structures.

4. REFERENCES

1. C. Chapados, R.M. Leblanc: Chem. Phys. Lett. 49, 180 (1977)
2. C. Chapados, D. Germain, R.M. Leblanc: Biophys. Chem. 12, 189 (1980)
3. C. Chapados, R.M. Leblanc: Biophys. Chem. 17, 211 (1983)
4. A. Désormeaux, R.M. Leblanc: Thin Solid Films 132, 91 (1985)
5. D. Frackowiak, S. Hotchandani, G. Bialek-Bylka, R.M. Leblanc: Photochem. Photobiol. 42, 567 (1985)

6. D. Frackowiak, J. Dutkcewicz, J. Grabowski, F. Fiksiński, H. Manikowski: Photosynthetica 13, 21 (1979)
7. D. Frackowiak, S. Hotchandani, R.M. Leblanc: Acta Univ. Lodz. Folia Biochim. Biophys. 5, 91 (1986)
8. D. Frackowiak, S. Hotchandani, R.M. Leblanc: Photochem. Photobiol. 42, 559 (1985)
9. D. Frackowiak, L. Lorrain, D. Wrobel, R.M. Leblanc: Biochem. Biophys. Res. Commun. 126, 254 (1985)
10. R. Popovic, R.M. Leblanc, M. Beauregard: In Progress in Photosynthesis Research, ed. by J. Biggins, Vol. 2 (M. Nijhoff Publishers, Dordrecht, 1987) p. 625
11. M. Beauregard, R.M. Leblanc, R. Popovic: In Progress in Photosynthesis Research, ed. by J. Biggins, Vol. 2 (M. Nijhoff Publishers, Dordrecht, 1987) p. 629
12. R. Popovic, M. Beauregard, R.M. Leblanc: Biochem. Biophys. Res. Commun. 144, 198 (1987)
13. R. Popovic, M. Beauregard, R.M. Leblanc: Plant Physiol. (in press)
14. V.V. Klimov, A.A. Krasnoskii: Photosynthetica 15, 592 (1981)
15. D.J. Murphy: Biochim. Biophys. Acta 864, 33 (1986)
16. M. Fragata, R. Popovic, E.L. Camm, R.M. Leblanc: Photosynthesis Res. (in press)
17. F. Boucher, L. Lavoie, A.F. Antippa, R.M. Leblanc: Can. J. Biochem. Cell. Biol. 61, 1117 (1983)
18. R. Carpentier, B. LaRue, R.M. Leblanc: Arch. Biochem. Biophys. 222, 411 (1983)
19. R. Carpentier, B. LaRue, R.M. Leblanc: Arch. Biochem. Biophys. 228, 534 (1984)
20. G. Bults, B.A. Horwitz, S. Malkin, D. Cahen: Biochim. Biophys. Acta 679, 452 (1982)
21. M. Havaux, O. Canaani, S. Malkin: Plant Physiol. 82, 834 (1986)
22. J.B. Hurley, T.G. Ebrey, B. Honig, M. Ottolenghi: Nature 270, 540 (1977)
23. F. Boucher, R.M. Leblanc: Photochem. Photobiol. 41, 459 (1985)
24. A. Cooper: Nature 282, 531 (1979)
25. F. Boucher, R.M. Leblanc, S. Savage, B. Beaulieu: Appl. Opt. 25, 515 (1986)
26. F. Boucher, R.M. Leblanc: Can. J. Spectrosc. 26, 190 (1981)

Inverse Yield Changes of Heat and Fluorescence During Photoinhibition of Photosynthesis

C. Buschmann[1] *and H. Prehn*[2]

[1]Botanisches Institut II, University of Karlsruhe, Kaiserstraße 12, D-7500 Karlsruhe, Fed. Rep. of Germany

[2]Institut für Biomedizinische Technik, FHGF, Wiesenstraße 14, D-6300 Gießen, Fed. Rep. of Germany

1. Introduction

Light much in excess of normal light present during the growth of a particular leaf is known to lead to a decrease of photosynthetic activity, which is termed photoinhibition (for a review see /1/). After photoinibition the yield of chlorophyll fluorescence is decreased (e.g. /2/). Although no direct measurements have been carried out up to now, it has been concluded that the surplus of energy absorbed by a photoinhibited leaf is given up by non-radiative deexcitation. The measuring of non-radiative deexcitation processes was made possible by photoacoustic spectroscopy and has already successfully been applied for detecting several photosynthetic parameters /3/, /4/, (review /5/).

In the experiment described here, fluorescence and photoacoustic signals were simultaneously measured during a photoinhibitory treatment of a radish leaf maintained in the photoacoustic cell. It has been previously checked /6/ that at the modulation frequency applied (279 Hz) the induction kinetic of the photoacoustic signal of radish leaves is determined only by heat emission. At this frequency the addition of non-modulated light saturating photosynthesis leads to a strong increase of the photoacoustic signal. 279 Hz was chosen as it yielded the highest heat induction kinetic. The "oxygen effect" /7/ may determine the signal of radish leaves only at frequencies below 150 Hz because the photoacoustic signal then decreases upon addition of non-modulated light /6/.

2. Material and Methods

Radish seedlings (Raphanus sativus) were grown on peat under white light (50 $\mu E \cdot m^{-2} \cdot s^{-1}$, 10 h per day). A 7 x 10 mm section of a leaf taken from a 2-week-old plant was placed into the closed photoacoustic cell with the lower side facing the excitation light. For the excitation the 632.8 nm light of a 5 mW He/Ne laser was mechanically chopped at a modulation frequency of 279 Hz. The chopped light passed to the photoacoustic cell via a fiber optic. The chlorophyll fluorescence was transmitted by another glass fiber and measured by means of a photodiode through a 665 nm cut-off filter and a 685 nm interference filter. The fluorescence signal was recorded after having been processed by a lock-in amplifier (time constant 1 s). The heat emission was measured with the microphone of the photoacoustic cell connected to a two-phase lock-in amplifier (time constant 1 s). Non-modulated white light was used as additional light (a) for saturating photosynthesis (2000 $\mu E \cdot m^{-2} \cdot s^{-1}$) or (b) for the photoinhibitory treatment (9000 $\mu E \cdot m^{-2} \cdot s^{-1}$). As light source a "cold light" (KL 150B, Schott, Mainz, F.R.G.) was used. For further details see /6/ and /8/.

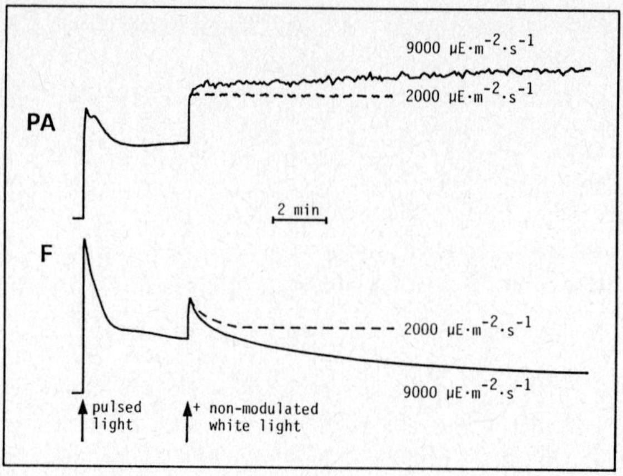

Fig. 1: Kinetic of the photoacoustic signal (PA) and of the chlorophyll fluorescence at 685 nm (F) measured simultaneously with a green radish leaf at 279 Hz. When the kinetics had reached a steady state, additional non-modulated white light was applied (a) saturating photosynthesis (2000 $\mu E \cdot m^{-2} \cdot s^{-1}$, signals: dashed line) or (b) leading to photoinhibition of photosynthesis (9000 $\mu E \cdot m^{-2} \cdot s^{-1}$, signals: solid line).

3. Results

The leaf excited with the 632.8 nm laser light pulses exhibits a transient (Fig. 1), which is similar to that for the chlorophyll fluorescence (F) and for the heat emission measured as photoacoustic signal (PA). From the kinetic maximum at the onset of illumination both signals decline to a low steady state after about 4 min of excitation. Non-modulated white light of 2000 $\mu E \cdot m^{-2} \cdot s^{-1}$ (Fig. 1, dashed line) added at the steady state of the transient results in a second kinetic of the fluorescence signal and a rise in the heat emission to a constant level. With a non-modulated white light of 9000 $\mu E \cdot m^{-2} \cdot s^{-1}$ (Fig. 1, 4 min after onset of excitation, solid line) the fluorescence rises first. Within 15 min the fluorescence then decreases to a value below the steady state reached before the addition of the non-modulated light. During these 15 min the photoacoustic signal increases further from the high value attained directly after the addition of the non-modulated light of 9000 $\mu E \cdot m^{-2} \cdot s^{-1}$.

4. Discussion

4.1 Induction Kinetic without Photoinhibition

The kinetic transients of fluorescence and heat emission measured with the radish leaf represent an induction kinetic of photosynthesis. The antiparallel relationship of the fluorescence yield and the rate of photosynthesis is well established since the first experiments of Kautsky and Hirsch in 1931 /9/. The kinetic maximum at the onset of the excitation is characterized by a low photosynthetic activity. Subsequent to the kinetic maximum the fluorescence and heat emission decline parallel to the increase of the rate of photosynthesis. After about 4 min the photosynthetic rate has reached its maximum and the fluorescence and heat emission are at a low steady state. The gradual appearance of the induction kinetics of fluorescence and heat emission during the greening in the light, which is connected to the establishment of photosynthetic activity has already been demonstrated for radish leaves /8/.

A simultaneous detection of chlorophyll fluorescence and of the photoacoustic signal measured with leaves was first decribed by Inoue et al. /10/. The induction kinetic of the photoacoustic signal was found to be in tendency antiparallel to the fluorescence emission. Later on it was demonstrated that this induction kinetic of the photoacoustic signal is detected only at low chopping frequencies and probably may arise from pulsed evolution of photosynthetic oxygen superimposed on the heat emission /7/. At high modulation frequencies the induction kinetic of the photoacoustic signal is determined only by the yield changes of heat emission. For our measuring system it has been checked before /6/ that for the radish leaves the "oxygen effect" can be excluded at modulation frequencies above 150 Hz. This has been carried out by measuring the effect of additional non-modulated white light, which is positive when only heat emission is sensed (see below) but which is negative when pulsed oxygen evolution contributes to the photoacoustic signal.

The modulated laser light wich is absorbed by the leaf is mainly used for photosynthesis. Part of the absorbed light energy is emitted as fluorescence and heat. When non-modulated light is added which saturates photosynthesis, the modulated light is fully transferred into heat and fluorescence and thus leads to an increase of the fluorescence and of the photoacoustic signal (Fig. 1, dashed line). The decrease of the fluorescence during the addition of non-modulated light can either be explained by an increase of "energy quenching" /11/ or by a "state 1 - state 2 transition" /12/. Both phenomena are used to interpret the decrease of fluorescence by means of an energy loss not directly involved in the photosynthetic reaction (a detailed discussion can be found in /6/).

4.2 Yield Changes during Photoinhibition

Photoinhibition is defined as the reduction of photosynthetic activity induced by illumination with light of intensities much in excess of that needed for photosynthetic light saturation /1/. Light saturation of photosynthesis for a particular leaf depends on the light intensity present during leaf expansion. Leaves grown in the shade exhibit photosynthetic light saturation at lower light intensities than leaves grown in full sun light /2/. During photoinhibition the functioning of specific photosynthetic components (the 32 kD protein and the reaction center of photosystem II, see review /1/) is disturbed. Photodamage of pigments occurs only later or at even higher light intensities.

During the photoinhibitory treatment of the radish leaf the fluorescence decreases gradually (Fig. 1, 4 min after onset of excitation, solid line). The steady state reached after about 15 min is much lower than the value attained with the non-modulated light of 2000 $\mu E \cdot m^{-2} \cdot s^{-1}$ and even lower than the steady state after 4 min of excitation with modulated light only. This strong decrease of fluorescence yield is already well documented as a result of photoinhibition (e.g. /2/). It was always concluded that this decrease of fluorescence is paralleled by an increase of non-radiative deexcitation. And this is what the increase of the photoacoustic signal during the 15 min of photoinhibitory treatment suggests.

5. Conclusion

Our experiment provides evidence for the hypothesis that photoinhibition leads to an increase of non-radiative deexcitation. By means of a simultaneous measurement of the fluorescence and of the photoacoustic signal at 279 Hz it could be demonstrated for the first time that during photoinhibition of a plant the decrease of chlorophyll fluorescence yield is paralleled by an increase of the yield of heat emission.

6. Literature

1. S.B. Powles: Annual Review of Plant Physiology 35, 15 (1984)
2. H.K. Lichtenthaler, R. Burgstahler, C. Buschmann, D. Meier, U. Prenzel, A. Schönthal: In Effects of stress on photosynthesis, ed. by R. Marcelle, H. Clijsters, M. van Poucke (M. Nijhoff / Dr. W. Junk, The Hague 1983) p. 353
3. C. Buschmann, H. Prehn: Photobiochemistry Photobiophysics 2, 209 (1981)
4. C. Buschmann, H. Prehn: Photobiochemistry Photobiophysics 5, 63 (1983)
5. C. Buschmann, H. Prehn: In Biological control of photosynthesis, ed. by R. Marcelle, H. Clijsters, M. van Poucke (M. Nijhoff, Dordrecht 1986) p. 83
6. C. Buschmann: Photosynthesis Research (in press)
7. G. Bults, B.A. Horwitz, S. Malkin, D. Cahen: Biochimica Biophysica Acta 679, 452 (1982)
8. C. Buschmann, H. Prehn: In Technical digest of the 4th topical meeting on photoacoustics, thermal and related sciences (Esterel 1985) p. ThA2.14
9. H. Kautsky, A. Hirsch: Die Naturwissenschaften 19, 964 (1931)
10. Y. Inoue, A. Watanabe, K. Shibata: FEBS Letters 101, 321 (1979)
11. U. Schreiber, U. Schliwa, W. Bilger: Photosynthesis Research 10, 51 (1986)
12. D.C. Fork, K. Satoh: Annual Review of Plant Physiology 37, 335 (1986)

Laser-Induced Optoacoustic Spectroscopy (LIOAS) of Photosynthetic Systems

C. Nitsch, S.E. Braslavsky, and G.H. Schatz

Max-Planck-Institut für Strahlenchemie,
D-4330 Mülheim a.d. Ruhr, Fed. Rep. of Germany

1. Introduction

To get information about the product stabilization within the photosynthetic electron transport chain, it is of interest to measure the heat evolved in the system upon irradiation. The cyanobacterium Synechococcus sp, containing both photosystems PS-I and PS-II, was chosen as the subject of investigation. LIOAS-measurements were performed a) in vivo with intact cells and b) in vitro with isolated PS-II and PS-I particles. In all cases the internal reference for maximal heat production [1] was the same system with inhibited photosynthetic activity. Compared to the conventional photoacoustic spectroscopy (PAS), LIOAS has a better time resolution (< 2 μs) and excludes contributions from modulated photosynthetic oxygen-evolution.

2. Materials and Methods

Apparatus: The LIOAS set-up was similar to the one described [2], except for the following changes. A 15 ns pulse from an excimer-pumped dye laser (Lambda Physics) at a repetition rate of 1 Hz was used for excitation at 628 and 677 nm (DCM and Pyridine 1 in MeOH, Lambda Physics). Laser beam diameters of 0.4, 0.9, 1.5 and 2.0 mm, corresponding to acoustic transit times τ_a [3] of 270 ns to 1.4 μs, were obtained with different pin-holes. The energy density was varied from 30 mJ/cm^{2}pulse down to 2.5 μJ/cm^{2}pulse (laser output between 35 μJ/pulse at 0.4 mm and 70 nJ/pulse at 2 mm beam diameter). To increase the S/N-ratio at the low energies, we employed a combination of two 400 kHz PZT detectors (Vernitron), attached to opposite walls of the flow cuvette. The signals were added and then amplified, averaged and energy-normalized as described [2]. The samples were thermostated at 17 °C in an external reservoir and pumped through the cuvette. To close the RC's, the reservoir was illuminated by cw fiber optic background light (Oriel 77501, Schott KG 3 for PS-II, KG 3 and RG 695 for PS-I).

Samples: Synechococcus cells were grown according to [4]. Oxygen-evolving PS-II particles of about 80 chlorophyll (chl) molecules per RC were prepared according to [5] and PS-I particles of about 85 chls/RC were obtained by further detergent treatment (G.H.S., in preparation). Algal cultures with a cell packed volume of ca. 1 μl/ml were diluted with 25% (v/v) ethylene glycol (EG), which produces larger LIOAS signals due to its thermoelastic properties. The PS-II and PS-I particles were resuspended in buffer solutions containing 25 % (v/v) EG.

The chl concentration was about 5-6 μM, and the absorbance at 677 nm was 0.4 to 0.5 per cm. With intact cells (size ca. 5 μm) an additional signal due to scattered-light absorption by the detector was observed as a small shoulder before the first peak. Scattered-light signals were simulated, by irradiating lipid emulsions in water/EG, and subtracted from the original signals in order to obtain pure absorption signals. PS-II reaction centers were either kept open by adding 1 mM $K_3Fe(CN)_6$ and dark adaptation or kept closed by adding 50 μM DCMU and additional background illumination. PS-I reaction centers were kept open by adding 1 mM Na-ascorbate and dark adaptation or kept closed by adding 1 mM $K_3Fe(CN)_6$ and background illumination. Furthermore, measurements were performed with PS-I particles adding 2 mM Na-dithionite to reduce the ferredoxin acceptors F_A, F_B of PS-I. Background illumination (KG 3, RG 695) served to reduce the acceptor F_X photochemically.

3. Results and Discussion

To obtain the fraction α of the absorbed light energy released as heat within our experimental time window, the first signal amplitude from an intact sample (open RC's),H_s, was compared to that from a sample with inhibited photosynthetic activity (closed RC's),H_r. In (1) and (2) α is related to the product of the quantum yield Φ_r of the primary photoreactions and to the molar energy content of the storing species, ΔE_r, by simple energy balance considerations [6]:

$$\alpha = (1-\Phi_{fr})H_s/H_r \quad (1) \quad ; \quad \Phi_r \Delta E_r / N_A h\nu = 1 - \alpha - \Phi_{fs}. \quad (2)$$

Φ_{fs} and Φ_{fr} denote the fluorescence quantum yields for open and closed RC's, respectively.

Signal energy dependence: Signals were measured as a function of the incident laser energy E (μJ/pulse) exciting at 628 and 677 nm, the absorption maxima of phycocyanin (PC) and chl-a, respectively. A reference should give a linear relationship between H_r and E, whereas for a system in which part of the energy is stored an initial small slope H/E should be observed, changing into a steeper one, due to the beginning of closure of the RC's at higher energies by the laser pulse itself. Correspondingly, the energy-normalized signal H_r/E should be independent of E (or even log E), whereas a plot of H_s versus log E should exhibit an inflexion point separating two H/E-levels, a lower one representing the thermal loss at open RC's, and a higher one representing the reference value. This was indeed observed for cells and PS-I particles for λ_{exc} = 677 nm (Fig.1). Varying the laser beam diameter from 0.4 to 2.0 mm, i.e., decreasing the energy density by a factor of 25, resulted in an equivalent shift of the inflexion point on the log E-axis (from -7.0 to -5.7; corresponding to 0.1 to 2.5 μJ/pulse). α values were estimated at energies $\leq$ 0.1 μJ and beam diameters $\geq$ 1.0 mm, corresponding to an energy density of less than 1 photon per RC in the particles.

For cells (λ_{exc} = 628 nm) and PS-II particles (λ_{exc} = 677 nm) a strong increase of the energy-normalized signals from intact samples up to the values for inhibited samples was

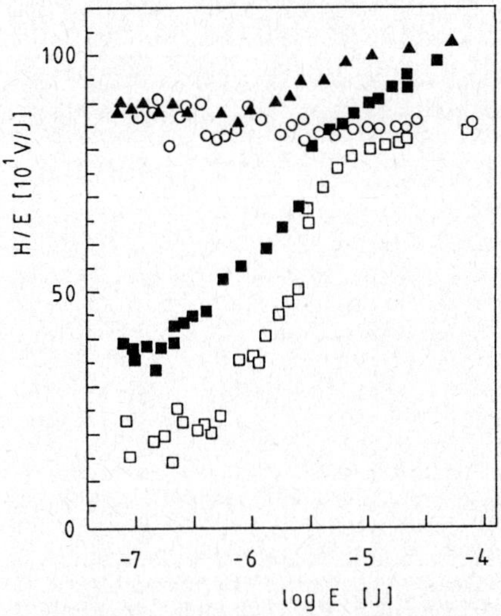

Figure 1

Semilogarithmic plot of the energy-normalized signals H/E versus the incident laser energy E; λ_{exc} = 677 nm

PS-I particles :

(o) inhibited
(□) intact

PS-II particles :

(▲) inhibited
(■) intact

observed with increasing energy. However, no plateau at high laser energies is reached. Correspondingly, the curve for inhibited samples shows constant signals H_r/E at low energies but an increase of the signals at higher energies (Fig.1). The inflexion point shifts towards higher energies at lower energy density. This effect can be explained by an increased singlet-singlet annihilation within the flat PS-II antenna, leading to a decrease of the fluorescence quantum yield and an enhanced heat production. This was already studied in purple bacteria and shown to be more pronounced for closed than for open RC's [7]. Nevertheless, correct α values can be obtained from (1) at sufficiently low excitation energies, where the fluorescence quantum yields have their maximum values.

Calorimetric evaluations: For cells, H_s/H_r = 0.4 and 0.65 ($\pm$10%) at 628 and 677 nm, respectively, were obtained. These data indicate a higher proportion of energy absorbed at 628 nm, used for PS-II photochemistry, than at 677 nm. This is consistent with findings that the PC in the phycobilisomes is preferentially coupled to PS-II. For isolated PS-II particles, containing only trace amounts of PC and allophycocyanin (APC), H_s/H_r = 1.0 and 0.35 at 628 and 677 nm, respectively, were obtained. The invariant value of H for open and closed PS-II at 628 nm indicates that the pigments absorbing at this wavelength (PC and APC) are not coupled to PS-II. The fact that the values obtained for isolated PS-II at 677 nm and for intact cells at 628 nm are identical again supports the idea of an efficient energy transfer from PC to the antenna chlorophyll in PS-II in vivo.

Using H_s/H_r = 0.4, Φ = 0.95, we obtained 0.65 for the term $\Delta E_r/N_A h\nu$. From this it follows that 35 % of the absorbed light energy is released by fast stabilization processes within 1.4 μs (upper limit for a beam diameter of 2.0 mm) whereas 65 % remains stored in the system. According to well known kinetic data of the electron transport chain in PS-II this time window includes the charge separation between P680 and the first acceptor I and the charge stabilization by the first quinone acceptor Q_A and the donor Z :

$$ZP^{*}IQ_A \rightarrow ZP^{+}I^{-}Q_A \rightarrow ZP^{+}IQ_A^{-} .$$

For PS-I particles excited at 677 nm, $H_s/H_r = 0.2$ is obtained. Assuming $\Phi_{fs} = \Phi_{fr} = 0.002$ and $\Phi_r = 0.95$ we obtain $\Delta E_r/N_A h\nu = 0.85$. This means that the energy loss in forward electron transfer in PS-I within our time window, which includes at least the charge separation between P700 and the first acceptor A_0 and the charge stabilization by a quinone acceptor A_1 [8], is significantly smaller than in PS-II.

$$P^{*}A_0A_1 \rightarrow P^{+}A_0^{-}A_1 \rightarrow P^{+}A_0A_1^{-} \rightarrow F_X \rightarrow F_A, F_B .$$

With PS-I particles under conditions where the secondary acceptors F_X, F_A and F_B are reduced, information about the back reaction of the radical pair $P^{+}A_0$ was obtained. With 2 mM Na-dithionite, a PS-I sample with open RC's (in the presence of 1 mM Na-ascorbate at an energy density of 0.3 photons/RC) exhibited a signal decrease of ca. 50 % . An H_s/H_r value of about 0.1 could be estimated. $\Phi_r \Delta E_r/N_A h\nu = 0.9$ results, which means that 90 % of the energy remains stored in the system. The formation of a triplet P700 is postulated, which would be the only possible reaction except recombination to $P^{*}700$ and giving up all energy as prompt heat. Its energy content should be close to that of the singlet and the quantum yield of its production, Φ_{isc} close to 1.0 under these conditions.

Acknowledgements: C.N. is the recipient of a fellowship award from the Alfried Krupp von Bohlen und Halbach-Stiftung, Essen. We are grateful for the support of Professor K. Schaffner.

4. Literature

1. S. Malkin, D. Cahen: Photochem. Photobiol. 29, 803 (1979)
2. M. Jabben, K. Heihoff, S.E. Braslavsky, K. Schaffner: Photochem. Photobiol. 40, 361 (1984)
3. K. Heihoff, S.E. Braslavsky: Chem. Phys. Lett. 131, 183 (1986)
4. T. Yamaoka, K. Satoh, S. Katoh: Plant Cell Physiol. 19, 943 (1978)
5. G.H. Schatz, H.T. Witt: Photobiochem. Photobiophys. 7, 1 (1984)
6. A.N. Garcia, G. Rossbroich, S.E. Braslavsky, H. Dürr, C. Dorweiler: J. Photochem. 31, 297 (1985)
7. J.G.C. Bakker, R. Van Grondelle, W.T.F. Den Hollander: Biochem. Biophys. Acta 725, 508 (1983)
8. A.W. Rutherford, P. Heathcote: Photosynth. Res. 6, 295 (1985)

Kinetic Analysis of the Photoacoustically Measured Photosynthetic Oxygen Response to Additional Blue and Far-Red Light

H. Dau and U.-P. Hansen

Institut für Angewandte Physik, Leibnizstraße,
D-2300 Kiel, Fed. Rep. of Germany

The processes of adaptation and regulation of the photosynthetic apparatus induced by changes in light intensity of broadband blue/green light or far red light (state transitions) were investigated by analysing the kinetics of the oxygen evolution as measured by the photoacoustic signal (PAS) [1,2]. In order to enable the quantitative evaluation of characteristic parameters, the responses from leaves of spinach were linearized by using an adequate light program [3] as given by the equation

$$I(t) = I_{m30} + I_{m300} + I_a(1+ 0.65\sin 2\pi ft). \qquad (1)$$

Two yield measurements could be performed simultaneously since the signals induced by the two different measuring lights I_{m30} (30 Hz, 2 Wm^{-2}) and I_{m300} (300 Hz, 4 Wm^{-2}) were measured by two separate lock-in amplifiers. A sequence of 22 frequencies f of the actinic light (I_a = 6 Wm^{-2}) ranging from 1 Hz to 1 cycle/2h was applied for the creation of a Bode-plot (Fig. 1) which gives the amplitudes (A) and phases (φ) of the induced photoacoustic signals vs. frequency f.

The usage of the low-frequency PA-signal as a measure of oxygen evolution could be justified by a comparison of frequency responses obtained from the 30-Hz yield signal with the theoretical frequency responses of the oxygen evolution as calculated from a combination of the 30-Hz signals with the 3oo Hz signals [1]. For t greater than 5s no significant difference between the kinetics of the computed oxygen evolution and the 30-Hz signal was found. For shorter times a reliable result was not obtained since the high-frequency signals became too noisy.

From the theory of linearized systems [3] it is known that the characteristic parameters (time-constants T_i, and amplitude factors a_i) can be obtained from curve-fitting of frequency responses (Fig. 1, Eq. 2a) or step-responses (Fig. 2, Eq.2b). With $p=2\pi\sqrt{-1}\,f$, n=3:

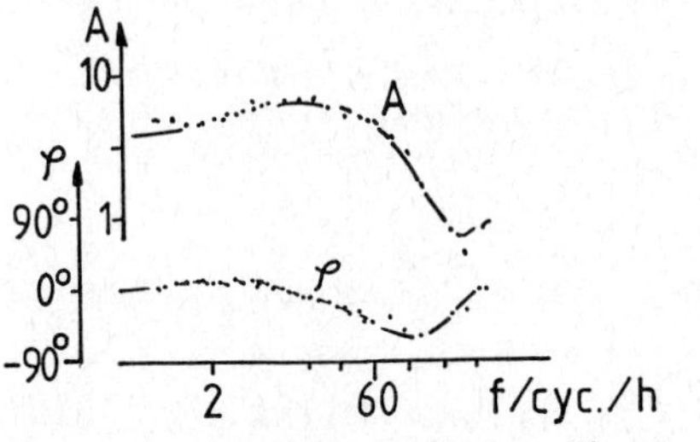

Fig. 1: Bode-plot of the amplitudes and phases of the yield of the PAS induced in spinach by light of sinusoidally modulated intensity (Eq.1).

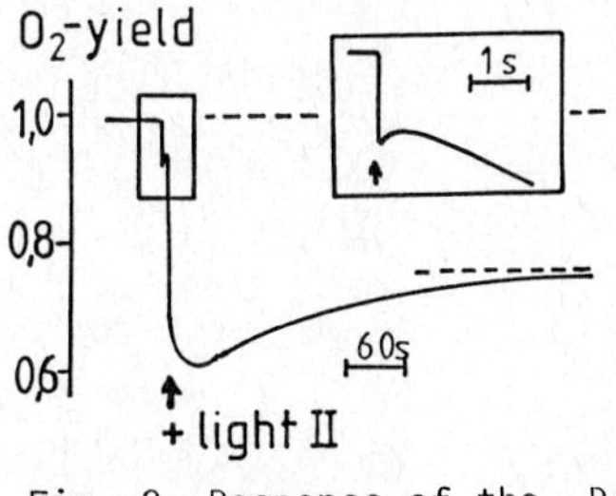

Fig. 2. Response of the PAS to a step-wise increase in intensity calculated from the data of table 1 by means of Eq. 2.

$$H(f)= H_0 \exp(\sqrt{-1}\,\varphi)=a_h+\sum_{i=1}^{n} a_i/(1+pT_i), \qquad g(t)=a_1+\sum_{i=1}^{n} a_i(1-\exp(t/T_i)). \quad (2a,b)$$

Typical values of T_i and a_i are shown in table 1. The time constants depend strongly on experimental conditions, thus the ranges of the most frequent values are given. A positive sign of a_i indicates that an increase of the actinic light (I_a, blue or 720 nm) leads to an increase of the oxygen evolution (chlorophyll-fluorescence) with the time constant T_i. For the sake of the comparison with the more familiar induction curves, the response to a step-wise increase in light intensity is calculated from the data in the table (Eq. 2b) and displayed in Fig. 2.

Table 1. Parameters of Eq. 2a obtained from curve-fitting (Fig. 1).

index i of component in Eq. 2a	h	1	2	3
time constant T_i/s	--	1...2	5..30	100..300
a_i of oxygen-yield, blue	-0.5	+0.8	-1.8	+0.4
a_i of oxygen-yield, 720 nm	0	+0.6	-0.1	+0.5
a_i of chl-fluorescence, 720 nm	0	-3.3	+0.15	+4.2

The assignment of the components of the oxygen response in the table to certain photosynthetic reactions is enabled by the usage of far-red (720 nm) actinic light (I_a modulated between 0.05 and 0.75 Wm^{-2}, Eq. 1). Table 1 shows that component no. 2 is scarcely stimulated according to the low modulation depth of the over-all intensity $I_{m30}+I_a$=10.05 to 10.75 W^{-2}. However, the components no. 1 and no. 3 are strongly stimulated. In accordance with the results of [2], and the positive sign of a_1 and a_3, this shows that T_1 and T_3 have to be assigned to the discharge of the plastoquinone pool and to the stimulation of the state-transition controller [4], respectively. The small coefficient of a_2 in the far-red induced responses of PAS and chlorophyll-fluorescence indicates the relationship to electron-flow. It could be shown that this component is related to the energy-quench [5]. The interference of the LHC-controller [2, 4] with the apparent absorption cross section of PS II leads to a simple non-linear model which enabled the correct prediction of the non-linear responses to step-wise increase in blue actinic light.

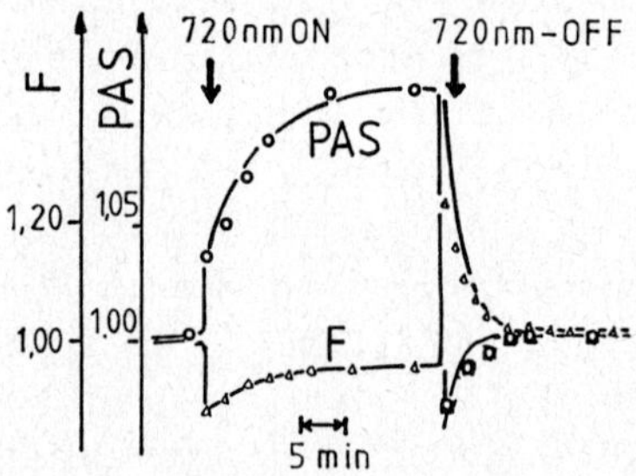

Fig. 3: Responses of PAS and fluorescence computed from a nonlinear kinetic model. The circles give experimental data.

1. G.Bults, B.A. Horwitz, S. Malkin, D. Cahen: BBA 679, 452 (1982)
2. O.Canaani, S. Malkin: BBA 766, 513 (1984)
3. U.P. Hansen: Ber. Deutsch. Bot. Ges. 98, 105 (1985)
4. J.F.Allen,J.Bennet,K.E.Steinback,C.J.Arntzen: Nature, 291, 25 (1981)
5. U.P. Hansen, J. Kolbowski, H. Dau: J. Exp. Bot. 38 (in press)

Applications of Power Optoacoustic Effects in Medicine and Biology

V.P. Zharov[1], *A.V. Kilpio*[2], *Y.O. Litvinov*[1], *V.I. Lotchilov*[1], *Y.S. Suchkov*[2], *and E.V. Shashkov*[1]

[1]Moscow Higher Technical School, Department of Biomedical Instruments, 107005-Moscow, USSR

[2]General Physics Institute of the Academy of Sciences of the USSR, 117942-Moscow, USSR

Conventional optoacoustics is primarily based on the application of relatively low-power lasers. When high-power laser radiation interacts with matter, very large acoustic waves can be produced via different physical mechanisms such as thermal expansion, evaporation, boiling, dielectric breakdown, etc. [1]. It should be noted that not all these mechanisms have been studied in detail yet. Nevertheless, some power optoacoustic (OA) effects with the formation of shock waves or sample destruction can be useful for high-pressure medical and biological investigations. These effects can be classified in the following way: 1) the OA "quasi-hydrostatic" effect in an acoustically closed volume with liquid; 2) the OA "hydrodynamical" effect in open space with the formation of shock waves initiated by optical breakdown; 3) the "mechanical" OA effect which involves mechanical motions of materials; 4) high efficiency "light-sound" transformation for ultrasound diagnostics.

The first effect was used to fracture gallstones. The experiments in vitro were carried out with a YAG:Nd laser with a pulse energy up to 12 J and 4 ms duration. The laser radiation with $\lambda = 1.06\,\mu$m was focused into the quartz light-guide of 0.4–1 mm diameter and 1.5–3 m length. The interaction of a laser pulse with a gallstone suspended in water at the light-guide output produced only small holes in the sample. The gallstones were fractured by applying many laser pulses with energy 12 J. The main cause of destruction was high thermal stress in the sample. A more economical method that is safe for medical procedures is shown in Fig. 1. This method includes grasping the stone in a Dormie trap and forming a narrow channel in the stone with a train of 5–10 low-power laser pulses with a repetition rate of 1 Hz and the energy in a single pulse equal to 0.1–0.3 J (Fig. 1a–b). After filling the channel with fluid, a high-power laser pulse is used to produce the OA hydrostatic effect inside the stone (Fig. 1c). As a result, the gallstone is fractured into several fragments at a laser pulse energy of 2.5–3 J (Fig. 1d). To improve the fracturing efficiency and to avoid damaging the bile duct wall with laser radiation, a strongly light-absorbing liquid is added to the interaction zone through a hollow guide. The best result was achieved using activated charcoal in water. The temperature change at the stone's surface during such a procedure was not more than 0.5°–1°C.

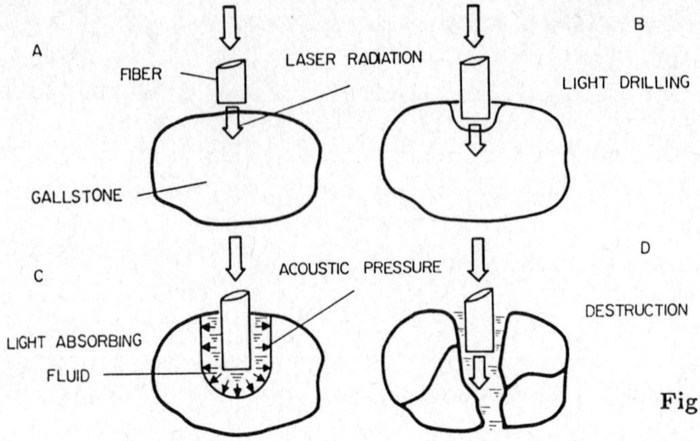

Fig. 1

For remote power action on biological samples the shock wave technique can be very useful. For example, extracorporal shockwave lithotrispy is the newest and most promising treatment for kidney-stone disease [2]. While this method uses electric arcs to create spherical blast waves, it is possible, instead, to use optical breakdown under focused laser radiation. In comparison with early experiments in this field [3] our research was carried out with a more powerful laser installation, "Kammerton", designed in the General Physics Institute of the USSR Academy of Sciences. In this setup a single laser pulse with $\lambda = 1.06\,\mu$m was transformed into the second harmonic with $\lambda = 0.53\,\mu$m and after optical amplification had an energy value up to 80 J at 3.6 ns duration. Radiation at the second harmonic was focused through a flat window in the first focus of the reflector itself attached to a $30 \times 30 \times 60$ cm water tank made of Plexiglas (Fig. 2). The reflector was a hollow half-ellipsoid made of aluminum with minor and major axes 7 cm and 14 cm, respectively. The shock waves initiated by laser radiation in the first focus of the ellipsoidal reflector were focused onto the second one where a piezoelectric gage or kidney

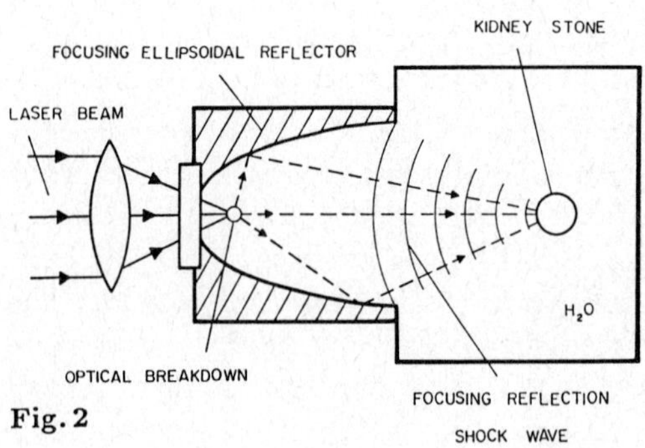

Fig. 2

Fig. 3

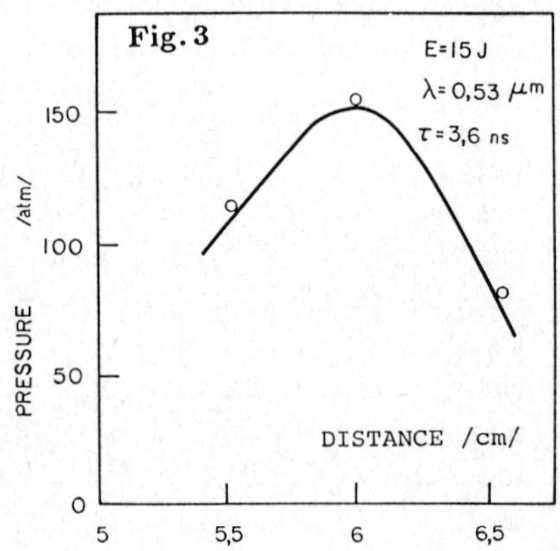

stone was placed. The shock waves' focusing zone was determined by the peak pressure registered by a piezoelectric transducer. Figure 3 presents the shock waves' pressure dependence on the distance along the axis from the output plane of the reflector at the pulse energy of 15 J. A pressure of 150 atm at 12 cm distance from the area of breakdown was achieved, while 50 atm pressure is quite enough for full fracturing of kidney stones after hundreds of separate shocks [2]. Fracturing experiments were carried out by suspending stone samples on a thread at the experimentally determined second focal point. In the monopulse regime, exfoliation of the small stone fragments was observed. But optimal application of this technique uses lasers in the pulse repetition regime. The experimental results confirm the possibility in principle of remote fracturing of kidney stones using the power OA effect for the formation of shock waves.

The mechanical OA effect causes liquid to be ejected through the nozzle of a small volume irradiated by short laser pulses. This effect was demonstrated in [4] for drop-on-demand ink jet printing. Our experiments show the usefulness of this effect for medical purposes. Figure 4 presents the OA medical injector which was made by inserting a light fiber in the needle of a medical syringe.

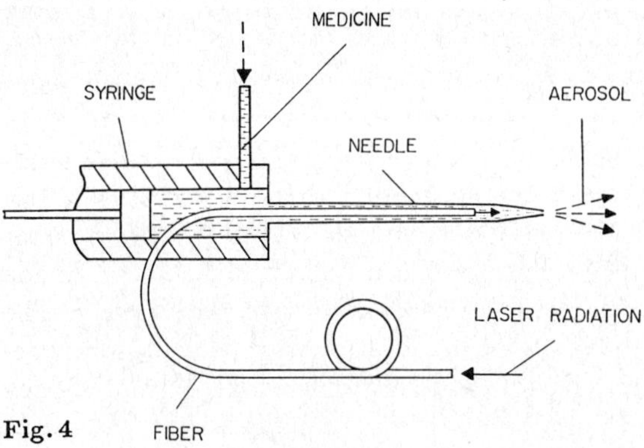

Fig. 4

This device is based on the formation and ejection of droplets of medicine being produced by the thermal energy deposited by a laser pulse in a small volume between the end of the fiber and the nozzle of the needle. The medicine can be released as an aerosol in the form of several drops or as one drop. This depends on the laser pulse energy and the absorption of the liquid. Hence, to achieve optimum ejection the optimum laser pulse energy and wavelength should be used. Different regimes were investigated using a YAG:Nd laser with a small energy ($10^{-4} - 10^{-2}$ J) and 4 ms duration. At strong absorption of the medicine only 1–10 μJ is needed for single-drop ejection. Such a device may be useful for dosing medicaments with a high accuracy and for aerosol treatment of a wound.

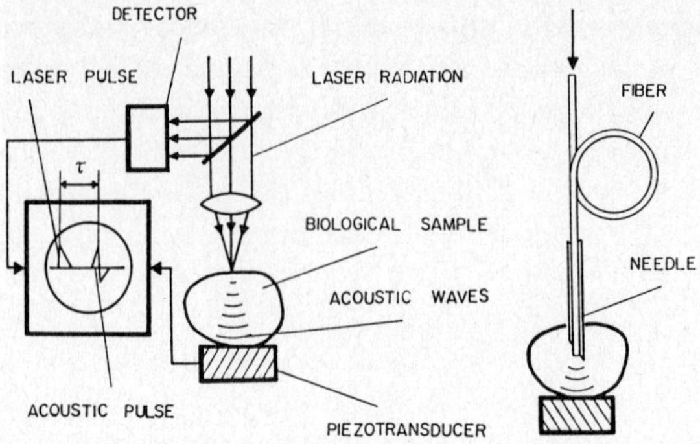

Fig. 5

For diagnostic purposes the laser OA probe was designed (Fig. 5). The principle of operation of this probe is based on the transformation of laser light into sound inside the special OA transformer at the end of the optic fiber. In some cases OA transformation can be realized in the biological sample itself (Fig. 5a). The OA probe can be inserted in a sample with the help of a syringe needle. The detection OA signals may be obtained in two ways: 1) by "transmission" with a piezotransducer which is in contact with the sample either directly or through an immersion adapter, for example, water; 2) by "reflection" when OA signals are transmitted through the fiber to the transducer in contact with this fiber at the input. In the second case it is necessary to use a fiber with a large diameter (up to 1–2 mm) because a strong attenuation of the OA signal in the fiber was registered. The possibilities of the laser OA probe were shown in its application for measurement of the acoustic charcteristics (speed of ultrasound, etc.) of kidney stones. Sound velocities were determined by measuring the time of the acoustic pulse delay in the sample with respect to the laser pulse. The experiments in vitro were carried out with a YAG:Nd laser with pulse energy up to 0.1 J and 10^{-7} s duration. The sound velocity in a kidney stone with average density 1.6 g/cm^3 was approximately (1.8–2) $\times 10^5$ cm/s. Another promising application of this technique is ultrasound diagnostics, using conventional ultrasound instruments to detect the OA signals, or diagnostics of vessels suffering from atherosclerosis, in combination with the fluorescence method.

To predict the influence of shock waves on biological tissue in laser ophthalmology or laser stone fracturing, experiments were carried out on OA formation of short acoustic pulses near cell structures of the onion-peel type in water [5]. Focused pulses of a YAG:Nd laser with 10 ns and 10 ps duration and 80 and 50 mJ energy, respectively, were used. The pressure of about 2×10^3 atm near the breakdown area was recorded by the shlirien technique. No cell destruction noticeable in the microscope was observed. This result can be explained by the fact that cells (and tissue also) have similar transmission acoustic characteristics to water [5].

In conclusion, we have presented the results of preliminary experiments on some new applications of power OA effects which indicate future perspectives of this technique for medical and biological purposes.

References

1. V.P. Zharov, V.S. Letokhov: *Laser Optoacoustic Spectroscopy*, Springer Ser. Opt. Sci., Vol. 37 (Springer, Berlin, Heidelberg 1986)
2. Ch. Chaussy et al.: *Extracorporeal Shock Wave Lithotripsy*, ed. by Ch. Chaussy (Karger Verlag, Munich 1982)
3. D.A. Russel: In Proceedings of 15th Int. Symp. on Shock Waves and Shock Tubes, ed. by P. Bershader (Stanford University 1985)
4. A.C. Tam, W.D. Gill: Appl. Opt. **21**, 1891 (1982)
5. E.N. Beilin, Y.A. Buyanov, V.P. Zharov et al.: Akust. Zh. (Russian) **32,** 240 (1987)

Photoacoustic Spectroscopy of Human Blood: Oxidation and Sedimentation Studies

P. Poulet, M. Ouzafe, and J. Chambron

Institut de Physique Biologique, Faculté de Médecine,
4, rue Kirschleger, F-67085 Strasbourg Cedex, France

1 INTRODUCTION

The blood is a suspension of cells, mainly red cells or erythrocytes, dispersed in a fluid, the plasma. Erythrocytes contain the hemoglobin, the protein which assures oxygen transport from the lungs to the tissues and the elimination of carbon dioxide.

We demonstrate that the oxygen-hemoglobin interaction can be studied on a red cell suspension using photoacoustic detection. In addition it is possible to measure the sedimentation of red cells in the plasma in a very short time by recording the time-evolution of the photoacoustic signal produced by a whole blood sample.

2 PHOTOACOUSTIC SPECTROSCOPY OF HEMOGLOBIN

Erythrocyte suspensions were prepared from whole blood : a test tube of heparinized blood was separated into two parts, one being reduced with sodium dithionite in slight excess. The two samples were then centrifuged and red cell suspensions were pumped in a syringe through the plasma. The photoacoustic cell was purged with nitrogen prior to the study of deoxygenated hemoglobin (Hb) and the sample was introduced anaerobically through a needle across the gas valve of the cell. The study of oxidized hemoglobin (HbO_2) was realized with the cell filled with air.

Photoacoustic spectra of Hb and HbO_2 were analyzed by the RG theory applied to thermally thick samples /1/. The absorption spectra calculated from photoacoustic experiments are presented in Fig. 1. They exhibit the Soret band in the near ultraviolet and the Q bands in the visible range.

A more precise verification of the validity of the theoretical model used, which neglects scattering effects, has been performed. The photoacoustically measured optical absorption spectra, using a thermal diffusivity of the red cell suspension of about $1.37\ 10^{-3}\ cm^2 s^{-1}$, agree well with the absorption spectra calculated from the estimated concentration of hemoglobin in red cells and the molar extinction coefficients of Hb and HbO_2. A more complete discussion of the above agreement and the use of the RG theory was reported in /2/. It justifies the use of theoretical models neglecting scattering effects for the study of hemoglobin oximetry and blood sedimentation.

3 HEMOGLOBIN OXIMETRY

The hemoglobin affinity for oxygen is described by the saturation curve of hemoglobin which shows the ratio $HbO_2/(Hb + HbO_2)$ as a function of the partial oxygen pressure PO_2.

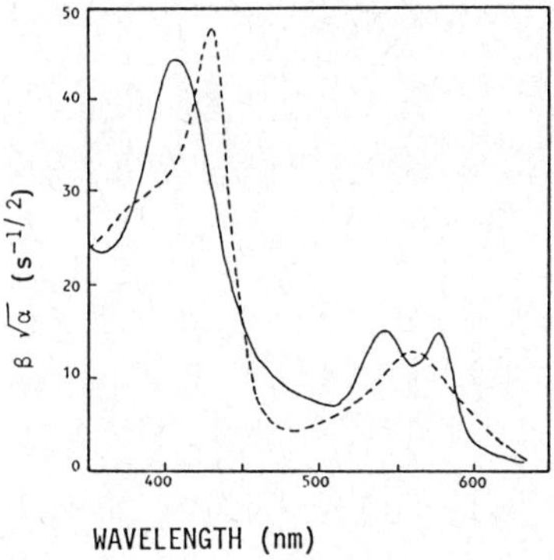

Fig. 1 : Photoacoustic absorption spectra of hemoglobin (---) and oxyhemoglobin ( ). β : optical absorption coefficient, α : thermal diffusivity

A modified photoacoustic cell containing a polarographic electrode enables the simultaneous measurement of PO_2 and of the photoacoustic signal at 470 nm (where the difference between the absorption of hemoglobin and oxyhemoglobin is maximum) during a slow increase of the PO_2 from zero (cell purged with 95 % N_2, 5 % CO_2) up to 400 mmHg, by oxygen diffusion through a Millipore filter placed between the photoacoustic cell and a second gas volume filled with oxygen and water. This arrangement represents a Helmholtz resonator with a resonance frequency of 150 Hz and a gain of about 3. The recorded photoacoustic signals were then linearized (versus optical absorption coefficient) by using the RG theory applied to a thermally thick sample and the saturation curve was calculated (Fig. 2).

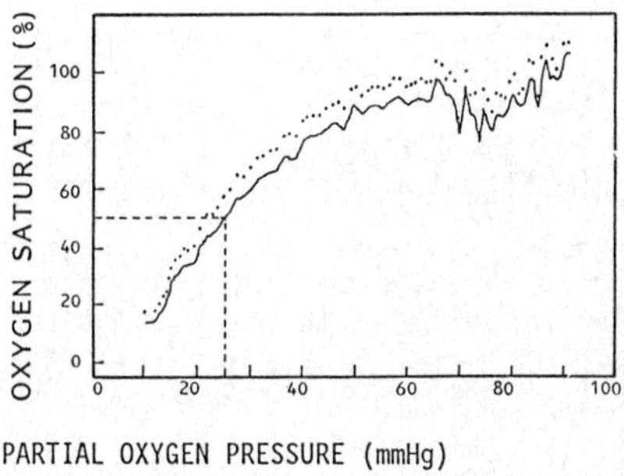

Fig. 2 : Saturation curve of adult human hemoglobin (5 % CO_2, 37°C. Signal amplitude as measured (...) and linearized (——).

The sigmoid shape of the saturation curve is due to the cooperative effects between the four globin chains of the hemoglobin molecule /3/. The effect of saturation on the PA signal shifts the saturation curve to the left. For linearized values, P50, the pressure at which 50 % of hemoglobin is oxidized, is about 26 mmHg (standard value 26.6) at 37°C. The Hill number /2,3/, expressing the cooperative effects is about 2.5 (standard value 2.8).

4 BLOOD SEDIMENTATION

Photoacoustic experiments reported above were performed on red cell suspensions. Measurements on whole blood exhibit the sedimentation of red cells in plasma /4,5/. The sedimentation rate of disk-shaped erythrocytes of thickness ϵ and diameter Ø is given by /2/

$$v = \epsilon\, \emptyset\, (\rho - \rho_0) / 7.65\, \mu \; ; \quad (1)$$

ρ_0 is the plasma density, ρ the red-cell density. μ the viscosity. The thickness of the plasma layer after a sedimentation time t is equal to vt. The sample can thus be considered as a two-layer stratified sample with an upper non-absorbing layer of plasma and a deeper, absorbing, thermally thick layer containing red cells. The evolution of the photoacoustic signal is given by /2, 6/

- amplitude $P(t) = P(o) \exp(-vt/\mu_s)$, (2)

- phase $\emptyset(t) = \emptyset(o) - vt/\mu_s$; (3)

μ_s is the thermal diffusion length in plasma.

The time evolution of the PA amplitude and phase are displayed in Fig. 3 and Fig. 4. Signals were recorded at various frequencies on samples from the same female blood kept on citric acid.

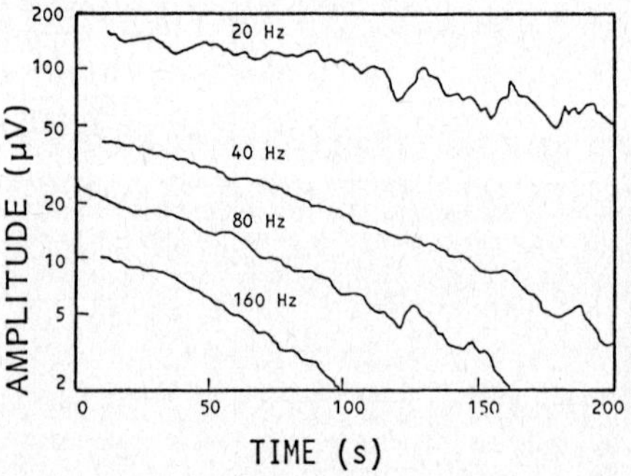

Fig.3 : Time evolution of PA amplitude at various chopping frequencies

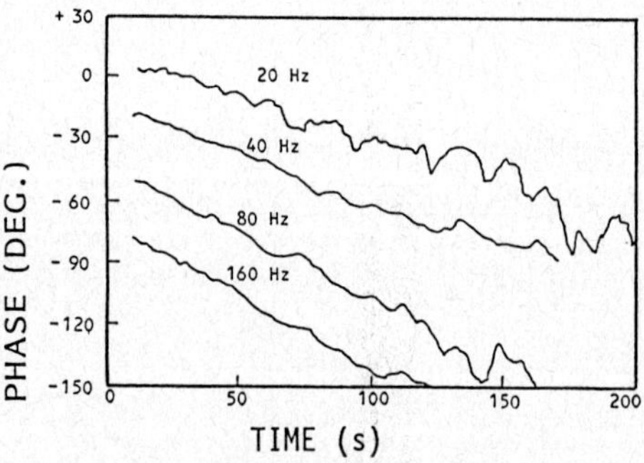

Fig. 4 : Time evolution of PA phase at various chopping frequencies

The mean sedimentation rates determined from the amplitude : $3.8 \times 10^{-7} \pm 0.5 \times 10^{-7}$ m/s and from the phase : $2.8 \times 10^{-7} \pm 0.6 \times 10^{-7}$ m/s are in good agreement with the expected value (1.5 to 4.2×10^{-7} m/s depending on the plasma viscosity) /2/. The sedimentation rate of a blood sample from the same donor is 14×10^{-7} m/s as measured with a conventional WESTERGREN method during the first hour (5mm). This greater value is due to the aggregation of red cells during the measuring time.

5 CONCLUSION

Photoacoustic detection is a potential diagnostic tool in blood analysis. Two examples have been considered :

- 1) The oxygen-hemoglobin interaction can be studied in red cells. PA detection presents various advantages over other methods using transmission spectroscopy on dilute solutions - concentration variations of various chemical agents - or on thin films - biocompatibility problems.

- 2) The sedimentation rate of red cells in blood can be measured in a very short time (less than 2 minutes). The results are not affected by aggregation : photoacoustic detection offers a means to study the size of red cells and the blood viscosity rather than the aggregation of erythrocytes as observed by the usual sedimentation methods.

6 REFERENCES

1. P. Poulet, J. Chambron, R. Unterreiner : J. Appl. Phys. <u>51</u>, 1738 (1980)
2. P. Poulet : <u>Spectroscopie Photoacoustique et Sciences Biomédicales : Contribution aux études du sang, de la peau et de la photosynthèse</u>, Thèse ULP Strasbourg, 1985
3. M. J. Perutz : Nature <u>228</u>, 726 (1970)
4. P. Poulet, J. Chambron, R. Unterreiner : in Photoacoustic Spectroscopy, Proceedings of the Institute of Acoustics, Chelsea College (1981)
5. P. Helander, I. Lundstrom : J. Photoacoust. <u>1</u>, 203 (1982)
6. N. C. Fernelius : J. Appl. Phys. <u>51</u>, 650 (1980)

Application of Photoacoustic Spectroscopy to Human Blood

Qing Pan[1;4], *Shu-ye Qiu*[2], *Shu-yi Zhang*[2], *Ji-bin Zhang*[3], *and Si-ming Zhu*[4]

[1]Department of Information Physics, Nanjing University, Nanjing, People's Republic of China
[2]Institute of Acoustics, Nanjing University, Nanjing, People's Republic of China
[3]Hematology Lab., The People's Hospital of Jiangsu Province, Nanjing, People's Republic of China
[4]Nanjing Medical College, Nanjing, People's Republic of China

We have measured the different types of whole blood and their separated components from adults and fetuses, normal persons as well as patients. The characteristics of the PAS spectrum of blood depend upon the molecular structure of heme. The spectrum varies with the density of hemoglobin, that is, the peak value ratio γ/β or γ/α becomes greater when the density is lower. There are differences among the PAS spectra of blood from some patients, such as anemia, leukemia and methemoglobin. Some simulated experiments have also been performed. It is believed that PAS technique will be a valuable tool in clinical diagnosis in hematology.

1. INTRODUCTION

As a valuable technique, photoacoustic spectroscopy (PAS) is increasingly applied in biology and medicine /1-4/. One of the principal advantages of PAS is that it enables one to obtain spectra, similar to optical spectra, of any type of solid, semisolid or liquid material. This capability is based on the fact that the absorbed light is converted to sound via nonradiative deexcitation. PAS is suitable to measure samples with light-scattering or amorphous properties, which cause so many difficulties in conventional spectroscopic techniques. Blood, as an example of a strong light-scattering sample, has been studied by the PAS technique /1/. But more work must be done before PAS can be used for hematological studies and clinical diagnosis. The aim of this study is to record the characteristic PAS spectra of different types of blood obtained from some patients and to provide information for the purposes mentioned above.

2. EXPERIMENTAL RESULTS AND DISCUSSIONS

A home-made experimental apparatus, a single-beam photoacoustic spectrometer is used /5/. In order to develop the application of PAS to hematology, we have measured the different types of whole blood and their separated components from adults and fetuses, normal persons as well as patients.

2.1 Normal Blood

As is well known, for normal blood, similar PAS spectra from whole blood, red blood cells and hemoglobin are obtained in the visible wavelength region. Three absorption peaks, called α , β and γ , appear at wavelengths of about 575 nm, 540 nm and 410 nm respectively. In addition, the PAS spectra of other separated components of blood such as heme, globin, membranes of red blood cells, white blood cells, platelets and plasma are also obtained. Except from heme, no characterized absorption peaks are

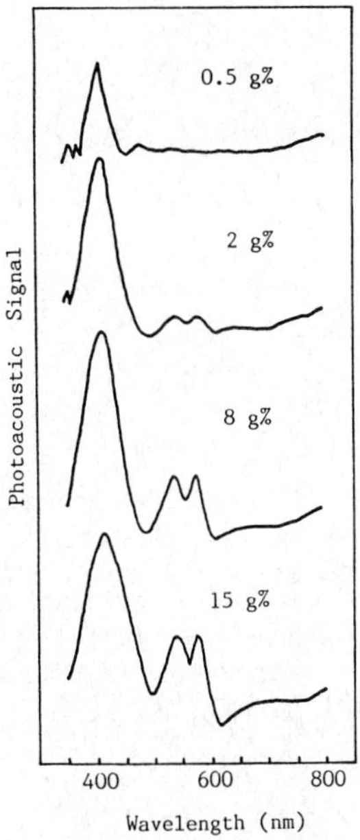

Fig.1 PAS spectra of hemoglobin with the density of 0.5, 2, 8 and 15 g%

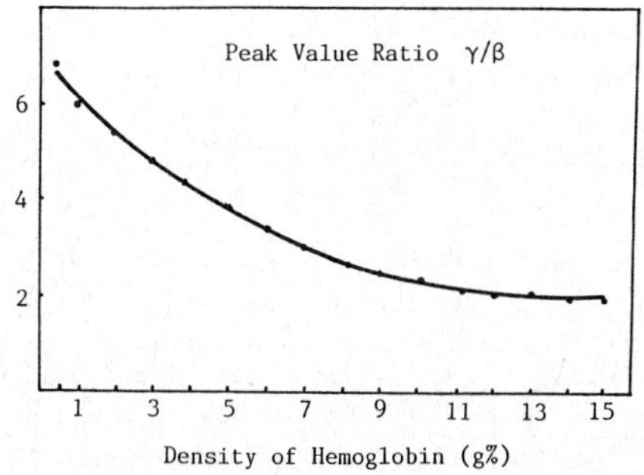

Fig.2 The γ/β of the spectra of hemoglobin with the density from 0.5 to 15 g%

displayed in the given region. On the other hand, the PAS spectrum of hemoglobin from adults is extremely similar to that from fetuses when both have the same density. Although the two kinds of hemoglobin include the heme with the same molecular structure, the globin, another component of hemoglobin, consists of different multipeptide chains. Thus, the results indicate that the PAS spectrum of normal whole blood essentially implies the optical absorption characteristics of hemoglobin inside the red blood cells, and the spectrum of hemoglobin displays the strong and characteristic absorption spectrum of heme. Adams et al. /2/ have shown the intensive peak γ is attributed to an allowed $\pi - \pi^*$ transition in a region remote from iron atom. Peak α and peak β, weaker than peak γ, are due to electronic transitions associated with charge-transfer interaction of iron orbitals with the ligand states.

The PAS spectral variation of hemoglobin in different densities from 0.5 to 15 g% is shown in Fig.1 and Fig.2. It can be seen that the three absorption peaks of hemoglobin with the varied density have no wavelength shift because of their same molecular structure of heme. However, the peak value ratio γ/β or γ/α increases monotonically with the decrease of the density of hemoglobin. In general, as the density is lower than 8 g%, peaks γ, β and α weaken and the γ/β goes up linearly with the decrease of density. When the density is higher than 8 g%, peak γ appears saturated and γ/β decreases slowly with the increase of density, and as the density lies between 12 and 15 g%, the ratio γ/β approaches a value of about 2.

2.2 Blood from Some Patients

Trying to find out the characteristic PAS spectra of hematonosis and hemoglobinopathy and to apply the PAS technique to hematology in clinical diagnosis, we investigated the PAS spectra of blood from some patients

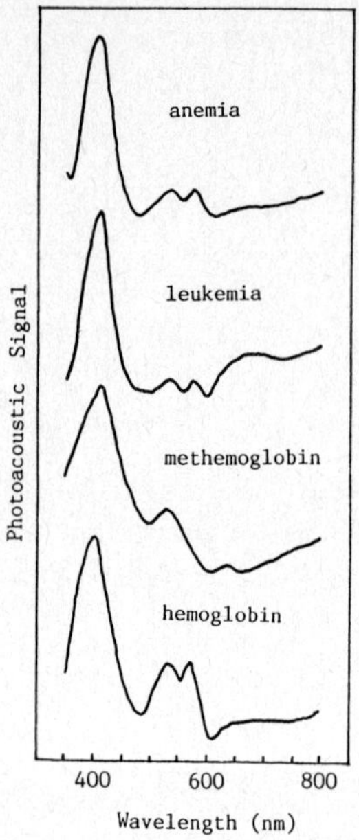

Fig.3 PAS spectra of blood from anemia, leukemia and methemoglobin patients, and the hemoglobin spectrum

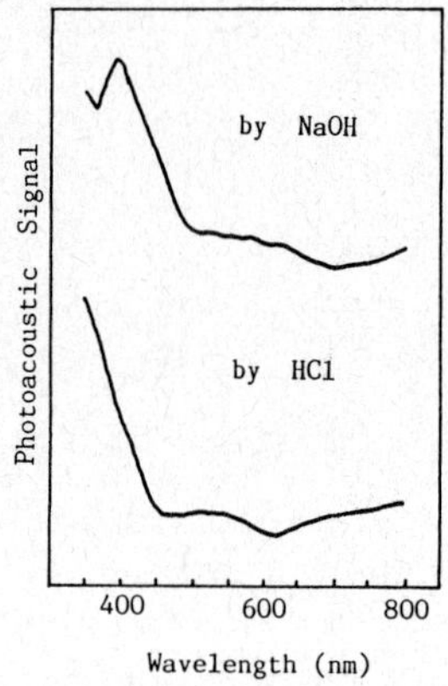

Fig.4 PAS spectra of hemoglobin affected by NaOH and HCl

suffering from anemia, léukemia and methemoglobin, which are clinically diagnosed, as shown in Fig.3. The spectra of anemia and leukemia show a similarity in three chracteristic absorption peaks and a greater ratio γ/β or γ/α than normal spectra, which are similar to that of the lower density hemoglobin and agree with the clinical results. On the other hand, a wide absorption peak located at the wavelength region longer than that of peak α appears in the spectrum of leukemia but is not seen in the spectrum of anemia. In the case of methemoglobin, the shape of the PAS spectrum is varied apparently and only two absorption peaks appear, one at a wavelength of about 410 nm and another at 530 nm. It is considered that methemoglobin has Fe^{3+} substituting for Fe^{2+} in the normal molecular structure of heme. This substitution changes the charge-transfer interaction of iron orbitals with the ligand states and then peak α and peak β merge into a peak located at 530 nm /6/. This variation of the structure of heme induces the change of the spectrum of blood.

Some simulated experiments have also been performed. The PAS spectra of hemoglobin affected by chemicals, such as NaOH and HCl,are shown in Fig.4. In the spectrum affected by NaOH, there is only one absorption peak which is a γ blue-shifted peak, with a wavelength of about 390 nm. Peak α and peak β disappear in the visible region. For the spectrum affected by HCl, peak γ shifts to a shorter wavelength than for NaOH; the other two peaks are also not observed. It is believed that OH^- or Cl^- is combined with Fe^{2+} in the heme in some way and changes the configuration of heme /7/. The PAS spectral features of hemoglobin depend upon the molecular structure of heme.

3. SUMMARY

In summary, according to the results of our experiments on the application of PAS to human blood we may conclude: (a) The PAS spectrum of blood shows the absorption properties of hemoglobin, and the spectral characteristics depend upon the molecular structure of heme. (b) The PAS spectrum of blood varies with the density of hemoglobin, that is, the peak value ratio γ/β or γ/α becomes smaller when the density is higher. As the density is higher than 8 g%, a PAS signal saturation appears. (c) There are some distinct differences in the PAS spectra from the blood of patients suffering from anemia, leukemia and methemoglobin, and that from healthy people. The characteristic PAS spectra of blood from the above patients can be obtained.

Thus, the PAS technique can be used conveniently and quickly to diagnose the hemotonosis with the lower density of hemoglobin and the hemoglobinopathy with the abnormal structure of heme. PAS will be a valuable tool to provide information in clinical diagnosis in hematology.

4。REFERENCES

1. A.Rosencwaig: Photoacoustics and Photoacoustic Spectroscopy, John Wiley & Sons, New York (1980)
2. M.J.Adams, B.C.Beadle, A.A.King and G.F.Kirkbright, Analyst, 101, 553 (1976)
3. G.M.Alter, J. Biol. Chem., 258, 14960 (1983)
4. G.M.Alter and S.J.McKee, Anal. Biochem., 132, 312 (1983)
5. S.Y.Que, S.Y.Zhang, C.N.Hu and L.H.Wei: J. Appl. Sci., 4, 108, (1986) (in Chinese)
6. W.B.Qin: Hemoglobinopathy, People's Health Publishing House,(1984) (in Chinese)
7. G.Z.Chen: Ultraviolet-Visible Spectroscopy, Atomic Energy Publishing House, (1983) (in Chinese)

Effect of Hematological Parameters on the PTR Signal During Blood Sedimentation

J. Marx[1], *C. Droullé*[2], *M. Egée*[1], *E. Van Schel*[1], *F. Potier*[1], *and G. Potron*[2]

[1]Service Universitaire d'Energétique, BP 347,
F-51062 Reims Cedex, France

[2]Laboratoire d'Hématologie, CHU, F-51100 Reims Cedex, France

Blood is essentially composed of a suspension of corpuscules (red and white blood cells, platelets) in a viscous liquid, the plasma. In the case of many human diseases, the suspension stability of blood is altered and the speed of sedimentation is accelerated, which can be measured with a standing test evaluating the evolution occurring after one or two hours.

When a blood sample is submitted to a visible laser excitation, this excitation is essentially absorbed by the red blood cells (the plasma is a transparent liquid which can be seen) when the sedimentation is in progress, the absorption coefficient keeps diminishing, which leads to a decrease of the photothermal signal (PTR) emitted by the sample surface (the plasma is opaque to infrared rays). Besides, some other poorly known phenomena - such as the agglutination of cells - result in a change in

- the thermal properties of the medium (diffusivity, concentration profile)
- the outline of the decrease chart of the PTR signal.

The first results /1,2/ using a sinusoïdal laser excitation and an approximate theoretical model /3/ allowed us to point out these phenomena. Several conditions such as the rapidity of the phenomena, the relative sample fragility and the necessity to study numerous cases for medical purposes, have led us to develop an automatic device for the recording and plotting of the PTR signal (amplitude and phase shift).

In this study, we are able to show, for the very first time, that there is indeed a connection between the time evolution of the PTR signal and some parameters which are known to influence the speed of sedimentation.

Among the different factors which act on the sedimentation rate of the red cells, the relative concentration of the cells and of the fibrinogen molecules, is a main one. On a given blood sample, this variation rate can be produced by adding either plasma of the same origin, or physiological serum. Such tests have been carried out on blood samples coming from different patients. Figure 1 shows the time dependence of the blood PTR signal, before (curve 1) and after adding different quantities of plasma (curves 2 and 3). Initially, we observe that the PTR amplitude directly depends on the absolute concentration of the red cells (so on the absorption coefficient) and keeps on diminishing very slowly as time passes. Besides, after the "rouleau" formation (agglutination of red cells) there occurs a quicker fall of the signal : this one comes earlier when the red cell concentration decreases compared to the fibrinogen one. After the addition of physiological serum in the same proportion as plasma, which gives the same absolute concentration of red cells (and so the same initial PTR signal), but leads to a diminution of the relative fibrinogen concentration, the sedimentation speed clearly increases (curve 4). Finally, the PTR signal immediately falls when the red cell agglutination is obtained by adding beff fibrinogen, the relative concentration of the red cells remaining constant (curve 5).

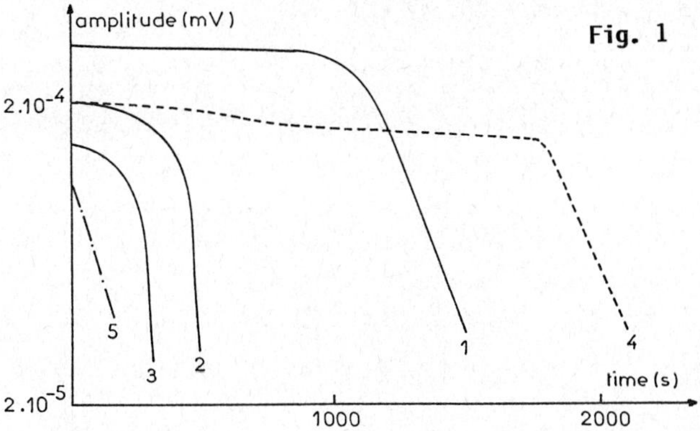

Fig. 1

Thanks to such studies, we will be better able to come near to and understand the relationships existing between the kinetics of the PTR signal and various biological mechanisms in the field of haematology and finally to result in a satisfying modelisation.

Bibliography :

1. J. Marx, M. Egée, R. Dartois, C. Droullé, G. Potron : in <u>Nouv. Rev. Fr. Hematol</u>. 27, 2 (Springer, Berlin, Heidelberg 1985) p. 112.
2. R. Dartois, <u>Thesis</u>, Reims, 1986.
3. R. Dartois, M. Egée, J. Marx, E. Van Schel : In <u>Rev. Gen. Therm. Fr.</u> 301, (1987) p. 22-32.

Penetration of Topically Applied Drugs Through Human Skin Investigated by Photoacoustic Spectroscopy

B. Sennhenn[1], *M. Rohr*[1], *K. Giese*[1], *and K. Kölmel*[2]

[1]Institut für Medizinische Physik und Biophysik der Universität, Gosslerstraße 10f, D-3400 Göttingen, Fed. Rep. of Germany
[2]Universitäts-Hautklinik, von-Siebold-Straße 3, D-3400 Göttingen, Fed. Rep. of Germany

1. INTRODUCTION

In vivo methods employed for the investigation of percutaneous absorption of drugs observe the radioactivity in excreta or blood after topical application of radiolabeled agents, the surface recovery, surface disappearence, and the biological or pharmacological response [1]. Most methods yield information about the amount of drug penetrating completely through full-thickness skin. The photoacoustic technique is unique in that it offers the opportunity for a detailed analysis of the penetration of topically applied drugs through the horny layer, which, as the outermost layer of the skin, forms the main diffusion barrier of the body. Experimental access to the distribution of drugs in this layer, and to its diffusion controlled variations, is of great importance for developing transcutaneous pharmaco-therapy.

In this paper we report on the penetration behaviour of the UV-absorbing drug Tizanidine, a muscle relaxant, applied to the skin surface in different vehicles. Differentiation between the amount of drug within the superficial corneal layers and within the total horny layer was achieved by suitable choice of the thermal diffusion length [2]. After application of a thin film of the preparation under test to the skin surface, the drug penetrates into the horny layer and reaches the underlying viable tissue, where the blood circulation causes a much faster transport.

2. DRUG DIFFUSION

To describe the diffusion of a light absorbing drug through the horny layer we proceed from the one-dimensional model of a vehicle film of thickness d applied to the horny layer of thickness L at time t = 0. The diffusion coefficients of the drug in vehicle and stratum corneum are given by D_F and D_S, respectively, and the partition coefficient stratum corneum/vehicle is denoted by K. If the drug is dissolved in the vehicle at initial concentration c_o, the concentration depends on time t and depth coordinate x according to

$$c_F(x,t) = \frac{c_o}{A}\sum_{n=0}^{\infty} \frac{\sin(u\cdot r_n)\,\cos(u\cdot r_n\cdot x/d)}{u\cdot r_n}\, e^{-(r_n^2 D_S t/L^2)} \qquad (1)$$

in the vehicle, $0 \leq x \leq d$, and

$$c_S(x,t) = \frac{K\cdot c_o}{2A}\sum_{n=0}^{\infty} \frac{\sin(2u\cdot r_n)\,\sin(r_n(L+d-x)/L)}{u\cdot r_n\cdot \sin(r_n)}\, e^{-(r_n^2 D_S t/L^2)} \qquad (2)$$

in the horny layer, $d \leq x \leq d+L$. In these relations r_n are the positive roots of the transcendental equation

$$\tan(u \cdot r_n)\ \tan(r_n) = u \cdot v \quad , \tag{3}$$

with the quantities u and v defined by

$$u = (d/L) \cdot \sqrt{D_S/D_F} \qquad \text{and} \qquad v = K \cdot L/d \quad , \tag{4a,b}$$

and

$$A = \sum_{n=0}^{\infty} \sin^2(u \cdot r_n)/(u \cdot r_n)^2 \quad . \tag{5}$$

With $\varepsilon(\lambda)$ being the specific absorption coefficient of the drug at wavelength λ, the increase of the absorption coefficient due to the drug, $\Delta\beta(\lambda, x,t)$, depends on concentration according to

$$\Delta\beta(\lambda,x,t) = \varepsilon(\lambda) \cdot c(x,t) \quad , \tag{6}$$

where the subscripts F and S denoting film and stratum corneum are omitted.

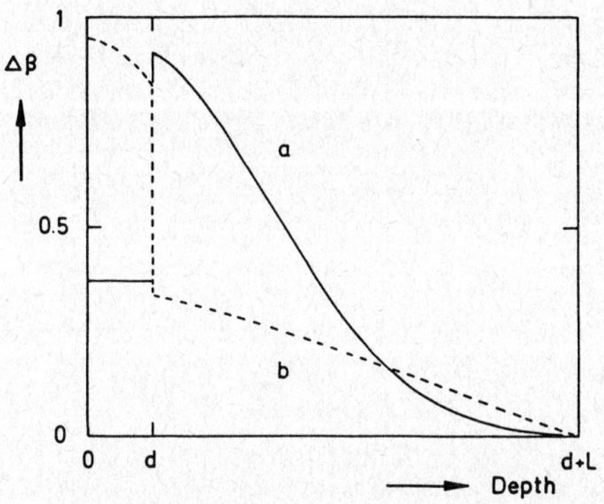

Fig. 1: Increase of the optical absorption coefficient due to drug distribution as a function of depth for two different combinations of diffusion coefficients in film and stratum corneum and stratum corneum/ vehicle partition coefficient. (See text)

Figure 1 shows the increase of the optical absorption in dependence on depth obtained as the result of an evaluation of (1) to (6) with d=2 µm and L=13 µm and two different combinations of diffusion coefficients D_F and D_S and partition coefficient K. Curve (a) represents a typical diffusion profile, which is determined by the diffusion of drug in the horny layer, curve (b) shows a diffusion profile with the release of drug from the vehicle being the rate determining step. For both profiles the diffusion coefficient of the drug in the stratum corneum was assumed to be $D_S=10^{-10}$ cm^2/s [3]. Curve (a) was evaluated with a larger diffusion coefficient in the vehicle, $D_F/D_S=6.3$, and a partition coefficient K=2.5, curve (b) with a smaller diffusion coefficient in the vehicle, $D_F/D_S=0.16$, and a partition coefficient K=0.4. The profiles are shown at times chosen such that the amount of drug up to a depth x=2d, containing the film and the superficial corneal layers, is the same in both cases.

3. EXPERIMENTAL RESULTS AND DISCUSSION

The photoacoustic spectrometer used for the in vivo measurements on skin has been described elsewhere [4,5]. Briefly, we employ a double-beam spec-

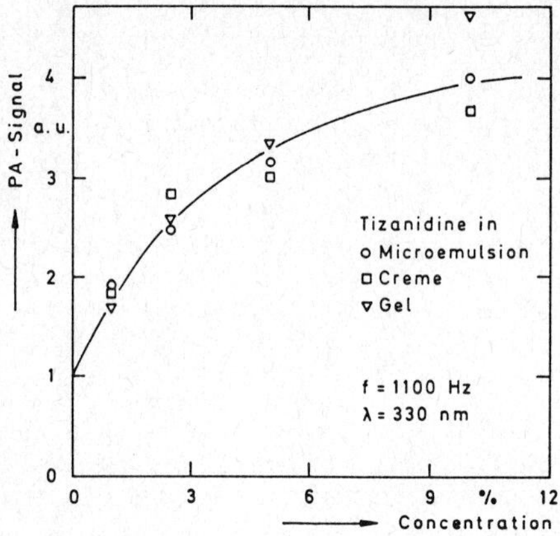

Fig. 2: Concentration dependence of the PA-signal of skin treated with Tizanidine. The drug was incorporated in 2 μm thick films of different vehicles

trometer with an open-ended differential detector which yields an effective reduction of bodily noise.

For the investigation of the penetration behaviour of topically applied drugs through human skin by photoacoustic spectroscopy it is necessary to apply thermally and optically thin films of the preparation under test. In fig. 2 the PA-signal of human skin treated with Tizanidine incorporated in 2 μm thick films of different vehicles is shown as a function of drug concentration. With the higher drug concentrations the applied film is no longer optically thin and the PA-signal becomes saturated. Only at the lower concentrations the PA-signal shows a linear dependence on drug concentration. In this range the photoacoustic difference-signal, which is the difference between the PA-signals of skin treated with the drug incorporated in a vehicle film and with the same vehicle as a placebo, sensitively detects the amount of light absorbing drug up to a depth of the thermal diffusion length.

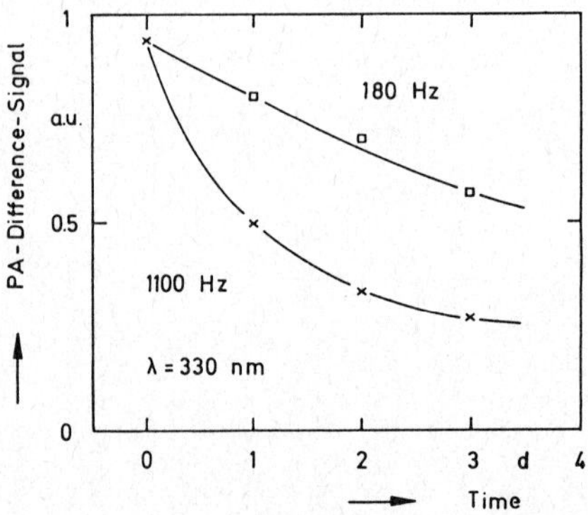

Fig. 3: Time dependence of the photoacoustic difference-signal of skin treated with Tizanidine incorporated in a gel. Chopping frequencies 180 Hz and 1100 Hz, drug concentration 2 %. The rate determining step is the drug diffusion in the horny layer

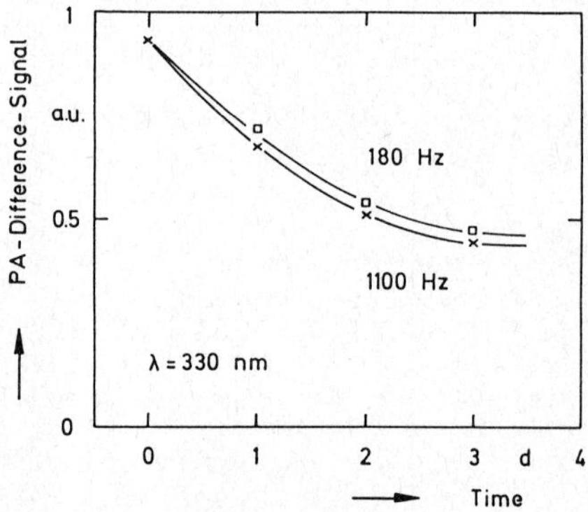

Fig. 4: Time dependence of the photoacoustic difference-signal of skin treated with Tizanidine incorporated in a cream. Chopping frequencies 180 Hz and 1100 Hz, drug concentration 2 %. The rate determining step is the release of the drug from the cream

In fig. 3 and fig. 4 temporal courses of the difference-signal are shown. The measurements have been carried out on six subjects over a time period of four days. Plotted are the mean values of the photoacoustic difference-signal. The penetration of Tizanidine has been measured at a wavelength of 330 nm, where the self-absorption of the skin is comparatively small. Fig. 3 shows the results obtained with Tizanidine incorporated in a gel. Here the 1100 Hz signal, which depends on the amount of drug within the film and the superficial corneal layers, decreases much faster with increasing time than the 180 Hz signal, which reflects the amount of drug within the film and the total horny layer. In the example represented in fig. 3 the time necessary for the release of drug from the vehicle is comparatively short and the drug diffusion in the stratum corneum is the rate determining process ($D_F >> D_S$, $K \gtrsim 1$). Curve (a) in fig. 1 shows a typical diffusion profile, evaluated for such conditions by use of (1) to (6). Fig. 4 represents the corresponding results obtained with Tizanidine incorporated in a cream. Here the difference-signals at both chopping frequencies show almost the same temporal decrease. In this case the release of drug from the vehicle is the rate determining process, lasting much longer than the diffusion of the drug through the horny layer. A representative diffusion profile evaluated for these conditions ($D_F << D_S$, $K \lesssim 1$) is given by curve (b) in fig. 1. As one can see, most of the drug is found within the film and the superficial corneal layers where it contributes to the photoacoustic signal at both chopping frequencies.

ACKNOWLEDGEMENTS

The authors thank the Deutsche Forschungsgemeinschaft, Bonn-Bad Godesberg, FRG, and the Paul G. Unna-Stiftung, Düsseldorf, FRG, for supporting this work.

References

[1] R. C. Wester, H. I. Maibach, in:Percutaneous Absorption, R. L. Bronaugh, H. I. Maibach (Eds.), Marcel Dekker, New York, 1985, pp. 245-249

[2] K. Giese, K. Kölmel, J. Physique, Colloque C6(1983)373-378

[3] R. J. Scheuplein, J. Invest. Derm. 48(1967)79-88

[4] K. Kölmel, A. Nicolaus, B. Sennhenn, K. Giese, J. Soc. Cosmet. Chem. 37(1986)375-385

[5] K. Giese, A. Nicolaus, B. Sennhenn, K. Kölmel, Can. J. Phys. 64(1986) 1139-1141

Photoacoustic In Vitro Investigation of Human Skin

U. Haas[1], *J. Franz*[2], *and F. Nimmerfall*[2]

[1]Institut für Physik, Universität Hohenheim, Garbenstraße 30, D-7000 Stuttgart 70, Fed. Rep. of Germany
[2]Pharma-Entwicklung, Sandoz AG, CH-Basel, Switzerland

Depending on modulation frequency, photoacoustic in vitro investigation of human cadaver skin reveals a change of water absorption at different depths of the epidermis of skin.

1. Introduction

The outermost part of human skin, the epidermis, provides a barrier which separates the interior of the organism from the environment. This barrier function is necessary to prevent excessive loss of water through the skin, as well as entrance of noxious substances. On the other hand, there exists a great interest in making use of the skin as an entrance for properly designed drug formulations. The human epidermis is a tissue layer, roughly 100 - 500 µm in thickness composed of, from inside to outside, the stratum germinativum, the stratum granulosum, and the stratum corneum. The stratum corneum is responsible for the barrier function of the skin. It is a horny layer, roughly 6 - 20 µm thick, consisting of flattened keratinized cells.

To get insight into the mechanism of skin penetration by drugs we have started to study first untreated human skin by photoacoustic (PA) spectroscopy. Because of its unique ability to yield a depth-profile analysis of the optical and thermal properties of the skin, PA spectroscopy should provide valuable experimental information on drug influence on skin constituents /1 - 3/.

2. Experimental

The in vitro study of human cadaver skin was performed in the UV, VIS, and NIR spectral region of light using an EG & G, Princeton Applied Research Model 6001 photoacoustic spectrometer as described previously /4/. We used a frequency range of light modulation from 10 Hz to 2 kHz, i.e., the thermal diffusion length μ_s varies between about 40 µm to about 4 µm, resp., in mammalian tissue, applying a thermal diffusivity for dry horny skin of $a = 7 \cdot 10^{-8}$ m²/s /5/. The optical absorption length $\mu_\beta = 1/\beta$ (β: optical absorption coefficient) of human skin is wavelength dependent: it is maximal at 750 nm ($\mu_\beta \simeq$ 80 µm) and decreases to smaller and longer wavelengths ($\mu_\beta \simeq 4$ µm at 250 nm, $\mu_\beta \simeq 40$ µm at 1400 nm) /6/. Therefore, the stratum corneum of human skin is translucent in the NIR, VIS, and in the UV region for wavelengths greater than 250 nm. For PA-investigations the stratum corneum of the human skin represents the case of optically thin and thermally thick material ($\mu_s < \mu_\beta$).

The experiments were performed on human cadaver skin obtained less than 24 h post mortem. Immediately upon biopsy the skin is frozen at -20 °C. Sections of 1 mm in thickness were prepared using a cooled microtome and stored at -70 °C. Specimens (4 mm x 8 mm) were punched from frozen sections and investigated by PAS after slowly warming to room temperature (about 23 °C).

3. Results

The PA-signal obtained in the UV and NIR region shows the absorption bands of intact human stratum corneum, corresponding to the absorption of keratin /1 - 3/ and water /4, 7/. The absorption bands of keratin were further validated by a comparison with the PA-spectrum of powdery keratin. The water content was examined in the NIR by observing the PA-signal of the first overtone of the hydrogen stretching vibration, as well as of the combination vibrations of the hydroxyls. The remarkable result of these investigations is that for a modulation frequency of f = 13 Hz the absorption peak of the first overtone is observed at a wavelength of 1456 nm corresponding to the absorption of water hydroxyls associated by hydrogen bonds. But for f = 1500 Hz the absorption peak appears at about 1367 nm corresponding to the absorption of "free" hydroxyls of water /4, 7/. The conclusion is that in the outermost layers of the stratum corneum the water molecules are adsorbed to the relatively dense keratin matrix, giving rise to the monomeric hydroxyl absorption, whereas in the innermost parts of stratum corneum, as well as in the other parts of the epidermis, water is associated. This result may be of great importance for a carrier function of water through the skin barrier.

In addition, we have started studies on the penetration of dimethylsulfoxide (DMSO) in human cadaver skin treated with DMSO before freezing. Typical absorption peaks allowed us to detect the deep penetration of DMSO into the epidermis of skin. Furthermore, we observed that DMSO seems to alter the keratin content in different depths beneath the skin surface compared to untreated skin.

Acknowledgements. The authors wish to thank Prof. Dr. H. Seiler (Inst. f. Physik), and Dr. T. Kissel (Sandoz AG) for their great interest in the progress of this work. The financial support by Sandoz AG is gratefully acknowledged.

1. A. Rosencwaig: Photoacoustics and Photoacoustic Spectroscopy, 1st. ed., Chemical Analysis, Vol. 57. Ed. by P.J. Elving and J.D. Winefordner. (John Wiley & Sons, Inc. New York, NY., 1980).
2. D. Deffond, J.L. Lévêque, J. Scott and D. Saint-Léger: J. Photodermatology 2, 279 (1985).
3. K. Giese, A. Nicolaus, B. Sennhenn, and K. Kölmel: Can. J. Phys. 64, 1139 (1986).
4. U. Haas and H. Seiler: Z. Naturforsch. A, 39A, 1242 (1984).
5. A.M. Stoll: In Advances in Heat Transfer 4, p. 117. Ed. by J.P. Hartnett and T.F. Irvine. (Academic Press, New York, NY., 1967).
6. G. Lischka und E.G. Jung: Lichtkrankheiten der Haut. Beiträge zur Dermatologie, Band 4, 2. Aufl. Ed. J. Metz. (Perimed Fachbuch-Verlagsgesellschaft mbH, Erlangen, FRG, 1982).
7. U. Haas: Can. J. Phys. 64, 1063 (1986).

A New Electronic Transition Observed in Concentrated Aqueous Solutions of Hematoporphyrin IX, as Detected by Photoacoustic Spectroscopy

A. Lachaîne[1] *and R. Pottier*[2]

[1]Department of Physics, Royal Military College of Canada, Kingston, Ontario K7L 5L0, Canada

[2]Department of Chemistry and Chemical Engineering, Royal Military College of Canada, Kingston, Ontario K7L 5L0, Canada

The nature of the mechanism responsible for the selective biodistribution of hematoporphyrin (Hp) related molecules towards tumor tissue is not yet clear. One physiological factor that has been implicated is the tissue pH [1]. We report here a new, pH-dependent spectroscopic transition, observable by photoacoustic (PA) spectroscopy, and discuss its implications in the selective biodistribution of porphyrin-type photochemotherapeutic agents.

Concentrated aqueous solutions of Hp show a new electronic transition, appearing as a shoulder or new peak on the red side of the Soret band, having a maximum intensity at or near 440 nm, as shown in Fig. 1. The intensity of this 440 nm band is very sensitive to pH, as indicated in Fig. 2. It is also sensitive to the concentration of Hp, the ratio of the intensity of the 440 nm band to that of the 385 nm band decreasing exponentially from a value of 0.88 to 0.20 for a corresponding change in the Hp concentration of 100 to 1500 μM. Increased concentration of Hp is associated with the conversion of dimers to aggregates [2–4]. On the other hand, for a fixed concentration (500 μM), this same ratio (440 nm/385 nm) increases from 0.26 to 0.48 for a corresponding change in chopping frequency from 100 to 160 Hz, and remains constant with further increase in chopping frequency. Taking 160 Hz as the characteristic frequency, and the thermal diffusivity of water as representative of the solution, one can obtain an approximate value for the thermal diffusion length (μ_s) as 0.0017 cm. Under these conditions, the optical penetration depth (μ_β) is of the order of 0.08 cm. Thus, the ratio of the peak intensities may be safely considered as an indication of the relative amplitude of both peaks.

Hp and other dyes adhere strongly to glass surfaces, especially under conditions that favor the formation of dimers and/or aggregates. It is thus possible that hydrophobic Hp dimers and/or aggregates become concentrated at the air/water interface of Hp solutions in the photoacoustic cell. The above data are thus consistent with a transition originating from a dimeric form of Hp. When 0.4% sodium dodecyl sulfate surfactant (SDS) was added to the Hp solution, no band at 440 nm could be observed by PA spectroscopy, regardless of the pH, Hp concentration or chopping frequency used.

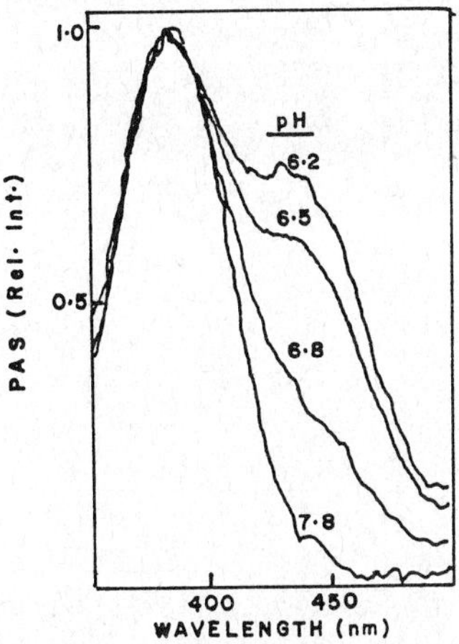

Fig. 1. Photoacoustic signal (PAS) vs excitation wavelength

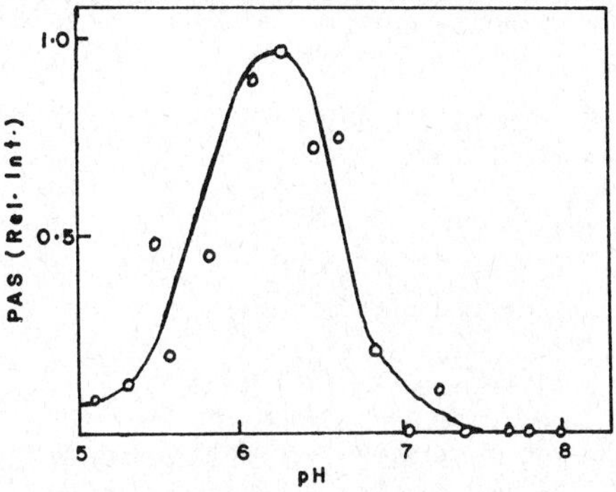

Fig. 2. PAS vs pH

Transmittance, fluorescence and volumetric titration data of Hp solutions indicate that the ionic species present near pH 6.3 is predominantly a neutral species, having no protons on the imino nitrogens and unionized carboxylic acids (no net charge). Since this pH roughly corresponds to that of fast-growing tumor tissue [5], the 440 nm species monitored by PA spectroscopy could well be the one that is responsible for the selective accumulation in tumor tissue. Thus, selective biodistribution could be explained on the basis of a relatively simple dynamic equilibrium between different ionic species, this equilibrium being sensitive to changes in the pH of the medium.

References

1. J. Moan, S. Sommer: "Uptake of the components of hematoporphyrin derivative by cells and tumors". Cancer Lett. **21**, 167–174 (1983)
2. R.F. Pasternack: "Aggregation properties of water-soluble porphyrins". Ann. N.Y. Acad. Sci. **206**, 614–630 (1973)
3. S.B. Brown, M. Shillock, P. Jones: "Equilibrium and kinetic studies on the aggregation of porphyrins in aqueous solutions". Biochem. J. **153**, 279–285 (1976)
4. G.A. Karns, W.A. Gallagher, W.B. Elliot: "Dimerization constants of water-soluble porphyrin in aqueous alkali". Bioorg. Chem. **8**, 69–81 (1975)
5. J.L. Wike-Hooley, J. Haveman, H.S. Reinhold: "The relevance of tumor pH to the treatment of malignant disease". Radiotherapy and Oncology **2**, 243–366 (1984)

Photoacoustic and Nuclear Magnetic Resonance Detection of Manganese III Hematoporphyrin in Mouse Organs

M. Ouzafe, J. Steibel, Y. Mauss, P. Poulet, and J. Chambron

Institut de Physique Biologique, Faculté de Médecine,
4, rue Kirschléger, F-67085 Strasbourg Cedex, France

Hematoporphyrin derivatives are used in the photodynamic treatment of various type of tumors. These photosensitizers bind preferentially to tumor tissues and to some organs like the liver and the kidneys. Two or three days after injection, the tumors are illuminated with red light. A partial or total regression of the tumors is observed in the few days following the irradiation /1/ . Tissue properties (oximetry, pH, optics...) and the biodistribution and the physical state of porphyrins determine the success of the therapy. Paramagnetic metalloporphyrins can also be considered as potential contrast agent for tumors /2/ . We used a manganese III - hematoporphyrin and report the preliminary results about its biodistribution obtained by magnetic resonance relaxometry and imaging and by photoacoustic spectroscopy.

Hematoporphyrin IX was purchased from Aldrich (Strasbourg). The synthesis of hematoporphyrin-manganese (Hp-Mn) was performed according to YONETANI and ASAKURA /3/. Female hybrid F1 mice were injected intraperitoneally either with 0.2 mmole/kg of bodyweight or with 1 mmole/kg. F1 mice were sacrificed and heart, kidney, liver, skin and brain were excised prior to relaxivity or photoacoustic analysis. The relaxation rates R_1 (longitudinal) and R_2 (transversal) were measured at 37°C on a Minispec PC20 Bruker spectrometer. Photoacoustic spectra of excised organs were recorded on a home-built photoacoustic spectrometer.

Relaxation evolution kinetics of Hp-Mn in various organs (Fig. 1) were similar in shape. A relaxation rate maximum was observed in the first hours after injection, thereafter the relaxation rate decreased to its initial value. Differences in the affinity of organs for Hp-Mn are displayed by their maximum relaxation rates, which decrease in the following order : kidney (2.5-23.3), liver (2.8-9.3), heart (1.4-3.8), skin (3.2-5.4), brain (2.0-2.0). The first numbers in brackets indicate $R_1/s^{-1}/$ of the untreated organs, the second numbers are R_1 6 hours after injection of 1 mmole/kg Hp-Mn.

The post-mortem evolution of the relaxation rate R_1 of the organs studied was found to be a very sensitive indicator of residual amounts of Hp-Mn : contrary to non-injected organs whose relaxation rates were stable after the death of the mouse, R_1 of kidneys of injected animals exhibited a rapid increase during the early hours following the sacrifice (Fig. 1).

The results obtained with photoacoustic spectroscopy strengthened this observation. Spectra of kidneys before and at different times after injection are presented on Fig. 2. The blank spectrum exhibits the absorption of hemoglobin. A short time after injection (1 hour) the absorption bands of Hp-Mn at 365 and 480 nm began to superpose to the Soret band of hemoglobin at about 415 nm. At longer times after injection, only the Hp-Mn absorption bands were observable, demonstrating an accumulation of hematoporphyrins in the kidneys.

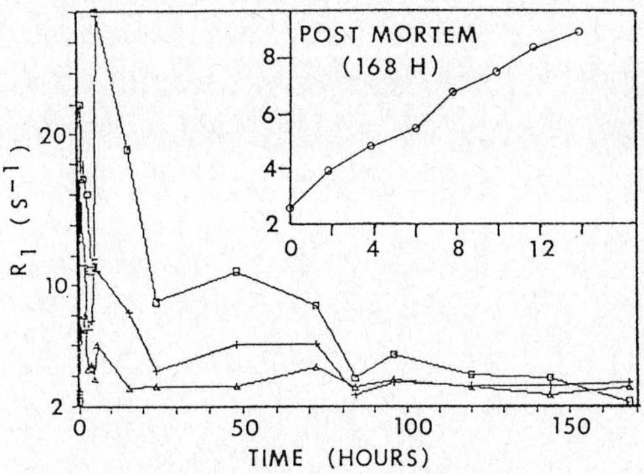

Fig 1 : Relaxation rate R_1 of kidney (–□–), liver (–+–) and skin (–△–) versus time after injection of Hp-Mn 1 mmole/kg. R_1 post mortem evolution of kidney

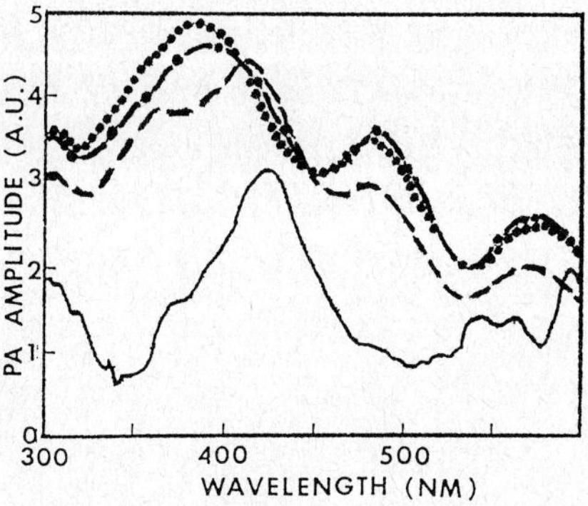

Fig 2 : Photoacoustic spectra of kidney : untreated (——), 1 hour (– – –), 2 hours (–•–) and 48 hours (•·····) after injection of Hp-Mn 0.2 mmole/kg

The relaxation rates R_1 - by their post-mortem evolution - as well as the photoacoustic spectra of organs demonstrate that Hp-Mn is always present in kidneys and, at a lower level, in the liver and the heart in the few weeks following injection. Hp-Mn is in a state where it does not contribute to the relaxation of water protons. Aggregation or crystallization of Hp-Mn seems to be responsible for the decrease of the relaxivity, rather than the excretion of the product. These results agree with measurements performed by extraction of Hp-Mn and absorption spectroscopy.

The results reported demonstrate that Hp-Mn is a potential candidate as MRI contrast agent : its relaxivity is high and the complex is very stable, however its toxicity appears to be a limiting factor. Photoacoustic spectroscopy appears to be more sensitive than relaxometry to monitor the amount and state of porphyrins in organs.

1. D Kessel, T.S. Dougherty : Porphyrin Photosensitization (Plenum Press, New York and London, 1983)

2. C Cohen, J.S. Cohen, C.E. Myers, M. Sohn : FEBS Lett. 168,70(1984)

3. T. Yonetani, T. Asakura : J. Biol. Chem. 244,4580 (1969)

Photoacoustic Imaging Immunoassay for Biological Component Microanalysis

T. Masujima[1], *Y. Munekane*[1], *C. Kawai*[1], *H. Yoshida*[1], *H. Imai*[2], *L. Juing-Yi*[3], *and Y. Sato*[3]

[1]Institute of Pharmaceutical Sciences, Hiroshima University, School of Medicine, Kasumi 1-2-3, Hiroshima 734, Japan
[2]Faculty of Pharmaceutical Science, Fukuyama University, Higashimura-cho, Fukuyama 729-02, Japan
[3]Research Institute for Nuclear Medicine and Biology, Hiroshima University, Kasumi 1-2-3, Hiroshima 734, Japan

A computerized imaging analyzer for laser photoacoustic microscopy has been developed and applied to the microanalysis of biological components in tissues or cell suspensions. Photoacoustic immunoassay was developed for the analysis of human λ-chain protein (λ type Bence Jones Protein), which is a principal indication of malignant lymphoreticular disease, particularly multiple myeloma [1]. In this paper, we report the use of enzyme immunoassay to perform more convenient and sensitive detection in microregions of the samples.

Imaging Analyzer

Figure 1 shows the system of the imaging analyzer. A He-Ne laser (25 mW) beam was introduced into the microscope with

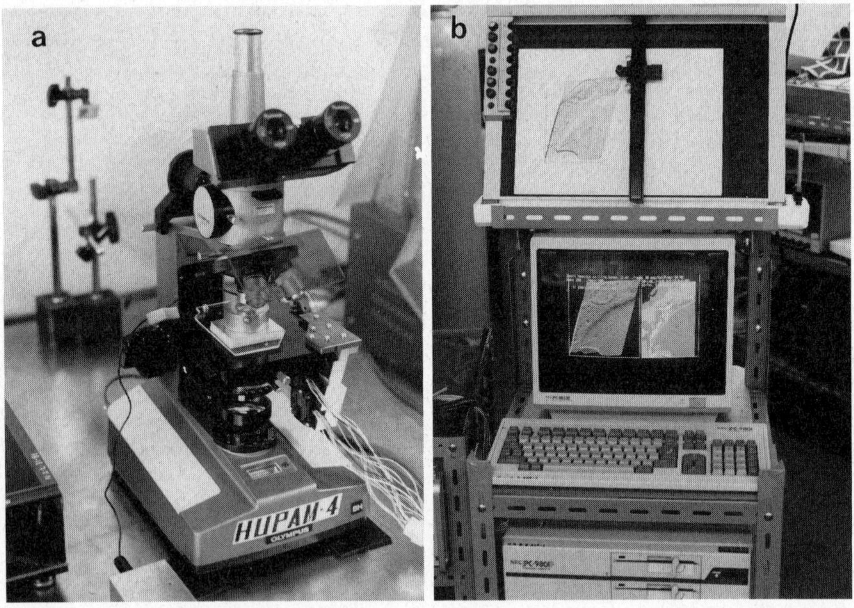

Fig. 1 Imaging analyzer for laser photoacoustic microscopy. a) Microscopy system and b) data processor

mirrors and a beam splitter and focused on the sample in the photoacoustic cell, which was mounted on an insulation plate on the microscope stage. This stage was modified to be scanned in 2-dimensions by pulse motors. The photoacoustic signal from a microregion of the sample was fed to a lock-in amplifier, and its output DC signal was introduced to an A/D converter in the microcomputer. The average of the signal corrected for the laser power fluctuation was calculated and the pulse motors were regulated to scan the sample in the desired direction. The final 3-dimensional(z-axis is the photoacoustic signal) figures and contour diagrams of the samples were displayed on a monitor.

The data thus obtained were processed to get the integration of the signal and the subtraction of the background. Finally, the PA-signal image was converted to the image of the amount of the target proteins using the appropriate standard curve.

Experimental

Samples (thin section of tissue or cell suspension) or standard human λ-chain were adsorbed on the nitrocellulose membrane filter by trapping on the membrane (for tissue), by spotting (for suspension), or by filtration (for standard solutions). The filter system with 46 holes (3 mmϕ) gave homogeneously distributed standard protein in the 3 mmϕ region on the membrane. The membranes with samples were dipped into a bovine serum albumin solution(in pH 5 acetate buffer) for 30 min to block the residual adsorbing sites on the membrane. These membranes were washed with phosphate buffered saline(PBS) solution and incubated in a first antibody (rabbit anti-human λ-chain IgG) solution for 30 min, then placed in a solution of the second antibody (goat anti-rabbit IgG), which was labeled with horse radish peroxidase, and left for 1 h. The area of antigen was stained by enzyme reaction in the substrate solution (for 10 min) which contained diaminobenzidine, H_2O_2, and the Ni^{2+} ion. The stained blue-brown spots of samples were measured by the imaging analyzer.

Results and Discussion

Figure 2 shows one example of the standard sample. The homogeneously dispersed circle is the region where 9.35 ng λ-

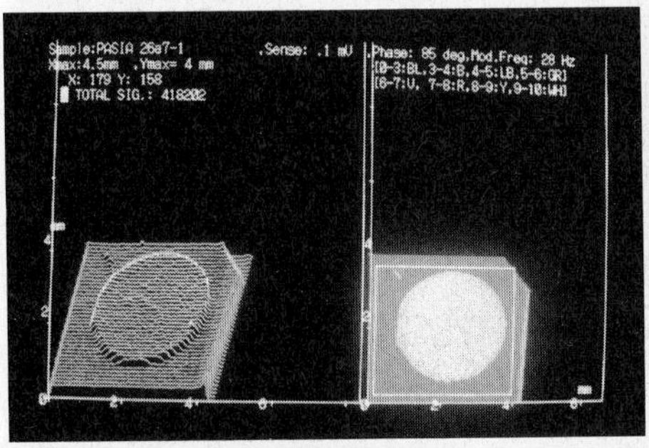

Fig. 2 PAM image of a spot of standard sample of human λ-chain (total 9.35ng)

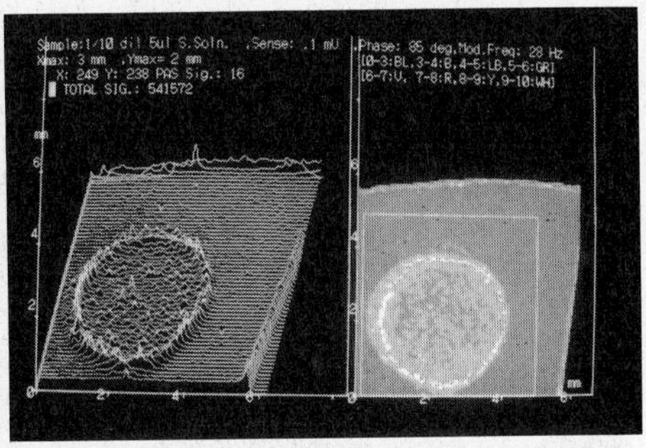

Fig. 3 PAM image of a spot of diluted human bone marrow cell suspension (5 µl)

chain was stained. The total signal(the background was subtracted) and the total amount was divided by the total number of spots (1 spot; 40x40 μm area). A curvilinear calibration plot for each microregion was obtained with a detection limit of 0.1 pg/40 μm area. The coefficient of variation was 5.9 % (n=5) at 3.25 pg/40 μm area. This calibration curve was fit to a allosteric function by the least squares method. Figure 3 shows the image of spotted bone marrow cell suspension (10 times dilution, 5μl). Each cell was stained as a dot and was shown in the image as a peak. The total amount of λ-chain in this sample was calculated to be 18 ng λ-chain/0.5 μl human bone marrow suspension.

Reference

1.T.Masujima,K.Wada, H.Yoshida, H.Imai, J. Photoacoustics,1,347(1982-83).

Basic Study of Photoacoustic Immunoassay and Determination of Trace Rheumatoid Factor

T. Kitamori[1], *K. Suzuki*[1], *T. Sawada*[2], *and Y. Gohshi*[2]

[1]Energy Research Laboratory, Hitachi, Ltd., 1168 Moriyama, Hitachi, Ibaraki 316, Japan

[2]Department of Industrial Chemistry, Faculty of Engineering, University of Tokyo, 7-3-1 Hongo, Bunkyo, Tokyo 113, Japan

The latex agglutination method has been developed for turbidimetric immunological determinations. In this method, latex microparticles which are coated by antibody/antigen are added to the sample, and they agglutinate with each other due to an immunoreaction as shown in Fig. 1, allowing the antigen/antibody to be determined on the basis of turbidimetric determination of the agglutinations [1]. On the other hand, trace analysis with photoacoustic spectroscopy (PAS) can be applied not only to true solutions but to turbid and colloidal solutions, and it has been stated that PAS sensitivity (the slope of the calibration curve) is independent of particle size, because the PA signal magnitude is proportional to the optical energy absorbed by the solution [2]. However, in the present study, a remarkable size dependence for sensitivity is found when the particle size is in a region close to that of the excitation beam wavelength. We propose a photoacoustic immunoassay (PIA) which determines latex agglutinations specifically and ultrasensitively with PAS using this observed sensitivity dependence on particle size. The principle of PIA is tested using rheumatoid factor (RF) as a sample antibody (normally, RF is an antigen, but it is an antibody for denatured γ-globulin), and the possibility of cancer diagnosis with PIA is discussed.

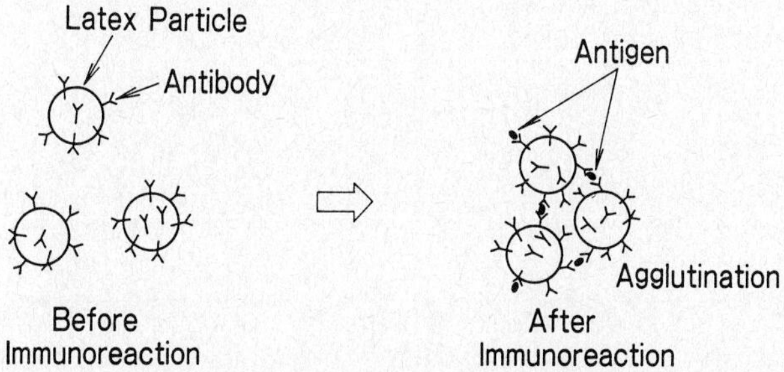

Fig. 1. Schematic illustration of latex agglutination by immunoreaction

Turbid solutions of monodisperse polystyrene latex particles of uniform size (0.10 - 5.0 μm) were prepared. For each particle size, the PA signal magnitude at four concentrations in the range of 0.1 - 1.0 ppb was measured with PAS (excitation wavelength: 488 nm, power: 1.8 W, modulation frequency: 181 Hz), and calibration curves were obtained as a regression line. The size dependence of sensitivity obtained as the slope of the calibration curves is plotted in Fig. 2. The sensitivity remarkably depends on the size between 0.3 - 0.8 μm, and the maximum sensitivity appears at 0.5 μm where the particle size coincides with the excitation beam wavelength. The physical meaning of this size dependence of the sensitivity is not clear at present, however, it seems to show a change in absorption cross-section due to dielectric loss which accompanies resonance scattering.

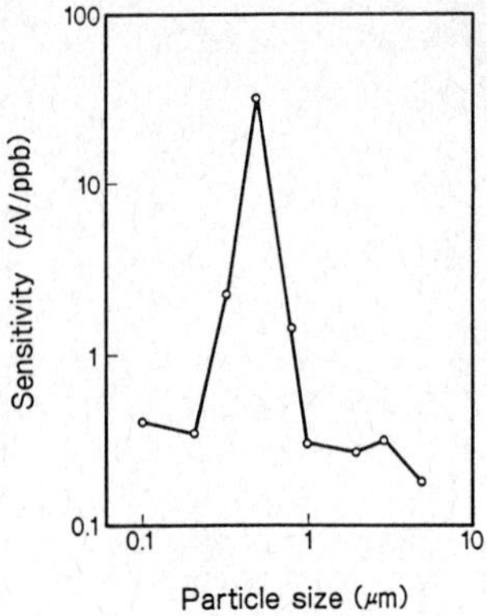

Fig. 2. Sensitivity size dependence of PAS

Fig. 3. Calibration curve of rheumatoid factor ▶

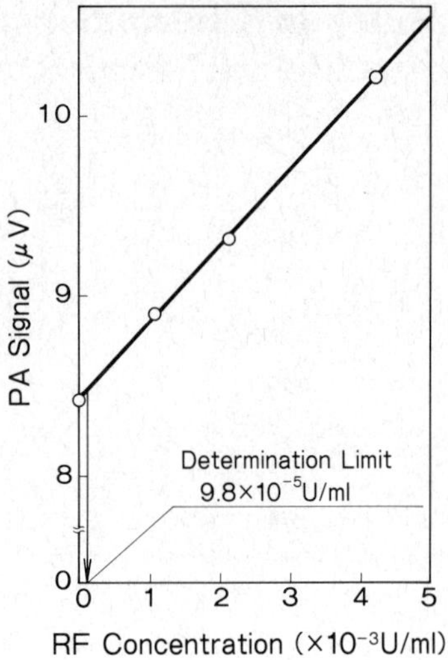

Using the size dependence of sensitivity shown in Fig. 2, microparticles whose size coincides with the wavelength of the excitation beam can be specifically determined with PAS. In the present study of PIA, latex particle agglutinations of excitation wavelength size, which were in a mixture of unreacted latex particles and various size agglutinations, were selectively determined.

Rheumatoid factor solutions of 21, 42 and 84 U(international unit)/mL were prepared by dissolving freeze-dried RF in ultrapurified water. Denatured γ-globulin (antigen of RF) coated 0.2 μm polystyrene latex particles were used as the reagent. After adding 3.5 mL of glycine buffered saline to the RF solutions of 30 μL and a blank solution, they were kept at 37 °C for five minutes, and 1 mL of the 0.03 w/w % reagent was added to each solution. The immunoreaction proceeded sufficiently at the chosen temperature within five minutes. After the immunoreaction, each solution was diluted by 1/20000, resulting in agglutinated turbid solutions corresponding to RF concentration of the order of 10^{-3} U/mL. The calibration curve of RF was obtained by PAS, and an example is shown in Fig. 3. The calibration curve of RF shows a good linearity for RF concentrations of 1.1 X 10^{-3}, 2.1 X 10^{-3} and 4.2 X 10^{-3} U/mL. The determination limit of RF, defined as double the standard deviation, is 9.8 X 10^{-5} U/mL. This value is at least three orders of magnitude lower than that of the conventional immunoassay using turbidimetry and the latex agglutination method. Furthermore, in comparison with radio immunoassay (RIA) and enzyme immunoassay (EIA), PIA is one to two orders more sensitive. Hence PIA is expected to determine some kinds of cancer marker antigens such as carcinoembryonic antigen, of whose concentrations in serum are at sub-ng/mL levels. In addition to sensitivity, in PIA no radioisotopes are used and just a few minutes are required for the immunoreaction and photoacoustic determination, providing advantages for medical screenings.

[1] J. M. Singer, et al., Am. J. Med. 21, 888 (1956)
[2] S. Oda, et al., Anal. 52, 650 (1980)

A Two-Photon Thermal Lensing Study of All-*trans* Retinal from 25,000 to 34,000 cm^{-1}

J.K. Rice[1*] *and R.W. Anderson*[2]

[1]Chemistry Department, University of Southern California, Los Angeles, CA 90089, USA
[2]Chemistry Department, University of California, Santa Cruz, CA 95064, USA

1 Introduction

In recent years two-photon spectroscopy has been applied to the determination of the singlet state ordering in visual chromophores. In long chain linear polyenes, model systems of visual chromophores, the lowest excited state has A_g^- symmetry[1]. BIRGE et al.[2,3] have observed the lowest excited state in retinol and retinal to have A_g^- symmetry using two-photon fluorescence excitation. BIRGE et al.[4] have also used two-photon thermal lensing to observe the lowest excited A_g^- state in the Schiff's base of all-trans retinal and in the 11-cis-locked conformation of retinal bound to opsin. In the Schiff's base, the A_g^- state is observed to be approximately 600 cm^{-1} below the B_u^+ state while in the 11-cis-locked chromophore the A_g^- state is observed 2000 cm^{-1} above the B_u^+ state. The symmetry classifications presented for visual pigments are approximate since they do not strictly belong to the C_{2h} symmetry group of polyenes. They do however have many common spectroscopic features making the C_{2h} symmetry assignments useful.

Due to the low quantum yield of fluorescence in retinal (<1.0 x 10^{-4}) and related molecules, there has been a shift to using non-fluorescing techniques such as thermal lensing. Accompanying this shift, the optimum solvents used in these studies have changed. CCl_4 is well suited as a solvent since it lacks C-H stretching overtones from hydrocarbon solvents which are observed in this energy region. The three bands observed in the one-photon absorption spectrum of retinal are labeled alpha, beta and gamma and appear at 27,000, 35,715 and 40,000 cm^{-1} respectively. The alpha band originates from the strong $B_u^+ \leftarrow A_g^-$ transition. The beta band is the only one of the three bands which exhibits vibrational structure. The absence of vibrational structure in

*Present Address: Naval Research Laboratory, Code 6111, Washington, D.C. 20375-5000

the alpha band has been addressed experimentally and theoretically[5,6]. It has been proposed that an unusually broad torsional potential for twisting about the C_6-C_7 single bonds is responsible for the lack of structure[7].

2 Experimental

The pulsed excitation beam, generated from a Molectron DL200 dye laser and UV-1000 nitrogen laser system, was aligned co-linearly with a 2mW Spectra Physics He-Ne monitoring laser. The dye laser output ranged from 150-500 microjoules. The dye laser and probe laser beams were focussed with 90mm and 110mm focal length lenses respectively into a 1cm sample cell. The dye laser intensity was averaged with a PAR boxcar averager. The thermal lensing signal intensity for the retinal in CCl_4 was obtained from 25 shots averaged on a storage oscilloscope. Using the solvent 3-methylpentane, 100-500 thermal lensing signals were averaged at each wavelength on a Model 6500 Biomation digitizer. Retinal, obtained from Tridom/Fluka (95% all trans and <5% 13-cis vitamin-A-aldehyde) was recrystallized twice from hexane. Spectrograde CCl_4 was used as the solvent and no thermal lensing absorptions were seen from the solvent throughout the spectral range.

3 Results

We report a structured two-photon absorption spectrum of retinal in the energy region corresponding to the $B_u^+ \leftarrow A_g^-$ transition. These features shown in Fig. 1 can be attributed to either 1) the

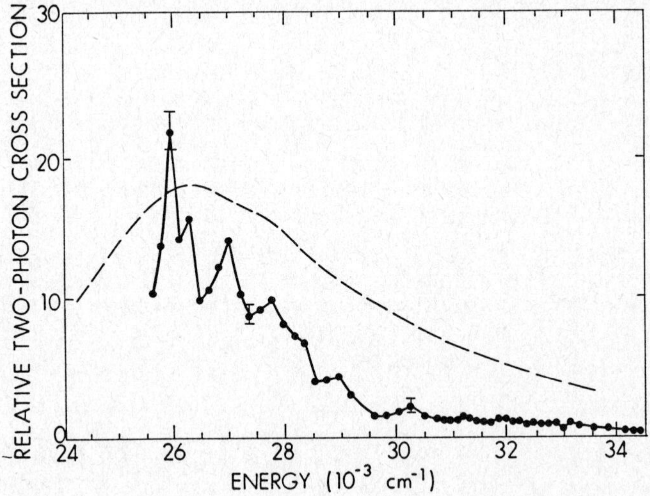

Fig. 1. The two-photon thermal lensing spectrum of 0.01M retinal in CCl_4 at 300K (solid) plotted at twice the photon energy. The corresponding one-photon spectrum (dashed).

two-photon forbidden $B_u^+ \leftarrow A_g^-$ transition with a maximum near 26,315 cm^{-1} seen through vibronic coupling, or 2) the high energy region of the lowest $A_g^- \leftarrow A_g^-$ transition with a maximum near 24,100 cm^{-1}. In the INDO-CISD calculations of BIRGE et al.[3], these two states have nearly equal two-photon cross sections in this energy range. Due to experimental limitations, the lowest energy obtainable was 25,640 cm^{-1}.

In an effort in ensure the solvent, CCl_4, was not responsible for the structural features, we prepared retinal samples in 3-methylpentane (3MP) and collected spectra from 25,500 to 29,000 cm^{-1} at 300K and at 77K. The spectra are shown in Fig. 2. A one-

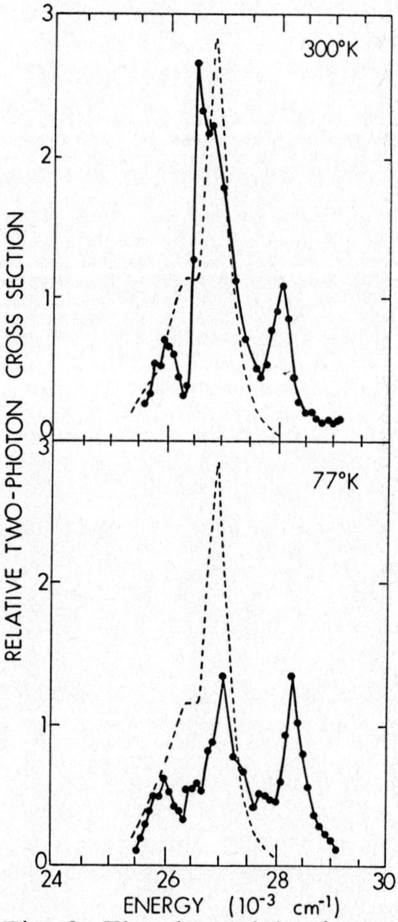

Fig. 2. The thermal lensing spectrum of retinal at 300K and 77K in 3-methylpentane (solid). This spectrum is normalized linearly with laser intensity below 27,600 cm^{-1} and to the square of the laser intensity above 27,600 cm^{-1}. The one-photon vibration overtone of the solvent (3MP) only (dashed).

photon C-H stretching overtone from 3MP is observed at 26,850 cm^{-1} at both temperatures. In addition, a two-photon dependent vibrational feature from retinal is seen at 28,050 cm^{-1} in the 300K sample and at 28,350 cm^{-1} in the 77K sample. These two-photon features seen in 3MP are not necessarily the same vibrational band. Transition energy shifts due to temperature can be small for $A_g^- \leftarrow A_\eta^-$ transitions and as large as 1000 cm^{-1} for $B_u^+ \leftarrow A_g^-$ transitions.

4 Conclusion

We conclude that the vibrational structure in the retinal spectrum is not solvent dependent but a physical feature of either the $B_u^+ \leftarrow A_g^-$ or $A_g^- \leftarrow A_g^-$ transition observable only in two-photon spectroscopy. We also speculate that structural features have not been observed in the two-photon fluorescence (TPF) excitation spectra of retinal and other visual chromophores because of the low fluorescence quantum yield.

There is a precedent for the observation of vibrational structure in a two-photon spectrum when none is seen in the corresponding one-photon spectrum. SMIRNOVA et al.[8] studied three organic dyes at room temperature using the TPF excitation technique. They conclude that the appearance of structure in the two-photon absorption spectrum is a result of preferential excitation of certain asymmetric molecular vibrations due to the two-photon selection rules.

In the two-photon retinal spectrum shown in Fig. 1, the energy separations between vibrational features from low to high energy are approximately 320, 725, 725, 485, 710, 1275, 950, and 800 cm^{-1}. No correspondence is found when these frequencies are compared to the normal vibration modes of all-trans retinal in the IR or Raman spectra of CURRY et al.[9]. The CCl_4 solvent has Raman active lines at 792 and 790 cm^{-1}. However, we cannot suggest any mechanism for observing these absorptions in the two-photon spectrum.

It may be difficult to assign the vibration features without knowing the location of the origin of the electronic transition. Pursuit of this research area to include the origin may lead to an understanding of the potential energy surface of the lowest excited B_u^+ or A_g^- state and the corresponding vibrational information may be available only through the use of multiphoton processes.

1. B.S. Hudson, B.E. Kohler, and K. Schulten: In Excited States ed. by E.C. Lim, Vol. 6, p. 1, (Academic Press: New York, 1982)
2. R.R. Birge, J.A. Bennett, B.M. Pierce and T.M. Thomas: J. Am. Chem. Soc. 100, 1533 (1978)

3. R.R. Birge, J.A. Bennett, L.M. Hubbard, H.L. Fang, B.M. Pierce, D.S. Kliger, and G.E. Leroi: J. Am. Chem. Soc. 104, 2519 (1982)
4. R.R. Birge, B.M. Pierce and L.P. Murray: Nato ASI, Ser. C, Spect. Bio. Mol. 139, 473 (1984)
5. B. Honig, A. Warshel and M. Karplus: Acc. Chem. Res. 8, 92 (1975)
6. A. Warshel and M. Karplus: J. Am. Chem. Soc. 96, 5677 (1974)
7. W. Hemley and B. Kohler: Biophys. 20, 377 (1977)
8. T.N. Smirnova, E.A. Tikhonov, and M.T. Shpak: JETP Lett. 29, 411 (1979)
9. B. Curry, A. Brock, J. Lugtenburg, and R. Mathies: J. Am. Chem. Soc. 104, 5274 (1982)

Photoacoustic Spectra of Green Leaves and of White Leaves Treated with the Bleaching Herbicide Norfluorazon

E.M. Nagel and H.K. Lichtenthaler

Botanisches Institut II, Universität Karlsruhe,
Kaiserstraße 12, D-7500 Karlsruhe, Fed. Rep. of Germany

Photoacoustic (PA) spectra (380-720 nm) of normal green leaves were compared with the corresponding absorption and reflectance spectra in order to obtain further information about the origin of the PA-signals. The spectra of the green leaves were also contrasted with spectra of white leaves, which were formed under the influence of the bleaching herbicide norfluorazon (10^{-4} molar; application via the root). These white leaves have a similar structure to the green leaves, but possess no photosynthetic pigments /1/ and /2/ and thus can help to decide which parts of the photoacoustic signal originate from the photosynthetic pigments and which are determined by the leaf structure or other pigments. The spectra of the 12-day-old plants were recorded with 22 and 515 Hz /3/, where the signals emanate from deeper inside (22 Hz: mesophyll and epidermis) or more from the surface (515 Hz: epidermis) of the leaves.

Green leaves (22 Hz): At a low chopping frequency the photoacoustic signal of the leaves is mainly determined by the absorption properties of the sample, as seen from the principal similarity between PA-spectra and absorption spectra (Figs. 1 and 2). In green radish leaves (Raphanus sativus) the red light maximum of the PA-spectrum (Fig. 1a) is higher than that of the blue light region. This is in contrast to the absorption spectra, where the blue light maximum is higher (Fig. 1b). In the green maize leaves (Zea mays) the shape of the 22 Hz PA-spectrum (Fig. 2a) is different from that of the green radish leaves. In maize the blue light plateau is higher than the red light maximum, as is the absorption spectrum (Fig. 2b). These differences may be caused by the different morphological structure of the leaves. In contrast to the bifacial radish leaf (palisade and spongy parenchyma), the maize leaves have the "Kranz-type" syndrome of the C_4-plants with two concentric mesophyll cell layers around the vascular bundle. The differences in the height of the PA-spectra in the blue light thus seems mainly to be due to the different arrangement of the cells and the aerial interspaces in the two leaf types. Another reason for these differences may be the following point: there is evidence that at low chopping frequencies (e.g. 22 Hz) the PA-signal of the photosynthetically produced oxygen is superimposed on the PA-signal due to the heat emission /4/. In view of this finding it may be that the contribution of the oxygen signal in the blue light region is higher for maize than for radish leaves. The third possibility, namely that in the maize leaves the blue light absorbed by carotenoids is transferred to a lower degree to the chlorophylls than in radish leaves, which would increase the PA-spectrum in the blue light, seems to be of less importance.

White leaves (22 Hz): In the white radish and maize leaves a small photoacoustic signal at 22 Hz remains, which is of constant height between 450 and 720 nm (Figs. 1b and 2b). Below 450 nm the PA-signal is higher

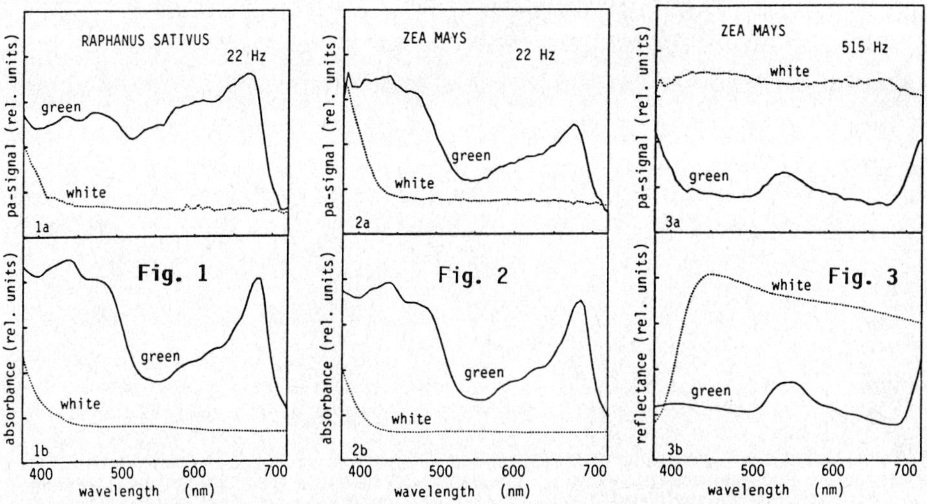

Figs. 1-3. Photoacoustic spectra at 22 and 515 Hz and absorption or reflectance spectra of radish and maize leaves. Mean of three spectra.

than above 450 nm in both plants. Below 450 nm the PA-spectrum of the white leaves is determined by the absorption of other pigments, possibly yellowish flavonoids, which are known to absorb below 450 nm.

PA-spectra at 515 Hz: At chopping frequencies of 515 Hz the PA-signals emanate from the epidermis region /3/. The spectra of radish and maize leaves are similar and are given here only for maize (Fig. 3a). The PA-spectrum of the **green leaf** resembles the reflectance spectrum (Fig. 3b) with a maximum at 550 nm. This indicates that the PA-spectrum at 515 Hz is determined by scattered green light from deeper leaf layers, which may be unspecifically absorbed in the chlorophyll-free epidermis. In contrast to the reflectance spectrum the PA-signal increases below 450 nm, which may be due to light absorbed by flavonoids, which are known to occur in the vacuoles of the epidermis cells and which absorb below 450 nm. The PA-signals of **white leaves** do not possess a specific maximum at 550 nm as the green leaves do. The signals are of the same level over the whole spectrum and are higher than the signals of the green leaves. This is similarly found in the reflectance spectra, except for the region below 450 nm.

Conclusion: The results obtained with radish and maize leaves indicate that the PA-spectra between 400 and 720 nm are primarily determined by the photosynthetic pigments and to a lesser degree by the leaf structure.

References:

1. S.M. Ridley: In Carotenoid Chemistry and Biochemistry, ed. by G. Britton and T.W. Goodwin (T.W. Pergamon Press, Oxford 1981) p. 353
2. H.K. Lichtenthaler, H.K. Kleudgen: Z. Naturforsch. 32c, 236 (1977)
3. E.M. Nagel, C. Buschmann, H.K. Lichtenthaler: Physiol. Plant. 70, 427 (1987)
4. G. Bults, B.A. Horwitz, S. Malkin, D. Cahen: Biochim. Biophys. Acta 679, 452 (1982)

Index of Contributors